Instrumental Methods of Analysis

B. Sivasankar

Visiting Professor
Department of Chemistry
College of Engineering, Guindy
Anna University, Chennai

OXFORD

UNIVERSITY PRESS

OXFORD
UNIVERSITY PRESS

Oxford University Press is a department of the University of Oxford.
It furthers the University's objective of excellence in research, scholarship,
and education by publishing worldwide. Oxford is a registered trade mark of
Oxford University Press in the UK and in certain other countries.

Published in India by
Oxford University Press
22 Workspace, 2nd Floor, 1/22 Asaf Ali Road, New Delhi 110002, India

First published in 2012

Digitally Printed in 2023

ISBN-13: 978-0-19-807391-8
ISBN-10: 0-19-807391-7

Typeset in Times New Roman
by Shubham Composers, New Delhi
Printed at Manipal Technologies Limited, Manipal

Preface

Analytical chemistry, a branch of chemistry dealing with identification of the chemical nature of materials (qualitative analysis) and quantification of the material content (quantitative analysis), is based on well-developed and established theoretical principles of chemical reactions. The subject has its roots in the other different branches of chemistry, namely physical chemistry, inorganic chemistry, and organic chemistry. Analytical methods developed in the first few decades of the twentieth century were essentially based on wet chemical methods involving volumetric and gravimetric methods. Developments in electronics and instrumentation have facilitated instrumental methods of analysis in gaining importance over the years. The advantages of instrumental methods include their detection capabilities of even trace quantities of substances and automation, especially suitable for quality control and process control in industries.

Analytical chemistry is a growing science as newer techniques emerge to cater to the needs of a variety of fields and is thus indispensable in the present technology-driven world. With the growing requirements in emerging and developing areas of electronics and semiconductors nanotechnology, biotechnology, clinical diagnostics, forensic science, environmental management, remote sensing, and space science and technology, the importance of analytical chemistry is bound to grow.

About the Book

This book—*Instrumental Methods of Analysis*—is written especially for both undergraduate and postgraduate students of chemical engineering, biochemical engineering, biotechnology, and chemistry. The philosophy of the book is to project the essentials of analytical chemistry, particularly for quantitative analysis so that the reader will have a comprehensive and thorough understanding of the subject.

All the spectroscopic methods involving the interaction of different regions of the electromagnetic spectrum with atoms and molecules have been included in a logical sequence in chapters 6–9. Similarly, the different types of separation methods have been arranged in sequence in chapters 12–15. Most other topics are independent of each other and have been dealt with in individual chapters.

Key Features

The special features of the book include the following.

Meets the Needs of Varied Streams The book can be used as a textbook for the core course of analytical chemistry/instrumental methods of analysis for postgraduate and undergraduate programmes in chemistry, material science, chemical engineering, biochemical engineering, and biotechnology.

Applications in Biotechnology and Chemical Engineering A good number of topics of interest and importance to both chemical engineering and biology/biotechnology, and chemistry students have been included in the book. These topics include the wet chemical methods, spectroscopic methods, various imaging techniques of optical microscopy, different centrifugation techniques for cell fractionation, methods of radioimmunoassay, and enzymatic methods in clinical diagnostics, among others.

Comprehensiveness The major components of the instruments and their functional aspects have been described so as to enable the reader to select an appropriate need-based method for analysis.

Illustrated Presentation Every important technique in the book is presented with a neat sketch of the instrument or the whole experimental set-up. Special care has been taken to reduce the complexity of figures that makes it easier for the students to reproduce them in the examinations.

In line with the philosophy of providing comprehensive information on various analytical methods, the book projects the essentials of analytical chemistry spread over nineteen chapters, covering the basics, characteristic features, and practical aspects of the techniques.

A brief introduction to each chapter gives an overview of the information and material content presented in the chapter. Each chapter is almost independent of other chapters and provides a complete picture of the basic concepts, principles involved in individual techniques, essential features of instrumentation, practical aspects, and a few case studies. Since the process of reading a textbook becomes a passive exercise, a few worked-out examples have been provided to enhance the comprehension of the subject at the end of each chapter. The review questions provided at the end of each chapter will further reinforce the learning.

Content and Structure

Chapter 1 introduces the concept of analytical chemistry and provides the scope and applications of analytical chemistry, the major steps involved in analytical process, a broad classification of analytical methods, and the basic principles of chemistry in terms of chemical equilibria. A detailed discussion on kinetic methods of analysis and their use mostly in clinical diagnostic applications is also included.

Chapter 2 discusses the importance and methods of assessment of analytical data for detecting different types of errors that are likely to be encountered and the procedures to be adopted for eliminating or minimizing such errors so that the final report is reliable and validated.

Chapter 3 discusses exclusively the use of wet chemical methods of analysis involving volumetric and gravimetric methods. This chapter has been included not only because these classical methods which have been in vogue for more than a century are simple, but also because they are relevant and will continue to be important in future as they are accurate and reliable, besides being simple and user-friendly even to beginners.

Chapter 4 deals with the principles, practice, and applications of refractive index measurements, polarimetry, circular dichroism, and optical rotatory dispersion; the latter three techniques are meant for the study of optically active compounds.

Chapter 5 on optical and electron microscopy has been included mainly to cater to the needs of biology, biochemistry, and biotechnology students as well as for material science students for studying the morphological characteristics of materials.

Chapters 6–9 deal with spectroscopic methods of analysis for qualitative and quantitative purposes as well as elucidation of molecular structure. *Chapter 6* discusses the energy level structures of atoms and molecules, basic principles, configuration and components of spectroscopic instruments, and the development of spectroscopic methods from visual colorimetry.

Chapter 7 deals exclusively with atomic spectroscopic techniques of atomic absorption spectrometry, atomic emission spectroscopy, and atomic fluorescence spectroscopy.

Chapter 8 discusses the different molecular absorption spectroscopic techniques of ultra-violet and visible spectroscopy, infrared spectroscopy, Raman spectroscopy, microwave spectroscopy, molecular fluorescence and phosphorescence spectroscopy, and turbidity and nephelometry. The chapter explains the basic principles, instrument details, practical aspects, and analytical applications of each of these methods.

Chapter 9 discusses the basic principles, instrumentation, and applications in understanding the structures of chemical compounds involving the magnetic resonance spectroscopic techniques of nuclear magnetic resonance spectroscopy and electron spin resonance spectroscopy.

Chapter 10 on mass spectrometry discusses the basics of the technique and its application in the study of molecular structure from simple organic molecules to complex biomolecules through the use of different ionization methods and instrumental techniques.

Chapter 11 on x-ray methods of analysis includes the topics of production of x-rays and their detection, analytical techniques of x-ray absorption spectroscopy, and x-ray fluorescence spectroscopy as investigative tools of materials and their surface characteristics.

Chapters 12–15 deal with all the separation techniques that find use in analytical, biological, and biochemical laboratories as well as in industries for separation and isolation of required substances and their purification.

Chapter 12 gives an overview of the different separation methods along with the preparative scale separation methods of liquid–liquid extraction, ion-exchange, and membrane separation methods used in chemical and biotechnological process industries.

Chapter 13 deals with the analytical separation techniques based on chromatography, the mechanisms of separation such as adsorption, partition, ion-exchange, size exclusion, and affinity. It also elaborates upon different types of equipment such as HPLC, GC, and supercritical fluid chromatography and their uses in qualitative and quantitative analyses. The chapter also gives an insight into the working of more advanced hyphenated techniques such as HPLC-MS, GC-MS, and GC-IR. The chapter also discusses planar chromatographic techniques of paper chromatography, TLC, and HPTLC and the recent developments in planar chromatographic techniques.

Chapter 14 discusses in detail the electrokinetic methods of separation such as electrophoresis and its variation of capillary electrophoresis. The unit also discusses the related techniques of isoelectric focusing and isotachophoresis which are exclusively used by biochemists/biotechnologists for the separation of biomacromolecules.

Chapter 15 deals with the principles of centrifugal separation as practised in bioprocess industries and biotechnology laboratories for the separation of biological products. Different centrifugation techniques such as differential centrifugation and density gradient centrifugation are included in this chapter. The chapter also delves upon analytical ultracentrifugation for the study of macromolecules and biopolymers.

Electroanalytical chemistry involves the use of simple as well as sophisticated methods based on the use of electrical energy in the study of reactions for analytical purposes.

Chapter 16 discusses the use of conductometry, potentiometry, electrogravimetry, coulometry, and voltammetry and newer developments as applicable to quantitative analyses of different analytes.

Chapter 17 deals with thermal methods of analysis that include thermogravimetry, differential thermal analysis, differential scanning calorimetry, thermomechanical analysis, dynamic mechanical analysis, and evolved gas analysis. These techniques find extensive use in the characterization of a wide variety of materials on the basis of the physical and chemical changes exhibited during programmed heating.

Chapter 18 discusses radiochemical methods of analysis that make use of radioisotopes in the characterization of materials, study of reaction mechanisms, and quantitative analysis. It also discusses neutron activation analysis, which is a highly sensitive non-destructive analytical technique for the study of archeological specimens and other materials. Radioimmunoassay, which is a highly sensitive and specific technique that finds use in the study of antigen–antibody interactions and separation and purification of biomolecules, is also discussed in this chapter.

Chapter 19 describes the procedure and applications of various analytical methods for the study of surfaces of solids. Some of these include methods based on adsorption of probe molecules, vibrational spectroscopic techniques of EELS, RAIRS for the study of surface adsorbed species, and electronic absorption spectral techniques of XPS and AES, and related techniques of ISS and SIMS. The chapter also discusses the use of thermal desorption technique of TPD and the related TPR in the study of catalysts.

B. Sivasankar

Acknowledgements

I wish to thank all my colleagues in the Department of Chemistry, Anna University, Chennai, particularly Professor V. Sadasivam, Head of the Department, Professor K. Rengaraj (retd.), and Dr. G.R. Rajarajeswari and well-wishers for their encouragement. I wish to thank the authorities of Anna University for the academic ambience and their support in academic pursuits. My family members, particularly my wife, Santha Kumari, deserve special mention for patience and understanding. I thank the editorial team of Oxford University Press, India, for their unstinting assistance and care at all stages of production of the book. Finally, I would like to dedicate this book to my teachers.

Any constructive criticism/comments from the readers of the book can be sent to sivasankar.bomma@gmail.com.

B. Sivasankar

Features of the Book

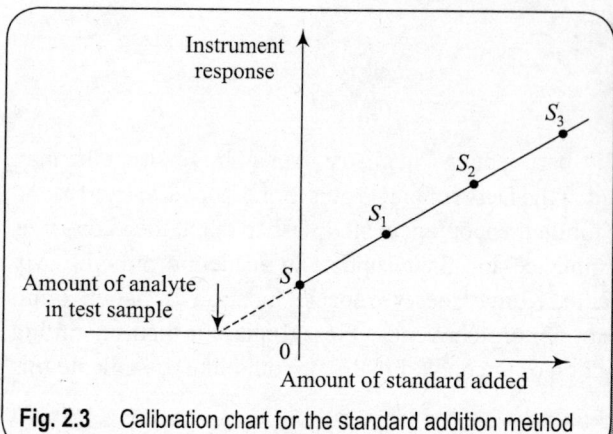

Fig. 2.3 Calibration chart for the standard addition method

Graphical Interpretation of Concepts

The text is interspersed with graphical illustrations in all chapters. These help to understand better the related mathematical deductions.

Neat Schematic Figures

Numerous schematic figures, help students to visualize related processes leading to the end result.

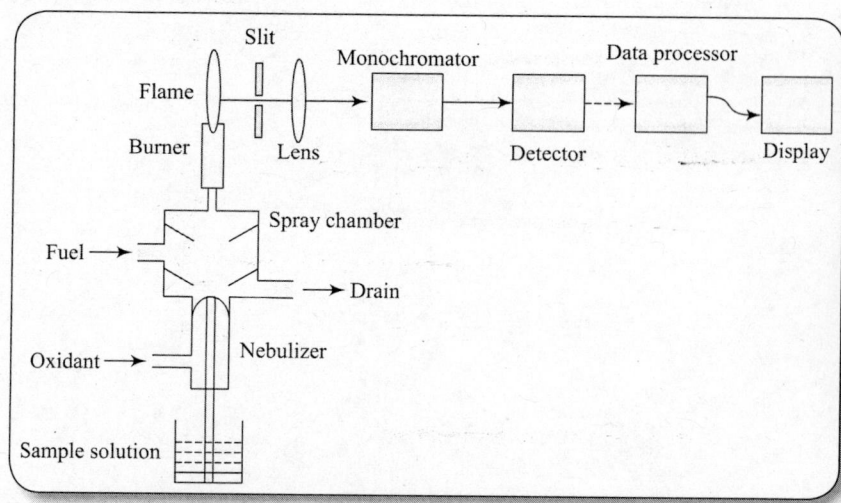

Table 7.1 Flames of different fuel gas–oxidant mixtures and their characteristics

Fuel	Oxidant	Temperature range (°C)	Burning velocity (cm s^{-1})
Natural gas/LPG	Air	1700–1900	40–43
Natural gas/LPG	Oxygen	2700–2800	370–390
Hydrogen	Air	2000–2100	300–440
Hydrogen	Oxygen	2500–2700	900–1400
Acetylene	Air	2100–2400	160–270
Acetylene	Oxygen	3000–3200	1100–2500
Acetylene	Nitrous oxide	2600–2800	280–300

Tabular Presentation

Tabular presentation of data helps to quickly grasp the information and also serves as a ready reckoner.

15.9.2 Applications of Analytical Ultracentrifuge

Analytical ultracentrifugation is used for accurate calculation of molecular weights of biopolymers such as polysaccharides, proteins, lipoproteins, nucleic acids, sub-cellular particles, cells, and viruses in their native state. In contrast, other techniques such as gel electrophoresis and size exclusion chromatography are applicable only to denatured samples. Analytical ultracentrifugation is also useful for the determination of molecular size and shape, density, sedimentation coefficient and frictional coefficient of analyte molecules, study of macromolecular interactions, conformation changes in enzymes, replication of DNA, and so on. The technique requires the use of high purity samples.

Applications

In addition to detailing the techniques, the book also outlines the relevant applications of the techniques, wherever appropriate.

Numerical Examples

A few simple and relevant numerical examples at the end of every chapter, augments the understanding gained in the relevant theoretical discussion.

SOLVED PROBLEMS

EXAMPLE 1 *Calculate the retention volumes, capacity factors, and relative retention of two components separated on a column with retention times of 1.8 and 3.2 min respectively on an adsorption column at a flow rate of 1.5 mL/min. The t_o value is 0.8 min.*

SOLUTION

$$V_{R1} = t_{R1} \times F = 1.8 \times 1.5 = 2.7 \text{ mL}$$
$$V_{R2} = 3.2 \times 1.5 = 4.8 \text{ mL}$$

SUMMARY

- Chromatography involves selective distribution of the components of a mixture between the immiscible stationary and mobile phases to bring about their separation by differential migration aided by the mobile phase as it percolates through a bed of the stationary phase.

- Separation of the components of a mixture takes place through any one of the five mechanisms of (i) adsorption,

Summary

The summary at the end of each chapter helps to recapitulate the concepts learnt in the chapter.

REVIEW QUESTIONS

Review Questions

Chapter-end review questions serve to test the understanding gained in that chapter.

1. Give an outline of the different types of thermal analytical methods.
2. Explain the principle involved in thermogravimetry with a suitable example.
3. What is a thermogram? How is it obtained?
4. What is DTG? What is its use?
5. Give a neat sketch of the block diagram of a thermobalance and describe the functions of its components.
6. How is TG useful in the analysis of solid fuels?

Brief Contents

Detailed Contents

1 Introduction to Analytical Chemistry

1.1 SCOPE AND APPLICATIONS OF ANALYTICAL CHEMISTRY

Analytical chemistry or, in a broader sense, analytical science aims at answering the two fundamental questions of 'what' and 'how much' as a part of characterization of materials. The first question is answered through *qualitative analysis* to identify a material through the chemical composition in the form of elements and compounds present, while the second question is answered through *quantitative analysis* of the exact composition of the constituents in the material. Apart from answering these basic questions, analytical chemistry also provides techniques and tools for *structural analysis* to understand the arrangement or structure of the constituents within a material and their interrelationship in the characterization of materials.

Analytical chemistry has become an indispensable discipline of daily life as its applications cover a variety of fields across various industries such as agriculture, medicine, and environmental monitoring.

In industry, analytical chemistry forms the basis of quality control and quality assurance in the production of high quality products and consistent maintenance of the required quality. The wide range of industries includes chemical, petroleum processing, petrochemical, polymer, rubber and plastics, metallurgical, electronics, food processing, pharmaceutical, construction, textile, and biochemical process industries. Quality control and assurance are based on chemical analytical methods exclusively devised to check the quality of raw materials, intermediates, and finished products. In addition, analytical methods find extensive use in process control, new product development, in-house research and development activities, and in effluent treatment and disposal. Besides these uses, analytical chemistry is also essential in determining the monetary or commercial value of a variety of materials such as process raw material (e.g., ores) and finished products (e.g., alloys, pharmaceuticals, and processed foods).

In agriculture, the nature of soil, its nutrient and mineral content for crop selection, the supply and standardization of fertilizers and crop-protecting chemicals, the nutritious quality of food crops and plant products, preservation and long-term storage of agricultural products, etc. require exclusive and specific analytical methods.

Medical and clinical diagnosis entirely depends on a variety of sophisticated analytical techniques and equipment in diagnosing physiological disorders and in health care management.

Analytical chemistry is an integral and indispensable part of monitoring of the environment for detecting the nature of pollutants and hazardous materials, determining their amounts and distribution over land and in water and atmosphere. Domestic and industrial waste management rely on analytical data for designing treatment and disposal methodologies.

1.2 ANALYTICAL PROCESS

The analytical process in the characterization of materials for qualitative and quantitative studies essentially involves six major steps. These include (i) identification and definition of a specific problem, (ii) selection of the method of analysis, (iii) sampling, (iv) separation from undesirable interference, (v) quantitative estimation, and (vi) assessment of the data generated and validation of the method.

The first step of an analysis involves identification and definition of the problem. It spells out the broad objective as well as the specific aim of an analysis, taking into consideration the qualitative nature of the analyte or test sample, the history of its availability, its procurement, and end use.

The selection of the method of analysis depends on the nature of the sample and its history as well as experimental parameters. The chosen method should be best suitable for the given situation to get maximum advantage with respect to accuracy, sensitivity or detection limit and selectivity, speed, cost-effectiveness, and legal acceptance.

Sampling aims at getting a true representative sample of the bulk material.

Separation from interferences is an important step aimed at isolating the desired analyte from interfering substances for the success of the analysis.

Quantitative estimation or quantitation is the actual measurement of the signal which is defined as

$$S = kA \tag{1.1}$$

where S is the signal generated by the analyte A and k is the proportionality constant. The proportionality constant is a measure of the sensitivity of the method chosen, which means that if k is large the value of S will also be large and vice versa. Different signals may be generated during analysis and the various analytical techniques may be classified on the basis of signals generated during an analysis (Table 1.1).

Assessment of the data generated in the previous step is important to check the reliability of the whole analysis and define the confidence limits on the basis of statistical analysis of the experimental data. Validation of the chosen analytical method involves comparing the results with those obtained using internationally accepted methods with regard to accuracy and precision.

1.3 SELECTION OF CHEMICAL REACTIONS FOR ANALYSIS

Two factors are important in choosing chemical reactions for qualitative as well as quantitative analysis: (i) Since chemical reactions are reversible or equilibrium reactions, which take place in both the directions, it is necessary to have the knowledge of the extent to which the reactions can proceed in the forward direction to form the desired product(s). In other words, one should know whether the reaction will reach the completion stage by transforming 100% of the reactants into products. This is the domain of thermodynamics. (ii) The second factor is the rate at which the reaction will proceed under a given set of conditions, particularly temperature and pressure. Thus, both thermodynamics and chemical kinetics are important in choosing a reaction for analytical purposes.

The progress of a chemical reaction between the analyte (reactant) A and a reagent R to form the product P may be traced as a function of time by determining either the concentration of the analyte or the product as a function of time. Figure 1.1 shows the progress of a reaction in terms of the concentration profiles of the reactant (Fig. 1.1a) and the product (Fig. 1.1b) respectively as a function of time.

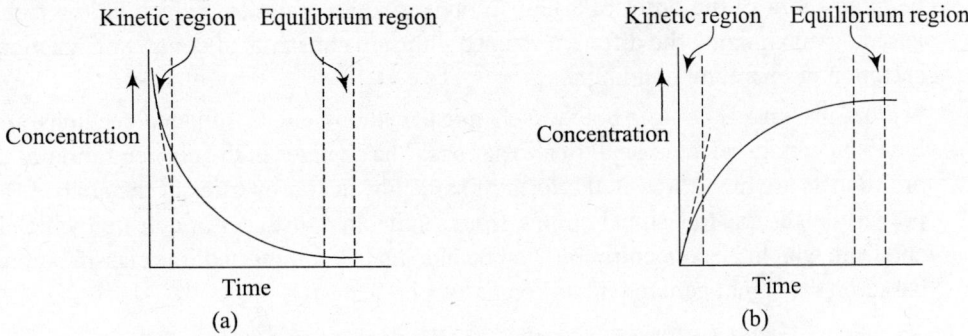

Fig. 1.1   Variation of concentration with time for (a) reactant and (b) product

The analyte concentration decreases linearly with time in the initial stages of the reaction and as the reaction progresses, the rate of change of concentration slows down and almost levels off. When this happens, the reaction is said to have attained equilibrium. The time required for attaining equilibrium varies with the nature of reaction and the equilibrium constant for the reaction which is independent of concentration at a given temperature. The thermodynamic equilibrium is a dynamic one with the rates of the forward and reverse reactions being equal. Once the equilibrium has been attained, no further change occurs in the concentration of the reactant.

The increase in the product concentration is also linear in the initial stages of the reaction and finally levels off as the reaction attains equilibrium. Thus, at equilibrium, the bulk concentration of the analyte or the product becomes constant as a function of time. However, all the reactions do not attain equilibrium at the same rate, and hence, the second important factor to be considered in choosing a reaction for analytical purposes is the kinetic aspect, that is, the rate of the reaction.

Based on the reaction time at which the concentration of the analyte is determined experimentally, the different analytical methods may broadly be classified into two major groups: (i) equilibrium methods and (ii) kinetic methods.

1.4   EQUILIBRIUM METHODS

Equilibrium methods based on thermodynamic considerations of reversible reactions involve measuring concentrations of the reactant or product (analyte in general) after the reaction has attained equilibrium, that is, when the bulk concentrations become constant. Chemical reactions chosen for quantitative analysis of materials should have high values for equilibrium constants and once the equilibrium has been attained, the concentrations of the reactants become almost negligible. The reactions which attain 99% or more completion may be called *irreversible reactions* as the rates of reverse reactions are negligible.

However, since equilibrium in a chemical reaction is dynamic one, reverse reaction does take place, though at a negligible rate, even in the so-called irreversible reactions. Hence, it is better to call such reactions as *analytical reactions*.

Analytical reactions are, thus, essentially equilibrium or reversible reactions whose equilibrium constants are greater than 10^4, and hence, reach completion (>99.9%) at a chosen temperature (mostly ambient temperature or at slightly elevated temperatures).

In addition, it is essential to select chemical reactions which attain thermodynamic equilibrium or reach completion almost instantaneously. The average reaction time to attain equilibrium should be quite short, of the order of a few seconds to a few minutes. Selectivity in such methods is based on maximizing the differences in equilibrium constants of competing reactions by proper selection of operating conditions.

Equilibrium reactions, in both wet chemical methods and instrumental methods, are essentially based on certain selected analytical reactions. The changes in the concentration of the reactants or products are monitored in the form of some physical properties of the analyte. The physical property gives rise to a signal output, from a suitable transducer or detector, which is displayed and from which the concentrations can be obtained. A summary of the classification of methods based on the signal generated has been given in Table 1.1.

Table 1.1 Classification of analytical techniques on the basis of signal generated

Signal generated	Analytical technique	Amount of analyte required
Mass	Gravimetry	10^{-1} mol L^{-1}
	Thermogravimetry	10–50 mg
Volume	Volumetry or titrimetry	10^{-1}–10^{-2} mol L^{-1}
Electrical potential	Potentiometry	10^{-1}–10^{-5} mol L^{-1}
Electrical current	Amperometry, gas chromatography with ECD	10^{-3}–10^{-10} mol L^{-1}
Electrical current and voltage	Voltammetry	10^{-3}–10^{-7} mol L^{-1}
Electrical current (or voltage) and mass	Electrogravimetry, coulometry	10^{-1}–10^{-4} mol L^{-1}
Electrical conductance	Conductometry, ion chromatography	10^{-1}–10^{-2} mol L^{-1}
Heat evolved or absorbed or temperature change	Thermal analysis, DTA, DSC	10–50 mg
Electromagnetic radiation absorbed or emitted	Spectroscopic techniques and spectral detectors used in GC and LC techniques	10^{-3}–10^{-6} mol L^{-1}
Electrical resistance	Gas chromatography with TCD or FID	10^3–10^{-6} mol L^{-1}
Ionizing radiation	X-ray, radiochemical methods and surface methods	10^3–10^{-10} mol L^{-1}

1.5 CONCEPTS OF CHEMICAL EQUILIBRIUM

Equilibrium reactions in both gas and liquid phases are governed by the *law of mass action*. The solvent used for most of the reactions in an analytical laboratory is water. Hence, it is necessary to have an understanding of the basic concepts involved in the analysis of equilibrium reactions and also the different types of chemical equilibrium used in aqueous media, e.g., acid–base or complex formation that one comes across.

Law of mass action, as proposed by Guldberg and Waage in 1863, states that 'the rate of a chemical reaction is proportional to the product of the active masses of the reacting substances'. The concept of 'active mass' is a thermodynamic term related to the concentration of the reactant in liquid phase (and partial pressure in vapour phase) as expressed by the relationship

$$\text{Activity} = \text{concentration} \times \text{activity coefficient} \tag{1.2}$$

In dilute solutions the value of activity coefficient is close to 1, and hence, the term 'activity' is replaced by 'concentration', since the latter can be determined experimentally.

A reversible reaction involving the reactants A and B whose concentrations are a and b respectively which give rise to products C and D of concentrations c and d respectively may be represented as

$$a\text{A} + b\text{B} \rightleftarrows c\text{C} + d\text{D}$$

The rate r_f of the forward reaction in which the reactants A and B are converted depends on their concentrations as expressed by the equation

$$r_f = k_f\,[\text{A}]^a\,[\text{B}]^b \tag{1.3}$$

where k_f is called the rate constant or rate coefficient and depends on factors such as temperature, pressure, and presence of a catalyst. The square brackets used in the equation represent the concentrations expressed in mol L^{-1}.

In a similar manner, the rate r_b of the back or reverse reaction in which the products C and D of the reaction recombine to form the original reactants A and B may be given as

$$r_f = k_b\,[\text{C}]^c\,[\text{D}]^d \tag{1.4}$$

where k_b refers to the rate constant of the backward reaction.

At equilibrium, the rates of the forward and the backward reactions are equal in magnitude, and hence,

$$k_f\,[\text{A}]^a\,[\text{B}]^b = k_b\,[\text{C}]^c\,[\text{D}]^d \tag{1.5}$$

which may be rearranged to give

$$[\text{C}]^c\,[\text{D}]^d/[\text{A}]^a\,[\text{B}]^b = k_f/k_b = K \tag{1.6}$$

The ratio of the rate constants for the forward and the backward reactions, k_f/k_b, is also a constant, called *equilibrium constant* and is given the symbol K. The equilibrium constant also depends on temperature and the presence of a catalyst in liquid and vapour phase reactions. The value of K can be calculated from the concentrations of the reactants and products which can be determined experimentally.

The greater the value of K at a given temperature, the farther right is the reaction at equilibrium, which may be represented as

$$a\text{A} + b\text{B} \rightleftarrows c\text{C} + d\text{D}$$

In general, the equilibrium constant K should be very large ($>10^4$) for the reaction chosen for analytical purposes so that the reaction is at least 99.99% to the right at equilibrium.

The thermodynamic relevance of the equilibrium constant is based on the change in Gibbs free energy function in the standard state

$$\Delta G^0 = -RT \ln K \tag{1.7}$$

where R is the gas constant ($8.314 \text{ J K}^{-1} \text{ mol}^{-1}$) and T is the absolute temperature. For any spontaneous reaction ΔG has to be negative, and hence, K should be positive. The change in the free energy depends on the changes in the enthalpy (ΔH^0) and entropy (ΔS^0) as given by

$$\Delta G^0 = \Delta H^0 - T \Delta S^0 \tag{1.8}$$

where the terms ΔG^0, ΔH^0, and ΔS^0 represent the thermodynamic quantities for one mole of the substance at the standard state of one atmospheric pressure and absolute temperature of 298 K.

Le Châtelier's principle predicts the effects of experimental parameters, such as temperature, pressure, and concentration of one of the reactants on the equilibrium constant K, and hence, on the equilibrium concentrations of the various species. Increase in temperature increases the rate of endothermic reactions and conversely decreases the rate of exothermic reactions. In addition to enthalpy effect, increase in temperature also favours an increase in the disorder of the system. Pressure affects the position of the equilibrium only in gas phase reactions and its effect is negligible in analytical reactions carried out mostly in aqueous medium. Though the equilibrium constant is independent of the concentrations of the reactants and products, the position of the equilibrium is however affected. For example, if the concentration of the reactant B is increased, the equilibrium shifts to the right and re-establishes itself. Catalysts do not affect the position of the equilibrium or the equilibrium constant, but they enhance the rate at which the equilibrium is attained.

1.6 TYPES OF EQUILIBRIA IN AQUEOUS MEDIA

The different types of equilibria that need to be considered in aqueous media from the analytical chemistry point of view include (i) self-dissociation of water, (ii) dissociation of acids and bases, (iii) hydrolysis of salts, (iii) metal–ligand complex formation, (iv) solubility of sparingly soluble salts, and (v) oxidation–reduction reactions.

1.6.1 Self-dissociation of Water

The small electrical conductivity of pure water was explained on the basis that it undergoes self-dissociation or self-ionization to generate a small amount of hydrogen and hydroxide ions as per the equation

$$H_2O \rightleftharpoons H^+ + OH^-$$

Since a free proton exists in water as a hydrated ion, also known as the *hydronium ion* H_3O^+, the above equation may be represented as

$$2\,H_2O \rightleftharpoons H_3O^+ + OH^-$$

The equilibrium constant K for the reaction is given as

$$K = [H^+][OH^-]/[H_2O] \tag{1.9}$$

Since the extent of ionization is very small, the concentration of un-dissociated water is considered as constant and Eq. (1.9) simplifies to

$$K_w = [H^+]\,[OH^-] \tag{1.10}$$

where K_w is called the *ionic product of water*, a special name given to the equilibrium constant expressing the self-dissociation of water. The value of ionic product of water at a temperature of 25°C is 1×10^{-14}. The value varies with temperature—it is 0.68×10^{-14} at 20°C and 1.47×10^{-14} at 30°C.

In pure water the concentrations of H^+ and OH^- are exactly equal, and hence,

$$[H^+] = [OH^-] = \sqrt{K_w} = 10^{-7} \text{ mol L}^{-1} \text{ at } 25°C$$

When the concentration of hydrogen ions is greater than 10^{-7} mol L^{-1}, the solution is said to be acidic and when the concentration of hydroxide ion is greater than 10^{-7} mol L^{-1}, the solution is said to be alkaline or basic.

It is more convenient to use 'p' notation for expressing the quantities which vary over a wide range of many powers of ten. The p function of any quantity X is defined as the logarithm to the base ten of the reciprocal of the quantity $(1/X)$ with a negative sign. Thus,

$$pX = \log_{10}(1/X) \quad \text{or} \quad pX = -\log_{10} X$$

The ionic product of water K_w is conveniently expressed as pK_w, the value being—$\log_{10}(1 \times 10^{-14}) = 14$.

1.6.2 Acid–base Equilibria

Lowry–Brønsted theory defines an acid as a proton donor and a base as a proton acceptor and the reaction between the two may be represented as

$$AH + B \rightleftharpoons A^- + BH^+$$

where AH and B represent the acid and the base respectively and A^- and BH^+ represent the *conjugate base* (of AH) and *conjugate acid* (of B) respectively.

Examples of Brønsted acids include neutral molecules (HCl, HNO_3, H_2SO_4, CH_3COOH, etc.), anions (HSO_4^-, $H_2PO_4^-$, etc.), cations (NH_4^+, $C_6H_5NH_3^+$), and aquated metal ions $[M(H_2O)_6]^{n+}$, etc. Similarly, examples of bases include NH_3, $C_6H_5NH_2$, OH^-, HPO_4^{2-}, etc.

An acid can donate a proton only when a base is ready to accept the proton. The above reaction may be considered to involve two conjugate acid–base pairs. The reaction will proceed to the right only when AH (acid 1) is relatively stronger than BH^+ (acid 2). In general, a stronger acid will lose its proton more readily than a weaker acid, and similarly a stronger base (B) will accept a proton more readily than a weaker base (A^-). Thus, in general, the conjugate base (A^-) of a strong acid (AH) is always weak, and similarly the conjugate acid (BH^+) of a strong base (B) is always weak. Conversely the conjugate base (A^-) of a weak acid (AH) is strong, and similarly the conjugate acid (BH^+) of a weak base is also strong.

Water is capable of accepting as well as donating protons and hence, called *amphiprotic*. In aqueous solutions, the solvent water acts as a base towards an acid and the reaction resulting in the formation of the conjugate acid, hydronium ion H_3O^+, may be represented as

$$AH + H_2O \rightleftharpoons A^- + H_3O^+$$

The equilibrium constant for the above reaction is given as

$$K_a = [A^-][H_3O^+]/[AH][H_2O] \tag{1.11}$$

The equilibrium constant K_a is called the *dissociation constant* or *ionization constant* of the acid at a given temperature, as the above reaction may be considered as *dissociation* or *ionization* of the acid in water. For example, the dissociation constant K_a of acetic acid at 25°C is 1.75×10^{-5} and it is expressed in terms of the p notation as $pK_a = -\log K_a = 4.76$.

The magnitude of the dissociation constant is useful in differentiating between *strong acids*, for example, hydrochloric acid and nitric acid (also strong electrolytes), and *weak acids* such as carboxylic acids. The equilibrium is to the far right in a strong acid–water system with almost 100% ionization of the strong acid in dilute solutions. The pK_a values are mostly less than zero (e.g., pK of $HNO_3 = -1.4$). In the case of polyprotic acids, such as sulphuric acid and phosphoric acid, the ionization occurs in successive stages, known as primary, secondary, and tertiary ionizations.

$$H_2SO_4 + H_2O \rightleftharpoons HSO_4^- + H_3O^+ \text{ (primary ionization)}$$
$$HSO_4^- + H_2O \rightleftharpoons SO_4^{2-} + H_3O^+ \text{ (secondary ionization)}$$

Primary ionization is complete ($pK_1 = -1.96$) but the secondary ionization is only partial and complete ionization to release two protons or hydronium ions occurs only in very dilute solution. Phosphoric acid (H_3PO_4) undergoes dissociation in three stages with pK values for the primary, secondary, and tertiary stages being 2.17, 7.21, and 12.36 respectively.

Water acts as an acid towards a base and the dissociation or ionization of the base and the corresponding equilibrium constant, called dissociation constant or ionization constant of the base at a given temperature, may be written as

$$B + H_2O \rightleftharpoons BH^+ + OH^-$$
$$K_b = [BH^+] [OH^-]/[B]] [H_2O] \tag{1.12}$$

The concentration of the solvent (water) varies negligibly during the dissociation of an acid or a base, and hence, dissociation constants may be written as

$$K_a = [A^-] [H_3O^+]/[AH] \tag{1.13a}$$

and

$$K_b = [BH^+] [OH^-]/[B] \tag{1.13b}$$

The product of the two dissociation constants is

$$K_a K_b = K_w \tag{1.14}$$

Since the concentrations of the reactants are expressed in $mol\ L^{-1}$, the units of K_a and K_b are also the same.

Lewis definition of acids and bases is based on the acceptance or donation of electron pairs by substances. An acid is an electron pair acceptor while a base is an electron pair donor, the interaction between the two results in the formation of a covalent bond. Examples of Lewis acids include BF_3, $AlCl_3$, SO_2, etc., and examples of Lewis bases include ammonia, organic amines, etc.

1.6.3 The pH Scale

Since the concentrations of H^+ and OH^- ions in aqueous solutions vary over a wide range (10^{-14} $mol\ L^{-1}$ to $1\ mol\ L^{-1}$), a convenient method of expressing the concentrations was introduced by Sorensen in terms of pH defined as the logarithm of the reciprocal of the H^+ ion concentration or the logarithm of H^+ ion concentration with a negative sign.

$$pH = \log_{10} (1/[H^+]) \quad \text{or} \quad -\log_{10} [H^+] \tag{1.15}$$

In neutral solutions the concentration of H^+ ion $= 1 \times 10^{-7}$, and hence, pH = 7. In acidic solutions the concentration of H^+ ion is greater than 1×10^{-7}, and hence, pH is less than 7 while in basic solutions pH is greater than 7.

Similarly, the OH^- concentration can be expressed in terms of pOH

$$pOH = \log_{10} (1/[OH^-]) \quad \text{or} \quad -\log_{10} [OH^-] \tag{1.16}$$

In pure water the ionic product of water $K_w = [H^+] [OH^-] = 1 \times 10^{-14}$ at 25°C. Hence, log $K_w = -14$ and it may be written as pH + pOH = pK_w = 14. Thus, a pH scale ranging from

0 to 14 may be conveniently used to express the concentrations of the H^+ and OH^- ions in dilute aqueous solutions at 25°C. When $[H^+] = 1$ mol L^{-1}, pH = 0 (pOH = 14) and when $[H^+] = 1 \times 10^{-14}$ mol L^{-1}, pH = 14 (pOH = 0).

1.6.4 Hydrolysis of Salts and pH of Salt Solutions

Acid–base reactions in aqueous medium always result in the formation of salt solutions of the acid–base pair (e.g., sodium chloride, sodium acetate, etc.). Salts are strong electrolytes as they exist as completely dissociated ions in solutions. The salt of a strong acid–strong base pair (e.g., sodium chloride or potassium nitrate) does not undergo any further reaction, and hence, the pH of the solution remains unchanged. However, if the salt is formed by the interaction of (i) a weak acid with a strong base (e.g., sodium acetate), (ii) a strong acid with a weak base (ammonium chloride), or (iii) a weak acid with a weak base (ammonium acetate), it undergoes hydrolysis in aqueous solutions, and hence, the pH of the salt solution is not neutral.

Salts of the type (MA) derived from the neutralization of weak acids represented as (HA) with strong bases represented as (MOH) are completely ionized and the anion undergoes hydrolysis forming undissociated acid and hydroxyl ions resulting in an increase in pH of the solution.

$$A^- + H_2O \rightleftharpoons HA + OH^-$$

The equilibrium constant for the hydrolysis reaction, called *hydrolysis constant* K_h, is the ratio of the ionic product of water K_w and the dissociation constant of the acid K_a.

$$K_h = K_w/K_a = [HA][OH^-]/[A^-] \tag{1.17}$$

The amounts of HA and OH^- generated by the hydrolysis of the salt are equal. Based on the assumption that hydrolysis occurs only to a small extent, the initial concentration of the salt is taken the same as that of the anion A^-. It can be written as

$$K_w/K_a = [OH^-]^2/C \tag{1.18}$$

where C is the initial concentration of the salt. Hence,

$$[OH^-] = (C K_w/K_a)^{1/2} \tag{1.19}$$

Since $[H^+] = K_w/[OH^-]$

$$[H^+] = (K_w K_a/C)^{1/2} \tag{1.20}$$

Using the p notation

$$pH = \tfrac{1}{2} pK_w + \tfrac{1}{2} pK_a + \tfrac{1}{2} pC \tag{1.21}$$

The above equation is useful for calculating the pH of the solution of the salt derived from a weak acid and a strong base.

Salts of the type (BX) derived from the neutralization of a strong acid represented as (HX) with a weak base represented as (BOH) are also completely ionized and the cation undergoes hydrolysis forming the undissociated base and the free hydrogen ions resulting in a decrease of the pH of the solution.

$$B^+ + H_2O \rightleftharpoons BOH + H^+$$

Similarly, based on the arguments detailed in the above paragraph, the pH of a solution of the salt derived from a strong acid and a weak base can be calculated using the formula

$$pH = \tfrac{1}{2} pK_w - \tfrac{1}{2} pK_b - \tfrac{1}{2} pC \tag{1.22}$$

where pK_b represents the dissociation constant of the base.

Salts derived from the neutralization of weak acids with weak bases are susceptible to hydrolysis in aqueous solutions and the pH of the solution depends on the relative magnitudes of the pK_a and pK_b but is independent of the concentration of the salt. The pH of the solution can be calculated using the formula

$$pH = \frac{1}{2}\,pK_w + \frac{1}{2}\,pK_a - \frac{1}{2}\,pK_b \tag{1.23}$$

1.6.5 Buffer Solutions

A buffer solution is the one which exhibits buffer action, that is, resists changes in pH when the solution is diluted or when a small amount of acid or base is added to it. Thus, a buffer solution is useful for controlling the pH of a reaction mixture in which pH changes occur due to the generation or absorption of hydrogen ions. The reaction mixture is said to be buffered. Buffer solutions are prepared from a mixture of the conjugate acid–base pair consisting of a weak acid (HA) and its sodium or potassium salt (A^-); for example, acetic acid and sodium acetate. A mixture of a weak base B and its salt BH^+, for example, ammonia and ammonium chloride also constitutes a buffer solution.

The buffer action of the mixture of a weak acid and its salt is explained by the Henderson–Hasselbalch equation according to which the pH of the buffer solution is determined by the dissociation constant of the weak acid and the ratio of the conjugate acid–base pair concentration.

$$pH = pK_a + \log_{10}([salt]/[acid]) \tag{1.24}$$

When the solution is diluted, the ratio of ([salt]/[acid]) remains constant, and hence, the pH remains constant. However, when a small amount of acid is added, it combines with the salt (A^-) to form the acid HA and the equilibrium, $HA = H^+ + A^-$, shifts to the left. Thus, the change in the ratio ([salt]/[acid]) is quite small and consequently, the change in pH also is negligible.

On the other hand, the equilibrium shifts to the right when a small amount of the base is added to the buffer solution, once again maintaining the ratio ([salt]/[acid]) constant and thereby the pH.

The buffering capacity of a buffer solution is maximum when the $pH = pK_a$ and is limited to a pH range as given by

$$pH = pK_a \pm 1 \tag{1.25}$$

The buffering capacity of a buffer solution consisting of a mixture of the base and its salt is similarly given by the relationship $pH = pK_b \pm 1$.

Examples of buffers commonly used in the analytical laboratory (the pH range is given in parentheses) include phthalic acid and potassium hydrogen phthalate (2.2–4.2), acetic acid and sodium acetate (3.8–5.8), sodium dihydrogen phosphate and disodium hydrogen phosphate (6.2–8.2), ammonia and ammonium chloride (8.2–10.2), and borax and sodium hydroxide (9.2–11.2).

1.6.6 Complexation Equilibria

Complexes or coordination compounds are formed by the reaction of metal ions (Lewis acids) and neutral or anionic ligands (Lewis bases). Ligands are classified into *monodentate* and *multidentate* ligands based on the number of coordination sites (also called donor or ligating sites) of the ligand involved in binding the metal. Examples of monodentate ligands include H_2O, NH_3, halide ions, etc. Multidentate ligands are also called *chelating ligands* which form coordination compounds with ring structures known as *metal chelates* or simply *chelates*. Chelating ligands which bind the metal simultaneously through two ligating sites are called *bidentate* ligands [e.g., ethylenediamine (en), oxalate (ox), acetylacetone (acac), etc.]. Similarly, chelating ligands which bind or coordinate the metal simultaneously with three, four, five, and six ligating sites are called tridentate, tetradentate, pentadentate, and hexadentate ligands respectively. Ethylenediamine tetraacetic acid (EDTA) is a hexadentate ligand which binds a metal with four oxygen and two nitrogen atoms (ligating sites) and finds extensive use in analytical chemistry.

In aqueous solutions, metal ions exist as aquated ions (aquo complexes) and the reaction between the metal ion and ligands may be considered as the replacement of the coordinated water molecules with the ligands. The complex formation reaction may be represented in a simple manner omitting the charges on the metal ions and the ligands.

$$[M (H_2O)_n] + L \rightleftharpoons [M (H_2O)_{n-1} L] + H_2O$$

where L represents a monodentate ligand. The number of coordinated water molecules being replaced depends on the denticity of the ligand. A bidentate ligand will replace two coordinated water molecules, a tridentate ligand will replace three coordinated water molecules, while a hexadentate ligand will replace six coordinated water molecules from the aquo complex. Multidentate ligands form more stable complexes compared to monodentate ligands due to *chelate effect* which is attributed to a large positive change in the entropy $(+\Delta S)$.

The reaction between a metal and a monodentate ligand L may be considered to occur in successive steps and the equilibrium constants (called *stability constants* or *formation constants*) for the successive reactions may be represented in a simple form as

$$M + L \rightleftharpoons ML; \qquad K_1 = [ML]/[M][L] \tag{1.26}$$
$$ML + L \rightleftharpoons ML_2; \qquad K_2 = [ML_2]/[ML][L] \tag{1.27}$$
$$ML_{(n-1)} + L \rightleftharpoons ML_n; \quad K_n = [ML_n]/[ML_{(n-1)}][L] \tag{1.28}$$

The equilibrium constants for the individual steps, $K_1, K_2, \ldots, K_n$ are called *successive* or *stepwise stability constants*. Complex formation may also be considered to involve the coordination of more than one ligand in a single step represented as

$$M + L \rightleftharpoons ML; \qquad \beta_1 = [ML]/[M][L] \tag{1.29}$$
$$M + 2L \rightleftharpoons ML_2; \quad \beta_2 = [ML_2]/[M][L]^2 \tag{1.30}$$
$$M + nL \rightleftharpoons ML_n; \quad \beta_n = [ML_n]/[M][L]^n \tag{1.31}$$

The equilibrium constants $\beta_1, \beta_2, \ldots, \beta_n$ are called *overall stability constants*. The overall stability constants are related to successive stability constants as

$$\beta_1 = K_1; \ \beta_2 = K_1 K_2; \ \beta_n = K_1 K_2 \cdots K_n \tag{1.32}$$

Volumetric titrations involving complex formation reactions are essentially based on the use of EDTA as the chelating agent as it reacts with almost all metals. This aspect is discussed in detail in Chapter 3.

1.6.7 Solubility Equilibria

Dissolution of substances (analytes and reagents) is an essential step in qualitative and quantitative analysis. The solubility of a substance depends on the nature of the substance, nature of the solvent, and other parameters such as temperature and the presence of other solutes. The substance may be a crystalline or amorphous solid or liquid or gas. Solids and liquids may be polar or non-polar and dissolve depending on the nature of the solvent. Ionic solids are relatively more soluble in polar or ionizing solvents as the solvation energy (hydration energy in water) overcomes the lattice energy due to favourable entropy factor. The dissolution of non-polar substances in polar solvents is due to a combination of forces such as van der Waals' forces, dipole–dipole interactions, hydrogen bond formation, and so on.

The solubility of an ionic compound in an ionizing solvent such as water may be represented as

$$AB + H_2O \rightleftharpoons A^+ (aq) + B^- (aq)$$

If AB is highly soluble, the equilibrium will be far to the right. However, if the ionic compound is sparingly soluble, the equilibrium will be far to the left. The solubility of such a sparingly soluble salt may be expressed in terms of the equilibrium constant called *solubility product constant*, K_{sp} as

$$K_{sp} = [A^+] [B^-]/[AB] [H_2O] \tag{1.33}$$

Since the substance is sparingly soluble, the concentrations of AB and the solvent water will remain unchanged, and hence, the equation may be simplified as

$$K_{sp} = [A^+] [B^-] \tag{1.34}$$

The solubility product constant is a measure of the solubility of the sparingly soluble substance and depends on the stoichiometric composition of the substance. It is expressed as a dimensionless quantity, but the solubility is usually expressed in g L^{-1} or mol L^{-1}. The solubility S of AB type compounds [e.g., AgX (X = Cl, Br, I, and CN), $BaSO_4$, $PbCrO_4$, PbS, CuS, etc.] is related to the solubility product constant as $K_{sp} = S^2$. Similarly, for AB_2 or A_2B type compounds (e.g., Ag_2CrO_4, CaF_2, etc.) the relation is $K_{sp} = 4 S^3$ and for A_2B_3 type compounds (e.g., As_2S_3, Sb_2S_3, etc.) the relation is $K_{sp} = 108 S^5$. The magnitude of K_{sp} depends on the temperature, generally increasing with increasing temperature.

The solubility of a sparingly soluble substance is affected by the presence of strong electrolytes. The effect may be either due to the presence of an ion common with the sparingly soluble substance (*common ion effect*) or due to different ions (*diverse ion effect*). The solubility of a sparingly soluble salt is also affected by the pH of the medium. These aspects are discussed in more detail with reference to gravimetry in Chapter 3.

The separation of metal ions into groups in qualitative inorganic analysis is based on two factors: (i) solubility product constant and (ii) common ion effect. Metal ions are precipitated by the addition of group reagents, for example, hydrogen sulphide is used to precipitate metal ions such as lead, mercury, copper, bismuth, tin, arsenic, and antimony in strongly acidic medium, while cadmium is precipitated under less acidic conditions by the same reagent. Metal ions, such as zinc and nickel, are precipitated as sulphides under alkaline conditions as they are soluble in acidic conditions. The control of the concentration of the precipitating agent, sulphide ion, is brought about through the common ion effect by controlling the pH of the medium.

The presence of complexing ligands and the nature of the solvent also affect the solubility of a sparingly soluble substance. For example, silver chloride completely dissolves in ammonia solution because of the formation of silver–ammine complex due to the coordinating ability of ammonia.

1.6.8 Redox Equilibria

Oxidation–reduction reactions involve the reaction between an oxidizing and a reducing agent which may be represented as

$$Ox_1 + Red_2 \rightleftarrows Red_1 + Ox_2$$

In the above representation, the oxidizing agent Ox_1 is reduced to Red_1 while the reducing agent Red_2 is oxidized to Ox_2. The equilibrium constant K for this reaction is related to electromotive force (emf) E^0_{cell} of redox system called the cell based on the thermodynamic relationships

$$\Delta G^0 = -RT \ln K \tag{1.35}$$

and

$$\Delta G^0 = -nFE^0_{cell} \tag{1.36}$$

and hence,

$$RT \ln K = nFE^0_{cell} \tag{1.37}$$

where ΔG^0 is the free energy change of the reaction at standard state and is equal to zero at equilibrium, R is the gas constant, T is absolute temperature, n is the number of electrons involved in the redox reaction, and F is the Faraday constant. The magnitude and sign of the E^0_{cell} indicate the spontaneity of the redox reaction and for a spontaneous reaction the E^0_{cell} should be positive. The equilibrium constant K for the reaction can also be calculated from the known values of E^0_{cell}. Redox reactions are useful in quantitative analysis because E^0_{cell} can be related to the concentration of the species involved in the reaction.

The overall redox reaction involving the two species, namely the oxidant and the reductant, may conveniently be considered as the sum of two half-reactions represented as

$$Red_2 \rightleftarrows Ox_2 + ne^-$$

and

$$Ox_1 + ne^- \rightleftarrows Red_1$$

The tendency of a chemical species to undergo oxidation or reduction depends on the standard reduction potential (also known as Latimer or standard oxidation potential) taking the half reaction

$$H^+ + e^- \rightleftarrows \tfrac{1}{2} H_2$$

the potential E^0 being arbitrarily taken as 0.000 V under standard conditions of 1 atmosphere pressure and unit activity at 25°C.

The standard reduction potential for the half reaction is related to the ratio of the concentrations of oxidized and reduced species as given by Nernst equation

$$E = E^0 + 0.0591/n \log ([Ox]/[Red]) \text{ at } 25°C \tag{1.38}$$

where E refers to the potential of the half reaction for activities of species other than unit activity. Application of Nernst equation for computing the potentials and theoretical titration curves for volumetric experiments are discussed in detail in Chapter 3.

1.7 KINETIC METHODS OF ANALYSIS

Kinetic methods of analysis are based on monitoring the concentrations of the reactant or the product at different time intervals—much below the time required for attaining equilibrium—usually within 30% change of the initial concentrations (Fig. 1.1). The conditions are dynamic with the concentrations varying continuously with time, almost linearly in the initial stages of the reaction. Selectivity in kinetic methods is achieved by the appropriate selection of reagents and reaction conditions so as to alter the rates at which the analyte and any potentially interfering substance can react. Kinetic methods are mostly based on catalysed reactions involving acid–base complexation, or redox reactions. Two types of catalysed reactions are used for analytical purposes: (i) reactions in which the catalyst itself is the analyte and (ii) reactions involving the use of a catalyst to enhance the rate of the reaction between the analyte and the reagent.

Kinetic methods are particularly useful when equilibrium methods cannot be used, because either the analytical reactions are too slow or do not proceed to 100% completion. Kinetic methods of analysis are especially suitable for most enzyme-catalysed reactions of biochemical and clinical diagnostic applications.

Kinetic methods are based on the determination of the order of the analytical reaction and the rate laws. *Order* of a reaction is defined as the total number of reacting species whose concentrations actually alter during the course of the reaction. Thus, a reaction involving the reactants A, B, and C giving rise to product(s) P is represented as

$$xA + yB + zC \rightarrow P$$

where x, y, and z represent the concentrations of the reactants A, B, and C respectively. The *rate* of the reaction is the derivative of the concentration of any reactant A, B, or C or the product P with respect to time.

$$\text{rate} = -d[A]/dt = -d[B]/dt = -d[C]/dt = d[P]/dt \tag{1.39}$$

The overall rate of the reaction R is

$$R = k\,[A]^x\,[B]^y\,[C]^z \tag{1.40}$$

where k is a proportionality constant also known as *rate constant*. The equation is called the *rate equation* or *rate law*; a mathematical expression depicting the variation of the rate of a chemical reaction as a function of time. The exponent or the power (x, y, or z) of each concentration term in the rate equation is called the order of the reaction with respect to the particular reactant. Thus, x is the order of the reaction with respect to reactant A and similarly, y is the order of the reaction with respect to the reactant B, and so on. The overall order n of the reaction is the sum of the exponents of the concentration terms in the rate equation ($n = x + y + z$). The order of a reaction may be zero, a fraction, or an integer also.

The order of a reaction depends on the nature of the reactants and has to be determined experimentally because it cannot be predicted on the basis of the stoichiometric equation. The order of a reaction is determined by determining the rate of the reaction under specified conditions and fitting the experimental data to theoretical rate expressions for different orders.

Quantitative analysis by kinetic methods essentially use chemical reactions involving two reactants, namely an analyte A and a reagent Z to give a product P which may be considered as a second-order reaction and the rate expression involves the concentrations of both the analyte and the reagent.

$$A + Z \rightarrow P; \quad \text{rate} = -d[A]/dt = d[P]/dt = k[A]\,[Z] \tag{1.41}$$

However, the mathematical analysis of a second-order rate expression is simplified by carrying out the reaction under *pseudo first-order* conditions by maintaining the concentration of one of the reactants (usually the reagent Z) quite large ($[Z] \gg [A]$), so that its concentration can be assumed to remain constant and does not vary with time during the course of the reaction. The rate of the reaction under such conditions is proportional to the concentration of the reactant or analyte A alone.

$$-d\,[A]/dt = k'[A] \qquad (1.42)$$

where k' represents the pseudo first-order constant.

1.7.1 Experimental Methods for the Determination of Rate of Reaction

A convenient method for the determination of the rate of a reaction under pseudo first-order conditions is known as the *differential method*. The method involves preparing a calibration chart and a graphical plot of rates of the reaction for different initial concentrations of the standard analyte. A typical experimental procedure involves the determination of the remaining concentration of the analyte or the product formed through the use of a suitable analytical method, either by a wet chemical method or by an instrumental method (e.g., spectrophotometry, potentiometry, chromatography, etc.) as a function of reaction time. In practice, a small portion of the reaction mixture is periodically withdrawn into separate sample tubes after specified time intervals (say every 5 or 10 minutes after initiating the reaction) and the reaction is arrested by quenching in an ice bath or freezing mixture. Alternatively, the reaction is stopped by the addition of a suitable reagent which combines with one of the reactants. The concentration of the analyte (reactant or product) in the withdrawn portions is determined by a chosen analytical method to give a series of data on the concentration of the analyte at different time intervals.

Alternatively, instrumental methods, such as spectrophotometry, conductometry, or potentiometry, can be used to monitor the changes in the concentration of the reactant or product continuously. The decrease in the concentration of the reactant or the formation of the product as a function of time for a given initial concentration of the reactant gives graphical plots as shown in Figs 1.1(a) and (b) respectively. The rate of the reaction for a given initial concentration of the reactant is obtained from the linear portions of the curves close to $t = 0$. The rate of the reaction is similarly determined for different initial concentrations of the reactant A. A plot of the variation of the rate of reaction for different initial concentrations of the reactants will be linear and this graphical plot serves as the calibration chart or graph for the determination of the concentration of reactant A in the test sample. The reactants in the test sample are subjected to the same reaction under identical conditions as used for preparing the calibration chart and the rate is determined. The concentration of analyte A, in the test sample is determined from the calibration based on the measured rate as in Fig. 1.2.

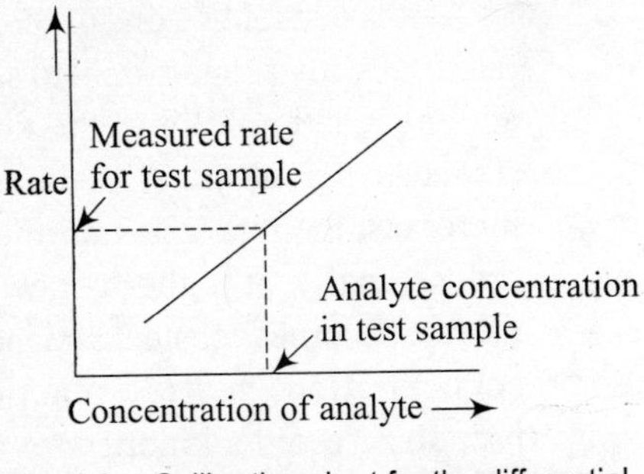

Fig. 1.2 Calibration chart for the differential method

Another graphical method is based on the equation

$$\ln [A]_t = -\,kt + \ln [A]_0 \qquad (1.43)$$

where $[A]_t$ represents the concentration of analyte A at reaction time $= t$ and $[A]_0$ at zero time, that is, initial concentration respectively. A graphical plot of the varying values of $[A]$ measured at different time intervals during the course of the reaction as a function of time gives a straight line with a slope of $-k$ and the intercept on the y-axis obtained by extrapolating the straight line to $t = 0$ gives the initial concentration of the analyte $[A]_0$.

1.7.2 Analytical Applications of Kinetic Methods

Kinetic methods for analytical purposes have been used mostly for catalysed reactions and for a few uncatalysed reactions. A variety of inorganic species, such as metal ions (e.g., copper, iron, cobalt, molybdenum, mercury, etc.), anions (e.g., iodide, bromide, cyanide, etc.), organic substances, and biochemicals, including metabolites and proteins in body fluids, have been analysed quantitatively using enzyme-catalysed reactions. Enzyme-catalysed reactions have gained importance over the years, particularly in clinical diagnostic applications.

1.8 ENZYME-CATALYSED REACTIONS

Enzymes are proteins with the specific biological function of catalysis. They catalyse a variety of reactions in living organisms and are also capable of functioning as catalyst *in vitro* (outside the living system). The characteristic features of enzymes include their high specificity towards substrates, inhibitors, activators, and products. They also exhibit region-specificity and stereo-specificity. Many enzymes require a cofactor or coenzyme to catalyse a reaction. For example, many dehydrogenase enzymes, which catalyse redox reactions, require nicotinamide adenine dinucleotide (NAD^+) as the hydrogen acceptor to form NADH. A variety of experimental parameters influences the rates of enzyme-catalysed reactions. These include temperature, pH, enzyme concentration, and substrate concentration. Each enzyme has a range of temperature over which it shows activity and an optimum temperature at which its catalytic activity is at maximum. Similarly, each enzyme has an optimum range of pH over which its catalytic activity persists and a pH at which the rate of the reaction is maximum. Hence, it is necessary to select proper experimental conditions for analytical determinations.

1.8.1 Mechanistic and Kinetic Aspects of Enzyme-Catalysed Reactions

A distinguishing feature of enzymatic reactions is the dependence of the rate on substrate concentration as shown in Fig. 1.3.

At low concentrations of the substrate, the reaction rate is directly proportional to substrate concentration and the reaction follows the first-order kinetics with respect to the substrate. As the substrate concentration increases, the rate falls and the reaction is of mixed order. Finally, at higher concentrations of substrate, the rate becomes constant and independent of substrate concentration, and the reaction is said to follow pseudo zero-order kinetics for a given concentration of the enzyme. Under these conditions the enzyme is saturated with the substrate and the rate is dependent only on the concentration of the enzyme.

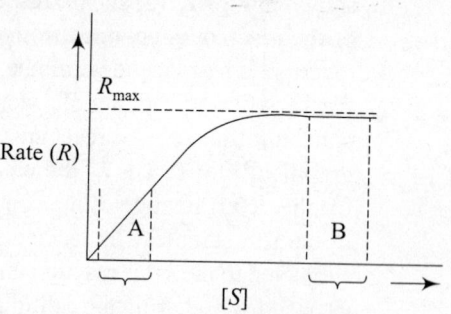

Fig. 1.3 Rate of enzyme-catalysed reaction on substrate concentration (regions A and B—analytical regions for substrate and enzyme respectively)

Michaelis and Menten explained the characteristic features of enzyme-catalysed reactions through a two-step mechanism known as Michaelis–Menten mechanism. In the first step, the enzyme E reacts reversibly with the substrate (reactant) S to form an enzyme–substrate complex or activated complex ES. In the second step, the complex decomposes irreversibly to form the product(s) and regenerate the enzyme.

$$E + S \underset{k_2}{\overset{k_1}{\rightleftarrows}} ES \overset{k_3}{\rightarrow} P + E$$

The rate equation, also known as Michaelis–Menten equation, for enzyme-catalysed reactions is given as

$$R = \frac{R_{max}\,[S]}{K_M + [S]} \tag{1.44}$$

where R is the overall rate and the Michaelis constant $K_M = (k_2 + k_3)/k_1 = [E][S]/[ES]$. The value of K_M is also equal to $[S]$ when $R = \frac{1}{2}\,R_{max}$.

The kinetic parameters K_M and R_{max} are useful to describe the enzyme-catalysed reaction. The value of K_M depends on the substrate and other experimental parameters, such as pH, ionic strength, temperature, and nature of the solvent used. The values of K_M are in the range of 10^{-1} to 10^{-6} M for most single substrate reactions. The smaller the K_M value, the greater is the activity of the enzyme for a given substrate. Each enzyme is characterized by the *turnover number*, that is, the number of substrate molecules converted into product per unit time. The turnover number of the enzyme is indicated by the maximum rate R_{max}. The values of K_M and R_{max} are obtained from the linear form of the equation known as Lineweaver–Burk equation.

$$1/R = (K_M/R_{max})\,[S] + 1/R_{max} \tag{1.45}$$

The plot of $1/R$ versus $1/[S]$ gives a straight line with a slope $= K_M/R_{max}$ and an intercept of $1/R_{max}$ as shown in Fig. 1.4.

For analytical purposes, the overall rate of the reaction R is proportional to the concentration of enzyme–substrate complex ES, and hence, may be related to the concentrations of the substrate and the enzyme.

$$R = k_3\,[ES] = k\,[S]\,[E] \tag{1.46}$$

When the substrate concentration is small compared to the enzyme concentration and the turnover number is not exceeded, the rate of the enzyme-catalysed reaction is dependent only on the concentration of the substrate for a given concentration of the enzyme, and hence, the reaction is of first order with respect to the substrate. Under these conditions, k [E] in Eq. (1.46) is a constant and the amount of the substrate can be quantitatively determined from the rate of the reaction (region A in Fig. 1.3).

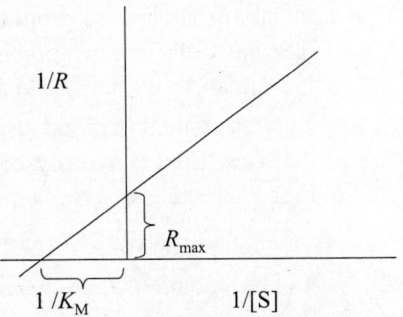

Fig. 1.4 Lineweaver–Burk plot

At large concentrations of the substrate, when the enzyme is saturated, the $k[S]$ in Eq. (1.46) becomes constant and the overall rate of the reaction is of first order with respect to the enzyme concentration. Under these conditions, the concentration of the enzyme can be determined quantitatively (region B in Fig. 1.3).

1.8.2 Applications of Enzymatic Analysis

Enzyme-catalysed reactions find use in diagnostic applications in two different ways. (i) Enzymes are useful as analytical reagents for the determination of a wide variety of biomolecules and metabolites. In such analytical determinations of an analyte (substrate, inhibitor, or activator), the enzymes function as catalysts, because of their high selectivity towards a particular analyte (substrate) which undergoes a reaction, inhibitor which decreases or inhibits the catalytic activity of the enzyme, or an activator which enhances the enzyme activity. Many enzymes have been immobilized onto inert supports, such as glass beads, polymer membranes, and walls of plastic tubes, and are commercially available for analytical purposes. Such immobilized enzymes have greater stability and have the advantage of reusability for hundreds of analysis. (ii) Enzymes themselves are analytes as their enhanced activity is helpful in the diagnosis of diseases.

1.8.3   Substrates as Analytes

Substrates, such as glucose, galactose, urea, uric acid, and alcohol are determined by enzyme-catalysed reactions for diagnostic purposes.

Glucose in blood is determined for diagnosing diabetes. The method of determination involves glucose oxidase-catalysed reaction in which glucose is oxidized to gluconic acid with the simultaneous formation of hydrogen peroxide.

$$C_6H_{12}O_6 + H_2O + O_2 \rightarrow C_6H_{12}O_7 + H_2O_2$$

The hydrogen peroxide formed is coupled with o-toluidine in the presence of a second enzyme, horseradish peroxidase, to give a coloured dye, and the absorbance of which is measured spectrophotometrically at 625 nm. The amount of glucose is determined by using calibration chart prepared from standard solution of glucose.

Galactose is determined to diagnose galactosemia, a rare metabolic disorder of inability to metabolize galactose properly, which leads to enlarged liver, cirrhosis, renal failure, and brain damage. Galactose in blood is determined based on the amount of hydrogen peroxide produced by the oxidation of galactose to form galactohexodialdose using galactose oxidase-catalysed reaction.

Urea content in blood and urine is determined to diagnose liver and kidney malfunction and diseases. Urea is hydrolysed by urease to form ammonia and carbon dioxide and urea content in the sample is determined from the amount of ammonia formed.

$$CO(NH_2)_2 + H_2O \rightarrow 2\,NH_3 + CO_2$$

The amount of ammonia is determined spectrophotometrically by modified Berthelot method in which ammonia forms a green coloured indophenol product on reacting with hypochlorite and salicylate in the presence of sodium nitroprusside. The absorbance measured at 630 nm is directly proportional to the amount of ammonia and is usually determined from the calibration chart prepared with standard solution of ammonia. A rapid method for the estimation of ammonia involves the use of specific ion-selective electrode for monitoring ammonium ion potentiometrically.

Uric acid ($C_5H_4N_4O_3$) content in blood plasma at levels higher than the reference range of 3.6–8.3 mg dL^{-1} and in urine is associated with the formation of kidney stones, leukaemia, renal dysfunction, and a form of arthritis known as gout. The amount of uric acid may be determined by measuring the absorbance at 292 nm. However, the absorption is not specific and the amount of uric acid in blood is quite small for accurate determination. Hence, uric acid is oxidized to allantoin by the enzyme uricase (specific to uric acid alone) and the absorbance measured once again. The difference in the absorbance is due to the uric acid in the sample.

Blood alcohol is determined to diagnose alcoholism by law enforcement agencies. Alcohol content in the sample is determined enzymatically by oxidizing alcohol to acetaldehyde by the enzyme alcohol dehydrogenase (ADH) with simultaneous reduction of nicotinamide adenine dinucleotide (NAD$^+$) to NADH. The increase in the absorbance at 340 nm is directly proportional to the concentration of NADH and in turn is directly proportional to the alcohol concentration in the sample.

1.8.4 Enzymes as Analytes

A variety of enzymes occurring in certain tissues are found in blood plasma at enhanced levels. Enzymes are usually measured by determining the rate at which they convert a substrate into the end product. Examples of enzymes useful for diagnostic purposes include serum glutamate–pyruvate transaminase (SGPT), serum glutamate–oxaloacetate transaminase (SGOT), lactate dehydrogenase (LDH), creatine kinase (CK), and alkaline phosphatase.

Enzymes generally known as transaminases catalyse the inter-conversion of energy storage molecules and in the synthesis and breakdown of amino acids in the liver. The concentrations of the different transaminases in the blood serum is usually low in the normal condition, but they increase when the liver is injured due to toxic substances and diseases such as viral hepatitis. Liver diseases are detected by determining the concentrations of two transaminases, alanine transaminase (ALT) and aspartate transaminase (AST), also known as serum glutamate–pyruvate transaminase (SGPT), and serum glutamate–oxaloacetate transaminase (SGOT) respectively. AST is also found in cardiac and skeletal muscles besides liver, and hence, measurement of its concentration has been used to diagnose heart attack. SGPT is determined using the reaction between α-ketoglutarate and L-alanine to produce glutamate and pyruvate. The pyruvate formed is converted to lactate by LDH in the presence of NADH. The decrease in the concentration of NADH, as measured from the absorbance at 340 nm, is directly proportional to the concentration of pyruvate which depends on the concentration of the enzyme SGPT. SGOT is similarly determined using the reaction between α-ketoglutarate and aspartate to produce glutamate and oxaloacetate, followed by the conversion of oxaloacetate to malate by LDH in the presence of NADH.

Lactate dehydrogenase (LDH) catalyses the oxidation of lactic acid to pyruvic acid in the presence of NAD^+ which gets reduced to NADH. It also catalyses the reduction of pyruvic acid to lactic acid in the presence of NADH. The increased level of LDH in blood serum is indicative of a variety of diseases such as haemolysis, tissue breakdown, cancer, meningitis, acute pancreatitis, encephalitis, and HIV. Lactate dehydrogenase is determined using L-lactate as the substrate to produce pyruvate in the presence of NAD^+ and the concentration of NADH formed being monitored spectrophotometrically at 340 nm.

Creatine kinase (CK), also known as creatinine phosphokinase (CPK), catalyses the conversion of creatine to phosphocreatinine and consumes adensosine triphosphate (ATP) during the reaction. It also catalyses the reverse reaction. The enzyme level in the blood serves as a marker for myocardial infarction, muscular dystrophy, and acute renal failure. Creatine kinase is determined using three enzyme-catalysed reactions in as many steps. In the first step, the substrate creatinine phosphate is converted to creatine by CK with the simultaneous phosphorylation of adenosine diphosphate (ADP) to adenosine triphosphate (ATP). In the second step, the ATP reacts with glucose in the presence of the enzyme hexokinase to form glucose-6-phosphate. In the final step, glucose-6-phosphate is oxidized to 6-phosphogluconate by the enzyme glucose-6-phospho dehydrogenase in the presence of NAD^+. The concentration of NADH formed is determined by measuring the absorbance at 340 nm.

Alkaline phosphatase (ALP) is a hydrolytic enzyme that catalyses dephosphorylation, that is, removal of phosphate group from a variety of molecules such as proteins, nucleic acids, and alkaloids, in alkaline medium. The enzyme exists in almost all cells in the human body and at higher levels in liver, bile duct, kidney, bone, and placenta. Higher than normal levels of ALP is indicative of blocked bile duct and active bone formations, whereas lower than normal levels is indicative of a variety of diseases such as anaemia, Wilson's disease, chronic myelogenous leukaemia, and so on.

Alkaline phosphatase is used in molecular biology for removing and introducing radio-labelled phosphate groups into DNA and for enzyme immunoassays. In the dairy industry, it is used as a marker or processing aid to determine whether pasteurization of cow's milk has been proper. Raw milk and improperly pasteurized milk contain ALP, which gives a yellow colour with *p*-nitrophenylphosphate within a couple of minutes. Alkaline phosphatase is heat denatured in properly pasteurized milk, and hence, will not answer the test. Alkaline phosphatase is determined quantitatively from the intensity of blue colour of thymolphthalein formed from the substrate thymolphthalein monophosphate over a specified time.

1.9 STOICHIOMETRIC CALCULATIONS

Quantitative analysis involves the measurement of weights of solids, volumes of liquids, and concentrations of solutions from which the masses of chemical species are calculated. *Mass* of an object is invariant as it refers to the quantity of the material of the object. In contrast, the *weight* of an object is defined as the force of attraction exerted on the object due to earth's gravity. Thus, weight w and mass m of an object are related as $w = mg$, where g represents the acceleration due to gravity of the earth. Generally, the terms *mass* and *weight* are used synonymously in quantitative analysis. The weight of the substance is measured and expressed in SI unit (International System of Units) as *kilogram* (kg); smaller units being gram (g), milligram (mg), microgram (μg), and even smaller units of nanogram (ng), picogram (pg), and femtogram (fg). The volume of liquid is measured and expressed in SI unit as *litre* (L); smaller units being millilitre (mL) and microlitre (μL).

The calculation of the mass or amount of chemical species is based on *stoichiometry*. The term *stoichiometry* refers to the quantitative relationship between reacting chemical species in terms of the combining ratios expressed in units of *moles*. One mole of a chemical substance contains Avogadro's number (6.022×10^{23}) of atoms, ions, or molecules. Stoichiometric calculations make use of atomic weights of elements and molecular weights or formula weights of substances. The atomic weights of elements are *relative masses* expressed as *atomic mass units* (amu) based on the assigned mass of 12 amu for the ^{12}C isotope. The *molar mass* (or gram atomic weight) is defined as the mass of one *mole* of the carbon, that is, Avogadro's number of carbon-12 atoms expressed as 12 g. All naturally occurring elements consist of mixtures of isotopes, and hence, the atomic weight of an element is actually the average of the isotope weights of the element, taking into account the natural abundance of the isotopes. For example, the atomic weight of carbon is given as 12.01115, taking into account the relative abundances of the naturally occurring isotopes of masses 12 and 13 in the ratio 100:1.08 and that of chlorine is given as 35.453 which occurs in nature as a mixture of isotopes of masses 35 and 37 at relative abundances of 100 and 32.5. The term *dalton* (Da) is used by biochemists to refer to the masses of macromolecules and substances such as proteins, mitochondria, ribosomes, chromosomes, viruses and bacteria. The mass of a single carbon-12 atom is taken as 12 Da, and hence, 1 Da = the reciprocal of Avogadro's number = 1.661×10^{-24} g. The molar mass of molecular substances is simply the *molecular weight* (the sum of the atomic weights of atoms in the molecular formula of a compound or gram molecular weight, that is, weight of one mole of the substance in grams). A better term for ionic substances, which do not exist as molecules, is *formula weight*.

The stoichiometry of a chemical reaction is usually expressed in the form of a balanced equation on the basis of the relative masses of reacting atoms, ions, or molecules. For example, the reaction between silver nitrate and sodium chloride in aqueous medium to precipitate silver chloride is represented as

$$Ag^+ NO_3^- + Na^+ Cl^- \rightarrow AgCl + Na^+ NO_3^-$$

From the balanced equation, one can understand that the actual reaction occurs between one silver ion and one chloride ion or, using the *mole* concept, between one mole of silver ions consisting of Avogadro's number of ions and one mole of chloride ions. Numerically, the mole is the atomic, molecular, or formula weight of a substance expressed in grams.

Similarly, the reaction between sulphuric acid and sodium hydroxide represented by the balanced equation indicates that two moles of sodium hydroxide are required to neutralize one mole of sulphuric acid.

$$H_2^{2+} SO_4^{2-} + 2 Na^+ OH^- \rightarrow 2 H_2O + Na_2^{2+} SO_4^{2-}$$

For determining the weights of reactants and products in a chemical reaction, stoichiometric calculations involve three steps: (i) conversion of the weight in grams or milligrams to the corresponding number of moles of substance, (ii) multiplication by a factor as indicated by the balanced stoichiometric equation, and (iii) re-conversion of the moles into the metric unit of grams or milligrams. When dealing with small quantities of substances, the unit *millimole* (mmol—one thousandth of a mole) is more convenient.

For example, to determine (a) how many grams of sodium chloride is required to convert 2.5 g of silver nitrate into silver chloride or (b) to calculate how many grams of silver chloride will be obtained from 2.5 g of silver nitrate, the three-step calculation procedure is adopted. According to the balanced stoichiometric equation, one mole (169.9 g) of silver nitrate reacts with one mole of sodium chloride (58.35 g) to form one mole (143.32 g) of silver chloride. In the first step, the number of moles in 2.5 g of silver nitrate is calculated as

$$\text{Number of moles} = \text{weight of the substance (g)/formula weight (g mol}^{-1})$$
$$= 2.5 \text{ g}/169.9 \text{ g mol}^{-1} = 0.0147 \text{ mol} = 14.71 \text{ mmol}$$

In the second step, the number of moles of sodium chloride required to react with 14.71 mmol of silver nitrate as per the stoichiometric equation is 14.71 mmol (multiplication factor = 1).

In the final step, the number of mmol of sodium chloride is reconverted to grams.

$$\text{Weight of substance} = \text{number of moles or mmol} \times \text{formula weight (g mol}^{-1})$$
$$= 14.71 \times 58.35 = 0.858 \text{ g}$$
$$\text{Weight of silver chloride formed} = 14.71 \text{ mmol} \times 143.32 \text{ g mol}^{-1} = 2.108 \text{ g}$$

Complexation reactions mostly involve use of ethylenediamine tetraacetic acid (EDTA) as the chelating agent to form complexes with metals. EDTA being a hexadentate ligand forms only 1:1 complex of the type ML with metal ions. Hence, it is convenient to state that one mole of EDTA reacts with one mole of the metal which can be represented by the stoichiometric equation as

$$M^{n+} + H_2Y^{2-} \rightarrow MY^{(n-2)+} + 2 H^+$$

where H_2Y^{2-} represents the dianion of EDTA.

A typical redox reaction is the oxidation of ferrous sulphate by potassium permanganate. The reaction involves the oxidation of Fe^{2+} to Fe^{3+} through the loss of one electron. The permanganate ion (MnO_4^-) containing Mn^{7+} undergoes reduction to Mn^{2+} involving the gain of five electrons. Hence, the balanced stoichiometric equation is written as

$$5\ Fe^{2+} + MnO_4^- + 8\ H^+ \rightarrow 5\ Fe^{3+} + Mn^{2+} + 4\ H_2O$$

Thus, it is obvious that five moles of ferrous ion react with one mole of permanganate ion.

Stoichiometric calculations may also be carried out using the unit of *equivalent weight* to express the quantitative relationship between reacting chemical species. The term *equivalent* represents the mass of the substance providing Avogadro's number of *reacting units*. The reacting unit may be an atom, ion, molecule, or an electron depending on the nature of the reaction. An example of the reacting unit in neutralization reactions is the hydrogen ion or *proton*, because according to Brønsted theory neutralization reactions involve the transfer of proton(s) between an acid and a base. Thus, monoprotic acids such as HCl, HNO_3, and CH_3COOH have one replaceable proton or one reacting unit, while diprotic acids (H_2SO_4 and COOH.COOH) have two replaceable protons or reacting units. The equivalent weight is obtained by dividing the molecular weight (or formula weight) by the number of reacting units. The equivalent weight of monoprotic acids is the same as the molecular or formula weight, whereas in the case of diprotic acids the equivalent weight is half the molecular weight.

In the case of oxidation–reduction reactions, the reacting unit is the *electron*, because these reactions involve the transfer of electron(s) between the oxidant and the reductant. Thus, for example, the redox reaction involving ferrous ion and permanganate involves the oxidation of ferrous ion to ferric ion with a change of one electron or reacting unit per ion, and hence, the equivalent weight of iron is the same as its atomic weight. The reduction of Mn^{7+} to Mn^{2+}, however, involves a change of five electrons per manganese, and hence, the equivalent weight of permanganate is one-fifth of the molecular weight or formula weight.

In precipitation reactions involving silver nitrate for the analysis of halides, the mole and equivalent weight are the same as represented in stoichiometric equations, and hence, the equivalent weight is the same as molecular weight (or formula weight). A similar relationship holds good for complexation reactions involving EDTA as the chelating ligand.

1.10 EXPRESSION OF CONCENTRATIONS OF SOLUTIONS

Chemical reactions in the analytical laboratory are mostly carried out using dilute aqueous solutions of reactants. Hence, it is necessary to use uniformly acceptable or standard units for expressing the amount of reacting substance in solution. The term *concentration* refers to the amount of the reactant present in the solution. Concentrations of solutions are commonly expressed in terms of *molarity* (M) as moles per litre (mol L^{-1}). One molar solution of a substance, say sodium chloride, represented as 1 M contains one gram molecular weight or formula weight of sodium chloride which is 58.35 g in one litre of the solvent. Similarly, a 0.1 M solution of $KMnO_4$ contains 15.8 g (one-tenth of the formula weight) of the substance dissolved in one litre of water.

Concentrations of the solutions are also expressed in terms of *normality* (N) which refers to the amount as equivalents per litre (eq L^{-1}). Thus, 1 N solution of sodium hydroxide contains one equivalent weight (same as the molecular weight), that is, 40 g of the substance dissolved in one litre of solvent. Similarly, 1 N solution of oxalic acid (diprotic acid) contains one equivalent weight of the substance which is one-half of the molecular weight since there are two equivalents (reacting units) per mole of the substance and in terms of molarity the concentration is 0.5 M.

Other units of concentrations used include molality and weight or volume percentage. *Molality* refers to the concentration of a substance in terms of one mole of the substance dissolved in 1000 g of the solvent. Concentrations of solutions, particularly of reagents, are expressed in terms of percentage as weight/volume (% w/v), that is, weight of the substance in grams dissolved in 100 mL of the solvent. Alternatively, volume/volume (% v/v) indicates the volume of the liquid substance in mL dissolved in 100 mL of the solvent.

1.11 REPORTING OF RESULTS

Analytical results are reported in a variety of ways. In the case of solid samples, the amount of analyte in a given test sample is reported in weight-by-weight per cent (% w/w). In the case of solutions containing dissolved solids, the solid content is reported in terms of weight-by-volume per cent (% w/v). In addition, the amount of dissolved analyte may be reported in terms of concentration as mol L^{-1}. Trace amounts of analytes are reported as parts per million (ppm) as mg kg^{-1} or mg L^{-1} and parts per billion (ppb) as μg kg^{-1} or μg L^{-1} depending on the nature of the test sample.

SUMMARY

- Analytical chemistry deals with qualitative analysis to identify the nature of chemical species and quantitative analysis to determine the amounts of chemical species. Analytical chemistry finds extensive use in a variety of fields.
- Analytical process involves six steps which include (i) identification and definition of a specific problem, (ii) selection of the method of analysis, (iii) sampling, (iv) separation from undesirable interference, (v) quantitative estimation, and (vi) assessment of the data generated and validation of method.
- Thermodynamic and kinetic considerations are important in choosing a reaction for analytical purposes because for quantitative analysis the chosen reaction should go to completion to almost 100% within a short time.
- Equilibrium methods of analysis allow sufficient time for reaction to attain equilibrium and determine the concentrations of the species.

- Equilibrium methods include a variety of instrumental methods which determine the concentration of the analyte based on some signals generated.
- Quantitative analyses in aqueous media involve four different types of equilibria, namely (i) acid--base, (ii) precipitation, (iii) complex formation, and (iv) redox.
- Kinetic methods of analysis are based on determining the rate of a reaction to determine the concentration of the analyte within the first 30% completion of reaction.
- Kinetic methods involve mostly catalysed reactions, particularly those involving enzymes as catalysts.
- Enzyme-catalysed reactions find use in clinical diagnostic applications to determine the metabolites in body fluids and enzymes.

REVIEW QUESTIONS

1. Write a brief note on the scope of analytical chemistry in industry.
2. Outline the steps involved in the analytical process.
3. What are the factors that need to be considered in the selection of a chemical reaction for analytical purposes?
4. How are equilibrium and kinetic methods of analysis distinguished?
5. What are the characteristic features of analytical reactions?
6. Classify the different analytical methods.
7. What is the significance of law of mass action?

8. Give a detailed note on the acid–base equilibria encountered in analytical chemistry.

9. Explain the terms: (a) ionic product of water, (b) dissociation constant of an acid, (c) stability constant, and (d) solubility product constant.

10. Explain the concept of pH and the action of buffers.

11. Discuss the significance of E_{cell} in redox reactions.

12. Explain the terms order and rate expression. How is the rate of a reaction determined experimentally?

13. What are enzymes? How are they useful in analysis?

14. Explain the kinetics and mechanism of enzyme-catalysed reactions.

15. Give an account on the use of enzymes in the analysis of metabolites in body fluids.

16. How are enzymes useful in clinical diagnosis? Give examples.

17. Explain the term stoichiometry and concepts by which it is expressed.

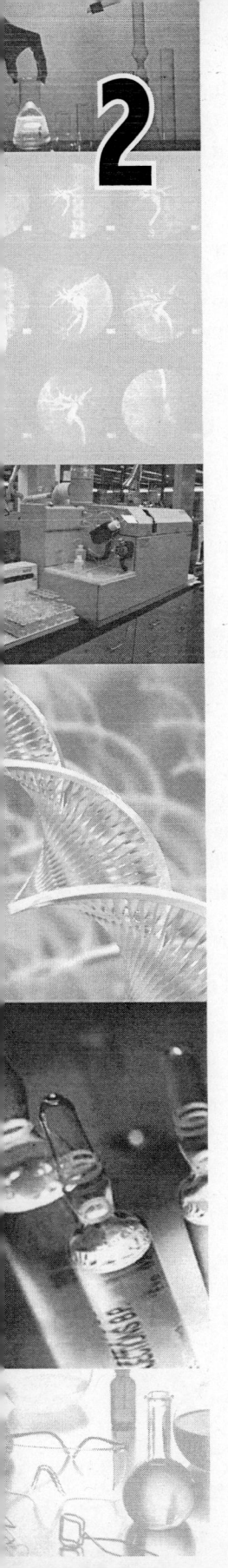

2 Assessment of Analytical Data

2.1 INTRODUCTION

The aim of quantitative analysis is to obtain a result as close to the true value as possible through the use of an appropriate analytical method in a correct manner. Quantitative analysis requires a sound knowledge on the part of the analyst about the selection of an appropriate analytical method, reliability, and accuracy of the chosen method, possible interferences, possible sources of errors in the analysis, reliability, and limitations of the equipment used, reproducibility of data obtained, as well as statistical distribution of values obtained. Once an appropriate analytical method has been selected and analysis has been carried out, it is essential that statistical analysis of the experimental data be carried out in order to ensure that the analytical report is reliable and acceptable. This chapter deals with the different types of errors commonly encountered and the assessment of analytical data by statistical methods.

2.2 DEFINITIONS OF TERMS

In the assessment of analytical data a wide variety of terms, such as true value, precision, accuracy, error, standard deviation, variance, confidence level, and so on are used, which need to be defined and explained before the assessment methods are discussed in detail.

2.2.1 True Value

True value is the correct value of a measurement and is known only for a standard sample. For other samples it may be arrived at with varying degrees of precision depending on the experimental method used.

2.2.2 Precision

Precision indicates the variability of measurements and the repeatability (concordance) as well as reproducibility of results. Precision may also be described as the degree of agreement between numerical values of two or more replicate measurements that have been made in exactly the same way. It may be expressed as an absolute or relative quantity. The most valuable precision indicator is the standard deviation. Precision may also be described by two other terms, namely variance and coefficient of variation.

2.2.3 Accuracy

The reliability of data is usually expressed in terms of accuracy and precision. Accuracy is defined as the nearness or closeness of a measurement or result to the true value. It describes the correctness of the experimental result. The only type of measurement that can give an accurate result is the counting of objects. All other measurements involve errors and give only an approximation of the true value. Accuracy may be expressed in terms of *absolute error* or in a more convenient way, as *relative error* because accuracy depends on factors, such as the difficulty of the analytical problem or the requirements of an analytical scientist in a particular situation.

2.2.4 Error

Error is the difference between the true value and the measured value. The actual difference between the true value (x_t) and the measured value (x) in the same unit is known as the *absolute error* (E_a). The measured value may be a single value or the average (mean) value (x_{av}) of a set of replicate measurements. Thus,

$$E_a = x_{av} - x_t \qquad (2.1)$$

Alternatively, *relative error* (E_{rel}) is calculated and expressed as percentage or parts per thousand (ppt).

$$E_{rel} = \frac{x_{av} - x_t}{x_t} \times 100 \qquad (2.2a)$$

$$E_{rel} = \frac{x_{av} - x_t}{x_t} \times 1000 \qquad (2.2b)$$

A positive or a negative sign is usually prefixed in the error indicating that the measured result is greater or less respectively than the true value.

2.2.5 Mean and Median

In statistics the *population mean* or *limiting mean* (μ) is the arithmetic average of a set of results when the population or the number of measurements N approaches infinity. The population mean is considered as the true value of the quantity being measured in the absence of any bias. In contrast, the *sample mean* (x_{av}) is the average of a finite set of data and often differs from the true value μ. The middle value of a replicate set of results is called the *median*.

2.2.6 Spread

The numerical difference between the highest and the lowest results is called the spread. It is a measure of precision.

2.2.7 Deviation

The numerical difference between an individual result and the mean or median of the set with respect of sign is called the deviation. It is expressed as a relative or absolute value.

2.2.8 Population Standard Deviation

The *population standard deviation* provides a measure of the precision of a population of data. It is the root-mean-square of the individual deviations from the mean for the population and is represented by the symbol σ. It is derived from the normal error curve (Fig. 2.1) by the equation

$$\sigma = \left[\frac{\sum\limits_{i=1}^{i=N} (x_i - \mu)^2}{N} \right]^{1/2} \qquad (2.3)$$

where x_i is the measured result, μ is the true value, and N is the number of results in the set approaching infinity.

In most cases since μ is not known and since the set of data (N) is a finite number, the term *sample standard deviation* (s) is used instead of population standard deviation. The sample standard deviation (symbol s) is given by the equation

$$s = \left[\frac{\sum\limits_{i=1}^{i=N} (x_i - x_{av})^2}{N-1} \right]^{1/2} \qquad (2.4)$$

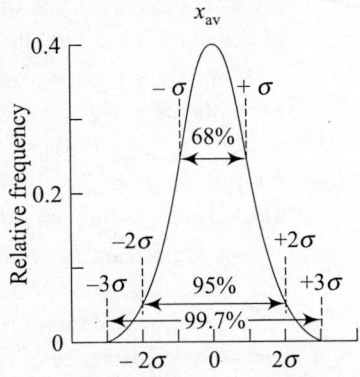

Fig. 2.1 Normal error curve

The population mean μ in Eq. (2.3) is replaced by x_{av}, which is the mean derived from the set of results. The number of degrees of freedom in a replicate set of results is equal to the number of the results in the set. It is an independent variable. The number of degrees of freedom decreases by one when another quantity, such as the mean, is derived from the set and also decreases by one for each derived value from the set of results. The number of degrees of freedom N [in Eq. (2.3) for population standard deviation] consequently decreases by one for sample standard deviation, and hence, N is replaced by $(N-1)$ in Eq. (2.4).

When the number of measurements N is large and approaches infinity, the arithmetic mean of the measurements will show less scatter from the true value μ and the scatter approaches zero. The precision improves as the square root of the number of measurements, the relationship being referred to as the *standard deviation of the mean* (s_{mean}) or as the *standard error*.

$$s_{mean} = s/\sqrt{N} \qquad (2.5)$$

The precision of a set of data is conveniently described in terms of standard deviation as it has the same units as the data itself. However, in statistical analysis precision is usually described in terms of variance (square of standard deviation).

2.2.9 Relative Standard Deviation and Coefficient of Variation

Relative standard deviation (RSD) is often used for comparison of precisions and is given as

$$RSD = s/x_{av} \qquad (2.6a)$$

RSD is expressed as a percentage called the *coefficient of variation* (CV).

$$CV = (s \times 100)/x_{av} \qquad (2.6b)$$

2.2.10 Variance

The square of the standard deviation is called variance (s^2 or σ^2). The values of variance are additive, i.e., $s^2_{a+b+c+\cdots} = s^2_a + s^2_b + s^2_c + \cdots$

2.2.11 Significant Figures

A significant figure is defined as the digit (any one of the ten numerals 0 to 9) which denotes the least number of digits that can be used to denote a given number. The digit of a number required for expressing the precision of the data or measurement is called the significant number. The number 0 is a significant digit only when it is not the first figure in a number. Thus, it is significant in the quantity 2.620 g or 1.082 g but not in 0.0032 kg, because the zeros in the last quantity indicate the position of the decimal and can be conveniently omitted by proper choice of the units. Thus, the quantity 0.0032 kg can be written as 3.2 g with two significant figures.

Significant figures must be such that the data will indicate only one uncertain figure. For example, the weight 1.8400 g indicates that the weight is accurate to ± 0.1 mg and should not be written as 1.84 g (indicating an accuracy of only ± 10 mg) or 1.840 g (indicating an accuracy of ± 1 mg). Similarly, a volume of 15.3 mL indicates that it is between 15.2 mL and 15.4 mL (accuracy of ± 0.1 mL) and should not be written as 15.30 mL as it indicates the accuracy to be ± 0.01 mL.

The significant figure, particularly in calculations, may be a rounded off depending on the last figure which has been retained or rejected. While rounding off, 1 is added to the last figure retained if the following figure to be rejected is 5 or more (e.g., 1.2787 is rounded off to 1.279). The sum or difference of two or more quantities cannot be more precise than the quantity having the largest uncertainty, and hence, during addition or subtraction each quantity should have only as many significant figures as available in the least accurately known quantity. For example, in the addition of numbers $212.3 + 14.567 + 3.4227$ the least accurately known quantity is the first one, and hence, the latter two quantities should be rounded off as 14.6 and 3.4 before the addition is carried out. Similarly, the percentage precision of a product of multiplication or a quotient (of division) cannot be greater than the percentage precision of the least precise factor used in the calculation. Hence, in multiplication or division one more significant figure is retained in each quantity than contained in the factor having the largest uncertainty. Thus, in multiplying the factors $1.23 \times 4.567 \times 0.2657$, the first factor has a large uncertainty, and hence, the two following factors should be taken as 4.567 and 0.266, retaining one more significant figure than the first one. However, when a calculator is used all the original factors are used and then the product is rounded off to three significant figures as 1.49.

2.3 TYPES OF ERRORS

In analytical experiments commonly encountered errors are classified on the basis of their origin into three major types, namely: (i) gross errors (E_g), (ii) determinate or systematic errors (E_s), and (iii) indeterminate or random (E_r) errors. The absolute error in a set of measurements is the sum of these three types of errors.

2.3.1 Gross Errors

Gross errors occur mainly due to the carelessness, or ineptitude on the part of the analyst carrying out the experiment. Examples of gross errors include spilling of the sample, accidental introduction of contaminants, reversing the sign of meter reading, or choosing a wrong scale on the meter and transposition of numbers while recording the data.

2.3.2 Systematic Errors or Determinate Errors

Systematic errors occur when the mean of a set of measurements differs from the true or the accepted value. These errors are measurable and have a definite value and sign (positive or negative). The errors may be *constant* errors when they are of the same magnitude for a set of measurements made in the same manner. In contrast, *proportional* errors increase with the magnitude of the measurement. Systematic errors may cause a bias in the measurement technique. The bias has either a positive or a negative sign and affects all the data in the set. When the mean value of a set of measurements is similar to the true value of the individual measurements spread over on either side of the true value, it gives rise to a distribution pattern of a normal curve as shown in Fig. 2.1; there is no bias in the set of data. On the other hand, a positive bias can be identified if all or a majority of the individual measurements are greater than the true value resulting in the mean value being greater than the true value.

Systematic errors have an assignable cause and it is possible to make corrections to minimize systematic errors by identifying the source of error. Systematic errors are caused by three different sources and these are: (i) method errors, (ii) instrument errors, and (iii) personal errors.

Method errors are inherent in the methods of analysis and are, hence, not easy to detect. These errors are the most serious in nature. Method errors arise due to non-ideal conditions employed such as the non-specificity of reagents, instability of species, and incompleteness or slowness of analytical reactions. For example, in volumetric titrations a small excess of titrant is necessary to cause a colour change in the indicator, thereby introducing a small error in the measured volume. Similarly, in gravimetry the precipitating reagent is usually non-specific for the analyte and an excess of precipitating reagent is necessary to complete the reaction, which causes the contamination of the analyte precipitate.

Instrumental errors include errors caused by all measuring devices such as volumetric apparatus, balances, and electronic instruments. For example, a pipette or burette does not hold exactly the same volume of liquid or deliver exactly the volume indicated on the apparatus due to changes in temperature, density of liquids, etc. Instrumental errors can be avoided or at least minimized by calibrating the apparatus or the instrument at regular intervals.

Personal errors occur due to wrong judgements as in the case of reading the graduation marks in a burette or pipette, visual detection of colour changes of indicators in titrations, or prejudices and preconceived ideas for attaining analytical values closer to the true or accepted value. Personal errors may also occur while recording the values in observation notes. Such errors are avoided by checking and rechecking the data as well notebook entries.

Minimizing systematic errors

Systematic errors can be minimized by taking appropriate measures, such as

(i) Calibration of apparatus and instruments.

(ii) Performing duplicate or parallel determinations to check for the precision.

(iii) Performing a blank determination (i.e., carrying out the analysis under exactly the same experimental conditions but without the sample).

(iv) Performing a control determination (i.e., carrying out the analysis using a standard substance under exactly the same experimental conditions).

(v) Selecting independent methods of analysis—for example, estimation of chloride in a sample can be carried out by volumetry and independently by gravimetry or potentiometry.

(vi) Adopting *standard addition method* in which a known amount of the substance being determined is added to the sample and then analysing the total amount of the substance. The difference between the results for samples with or without the added standard gives the recovery of the amount of the added standard and if the recovery is satisfactory the accuracy of the method is relatively high. The standard addition method finds extensive use in spectrophotometric and polarographic methods of analysis.

(vii) Adopting *internal standard method* in which a fixed amount of a reference substance (internal standard) is added to a series of known concentrations of the substance being determined. The ratios of the measured physical value (for example, the intensity of absorption or emission peak in spectroscopy or peak height or peak area in a chromatogram) of the internal standard and the series of known concentrations should linearly vary with the concentration values. The concentration of a test sample is obtained by adding the same amount of internal standard to the test sample and determining the position where the ratio obtained falls on the concentration scale.

2.3.3 Random Errors or Indeterminate Errors

A scatter of experimental results from a set of measurements carried out on the sample is known as random or accidental error and indicates the lack of precision of the data. The random errors cannot be predicted or estimated and are, hence, known as indeterminate errors. Such errors arise from unpredictable inaccuracies in the individual steps in a procedure. Typical examples of data with random errors include a set of replicate data for calibration of a burette, pipette or a weight, pH measurements or absorbance measurements on a sample even though the data has been collected by the same analyst under identical conditions. Random errors in experimental data cannot be avoided and occur with a random distribution. They are best treated by statistical techniques based on mathematical laws of probability so that some conclusion on the most probable result of a series of measurements can be arrived at.

2.4 STATISTICAL TREATMENT OF RANDOM ERRORS

The basis of statistical techniques is that a finite or small number of replicate experimental data is considered as a minute fraction of an infinite number of data that could be obtained with an infinite amount of sample in infinite time. The finite number of replicate experimental data is designated as *sample* and considered as a subset of the infinite set of data designated as *population* or *universe*. In the statistical treatment of data for analysing the random errors it is assumed that the sample is truly representative of the population. The standard deviation indicates the spread of data but the relative distribution of the results can be obtained only by frequency analysis.

2.4.1 Distribution of Random Errors

If the number of experimental data is large (more than 25) the deviations of the data from the mean will give rise to a smooth symmetrical Gaussian curve called the *normal distribution curve* or *normal error curve* as shown in Fig. 2.1. The normal error curve is described by the general equation

$$y = \frac{\exp\left[-(x-\mu)^2/2\sigma^2\right]}{\sigma(2\pi)^{1/2}} \tag{2.7}$$

where μ is the mean and σ is the standard deviation.

In general the experimental results of any replicate analysis may be fitted to a Gaussian curve. The characteristics of such a normal distribution curve include: (i) the mean of a set of data is the most frequently observed result, (ii) the experimental data form a cluster symmetrically distributed around the mean value, and (iii) a large number of experimental data are generally found to diverge from the central mean value to a small extent while the number of data with large divergences from the mean value is relatively small. Thus, the width of the curve is determined by σ, indicating the spread or precision of a set of experimental data and is unique for that set of data. This type of distribution will contain 68.3% of all the values within an interval of one standard deviation, i.e., $\mu \pm \sigma$, and 95.5% and 99.7% of all the values within $\mu \pm 2\sigma$ and $\mu \pm 3\sigma$ respectively, and (iv) in the absence of any systematic error the mean value of a large set of data approaches the true value. Hence, it is possible to minimize random error by performing a large number of replicate analyses.

2.5 EVALUATION OF EXPERIMENTAL RESULTS

A truly representative 'statistical sample' of results requires a large number of determinations (usually more than 25) which is not economical. Hence, it is necessary to use statistical methods to evaluate the reliability of the determinations made. Even before attempting a statistical approach it is necessary to check the set of data for occurrence of any possible gross error either in the experimental result or in arithmetic work. Such a gross error is revealed when a result deviates widely from the mean. When it is certain that gross errors are absent a two-step statistical approach is adopted to evaluate the experimental data collected by a smaller set of replicate measurements. The first step is to examine the reliability of the measurements themselves followed by the second step of assessment of the meaning or significance of the results.

2.5.1 Reliability of Measurements

Reliability of the set of measurements or experimental results is checked by a process of data rejection based on setting a *confidence level* of 90% or 95%. A higher confidence level will include all the experimental data with a risk of including the data affected by gross error. In contrast, a lower confidence level may exclude experimental data which are rightly the part of the statistical sample. A convenient criterion used for data rejection step is based on $x \pm 2\sigma$ interval which contains 95.5% of all the experimental data. If this criterion is considered to have a wide limit an alternate criterion based on *Q-test* at a 90% confidence level may be used for data rejection. The parameter Q is called the *rejection quotient* and defined as

$$Q = a/w \tag{2.8}$$

where a is the difference between the questionable data and the nearest value and w is the difference between the largest and the smallest value in a set of x_i. The questionable data is rejected if the calculated value of Q exceeds the critical value of Q listed in Table 2.1 of critical values of Q for different confidence levels.

Table 2.1 Critical values of rejection quotient Q at different confidence levels

Number of results	Q_{90}	Q_{95}	Q_{99}
3	0.941	0.970	0.994
4	0.765	0.829	0.926
5	0.642	0.710	0.821
6	0.560	0.625	0.740
7	0.507	0.568	0.680
8	0.468	0.526	0.634
9	0.437	0.493	0.598
10	0.414	0.466	0.568
20	0.300	0.342	0.425
30	0.260	0.298	0.372

For example, in a set of five results of iron estimation of 8.0, 8.2, 8.4, 8.5, and 9.6 ppm the questionable or suspect data is 9.6 ppm. The Q test is useful to determine whether the suspect value is due to some accidental error and can be rejected.

Thus, for $Q_{95} = (9.6 - 8.5)/(9.6 - 8.2) = 1.1/1.6 = 0.688$.

The calculated Q_{95} value of 0.688 is less than the critical value of 0.710 for a set of five results, and hence, should not be rejected for a confidence level of 95%. However, the calculated value exceeds the critical value of 0.642 for Q_{90}, and hence, should be rejected if the confidence level is taken as 90%.

2.5.2 Analysis of Data

After establishing the reliability of the replicate set of experimental results the second step of computing the experimental mean value is taken up. The experimental mean will be a true mean approaching the true or actual value only when an infinite number of experimental results have been accumulated. The true mean, on the other hand, remains unknown with a finite number of experimental results and the value of the standard deviation alone cannot indicate the extent of closeness between the sample mean and the true value. Under such circumstances statistical analysis of results is attempted using the *t-factor*. The *t*-factor involves the calculation of a *confidence interval* or *confidence limit* about the experimental mean within which there is a 90% or greater confidence (or probability) of finding the true mean. The confidence limit (CL) for μ or x_{av} for N replicate measurements is given as

$$CL = x_{av} \pm \frac{ts}{N^{1/2}} \tag{2.9}$$

where x_{av} is the experimental mean, t is a statistical factor derived from the normal error curve depending on the number of degrees of freedom and the confidence level (Table 2.2), s is the estimated standard deviation, and N is the number of results.

Table 2.2   Values of *t*-factor for various levels of confidence

Degrees of freedom	Confidence level (%)		
	90	95	99
1	6.31	12.7	63.7
2	2.92	4.30	9.92
3	2.35	3.18	5.84
4	2.13	2.78	4.60
5	2.02	2.57	4.03
6	1.94	2.45	3.71
7	1.90	2.36	3.50
8	1.86	2.31	3.36
9	1.83	2.26	3.25
10	1.81	2.23	3.17
11	1.80	2.20	3.11
12	1.78	2.18	3.06
13	1.77	2.16	3.01
14	1.76	2.14	2.98
15	1.75	2.13	2.95
16	1.75	2.12	2.92
17	1.74	2.11	2.90
18	1.73	2.10	2.88
19	1.73	2.09	2.86
20	1.73	2.09	2.85
Infinity	1.64	1.96	2.58

For example, if the experimental mean x_{av} of a set of eight results is 1.536 with a standard deviation of 0.008, the t value for 90% confidence limit with $(8 - 1 = 7)$ degrees of freedom is 1.90. Hence, the 90% confidence limit for the true value is given as

$$90\% \text{ CL for the true value} = 1.536 \pm (1.90 \times 0.008)/\sqrt{8} = 1.536 \pm 0.005$$

Thus, with 90% confidence the true value lies in the range of 1.531 to 1.541.

For the same data the t value for 95% confidence limit is 2.36. The true value at 95% confidence level is given as

$$95\% \text{ CL for the true value} = 1.536 \pm (2.36 \times 0.008)/\sqrt{8} = 1.536 \pm 0.007$$

With 95% confidence the true value lies in the range of 1.529 to 1.543.

In general the values of t and $s/N^{1/2}$ decrease with increasing the number of replicate results and the confidence interval is smaller.

2.6 COMPARISON OF RESULTS

A comparison of a replicate set of experimental measurements with a true value or another independent replicate set of data provides a means to determine the relative accuracy and precision of analytical procedures adopted. Two commonly used methods for such comparison include the F-test or variance ratio method and the Student's t-test.

2.6.1 *F-test*

F-test involves comparison of the precisions of two sets of data and is adopted when two separate replicate sets of data or results from different analytical methods or different laboratories need to be compared. The test involves determining the ratio of the variances of the two sets to determine any significant difference in precision. F is calculated using the formula

$$F = s_a^2/s_b^2 \tag{2.10}$$

taking the larger variance as numerator by convention, and hence, the value of F is always greater than unity. The calculated F value is compared with critical values of F in the F-table calculated from an F-distribution taking into account the numbers of degrees of freedom to determine the statistical significance. The critical values are calculated on probability basis that they will be exceeded only in 5% or 10% of the cases. The critical values for F at 5% and 10% probability are given in Table 2.3.

For example, comparison of the precision of two sets of experiments by the F-test is carried out as follows: The standard deviation of the first set of results of 7 determinations is 0.215 and for the second set of results of 11 determinations is 0.258.

The calculated value of $F = (0.258)^2/(0.225)^2 = 1.44$.

The calculated value of F of 1.44 is less than the tabulated value of F of 4.06 at 95% confidence level (for 10 degrees of freedom for the numerator and 6 degrees of freedom for the denominator) it can be concluded that there is no significant difference in the precisions of the two sets of results. The standard deviations are due to random error only and do not depend on the sample.

Table 2.3 Values of F at 95% and 90% confidence levels

Degrees of freedom (denominator)	Probability (%) 5* 10 **	3	4	5	6	7	8	9	10	15	Infinity
					Degrees of freedom (numerator)						
3	5	9.28	9.12	9.01	8.94	8.89	8.85	8.81	8.79	8.70	8.53
	10	5.39	5.34	5.31	5.28	5.27	5.25	5.24	5.23	5.20	5.13
4	5	6.59	6.39	6.26	6.16	6.09	6.04	6.00	5.96	5.86	5.62
	10	4.19	4.11	4.05	4.01	3.98	3.95	3.84	3.92	3.87	3.76
5	5	5.41	5.19	5.05	4.95	4.88	4.82	4.77	4.74	4.62	4.36
	10	3.62	3.52	3.45	3.40	3.37	3.34	3.32	3.30	3.24	3.10
6	5	4.76	4.53	4.39	4.28	4.21	4.15	4.10	4.06	3.94	3.67
	10	3.29	3.18	3.11	3.05	3.01	2.98	2.96	2.94	2.87	2.72
7	5	4.35	4.12	3.97	3.87	3.79	3.73	3.68	3.64	3.51	3.23
	10	3.07	2.96	2.88	2.83	2.78	2.75	2.72	2.70	2.63	2.47
8	5	4.07	3.84	3.69	3.58	3.50	3.44	3.39	3.35	3.22	2.93
	10	2.92	2.81	2.73	2.67	2.62	2.59	2.56	2.54	2.46	2.29
9	5	3.86	3.63	3.48	3.37	3.29	3.23	3.18	3.14	3.01	2.71
	10	2.81	2.69	2.61	2.55	2.51	2.47	2.44	2.42	2.34	2.16
10	5	3.71	3.48	3.33	3.22	3.14	3.07	3.02	2.98	2.85	2.54
	10	2.73	2.61	2.52	2.46	2.41	2.38	2.35	2.32	2.24	2.06
15	5	3.29	3.06	2.90	2.79	2.71	2.64	2.59	2.54	2.40	2.07
	10	2.49	2.36	2.27	2.21	2.16	2.12	2.09	2.06	1.97	1.76
Infinity	5	2.60	2.37	2.21	2.10	2.01	1.94	1.88	1.83	1.57	1.00
	10	2.08	1.94	1.85	1.77	1.72	1.67	1.63	1.60	1.49	1.00

Confidence levels: * 95% and ** 90%

2.6.2 Student's *t*-test

Student's *t*-test involves comparison of the mean value of a set of results with the true value or standard value if it is known. If the true value is not known, the mean value of a set of results may be compared with the mean value of a second set. The test is useful for comparing the results obtained by different methods of analysis, one of which is a standard or accepted method. The *t*-value of a set of data is calculated and compared with the tabulated value of a given number of tests at a desired confidence level. If the calculated *t*-value is less than the tabulated value, it is concluded that there is no significant difference between the two sets of data or between the standard method and a test method of analysis. On the other hand, if the calculated *t*-value exceeds the tabulated value it is concluded that there is a significant difference between the two sets of data for a given confidence level.

If the true value is known then *t*-value of a set of results obtained by a method of analysis can be calculated using the equation

$$t = \frac{(x_{av} - \mu)\sqrt{N}}{s} \tag{2.11}$$

For example, if the true value or an acceptable standard value of an analyte in a sample is 12.8 ppm and the mean of 6 determinations is 12.6 ppm with a standard deviation of ± 0.2 ppm the value of *t* can be calculated as

$$t = \frac{(12.6 - 12.8)\sqrt{6}}{0.2} = 2.45$$

Since the calculated value of 2.45 is less than the tabulated value of 2.57 at 95% confidence level for five degrees of freedom, it is concluded that the experimental data is acceptable with 95% confidence.

When the true value is not known the t-test can be used to compare the means of two sets of results. The method is also useful to compare the mean and the precision of a new analytical method developed with that of the accepted standard method or reference method. The value of t is calculated by replacing μ in the above expression with the mean of the second set and introducing the *pooled standard deviation* s_p to give

$$t = \frac{(x_1 - x_2)}{s_p \sqrt{1/N_1 + 1/N_2}} \tag{2.12}$$

The pooled standard deviation s_p is calculated from the two sample standard deviations s_1 and s_2 using the expression

$$s_p = \sqrt{\left[(N_1 - 1)s_1^2 + (N_2 - 1)s_2^2\right]\big/(N_1 + N_2 - 2)} \tag{2.13}$$

The method is applicable only when there is no significant difference between the precisions of the two methods. Hence, it is necessary to assess the results first by the F-test before using the t-test to compare the methods.

For example, if the mean, standard deviation, and number of sample results for the method A, used for the determination of chromium content of an alloy are 5.28%, ±0.08%, and 7 respectively and 5.42%, 0.075%, and 6 respectively for method B, the precisions of the two methods are evaluated by F-test as per Eq. (2.10).

$$F = (0.08)^2/(0.075)^2 = 1.14$$

The calculated value of 1.14 is less than the tabulated F-value of (Table 2.3) 4.95 for 6 degrees of freedom for s_a, and 5 degrees of freedom for s_b, hence, it is concluded that the precisions of the two methods are comparable and the t-test can be used. The pooled standard deviation s_p for the two methods is now calculated using Eq. (2.13).

$$s_p = \sqrt{\left[(7-1)(0.08)^2 + (6-1)(0.075)^2/(7+6-2)\right]} = \pm 0.077$$

For comparison of the results of the two methods the t-test is carried out for the given data using Eq. (2.12).

$$t = (5.28 - 5.42)\big/\left(0.077\left(\sqrt{1/7 + 1/6}\right)\right) = 0.14/(0.077 \times 0.556) = 3.27$$

The calculated value of t is 3.27 is greater than the tabulated value of t of 2.20 for 11 degrees of freedom ($N_1 + N_2 - 2 = 11$) at 95% confidence, it can be concluded that there is significant difference between the mean results of the two methods of analysis.

2.6.3 Paired t-test

Paired t-test is used particularly for evaluating the results of a new method of analysis in comparison with an accepted or a standard method. For example, the determination of a metabolite or biochemical may be performed over a wide range of compositions using a new procedure and the data obtained needs to be compared with the results obtained by a standard method or accepted method for determining the acceptability of the new procedure. The difference d_i between each of the paired measurements from the two methods on each sample is calculated. The average or mean difference D of all the individual differences is also calculated. The paired t-test involves calculating the t-value as given by the formula

$$t = (D/s_d)\sqrt{N} \tag{2.14}$$

where the standard deviation s_d is calculated using the formula

$$s_d = \sqrt{\Sigma(d_i - D)^2/(N-1)} \tag{2.15}$$

If the calculated value of t is less than the tabulated value for a given number of degrees of freedom, then it is concluded that there is no significant difference between the new test method and the standard method.

2.7 STANDARDIZATION OF INSTRUMENTAL METHODS OF ANALYSIS

All measurements by instrumental methods are subject to random errors with a distribution giving rise to a normal error curve. The measurements are associated with a degree of noise which is reflected in the precision of the background signal. It is necessary to distinguish between noise due to instrumental factors, the signal generated by a blank, and the signal due to the analyte sample by specifying the limits of detection and quantitation.

2.7.1 Limit of Detection and Limit of Quantitation

The *limit of detection* (LOD) or *detection limit* is the smallest concentration or the absolute amount of analyte that can be detected in the presence of a signal due to the blank and the background noise. Different methods are adopted to define the LOD. A simple method is to set the LOD as the concentration of the analyte which gives rise to a peak or signal which is twice that of the background signal. Another method defines the LOD as the concentration of the analyte that gives a signal three times the standard deviation of the background signal. The precision of a set of measurements at the LOD is about 33% and for quantitative measurements the concentration of the analyte, called the *limit of quantitation* (LOQ) should be at least ten times the LOD.

The International Union of Pure and Applied Chemistry (IUPAC) defines the LOD as three times the ratio (s/S) derived from a confidence level of 95% for a set of reasonable number of measurements (about 7–10 measurements), where s is the standard deviation of the measurements and S is the slope or sensitivity of the calibration curve at levels approaching the limit.

2.7.2 Calibration Chart or Curve

In most of the instrumental techniques a calibration chart or curve is employed to standardize the analytical methods. The procedure involves preparing a series of standards containing varying amounts of the analyte. The analyte content of the standards is determined under identical conditions by a chosen analytical method. The signal generated by the analyte in each of the standard is plotted on the y-axis against the amount (or concentration) of the analyte on the x-axis to give a straight line as shown in Fig. 2.2.

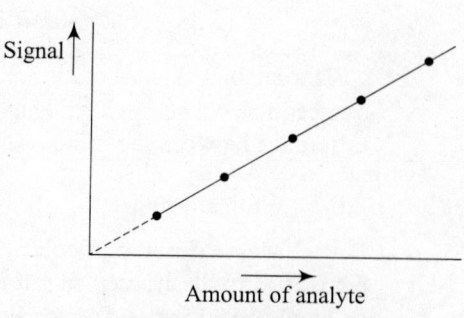

Fig. 2.2 A typical calibration chart

The signal generated by the test sample under the same specified analytical conditions is interpolated and from the intersection a perpendicular is drawn to the x-axis to determine the amount or concentration of the analyte in the test sample. The calibration curve obtained by plotting the absorbance (signal) versus concentration of the analyte using the Beer–Lambert equation in spectrophotometry is a typical example. Other examples include the plot of peak height (or peak area) versus concentration in chromatographic analysis and height of diffusion current plotted against amount of analyte in polarographic analysis.

Correlation coefficient (r) is useful to determine whether the variables x_i and y_i of a set of n data points are linearly related. It is calculated using the formula

$$r = (n\Sigma x_i y_i - \Sigma x_i \Sigma y_i)/\{[n\Sigma x_i^2 - (\Sigma x_i)^2][n\Sigma y_i^2 - (\Sigma y_i)^2]\}$$

The value of r lies between $+1$ and -1 and the nearer it is to ± 1, the greater is the probability of a linear relationship between the variables x_i and y_i. When the value of r tends towards zero the relationship between the variables x_i and y_i is no longer linear. However, the relationship between the two variables may be non-linear.

2.7.3 Method of Standard Addition

Calibration procedure using a series of standard solutions containing known concentrations of the analyte is beset with the problem of matching the matrix of the standard with that of the test sample. Matrix effects include the effects of the nature of the solvent, presence of inert or non-interfering components, and physical properties, such as viscosity and surface tension on the quality and reliability of analysis. It is necessary, at least, to minimize the matrix effects if not eliminate by preparing standard solutions of similar composition and concentration range as the test sample. However, in many types of analysis it is difficult to overcome the matrix effects. The method of standard addition is adopted particularly when interference from such matrix effects cannot be overcome.

The principle of standard addition method is based on the fact that the signal generated by the instrument must be a linear function of the analyte concentration with the signal showing zero reading in the absence of the analyte. The procedure involves 'spiking', that is, adding standard solutions of increasing concentration (S_1, S_2, S_3, etc.) of the pure analyte to the test sample solutions of constant amount taken in separate flasks. All the solutions are diluted to the same final volume and the instrument response is recorded under identical conditions for the spiked solutions as well as for an unspiked solution of the test sample diluted to the same total volume. A plot of the instrument reading as a function of the added analyte concentration on the x-axis gives a straight line as shown in Fig. 2.3. The unknown concentration of the analyte in the test sample is obtained from the point at which the extrapolated line intersects the x-axis when y = zero.

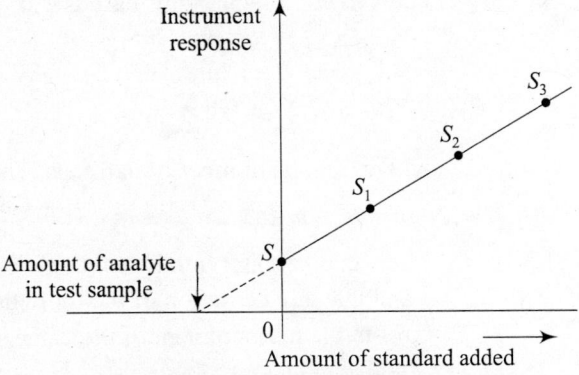

Fig. 2.3 Calibration chart for the standard addition method (S = unspiked test sample; S_1, S_2, and S_3 = spiked test samples)

The standard addition method may also be carried out using only two solutions, one containing a fixed amount of test sample and another containing the same amount of test sample and a known amount of the standard analyte. The instrument response for the spiked and unspiked sample solutions are recorded and the amount of analyte in the test sample calculated using the formula

$$C_X = R_X \, C_S / (R_S - R_X)$$

where C_X and C_S refer to the concentrations of the analyte in the test sample and in the standard solution, and R_X and R_S refer to the instrument response or detector signal for the test sample and the standard.

The method of standard addition finds extensive use in spectroscopic techniques, such as atomic absorption spectrometry, flame emission spectroscopy, and electroanalytical techniques, for example, polarography and anodic stripping voltammetry.

2.7.4   Method of Least Squares

The basis of many analytical methods is the calibration curve, obtained by plotting the measured quantity y as a function of an independent variable x such as concentration of an analyte. The calibration curve actually approximates to a straight line within certain range of values of x, which is desirable as it leads to a direct linear relationship between the independent and the dependent variables. However, all the data points may not fall exactly on the straight line. A best *straight line fit* of such two-dimensional data is obtained by a statistical technique called the *regression analysis* and the method of least squares is one such method of analysis of two-dimensional data.

The method of least squares is based on two assumptions: (i) that a linear relationship is valid between the measured variable y and (ii) the independent variable x.

(i) The linear relationship is expressed as

$$y = mx + c$$

where c is the intercept (the value of y when $x = 0$) and m is the slope of the straight line. The intercept and the slope are referred to as the *parameters* of the model or *regression coefficients*, which in this case is a straight line. The slope m is given by the equation

$$m = \frac{\sum (x_i y_i) - \sum x_i \sum y_i}{N \sum x_i^2 - \left(\sum x_i\right)^2}$$

where N is the number of data sets. The intercept c is given by the equation

$$c = y_{av} - m \, x_{av}$$

(ii) The second assumption is that there is no error in the values of the independent variable x and that any deviation of the individual data points from the straight line is due to the error in the measurement of the value of y.

The vertical deviation of any given point from the straight line is called the residual and the line generated by the least squares method is the one which minimizes the sum of the squares of the residuals for the data points.

● ● ● ● ● ● ● ● ● ● ● ● ● ● ● ● ● ●   **SOLVED PROBLEMS**

EXAMPLE 1   *Calculate the mean and the standard deviation of the following set of four results of 0.256 g, 0.268 g, 0.252 g, and 0.262 g obtained for the estimation of chloride by argentometric titration.*

SOLUTION

Mean of the four results $x_{av} = 0.2595$ g

(x_i)	$(x_i - x_{av})$	$(x_i - x_{av})^2$
0.256	0.0035	1.225×10^{-5}
0.268	0.0085	7.225×10^{-5}
0.252	0.0075	5.625×10^{-5}
0.262	0.0025	6.25×10^{-6}
$\Sigma = 1.038$		$\Sigma = 1.47 \times 10^{-4}$

Standard deviation as per Eq. (2.4) $= (1.47 \times 10^{-4}/(4-1))^{1/2} = 7.0 \times 10^{-3}$

EXAMPLE 2   *Calculate the standard deviation of the individual results and the standard deviation of the mean and express them in terms of absolute and relative values for the set of data obtained for the estimation of iron in a sample as 86.1 mg, 85.2 mg, 85.8 mg, 85.7 mg, 85.5 mg, and 85.3 mg.*

SOLUTION

The set of individual results (x_i) and the values of $(x_i - x_{av})$ and $(x_i - x_{av})^2$ are tabulated and the mean value, standard deviation, and standard deviation of the mean are calculated.

(x_i)	$(x_i - x_{av})$	$(x_i - x_{av})^2$
86.1	-0.5	0.25
85.2	-0.4	0.16
85.8	0.2	0.04
85.7	0.1	0.01
85.5	-0.1	0.01
85.3	-0.3	0.09
$\Sigma = 513.6$		$\Sigma = 0.56$

$x_{av} = 513.6/6 = 85.6$

Standard deviation (absolute) $= 0.335$ mg; CV $= (0.335/85.6) \times 100 = 0.39\%$

Standard deviation of the mean as per Eq. (2.5) (absolute) $= 0.335/\sqrt{6} = 0.137$ mg

      CV $= (0.137/85.6) \times 100 = 0.16\%$

EXAMPLE 3   *The following data was obtained in a set of replicate analysis of the nickel content of an alloy in per cent as 7.72, 7.86, 7.54, 4.58, 7.62, 7.66, and 7.05. Assess the data by Q-test for a confidence limit of 95%.*

SOLUTION

The value 4.58 may be rejected outright as having a gross error and in the remaining data the questionable data for which Q-test is to be performed for inclusion or rejection of the data is 7.05. As per Eq. (2.8), $Q = a/w$

    $Q = 7.54 - 7.05/7.86 - 7.05 = 0.605$

The calculated Q value of 0.605 is less than the Q critical value of 0.625 for a set of six results with a confidence limit of 95% (Table 2.1). Hence, the data 7.05 should not be rejected.

EXAMPLE 4   *The protein content of a sample determined from five analysis using a new method of analysis are 46.2%, 45.8%, 46.4%, 45.9%, and 46.3%. The protein content of the same sample was determined to be 46.4% by a standard method of analysis. Comment on the acceptability of the new method.*

SOLUTION

All the five values are retained and the mean and the standard deviation are calculated in the first step.

(x_i)	$(x_i - x_{av})$	$(x_i - x_{av})^2$
46.2	0.08	0.0064
45.8	−0.32	0.1024
46.4	0.28	0.0784
45.9	−0.22	0.0484
46.3	0.18	0.0324
$\Sigma = 230.6$		$\Sigma = 0.268$

$$x_{av} = 46.12 \text{ and } s = (0.268/4)^{1/2} = 0.259$$

Applying the student's t-test as per Eq. (2.11)

$$t = \frac{(x_{av} - \mu)\sqrt{N}}{s}$$

$$= \frac{(46.12 - 46.4)\sqrt{5}}{0.259} = 2.42$$

Since calculated value of 2.42 is less than $t_{tabulated}$ value of 2.57 at 95% confidence level, it is concluded that the new method is acceptable and there is no significant difference between the new method and the standard method.

SUMMARY

- Analytical results must be accurate, reproducible, reliable, and free from errors so that these can be accepted.

- Different types of errors are commonly encountered in the reported analytical data. These include gross errors, systematic errors, and random errors.

- Gross errors occur mainly due to carelessness of the analyst.

- Systematic errors include (i) method errors due to an improper selection or inherent limitation of the analytical method, (ii) instrumental errors, and (iii) personal errors due to prejudices.

- Systematic errors are determinable and hence can be minimized by simple methods such as calibration of apparatus and instruments, performing duplicate experiments, and adopting different analytical methods.

- Random errors are indeterminable errors as they are not predictable. Such errors indicate the lack of precision in the experimental results.

- Random errors are best treated by statistical analysis of the experimental results so that the distribution of such errors can be expressed in terms of standard deviation and the acceptability of the results at certain confidence levels.

- The experimental results are subjected to student's t-test, F-test, and paired t-test to determine the acceptability of the results and also to compare the relative accuracies of different analytical methods.

- The limits of detection and accuracy of the instrumental methods are usually standardized by calibration method or standard addition method.

REVIEW QUESTIONS

1. Explain the terms precision and accuracy.
2. How is error expressed?
3. Explain the terms population standard deviation, sample standard deviation, and standard error. What is their significance?
4. Classify and give a brief account on the different types of errors.
5. What are systematic errors? How are they minimized?
6. What are random errors? How are they treated?
7. What is Q-test? How is it useful?
8. What is confidence level of a set of results? How is it calculated?
9. Give an account on the comparison of results obtained by two different methods using F-test and student's t-test.
10. Give a detailed account on the methods of standardization of instrumental methods for quantitative analysis.

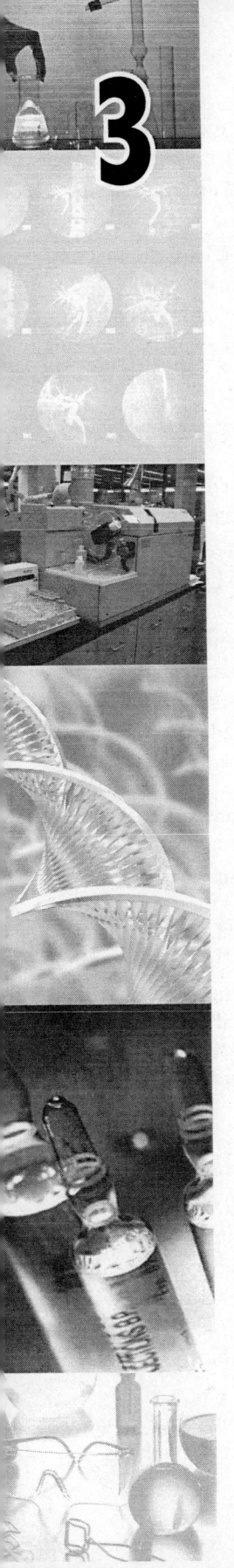

3 Wet Chemical Methods of Analysis

3.1 INTRODUCTION

Wet chemical methods of analysis are traditional or classical methods that have been in practice for more than a century. The methods include (i) *volumetry* and (ii) *gravimetry*. These are macroscopic methods requiring about 25–50 mg of analyte for each titration in volumetry and about 200–300 mg in the form of a solid in gravimetry. Both these methods have several advantages, such as (i) use of simple glassware and other common laboratory equipment for analysis of samples, (ii) better precision than most other instrumental methods, (iii) useful for calibrating and validating analyses carried out using instruments, (iv) non-requirement of frequent calibration, (v) amenable for automation, (vi) preferred for analysing small numbers of samples and carrying out one-off analyses, and (vii) relatively inexpensive. However, the wet chemical methods also have disadvantages such as being labour intensive. In addition, gravimetry, in particular, is time-consuming.

3.2 VOLUMETRY

Volumetry involves the measurement of volumes of the reactants taking part in the reaction used in quantitative analysis. The experimental procedure of measuring the volume of the reagent required to completely react with a known volume of the analyte solution is called *titration* and involves the addition of a standard solution of the reagent (titrant) from a burette to a known volume of the analyte solution (titrand) until the reaction is judged to be complete. The reaction is usually judged to be complete on the basis of a colour change of a suitable indicator or an instrument. Volumetry is also known more appropriately as *titrimetry* or *titrimetric analysis*.

The completion of reaction between the reagent and the analyte during titration is called the *equivalence point* which is a theoretical point when the amount of added titrant is chemically equivalent to the amount of the analyte taken. It is not possible to experimentally detect this equivalence point, but its position can be estimated by careful observation of a physical change associated with the theoretical equivalence point. Such a physical change associated with the equivalence point is called the *end point* which can be determined experimentally. It is necessary that the end point should exactly coincide with the equivalence point with respect to volume or mass of the reagents or the difference between the two is small for accurate results. Any difference between the equivalence and end points is called *titration error*.

A simple physical change that can be easily visualized and useful for the detection of end points in titrations is colour change of an indicator. Indicators are inorganic or organic substances which show a change of colour at or close to the theoretical equivalence point, and hence, useful for detection of end points. Different types of indicators applicable for the different types of volumetric titrations are discussed in the following sections.

For a satisfactory volumetric analysis the essential prerequisites include (i) simple reaction which goes to completion rapidly as per stoichiometric relationship or equation, (ii) availability of a suitable indicator which shows a detectable visual change, such as a sharp change in colour or formation of a precipitate, at the stoichiometric end point, and (iii) absence of any complication by additional or side reactions.

3.3   CLASSIFICATION OF VOLUMETRIC METHODS

Volumetric titrations are classified into four types based on the nature of chemical reaction as (i) acid–base or neutralization titrations, (ii) complexometric titrations, (iii) precipitation titrations, and (iv) oxidation–reduction titrations.

Acid–base reactions essentially involve the reaction of hydrogen ions with hydroxide ions to form water. In complexometric titrations, complex-forming or chelating agents (mostly EDTA) are used as reagents to form soluble complexes with metal ions for quantitative estimation of metal ions. In precipitation titrations, sparingly soluble substances or precipitates are formed as products. Important examples of precipitation titrations called *argentometry* are based on using silver nitrate as the reagent for the estimation of halides. Oxidation–reduction titrations involve electron transfer reactions between an oxidizing and a reducing agent. Oxidizing agents, such as potassium permanganate, potassium dichromate, ceric sulphate, iodine, etc., and reducing agents, such as iron(II), tin(II) compounds, sodium thiosulphate, and so on are used in oxidation–reduction titrations.

3.4   STANDARD SOLUTIONS AND STANDARD SUBSTANCES

Standard solutions are solutions of reagents whose concentrations are exactly known. The concentrations of solutions used in volumetric analysis are expressed either in terms of *molar* (or *molarity*) and given the symbol M or in terms of *normal* (or *normality*) with the symbol N. Molar concentration is defined as the number of moles of the substance dissolved in 1 L of the solution, whereas normality refers to the number of equivalents of the substance dissolved in 1 L of the solution.

The molar concentrations and normal concentrations of monobasic acids (e.g., hydrochloric acid, perchloric acid, acetic acid, etc.) and monoacidic bases (e.g., sodium hydroxide, potassium hydroxide, etc.) are the same as the molecular weights and equivalent weights are the same. Thus, a solution of HCl containing 36.5 g of the substance in 1 L of water is designated as 1 M or 1 N solution. Similarly, a 1 M or 1 N solution of NaOH contains 40 g of the substance in 1 L of water. In the case of a dibasic acid containing two protons [e.g., H_2SO_4 or oxalic acid—$(COOH)_2$] or a diacidic base (e.g., Na_2CO_3), the equivalent weight is equal to one-half of the molecular weight, and hence, the concentration of a solution of sodium carbonate containing 106 g of the substance in 1 L of water is expressed as 1 M or 2 N.

Standard substances are substances which are used for preparing standard solutions. Standard substances are classified into two types: (i) primary standards and (ii) secondary standards.

A substance is called a *primary standard* when it is available in pure form and can be directly weighed and dissolved in a known volume of solvent (usually water) to prepare a standard solution. A primary standard must satisfy certain prerequisites, such as (i) it must be available in the pure form or must be readily purified, (ii) it should be possible to store the substance in the pure form and its composition should not change during storage, (iii) the substance should not absorb moisture (hygroscopic) or carbon dioxide in air during weighing, and (iv) the reaction of the standard solution with the analyte should be almost instantaneous and stoichiometric. It is difficult to satisfy all the criteria by a single substance.

Commonly used primary standards in volumetric analyses include

- Sodium carbonate (Na_2CO_3), sodium tetraborate ($Na_2B_4O_7$), and potassium hydrogen phthalate ($KH(C_5H_5O_4)$) for acid–base titrations.
- Magnesium sulphate and zinc sulphate for complexometric titrations.
- Sodium chloride ($NaCl$) and potassium chloride (KCl) for argentometric titrations.
- Oxidizing agents, such as potassium dichromate ($K_2Cr_2O_7$), potassium bromate ($KBrO_3$), ammonium ceric sulphate ($(NH_4)_4Ce(SO_4)_4.2\,H_2O$) and reducing agents, such as sodium oxalate ($Na_2C_2O_4$), ferrous ammonium sulphate or Mohr salt ($FeSO_4\,(NH_4)_2SO_4.6H_2O$), for oxidation–reduction titrations.

A *secondary standard* is a substance that does not satisfy the prerequisites of a primary standard but still can be used for standardization, provided it has been standardized with a primary standard. Examples of secondary standards include hydrochloric acid, sulphuric acid, acetic acid, sodium hydroxide, disodium salt of EDTA, silver nitrate, potassium permanganate, iodine, sodium thiosulphate, etc.

3.5 NEUTRALIZATION TITRATIONS

Neutralization reactions are also known as *acidimetry* or *alkalimetry* as they basically involve the neutralization of a Lowry–Brønsted acid with a Lowry–Brønsted base in aqueous media. These acids and bases are either completely ionized as in the case of the strong acids and bases or partially ionized in the case of weak acids and bases. The actual reaction involves the neutralization of the protons (H^+) by the hydroxide (OH^-) ions to form neutral water. The counter ions (X^- or B^+) are not involved in the reaction, and hence, remain unaltered.

$$H^+X^- + B^+OH^- \rightleftarrows H_2O + B^+ + X^-$$

Although the reaction is reversible, the equilibrium is far to the right with the negligible rate of the reverse reaction. Hence, the forward reaction (neutralization) may be considered to go to completion as if it is an irreversible reaction. A more appropriate term would be to call the reaction as *analytical reaction* since the equilibrium constant for the reaction in most cases is greater than 10^4.

Some important neutralization titrations are discussed in this chapter: (i) strong acid versus strong base, (ii) strong acid versus weak base, (iii) weak acid versus strong base, (iv) weak acid versus weak base, (v) mixtures of strong and weak acids and bases, (vi) polybasic acids versus strong base, (vii) displacement reactions, and (viii) titrations in non-aqueous solvents.

Standard solutions of mostly strong acids or strong bases are used in neutralization titrations as these substances react completely with an analyte giving an easily detectable sharp end point. In contrast, in the case of weak acids or weak bases the end point is relatively less sharp. Dilute solutions of hydrochloric, sulphuric, or perchloric acid in the concentration range of 0.05–0.1 M are the commonly used strong acids for titrations. Nitric acid is not preferred because of its oxidizing properties in addition to its acidic properties. Hot concentrated sulphuric and perchloric acids are also potent oxidizing agents and hazardous to handle, but cold dilute solutions can be used without any detrimental effects. Standard solutions of bases that are commonly used include those of sodium hydroxide and potassium hydroxide. Barium hydroxide is also used sometimes. Standard solutions of acids may be prepared by titrating against a standard solution of sodium carbonate. Anhydrous sodium carbonate is a primary standard, and hence, a standard solution (~ 0.1 N) of it can be prepared by weighing the required quantity of the solid and dissolving it in a known volume of water. The titration may be carried out by pipetting out a known volume of the standard sodium carbonate solution into a conical flask and titrating with the acid solution which is to be standardized. Alternatively, the standard solution of sodium carbonate may be added from the burette to a known volume of the acid to be standardized. Methyl orange is used as the indicator. The reaction consumes two equivalents of the monobasic acid and may be represented as

$$2\,HX + Na_2CO_3 \rightarrow 2\,NaX + CO_2$$

The end points in neutralization titrations are detected by the use of suitable acid–base indicators or by instrumental methods such as conductometry, potentiometry, or coulometry. In classical methods of analysis, acid–base indicators are used. These are water-soluble weak organic acids or bases which display a colour change depending on the solution pH.

3.5.1 Theory of Acid–base Indicators

In aqueous solutions, acid–base indicators undergo dissociation readily to form conjugate acid–base pairs. An acid-type indicator may be represented as HIn which dissociates to form the conjugate acid–base pair H_3O^+ and In^- as represented by the equation

$$HIn + H_2O \rightleftarrows H_3O^+ + In^- \tag{3.1}$$

The equilibrium constant for the above reaction, called the *acid dissociation constant*, K_a, is given as

$$K_a = \frac{[H_3O^+][In^-]}{HIn} \tag{3.2}$$

The above equation may be rearranged to give

$$[H_3O^+] = K_a \frac{[HIn]}{In^-} \tag{3.3}$$

It is clear that the hydronium ion concentration will determine the ratio of concentrations of the conjugate acid–base pair of the indicator. The protonated and the deprotonated forms of the indicator have different colours. The colour of the solution is, in turn, determined by the ratio of the concentrations of HIn and In^- species. The human eye can detect colour difference of a solution containing both HIn and In^- species only when the ratio $[HIn]/[In^-] \geq 10$ when the acid colour predominates. The base colour predominates when the ratio $[HIn]/[In^-] \leq 0.1$. Substituting for the ratio of concentrations of the two species in Eq. (3.3), the hydronium ion concentration required to effect the complete indicator colour change can be arrived at as $[H_3O^+] = 10\,K_a$ for the acid colour and $[H_3O^+] = 0.1\,K_a$ for the base colour. Taking negative logarithms, the pH required for exhibiting the colour change is obtained as

$$\text{pH (acid colour)} = - \log (10\, K_a) = pK_a + 1 \qquad (3.4)$$
$$\text{pH (basic colour)} = - \log (0.1\, K_a) = pK_a - 1 \qquad (3.5)$$

So the pH range for the indicator to show a colour change $= pK_a \pm 1$.

For a base-type indicator the equilibrium reaction is represented as

$$\text{In} + H_2O \rightleftarrows HIn^+ + OH^-$$

and the range of pH for colour change can be obtained in a similar manner.

A few common acid–base indicators for different pH ranges are listed in Table 3.1.

Table 3.1   A few examples of acid–base indicators

Indicator	pK_a	pH range	Colour change
Acid-type indicators			
Thymol blue	1.65	1.2–2.8	Red–yellow
	8.90	8.0–9.6	Yellow–blue
Bromocresol green	4.66	3.8–5.4	Yellow–blue
Bromocresol purple	6.12	5.5–6.8	Yellow–purple
Bromothymol blue	7.10	6.2–7.6	Yellow–blue
Phenol red	7.81	6.8–8.4	Yellow–red
Phenolphthalein	9.60	8.3–10.0	Colourless–red
Thymolphthalein	9.30	9.3–10.5	Colourless–blue
Base-type indicators			
Methyl yellow		2.9–4.0	Red–yellow
Methyl orange	3.70	3.1–4.4	Red–orange
Methyl red	5.00	4.2–6.3	Red–yellow
Alizarin yellow GG		10–12	Colourless–yellow

Acid–base indicators are usually used as solutions of the indicator. The most commonly used phenolphthalein indicator solution is prepared by dissolving 1 g of the solid in 100 mL of ethanol and diluted with 100 mL of distilled water. Methyl orange indicator solution is prepared by dissolving 0.1 g of sodium salt in 200 mL of distilled water and acidified with about 3 mL of 0.1 N hydrochloric acid. The solution may be filtered if necessary. Methyl red indicator solution is prepared by dissolving 0.1 g of the solid in about 60 mL of ethanol and diluted to 100 mL with distilled water. Thymolphthalein is usually used as a 0.04% solution in water (40%)–ethanol (60%) mixture.

Mixed indicators may be used in certain cases when single acid–base indicators do not give sharp colour change or when a colour change is desired over a narrow selected range of pH. Mixed indicators are prepared by proper selection of indicators having pK values which are close to each other and the overlapping colours are complementary at an intermediate pH value. For example, a mixture of 0.1% ethanolic solutions of phenolphthalein and 1-naphtholphthalein in 3:1 ratio is useful for the titration of phosphoric acid to the dibasic stage ($K_2 = 6.3 \times 10^{-8}$) with an end point at pH 8.7. The mixed indicator shows a colour change from pale rose to violet at pH 8.9, close to the theoretical end point. Similarly, a mixture of thymol blue and cresol red (3:1 mixture of 0.1% aqueous solutions) shows a colour change at pH 8.3 and is useful for the titration of carbonate to hydrogen carbonate stage.

Screened indicators consist of a mixture of a pH sensitive dye and an acid–base indicator. The addition of the dye improves the colour change exhibited by the single indicator by producing the complement of one of the indicator colours. For example, the mixture of xylene cyanol FF (0.3%) and methyl orange (0.2%) in 50% ethanol shows a colour change of green–grey–magenta from the alkaline to the acid side. The grey colour corresponds to a pH of 3.8. Another example of a screened indicator is a mixture of methyl green and phenolphthalein (2:1 ratio of 0.1% ethanolic solutions) shows a colour change from grey to pale blue at a narrow pH range of 8.4–8.8.

3.5.2 Titration Curves

The titration curve is a graphical presentation of the changes in pH (or any other measurable physical parameter such as potential or conductance) of the analyte solution monitored as a function of the incremental addition of the reagent from the burette. A titration curve may be plotted from the data obtained based on experimental measurement of pH for each incremental addition of the standard solution of the base from the burette (titrant) to the known volume of acid (titrand) taken in the conical flask. A theoretical titration curve may also be constructed calculating the pH of the reaction medium from the known volumes and concentrations of the acid and the base. Strong acids and bases are completely ionized in aqueous media but in the case of weak acids and bases, additional information with regard to the dissociation constants are required to calculate the pH theoretically. Such theoretical titration curves are useful to understand the theoretical basis of detecting end (equivalence) points, the selection of suitable indicator for a given titration, and also in identifying titration errors. The equivalence point of a titration is characterized by an abrupt change in the relative concentrations of titrand and the titrant. In acid–base titrations, the neutralization of protons with the added hydroxide ions brings about an abrupt change in the pH of the reaction medium at the equivalence point of the reaction. A reasonably sharp end point can be visualized using an acid–base indicator if the pH change is within 4–10.

3.5.3 Titration of a Strong Acid with a Strong Base

Strong acids as well as strong bases exist as completely dissociated ions in solution and the acid–base reaction may be considered as the neutralization of hydronium ions and hydroxide ions.

The computation of a theoretical titration curve is best explained with the help of an example. The changes in the pH of a solution of 50 mL of 0.1 N HCl being titrated with 0.1 N NaOH are computed and shown below.

(i) In the first step prior to the addition of NaOH, the hydrogen ion concentration in the acid solution is essentially due to the completely dissociated HCl. The hydrogen ion concentration due to the self-dissociation of water is negligible, and hence, the analytical concentration of the HCl (0.1 N) is taken as the concentration of hydrogen ions and pH = 1.0.

(ii) After the addition of 10 mL NaOH the remaining H^+ concentration = (40 mL × 0.1 N)/ (total volume of 60 mL) = 0.067 N and the pH = $-\log$ (0.067) = 1.18 and pOH = pK_w – pH = 14 – 1.18 = 12.82.

The concentrations of H^+ and OH^- along with pH and pOH are given in Table 3.2 based on similar calculations for the addition of different volumes of NaOH up to the equivalence point.

(iii) At the equivalence point when all the H^+ ions have been neutralized by the addition of OH^-, the reaction mixture is an aqueous solution of the sodium chloride. Based on the self-dissociation of water $[H^+] = [OH^-] = 1 \times 10^{-7}$ and pH = pOH = 7.

(iv) After the equivalence point, only excess OH^- will be present due to the addition of NaOH and the concentration of OH^- is calculated from the known of volume of NaOH added after the equivalence point and the total volume. Thus, for the added volume of 50.1 mL of NaOH, the total volume of the solution being 100.1 mL, $[OH^-] = 0.1$ mL $\times 0.1$ N/100.1 $= 1 \times 10^{-4}$ N or the pOH = 4 and the corresponding pH = pK_w – pOH = 14 – 4 = 10.0.

Table 3.2 Computed data for the titration curve of HCl vs NaOH

S. No.	Volume of NaOH (mL)	Total volume (mL)	[H⁺]	[OH⁻]	pH	pOH
1.	0	50	0.1	1×10^{-13}	1.0	13.0
2.	10	60	0.067	1.51×10^{-13}	1.18	12.82
3.	20	70	0.043	2.34×10^{-13}	1.37	12.63
4.	30	80	0.025	2.5×10^{-12}	1.60	12.40
5.	40	90	0.011	1.1×10^{-12}	1.95	12.05
6.	49.5	99.5	5.03×10^{-4}	5.03×10^{-10}	3.30	10.70
7.	49.9	99.9	1×10^{-4}	1.0×10^{-10}	4.00	10.00
8.	50	100	1×10^{-7}	1×10^{-7}	7.00	7.00
9.	50.1	100.1	9.99×10^{-09}	9.99×10^{-5}	10.00	4.00
10.	51	101	9.9×10^{-09}	9.9×10^{-4}	11.00	3.00
11.	52	102	1.96×10^{-11}	1.96×10^{-3}	11.30	2.70
12.	55	105	4.76×10^{-11}	4.76×10^{-3}	11.68	2.32

Figure 3.1 shows the pH titration curve for the titration of hydrochloric acid against sodium hydroxide based on the data from Table 3.2.

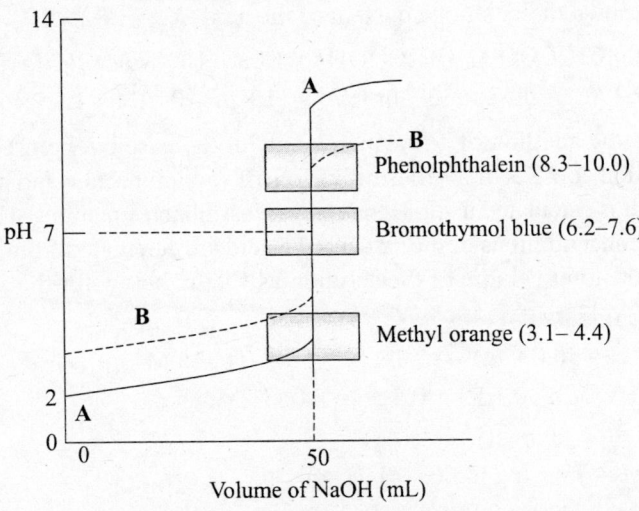

Fig. 3.1 Titration curve for HCl vs NaOH for different concentrations and selection of indicator for the titration. Curve A: 0.1N HCl vs 0.1 N NaOH. Curve B: 0.001 N HCl vs 0.001 N NaOH

The change in pH for incremental addition of base at the beginning of the titration is very small, giving rise to a plateau (buffer region). When almost all the available protons have been neutralized by the added hydroxide ions, the change in pH is large giving rise to the neutralization or inflection region which is vertical and almost parallel to the y-axis. The theoretical end point of the titration is at pH = 7. Further addition of the base beyond the equivalence point increases the pH only to a small extent, giving rise to a second plateau. From the figure, it is clear that phenolphthalein is the proper indicator for the titrations of a strong base against a strong acid as it shows a colour change closer to the theoretical equivalence point with a relatively negligible error. Methyl orange is not a suitable indicator as the colour change occurs much before the equivalence point that is before the neutralization region begins.

The concentrations of the analyte and the titrand affect the shape of the titration curves and also the detection of end point. Usually, strong acid versus strong base titrations are carried out with concentrations of the reagents being 0.05–0.1 N as the change in pH at the equivalence point is quite large (curve A). The titrations may be carried out even with very dilute solutions in the concentration range of 0.001–0.05 N with care as the change in pH at the equivalence point is markedly less as can be seen in Fig. 3.1 (curve B). Phenolphthalein or methyl orange is not suitable for detection of end point at such low concentrations of the reagents and only bromothymol blue provides a satisfactory end point.

3.5.4 Titration of a Weak Acid with a Strong Base

A typical example of the titration involving a weak acid being neutralized by a strong base is that of acetic acid versus sodium hydroxide. The reaction generates the sodium salt of the weak acid (sodium acetate) which is completely dissociated, but unlike the salt of the strong acid and strong base, such as sodium chloride, it undergoes hydrolysis. Hence, for computing the theoretical titration curve, four different types of calculations are required.

(i) In the first stage prior to the addition of any base, the solution is weakly acidic and the pH of the solution is calculated from the concentration of the acid and its dissociation constant. Thus, for an initial concentration of 0.1 N acetic acid (50 mL) the pH is calculated from the known dissociation constant of the acid ($K_a = 1.82 \times 10^{-5}$ at 25°C) as given below.

$$[H^+][CH_3COO^-]/[CH_3COOH] = 1.82 \times 10^{-5} \text{ since } [H^+] = [CH_3COO^-]$$
$$[H^+]^2/0.1 = 1.82 \times 10^{-6} \text{ or } [H^+] = (1.82 \times 10^{-6})^{1/2} = 1.35 \times 10^{-3} \text{ or pH} = 2.87$$

(ii) After the addition of a few increments of the base but before the equivalence point, the solution consists of a series of buffers of sodium acetate and acetic acid. The pH of each buffer is calculated using Henderson–Hasselbach equation: $pH = pK_a + \log ([salt]/[acid])$. The concentrations of the salt and the acid are calculated from the volume of alkali added and the total volume of the solution as shown below for the addition of 5 mL of NaOH (total volume = 55 mL).

$$[HA] = 50.0 \times 0.1 - (5.0 \times 0.1)/55 = 0.0818 \text{ M}$$
$$[A^-] = 5.0 \times 0.1/55 = 0.5/55 = 0.0091 \text{ M}$$
$$pH = pK_a + \log ([A^-]/HA])$$
$$= 4.74 + \log (0.0091/0.0818)$$
$$= 4.74 + (-0.95) = 3.79$$

The same procedure is adopted for calculating the pH after the addition of any other volume of NaOH before the equivalence point.

(iii) At the equivalence point, the solution contains only the salt (conjugate of the weak acid) whose concentration is 0.05 M. The amounts of HA and OH⁻ formed by the hydrolysis of the salt will be equal. Under such conditions

$$\frac{[HA][OH^-]}{[A^-]} = \frac{[OH^-]^2}{[A^-]} = K_b = \frac{K_w}{K_a}$$

$$[OH^-] = \sqrt{([A^-] K_w / K_a)}$$

Since $[H^+] = K_w / [OH^-]$ it may be written as

$$[H^+] = \sqrt{(K_w K_a / [A^-])}$$

or

$$pH = \tfrac{1}{2} pK_w + \tfrac{1}{2} pK_a + \tfrac{1}{2} \log[A^-] \hspace{2cm} (3.6)$$
$$= 7 + 2.37 + \tfrac{1}{2}(-1.30) = 8.72$$

(iv) After the equivalence point, the excess of the strong base added predominates over the salt formed in determining the pH of the solution, and hence, the pH can be calculated from the volume of the excess base added and the total volume. Thus, after the addition of 50.1 mL of NaOH (excess of 0.1 mL), the pH of the solution is calculated as

$$[OH^-] = 0.1 \times 0.1/100.1 = 9.99 \times 10^{-5} \text{ M}$$

$$pH = pK_w - pOH = 14.0 - 4.0 = 10.0$$

A table similar to Table 3.2 may be constructed for the addition of different volumes of NaOH based on the method of calculations shown above and using the data a theoretical titration curve can be drawn. The pH titration curve for the titration of 0.1 N acetic acid against 0.1 N sodium hydroxide is shown in Fig. 3.2.

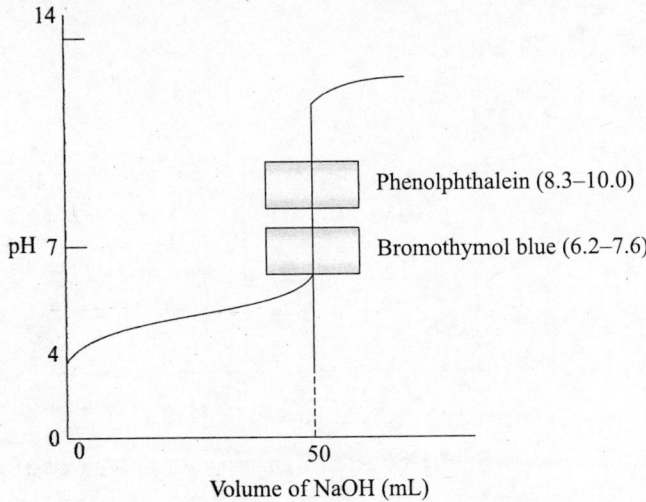

Fig. 3.2 Titration curve for acetic acid vs NaOH

It can be seen from the figure that bromothymol blue cannot be a suitable indicator for the titration and only an indicator which shows a colour change in the basic region, such as phenolphthalein, thymol blue, or thymolphthalein, is a suitable indicator. At lower concentrations of the titrand and titrant, the initial pH values before the addition of the base will be higher and the equivalence point pH will be lower.

Titrations of acids weaker than acetic acid ($K_a < 10^{-6}$) with a strong base will show pH titration curves with the smaller equivalence point regions. Very weak acids with $K_a < 10^{-8}$ cannot be determined by titration with a strong base in aqueous media as the reaction between the acid and the base is incomplete.

3.5.5   Titration of a Weak Base with a Strong Acid

A typical example of weak base is ammonia which can be titrated against a strong acid such as hydrochloric acid. The pH of the solution at the theoretical equivalence point for the titration can be predicted using the following equation.

$$pH = \tfrac{1}{2}\, pK_w - \tfrac{1}{2}\, pK_b - \tfrac{1}{2}\, pC \tag{3.7}$$

where pK_b is the dissociation constant of the base (1.85×10^{-5} for ammonia) and $pC = - \log [C]$ where C is the initial concentration of the base. Thus, for the titration of 50 mL of 0.1 N aqueous ammonia with 0.1 N hydrochloric acid, the pH at the equivalence point is 5.28 ($7.0 - 2.37 + \tfrac{1}{2}$ $(1.3) = 5.28$). The theoretical titration curve may be constructed by calculating the pH values for various amounts of acid added prior to the equivalence point using the equation

$$pH = pK_w - pK_b - \log([salt]/[base]) \tag{3.8}$$

After the equivalence point the excess of H^+ ions added suppress the hydrolysis of the salt and the pH can be calculated from the amount of acid added and the total volume in the titration cell. Figure 3.3 shows the titration curve for ammonia versus hydrochloric acid and the selection of indicator. Methyl orange, methyl red, bromophenol blue, or bromocresol green may be used as the indicator for this titration.

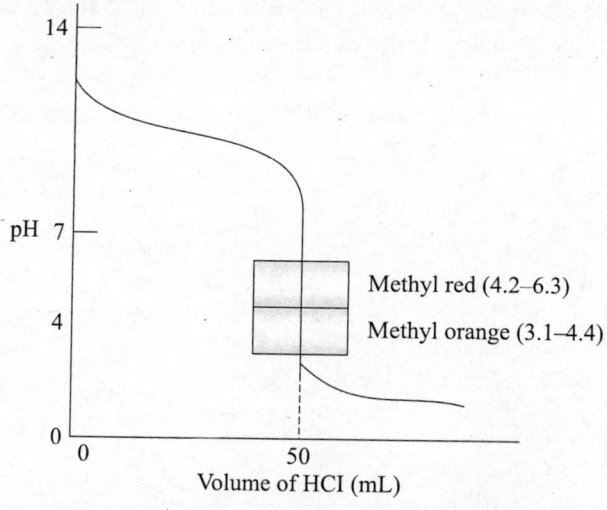

Fig. 3.3   Titration of ammonia with hydrochloric acid

In general, for titrations of weak bases having dissociation constant values greater than 10^{-6} with strong acids, bromophenol blue, or bromocresol green may be used as indicators. If the K_b of the weak base is less than 10^{-7}, bromophenol blue or methyl orange may be used as indicators.

3.5.6   Titration of a Weak Acid with a Weak Base

The titration curve for the neutralization of a weak acid, such as acetic acid by a weak base (e.g., aqueous ammonia), is shown in Fig. 3.4.

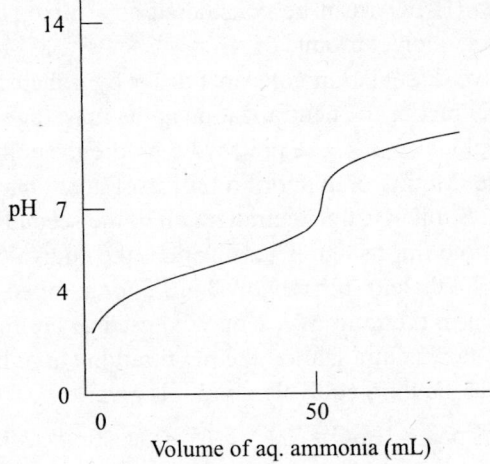

Fig. 3.4 Titration curve for acetic acid versus aqueous ammonia

The equivalence point pH is given by the following equation

$$pH = \frac{1}{2}\,pK_w + \frac{1}{2}\,pK_a - \frac{1}{2}\,pK_b \tag{3.9}$$

the value being 7.1 (7.0 + 2.38 − 2.37 = 7.1). The change near the end point is not sharp but gradual as seen in the figure, and hence, single indicator is not suitable for detection of the end point. However, a mixed indicator of neutral red and methylene blue may be used for the above titration.

3.5.7   Neutralization of Mixtures of Strong and Weak Acids or Strong and Weak Bases

A mixture of a strong monobasic (or monoprotic) acid and a weak monoprotic acid can be titrated against a strong base and two distinct end points can be obtained provided the dissociation constants of the two acids differ by more than 10^4 and their concentrations are of the same order. The first end point will be for the neutralization of the strong acid as the dissociation of the weak acid will not occur till all the protons of the strong acid have been completely neutralized. The dissociation of weak acid commences only after the complete neutralization of the strong acid as the titration progresses and a second end point for the neutralization of the weak acid can be detected. Thus, the individual concentrations of the two acids can be determined.

A mixture of a strong base and a weak base can similarly be titrated with a strong acid to give two end points; the first, indicating the complete neutralization of the strong base and the second, for the neutralization of the weak base.

3.5.8   Titration of Polybasic Acids with a Strong Base

Polyprotic acids can be considered as mixtures of strong and weak acids and multiple end points can be determined in the titration, provided the dissociation constants for the different stages differ by more than 10^4. Thus, for example, sulphurous acid (H_2SO_3) can be considered as a mixture of a strong and a weak acid as the two dissociation constants K_{a1} (1.7×10^{-2}) and K_{a2} (1.0×10^{-7}) differ widely and two end points can be detected in the titration. The first end point is quite sharp as the pH change is quite large but for the second, the change in pH is much less pronounced. On the other hand, for sulphuric acid the two dissociation constants differ only by about 100, and hence, only a single end point after the complete neutralization of both the protons can be detected.

Phosphoric acid (H_3PO_4) can be considered as a mixture of three monoprotic acids with different acid dissociation constants of $K_{a1} = 7.5 \times 10^{-3}$, $K_{a2} = 6.2 \times 10^{-8}$, and $K_{a3} = 5 \times 10^{-13}$. Since the successive dissociation constants differ by a factor of 10^5, three distinct end points can be detected. In practice the neutralization at the first stage is complete with the equivalence point predicted at $pH = (\frac{1}{2} pK_1 + \frac{1}{2} pK_2) = 4.6$ before the commencement of the neutralization of the second stage. Methyl orange or bromocresol green may be used as the indicator for the first stage titration. Similarly, the neutralization of the second stage also is complete before the commencement of the third stage neutralization, the equivalence point pH being $(\frac{1}{2} pK_2 + \frac{1}{2} pK_3) = 9.7$. Thymolphthalein (pH range 9.3–10.5) or a mixed indicator of phenolphthalein and 1-naphtholphthalein in the ratio of 3:1 may be used as an indicator for titrating up to second stage. In the third stage neutralization the pH titration curve is flat and no suitable indicator is available. The pH at the third equivalent point is given

$$\frac{1}{2} pK_w + \frac{1}{2} pK_3 - \frac{1}{2} pC = 7.0 + 6.15 - \frac{1}{2} (1.6) = 12.35 \text{ for } 0.1 \text{ M} H_3PO_4 \qquad (3.10)$$

The titration curve for the neutralization of phosphoric acid is shown in Fig. 3.5.

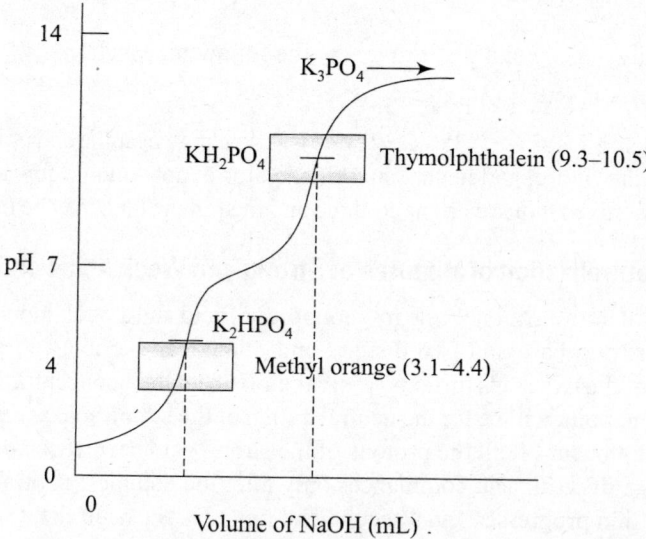

Fig. 3.5 Titration curve for the neutralization of phosphoric acid

3.5.9 Titrations in Non-aqueous Media

Titrations in non-aqueous solvents are helpful when the substances are less soluble in aqueous medium or when detection of end point becomes difficult in aqueous solution. The acid–base theory of Brønsted and Lowry can be applied to non-aqueous solvents, and hence, acid–base titrations can be carried out in non-aqueous media also. However, such titrations can be carried out provided the substances exhibit their ability to release or accept proton(s) in the chosen non-aqueous solvent and a suitable indicator is available to detect the end point. Non-aqueous solvents are classified into four types, namely aprotic, protophilic, protogenic, and amphiprotic.

Aprotic solvents are low dielectric solvents and chemically neutral or inert towards acids and bases and do not cause ionization or dissociation of solutes. Typical examples of aprotic solvents include toluene and carbon tetrachloride. These solvents are mostly used to dilute reaction mixtures as they do not participate in the chemical reactions.

Protophilic solvents have high affinity for protons and the reaction with protons may be represented as

$$S + HA \rightleftarrows SH^+ + A^-$$

The reversible reaction is influenced by the nature of the acid HA and the solvent S. Examples of protophilic solvents include liquid ammonia, ketones, and amines. These solvents enhance the acid strengths of weak acids because of their *levelling effect*. Such levelling solvents find use in quantitative analyses of very weak acids in the presence of strong acids as all acids will be levelled to the same strength.

Protogenic solvents are acidic in nature and readily donate protons. These solvents are also levelling solvents as they enhance the strength of weak bases and all bases are levelled to the same strength. Examples include anhydrous sulphuric acid and hydrogen fluoride.

Amphiprotic solvents exhibit both protophilic and protogenic nature as they are capable of accepting and also donating protons. They undergo self-ionization to a small extent and exhibit their acidic characteristics

$$HS \rightleftarrows H^+ + S^-$$

Examples include water, weak organic acids, and alcohols. These solvents behave as acids in the presence of a base by donating protons. They behave as bases in the presence of an acid by accepting protons. Thus, for example, water and a weak organic acid, such as acetic acid, accept protons and behave as bases in the presence of a strong acid such as sulphuric or perchloric acid.

$$2\,H_2O + H_2SO_4 \rightleftarrows 2\,H_3O^+ + SO_4^{2-}$$
$$CH_3COOH + HClO_4 \rightleftarrows CH_3COOH_2^+ + ClO_4^-$$

These solvents donate protons and behave as acids in the presence of a base such as ammonia.

$$H_2O + NH_3 \rightleftarrows OH^- + NH_4^+$$
$$CH_3COOH + NH_3 \rightleftarrows CH_3COO^- + NH_4^+$$

Commonly used non-aqueous solvents include glacial acetic acid (useful as a differentiating solvent for weak acids) together in acetonitrile and nitromethane; acetonitrile together with chloroform or phenol; and dioxan, dimethylformamide, and ethylene glycol together with isopropanol or 1-butanol.

Only a few indicators are effectively available for titrations in non-aqueous media. These include methyl red, thymol blue, crystal violet, quinaldine red, and 1-naphthol benzein. Methyl red is used in the form of a 0.2 per cent solution in dioxan and shows a colour change from yellow to red at the end point for acid–base titrations. Thymol blue (0.2% solution in methanol) shows a sharp colour change from yellow to blue at the end point, whereas quinaldine red (0.1% in ethanol) shows a colour change from purple red to pale green for determination of acids in dimethylformamide solution. For titrations of organic bases (e.g., pyridine) against perchloric acid in acetic acid medium, a 0.5 per cent solution of crystal violet in acetic acid is used as the indicator showing a colour change from violet through blue, green, and finally to greenish yellow. For titrations of weak bases against perchloric acid in nitromethane containing acetic anhydride 1-naphthol benzein indicator shows a colour change from yellow to green.

3.5.10 Applications of Acid–base Titrations

A few typical examples of the applications of acid–base titrations include the determination of types of alkalinity in water, dissociation constants of weak acids and amino acids, estimation of protein content and nitrogen content in organic and other complex substances, such as food and food products, pharmaceuticals, coal, etc., by Kjeldahl method, and acid value and saponification value of oils and fats are discussed in brief.

Determination of alkalinity of water is an important example of the application of acid–base titration from the viewpoints of public health and pollution control for determining the quality of drinking water and for the characterization of process water as well as wastewater in industries.

Natural water shows a characteristic alkaline nature which is attributed mostly to the presence of hydroxide, carbonate, bicarbonate ions, and to some extent to silicates such as $HSiO_3^-$ and SiO_3^{2-}. In addition, salts of weak organic acids undergo hydrolysis increasing the concentration of hydroxide ions. Hydroxide and bicarbonate ions cannot coexist as they react with each other forming carbon dioxide and water. Alkalinity of water is classified based on the type of anion present in water as bicarbonate (HCO_3^-), carbonate (CO_3^{2-}), and hydroxide (OH^-) alkalinity.

The total alkalinity of water and the different types of alkalinity in a sample of water are determined by titrating with a standard solution of hydrochloric acid. The titration is an example of a *displacement titration* involving the displacement of a weak acid (carbonic acid) from its salt (sodium carbonate) with a strong base by a strong acid. The titration is carried out in two stages. In the first stage a known volume of sample water (say 50 mL) is taken in the conical flask and a few drops of phenolphthalein indicator are added. The contents of the flask are titrated against standard solution of hydrochloric acid; the end point being indicated by the colour change from pink to colourless. The volume of acid consumed is noted as phenolphthalein end point V_{phe}. The phenolphthalein end point corresponds to the neutralization of hydroxide and neutralization of carbonate ions to form bicarbonate ions by the added hydrochloric acid.

$$OH^- + H^+ \rightarrow H_2O$$
$$CO_3^{2-} + H^+ \rightarrow HCO_3^- + H_2O$$

Thus, phenolphthalein end point indicates the neutralization of hydroxide alkalinity and half the carbonate alkalinity.

In the second stage of the titration, a few drops of methyl orange, or bromocresol green indicator are added to the contents of the flask and the titration is continued against the same standard solution of hydrochloric acid. The end point is indicated by the colour change from yellow to pink/red at pH 4.5 and the volume of the acid consumed is noted as the methyl orange end point V_M. Methyl orange alkalinity corresponds to the complete neutralization of all types of alkalinity, that is, total alkalinity. The neutralization of total alkalinity involves the two reactions mentioned above as well as the neutralization of bicarbonate ions.

$$HCO_3^- + H^+ \rightarrow H_2O + CO_2$$

The titration curve is shown in Fig. 3.6.

Thus, with a single two-stage titration the total alkalinity, the different types of alkalinity, and their concentrations can be determined with the help of the titre values at phenophathalein end point V_{phe} and the methyl orange end point V_M based on the following relationships.

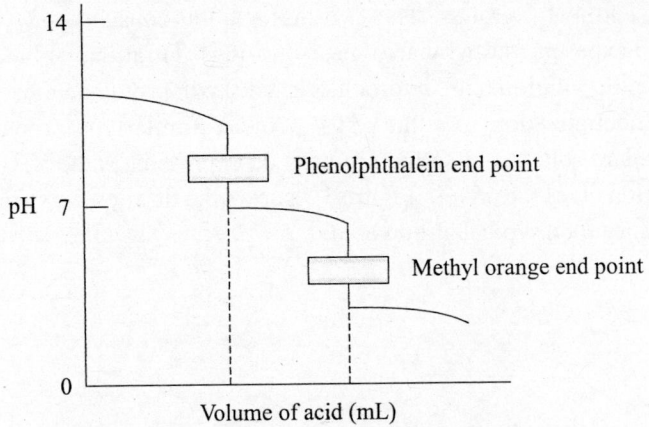

Fig. 3.6   Determination of alkalinity of a sample of water

(i)   Titre values of $V_{phe} = 0$ and $V_M > 0$ indicate the absence of both hydroxide and carbonate ions and any alkalinity is attributed to the presence of bicarbonate ions only.

(ii)   A titre value of $V_{phe} = V_M$ indicates the absence carbonate and bicarbonate ions and the alkalinity is entirely due to the hydroxide ions.

(iii)   When $V_{phe} = \frac{1}{2}\, V_M$ the alkalinity is due to the presence of carbonate ions only.

(iv)   When $V_{phe} < \frac{1}{2}\, V_M$ the alkalinity is due to the presence of both carbonate and bicarbonate ions the concentration of $[CO_3^{2-}] = 2V_{phe}$ and that of $[HCO_3^-] = (V_M - 2V_{phe})$.

(v)   When $V_{phe} > \frac{1}{2}\, V_M$ the presence of hydroxide and carbonate is indicated, the concentration of $[OH^-] = (2\,V_{phe} - V_M)$ and that of $[CO_3^{2-}] = 2\,(V_M - V_{phe})$.

Alkalinity of water may be expressed in terms of molarity of respective ions but is usually expressed in terms of equivalent amount of calcium carbonate in units of ppm or mg/L by using the conversion factor of 1 mL of N/50 acid = 1 mg of $CaCO_3$.

Different types of alkalinity are related to different types of hardness of water as summarized below:

> Non-carbonate hardness = total hardness − total alkalinity
>
> Calcium alkalinity = total alkalinity or calcium hardness whichever is smaller
>
> Magnesium alkalinity = total alkalinity − calcium alkalinity
>
> Sodium alkalinity = total alkalinity − total hardness
>
> Permanent magnesium hardness = magnesium hardness − magnesium alkalinity
>
> Permanent calcium hardness = calcium hardness − calcium alkalinity

Dissociation constants of weak acids and bases can be determined by monitoring pH change as a function of volume of titrant during acid–base titrations. The dissociation constant of a weak acid is exactly equal to the pH at which 50 per cent of the acid is neutralized by the added base. Similarly, the dissociation constant of a weak base is numerically equal to the pOH when the base is half neutralized. The value of $pOH = pK_w - pH$.

Dissociation constants of amino acids can also be determined by neutralization titrations. Amino acids contain both an acidic (carboxyl) and a basic (amino) group. In aqueous solution the amino acid usually exists as a zwitter ion (internally neutralized molecule) containing the COO⁻ and NH_3^+ groups. The two dissociation constants pK_{COOH} and the pK_{NH2} can be determined experimentally by acid–base titrations. Titrations of the zwitter ion solution with a standard acidic solution (e.g., hydrochloric acid) will protonate the weaker COO⁻ group and the pH at half neutralization gives the pK_{COOH} value. Similarly, titration of the zwitter ion solution with a standard solution of sodium hydroxide deprotonates the NH_3^+ group and the pH at half neutralization gives the pK_{NH2}. Figure 3.7 shows the titration curve and the determination of the two pK values for a typical amino acid.

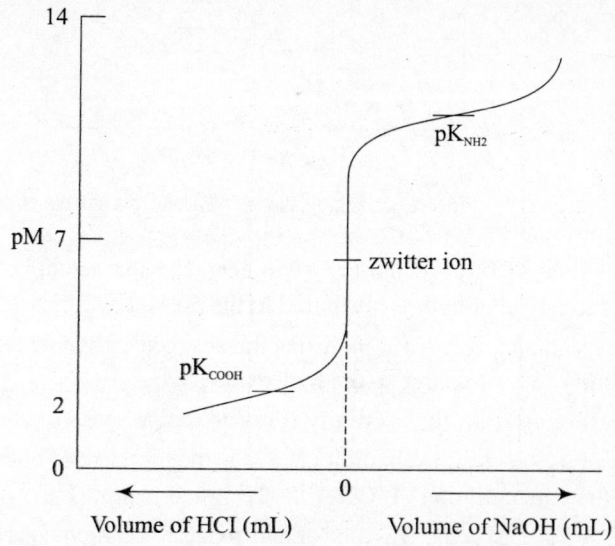

Fig. 3.7 Determination of the dissociation constant of an amino acid

Protein estimation by Kjeldahl method is based on determining the nitrogen content of the sample and calculating the protein content from the per cent nitrogen of the sample. The method is applicable to biological as well as organic compounds. The method involves converting organic nitrogen into inorganic ammonia and estimating the ammonia by titration. In the first step, the organic or protein sample is digested with concentrated sulphuric acid in the presence of selenium or copper salt as the catalyst and potassium sulphate to raise the boiling point of sulphuric acid. The organic nitrogen is converted into ammonia which gets fixed as ammonium sulphate. Ammonia is liberated from the ammonium sulphate solution by distilling it with a large excess of concentrated (50% w/v) solution of sodium hydroxide. The liberated ammonia is once again absorbed in a standard solution of sulphuric acid of known volume and the excess acid is back titrated with a standard solution of sodium hydroxide to determine the amount of ammonia. The nitrogen content of the sample is calculated from the amount of ammonia as determined by the titration. The per cent nitrogen is then multiplied by the factor 6.25 (because a large number of proteins contain very nearly the same per cent of nitrogen of 16%) to get the protein content of the sample.

Acid value and saponification value of oils and fats are important parameters in the characterization of oils and fats. Acid value of oil is defined as the number of milligrams of KOH required to neutralize the free fatty acid in 1 g of oil and is significant in determining the free fatty acid content of oil which is related to the extent of rancidity of the oil. Acid value of oil/fat is determined by titration of an ethanolic solution of the oil/fat with a standard solution of sodium hydroxide using phenolphthalein as indicator. Saponification number of an oil is defined as the number of milligrams of potassium hydroxide required to neutralize the free fatty acid obtained by the hydrolysis of 1 g of oil/fat. The procedure involves hydrolysing the weighed sample of oil or fat with a known excess of potassium hydroxide solution under reflux conditions to liberate and neutralize the fatty acids and determining the excess unreacted potassium hydroxide by back titration with a standard solution of hydrochloric acid using phenolphthalein indicator.

3.6 PRECIPITATION TITRATIONS

Precipitation titrations involve reactions in which the analyte is converted into a sparingly soluble substance. Such reactions are based on the principle of solubility product constant which may be represented as

$$A_xB_y \rightleftharpoons x A^{m+} + y B^{n-} ; \quad K_{sp} = [A^{m+}]^x [B^{n-}]^y \tag{3.11}$$

where A_xB_y represents the sparingly soluble salt undergoing dissociation (dissolution) into its constituent ions A^{m+} and B^{n-} and K_{sp} is the equilibrium constant at a given temperature for the above reaction and is called *solubility product constant* (the product of the concentrations of the constituent ions). The solubility product relationship is valid only in very dilute solutions where the analytical concentrations of the species may be taken as equal to the activities. At moderately higher concentrations of salts, the activity coefficients of the ions become smaller due to an increase in ionic strength. This effect called the *salt effect* increases the solubility of the sparingly soluble substance.

3.6.1 Argentometry

Argentometry is the most widely used precipitation reaction and involves the titration of silver nitrate for the estimation of halides (except fluoride) and thiocyanate. The detection of end points in argentometric titrations is achieved both by the use of chemical indicators and instrumental methods (e.g., conductance and potentiometric titrations). The precipitation reaction may be represented as

$$Ag\,NO_3 + X^- \rightleftharpoons AgX \downarrow + NO_3^-$$
$$\text{(precipitate)}$$
$$(X^- = Cl^-, Br^-, I^-, \text{ and } SCN^-)$$

The equilibrium of the above reaction is to the far right with the rate of the forward reaction (precipitation) being very large compared to the negligible rate of the reverse reaction (solubilization). The *solubility product constant* K_{sp} is given by the expression

$$K_{sp} = [Ag^+][X^-]/[AgX] \tag{3.12}$$

and since the activity of the solid AgX is unity the expression simplifies to

$$K_{sp} = [Ag^+][X^-] \tag{3.13}$$

The titration curve for a typical precipitation reaction exemplified by the titration of chloride by silver nitrate may be constructed theoretically based on the solubility product principle. For a titration of 50 mL of 0.1 M sodium chloride against 0.1 M silver nitrate, the variation of ionic concentrations of the added silver and remaining chloride ions as the titration progresses may be calculated from the known solubility product constant value K_{sp} ($[Ag^+][Cl^-]$) of 1.2×10^{-10}

for silver chloride at room temperature. The corresponding p-values (negative logarithms) are $pAgCl = pAg^+ + pCl^- = pK_{sp} = 9.92$. The initial concentration of Cl^-, before the addition of any silver nitrate, is 0.1 mol/L and $pCl^- = 1$. Before the end point but after the addition of 20 mL of silver nitrate, the ionic concentrations of the remaining chloride and silver are given as $[Cl^-] = 0.1$ mol/L $\times$ 30 mL / 70 mL $= 0.043$ or $pCl^- = 1.37$ and $pAg^+ = 9.92 - 1.37 = 8.55$.

At the equivalent point $pAg^+ = pCl^- = 9.92/2 = 4.96$.

After the end point at the addition of 50.1 mL of silver nitrate the ionic concentration of Ag^+ will be $[Ag^+] = 0.1$ mol/L $\times$ 0.1 mL / 100.1 mL $= 1 \times 10^{-4}$ or $pAg^+ = 4$ and $pCl^- = 9.92 - 4.0 = 5.92$.

The calculated values differ from the actual concentrations of silver and chloride ions ($\sim 1 \times 10^{-5}$) just after the equivalence point due to the fact that the precipitated silver chloride undergoes dissociation. The p-values of silver and chloride ions (and similarly for bromide and iodide) for various volumes of silver nitrate added during the titration computed in the above manner are given in Table 3.3.

Table 3.3 Values of pAg^+ and pCl^-, pBr^-, and pI^- for the individual titrations of chloride, bromide, and iodide (50.0 mL each) with silver nitrate (0.1 M). (K_{sp} AgCl $= 1.2 \times 10^{-10}$, K_{sp} AgBr $= 3.5 \times 10^{-13}$, and K_{sp} AgI $= 1.7 \times 10^{-16}$)

Vol. of AgNO$_3$ (0.1 M) added (mL)	Titration of chloride (50 mL of 0.1 M)		Titration of bromide (50 mL of 0.1 M)		Titration of iodide (50 mL of 0.1 M)	
	pCl^-	pAg^+	pBr^-	pAg^+	pI^-	pAg^+
0	1.0	–	1.0	–	1.0	–
10	1.18	8.74	1.18	11.28	1.18	14.59
20	1.37	8.55	1.37	11.08	1.37	14.40
25	1.48	8.44	1.48	10.97	1.48	14.29
30	1.60	8.32	1.60	10.84	1.60	14.17
40	1.95	7.96	1.95	10.5	1.95	13.82
49	3.00	6.92	3.00	9.45	3.00	12.77
49.5	3.30	6.62	3.30	9.15	3.30	12.47
49.9	4.00	5.92	4.00	8.45	4.00	11.77
50.0	4.96	4.96	6.23	6.23	7.89	7.89
50.1	5.92	4.00	8.45	4.00	11.77	4.00
50.2	6.22	3.70	8.75	3.70	12.07	3.70
50.5	6.62	3.30	9.15	3.30	12.47	3.30
51.0	6.92	3.00	9.45	3.00	12.77	3.00
52.0	7.22	2.70	9.75	2.70	13.07	2.70
55.0	7.00	2.32	10.13	2.32	13.45	2.32

The theoretical titration curve for the precipitation titration may be drawn by plotting the pAg^+ values as a function of the volume of silver nitrate added (Fig. 3.8). A similar titration curve for the titration of bromide or iodide ions against silver nitrate is also shown in the same figure and the values of pAg^+, pBr^-, and pI^- are listed in Table 3.3.

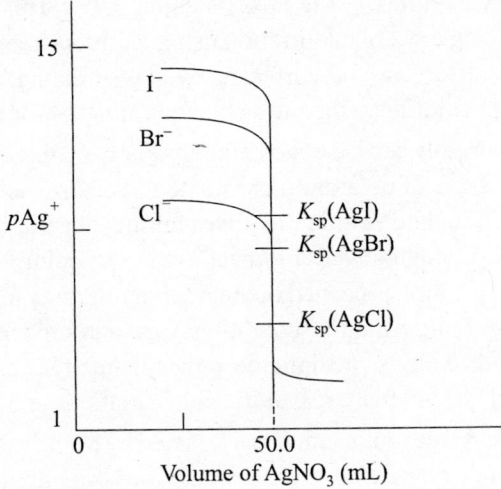

Fig. 3.8 Precipitation titration curves for halides with silver nitrate

3.6.2 Detection of End Points

Chemical indicators used in the precipitation titrations should in general show a colour change over a narrow or limited range of p-function of the analyte or the precipitating reagent and colour change should occur within the inflection zone of the titration curve. Three types of chemical indicators find use in argentometric titrations. These include (i) the formation of a coloured precipitate in Mohr's method, (ii) the formation of a soluble coloured compound in Volhard's method, and (iii) an adsorption indicator in Fajans' method.

Mohr's method

Mohr's method is based on the appearance of a coloured precipitate at the end point. Chloride, bromide, and cyanide ions can be estimated quantitatively by titrating with a standard solution of silver nitrate in the presence of a small amount of potassium or sodium chromate as indicator. At the end point, the chromate ion (of the yellow-coloured indicator solution) reacts with the first drop of excess of silver ion added from the burette to form a brick-red precipitate of silver chromate. The titration involves a fractional precipitation, silver chloride being precipitated first due to its lower solubility compared to silver chromate, and also due to the high initial concentration of chloride. At the equivalence point, both the salts are in equilibrium with the solution. The concentration of $[Ag^+]$ at the equivalence point will be $K_{sp}^{1/2}(AgCl) = 1.1 \times 10^{-5}$ M. For precipitating silver chromate at this concentration of Ag^+ ion, the chromate ion concentration required will be about 0.014 M (about 150 times the silver or chloride concentration) as shown by the calculation:

$$[CrO_4^{-2}] = \frac{K_{sp}(Ag_2CrO_4)}{[Ag^+]^2} = \frac{1.7 \times 10^{-12}}{(1.1 \times 10^{-5})^2} = 1.4 \times 10^{-2} \text{ M}$$

However, in practice the concentration of potassium chromate is maintained at much lower level of 0.003–0.005 M as higher concentrations impart a deep yellow–orange colour impairing the detection of the end point clearly. The use of dilute chromate solution requires slightly more of silver ion to be added to induce the precipitation of silver chromate. The error so introduced is negligible when the concentration of silver nitrate titrant is in the range of 0.1 M but at 0.01 M concentration of silver nitrate, the error is about 0.4%. This systematic positive error can be corrected by carrying out a blank titration using chloride-free suspension of calcium carbonate. Alternatively, the error may be corrected by standardizing the silver nitrate with a standard solution of sodium chloride using the same conditions as for the analyte.

The titration is usually carried out by taking a known volume of the halide solution in a conical flask to which 1–2 mL of potassium chromate indicator is added and the contents of the flask are titrated with a standard solution of silver nitrate taken in the burette. The indicator forms a red-coloured precipitate of silver chromate at the end point with the first drop of excess silver nitrate added. The titration is carried out in neutral or faintly alkaline conditions in the pH range of 6.5–9 (neutral or faintly alkaline conditions) and at room temperature. In alkaline solutions at pH > 9, silver hydroxide is precipitated rather than silver chromate. On the other hand, the solubility of silver chromate increases in acidic solutions and also at higher temperatures.

Iodide and thiocyanate ions cannot be estimated by Mohr's method as the precipitates of silver iodide and silver thiocyanate adsorb chromate ions strongly resulting in a false and indistinct end point.

Volhard's method

Volhard's method involves titration of silver in nitric acid medium against a standard solution of potassium thiocyanate or ammonium thiocyanate using ferric nitrate or ferric alum as the indicator. Silver thiocyanate is precipitated quantitatively during the titration ($K_{sp} = 7.1 \times 10^{-13}$) and after complete precipitation of silver, a slight excess of thiocyanate added from the burette reacts with the indicator to form a red-coloured complex species ($[Fe(SCN)]^{2+}$), indicating the end point of the titration.

The method is applicable for the determination of chloride, bromide, and iodide in acidic medium. In practice, a known excess of a standard solution of silver nitrate is added to precipitate the halide from a known volume of the analyte solution and the excess of silver nitrate is back-titrated with a standard solution of thiocyanate. In the estimation of chloride by this method, it should be noted that silver chloride is more soluble compared to silver thiocyanate and during the back titration the added thiocyanate will displace the chloride from the silver chloride precipitate thereby dissolving the silver chloride precipitate. Hence, the silver chloride precipitate formed by the added excess of silver nitrate should be boiled to coagulate the silver chloride precipitate and filtered-off before the cold filtrate is back-titrated against potassium thiocyanate. Alternatively, nitrobenzene (~1–2 mL) may be added to the silver chloride precipitate before the back titration is carried out. Nitrobenzene forms a protective film surrounding the silver chloride particles from interacting with thiocyanate. A few drops of ferric alum indicator are added to the contents of the conical flask and the mixture is then back-titrated against potassium or ammonium thiocyanate. At the end point, after all the silver has been precipitated, the added thiocyanate reacts with iron to form a red coloured solution indicting the end of the titration.

The estimation of bromide or iodide by Volhard's method does not require the addition of nitrobenzene prior to the back titration of the excess silver nitrate with thiocyanate. This is because the solubilities of silver bromide and silver thiocyanate are almost comparable [$K_{sp}(AgBr)$ = 3.5×10^{-13} and (K_{sp} (AgSCN) = 7.1×10^{-13}]. The titration error is negligibly small, even though silver bromide is slightly more soluble. Silver iodide is much less soluble [$K_{sp}(AgI)$ = 1.7×10^{-16}], and hence, does not interfere in the back titration. However, the iron(III) indicator must be added only in the presence of excess of silver ions (just before the back titration) as iodide ions tend to reduce iron(III).

Fajans' method

Fajans' method makes use of adsorption indicators to detect the end point in the titration of halides with silver nitrate. The adsorption indicator ion gets adsorbed onto the colloidal precipitate strongly at the end point showing a different colour from that of the free indicator. In the titration of chloride or bromide with silver nitrate in the pH range 4.5–7, dichlorofluorescein (exists as dichlorofluoresceinate anion) is used as an adsorption indicator. In the initial stages of the titration when silver nitrate is added from the burette, the colloidal silver chloride precipitate is surrounded by excess of chloride ions in the primary coordination sphere. The anionic dichlorofluoresceinate cannot adsorb onto the silver chloride precipitate and remains free, showing a pale green or yellow green colour. Immediately after the complete precipitation of all the available chloride, the silver chloride precipitate is surrounded by silver ions from the first drop of excess silver nitrate added. The change of the ionic atmosphere surrounding the silver chloride precipitate facilitates the strong adsorption of the anionic indicator. On adsorption, the indicator changes its colour to red indicating the end point.

Other adsorption indicators for argentometric titrations include (i) fluorescein in the titration of chloride, bromide, and iodide with silver nitrate in neutral or weakly basic solutions showing a colour change from yellow green to pink, (ii) eosin (tetrabromofluorescein) in the titration of bromide, iodide, and thiocyanate with silver nitrate in acetic acid solution (pink to red violet), (iii) bromocresol green for the titration of thiocyanate with silver nitrate at pH 4–5, (iv) methyl violet for titrating Ag^+ with chloride in acid solution, (v) rhodamine 6G for the estimation of Ag^+ with bromide in dilute nitric acid medium (< 0.3 M), and (vi) tartrazine in the titration of silver nitrate with iodide or thiocyanate. Tartrazine can also be used for the titration of a mixture of iodide and chloride with silver nitrate and for the back-titration of silver nitrate with iodide (colourless to green). The use of adsorption indicators requires stringent conditions in that the precipitate should be colloidal and should not coagulate. The pH should be suitable for the indicator to be predominantly in the ionic form. The titration should be carried out in subdued light.

Dichlorofluorescein

Precipitation titrations using adsorption indicators may also be carried out for the (i) estimation of sulphate by titrating with barium chloride in the pH range of 1.5–3.5 using thorin as indicator, (ii) estimation of lead with potassium chromate in neutral pH using orthochrome T as indicator, and (iii) estimating mercury with potassium chloride using bromophenol blue as indicator.

3.7 COMPLEXATION TITRATIONS

Complexation or complex formation reactions used in volumetric analysis involve titrations with the chelating agents (also called complexones). The titrations are also known as *chelometric titrations*. Chelating agents include ethylenediamine tetra acetic acid (EDTA), nitrilotriacetic acid (NTA), 1,2-diaminocyclohexane–*N,N,N′,N′*-tetra acetic acid (CDTA), ethylene glycol bis-(2-aminoehtyl ether)-N,N,N',N'-tetra acetic acid (EGTA), and so on. The structures of some of the chelating agents are shown below.

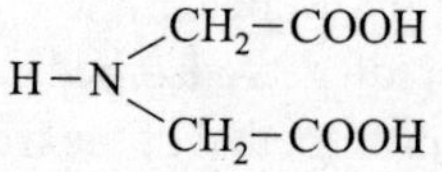

NTA

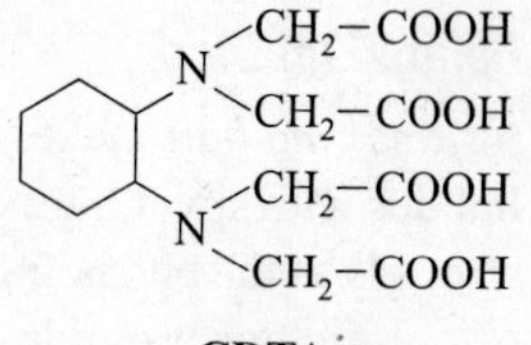

CDTA

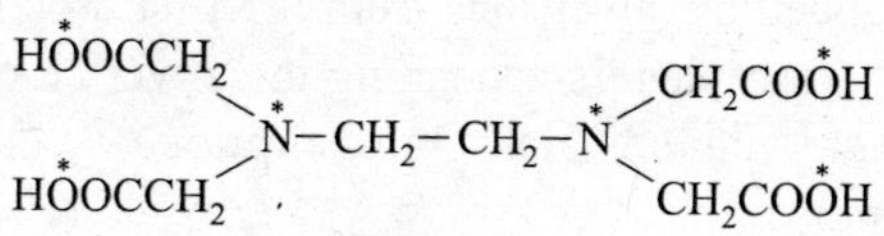

EDTA

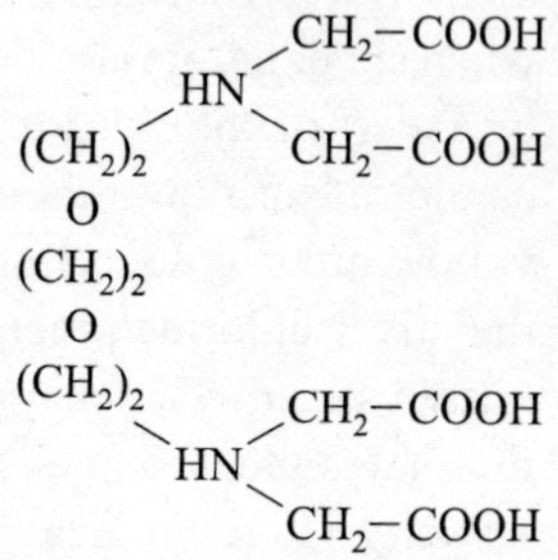

EGTA

EDTA is the most commonly used chelating agent. CDTA forms more stable complexes compared to EDTA but reacts with metal ions rather slowly, and hence, the end point in a titration is not very sharp. EGTA, on the other hand, is used mainly in the case of determination of calcium in a mixture of calcium and magnesium.

EDTA has four titratable protons and the stepwise dissociation constants may be represented as

$$H_4Y \rightleftarrows H_3Y^- + H^+ ; \quad K_{a1} = \frac{[H_3Y^-][H^+]}{[H_4Y^-]} = 1.0 \times 10^{-2} \tag{3.14}$$

$$H_3Y^- \rightleftarrows H_2Y^- + H^+ ; \quad K_{a2} = \frac{[H_2Y^{2-}][H^+]}{[H_3Y^-]} = 2.2 \times 10^{-3} \tag{3.15}$$

$$H_2Y^{2-} \rightleftarrows HY^{3-} + H^+ ; \quad K_{a3} = \frac{[HY^{3-}][H^+]}{[HY^{3-}]} = 6.9 \times 10^{-7} \tag{3.16}$$

$$HY^{3-} \rightleftarrows Y^{4-} + H^+ ; \quad K_{a4} = \frac{[Y^{4-}][H^+]}{[HY^{3-}][H^+]} = 5.5 \times 10^{-11} \tag{3.17}$$

EDTA is a hexadentate ligand capable of forming stable, colourless, and water-soluble complexes with most metal ions. The coordination sites in the ligand include four carboxyl oxygen atoms and two amino nitrogen atoms shown with * in the structural formula of EDTA. The chelating agent is commercially available in the form of disodium salt (represented as Na_2H_2Y) for volumetric analysis, the salt being relatively more soluble than the free acid. The availability of suitable indicators called *metal ion indicators* to detect the end point in volumetric titrations makes EDTA as an important reagent for the estimation of almost all the metal ions and also a few anions.

Metal–EDTA complex has an octahedral geometry as shown below. A simple representation of the complex is also shown below.

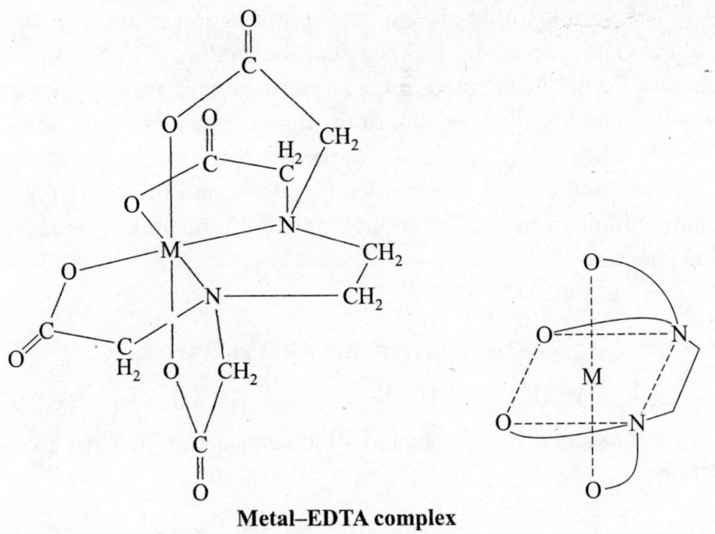

Metal–EDTA complex

Metal ions form stable 1:1 complexes with EDTA almost instantaneously, and hence, EDTA is quite suitable for quantitative analyses of metal ions by volumetry. Divalent metal ions mostly form stable complexes in the pH range of 8–10 (e.g., Ca, Mg, Sr, and Ba) or mildly acidic pH of 4–6 (e.g., Pb, Cu, Zn, Co, Ni, Mn, Fe, Cd, Sn, and Al^{3+}), whereas metal ions of higher oxidation states (e.g., Fe^{3+}, Bi^{3+}, Th^{4+}, Zr^{4+}, etc.) form stable complexes in acidic pH of 1–3.

3.7.1 Metal–EDTA Equilibrium

The complexation reaction may be represented as

$$M^{n+} + Y^{4-} \rightleftarrows MY^{(4-n)-}$$

The equilibrium constant for the above reaction is called the *stability constant* or *formation constant* of the complex and is given as

$$K_f = \frac{[MY^{(4-n)-}]}{[M^{n+}][Y^{4-}]} \qquad (3.18)$$

The logarithms of the stability constant values ($\log K_f$) of EDTA complexes with different metal ions are quite high in the range of 8–25. The $\log K_f$ values for some of the metal–EDTA complexes are given in parentheses:

Mg^{2+} (8.7); Ca^{2+} (10.7); Mn^{2+} (13.8); Fe^{2+} (14.3); Co^{2+} (16.3); Ni^{2+} (18.6); Cu^{2+} (18.8); Zn^{2+} (16.7); Sr^{2+} (8.6); Ba^{2+} (7.8); Cd^{2+} (16.6); Hg^{2+} (21.9); Pb^{2+} (18.0); Al^{3+} (16.3); Fe^{3+} (25.1); Cr^{3+} (24.0); La^{3+} (15.7); Ce^{3+} (15.9); Lu^{3+} (20.0); Th^{4+} (23.2); Ag^{+} (7.3); Na^{+} (1.7)

The metal–EDTA equilibrium is affected by the pH of the medium due to the competing reaction of protonation of the chelating anion ethylenediamine tetraacetate. The actual metal–EDTA equilibrium at any given pH may be represented as

$$M^{n+} + H_4Y \rightleftarrows MY^{(4-n)-} + 4\,H^+$$

As the hydrogen ion concentration increases, the above equilibrium will be shifted to the left as protons compete with the metal for the available coordination sites on the ligand, and at low pH values, protonation reaction predominates and complex formation with the metal will be negligible. The free ligand H_4Y exists as a mixture of Y^{4-}, HY^{3-}, H_2Y^{2-}, H_3Y^-, and H_4Y depending on the pH of the medium. Thus, the actual concentration of Y^{4-} depends on the pH of the medium and is only a fraction of the total concentration of the free (uncomplexed) ligand at a given pH. For calculating the K_f value of a metal complex, it is necessary to calculate the actual concentration of Y^{4-} at the given pH. The concentration of Y^{4-} is calculated at a given pH of the medium from the knowledge of the dissociation constants of the ligand using the following equation.

$$Y^{4-} = \alpha_4 . C_L \qquad (3.19)$$

where C_L is the total concentration of the free ligand given by

$$C_L = [Y^{4-}] + [HY^{3-}] + [H_2Y^{2-}] + [H_3Y^-] + [H_4Y] \qquad (3.20)$$

and α_4 is the fraction of the free ligand which exists as Y^{4-}. The value of α_4 is calculated using the equation

$$\alpha_4 = \frac{K_{a1} K_{a2} K_{a3} K_{a4}}{[H^+]^4 + K_{a1}[H^+]^3 + K_{a1} K_{a2}[H^+]^2 + K_{a1} K_{a2} K_{a3}[H^+] + K_{a1} K_{a2} K_{a3} K_{a4}} \qquad (3.21)$$

where K_{a1}, K_{a2}, K_{a3}, and K_{a4} are the dissociation constants of the free ligand.

The values of α_4 for EDTA at different pH values are 3.7×10^{-14} (pH 2.0), 2.5×10^{-11} (pH 3.0), 3.6×10^{-9} (pH 4.0), 3.5×10^{-7} (pH 5.0), 2.2×10^{-5} (pH 6.0), 4.8×10^{-4} (pH 7.0), 5.4×10^{-3} (pH 8.0), 5.2×10^{-2} (pH 9.0), 3.5×10^{-1} (pH 10.0), 8.5×10^{-1} (pH 11.0), and 9.8×10^{-1} (pH 12.0).

As can be seen from the above discussion, the equilibrium constant K_f for the metal–EDTA complex is dependent on the pH of the medium as the concentration of Y^{4-} varies with pH. Hence, for a given pH a *conditional formation constant* K_f* may be defined as

$$K_f* = K_f \alpha_4 = [MY^{(4-n)-}]/[M^{n+}] C_L \qquad (3.22)$$

Standard solutions of EDTA of the order of 0.05–0.1 M can be prepared from the commercially available free acid or the disodium salt. The free acid can be used as a primary standard after drying at 130–140°C and the standard solution is prepared by dissolving the appropriate weight in a minimum amount of base. Alternatively, the disodium salt $Na_2C_{10}H_{14}O_8N_2 \cdot 2H_2O$ (formula weight of 372.24 contains about 0.3% moisture) can be dissolved in distilled water for use as the moisture content does not vary much. For more accurate determinations, the disodium salt is dried at 80°C for several days in an atmosphere of 50% relative humidity prior to use. The EDTA solution may be standardized with nearly neutral solution of zinc chloride or zinc sulphate prepared from known weight of zinc pellets or with a standard solution of lead prepared from dried lead nitrate.

3.7.2 Titration Curves

The theoretical titration curves for metal–EDTA titration are obtained by plotting pM values as a function of volume of EDTA added during the titration. The construction of theoretical titration curve for the titration of 25 mL of 0.1 M of Mg^{2+} (log K_f = 8.7) and 0.1 M Zn^{2+} (log K_f = 16.7) each separately with 0.1 M EDTA at pH 10 is illustrated by calculating the pM values for incremental addition of specific volumes of EDTA in four stages.

(i) Before the addition of EDTA, the number of millimoles (mmol) of metal = 25 mL × 0.1 M = 2.5 mmol.

$$p\text{Mg}^{2+} \text{ or } p\text{Zn}^{2+} = -\log [\text{M}^{2+}] = -\log (0.1 \text{ M}) = 1.00$$

(ii) Before attaining the equivalence point, after the addition of 10 mL of EDTA. The conditional formation constant is quite large for both the metal–EDTA complexes, and hence, dissociation of the MY^{2-} to give M^{2+} may be neglected and the free metal concentration is given as

$$[\text{Mg}^{2+}] = (25 \text{ mL} \times 0.1\text{M}) - (10\text{mL} \times 0.1 \text{ M}) = 15.0 \text{ mL} \times 0.1 \text{ M/35 mL}$$
$$= 2.5 \text{ mmol} - 1 \text{ mmol} = 1.5 \text{ mmol/35 mL} = 0.0429 \text{ M}$$
$$p[\text{Mg}^{2+}] = -\log (0.0429) = 1.37 ; \text{ Similarly } p[\text{Zn}^{2+}] = 1.37$$

(iii) At the equivalence point, all the metal ions exist only as the metal–EDTA complex. The concentration of the metal is 2.5 mmol/50 mL = 0.05 M. The free metal concentration his calculated using Eq. (3.22). For Mg–EDTA titration, $K_f = 5.01 \times 10^8$ and α_4 at pH 10 = 0.35.

$$5.01 \times 10^8 \times 0.35 = 1.75 \times 10^8 = 0.05/[\text{M}^{2+}]C_L$$

Since at the equivalence point $[\text{M}^{2+}] = C_L$

$$[\text{Mg}^{2+}] = (0.05/1.75 \times 10^8)^{1/2} = 1.69 \times 10^{-5}$$
$$p[\text{Mg}^{2+}] = -\log (1.69 \times 10^{-5}) = 4.77$$

For the Zn–EDTA system $K_f = 5.01 \times 10^{16}$ and α_4 at pH 10 = 0.35.

$$5.01 \times 10^{16} \times 0.35 = 1.75 \times 10^{16} = 0.05/[\text{M}^{2+}] C_L$$

Since $[\text{M}^{2+}] = C_L$ at the equivalence point

$$[\text{Zn}^{2+}] = (0.05/1.75 \times 10^{16})^{1/2} = 1.69 \times 10^{-9}$$
$$p[\text{Zn}^{2+}] = -\log (1.69 \times 10^{-9}) = 8.77$$

(iv) After the equivalence point, the concentration of the ligand C_L is equal to the concentration of the ligand added as the liberation of free ligand by the dissociation of the complex MY^{2-} is negligible (and may be neglected). Thus, after the addition of 25.1 mL of EDTA, the concentration of MY^{2-} will be 0.0499 M and the free ligand concentration will be 0.1 mL $\times$ 0.1 M/50.1 mL = 2.0×10^{-4} M. Therefore, the concentration of Mg^{2+} is arrived at from Eq. (3.22) as

$$0.0499/([\text{Mg}^{2+}] \times 2.0 \times 10^{-4}) = 1.75 \times 10^8$$
$$[\text{Mg}^{2+}] = 1.42 \times 10^{-6} \text{ and } p[\text{Mg}^{2+}] = 5.84$$
$$\text{Similarly, } p[\text{Zn}^{2+}] = -\log (1.43 \times 10^{-14}) = 13.84$$

Similarly, after the addition of 30 mL of EDTA, the concentration of MY^{2-} will be 0.045 M and the free ligand concentration will be 5 ml $\times$ 0.1 M/55 mL = 9.09×10^{-3} M. Therefore, the concentration of Mg^{2+} is arrived at from Eq. (3.22) as

$$0.045/([\text{Mg}^{2+}] \times 9.09 \times 10^{-3}) = 1.75 \times 10^8$$
$$[\text{Mg}^{2+}] = 2.6 \times 10^{-8} \text{ and } p[\text{Mg}^{2+}] = 7.59$$
$$\text{Similarly, } p[\text{Zn}^{2+}] = -\log (2.83 \times 10^{-16}) = 15.55$$

A plot of the calculated values of pM against volume of EDTA in mL gives the theoretical titration curve (Fig. 3.9).

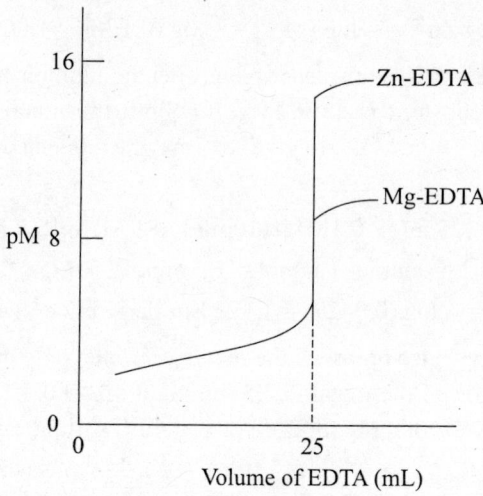

Fig. 3.9 Titration curve for complexation reaction

As can be seen from Fig. 3.9, the larger the value of log K_f (the more stable the chelate), the larger will be change in pM values at the end point. The titration may also be carried out at a lower pH for more stable chelates. The determination of a metal forming more stable complex can be carried out in the presence of another metal which forms a weak or less stable complex particularly at lower pH.

Based on the stability constant values, the metals may be brought under three groups. One group of metals (e.g., Fe^{3+}, Th^{4+}, Sc^{3+}, and Hg^{2+}) whose stability constant values are quite high (log K_f values in the range 20–25) can be titrated at low pH values in the range 2–4. The second group of metals (e.g., Cu^{2+}, Ni^{2+}, Zn^{2+}, Al^{3+}, Mn^{2+}, La^{3+}, and Fe^{2+}) with log K_f values in the range 12–20 can be titrated in the pH range 4–7. The third group of metals (e.g., Ca^{2+}, Sr^{2+}, and Mg^{2+}) form weak complexes (log K_f values in the range of 8–12) require a pH of ~ >10 for titrations with EDTA. Thus, the first group of metals can be titrated in the presence of others and similarly, the second group of metals can be titrated in the presence of the third group of metals. At higher pH range, all metals can be titrated. However, a few metals undergo hydrolysis and precipitate out in alkaline conditions (e.g., Fe^{3+} and Th^{4+}), and hence, cannot be titrated directly. These metal ions can be determined by back titration or by the use of *masking agents* to prevent precipitation.

During EDTA titrations, masking is made use of to prevent interference from a second metal that may be present in the analyte sample. Masking may be achieved by precipitation, complex formation, or oxidation or reduction, or kinetically by making use of a slow reaction between the interfering metal and EDTA. For example, in the titration of bismuth in the presence of lead, the latter can be precipitated as sulphate. Cyanide ions mask Cd, Co, Cu, Ni, and Zn so that calcium and magnesium can be determined by EDTA titration. Similarly, copper can be masked by complexing with ammonia during a titration with EDTA using murexide indicator. Chromium(III) reacts very slowly with EDTA, and hence, other metal ions can be titrated in its presence. Copper(II) can be masked by reducing it to Cu(I) with ascorbic acid or complexing with iodide.

3.7.3 Metal Ion Indicators

The use of EDTA in volumetric analyses depends on the availability of suitable indicators which show a colour change at the end point. Indicators used in complexation titrations are generally known as *metal ion indicators* because these indicators indicate the presence or absence of the analyte metal ion through an appropriate change in colour. The prerequisites of a substance to function as a metal ion indicator are as follows: (i) the indicator must be at least selective if not specific in forming a coloured complex with the metal, (ii) the indicator must form a stable complex with the metal but the metal-indicator complex should be less stable compared to the metal–EDTA complex so that the free indicator is liberated instantaneously at the equivalence point, (iii) the colour of the free indicator should be distinctly different from that of the metal-indictor complex, and (iv) the indicator should be very sensitive to even very low concentration of the metal ion so that the colour change between the metal-indicator complex and the free indicator is sharp and close to the equivalence point.

Dyes are usually used as metal ion indictors in EDTA titrations as they are also chelating agents. The dyes are also organic acids or bases and undergo protonation–deprotonation reactions depending on the pH of the medium. Hence, EDTA titrations have to be carried out in buffered solutions depending on the dye used as indicator. Typical examples of dyes used as metal ion indicators, the pH range, and the colour change from the metal-indicator complex to the free indicator are listed below:

(i) *Eriochrome black* T (EBT) also called *solochrome black* (1-(1-hydroxy-2-naphthylazo)-6-nitro-2-naphthol-4-sulphonate) is sensitive to Mg, Zn, Mn, Ca, Cd, Hg, and Pb showing a colour change from red (M-In complex) to blue (free In) at pH = 10.

(ii) *Murexide* (ammonium purpuriate) sensitive to Cu, Ni, Co, Ca, and lanthanides shows a colour change from yellow to blue violet (for Ni and Co), orange to blue violet (for Cu), and red to blue violet in the pH range 10–11.

(iii) *Xylenol orange* (3,3′[N,N-di(carboxymethyl)-aminomethyl]-O-cresol- sulphonphthalein) is sensitive to Mn, Ni, Cd, Pb, and Sn in the pH range of 4–6, to Bi, Co, Zn, and Th in the pH range 1–2, and shows a colour change from red to lemon yellow.

(iv) *Methylthymol blue* (thymolsulphonaphthalein di(methylimine diacetic acid)) is sensitive to Co, Zn, Cd, Hg, Al, Ni, and Mn in the pH range 4–6; Be, Ti, Zr, and Hf at pH 0–2 showing colour change from blue to yellow. It is also useful for titrations involving alkaline earth metals, Ca, Sr, Ba, and Mg at pH 12, the colour change at the end point being blue to colourless or grey.

(v) *Patton and Reeder's* indicator (HHSNNA; 2-hydroxy-1-(2-hydroxy-4-sulpho-1-naphthylazo)-3,6-disulphonic acid) is specific for Ca. Its working pH range is 12–14 where it shows a colour change from wine red to blue.

(vi) *Bromopyrogallol red* (dibromopyrogallol sulphonphthalein) is sensitive to many cations and selective to Bi, shows a colour change from blue to red at pH 2–3.

(vii) *Variamine blue* (4-methoxy-4′-aminodiphenylamine) is specific for Fe(III) at pH 3 showing a colour change from blue to yellow.

The intense colour is discernible to the eye even at very low concentrations of the metal ion in the range of 10^{-6} M. These indicators satisfy the other prerequisites, namely the colour of the free indicator is different from that of the metal ion indicator complex and the metal-indicator complex is usually less stable compared to the complexes formed by better chelating agents such as EDTA, NTA, etc. Most of the metal–EDTA complexes have stability constant values in the range of 10^8 to 10^{25}, whereas the metal ion-indicator complexes have stability constant values in the range of 10^4–10^6. Hence, EDTA is capable of displacing the indicator from the metal ion-indicator complex thereby liberating the free indicator at the end point during a titration.

3.7.4 Theory of Metal Ion Indicators

The principle involved in the use of a typical metal ion indicator during the titration of a metal ion with a chelating agent, such as EDTA, may be conveniently explained in the case of eriochrome black T (EBT) whose structure is shown below.

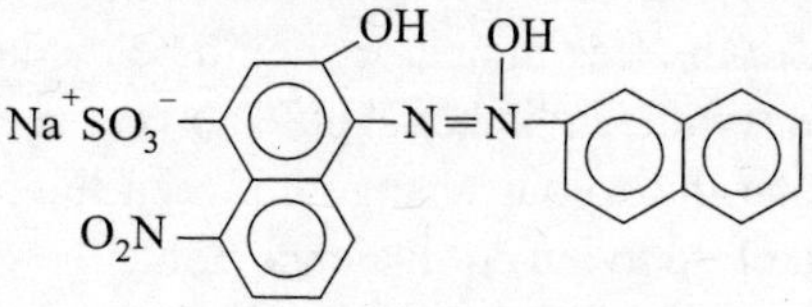

Eriochrome black T

The indicator is a weak acid with three replaceable protons. The self-ionization of the indicator may be represented as

$$H_2In^- + H_2O \rightleftharpoons HIn^{2-} + H_3O^+; \quad K_1 = 5 \times 10^{-7}$$
$$HIn^{2-} + H_2O \rightleftharpoons In^{3-} + H_3O^+; \quad K_2 = 2.8 \times 10^{-15}$$

The conjugate acid–base pairs have different colours and the indicator behaves as an acid–base indicator also.

The EBT–metal ion complexes of most of the divalent metal ions (MIn^-) are red in colour, whereas the free HIn^{2-} is blue in colour. If the pH of the reaction medium is distinctly alkaline, the free indicator exists as HIn^{2-}. The complexometric titration for the determination of the concentration of a given metal ion (analyte) involves titrating a known volume of the metal ion solution under buffered conditions against a standard solution of disodium salt of EDTA (HY^{2-}). The analyte is buffered to pH = 10 with ammonia–ammonium chloride buffer and a small amount 40–50 mg of the EBT indicator mixture (prepared by mixing and grinding about 100 mg solid EBT with about 1 g of NaCl) is added. The red-coloured solution is titrated against a standard solution of disodium salt of EDTA. As the titration progresses the added EDTA reacts with the free metal ion. Closer to the end point in the absence of any free metal ion, the added EDTA reacts with the metal ion-indicator complex, liberating the free indicator. Thus, the end point of the titration is indicated by a colour change from red (of the metal–EBT complex) to blue (of the free indicator). The change in the pM values at the equivalence point is $>10^4$ in order and the end point is close to the equivalence point.

The formation of the metal ion-indicator complex and subsequent liberation of the free indicator may be represented as

$$M^{n+} + HIn^{2-} \rightarrow MIn^- + 2\,H^+$$
$$MI^{n-} + H_2Y^{2-} \rightarrow MY^{2-} + HIn^{2-}$$

    (wine red)   (blue)

Eriochrome black T forms stable red-coloured complexes with more than twenty different metal ions, but only a few of the complexes have stability constants that are appropriate for end-point detection. The indicator is an ideal choice for titrating magnesium or zinc with EDTA but is not satisfactory for calcium because the calcium–EBT complex is very weak (about one-fortieth compared to that of magnesium–EBT complex) and the free indicator is liberated well before the end point.

3.7.5 Types of EDTA Titrations

EDTA titrations may be broadly classified into four types, depending on the mode of using EDTA for complexing the metal ion. These include (i) direct titration, (ii) back titration, (iii) substitution or replacement titration, and (iv) miscellaneous methods involving exchange or precipitation reactions.

In *direct titration*, the metal ion solution is buffered and titrated directly with disodium salt solution of EDTA at a desired pH (depending on the nature of metal ion indicator used). At the end of titration, the colour of the metal ion-indicator changes, indicating the end point of the titration. For example, in titration of zinc(II) or magnesium(II) ions buffered at pH = 10 (ammonia–ammonium chloride buffer) using eriochrome black T as the indicator the colour change at the end point is wine red to steel blue. Nickel(II) can be titrated directly with EDTA in the presence of murexide indicator.

Sometimes, the analyte metal ion cannot be directly titrated with EDTA. At other times, the reaction between EDTA and the metal ion is relatively slow or may be the detection in colour change is difficult (estimation of nickel using EBT as the indicator) or a suitable indicator is not available. In such cases, *back titration* is used. In this process, a solution of disodium salt of EDTA is added in excess (known amount) to the nickel solution (analyte) of known volume. Ammonia–ammonia chloride buffer is used to bring the pH of the solution to 10. After this, EBT indicator is added. As EBT is in its free form, the solution is blue in colour. A solution of zinc or magnesium sulphate is taken in a burette. This solution is used for titration against the reaction mixture containing excess EDTA. EDTA then reacts with the sulphate solution of zinc or magnesium. Once all the EDTA has exhausted, on addition of another drop of the metal solution a red-coloured metal ion complex is formed.

Chromium(III) and cobalt(III) react with EDTA very slowly, and hence, a direct titration is not suitable for the estimation of these metal ions. Hence, back titration method is adopted by adding a known excess of the EDTA solution to the analyte and determining the excess unreacted EDTA by back titration with a standard solution of zinc sulphate after buffering at pH =10 using EBT as the indicator.

Substitution or *replacement titration* is adopted when the analyte metal ion forms complexes with EDTA more stable than the complexes formed by magnesium or calcium and also when the analyte metal ion does not react or reacts unsatisfactorily with the indicator or when no suitable indicator is available. In the estimation of such metal ions a known excess of magnesium–EDTA complex is added to the analyte metal ion. The analyte metal ion displaces magnesium from the magnesium–EDTA complex as it forms a relatively more stable complex with EDTA.

$$M^{n+} + Mg\text{-EDTA} \rightarrow M\text{-EDTA} + Mg^{2+}$$

The liberated magnesium is estimated by direct titration with a standard solution of disodium salt of EDTA and the amount of magnesium is equivalent to the analyte metal ion present.

Another example of this type of titration is that of calcium which gives poor end point with EBT indicator. However, an improved end point is obtained if the titration is carried out in the presence of magnesium, as calcium displaces magnesium from its EDTA complex.

Miscellaneous methods include exchange reactions or precipitation reactions. For example, silver and gold cannot be titrated directly. Hence, an exchange reaction between tetracyanonickelate(II) and silver or gold liberates equivalent amount of nickel which is determined by titrating with a standard solution of disodium salt of EDTA using murexide as indicator.

$$[Ni(CN)_4]^{2-} + 2\,Ag^+ \rightarrow 2[Ag(CN)_2]^- + Ni^{2+}$$

Such exchange reactions are also helpful in the estimation of anions, such as chloride, bromide, iodide, and thiocyanate, which form sparingly soluble salts with silver. The anion is precipitated with silver nitrate and the silver salt is dissolved in a solution of tetracyanonickelate(II). The equivalent amount of nickel liberated is determined by EDTA titration.

Precipitation reactions are also used for determining the analyte indirectly by complexometric titrations. Thus, sulphate is precipitated as $BaSO_4$ or $PbSO_4$ and the separated precipitate is dissolved in an excess of standard solution of disodium salt of EDTA and excess of EDTA is back-titrated with a standard magnesium or zinc solution using EBT as indicator. Similarly, fluoride is precipitated as lead chlorofluoride, the separated precipitate is dissolved in dilute nitric acid. The liberated lead is titrated with a standard solution of EDTA after adjusting the pH to 5–6 and using xylenol orange as indicator. Phosphate is precipitated as $Mg(NH_4)$ $PO_4.6H_2O$ and the precipitate is dissolved in dilute hydrochloric acid to liberate magnesium. A known excess of standard EDTA solution is added to the reaction mixture, the pH of the solution is adjusted to 10, and the excess EDTA is back titrated with standard magnesium solution.

3.7.6   Applications of EDTA Titrations

EDTA titrations are used for the determination of almost all metal ions and a few anions, such as phosphate and sulphate, by direct or indirect titrations.

Estimation of metal ions is discussed here with a few typical examples.

Magnesium and zinc can be titrated with standard solution of EDTA using EBT indicator in ammonia–ammonium chloride buffer (pH = 10). The buffer is prepared by mixing 17.5 g of ammonium chloride and 142 mL of liquor ammonia (sp. gravity ~ 0.90) and diluting with distilled water to 250 mL. An aliquot of 20 mL (~ 0.05 M) of the analyte (Mg^{2+} or Zn^{2+}) is pipetted out accurately into a conical flask. Approximately 20 mL of the freshly prepared ammonia–ammonium chloride buffer is added, followed by about 50 mg of the EBT indicator mixture. The contents of the flask are titrated against a standardized (~ 0.05–0.1 M) solution of disodium salt of EDTA. The end point of the titration is indicated by the change of the wine red colour (of the metal-indicator complex) to steel blue (free indicator colour). From the volume and known strength of the EDTA solution, the analyte concentration is calculated using the formula

$$1 \text{ mol EDTA} \equiv 1 \text{ mol of } Mg^{2+}/Zn^{2+}$$

Nickel can be directly titrated against standard solution of EDTA (~0.05–0.1 M) by using murexide indicator. The indicator is prepared by mixing about 0.1 g of the solid indicator with about 10 g of solid potassium nitrate. Nickel reacts with EDTA rather slowly, and hence, the titration must be carried out slowly by dropwise addition of EDTA solution particularly near the end point. An aliquot of nickel solution (20 mL of ~0.05 M) is pipetted out accurately into a conical flask and diluted to approximately 100 mL with distilled water. About 50 mg of the murexide indicator mixture is added followed by about 10 mL of 1 M ammonium chloride solution (5.35 g/100 mL). A concentrated liquor ammonia solution is added dropwise to bring the pH to about 7 as shown by the yellow colour of the solution. The analyte solution is titrated with EDTA. When the end point is near, the solution is made distinctly alkaline by the addition of about 10 mL of liquor ammonia (pH ~10) and the titration is continued till the yellow colour changes to violet indicating the end point. If the pH of the final solution is less than 10, an orange–yellow colour develops, and it is necessary to add more of liquor ammonia to make the solution clear yellow.

Alternatively, nickel may be titrated directly with EDTA using bromopyrogallol red indicator in a buffer solution of ammonia (1 M)–ammonium chloride (1 M). The indicator is prepared by dissolving about 50 mg of bromopyrogallol red in about 100 mL of 50% ethanol. The analyte solution (20 mL) is diluted to about 100 mL with distilled water, mixed with about 10–15 drops of the indicator solution and buffered with 10 mL of the buffer solution and titrated with EDTA solution till the end point is indicated by a colour change from blue to claret red.

Determination of hardness of water is an important application of EDTA titrations.

Hardness is an important parameter for assessing the quality of water for domestic as well as industrial use. When water does not produce lather readily with soap, it is said to be *hard water*. The presence of dissolved salts of calcium, magnesium, and to a smaller extent of strontium and iron is responsible for the hardness of water. The anions associated with these metal ions include chloride, sulphate, and bicarbonate but these do not contribute to the hardness of water. When soap is applied to water, these metal ions form insoluble precipitates with fatty acid components (e.g., stearic, palmitic, oleic acids, etc.) of soap.

$$M^{2+} + 2\,C_{17}H_{35}COONa \rightarrow M(C_{17}H_{35}COO)_2 + 2\,Na^+$$

(M = Ca, Mg)   Sodium stearate   Calcium/Magnesium stearate
           (soluble)   (insoluble)

Thus, the soap applied to hard water is utilized initially to precipitate out the hardness-causing metal ions as fatty acid salts. Further addition of soap is required to initiate lather formation in the case of hard water. Hard water consumes more amount of soap to produce the same amount of lather compared to *soft water*.

Hard water has several domestic as well as industrial disadvantages.

- Due to its bad taste, it is not used as potable water.
- Consumption of hard water causes formation of calcium oxalate in kidneys leading to formation of stones.
- Organoleptic qualities (i.e., taste, smell, sight, and touch) of food cooked in hard water are affected due to dissolved salt deposition.
- Cooking time increases when pulses, beans, and cereals are cooked in hard water leading to increased consumption of fuel and time.
- Hard water affects the cleaning and lather-forming ability of soap, and thus, bathing and washing consumes more soap.
- Manufacturing industries, such as textile dyeing and printing, pharmaceutical, processed foods, fine chemicals and sugar do not accept hard water usage.
- The water used for boilers should have zero hardness as hard water may lead to boiler explosion.

Classification of hardness of water is based on whether the hardness is destroyed or not by boiling the water: (i) *carbonate hardness*, also known as *temporary hardness*, is destroyed by boiling the water, whereas (ii) *non-carbonate hardness or permanent hardness* is not destroyed by boiling the water. The *total hardness* of water is the sum of temporary and permanent hardness.

Presence of calcium and magnesium bicarbonates in water causes *temporary hardness* of water. When water is boiled, these salts decompose to form insoluble calcium carbonate and magnesium hydroxide which can be filtered. The filtered water is free from temporary hardness.

$$Ca(HCO_3)_2 \rightarrow CaCO_3 + H_2O + CO_2$$
$$Mg(HCO_3)_2 \rightarrow Mg(OH)_2 + 2\ CO_2$$

Permanent hardness cannot be removed by boiling the water and is caused mainly due to the presence of chlorides and sulphates of calcium and magnesium and to a minor extent of iron and other heavy metals.

Determination of different types of hardness of water is carried out by EDTA titrations. The older method of determining the hardness from the thickness of the lather layers formed by the test sample of water and 'standard' hard water under identical conditions, has been replaced by the more reliable and easy to perform EDTA titration. The basic principle involves the determination of the content of the hardness-causing metal ions (calcium and magnesium) by titrating a known volume of sample water with a standard solution of disodium salt of EDTA.

Determination of total hardness of water is carried out by taking a known volume of sample water in a conical flask. The analyte is buffered to pH = 10 with ammonia–ammonium chloride buffer and 40–50 mg of EBT indicator is added. The contents of the flask are titrated against a standard solution of disodium salt of EDTA taken in the burette. Initially, the indicator forms wine red-coloured metal ion-indicator complex with a small amount of the calcium and magnesium present in the sample water. As the titration progresses the added EDTA reacts with the freely available calcium and magnesium ions in the sample water. When all the freely available calcium and magnesium are exhausted, the added EDTA displaces the indicator from the metal ion-indicator complex liberating the blue-coloured free indicator HIn^{2-}. Thus, the colour change at the end point is red to blue in this titration. The reactions may be represented by equations given below where M is the mixture of calcium and magnesium present in the sample water.

$$M^{n+} + HIn^{2-} \rightarrow MIn^- + 2\ H^+$$
$$MIn^- + H_2Y^{2-} \rightarrow MY^{2-} + HIn^{2-}$$
$$\text{(wine red)} \qquad\qquad \text{(blue)}$$

The total amount of calcium and magnesium present in the sample water corresponds to the total hardness. The total hardness is calculated on the basis of the stoichiometric relationship

1000 mL of 1 M EDTA = 1 g atomic weight of metal

The total hardness of water is expressed in terms of equivalent amount of calcium carbonate in units of parts per million (ppm) or milligram per litre (mg/L) by using the formula

1 mL of 1 M EDTA = 100 mg of $CaCO_3$

Permanent hardness is determined after destroying temporary hardness of the sample water. Temporary hardness is destroyed by boiling the sample water of known volume for about 10 min. The bicarbonates of calcium and magnesium contributing to the temporary hardness undergo thermal decomposition and precipitate as calcium carbonate and magnesium hydroxide. The sample water is then cooled to room temperature and filtered through a Whatman No.1 filter paper to collect the precipitate. The precipitate collected on the filter paper is washed with distilled water so that any metal ions adhering to the precipitate are removed and the washings are added to the filtrate. The filtrate is buffered to pH = 10 with ammonia–ammonium chloride buffer and titrated against the same standard solution of disodium salt of EDTA using eriochrome black T indicator. The total amount of calcium and magnesium is calculated using the stoichiometric formula and expressed in terms of calcium carbonate in ppm unit. The hardness determined after destroying the temporary hardness corresponds to the permanent hardness of the sample water.

Temporary hardness or *carbonate hardness* of water cannot be determined directly by EDTA titration but calculated by subtracting the amount of permanent hardness from the value of total hardness. Carbonate hardness of water (responsible for the alkalinity of water) can be estimated by neutralization titration.

It is necessary to know the amount of calcium and magnesium contents separately for softening of hard water by lime–soda process. The amount of hardness of water due to calcium alone (calcium hardness) can be determined by EDTA titration. A known volume of the sample water is buffered to pH = 12 with potassium hydroxide and triethylamine. This leads to precipitation of magnesium as magnesium hydroxide. This solution is titrated against a standard solution of disodium EDTA using Patton and Reeder indicator. The end point of the titration is indicated by the colour change from wine red to blue. There is no interference from precipitated magnesium hydroxide during titration. The amount of calcium in the sample water is calculated using the stoichiometric relationship and the calcium hardness is expressed in terms of calcium carbonate in units of ppm. The amount of magnesium hardness can be obtained by subtracting the calcium hardness value from the value of total hardness.

3.8 REDOX TITRATIONS

Oxidation–reduction reactions which occur stoichiometrically and rapidly find use in volumetric titrations for quantitative analyses. The analyte may undergo oxidation or reduction when it reacts with a titrant and accordingly the titrant will be an oxidizing or a reducing agent. Oxidizing agents commonly used in redox titrations include ceric sulphate, potassium permanganate, potassium dichromate, and iodine. Iron(II) and sodium thiosulphate are used as reducing agents.

The oxidation–reduction reaction may simply be represented in general as

oxidant + $ne \rightleftarrows$ reductant

As the oxidation–reduction reaction involves the transfer of electron(s), the equilibrium constant for this reaction K is related to electromotive force (emf) or the redox potential E. The potential of the redox system may be expressed in terms of Nernst equation at 25°C as

$$E = E^0 + (0.0591/n) \log ([\text{oxidant}]/[\text{reductant}]) \tag{3.23}$$

where E and E^0 represent the redox potential at any given point during the reaction (titration) and standard redox potential of the system, respectively, and n represents the number of electrons involved in the reaction. A sudden change in the potential of the two redox systems at the equivalence point provides the means of detecting the end point. The end point of a redox titration may be determined by monitoring the potential developed in the reaction mixture as a function of the added volume of titrant using a suitable indicator electrode in conjunction with a potentiometric set-up. Alternatively, a redox indicator which shows a colour change at the end point may be used.

The progress of the titration and the equivalence point may be computed theoretically in the form of a titration curve as shown in the case of acid–base, precipitation, and complexation titrations. For example, the titration of 50 mL of 0.1 M solution of ferrous iron with 0.1 M ceric solution can be carried out in the presence of sulphuric acid. Ceric sulphate ($Ce(SO_4)_2$, MW = 333.25) and ammonium ceric sulphate ($(NH_4)_4[Ce(SO_4)]_4.2H_2O$, MW = 632.56) are primary standards. A standard solution of cerium(IV) can be prepared by weighing accurately the salt and dissolving in a known volume of distilled water.

The reaction between cerium(IV) and iron(II) may be represented as

$$Ce^{4+} + Fe^{2+} \rightleftarrows Ce^{3+} + Fe^{3+}$$

where ferrous iron gets oxidized to ferric and ceric is reduced to cerous. The standard redox potential for the Fe^{3+}/Fe^{2+} couple is 0.76 V and for the Ce^{4+}/Ce^{3+} couple it is 1.45 V.

The change in the potential for incremental addition of ceric solution may be calculated considering the Fe^{3+}/Fe^{2+} couple or the Ce^{4+}/Ce^{3+} couple. Before the addition of the titrant (ceric) the potential cannot be calculated as Fe^{3+} is totally absent. After the addition of 5 mL of ceric solution, the potential may be calculated using the Fe^{3+}/Fe^{2+} couple, the potential depending on the ratio of ($[Fe^{3+}]/[Fe^{2+}]$).

$$\begin{aligned} E_{Fe} &= E°_{Fe} + (0.0591/1) \log ([Fe^{3+}]/[Fe^{2+}]) \\ &= 0.76 + 0.0591 \log (5/45) \\ &= 0.76 - 0.057 = 0.69 \text{ V} \end{aligned}$$

The exact concentrations of Fe^{3+} and Fe^{2+} may also be calculated by taking into account the total volume.

$$[Fe^{3+}] = 5 \text{ mL} \times 0.1M/55 \text{ mL} = 0.00909 \text{ M}$$
$$[Fe^{2+}] = 45 \text{ mL} \times 0.1 \text{ M}/55 \text{ mL} = 0.0818 \text{ M}$$

Hence, $E_{Fe} = E°_{Fe} + (0.0591/1) \log (0.00909/0.0818) = 0.69$ V.

Similar calculations for the addition of increments of ceric solution before the end point can be carried out.

At the equivalence point (when 50 mL of 0.1 M ceric solution has been added), the concentrations of Fe^{3+} and Ce^{3+} are exactly equal and the species Fe^{2+} and Ce^{4+} are totally absent, and hence, no redox couple exists. Thus, at the equivalence point, the potential of the system is given by the relationship

$$(E°_{Fe} + E°_{Ce})/2 = (0.76 + 1.45)/2 = 1.105 \text{ V}$$

After the equivalence point, the system contains Ce^{4+}/Ce^{3+} couple as excess of ceric solution is added. The potential of the system is calculated by using the following equation.

$$E_{Ce} = E°_{Ce} + (0.0591/1) \log ([Ce^{4+}]/[Ce^{3+}])$$

Thus, when 50.1 mL of ceric solution has been added, the ratio of ($[Ce^{4+}]/[Ce^{3+}]$) 0.1/50, and hence,

$$E_{Ce} = 1.45 + 0.0591 \log 0.1/50 = 1.29 \text{ V}$$

Alternatively, the exact concentrations of Ce^{4+} and Ce^{3+} may be calculated as

$$[Ce^{4+}] = 0.1 \text{ mL} \times 0.1 \text{ M}/100.1 \text{ mL} = 0.00999 \text{ M}$$
$$[Ce^{3+}] = 50 \text{ mL} \times 0.1 \text{ M}/100.1 \text{ mL} = 0.0499 \text{ M}$$
$$E_{Ce} = 1.45 + 0.0591 \log (9.99 \times 10^{-5}/0.0499) = 1.29 \text{ V}$$

The theoretical titration curve obtained by plotting the potential of the redox system as a function of the volume of titrant is shown in Fig. 3.9.

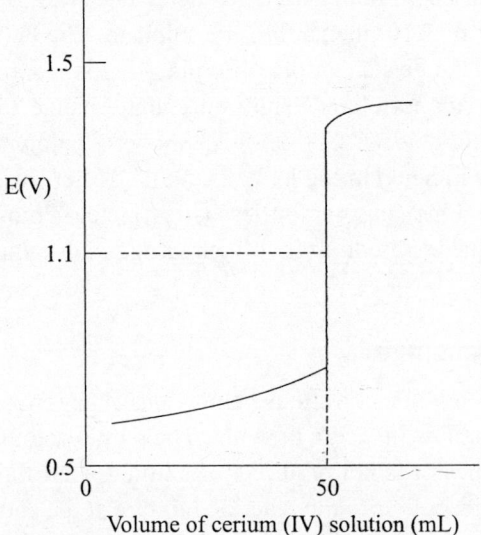

Fig. 3.10 Titration curve of ferrous vs ceric solution

As can be seen, from Fig. 3.10, the potential of the redox system increases gradually till almost the equivalence point is reached. At the end point, the potential changes abruptly and once again after the equivalence point, the change in the potential is gradual. The end point of the redox titration is indicated by the midpoint of inflection zone and the volume of titrant corresponding to the end point is obtained graphically by dropping a perpendicular to the x-axis.

The end point may be determined practically by adding a suitable redox indicator which shows a colour change at the end point.

3.8.1 Redox Indicators

Redox indicators are chemical species which exhibit different colours in their oxidized and reduced forms. At or close to the end point of the redox titration, the indicator also undergoes oxidation or reduction reaction which may be represented as

$$In_{ox} + ne^- \rightleftarrows In_{red}$$

where In_{ox} and In_{red} represent the oxidized and the reduced forms, respectively, of the indicator having different colours. The indicator changes colour at a potential as given by Nernst equation.

$$E = E^0_{In} + (0.059/n) \log ([In_{ox}]/[In_{red}]) \tag{3.24}$$

where E^0_{In} is the standard potential of the indicator. The colour change is detectable by the naked eye depending on the ratio of ($[In_{ox}]/[In_{red}]$). The colour of the In_{ox} species dominates when the ratio is 10 and that of the reduced form when the ratio is 0.1 provided the intensities of the colours of the two species are comparable. Under such circumstances, the potential range for a detectable colour change is given by the following equation.

$$E = E^0_{In} \pm (0.059/1) \tag{3.25}$$

A sharp colour change is observed only when the standard potentials of the indicator and the redox systems differ by at least 0.15 V.

Redox indicators commonly used for detecting the end points in redox titrations include ferroin a complex of 1,10-phenanthroline sulphate. The Fe(III) complex $[Fe(phen)_3]^{3+}$ is pale blue in colour and the Fe(II) complex $[Fe(phen)_3]^{2+}$ is red in colour. The ferroin indicator (E^0_{In} = 1.11 V in 1 M sulphuric acid) shows a colour change from pale blue (In_{ox}) to red (In_{red}), N-phenylanthranilic acid (E^0_{In} = 0.89 V) shows a colour change from purple red (In_{ox}) to colourless (In_{red}), diphenylamine [E^0_{In} = 0.76 V, violet (In_{ox}) to colourless (In_{red})], starch-I_3^-, KI [E^0_{In} = 0.53 V, blue (In_{ox})-colourless (In_{red})], etc. Almost all the redox indicator reactions taking place in aqueous medium involve hydrogen ions, and hence, the observed end point is pH dependent.

3.8.2 Permanganometry

Permanganometry involves titrations using potassium permanganate as the oxidizing agent for quantitative estimations in acidic medium. The purple colour of the permanganate is so intense that a separate indicator is not required. The titrant itself functions as a self-indicator giving a permanent pale pink colour to the reaction mixture at the end point of the titration. In sulphuric acid medium permanganate oxidizes reducing species, such as iron(II), and undergoes reduction to manganese(II). The reaction is quite fast and quantitative.

$$MnO_4^- + 8\ H^+ + 5\ e^- \rightarrow Mn^{2+} + 4\ H_2O\ ; E^0 = 1.51\ V$$

The reduction of Mn(VII) to Mn(II) involves five electrons, and hence, the equivalent weight of $KMnO_4$ is one-fifth of its molecular weight. Potassium permanganate is capable of oxidizing chloride ions (E^0 = 1.36 V), and hence, permanganometric titrations cannot be performed in aqueous medium containing chloride ions.

Potassium permanganate is not a primary standard, and hence, requires standardization. A standard solution of potassium permanganate may be prepared by titrating with a known volume of standard solution of sodium oxalate in sulphuric acid medium. The oxalate ion is converted into undissociated acid. The oxidation of oxalic acid to carbon dioxide by permanganate is represented by

$$2\ MnO_4^- + 5\ H_2C_2O_4 + 6\ H^+ \rightarrow 2\ Mn^{2+} + 10\ CO_2 + 8\ H_2O$$

The oxidation of oxalate at room temperature is slow, and hence, the titration is carried out by the addition of permanganate solution slowly to a hot solution ($\sim 80°C$) of the oxalate. Alternatively, the titration may be carried out initially at room temperature by adding the permanganate solution till about 90% of the oxalate is titrated. The reaction mixture is heated to about 60°C and the titration completed, the last 1 mL of permanganate being added dropwise. The end point is indicated by the appearance of a permanent pale pink colour.

Potassium permanganate solution may also be standardized using a standard solution of ferrous iron prepared from the primary standard ferrous ammonium sulphate (Mohr salt). The oxidation of iron(II) by potassium permanganate is rapid and quantitative in sulphuric acid medium, and hence, the standardized permanganate solution may be used for quantitative estimation of iron(II) in any sample.

$$MnO_4^- + 5\ Fe^{2+} + 8\ H^+ \rightarrow Mn^{2+} + 5\ Fe^{3+} + 4\ H_2O$$

3.8.3 Dichrometry

Dichrometry refers to the use of potassium dichromate as the oxidizing agent. The reduction of dichromate to chromium (III) in a strong acid medium may be represented as

$$Cr_2O_7^{2-} + 14\,H^+ + 6\,e^- \rightarrow 2\,Cr^{3+} + 7\,H_2O; \quad E^0 = 1.33\;V$$

Pure potassium dichromate of analytical grade is commercially available and it is a primary standard. A standard solution of the reagent can be prepared by weighing accurately the salt, dissolving it in distilled water and making up to a known volume in a volumetric flask. The equivalent weight of $K_2Cr_2O_7$ is one-sixth of its molecular weight. The detection of the end point of the titration requires an indicator such as diphenylamine.

The determination of the concentration of iron in a given sample solution is a typical example of dichrometry.

$$Cr_2O_7^{2-} + 6\,Fe^{2+} + 14\,H^+ \rightarrow 2\,Cr^{3+} + 6\,Fe^{3+} + 7\,H_2O$$

It is necessary that iron must be present in the +2 oxidation state, and hence, any iron (III) that may be present is reduced to iron(II) by treating with the reducing agent, tin (II) chloride, in hydrochloric acid medium. Any excess of the added reducing agent is destroyed by the addition of mercuric chloride. The reactions may be represented as

$$2\,Fe^{3+} + Sn^{2+} \rightarrow 2\,Fe^{2+} + Sn^{4+}$$
$$Sn^{2+} + 2\,HgCl_2 \rightarrow Sn^{4+} + Hg_2Cl_2(s) + 2\,Cl^-$$

The procedure involves adding a few mL of concentrated hydrochloric acid to a known volume of the iron(III) solution. The mixture is heated to about 80°C. The colour of the mixture is pale yellow due to presence of iron(III). To this pale yellow hot solution, tin(II) chloride is added dropwise with constant stirring till the pale yellow colour of iron(III) changes to faint green colour (free from any tinge of yellow). The solution is cooled to room temperature. After this 5–6 mL of mercuric chloride solution is added leading to the formation of a silky white precipitate of mercurous chloride. But sometimes, instead of white precipitate a grey-black precipitate of finely divided mercury is formed, which has the ability to reduce both dichromate and iron(III), thereby in the titration. This happens due to the presence of excess of tin(II) or addition of mercuric chloride to hot solution. Such a solution should be discarded. Eighty-five per cent phosphoric acid (approximately 5 mL) and a few drops of sodium diphenylamine sulphonate indicator are added to the iron(II) solution. Standard solution of potassium dichromate is used for titration of the reaction mixture. Appearance of violet–blue colour indicates the A end point. Fe (II)/Fe (III) couple and the In_{ox}/In_{red} couple of diphenylamine indicator have the same standard potentials (0.76 V). Thus, when phosphoric acid is not present, the end point is indicated much before the actual one. Addition of phosphoric acid leads to formation of a soluble complex with Fe(III). This complex formation alters the standard potential of Fe(II)/Fe(III) pair, and thus, helps in detection of the correct end point.

3.8.4 Iodometry

In *iodometry*, the iodide ion is oxidized which leads to the liberation of iodine. In contrast, *iodimetry* involves a direct titration with iodine solution. The redox reaction involving iodine may be represented as

$$I_2 + 2\,e^- \rightarrow 2\,I^-; \quad E^0 = 0.535\;V$$

Iodide ion is a moderate reducing agent. For analysing oxidizing agents, iodide ion is added in excess in the form of potassium iodide to a known volume of oxidant. The reaction between the iodide and the oxidant results in the liberation of free iodine equivalent to the amount of oxidant present. The liberated iodine is titrated against a standard solution of sodium thiosulphate using starch as the indicator. Sodium thiosulphate ($Na_2S_2O_3.5H_2O$) is not a primary standard, and hence, the reagent solution can be standardized iodometrically by titrating against a standard solution of potassium dichromate or a standard solution of Cu^{2+}.

The determination of copper(II) in a sample solution is a typical example of iodometric method. Copper(II) is reduced by the added iodide to form cuprous iodide. The optimum pH range for this reaction is 4–5.5.

$$2\,Cu^{2+} + 4\,I^- \rightarrow 2\,CuI(s) + I_2$$

The liberated iodine is titrated against a standard solution of sodium thiosulphate using starch as indicator.

$$I_2 + S_2O_3^{2-} \rightarrow 2\,I^- + S_4O_6^{2-}$$

Starch forms a blue-coloured adsorption complex with free iodine. At the end point, all the free iodine is consumed and the adsorption complex is destroyed discharging the blue colour.

The procedure for the estimation of copper(II) involves adjusting the pH of a known volume (~20 mL) of copper(II) solution to about 4.5–5.5 by the addition of a few drops of sodium carbonate. A faint bluish-white precipitate of copper hydroxide appears which is just dissolved by the addition of a few drops of dilute acetic acid to buffer the solution to the desired pH. A solution of 10% potassium iodide solution (~ 20 mL) is added to the copper solution to precipitate cuprous iodide and liberate iodine. The liberated iodine is then titrated against a standard solution of sodium thiosulphate till the dark brown colour fades to pale yellow towards the end of the titration. About 2 mL of a freshly prepared solution of starch (1%) is added to form the blue-coloured starch–iodine complex. The titration is continued till the end point is indicated by the disappearance of the blue colour.

3.8.5 Applications of Redox Titrations

Redox titrations find a variety of applications in analytical laboratory and industrial processes. A few typical applications are discussed in this section.

Dissolved oxygen (DO) content of water is an important parameter in determining the quality of water used for different purposes. Oxygen is soluble in water only to a small extent. At ambient conditions of temperature and pressure the maximum concentration of dissolved oxygen is about 8 mg/L. The presence of oxygen in water is useful for certain applications and at the same time undesirable for other applications. Dissolved oxygen is absolutely essential for sustaining aquatic life. Dissolved oxygen is required for the aerobic microorganisms involved in the treatment of water for municipal supply for domestic consumption as well as in the treatment of domestic sewage and effluents from industries. On the other hand, the presence of dissolved oxygen is detrimental for certain industrial purposes. For example, the dissolved oxygen is responsible for corrosion of boilers used in industry and power plants for generating steam, and hence, boiler feed water should be free from dissolved oxygen. The dissolved oxygen content also indicates the extent of pollution of water caused by organic and oxidizable impurities in water bodies as well as in wastewater.

The dissolved oxygen content in a sample of water is determined by *Winkler method*, a method based on iodometry. A known volume of the sample water is treated with a mixture of manganese sulphate and alkaline potassium iodide. Under the alkaline conditions, a gelatinous precipitate of manganese hydroxide is formed. The dissolved oxygen present in the sample is trapped by the $Mn(OH)_2$ precipitate resulting in the oxidation of Mn^{2+} to Mn^{4+} and the formation of a brown-coloured precipitate of manganic oxide, $MnO.(OH)_2$.

$$Mn^{2+} + 2\ OH^- \rightarrow Mn(OH)_2$$
$$Mn(OH)_2 + \frac{1}{2}\ O_2 \rightarrow MnO.(OH)_2$$

The precipitate is allowed to settle for a few minutes. The precipitate is then dissolved by adding 2–3 mL of concentrated sulphuric acid to liberate the oxygen which oxidizes the iodide ions of the added potassium iodide to iodine. The liberated iodine is proportional to the amount of dissolved oxygen present in the sample water. The liberated iodine is titrated against a standard solution of sodium thiosulphate using starch as indicator.

$$MnO.(OH)_2 + 4\ H^+ + 2\ I^- \rightarrow I_2 + Mn^{2+} + 3\ H_2O$$
$$I_2 + 2\ S_2O_3^{2-} \rightarrow S_4O_6^{2-} + 2\ I^-$$

From the above equations, it is clear that 4 moles of thiosulphate correspond to 1 mole of dissolved oxygen. Nitrite, if present (usually present in treated wastewater), interferes with the determination. Addition of sodium azide destroys the nitrite in the sample water when the latter is acidified.

$$2\ NaN_3 + H_2SO_4 \rightarrow 2\ HN_3 + Na_2SO_4$$
$$HNO_2 + HN_3 \rightarrow N_2 + N_2O + H_2O$$

In practice, water sample is collected by carefully filling a 200-mL bottle to the brim and closing with the stopper below the water surface so as to eliminate direct contact of the sample with atmospheric oxygen. About 2 mL of 50% $MnSO_4$ solution is introduced into the bottle with the help of a dropper pipette below the surface. About 2 mL of alkaline iodide–azide solution (prepared by dissolving 40 g NaOH, 20 g KI, and 0.5 g NaN_3 in 100 mL of water) is introduced into the bottle in a similar manner. The bottle is closed with the stopper and shaken thoroughly to mix the contents. The brown-coloured precipitate of manganic oxide formed is allowed to settle for about 15 min. Then concentrated sulphuric acid (2–3 mL) is added to the bottle and the bottle is turned upside-down two or three times to mix the contents. The brown-coloured precipitate completely dissolves and releases iodine in the solution. An aliquot of 100 mL of the solution is pipetted out and the liberated iodine is titrated against a standard 0.01 N solution of sodium thiosulphate. Freshly prepared starch solution of 1% (2 mL) is added towards the end of the titration when the iodine solution has become pale yellow in colour. The titration is continued till the disappearance of the blue colour of the starch indicator. The dissolved oxygen content of the sample water is calculated as $8 \times V$, where V is the volume of 0.01 N thiosulphate consumed in the titration and expressed as mg/L.

Chemical oxygen demand (COD) is an important characteristic of wastewater, particularly from industries. The extent of pollution or pollution strength is determined by COD in domestic sewage as well as in industrial wastewaters. The COD gives a quantitative measure of pollutants in the form of organic matter and oxidizable species in wastewater and is an important parameter for designing the subsequent pollution abatement methodology.

COD is defined as the amount of oxygen required under specific conditions for the oxidation of organic and oxidizable inorganic matter present in the wastewater. The COD is expressed in mg/L. Any chloride present in the wastewater is also oxidized under experimental conditions, and hence, the value of COD has to be corrected for the chloride content.

Determination of COD is based on the principle of complete oxidation of the organic and oxidizable matter with excess potassium dichromate and determining the excess dichromate. An accurately known volume (say 50 mL) of the sample wastewater is refluxed for about 3 h with an accurately known excess of potassium dichromate (~1.5 g) in the presence of a few mL of 50% sulphuric acid, silver sulphate as catalyst, and a small amount of mercuric sulphate to eliminate any interference from chlorides. Potassium dichromate oxidizes all the organic matter completely to water, carbon dioxide, and ammonia. Any other oxidizable species present in the sample is also oxidized. The excess remaining dichromate is determined by titrating the reaction mixture with a standard solution of ferrous ammonium sulphate using diphenylamine as indicator. A blank titration is carried out using distilled water instead of the sample. The COD of the wastewater sample is calculated using the formula

$$\text{COD} = \frac{(V_b - V_s)N \times 8 \times 1000}{X} \quad \text{or 1 mL of 1 N } K_2Cr_2O_7 = 8 \text{ mg of } O_2 \quad (3.26)$$

where V_b and V_s represent the volumes of ferrous ammonium sulphate of normality N used for the blank and test sample, respectively, and X is the volume of wastewater sample.

Biochemical oxygen demand (BOD) is a measure of oxygen required by aerobic microorganisms during the biochemical (biological) breakdown of organic matter present in the wastewater under aerobic conditions. It is an important parameter in characterizing (i) domestic sewage and (ii) industrial wastewaters. It is also useful for assessing the self-purification capability of flowing and stagnant water in rivers, lakes, etc. It is used as a guideline by regulatory authorities to check the quality of effluents discharged into water bodies.

The BOD of a sample is calculated from dissolved oxygen content values determined in the sample wastewater before and after incubation at 20°C for 5 days. The BOD is referred to as BOD_5. The DO content of the sample after incubation for 5 days will be less compared to that in the sample before incubation as it would have been consumed by aerobic organisms. Thus, the decrease in DO content is a measure of BOD of the sample. It is expressed in mg/L. The sample will have to be suitably diluted with distilled water when the organic load in the sample is too heavy. The sample may be supplied with oxygen if it does not contain enough oxygen.

BOD is indicative of the pollutants which are amenable for biodegradation. The BOD value of a wastewater sample is usually less than that of COD because COD value is indicative of the total amount of pollutants some of which may not be amenable for biodegradation. Such pollutants are oxidized by potassium dichromate during the determination of COD.

3.9 GRAVIMETRY

In gravimetry, an analytical balance is used for measuring the mass of an analyte or a substance derived from the analyte. In the simplest method, the sample is either dried or heated to determine the volatile and/or non-volatile components. The case of the determination of *total dissolved solids* (TDS) in a sample of water is a typical example. In this method, a known quantity of water is evaporated and the total dissolved solids are left behind which are then weighed on the balance. Metals may be deposited electrolytically and weighed. Alternatively, metals may be precipitated as sparingly soluble salts from aqueous solutions and the precipitates weighed after appropriate heat treatment. Liquid samples may be subjected to fractional distillation followed by weighing the residue and the fractions obtained separately.

The advantages offered by gravimetric analysis are:

(i) It is possible to achieve high level of accuracy and precision by using modern analytical balances. The repeatability of results can be within ± 0.3–0.5%.

(ii) Possible sources of error are readily checked since filtrates can be tested for completeness of precipitation and precipitates may be examined for the presence of impurities.

(iii) It is an absolute method as it involves direct measurement without any form of calibration being required.

(iv) Determination can be carried out with relatively inexpensive apparatus.

However, the major disadvantage associated with gravimetric analysis is that it is time-consuming besides being a macroscopic method involving relatively large samples.

Gravimetric methods may be broadly classified into (i) volatilization methods and (ii) precipitation methods.

3.10 VOLATILIZATION METHODS

The principal steps involved in volatilization methods include (i) theoretical/qualitative identification of the volatile product or residue that can be obtained from the analyte, (ii) subjecting the analyte to chemical or heat treatment to liberate the volatile matter in the form of a gas, (iii) separation and purification of the evolved gas, and (iv) absorption of the purified gas in a suitable reagent held in an absorption tube and weighing the tube. For example, the carbonate content of an ore sample (e.g., calcite or dolomite) or the bicarbonate content of an antacid may be determined by treating the analyte of known weight with an excess of dilute sulphuric acid to liberate carbon dioxide gas which is dried over fused calcium chloride and absorbed over anhydrous sodium or potassium hydroxide kept in an absorption tube. The increase in the weight of the absorption tube is equal to the amount of carbon dioxide liberated from the analyte sample. Alternatively, the weighed ore sample may be subjected to a heat treatment to expel the carbon dioxide and the residue (calcium oxide) weighed to calculate the carbonate content of the sample. Determination of carbon and hydrogen content of organic compounds is yet another example of gravimetric experiment in which a weighed organic sample is subjected to combustion process to convert the hydrogen and carbon into water vapour and carbon dioxide, respectively, the volatile gases being absorbed in separate absorption tubes of fused $CaCl_2$ and anhydrous KOH and weighed. Other examples of volatilization methods include the determination of sulphur content in sulphides by the evolution of hydrogen sulphide and by volatilizing off sulphur dioxide from sulphites.

3.11 PRECIPITATION METHODS

Precipitation methods are more commonly used for the quantitative analysis of inorganic samples, and hence, discussed in more detail. A quantitative precipitation reaction is brought about if the solubility product of the precipitate is small. Precipitates may be of different chemical types such as salts, complexes and double salts (e.g., chlorides, sulphides, hydroxides). The precipitation process is carried out by the addition of a solution of the precipitating agent under controlled conditions of certain experimental parameters such as pH and temperature.

3.11.1 Theoretical Principles of Precipitation Methods

Precipitation methods are based on four principal steps: (i) In the first step, an analyte of known weight is dissolved in a known volume of solution and reacted with a suitable reagent to form a precipitate. (ii) The precipitate is collected on a filter medium and washed to remove any impurities. (iii) The precipitate is given a heat treatment to convert it into a stoichiometric compound of known composition. (iv) In the final step, the compound is weighed using an analytical balance and the amount of analyte in the sample is calculated. For example, in the determination of lead content of a sample (e.g., alloy or ore), the first step involves the preparation of solution of the sample containing a known weight in a known volume and converting the soluble lead into an insoluble precipitate. Lead may be precipitated as lead chromate by reacting with potassium chromate. In the second step the lead chromate precipitate is collected over a sintered glass crucible of porosity G-4 and washed thoroughly to free it from impurities. The precipitate is dried in an air oven at 105–115°C for an hour and cooled to room temperature in a desiccator as part of the third step to yield the compound of known composition as $PbCrO_4$. The final step involves weighing the precipitate and calculating the lead content of the sample.

3.11.2 Criteria for an Ideal Gravimetric Estimation

There are certain prerequisites to be satisfied for successful quantitative analysis by precipitation method. These include: (i) The chemical reagent used for precipitating the analyte should preferably be specific for analyte or at least should be selective under specified conditions. No chemical reagent satisfies the requirement of specificity for a particular analyte, and hence, the obvious choice is restricted to a selective reagent. (ii) The precipitating reagent should react with analyte rapidly and quantitatively. (iii) The precipitate formed should have very low solubility; otherwise, significant loss of analyte will occur during subsequent filtering and washing steps. (iv) The precipitate formed should be granular forming large-sized particles so that it can be easily filtered and washed. (v) The precipitate should yield a compound of known and stable stoichiometric composition after heat treatment so that its weight can be related to the amount of analyte present. (vi) The precipitate after drying must be stable and should not absorb moisture or carbon dioxide from surrounding atmosphere.

3.11.3 Precipitating Agents

Commonly used precipitating reagents used in gravimetry for the estimation of cations and anions may be broadly classified into inorganic and organic precipitating agents.

Examples of *inorganic precipitating agents* include (i) hydrochloric acid for precipitating silver as AgCl, (ii) sulphuric acid for precipitating lead as $PbSO_4$ and barium as $BaSO_4$, (iii) potassium chromate for precipitating lead as $PbCrO_4$ and barium as $BaCrO_4$, (iv) liquor ammonia for precipitating metals, such as iron(III), aluminum(III), chromium(III) as hydroxides which on heat treatment yield oxides of composition M_2O_3, (v) ammonium thiocyanate for precipitating copper(I) as CuSCN, (vi) silver nitrate for precipitating anions, such as chloride, bromide, iodide as silver halides of composition AgX, (vii) ammonium molybdate for precipitating phosphate as the phosphomolybdate which on heating at 800–825°C gives the compound of composition $P_2O_5 \cdot 24MoO_3$, and (viii) barium chloride for precipitating sulphate as $BaSO_4$.

Commonly used *organic precipitating agents* include (i) dimethylglyoxime (DMG) for precipitating nickel and palladium as dimethylglyoxime complexes of composition $M(C_4H_7O_2N_2)_2$, (ii) ammonium oxalate for precipitating calcium as oxalate monohydrate $(CaC_2O_4.H_2O)$, (iii) 8-hydroxyquinoline (oxine -C_9H_7ON) for precipitating divalent metal ions Cu, Zn, Mg, Cd, and Pb as $M(C_9H_7ON)_2$, trivalent metal ions Al, Fe, Bi, and Ga as $M(C_9H_7ON)_3$ and tetravalent metal ions Th and Zr as $M(C_9H_7ON)_4$, and (iv) anthranilic acid for precipitating Cd, Zn, Ni, Co, and Cu as $M(C_7H_7O_2N)_2$.

3.11.4 Factors Affecting Solubility of Precipitates

The various factors affecting the solubility of a precipitate include (i) ionic strength of the medium or diverse ion effect, (ii) common ion effect, and (iii) pH. The solubility of a precipitate is expressed in terms of its solubility product constant, K_{sp}. For the solubility equilibrium

$$AB = A + B$$

The solubility product constant is given as

$$K_{sp} = [A][B] = \gamma_A C_A \gamma_B C_B \qquad (3.27)$$

where C is the analytical concentration of the species and γ the activity coefficient relating to the charge on the species Z and ionic strength μ at 298 K as

$$-\log_{10}\gamma_A = 0.51 \, Z_A^2 \, \mu^{1/2} \qquad (3.28)$$

The solubility product constant K_{sp} and the solubility of AB increase with increasing ionic strength of the solution at a given temperature. The effect of ionic strength or the diverse ion effect (potassium nitrate) on the solubility of barium sulphate is shown in Fig. 3.11.

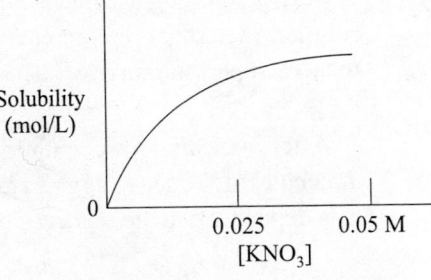

Fig. 3.11   Diverse ion effect on the solubility of barium sulphate

The solubility of sparingly soluble substances is also affected by the presence of common ions in solution. The solubility of a sparingly soluble salt in equilibrium with its own ions in solution decreases as the concentration of one of the ions increases. For example, the solubility of silver chloride precipitate decreases if a slight excess of either chloride or silver ions is present in solution at equilibrium. The decrease in the solubility of a precipitate is utilized in gravimetric analysis by adding a slight excess of the precipitating agent to complete the precipitation of the analyte ion. However, a large excess of the common ion may cause an increase in solubility of the precipitate either by neutral salt effect or by the formation of a soluble complex. Thus, the solubility of silver chloride increases with chloride ion concentration due to the formation of complex ionic species such as $[AgCl_2]^{2-}$ and $[AgCl_3]^{3-}$. The effect of chloride ion concentration (KCl) on the solubility of silver chloride is shown in Fig. 3.12.

The pH of the medium also affects the sparingly soluble substances as in the case of metal hydroxides and salts of weak acids such as insoluble carbonates, sulphides, phosphates, oxalates, and so on.

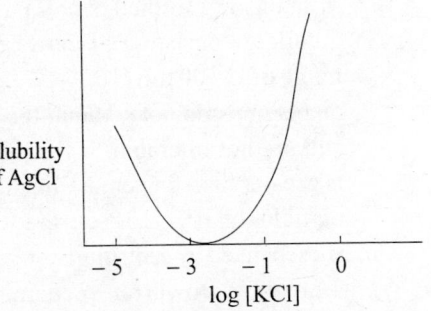

Fig. 3.12   Common ion effect on the solubility of silver chloride

3.11.5 Mechanism of Formation of Precipitates

Precipitate formation may be considered to take place in three stages, which include (i) supersaturation, (ii) nucleation, and (iii) precipitate particle growth. According to von Weimarn, precipitation cannot occur until the solution is supersaturated with respect to a given compound. A solution becomes supersaturated, when it contains a higher concentration of solute than that may exist in a saturated solution. Thus, when a reagent is added to an analyte solution forming a sparingly soluble substance, the solubility product for the sparingly soluble compound is immediately exceeded and the solution becomes supersaturated. The extent of supersaturation determines the precipitate particle size and filterability.

Nucleation may be defined as the formation of a more stable phase from the metastable phase of supersaturation. Nucleation, or more specifically, homogeneous nucleation involves the aggregation of small groups of ions or molecules to form primary nuclei of submicroscopic dimensions, but large enough for crystal lattice forces to become operative. The number of nuclei formed depends on the degree of supersaturation or relative supersaturation, R_s. Relative supersaturation is defined as

$$R_s = (Q - S)/S \tag{3.29}$$

where Q is the actual concentration of the solute and S the equilibrium concentration.

The rate at which the value of Q decreases to S determines the rate of precipitation. Nucleation may also be brought about when aggregation is initiated by particulate impurities within the solution (heterogeneous nucleation). Homogeneous nucleation depends exponentially on the relative supersaturation of the solution, whereas heterogeneous nucleation is independent of it.

After nucleation, precipitation continues by particle growth during which further ions or molecules are added to aggregates formed by nucleation. The rate of precipitate particle growth, v, is dependent on the relative supersaturation as given by

$$v = k R_s \tag{3.30}$$

where k is proportionality constant. The dependence of the rates of homogeneous and heterogeneous nucleation and particle growth is shown in Fig. 3.13.

The nature of precipitated particles will be determined by the relative rates of nucleation and particle growth. When nucleation predominates, that is, more number of nuclei are formed, small particles are produced and a colloid consisting of particles with diameters in the range of 1–$100\ \mu m$ (10^{-7}–10^{-5} cm) is formed. Colloidal dispersions do not settle at the bottom of the container and are not filterable. If the rate of the particle growth is greater than fine crystalline dispersions consisting of particles of 10^{-5}–10^{-3} cm are formed. Such dispersions are separated by centrifugation but difficult to filter. When the particle growth rate predominates, coarse precipitates (particle diameters greater than 10^{-3} cm) are formed which are readily collected on filter paper or filter crucible.

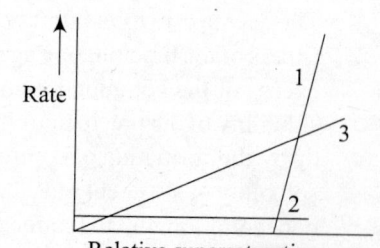

Fig. 3.13 Effect of relative supersaturation on the rates of nucleation and particle growth: 1. homogeneous nucleation 2. heterogeneous nucleation 3. particle growth

3.11.6 Colloidal Precipitates

Gravimetric analysis often encounters situations in which the rate of homogeneous nucleation predominates resulting in colloidal precipitates. Moreover, the precipitation process always involves the formation of colloidal particles in the initial stages. The hydroxide precipitates of most of the metals are colloidal in nature. It is necessary to treat such colloidal precipitates to facilitate their filtration. The characteristics of colloids include non-settling nature under gravity and their large surface area. The large surface area is responsible for adsorption characteristics of colloidal dispersions. Electrovalent colloids tend to adsorb ions, which are common to them. For example, colloidal silver chloride tends to adsorb either silver or chloride ions depending on the excess availability of silver or chloride ions during precipitation of silver from silver nitrate solution by the addition of sodium chloride. The adsorption results in a primary adsorption layer on the surface of the colloidal particle which is either positively charged (when silver ions are adsorbed) or negatively charged (when chloride ions are adsorbed in the presence of excess of chloride). The charge on the primary adsorption layer is balanced by a counter ion layer surrounding the particle consisting of the ions of opposite charge. The primary adsorbed layer is mono-atomic in depth and the ions are usually held relatively more strongly, while the secondary layer is diffuse and mobile consisting of the counter ions held loosely. The presence of this electrical double layer consisting of the primary adsorbed layer and the secondary counter ion layer is responsible for the stability and electrical properties of electrovalent colloids. Since the secondary layer on all the colloidal particles has the same charge, the particles repel each other and do not bind to form larger aggregates. The thickness of the electrical double layer is minimum when the supernatant liquid in contact with the precipitate does not contain an excess of either cations or anions. However, for a gravimetric precipitation to be quantitative and complete, it is necessary to have a slight excess of the precipitating agent resulting in an excess of charged species and a consequent increase in the thickness of the electrical double layer. The electrostatic repulsion between like-charged particles stabilizes the colloidal dispersions. Such colloidal dispersions do not settle and are difficult to filter. The stability of colloidal precipitates can be decreased by the addition of an electrolyte, stirring or heating. Addition of an electrolyte neutralizes the charge on the secondary layer, decreases the electrostatic repulsion between particles, and facilitates the binding of the individual particles into an amorphous mass, which settles down and is amenable for filtration. The conversion of a colloidal precipitate into a filterable solid is called coagulation or agglomeration. Heating and stirring the dispersion to remove the adsorbed ions and thereby decrease the thickness of the electrical double layer can also bring about coagulation of colloidal particles.

The analyte forming colloidal suspension is best precipitated from a hot solution containing sufficient electrolyte with constant stirring to yield filterable solid. Digestion improves the filterability of coagulated colloidal precipitate. During digestion, weakly bound water is expelled from the colloidal precipitate resulting in a dense mass, which is easy to filter.

3.11.7 Contamination of Precipitates

The sources of contamination of precipitates affecting the gravimetric analysis include post-precipitation and coprecipitation.

Post-precipitation involves the deposition of a sparingly soluble impurity of similar properties to the analyte precipitate on the surface of the latter after its formation. The classical example of post-precipitation is that of magnesium oxalate (impurity) depositing on the calcium oxalate precipitate. Calcium oxalate precipitates out in the presence of magnesium ions satisfactorily without any interference but if the precipitate remains in contact with mother liquor containing magnesium ions for a prolonged period, magnesium oxalate precipitates on top of calcium oxalate. Post-precipitation also takes place involving zinc and mercury sulphides. Post-precipitation is avoided by filtering the calcium oxalate within 1 or 2 h after precipitation.

Coprecipitation involves the inclusion of soluble substances in the precipitate during its formation. The major types of coprecipitation include (i) adsorption, (ii) mixed crystal contamination due to isomorphic inclusion, (iii) occlusion, and (iv) mechanical entrapment of non-isomorphic substances.

Adsorption depends on the surface area of the precipitate particles. Precipitate formation always involves the formation of colloidal particles in the initial stages, the particles growing to larger-sized coarse particles gradually. Colloidal particles with their large surface area facilitate adsorption of different types of ions/impurities in the primary layer. The ionic impurities of the primary layer as well as the counter ion layer precipitate along with the main precipitate. In the case of crystalline precipitates with large-sized crystals, contamination due to surface adsorption is less. Colloidal precipitates have to be washed to remove the adsorbed impurities.

Mixed crystal contamination of the precipitate occurs due to substitution of the precipitate lattice with impurity ions of similar crystallinity. Such mixed crystal contamination is possible in the precipitation of barium as barium sulphate in the presence of lead ions, strontium sulphate in the presence of barium, and cadmium sulphide in the presence of manganese. Isomorphic substitution of K^+ for NH_4^+ in $MgNH_4PO_4$ results in the coprecipitation of $MgKPO_4$. Mixed crystal contamination can occur both in colloidal as well as in crystalline precipitates. The only way this problem can be overcome is by separating the analyte from the contaminating ions prior to precipitation.

Occlusion occurs during the formation of the precipitate when foreign ions in the counter-ion layer get trapped within the rapidly growing crystal. The amount of the occluded material is greatest within that part of the crystal that formed first.

Mechanical entrapment occurs when several crystals growing together come closer to each other and trap a portion of the solution in pockets between the crystals. Occlusion and mechanical entrapment are at a minimum when the rate of precipitate formation is low and the degree of supersaturation is low. Digestion of precipitates decreases contamination due to occlusion and mechanical entrapment.

Contamination of precipitates by coprecipitation can result in positive as well as negative errors in gravimetric analysis depending on whether the contaminant is a compound of the ion being determined or not. For example, in estimation of chloride as silver chloride a positive error results due to adsorption of silver nitrate on the colloidal silver chloride precipitate. In contrast, in the determination of barium as barium sulphate, both positive and negative errors can result due to contamination. When barium chloride is the contaminant a negative error results due to the fact that the molar mass of the contaminant is smaller than that of barium sulphate. However, if the contaminant is barium nitrate a positive error is observed.

3.11.8 Practical Aspects

Precipitation (the first step in gravimetric analysis) is usually carried out by adding the precipitating reagent solution to a hot and dilute solution of the analyte. The addition is carried out slowly with constant stirring so as to produce gravimetric precipitates of relatively large-sized crystals or flocs of coarse particles. The *optimum conditions for the formation of large-sized crystalline precipitates* are discussed in the following paragraphs.

To obtain easily filterable precipitates of large particle size for gravimetric experiments, relative supersaturation must be low so that the rate of precipitation may also be reduced. For the best possible precipitate, the difference (Q–S) should be controlled and kept to a minimum. However, on mixing the reagent and the analyte solutions, a momentarily high (Q–S) is unavoidable particularly when S is small. A low value for S is necessary for quantitative precipitation. Hence, optimum conditions for gravimetric precipitations may be attained practically through the following techniques.

(i) *Dilution*: Precipitation should be carried out using dilute solutions of reagent and analyte. Dilution slows the nucleation process thereby aiding the particle growth and formation of large-sized particles. However, too dilute solutions will result in the loss of the precipitate due to increased solubility and a greater experimental error. Further, handling large volumes of solutions poses practical difficulty and increases the time required for analysis.

(ii) *Precipitation at maximum solubility of the sparingly soluble substance*: The maximum solubility (or the equilibrium solubility S) of the precipitate as determined by the solubility product constant, K_{sp} at a given temperature increases at higher temperatures. In a hot solution of the analyte, the value of S will be greater and the relative supersaturation is lower. Hence, precipitation may be carried out from a hot solution. Yet another method of precipitation at maximum solubility of the precipitate is achieved by control of pH, particularly when the precipitate is soluble in an acid solution. For example, the precipitation of calcium oxalate is usually carried from a hot acidic solution by the slow addition of ammonium oxalate followed by aqueous ammonia to increase the pH.

(iii) *Slow addition of precipitating agent with constant stirring*: The slow addition of a precipitating agent results in the formation of large-sized precipitate particles by keeping the value of Q low. Constant stirring during addition of the precipitating agent disperses the reagent in the bulk of the medium and prevents high local concentration.

(iv) *Digestion of the precipitate*: Digestion or aging of the precipitate involves allowing the precipitate to remain in contact with the mother liquor for about an hour or more. During this *aging* process, the larger sized precipitate particles grow at the expense of the smaller-sized particles, that is, the smaller particles dissolve and reprecipitate on the surface of the larger crystals. Hence, the filterability of the precipitate is enhanced. Digestion is also helpful in minimizing the problems associated with coprecipitation. In the case of curdy precipitates, such as silver chloride, agglomeration occurs enhancing the filterability of such precipitates.

Filtration and washing of precipitates aim to separate the precipitate from the mother liquor. Filter papers (ashless quantitative Whatman series filter papers) and sintered glass crucible with porous septum are conventionally used as filter media.

It is necessary to wash a gravimetric precipitate as it will be contaminated with non-volatile soluble compounds, such as precipitating reagent, added in excess. Ideally, a wash liquid should not dissolve the precipitate but only the contaminant. The wash liquid should have no chemical interaction with the precipitate resulting in the dispersion of the precipitate or formation of volatile or insoluble product. It should be volatile at the drying temperature and should not contain any components which may interfere with subsequent analysis of the filtrate. Pure water is used only when it is certain that it will not dissolve the precipitate. Most commonly used wash liquids are dilute aqueous solutions of electrolytes containing a common ion with the precipitate so as to minimize the solubility of the precipitate as well as to avoid *peptization*, that is, dispersion of the precipitate back to colloidal state which will easily pass through the filter medium. Hot solutions of ammonium salts are preferred as ammonium salts are volatile at the drying temperature. Thus, a dilute solution of ammonium oxalate is used to wash calcium oxalate precipitate in the estimation of calcium while a dilute solution of ammonium nitrate is used to wash iron hydroxide precipitate. Dilute nitric acid reduces the solubility of the silver chloride precipitate and also prevents its peptization.

If the precipitate is susceptible to air oxidation, as in the case of copper sulphide, acidified hydrogen sulphide solution may be used as a wash liquid. Washing of precipitates is carried out by using a number of small volumes of the wash liquid rather than washing once or twice with a large volume. The wash liquor is completely drained off the precipitate before adding the next portion of the wash liquid.

Heat treatment of precipitates is given at an appropriate level to convert the precipitate to a stoichiometric compound suitable for weighing. The heat treatment may be a simple drying operation in an air oven maintained at 110–130°C to remove the solvent water and other volatile species. Examples of precipitates subjected to drying include the calcium oxalate monohydrate, aluminum oxinate, lead chromate, nickel dimethylglyoximate, etc.

A rigorous heat treatment is given to precipitates which do not yield compounds of known composition by simple drying. In addition, certain precipitates may not be amenable for collection over a sintered glass crucible. Such precipitates are collected over Whatman ashless filter paper. For example, hydroxide precipitates of metals, such as iron, chromium, aluminum or sulphate precipitates of lead and barium are collected over filter papers. The precipitate is given the wash treatment as described earlier and finally transferred quantitatively to the filter paper and given a final wash. The collected precipitate is dried along with the filter paper in an air oven at 105–120°C. The filter paper containing the dried precipitate is folded and placed in a previously weighed silica crucible. The filter paper and precipitate are heated initially with a low flame of a Bunsen burner or heated in an electric Bunsen till all the filter paper is burnt off. The precipitate is then subjected to strong heating to temperatures up to 1000°C. After heat treatment, the crucible containing the precipitate is cooled to room temperature in a desiccator. For example, calcium oxalate monohydrate precipitate is heated to 500–600°C to form $CaCO_3$ which can be weighed. Alternatively, the precipitate may be heated to temperatures above 800°C to give CaO. The amount of calcium in the sample may be calculated from the weight of the $CaCO_3$ or CaO as they are stable compounds.

Weighing is the final step in a gravimetric analysis. The stoichiometric compound obtained by appropriate heat treatment of the precipitate is weighed using a good analytical balance. The balance should be capable of weighing the sample with an accuracy of ±0.1 mg and the weight of the precipitate should be at least 200 mg so that the error in weighing will be less than 0.1%. The crucible (either the sintered glass crucible or silica crucible) is weighed before collecting the precipitate accurately and once again after the precipitate has been subjected to heat treatment and cooling in a desiccator. The weight of the precipitate is calculated from the difference in weights. It is advisable to repeat the heat treatment of the precipitate for a short period of 10–15 min, cooling to room temperature in the desiccator and weighing to ensure the completeness of the heat treatment indicated by the constancy in the weight of the precipitate.

3.11.9 Homogeneous Precipitation

Homogeneous precipitation has the advantage over conventional precipitation in that the precipitating agent is generated *in situ* slowly under controlled conditions, and hence, the precipitate is relatively pure, dense, and readily filterable. Colloidal precipitate formation and consequent contamination of precipitate is eliminated. The technique involves the generation of the precipitating agent in a solution of the analyte through a slow chemical reaction such as hydrolysis of a precursor. For example, metal ions, such as Al, Ga, Fe, Th, Bi, and Sn, can be precipitated by hydroxide ions generated *in situ* by slow hydrolysis of added urea.

$$CO(NH_2)_2 + 3H_2O \rightarrow CO_2 + 2NH_4^+ + 2OH^-$$
$$M^{n+} + nOH^- \rightarrow M(OH)_n$$

Similarly, dimethyl sulphate on hydrolysis generates sulphate ions homogeneously for precipitation of Ba, Ca, Sr, and Pb as sulphates.

$$(CH_3)_2SO_4 + 2H_2O \rightarrow 2CH_3OH + 2H^+ + SO_4^{2-}$$

Other examples precipitating agents used in homogeneous precipitation include thioacetamide for generating sulphide ions to precipitate Sb, Mo, Cu, and Cd, trimethyl phosphate for generating phosphate ions for precipitating Zr and Hf as phosphates, ethyl oxalate for generating oxalate ions and precipitating Mg, Ca, and Zn as oxalates, 8-hydroxyquinoline in the estimation of Al, U, Mg, and Zn as oxinates or oxides, and trichloroacetic acid for the generation of carbonate ions for precipitating La and Ba as carbonates.

3.11.10 Examples of Gravimetric Estimations

Nickel content of a sample solution is quantitatively determined by precipitating it as its dimethylglyoxime (DMG) complex. A 1% aqueous–alcoholic solution of DMG is added to a hot mildly acidic solution of nickel of known volume and the reaction mixture is rendered ammoniacal by the addition of liquor ammonia solution. The precipitate is collected over a sintered glass crucible of G-4 porosity, washed, dried, and weighed as $Ni(C_4H_7O_2N_2)_2$.

The procedure for the determination of nickel involves dissolving the weighed sample of nickel (salt or alloy) in dilute hydrochloric acid and the volume of the solution is made up to a known value using a volumetric flask. After this, a known volume of this made up solution is taken in a beaker provided with a glass rod, a rubber policeman, and a watch glass. This solution is heated to approximately 70°C. A saturated aqueous alcoholic (1%) solution of dimethylglyoxime (H_2DMG) is added slowly with stirring to the hot nickel solution. An excess of liquor ammonia solution is added to the reaction mixture till it is ammoniacal as indicated by the smell. The cherry red-coloured nickel dimethylglyoximate complex $Ni (HDMG)_2$ precipitates. The precipitate is digested over a hot water bath for about 30 min and then allowed to stand for about an hour at room temperature. A test for completion of precipitation is carried out by adding a few drops of DMG solution over the side wall of the beaker. The appearance of white precipitate of DMG indicates completion of precipitation. The precipitate is collected over a previously weighed G-4 sintered glass crucible, washed with cold water until free from chloride and finally with aqueous alcohol and dried at 110–120°C. The dried precipitate $Ni (C_4H_7O_2N_2)_2$ is cooled to room temperature in a desiccator and weighed using an analytical balance. The nickel content of the sample is calculated from the known composition and weight of the precipitate.

Aluminum is precipitated by homogenous precipitation as 8-hydroxyquinolinate (oxinate) complex by the addition of a solution of 8-hydroxyquinoline (oxine) in acetic acid in the presence of urea. The precipitate is collected over a sintered glass crucible, washed, dried, and weighed as Al $(C_9H_6ON)_3$. Alternatively, the precipitate may be collected over a Whatman ashless filter paper and ignited in a silica crucible to about 1000°C to yield Al_2O_3, which can be weighed.

In practice, a weighed quantity of aluminum sample (e.g., bauxite ore) is baked with hot concentrated hydrochloric acid in a china dish. Aluminum is converted to soluble chloride and is dissolved in hot dilute hydrochloric acid. The reaction mixture is transferred quantitatively to a filter paper to filter off the insolubles (mostly silica) and washings are added to filtrate. The filtrate containing aluminum is made up to a known volume in the standard flask. An aliquot of known volume of the aluminum solution containing about 40–50 mg of the metal is pipetted out into a beaker and diluted to about 150 mL. The solution is made distinctly acidic by adding about 2 mL of concentrated hydrochloric acid. A 10% solution of 8-hydroxyquinoline in 2 M acetic acid (about 10 mL) and 5 g of urea are added to the solution. The mixture is heated to about 90°C and maintained in the hot condition for about 2 h. Urea undergoes hydrolysis and releases hydroxide ions which increase the pH of the solution. The precipitating agent is generated *in situ* and the greenish yellow supernatant becomes pale orange yellow on completion of precipitation. The mixture is cooled to room temperature and filtered through a previously weighed sintered glass G-4 crucible. The precipitate is washed initially with a little hot water and finally with cold water and dried at 130°C. The precipitate is cooled in a desiccator and weighed. The aluminum content of the sample is calculated from the weight of the precipitate Al $(C_9H_6ON)_3$.

The precipitate obtained by homogeneous precipitation of aluminum oxinate may be collected over a Whatman ashless quantitative filter paper, washed, dried, and finally ignited in a silica crucible at about 1000°C to yield Al_2O_3. The oxide is cooled to room temperature in a desiccator and weighed. The aluminum content can be calculated from the weight of Al_2O_3.

Chloride is estimated as silver chloride. The chloride solution is acidified by adding dilute nitric acid followed by the addition of a slight excess of 0.1 M aqueous solution of silver nitrate. The initially formed colloidal precipitate is coagulated by digestion and stirring after which the supernatant becomes almost clear. The precipitate is collected over a sintered glass G-4 crucible washed with dilute nitric acid, dried at 130–140°C, cooled to room temperature in a desiccator and weighed as AgCl. The silver chloride precipitate is photosensitive, and hence, it is advisable to carry out the estimation in subdued light without exposing to direct sunlight or cover the beaker with brown or black paper.

Sulphate solution is acidified with concentrated hydrochloric acid (about 0.5 mL) and heated to boiling. A 5% aqueous solution of barium chloride is added slowly dropwise to the hot solution with constant stirring to precipitate as barium sulphate. The precipitate is allowed to settle for a few minutes and test for completion of precipitation is carried out by adding a few drops of barium chloride solution. The settled precipitate is filtered through a Whatman ashless filter paper No. 40. The precipitate is washed with cold water and dried at 110–120°C. The dried precipitate along with the filter paper is placed in clean, weighed silica crucible and ignited to a final ignition temperature of 600–800°C. Barium sulphate is reduced to sulphide by the carbon of the filter paper, and hence, the reduction is avoided by first charring the paper slowly and then burning off the carbon slowly at low temperatures with free access to air. Any reduced precipitate may be reoxidized to sulphate by adding a drop or two of concentrated sulphuric acid to the cooled precipitate and reheating. The ignited precipitate is cooled to room temperature in a desiccator and weighed as $BaSO_4$.

Organic functional groups can be quantitatively analysed by precipitation methods. Thus, methoxy and ethoxy groups can be converted to methyl and ethyl iodides, respectively, by decomposing with hydrogen iodide and the alkyl iodides formed can be quantitatively precipitated as AgI by treating with silver nitrate. Similarly, carbonyl group can be precipitated to yield a compound of a stoichiometric formula of $RCH=NNHC_6H_5(NO_2)_2$, by reacting with 2, 4-dinitrophenylhydrazine. Aromatic carbonyl group can be decomposed by heating with copper carbonate at about 230°C to liberate carbon dioxide on distillation, which is absorbed in suitable reagent in an absorption tube and weighed. Sulphanilic acid is oxidized by nitrous acid to yield sulphate which is precipitated as $BaSO_4$ following the procedure for sulphate determination.

3.12 ANALYSIS OF ALLOYS, ORES, AND COMPLEX MATERIALS BY WET CHEMICAL METHODS

A few examples of analysis of alloys, ores, and complex materials by following the principles of wet chemical methods are discussed in this section.

3.12.1 Analysis of an Iron Ore

Iron exists in earth's crust mostly as its oxide ores—hematite (Fe_2O_3) and magnetite (Fe_3O_4). The weighed quantity of ore sample is decomposed by heating with concentrated hydrochloric acid in the presence of a small amount of tin(II) chloride. Iron goes into solution while the silicates present in the ores are not decomposed and remain as insolubles. The insolubles are filtered off using a Whatman quantitative filter paper. The filtrate is collected along with the washings of the insolubles and made up in a standard volumetric flask. A measured volume of the made up solution is pipetted out into a conical flask and the iron content estimated by dichrometry as per the procedure described in Section 3.8.3.

The iron content of the made up solution may also be determined by gravimetry. Iron is oxidized to iron(III) by an equivalent amount of potassium permanganate added exactly and the resulting iron(III) is precipitated as hydroxide by treating with ammonium chloride and ammonium hydroxide reagents. Alternatively, the iron (III) may be precipitated as oxinate by treating with a 2 M acetic acid solution of 8-hydroxyquinoline (oxine). The precipitated oxide or oxinate is collected over a Whatman ashless quantitative filter paper and ignited in a silica crucible at 850–900°C to get Fe_2O_3. The iron content of the sample is calculated from the weight of Fe_2O_3.

3.12.2 Analysis of Brass

Brass is an alloy primarily of copper and zinc. It may contain small amounts of lead and tin. The weighed quantity of the alloy is decomposed by heating with a few millilitres of 1:1 nitric acid. The solution is then treated with concentrated sulphuric acid and evaporated till white fumes of SO_3 evolve. The solution is cooled and made up in a standard volumetric flask. The copper content of the solution is determined by iodometric method as described in Section 3.8.4. For the estimation of zinc, interference from copper has to be eliminated by precipitating copper as copper sulphide. Hydrogen sulphide gas is passed through the analyte solution of known volume to precipitate the copper as copper sulphide which is filtered off. The filtrate and washings of the precipitate are collected and hydrogen sulphide is destroyed by boiling with dilute nitric acid.

The reaction mixture is cooled and the zinc content is determined by titrating with a standard solution of EDTA as described earlier.

3.12.3 Analysis of Solder

Solder is an alloy of lead and tin and both the metals are complexed by adding an excess of standard EDTA solution. The excess of EDTA is determined by titrating with standard lead nitrate solution using xylenol orange. The volume of EDTA solution consumed to complex both the metal ions corresponds to the total content lead + tin. The tin(IV)–EDTA complex in the reaction mixture is then decomposed by the addition of sodium fluoride and the liberated EDTA is once again titrated with the same standard solution of lead nitrate to determine the tin content of the sample. The lead content is calculated from the difference in the volumes of EDTA required to complex lead + tin and tin alone.

In practice, the weighed amount of solder is dissolved in a mixture of concentrated hydrochloric acid and concentrated nitric acid. The solution is boiled to expel nitrous fumes and chlorine. On cooling, some amount of lead chloride may get precipitated. The reaction mixture is treated with a known excess of standard EDTA solution and boiled for about 1–2 min to dissolve the lead chloride. The solution is diluted and made up in a standard volumetric flask. An aliquot of the solution is pipetted out into a conical flask, diluted, and mixed with hexamine solution. A few drops of xylenol orange indicator are added and the solution containing excess EDTA is titrated with a standard solution of lead nitrate till the end point is indicated by a colour change from yellow to red. The total amount of lead and tin together is determined from the volume of EDTA (V_1 mL) consumed for complex formation. To the contents of the same flask, approximately 2 g of sodium fluoride is added to liberate EDTA from the tin–EDTA complex. The solution acquires a yellow colour and the liberated EDTA is titrated against the same standard lead nitrate solution. The approach of the end point is indicated by a pink or red colour changing slowly to yellow colour. The titration is continued with dropwise addition of lead nitrate till a permanent red colour is obtained. The amount of tin in the sample is calculated from the volume of EDTA (V_2 mL) liberated. The amount of lead is calculated from the difference in the volumes ($V_1 - V_2$) of EDTA.

Alternatively, the weighed quantity of solder may be treated with concentrated nitric acid to precipitate tin as its hydrated oxide. The oxide is collected quantitatively over a Whatman ashless filter paper and ignited to high temperatures in a silica crucible to give the oxide SnO_2. The oxide is weighed and the tin content of the sample is calculated. The filtrate obtained by filtering off the hydrated tin oxide contains lead. The solution is made up to a known volume in a standard flask. The lead content is determined by precipitating lead as lead chromate from a known volume of the made up solution.

3.12.4 Analysis of Cement

Cement is a complex substance containing silicate of calcium as the main constituent and smaller amounts of silicates of aluminum, magnesium, and iron. The binding quality of cement depends on the calcium content and the setting and hardening properties on the amounts of other constituents. Hence, the analysis of cement involves the estimation of silica and calcium as the major constituents and the determination of the mixed oxides of aluminum and iron. Silica is estimated as the insoluble residue obtained by decomposing the cement to extract the metal ions into solution. The calcium, aluminum, and iron contents are determined from the solution.

In practice, a weighed quantity of cement (~1 g) is baked with about 5 mL of concentrated hydrochloric acid in a china dish to convert the metal ions into chlorides. The metal ions are extracted with hot water leaving behind the silica as the insoluble residue. The silica is collected quantitatively over a Whatman ashless filter paper and washed with water. The filtrate and washings are collected and made up to a known volume in a standard flask. The residue collected on the filter paper is dried in an air oven at 105–120°C. The dried filter paper along with the residue is placed in a previously weighed silica crucible. The filter paper is burnt off and the residue is ignited to temperatures 800–900°C. The crucible and the residue are cooled to room temperature in a desiccator. The residue is weighed as SiO_2.

Calcium in the made up solution is determined by pipetting out a known volume of the solution into a beaker. The solution is diluted to about 150 mL, heated to about 70–80°C, and the pH of the solution is adjusted to 4–5 by the addition of base. A 10% solution of ammonium oxalate is added slowly to the hot solution of the analyte followed by the addition of liquor ammonia to render the solution distinctly ammoniacal. The precipitate of calcium oxalate is digested over a hot water bath, and allowed to settle for about 30 min. The test for completion of precipitation is carried out by the addition of a few drops of ammonium oxalate through the walls of the beaker. The clear supernatant solution is filtered through a previously weight sintered glass G-4 crucible. The bulk of the precipitate is washed in the beaker with a dilute solution of ammonium oxalate and quantitatively transferred to the crucible. After a final washing of the precipitate in the crucible with distilled water, the crucible containing the precipitate is dried in an air oven at 110–120°C. The dried precipitate is cooled to room temperature in a desiccator and the precipitate is weighed as $CaC_2O_4.H_2O$. The amount of calcium is calculated from the weight and known formula of the precipitate.

The aluminum and iron content of cement is usually determined as mixed oxides. A known volume of the made up solution is pipetted out into a beaker, and the aluminum and iron are precipitated as hydroxides by the addition of liquor ammonia in the presence of ammonium chloride. The hydroxide precipitates are collected over a Whatman ashless filter paper and washed with dilute ammonia solution. The filter paper containing the precipitate is dried in an air oven at 110–120°C. The dried filter paper and precipitate are placed in a previously weighed silica crucible and ignited to 900–1000°C. The crucible containing the mixed oxides is cooled to room temperature in a desiccator and weighed. The mixed oxide content is reported in terms of percentage.

● ● ● ● ● ● ● ● ● ● ● ● ● ● ● ● ● ● **SOLVED PROBLEMS**

EXAMPLE 1   *Calculate the amounts of sodium carbonate and sodium hydroxide present in a solution if the solution on titration with 0.20 M HCl gave an end point with phenolphthalein at 21.4 mL and another end point with methyl orange at 32.3 mL.*

SOLUTION

Since on titration the end points indicated that $V_{phe} > \frac{1}{2} V_M$,

$\qquad [OH^-] = (2 V_{phe} - V_M)$ and $[CO_3^{2-}] = 2 (V_M - V_{phe})$

$\qquad [OH^-] = (2 \times 21.4 \text{ mL} - 32.3 \text{ mL}) = 10.5 \text{ mL of 0.2 M HCl}$

$\qquad \qquad = 2.1 \text{ mmol}$

Amount of NaOH = 2.1 mmol × 40 mg = 84 mg

$\qquad [CO_3^{2-}] = 2 (32.3 \text{ mL} - 21.4 \text{ mL}) = 21.8 \text{ mL of 0.2 M HCl}$

$\qquad \qquad = 4.36 \text{ mmol}$

Amount of Na_2CO_3 = 4.36 mmol × 53 mg = 231.08 mg

EXAMPLE 2 *A water sample of 100 mL consumed 15.7 mL of 0.11M HCl at the phenolphthalein end point and 37 mL of the same HCl at the methyl orange end point. Identify the different types of alkalinity present in a sample of water and express their amounts in terms of calcium carbonate equivalents.*

SOLUTION

Since $V_{phe} < \frac{1}{2} V_M$ the alkalinity is due to the presence of both carbonate and bicarbonate ions. The concentration of $[CO_3^{2-}] = 2 V_{phe}$ and that of $[HCO_3^-] = (V_M - 2 V_{phe})$.

The conversion factor for expressing the alkalinity in terms of calcium carbonate equivalents is 1 mL of N/50 acid = 1 mg of $CaCO_3$.

$$[CO_3^{2-}] = 2 \times 15.7 \text{ mL} = 31.4 \text{ mL of } 0.11 \text{ M HCl}$$

Carbonate alkalinity in terms of $CaCO_3$ equivalents = (31.4 mL × 0.11 M)/0.2 M

$$= 172.7 \text{ mg/L or ppm}$$

$$[HCO_3^-] = (37.0 - 31.4) = 5.6 \text{ mL of } 0.11 \text{ M HCl}$$

Bicarbonate alkalinity in terms of $CaCO_3$ equivalents = (5.6 mL × 0.11 M)/0.02 M

$$= 30.8 \text{ mg/L or ppm}$$

EXAMPLE 3 *A food sample of 0.56 g was subjected to Kjeldahl analysis for determining the protein content. The liberated ammonia consumed 26.2 mL of 0.1 M HCl. Calculate the protein content of the sample.*

SOLUTION

1 mmol of ammonia reacts with 1 mmol of HCl and 1 mmol of nitrogen is present in 1mmol of NH_3.

The number of mmols of HCl consumed = 0.1 M × 26.2 mL = 2.62 mmols

Nitrogen content of the sample = 2.62 mmol × 14 (at.wt. of N) = 36.68 mmol

Protein content of the sample = 36.68 mmol × 6.25 × 100/560 mg = 40.94%

EXAMPLE 4 *A sample water of 100 mL required 15.8 mL of 0.024 M EDTA solution with eriochrome black T as indicator and 9.6 mL of the same EDTA for 100 mL of water from the same source after boiling, cooling, and filtering off the precipitated calcium carbonate. Calculate the total, permanent, and temporary hardness in terms of calcium carbonate equivalents.*

SOLUTION

From the stoichiometric reaction 1 mL 1.0 M EDTA = 100 mg of $CaCO_3$

$$15.8 \text{ mL} \times (0.024 \text{ M}/1 \text{ M}) = 37.92 \text{ mg of } CaCO_3 \text{ in } 100 \text{ mL of sample}$$

Total hardness of the sample water = 379.2 mg/L

Permanent hardness of the sample water = 230 mg/L

Temporary hardness = Total hardness – permanent hardness = 149.2 mg/L

EXAMPLE 5 *A 25-mL solution of 0.1 M sodium chloride was titrated against 0.1 M silver nitrate by Mohr's method. Calculate the concentration the pAg values for the addition of 0 mL, 5 mL, 15 mL, 25 mL, and 26 mL of silver nitrate. The K_{sp} of AgCl = 1.2 × 10^{-10}.*

SOLUTION

Before the addition of silver nitrate $[Cl^-] = 0.1$ M

$$K_{sp} = 1.2 \times 10^{-10} \text{ or } pAgCl = pAg^+ + pCl^- = 9.92$$

The initial concentration of Cl^-, before the addition of any silver nitrate, is 0.1 mol/L and $pCl^- = 1$. Hence, pAg = 9.92–1.0 = 8.92.

Before the end point but after the addition of 5 mL of silver nitrate the ionic concentrations of the remaining chloride and the corresponding pAg are

$$[Cl^-] = 0.1 \text{ mol/L} \times 20 \text{ mL}/30 \text{ mL} = 0.067 \text{ or } pCl^- = 1.17; \text{ and}$$

$$pAg = 9.92 - 1.17 = 8.75$$

$$[Cl^-] = 0.1 \text{ mol/L} \times 10 \text{ mL}/40 \text{ mL} = 0.025 \text{ or } pCl^- = 1.60; \text{ and}$$

$$pAg = 9.92 - 1.60 = 8.32$$

At 25 mL [Cl$^-$] and [Ag$^+$] = 0; Hence, pAg = 9.92/2 = 4.96.

After the end point (26 mL of titrant addition) [Ag$^+$] = 1.0 mL × 0.1 M/51 mL = 1.961 × 10^{-3}

$$pAg = 2.71$$

EXAMPLE 6 *Two BOD bottles each containing 5.0 mL of sewage sample was diluted with distilled water to 250 mL. One 100 mL portion of the diluted sample in the first BOD bottle consumed 6.2 mL of 0.05 N thiosulphate in the Winkler's method for the determination of dissolved oxygen. A 100 mL portion of the second BOD bottle was incubated at 20°C for five days and consumed 1.4 mL of the same thiosulphate solution for the Winkler's method. Calculate the BOD content of the sample.*

SOLUTION

Difference in volume of thiosulphate solution required for the sample solution from the first and the second BOD bottles = 6.2 – 1.4 = 4.8 mL of 0.05 N thiosulphate solution.

Since 1 L of 1 N thiosulphate solution = 8 g of oxygen,

$$4.8 \text{ mL of } 0.05 \text{ N thiosulphate solution} = 8 \text{ g} \times 4.8 \text{ mL} \times 0.05 \text{ N} \times 100/(250 \times 1000)$$
$$= 7.68 \times 10^{-4} \text{ g oxygen/5 mL sample}$$

Oxygen required for 1000 mL of sample = 7.68 × 10^{-4} g × 1000/5 = 0.154 g

BOD = 154 mg/L

EXAMPLE 7 *A 10-mL sample of wastewater was refluxed with 25 mL of potassium dichromate solution and after refluxing the excess unreacted dichromate required 12.6 mL of 0.12 M ferrous ammonium sulphate (FAS) solution. A blank of 10 mL of distilled water on refluxing with 25 mL of dichromate solution required 24.2 mL of 0.12 M FAS solution. Calculate the COD value of the wastewater.*

SOLUTION

Difference in volumes of the FAS required for the blank and sample solution = 24.2 – 12.6 = 11.6 mL

Since 1 L of 1 M FAS = 8 g of oxygen,

$$11.6 \text{ mL of } 0.12 \text{ M FAS} = (8 \text{ g} \times 11.6 \text{ mL} \times 0.12 \text{ M})/1000$$
$$= 0.011 \text{ g of oxygen/10 mL of wastewater}$$

Oxygen requirement for 1000 mL of wastewater = 1.11 g

COD value of the wastewater sample = 1110 mg/L

EXAMPLE 8 *Calculate the percentage of iron in a sample of 1.25 g which on conversion yielded 0.864 g Fe$_2$O$_3$ (FW – 159.69).*

SOLUTION

$$0.864 \text{ g of Fe}_2\text{O}_3 = 0.864/159.69 = 5.41 \times 10^{-3} \text{ moles}$$

Amount of Fe in grams (2 moles of Fe in Fe$_2$O$_3$)

$$= (2 \times 5.41 \times 10^{-3} \times 55.85) = 0.6043 \text{ g}$$

Percentage of iron in the sample = (0.6043/1.25) × 100 = 48.34

SUMMARY

- Volumetric analysis, also known as titrimetric analysis, involves measuring the volumes of the reactants for quantitative purposes using the experimental procedure of titration. The volume of the standard solution required to completely react with a known volume of the analyte is determined by using a visual indicator which undergoes an abrupt colour change at or close to the theoretical equivalence point. The choice of the indicator depends on the nature of the reactants.

- In acid–base reactions the change in pH can be computed theoretically or monitored experimentally to obtain a titration curve. The titration curve is useful to select a suitable visual indicator to detect the end point. An abrupt change in pH occurs close to the end point.

- Precipitation titrations involving silver nitrate as one of the reactants make use of indicators which show formation of a coloured precipitate or formation of a coloured solution or an adsorption indicator to detect the end point.
- Complex formation reactions involving EDTA as the chelating agent to form complexes with metal ions use metal ion indicators to detect the end point.
- Red-ox titrations use indicators which change colour depending on the red-ox potential of the indicator and that of the reductant–oxidant couple.

- Gravimetric analysis involving precipitation reactions is based on converting the soluble analyte into an insoluble precipitate, which is collected quantitatively over a filter medium, washed, dried, and given appropriate heat treatment to convert the precipitate into a stoichiometric compound, which is then weighed. From the known weight of the compound and its formula, the amount of analyte in the sample is determined.
- Gravimetric analysis by volatilization is based on converting the analyte constituent of the sample into volatile matter or into a residue from the weight of which the amount of analyte in the sample is determined.

REVIEW QUESTIONS

1. Classify the wet chemical methods of analysis. What are the advantages of wet chemical methods?
2. Give the classification of volumetric analyses.
3. What are standard solutions? How are they prepared?
4. Distinguish between primary standards and secondary standards. Give two examples each of primary and secondary standards.
5. What are the important characteristics of primary standards?
6. Explain how an acid–base indicator functions.
7. How are indicators selected for different acid–base titrations?
8. List the applications of acid–base titrations.
9. What is alkalinity of water due to? What are the different types of alkalinity present in water?
10. How are hydroxide, carbonate, and bicarbonate alkalinities of water determined?
11. What are complexometric titrations?
12. What are metal ion indicators? Give examples.
13. Explain with a suitable example the functioning of a metal ion indicator.
14. Explain the terms direct titration and back titration with respect to EDTA titrations.
15. What is hard water? What are its characteristics?
16. What are the disadvantages of hard water?
17. Explain the terms total hardness, temporary hardness, and permanent hardness of water.
18. How is total hardness of water determined by EDTA titration?
19. Discuss the principle and procedure involved in the determination of permanent and temporary hardness of water.
20. Describe the procedure involved in the determination of chloride content of water by Mohr's method.
21. What is Volhard's method? How is it carried out?
22. What is an adsorption indicator? How does it function?
23. Explain the principles involved in oxidation–reduction titrations.
24. What are redox indicators? How do they function? Give two examples of redox indicators.
25. Explain the principles involved in permanganometry, dichrometry, and iodometry.
26. How is iron determined by dichrometry?
27. How is copper determined by iodometry?
28. Explain the terms BOD and COD. What is their significance?
29. How is dissolved oxygen content of water determined?
30. Describe the procedure for the determination of COD.
31. How is BOD determined?
32. Give the classification of gravimetric methods.
33. Give examples of volatilization methods of analysis.
34. Discuss in detail the principles involved in precipitation methods of analysis.
35. What are precipitating agents? Give two examples.
36. How is nickel determined gravimetrically?
37. Give an account of the principle and procedure involved in the analysis of solder.
38. Describe the experimental procedure for the analysis of brass.
39. Write a note on the analysis of hematite.
40. What are the important constituents of cement? How are they determined?

4 Optical Methods

4.1 INTRODUCTION

Optical methods of analysis make use of visible light for characterizing substances on the basis of their physical interaction with radiation. The methods include refractometry, polarimetry, optical rotatory dispersion, and circular dichroism. This chapter discusses the principles, techniques, and applications of these methods.

4.2 REFRACTION

The phenomenon of *refraction* of light refers to the bending of a ray of light as it passes obliquely from one medium to another of different optical density. If the ray passes from vacuum or air to an optically denser medium, the ray bends towards the normal (Fig. 4.1). The velocity of light also changes as it crosses the boundary from one medium to another. Light travels in a straight line in vacuum or in any medium such as air. Its velocity in vacuum is constant at 3×10^8 ms^{-1} and decreases by only about 0.03% in air. However, light travels slowly through any other relatively denser medium due to its interaction with the constituent atoms and molecules of the medium.

4.3 REFRACTIVE INDEX

The ratio of the velocity of light in vacuum to the velocity in another medium is defined as the *index of refraction* or *refractive index* (η) of the substance. It is not necessary to measure the velocity of light in different media to determine the refractive index of a substance. Refractive index of any medium is calculated from the angles of incidence and refraction as shown in Fig. 4.1 using Snell's formula

$$\eta = \frac{\eta_{air}}{\eta_{medium}} = \frac{\sin i}{\sin r} \tag{4.1}$$

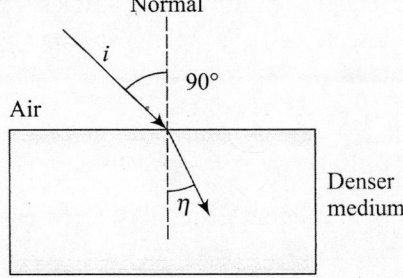

Fig. 4.1 Representation of the phenomenon of refraction

The angle of refraction will be greater than the angle of incidence when light passes from a denser medium to a rarer one, as shown in Fig. 4.2. Since the ratio (sin i/sin r) is a constant for the same wavelength and temperature, the angle r will increase with increasing values of the angle i. When the angle i is increased to a value so that the angle $r = 90°$, the refracted ray of light just grazes the surface and passes along the surface of the denser medium as shown in Fig. 4.2. The refracted ray of light when $r = 90°$ is called the *critical ray* and the angle of incidence which results in the angle of refraction $r = 90°$ is called the *critical angle*. The critical ray serves as a reference line in refractometer for measuring the refractive index of a medium. Any ray of light passing from a denser medium to a rarer one with an angle of incidence greater than the critical angle will not pass through to the rarer medium but will be totally reflected back from the surface into the denser medium resulting in total reflection of the light.

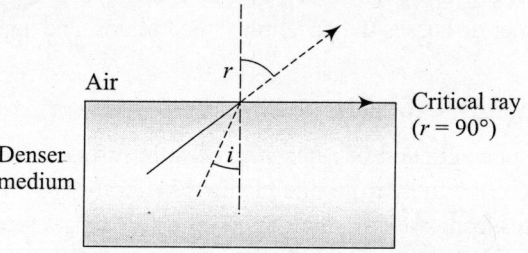

Fig. 4.2 Critical ray of light due to refraction

The refractive index of a transparent medium depends on the wavelength of light passing through it and also on the temperature of the medium. In the case of gases, it varies with pressure also. The variation of the angle of refraction with wavelength is known as *dispersion*. In general, refractive index of a medium decreases gradually with increasing wavelength except at regions of absorption where the refractive index changes abruptly. Hence, it is necessary to specify the wavelength when the refractive index of a medium is quoted. The most commonly used reference wavelength is that of sodium D line (the doublet at 589 nm and 589.6 nm) and the refractive index is specified as η_D^{20} to indicate the wavelength and the temperature of measurement as 20°C.

Refractive index is a characteristic constant of the medium, provided it has been measured at specified conditions of wavelength of the incident ray and temperature and also pressure in the case of gases. The values of refractive index of organic liquids vary between 1.2 and 1.8, whereas those of organic solids vary between 1.3 and 2.5.

Refractive index of a liquid varies with temperature and pressure. However, the *specific refraction* (η_D) is independent of these variables. The Lorentz–Lorenz equation (also known as the Clausius–Mossotti relation) relates the specific refraction, the refractive index, and the density ρ of a liquid as

$$\eta_D = \frac{\eta^2 - 1}{\eta^2 + 2} (1/\rho) \tag{4.2}$$

Since the density of a liquid generally decreases with increasing temperature, the index of refraction also decreases as the temperature increases. For most organic liquids, the decrease in refractive index is approximately 0.0005 for an increase in temperature by 1°C, whereas for water the decrease is much less (~0.0001).

The specific refraction is characteristic of the substance, and hence, useful for the identification of the substance and also as a criterion of its purity. The specific refraction increases linearly with increasing length of the carbon chain of a homologous series of compounds.

The *molar refraction* $M\eta_D$ is calculated by multiplying the specific refraction with molecular weight of the substance. The molar refraction is an additive property of the atomic or group refractions of the constituent elements or groups in chemical compounds. Hence, molar refraction can be computed from known values of atomic and group refractions listed in Table 4.1.

Table 4.1 Atomic and group refractions

Atoms and groups	Refraction	Atoms and groups	Refraction
H	1.100	S (C=S) thiocarbonyl	7.97
C	2.418	N (primary aliphatic amine)	2.322
F	1.0	N (sec. aliphatic amine)	2.499
Cl	5.967	N (tert. aliphatic amine)	2.840
Br	8.865	N (primary aromatic amine)	3.21
I	13.900	N (sec. aromatic amine)	3.59
O (C=O) carbonyl	2.211	N (tert. aromatic amine)	4.36
O (O–H) hydroxyl	1.525	N (amide)	2.65
O (C–O) ether, ester	1.643	NO_2 aromatic nitro	7.30
S (S–H) mercapto	7.69	$C \equiv N$	5.459

Refractive index measurement is useful in the elucidation of molecular structure based on molar refraction. Molar refraction is related to the polarizability of molecules due to shifting of electron density between nuclei linked through chemical bond.

4.3.1 Measurement of Refractive Index

Refractive index measurement of crystalline solids is carried out using a microscope by the *Becke line method*. A specimen of the crystalline sample is mounted on a microscope slide and placed under an optical microscope with suitable magnification in the range 10–$100\,x$. The solid specimen is immersed in a drop of a liquid of known refractive index and observed through the microscope. As the microscope is focused up, a hollow moves away from the centre or towards the centre of the crystal, depending on whether the liquid has a higher refractive index than the crystal or vice versa. A set of standard liquids of known refractive indices are used for determining the refractive index of the crystalline solid. If the refractive indices of the solid and the chosen liquid standard are identical, no boundary can be seen between the liquid and the solid. An important industrial application of the Becke line method is to determine the quality of bauxite ore.

Refractive index of any liquid medium denser than air is determined by an instrument called *refractometer* based on the principle of refraction. Refractometer determines the refractive index of a medium by measuring the position of the critical ray. Commonly used refractometers for measuring the refractive index of liquids include those of Abbe, Pulfrich, and Immersion type.

4.3.2 Abbe Refractometer

Abbe refractometer was the first laboratory refractometer developed by Ernst Abbe during the late 1800s. It is the most popular one as it needs less quantity of sample and uses white light. In contrast, Pulfrich refractometer uses a monochromatic light. Modern Abbe refractometers have built-in digital display and also use solid-state Peltier effect to heat and cool the sample so that the variation of refractive index of the sample as a function of temperature can be determined. They are also capable of measuring refractive indices at different wavelengths of incident light.

The working principle of Abbe refractometer remains the same as in the original instrument of Abbe, in spite of the sophistication of the instrument. For measuring the refractive index of liquids, a drop of the sample is kept sandwiched as a thin layer between an illuminating prism and a refracting prism made of glass in the instrument (Fig. 4.3). The bottom surface of the illuminating prism is roughened by grinding so that when light from a source is passed through the prism, each roughened point on the surface scatters the light in all directions. The refracting prism is made of glass of a high refractive index (1.75) so that refractive indices of samples less than 1.7 can be determined by the instrument. The working range of Abbe refractometer is 1.3000–1.7000 with a precision of 0.0001. The ray of light travelling from point A to point B through the sample and entering the refracting prism will have the maximum possible angle of incidence, and hence, will be refracted to point C with the maximum possible angle of refraction. All the other rays of light (say point D) passing through the sample will have angles of incidence less than the maximum angle, and hence, will be refracted to the left of point C. A detector placed at point C will show a bright region to the left of point C and a dark region to the right of point C.

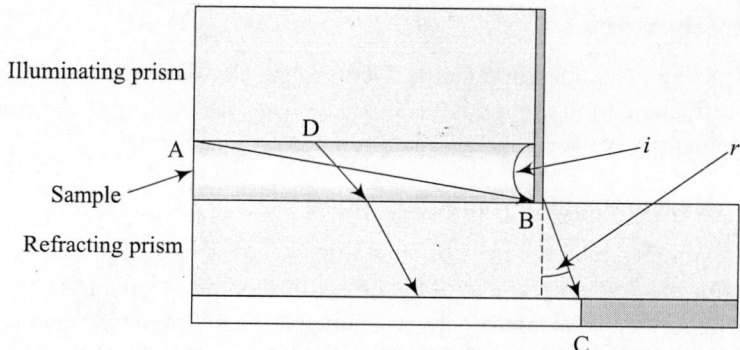

Fig. 4.3 Optical path in Abbe refractometer

The borderline between the bright and dark regions will not be sharp because of dispersion, when white light (polychromatic radiation) is used, as refractive index varies with the wavelength of the incident beam of light. Hence, in Abbe refractometer, a set of compensating prisms are introduced into the optical path after the refracting prism to compensate the dispersion effect and make the borderline sharp similar to that obtained when monochromatic light of 589 nm is used. The block diagram and the various components of the instrument are shown in Fig. 4.4(a) and the view of the borderline at the cross-wires and the refractive index scale in Fig. 4.4(b).

Samples with different refractive indices produce different angles of refraction and thereby shift the position of the borderline between the light and dark regions. The position of the borderline between the light and dark regions on a calibrating scale serves to indicate the refractive index of the sample.

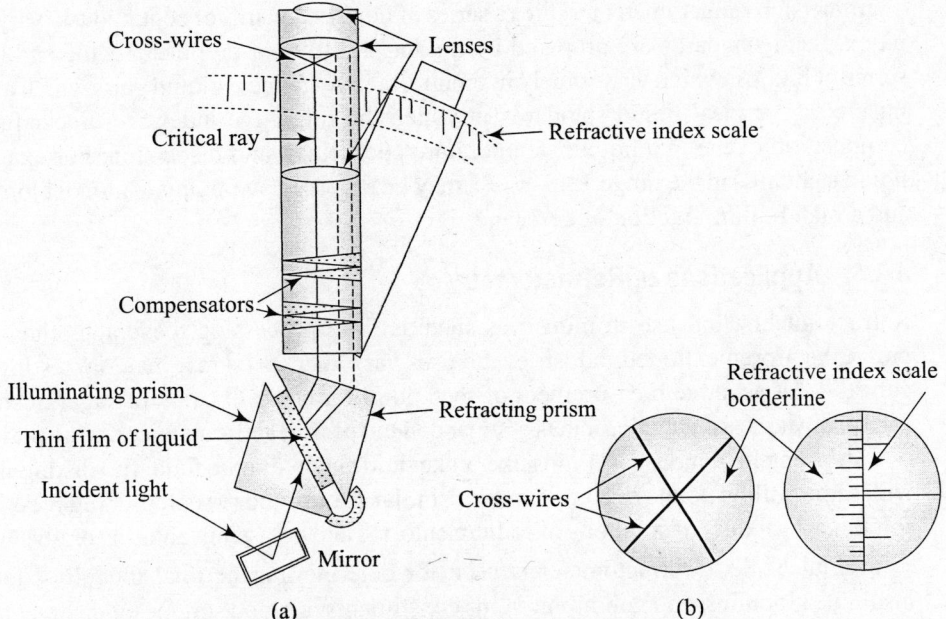

Fig. 4.4 (a) Block diagram of Abbe refractometer and (b) border line for total internal
reflection at intersection of cross-wires and view of scale

4.3.3 Immersion Refractometer

Immersion refractometer consists of a single prism mounted on a telescope with a scale placed
below the eyepiece. As the name implies, the instrument is dipped directly into the sample
liquid of about 15 mL contained in a small beaker. The beaker is surrounded by a water bath for
temperature control. The temperature of the sample has to be maintained within $\pm 0.1°C$ for an
accuracy of 0.02 divisions in the scale which corresponds to a refractive index of 0.000074 with
the sodium D line. The optical path, the position of the critical ray, and the view of the scale
through the eyepiece are shown Fig. 4.5.

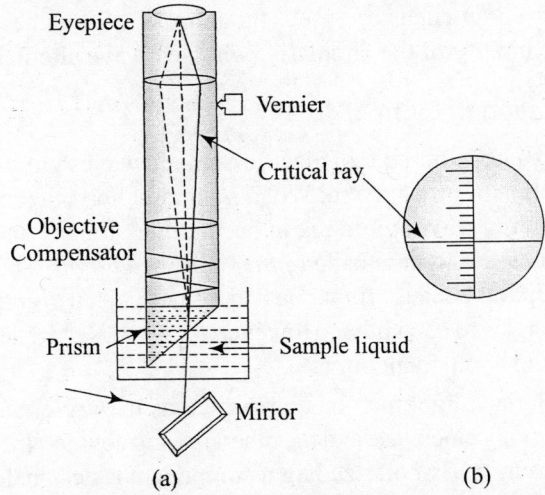

Fig. 4.5 (a) Optical path in immersion refractometer and (b) view through the eyepiece

Immersion refractometer requires a series of liquid standards of continuously varying refractive index. Such standards are prepared by mixing in different proportions, the end members of a series of liquids, which vary widely in their refractive indices. Liquid pairs which are completely miscible give ideal liquid standards in which the refractive index is a linear function of the composition of the two liquids, temperature coefficient, and dispersion. For example, a set of liquid standards in the range 1.45–1.632 may be prepared by mixing α-monochloro naphthalene and a high boiling fraction of kerosene.

4.3.4　Applications of Refractometry

Refractometers find use in industries such as food processing, beverage, chemicals, and so on. In the aforementioned industries the uses vary such as (i) raw material testing, (ii) process control—for accurate measurement of seed point for crystallization of sugar in sugar industry, (iii) quality assessment of products—sugar content of fruit and fruit juices, wort, and finished beer or water content of industrial fluids like brake fluid and hydraulic fluid, (iv) distinguishing liquids with same boiling points or compounds of similar nature, and (v) determining the composition of solutions—mixture of solutions of sodium chloride and potassium chloride or glycerol and water.

In clinical field, refractometer is used for determining the total globulins, total albumen, insoluble globulins, and non-albuminous constituents in blood serum, solid content of urine, etc.

4.4　POLARIMETRY

The term *polarimetry* refers to the measurement of the polarization of electromagnetic waves, particularly of radiation in the visible and radiofrequency regions. In the visible region of electromagnetic radiation, polarimetry involves the measurement of the change in the direction of vibration of polarized light as it passes through an optically active medium. It is commonly used to measure optical properties, such as birefringence, optical rotation and dichroism of materials. Polarimetry of thin films and surfaces is known as *ellipsometry*.

Polarimetry is useful in the computational analysis of waves as in the case of radars for improving the characterization of targets. This is achieved by polarimetry to estimate the fine texture of a material and help resolve the orientation of small structures in the target. When circularly polarized antennas are used, the number of bounces of the received signal can be resolved as the chirality of the circularly polarized wave alternates with each reflection.

4.4.1　Polarization of Light

Light consists of a large number of electromagnetic waves vibrating in all possible orientations around the direction of propagation. When such radiation interacts with materials, it is possible to separate or sort out waves vibrating in one particular plane only. This phenomenon is known as *polarization* and the light is called *plane polarized light*. The term *plane polarized light* is not an apt description because light consists of both electric and magnetic field vectors which vibrate at right angles to each other. However, the term is used taking into account the vibration of the electric field component only.

The plane polarized light may be represented as the vector sum of *two circularly polarized light rays*, one moving clockwise and the other moving counterclockwise with the same amplitude of vibration. The circularly polarized light component is designated as *d* or dextrorotatory (right handed) when it rotates clockwise around the direction of propagation and as *l* or levorotatory (left handed) when it rotates counterclockwise.

Polarization of electromagnetic waves, particularly of radiation in the visible and radiofrequency regions, is well documented and has many applications. Polarization of electromagnetic radiation may be brought about by reflection, refraction, or diffraction as the radiation interacts with materials.

Polarization by reflection was first noticed by Brewster in 1912. When light is incident on a reflecting surface, such as glass, so that the angle of incidence i satisfies the condition $\tan i = \eta$ (where η is the refractive index of the reflecting material), then only the component of light vibrating parallel to the mirror surface (that is perpendicular to the plane of incidence) will be reflected. The critical angle is called *Brewster angle*. Polarization by reflection is shown in Fig. 4.6.

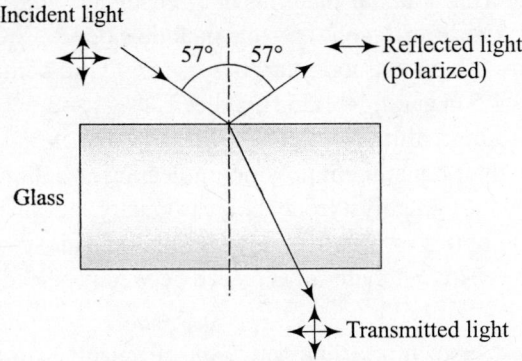

Fig. 4.6 Polarization by reflection ($\tan i = \eta$)

Polarization by absorption occurs when certain crystals, called *dichroic crystals*, such as tourmaline or iodocinchonidine sulphate, absorb strongly the light vibrating in one direction, but only weakly absorb the light vibrating in the perpendicular direction. The phenomenon can be demonstrated using two plastic sheets embedded with such crystals to act as polaroid filters. When the two sheets are aligned parallel to each other with the crystals having the same orientation, light absorbed by one polarizing filter passes unimpeded through the second filter. However, if the second filter is rotated, the intensity of polarized light decreases. This is because the plane of the polarized light corresponds to one of the absorption planes in the second filter. The polarized light from the first filter is totally absorbed by the second filter when the two filters are perpendicular to each other. Polaroid filters are effective only in the wavelength region of 500–680 nm.

Polarization by double refraction or *birefringence* involves the splitting of a ray of light into two rays called the *ordinary ray* (O-ray) and *extraordinary ray* (E-ray) when light passes through certain materials called *birefringent* materials. In most materials, such as glass or thin wafers of sodium chloride crystal, the refractive index of light is independent of the direction of incidence of the ray of light. Such materials are called *isotropic* and their optical properties remain the same in all the three Cartesian axes or crystallographic directions and the refractive index of the material is a constant. *Snell's law* ($\sin i/\sin r = \eta$) holds good for the angles of incidence other than zero degree. However, in certain materials (e.g., calcite or boron nitride crystals), the refractive index is not the same for all directions of incidence of light. Such crystals are *anisotropic* (because their structure is anisotropic or directionally dependent) and exhibit *birefringence*. Anisotropic crystals posses at least one direction along which the incident ray of light is not split. If the y- and z-directions are equivalent in terms of the crystalline forces, then the x-axis is unique and is called the *optic axis* of the material. The propagation of light along the optic axis would be independent of its polarization; its electric field is everywhere perpendicular to the optic axis; and it is called the *ordinary ray* or *O-ray*. When a ray of light is incident on such crystals in any other direction, the light is split into the ordinary ray which obeys the Snell's law and the extraordinary ray or E-ray which does not conform to Snell's law. The extraordinary ray has the electric field parallel to the optic axis. Both the rays are linearly polarized.

The reason for birefringence is the fact that in anisotropic media the electric field vector and the dielectric displacement can be non-parallel (for the extraordinary polarization), though being linearly related. When the material has a single axis of anisotropy (optical axis), it is called *uniaxial* material. The uniaxial material is birefringent with two different refractive indices. Examples of uniaxial birefringent crystals include calcite ($CaCO_3$) with the refractive indices of the ordinary ray η_O and extraordinary ray η_E at 590 nm being 1.658 and 1.486, respectively, quartz (SiO_2, $\eta_O = 1.544$ and $\eta_E = 1.553$), rutile (TiO_2, $\eta_O = 2.616$ and $\eta_E = 2.903$), ice ($\eta_O = 1.309$ and $\eta_E = 1.313$), sodium nitrate ($NaNO_3$, $\eta_O = 1.587$ and $\eta_E = 1.336$), etc. Plastics on moulding or extrusion have their constituent polymer molecules in a stretched conformation and thereby become birefringent (e.g., polystyrene, polycarbonate, cellophane, etc.). Birefringent materials are used widely in optics to produce polarizing prisms and retarder plates (e.g., quarter-wave plates). When a birefringent material is placed between crossed polarizers, interference colours can be observed.

Certain crystals have more than one axis of anisotropy, and hence, give rise to biaxial birefringence, also known as *trirefringence*. For such a material, the refractive index tensor is given in terms of three distinct eigen values. Examples of biaxial materials include epsom salt ($MgSO_4{\cdot}7H_2O$, $\eta = 1.433$, 1.455, and 1.461), white mica (Muscovite-KAl_2 (OH)$_2$ Si_3AlO_{10}, $\eta = 1.563$, 1596, and 1.601), olivine (($Mg, Fe)_2SiO_4$, $\eta = 1.640$, 1.660, and 1.680), perovskite ($CaTiO_3$, $\eta = 2.300$, 2.340, and 2.380), etc. Birefringence and related optical effects of optical rotation and circular dichroism can be measured by polarimetry.

Birefringence can also be created in optically isotropic materials by (i) stretching or bending to form mechanically deformed materials, e.g., plastics become birefringent when moulded or extruded as the molecules are aligned in a stretched conformation, (ii) applying an electric field to induce molecules to align in a given direction or align asymmetrically, and (iii) applying a magnetic field to make the material circularly birefringent.

Birefringence of a material can be measured by measuring the changes in polarization of light passing through the material using a polarimeter or a polarizing microscope.

The phenomenon of birefringence of the retinal nerve fibre layer is used to monitor and assess glaucoma (pressure within eye) and in optical mineralogy to determine the chemical composition and history of minerals and rocks. A different form of birefringence in elastic materials, known as *elastic birefringence*, causes the splitting of shear waves, and hence, is useful in seismology. Birefringent materials find extensive use in optical devices, such as liquid crystal displays, colour filters, wave plates, light modulators, optical axis gratings, polarizing prisms, birefringence filter, in electronic and television cameras.

4.4.2 Polarizers

Polarization of visible light by double refraction has been used for the production of polarizers and analysers. Calcite crystal is birefringent and both the ordinary and extraordinary rays are linearly polarized. A plane polarized beam of light can be obtained by removing one of the rays. William Nicol constructed a polarizer made of calcite crystal to get plane polarized light in 1928. The Nicol prism is constructed from Iceland spar, a naturally occurring calcite mineral, because of its transparency and availability in large-sized crystals. The prism is constructed so as to transmit only the extraordinary ray and remove the ordinary ray by total internal reflection within the crystal. The normal angle of 71° between the two end-faces of the calcite crystal is cut down to 68° and the crystal is diagonally cut along the line A–B as shown in Fig. 4.8. The cut surfaces are polished and pasted with Canada balsam. The refractive index of Canada balsam (1.55) is intermediate between the indices of the calcite crystal for the ordinary and extraordinary rays (1.658 and 1.486, respectively). The ordinary ray of light is incident on the balsam at an angle greater than the critical angle, and hence, is totally reflected. The extraordinary ray with a direction of vibration perpendicular to the surface of the balsam alone is transmitted through the crystal as shown in Fig. 4.7.

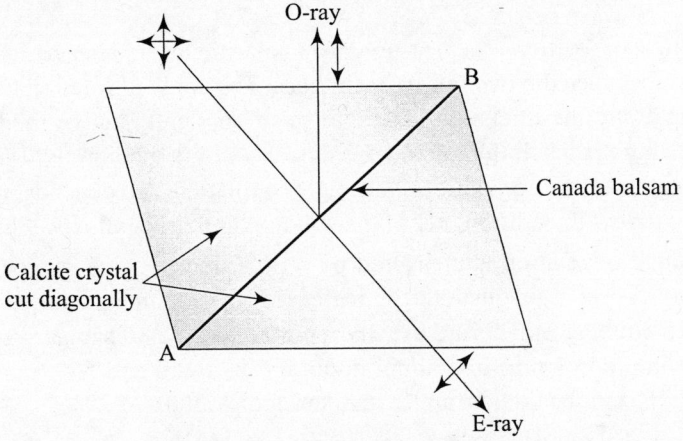

Fig. 4.7 Nicol prism transmitting the E-ray

The Nicol prism requires smaller pieces of calcite crystal and is widely used as it is relatively cheaper. However, it has certain disadvantages in that the light transmitted through the Nicol prism is displaced slightly from the original beam of light and revolves in a circle as the prism is rotated. In addition, the detection of balance point in a polarimeter is somewhat difficult as the combination of two Nicol prisms does not produce total extinction at any point in the field of vision. A better polarizer is Glan–Thomson prism which is also constructed from calcite crystals but the angles of the prism faces and the angle of the diagonal cut are different.

Polarizers for ultraviolet light are made of quartz crystals pasted with glycerine or castor oil as Canada balsam absorbs ultraviolet light.

4.4.3 Polarimetry Theory

The most commonly measured optical property of chemical/biochemical substances is optical rotation. Solutions of optically active substances exhibit circular birefringence, that is, rotate the plane of polarization of polarized light as it passes through the sample. As mentioned earlier, plane polarized light may be considered as the vector sum of the two components, the d and l circularly polarized light rays. An optically active medium will show different indices of refraction for d and l components of the plane polarized light. Hence, either d or the l component would be slowed down on passing through the optically active medium, resulting in the rotation of the plane polarized light. For example, if the left circularly polarized light component is slowed down compared to the right circularly polarized light component, the resultant would be a plane polarized light rotated somewhat to the right from its original position. The angular phase difference φ between the two circular components is given by the relationship

$$\varphi = \frac{2\pi d}{\lambda} (\eta_d - \eta_l) \tag{4.3}$$

where d is the thickness (or length) of the optically active medium through which the light ray passes, λ is the wavelength of the light ray, and η_d and η_l refractive indices, respectively, of the d and l components of light. The angle of rotation α is one-half of the angular phase difference φ of the two circular components, and hence, the above equation can be written as

$$\alpha = \frac{\pi d}{\lambda} (\eta_d - \eta_l) \text{ or } (\eta_d - \eta_l) = \alpha\lambda/\pi d \tag{4.4}$$

The angle of rotation α is measured by a polarimeter, and hence, it is possible to calculate the difference between the two refractive indices. The angle of rotation depends on the wavelength of the light and the thickness or length of the optically active medium. When the optically active medium is a solution, then the thickness of the medium will have to be combined with the concentration of the solution. The angle of rotation also depends on temperature as it affects refractive index. In addition, pH of the solution and time influence the angle of rotation.

The angle of rotation is measured by a polarimeter using monochromatic radiation and a constant thickness or length of the optically active medium maintained at a constant temperature. Polarimetric measurements are usually reported for a set of standard conditions using the green mercury line (546.1 nm) or sodium doublet (589.0 nm and 589.6 nm), a length of 1 mm for solids and 10 cm for liquids, and a standard temperature of 20°C.

The *specific rotation* or *rotation under standard conditions*, [α], of an optically active substance is given as

$$[\alpha] = 100 \, \alpha_\lambda^t/dC \tag{4.5}$$

where α^t is the angle of rotation measured at temperature t using a light of wavelength λ, d the thickness of the medium expressed in decimeters (dm), and C its concentration in terms of grams of solute per 100 mL of solution. The specific rotation is characteristic of the chemical substance, and hence, useful for identification of the substance. Thus, the specific rotation $[\alpha]_D^{20}$ (sodium doublet line) of sucrose, glucose, and fructose, respectively, are $+66.5°$, $+52.5°$, and $-93°$.

The *molecular rotation* [Φ] of a substance may be calculated using the formula

$$\Phi = \alpha \times MW/100 \tag{4.6}$$

where MW is the molecular weight of the substance.

4.4.4 Polarimeter

Polarimeter is an instrument used to measure the angle of rotation of plane polarized light passing through an optically active medium. A simple instrument consists of polarizing filters, whereas a more sensitive instrument is based on interferometer. The block diagram of a simple polarimeter is shown in Fig. 4.8. The main components include (i) a source of light, (ii) sample holder, (iii) polarizer and analyser prisms, (iv) half-shade device, (v) eyepiece, and (vi) scale.

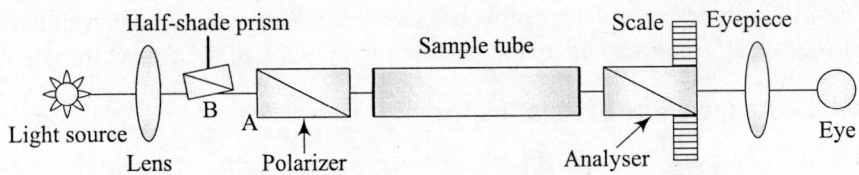

Fig. 4.8 Optical parts of a polarimeter

The light source is usually a sodium vapour lamp emitting light of wavelengths 589.0 nm and 589.6 nm. The continuous background of the sodium lamp can be eliminated with a 6 cm thick filter of a solution of 7 % potassium dichromate. Alternatively, mercury vapour lamp which emits light of several wavelengths in the visible region at 435.8, 491.6, 546.1, 577.0, and 579.1 nm can also be used with a proper choice of filters to isolate each wavelength. Continuous light sources, such as sunlight or tungsten filament lamp, can also be used.

The most commonly used polarizer in polarimeters is called *Lippich polarizer* with associated half-shade prism. It is referred to as a two-prism polarizer since it consists of two Nicol prisms A and B. The smaller prism B covers only one-half of the thickness of prism A. The arrangement of the two prisms facilitates the rotation of one-half of the field viewed in the eyepiece and also divides the view into two halves by a sharp line. The line is the image of the edge of the smaller prism B. Thus, one-half of the light from the source passes through both the polarizer prisms A and B while the other half passes only through the prism A. The principal planes of the two prisms are inclined at a small angle. When the principal plane of the analyser prism is at right angle to polarizer prism A, one-half of the field in the eyepiece is dark and when it is at right angle to the optical plane of B, the other half of the field is dark. When the analyser prism is rotated from one position to another, the intensity of one-half of the field in the eyepiece increases from zero to maximum while the intensity of the other half decreases from maximum to zero. The intermediate position when both the halves in the eyepiece are of equal brightness is taken as the balance point. The angle between the two Lippich prisms can be varied to alter the sharpness of the balance point for easy detection. Sharpness increases with decreasing angle between the prisms.

The sample holder is a glass tube of 10 cm (1 dm) or 20 cm (2 dm) in length (the length is expressed more commonly in decimeters dm) with its ends covered by strain-free colour corrected glass disks to prevent even slight circular polarization of light. The glass disks are used for visible region, whereas quartz disks are used for ultraviolet region. The exact length of the polarimeter tube is calibrated by measuring the rotation of polarization using a standard solution of nicotine in ethanol at a specific temperature. Analyte solutions are prepared by dissolving weighed quantities of the substance and dissolving in known volume of distilled water. The intensity of light transmitted through the polarizer prism, sample solution, and the analyser prism is given by Malus' law as

$$I_t = KI_0 \cos^2\theta \tag{4.7}$$

where I_t and I_0 are, respectively, the intensities of the light transmitted though the analyser prism and incident on the analyser, K is a constant approximately equal to 1, and θ is the angle between the directions of transmission of the polarizer and analyser prisms.

A graduated circular scale fitted with a vernier scale measures the angle through which the analyser prism is rotated to an accuracy of 0.002°.

In sugar industry, polarimeter is used as a *saccharimeter* for determining the concentration of sucrose by using a scale graduated directly in terms of the concentration of sucrose.

4.4.5 Applications of Polarimetry

The following are the various applications of polarimetry.

(i) Qualitative analysis for identification of the sample by determining its specific rotation as it is indicative of the nature of the sample. Sugars and amino acids and a variety of organic substances can be identified through specific rotation.

(ii) Quantitative analysis of optically active substances for the determination of the concentration of the analyte based on the principle that optical rotation of a substance is directly proportional to its concentration, provided the path length is kept constant. A series of standard solutions, usually in the concentration range of 0–25% (w/v), is prepared and a calibration or working graph of optical rotation α, as a function of known concentrations of the standards, is prepared. The optical rotation of the analyte solution is determined (under identical conditions as determined for the standards) and the concentration of the analyte is determined graphically from the calibration graph as shown in Fig. 4.9.

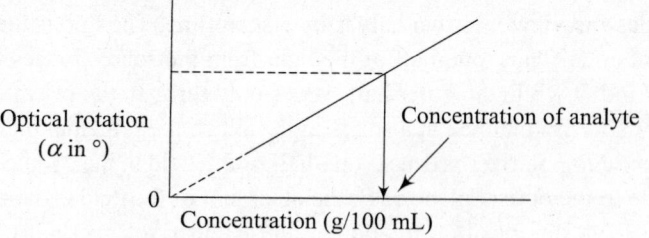

Fig. 4.9 Quantitative analysis by polarimetry

Alternatively, the concentration of the optically active substance can be calculated by using Eq. (4.5) if the specific rotation of the substance is known and the analyte contains no other optically active substance.

(iii) Estimation of individual analytes in a mixture is yet another application, the typical example being the estimation of glucose and sucrose in a mixture of the two. The procedure involves the determination of the optical rotations of known volumes of the analyte mixture and that of the completely hydrolysed mixture (hydrolysed with 5 mL of concentrated HCl) by a polarimeter. The individual amounts of sucrose (S) and glucose (G) are calculated as follows:

$$A = d\,(S\alpha_{\text{sucrose}} + G\alpha_{\text{glucose}}) \tag{4.8a}$$
$$B = d\,[(S1.052\alpha_{\text{invert}}) + (G\alpha_{\text{glucose}})] \tag{4.8b}$$
$$(A-B) = dS\,(\alpha_{\text{sucrose}} - 1.052\,\alpha_{\text{invert}}) \tag{4.8c}$$

Rearranging Eq. (4.8c) gives

$$S = (A-B)/d\,(\alpha_{\text{sucrose}} - 1.052\,\alpha_{\text{invert}}) \tag{4.8d}$$

where A and B are, respectively, the optical rotations of the analyte mixture and the completely hydrolysed mixture; α_{sucrose}, α_{glucose}, and α_{invert} the specific rotations of sucrose, glucose, and invert sugar, respectively; and d the tube length (optical path) in the polarimeter. The factor 1.052 is the ratio of sucrose to invert sugar because 342.2 g of sucrose yields 360.2 g of invert sugar. The rotation due to S grams of sucrose alone in the analyte mixture is calculated from the known value of its specific rotation [Eq. (4.5)] and this amount is subtracted from the optical rotation A of the analyte mixture to get the optical rotation due to glucose alone in the analyte mixture. The amount of glucose present (G) in the analyte mixture is obtained by using Eq. (4.5).

(iv) Comparison of strengths of HCl and H_2SO_4 can be performed by polarimetry. The strength of an acid is a measure of its ionization or H^+ ion concentration. Although the strengths in terms of normality may be the same, the acids may not ionize to the same extent, and hence, the H^+ ion concentrations will vary. The principle of polarimetric determination of the strengths of the two acids is based on the fact that H^+ catalyses the inversion of sucrose, and by monitoring the progress of the reaction, H^+ ion concentrations of the two acids, that is, their relative strengths can be determined.

Sucrose is dextrorotatory, and on hydrolysis yields an equimolar mixture of glucose and fructose. The levorotation of fructose is more than the dextrorotation of glucose, and hence, the equimolar mixture obtained by complete hydrolysis of sucrose has a net levorotation. The hydrolysis reaction is known as *inversion* because of the inversion of optical rotation. The experiment is carried out by mixing a known volume of 20% (w/v) solution of sucrose with an equal volume of 0.5 N HCl and recording the angle of rotation at regular time intervals of 5 or 10 min to approximately 100 min, starting from zero (immediately after mixing the solutions). The reading after 48 h is taken as at infinite time indicative of complete hydrolysis of sucrose. The procedure is repeated separately with sulphuric acid of same strength. The rate constant, k, for the acid catalysed hydrolysis of sucrose by the two acids is calculated individually using the first order rate equation [$k = (2.303/t)\ \log(a/(a - x))$], where a and $(a - x)$ represent the concentration of sucrose at zero time and individual time intervals t, respectively. The ratio of the k values gives the relative strengths of the two acids.

4.5 OPTICAL ROTATORY DISPERSION AND CIRCULAR DICHROISM SPECTRA

Optical rotatory dispersion (ORD) and circular dichroism (CD) are related to the rotatory power and light absorption of optically active compounds. The change in optical activity as a function of wavelength of incident light is called *optical rotatory dispersion*. On changing the wavelength of the incident light to shorter wavelengths in the vicinity of electronic absorption spectral band of an optical isomer of a compound, the rotatory power of the compound (specific rotation $[\alpha]$) first increases strongly, then decreases to zero, and changes sign with decreasing wavelength of the incident radiation. Within the electronic absorption spectral band, the refractive indices of the right- and left-circularly polarized light components are different ($\eta_l \neq \eta_d$).

Similarly, the molar absorptivity values of the right- and left-circularly polarized light components (ε_r and ε_l respectively) are also different as they are absorbed by the optically active medium to different degrees, and hence, $\varepsilon_l - \varepsilon_d \neq 0$. This effect changes the linearly polarized light into an elliptical polarized light and the phenomenon is known as *circular dichroism*. The vector diagram of elliptically polarized light with an ellipticity of θ is shown in Fig. 4.10. The ellipticity is defined as the angle whose tangent is the ratio of the minor axis and the major axis of the ellipse. Circular dichroism is a plot of θ as a function of wavelength λ. The angle θ is given as $\theta = (\varepsilon_l - \varepsilon_d)\ 3305$.

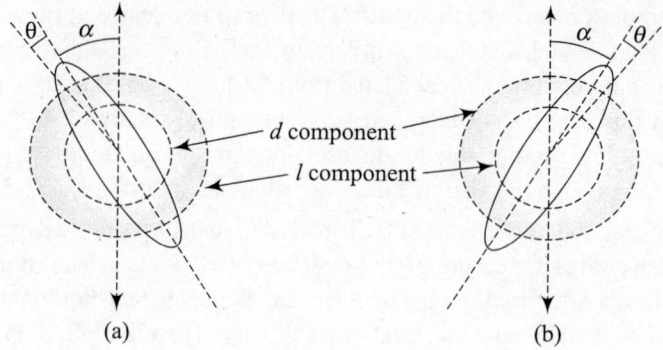

Fig. 4.10 Elliptically polarized light when (a) $\eta_d > \eta_l$ and $\varepsilon_l > \varepsilon_d$ and (b) $\eta_d < \eta_l$ and $\varepsilon_l < \varepsilon_d$

The combined effect of optical rotatory dispersion and circular dichroism is known as *Cotton effect* after the name of the discoverer. Cotton effect by the two enantiomers of an optically active compound known as the positive and negative Cotton effect in the region of absorption band is shown in Fig. 4.11(a) and Fig. 4.11(b), respectively. The figures are obtained by superimposing the absorption spectrum, ORD spectrum, and the CD spectrum on the same graph.

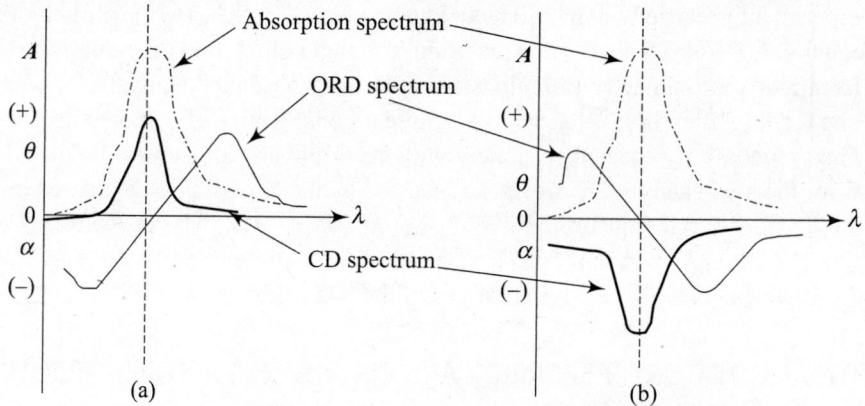

Fig. 4.11 (a) Positive Cotton effect by the levorotatory isomers and
(b) Negative Cotton effect by the dextrorotatory isomers

In Fig. 4.11, the y-axis symbols are A, absorption in the absorption spectrum, $\theta = (\varepsilon_1 - \varepsilon_d)$ for CD spectrum, and α, angle of rotation for the ORD spectrum, whereas the x-axis represents wavelength.

The ORD and CD spectra of compounds can be recorded by spectropolarimeters which have the required optics for measuring the optical activity, as well as the absorbance at different wavelengths of incident beam of light. The incident light first passes through a monochromator and then through a polarizer before it passes through the sample solution. The plane polarized light emerging from the sample passes through an analyser before reaching the photomultiplier tube detector. For measuring the circular dichroism, the plane polarized light is split into the left- and right-circularly polarized light components by a quarter wave plate and the intensities of the different components are measured by photomultiplier tube detector. The ORD and CD spectra are useful in assigning the absolute configuration of optically active compounds. The midpoint of the ORD curve matches with the peak maxima of the absorption spectrum and the

CD spectrum as shown by the dotted vertical line in Fig. 4.11. The mirror image enantiomer shows an exactly opposite behaviour with respect to its rotatory power as a function of decreasing wavelength of the incident light as shown in Fig. 4.11(b).

ORD and CD are useful in the determination of structure of optically active substances such as amino acids, proteins, antibiotics, terpenes, steroids, and coordination compounds. The sign of the Cotton effect exhibited by these compounds is useful in determining the stereochemistry at the asymmetric centre.

SOLVED PROBLEMS

EXAMPLE 1 *Calculate the molar refraction of ethyl acetate if the refractive index and density are 1.3701 and 0.901 g cm^{-3}, respectively. Compare with the value of molar refraction obtained from the data provided in Table 4.1.*

SOLUTION

From Eq. (4.2)

$$\eta_D = \frac{\eta^2 - 1}{\eta^2 + 2} (1/\rho)$$

$$= (0.8772 / 3.8772)(1/0.901 \text{ g cm}^{-3}) = 0.2511 \text{ cm}^3 \text{ g}^{-1}$$

$M\eta_D$ = specific refraction × molecular weight

$$= 0.2511 \text{ cm}^3 \text{ g}^{-1} \times 88 = 22.10 \text{ cm}^3 \text{ mol}^{-1}$$

$M\eta_D$ is calculated based on the molecular formula $C_2H_5OOCCH_3$.

Carbon	2.418 × 4 = 9.672
Hydrogen	1.10 × 8 = 8.800
Carbonyl (C=O)	2.211 × 1 = 2.211
Ether (C–O)	1.643 × 1 = 1.643

	= 22.326 cm^3 mol^{-1}

EXAMPLE 2 *Calculate the concentration of an optically active substance of specific rotation of 130° if a 20 cm solution gave a polarimeter reading of 11.2° at 589 nm.*

SOLUTION

Using Eq. (4.5)

$$[\alpha] = 100\alpha'_\lambda/dC$$

$$C = 11.2° \times 100/130° \times 2 \text{ dm} = 4.30 \text{ g L}^{-1}$$

EXAMPLE 3 *Calculate the value of specific rotation of an optically active substance if a 5 gL^{-1} solution in a 10 cm tube showed an angle of rotation of 8.2°.*

SOLUTION

Using Eq. (4.5)

$$[\alpha] = 100 \, \alpha'_\lambda/dC$$

$$[\alpha] = 100 \times 8.2/1 \times 5 = 164°$$

SUMMARY

• Refractive index of the substance is the ratio of the velocity of light in vacuum to that through the substance. It is calculated using Snell's formula of sin i/sin r.

• Refractive index of a substance is a characteristic constant and depends on the wavelength of incident light and temperature in the case of liquids and in addition on pressure in the case of gases.

- Internationally sodium D line (589 and 589.6 nm) is used as the reference wavelength to measure the refractive index at 20°C. Most organic liquids have refractive index values in the range 1.2–1.8 and solids have values between 1.3 and 2.5.

- The specific refraction is a constant for a given substance and hence useful for the identification of the substance and also for the determination of its purity. Molar refraction is the product of the specific refraction and molecular weight of the substance. It is an additive property and can be calculated from known values of atomic and group refractions.

- Becke line method uses a microscope for determining the refractive index of crystalline solids, whereas a refractometer is used in the case of liquids denser than air. Refractometers such as Abbe or Immersion-type refractometers determine the refractive index of a medium by measuring the position of the critical ray when the angle of refraction of the light passing through the medium is 90°.

- Polarization of light to yield plane polarized light is said to occur when light waves vibrating in one plane only are separated out on passing through a medium. The plane polarized light is considered as a vector sum of two circularly polarized light rays of same amplitude, the right circularly polarized dextrorotatory ray moving clockwise, and the left circularly polarized ray (levorotatory) moving counter clockwise. Polarization of visible light is brought about by double refraction when light is allowed to pass through birefringent materials.

- Optically active substances have different refractive indices for the d and l components of plane polarized light and hence rotate the plane of plane polarized light. The angle of rotation of the plane polarized light of a given wavelength is related to the difference $(\eta_d - \eta_l)$ as it passes through an optically active medium and is useful for analytical purposes.

- Polarimeter provided with a sodium vapour lamp is the most commonly used instrument to measure the angle of rotation of liquids and solutions. Polarimeter uses the principle of double refraction. Monochromatic radiation from the sodium vapour lamp passes through the polarizer prism made of calcite crystal to generate plane polarized light. The plane of plane polarized light is rotated through an angle as it passes through an optically active medium. The angle of rotation is measured by rotating the analyser prism also made of calcite crystal.

- The angle of rotation depends on the nature of the optically active substance, its concentration, and the path length of the medium. The specific rotation calculated from the angle of rotation is characteristic of a substance and hence useful in qualitative analysis. Since the angle of rotation depends on the concentration of the optically active substance in a solution, polarimeter is useful for quantitative analysis of optically active substances.

- Optical rotatory dispersion and circular dichroism depend on the change in the angle of rotation and absorption coefficient characteristics respectively of optically active substances as a function of the wavelength of the incident light. These techniques find use in the elucidation of molecular structure and configuration of optically active substances.

REVIEW QUESTIONS

1. What is molar refraction? How is it useful in identification of the sample?

2. How is refractive index of a liquid determined?

3. Draw a neat sketch of Abbe refractometer and explain the principle and procedure involved in the determination of refractive index.

4. Write a note on the different methods of polarization of electromagnetic radiation in the visible region.

5. Discuss the theoretical principles of polarimetry.

6. Explain the terms 'plane polarized light' and 'angle of rotation'.

7. Draw a sketch of polarimeter and explain the functions of the components.

8. Write a note on the applications of polarimetry.

9. Explain the terms 'optical rotatory dispersion' and 'circular dichroism'. What is their importance?

10. Give an account of the use of ORD and CD spectra in the analysis of optically active compounds.

5 Microscopy

5.1 INTRODUCTION

The name *microscope* (Greek: *micro* = small and *scope* = view) was coined by Giovanni Faber for the microscope used by Galileo Galilei in 1625. Though, microscope in the form of magnifying glass has been in use for about 1000 years (since about 1011–1020), the first instrument to facilitate the viewing of objects that are too small to be seen by the naked eye came into existence only in 1595. The technique of viewing small objects using such an instrument is called microscopy. The *optical microscope* containing one or more lenses to view an enlarged image of an object placed in the focal plane of lenses was invented much later.

Microscopes used at present may be broadly classified into three groups: (i) optical or light microscope, (ii) electron microscope, and (iii) scanning probe microscopes.

5.2 OPTICAL MICROSCOPE

The operation of optical microscope, also known as light microscope, is based on the optical theory of lenses and magnifies the image generated by the passage of waves of visible light through the sample. Two basic configurations of optical microscopes are known as (i) simple (one lens) microscope and (ii) compound microscope containing many lenses. Early microscopes were called *simple microscopes* because they had only one lens and worked like magnifying glasses. These early microscopes had limited magnification to the extent of 266 times the original size of the object expressed as 266x.

5.2.1 Compound Light Microscope

Compound microscope is a better version of the optical microscope using visible light and is the most commonly used one. The main components of the compound light microscope include (i) ocular lens or the eyepiece located at the top of the body tube, (ii) objective lenses mounted on a circular nose piece connected to the body tube, (iii) mechanical stage or object holder, and (iv) light source and optical devices. Commercially, two basic models of microscopes are available, *monocular microscope* with a single viewing tube and eyepiece and the more popular *binocular microscope* which enables viewing with both eyes for a better field of view. Figures 5.1(a) and (b) show the two models.

The optical assembly is fixed to a rigid arm attached to a rigid U-shaped foot or base to provide necessary stability when placed on the work table. The arm can be pivoted on the base to adjust the viewing angle. Two focusing wheels, one for coarse focusing and another for fine focusing are mounted on the arm.

The glass eyepiece is mounted at the top end of the eyepiece tube. Eyepieces are interchangeable and different eyepieces with different degrees of magnification can be inserted into the body tube. Typical magnification values for eyepieces include 5x and 10x. The objective lens consisting of one or more lenses, made of glass, are screwed into a circular nose piece which may be rotated around the cylindrical microscope body tube to select the required objective lens. Most compound microscopes have three objective lenses—a scanning lens (4x), a low power lens (10x), and a high power lens (ranging from 20x to 100x). In some microscopes a fourth objective lens called an oil immersion lens is also provided. When this lens is to be used, a drop of oil is placed on the cover slip covering the object and the lens is lowered carefully till the objective lens is just immersed in the oil film. The refractive indices of the oil and the lens are closely matched so that the light is transmitted from the object with minimum refraction. The oil immersion lens has a magnification of about 50x to 100x.

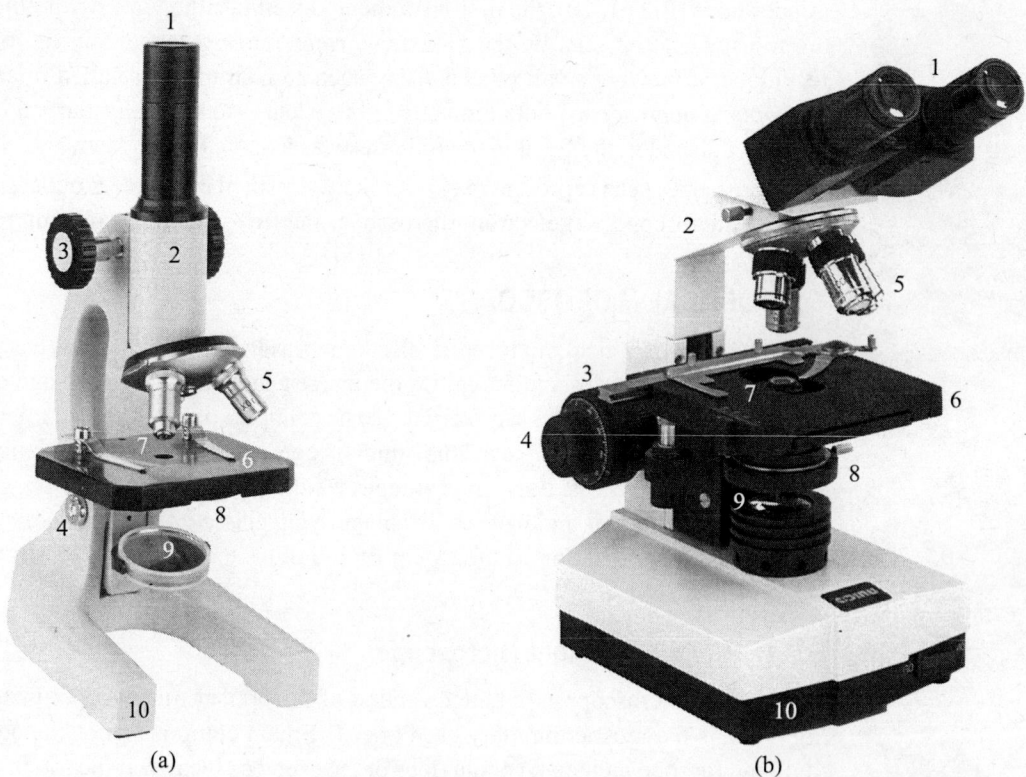

(a)   (b)

Fig. 5.1 Microscopes: (a) monocular and (b) binocular models. 1. Eyepiece, 2. Body tube, 3. Coarse focus, 4. Fine focus, 5. Objective, 6. Stage, 7. Specimen slide, 8. Diaphragm, 9. Mirror (or light source), and 10. Base

The nose piece containing the objective lens is positioned at the lower end of the microscope body tube. The objective lens collects light from the specimen. The stage is a platform located just below the objective. The sample to be viewed is placed on a rectangular glass plate (slide) which is then mounted on the stage. The object is covered with a thin glass sheet called cover slip to prevent any dust particles coming into contact with the object. The stage has a hole in the centre through which light passes to illuminate the object. The source of light may be daylight, which is directed through a mirror on to the object, or artificial light. The optical path of the compound microscope is shown in Fig. 5.2.

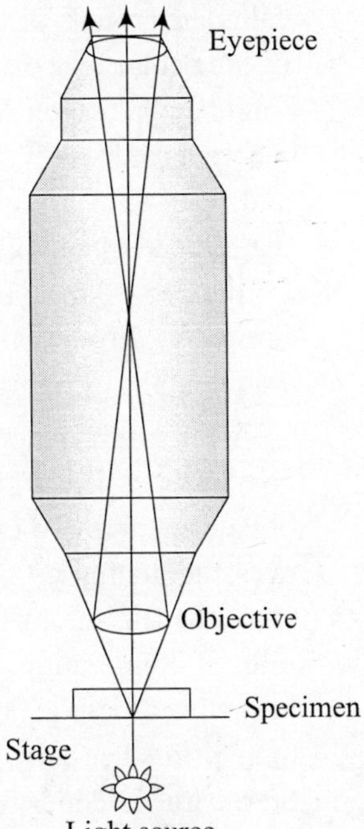

Microscopes are also provided with their own light sources which are focused by an optical device consisting of condenser, diaphragm, and filter to control the intensity and quality of light. August Köhler of Germany developed a configuration of illumination which is known as *Köhler illumination* to provide an evenly illuminated field of view with a reduced optical glare from the light source. The configuration consists of a collector lens which focuses the light onto the front aperture of a condenser. The condenser in turn focuses the light onto the specimen using a diaphragm and a condenser focus control. Köhler illumination is considered as an important configuration for achieving the best optical resolution on a light microscope and forms the basis for modern microscopes of phase contrast, confocal, and epifluorescence types.

The objective lens has a very short focal length and functions as a magnifying glass. The lens is brought very close to the object to be viewed so that the light transmitted through the object is focused inside the microscope tube. A real, enlarged, and inverted image of the object is formed at the focal point. The eyepiece of the microscope is a compound lens consisting of two component lenses (air separated couplet) fixed at the ends of the eyepiece tube. The virtual image of the object is brought to focus between the two component lenses of the eyepiece enabling the eye to focus on the virtual image. The image is viewed with the eyes focused at infinity to avoid tired eyes and headache.

Fig. 5.2 Optical path in a compound microscope

The actual *power of magnification* of the compound light microscope is the product of the powers of the eyepiece and the objective lens. Modern compound light microscopes, under optimal conditions, can magnify an object from 1000x to 2000x but have low resolution.

The *resolving power* of a microscope is defined as the ability to distinguish two points apart from each other. The resolution of a microscope depends on a number of factors, such as an inherent theoretical limit to resolution, imposed by the wavelength of visible light (400–700 nm), the refractive index of the objective lens, and the numerical aperture (NA) of the objective lens. The theoretical limit of resolution is calculated from the known wavelength (λ) of light used and the numerical aperture by the formula

Resolution = $0.61\lambda/\text{NA}$

The *theoretical resolution limit* of most of the microscopes is about 0.2 μm (200 nm). At high magnifications the transmitted light forms diffraction rings surrounding point objects. Since compound light microscopes have a limited depth of field, only thin slices of objects can be viewed. The image seen with these types of microscopes is two dimensional.

Microscopes provide an image which appears larger than the original object through magnification to facilitate the viewing of small details. The magnification of an object can be calculated roughly by multiplying the magnification of the objective lens times the magnification of the ocular lens. Even though magnification can be achieved without any limitation there is a limit beyond which details do not become clearer. Such a limitation is called *empty magnification* wherein the images are much bigger but their details do not become clearer. Thus, magnification and resolution are equally important to the quality of the information in an image. Greater magnification and higher resolution than that is possible in light microscopes may be achieved by the use of specialized techniques (e.g., confocal microscopy). However, the resolution is limited by diffraction. The resolving power of the microscope may be improved by using shorter wavelengths of light such as the ultraviolet.

A three-dimensional view of an object can be projected onto the eyes by using a *stereo microscope*. The stereo or dissecting microscope uses two separate optical paths with two objectives and two eyepieces to provide slightly different viewing angles to the left and right eyes facilitating a three-dimensional view. Stereo microscopes make use of reflected light to view the surface of the object. However, the instrument is also capable of using transmitted light of illumination. The stereo microscope with a magnification up to 100x is mostly used for low-power magnification on large subjects, to study the surface characteristics of solid surfaces (e.g., printed circuit boards), for dissection purposes (e.g., microsurgery), sorting of objects, etc. Stereo microscopes with attached digital cameras controlled by suitable software are useful for projecting the three-dimensional images onto LCD monitors.

Microscopes with additional features, such as fluorescence microscope, phase contrast microscope, and differential interference contrast microscope with provision for automation and digital imaging, have emerged into vogue. Ultraviolet light is used to enable the resolution of smaller features as well as to image samples that are transparent to the eye. Near-infrared light is used to image circuitry embedded in bonded silicon devices as silicon is transparent in this region. Many wavelengths of light, ranging from the ultraviolet to the visible are used to excite fluorescence emission from objects for viewing by eye or with sensitive cameras.

5.3 IMAGING TECHNIQUES

The light microscope is an important tool for studying the surface characteristics of a wide variety of materials, including biological samples. The image formed in a conventional light microscope is improved by different illumination and imaging techniques. The microscopic techniques based on different illuminating and imaging techniques that find extensive use in the study of biological samples include bright-field, dark-field, phase-contrast, fluorescence, and confocal microscopic techniques.

5.3.1 Bright-field Microscopy

Bright-field microscopic imaging technique is the simplest illumination method commonly used with a conventional microscope for viewing specimens, where the image formed by the deviated light is seen against the background of the undeviated light (bright-field). A visible or white light from the source is aimed at a condenser lens below the stage and shaped into a cone whose apex is focused at the plane of the specimen. Specimens are seen as they have the ability to refract light passing through them. Hence, the light is transmitted through the sample mounted suitably on a glass microscope slide, through the objective lens, and finally on to the eye through a magnifying lens. The specimen is seen as the different parts change the light to different extents due to differences in colour or thickness thereby producing a contrast. If the specimen and the surrounding medium have the same refractive index the specimen cannot be seen. Hence, biological materials must have either inherent contrast or should be artificially stained.

This technique has the advantages of using a simple equipment and non-requirement of any preparation of the sample. However, the technique is associated with the disadvantages of low contrast, particularly of biological samples, and low resolution due to blur caused by out-of-focus image. The disadvantages may partly be overcome by (i) staining the biological samples with simple stains such as methylene blue, crystal violet, or safranin or differential stains (e.g., negative stains, flagellar stains, endospore stains, etc.), (ii) varying the intensity of light from the source through an iris diaphragm, (iii) using blue filter or a polarizing filter for illuminating the sample to highlight the features not visible under white light, and (iv) using an oil immersion objective lens in conjunction with an immersion oil (which has the same refraction as glass) placed on a glass cover over the sample for improving the resolution.

Bright-field microscopy is usually used for viewing cell suspensions, water samples, algae, plant material, naturally pigmented specimens or stained specimens, thick sections of tissue, thin sections of organelles, smears, blood, negative stained bacteria, etc.

5.3.2 Dark-field Microscopy

Dark-field microscopic imaging technique, as the name indicates, has a dark background field with light radiating from the sample (the sample is bright) thereby providing a contrast. The technique was introduced first as oblique illumination (illumination of the specimen from the sides rather than from bottom) for viewing striae of diatoms which could not be resolved by normal illumination in bright-field microscopy. The technique became popular in medicine and biology after the discovery of syphilis spirochete in 1905, because the structures of many biological specimens are of low contrast that cannot be revealed by the bright-field compound microscopes.

Dark-field microscopy uses a different illumination system in which a specialized condenser is designed to form a hollow cone of light (rather than a filled cone of light as in bright-field microscopy) with its apex focused at the plane of the specimen. As the light moves past the plane of the specimen it spreads again into a hollow cone surrounding the objective lens. The objective lens lies within the dark hollow of this cone, and hence, the entire field of view appears dark in the absence of a sample on the mechanical stage of the microscope. When a specimen is placed on the stage the apex of the light cone strikes it providing an oblique illumination of 360 degrees around the specimen. An image of the specimen is formed by light rays scattered by the specimen and captured by the objective lens. The image appears bright against a dark background field, and hence, the name dark-field microscopy.

Figure 5.3 shows the bright-field and dark-field microscopic illumination. A solid cone of bright light illuminates the object and enters the objective lens in bright-field imaging while in the dark-field an opaque disk placed below the condenser blocks the centre of the light beam to produce a hollow cone of light. This light does not directly enter the objective lens.

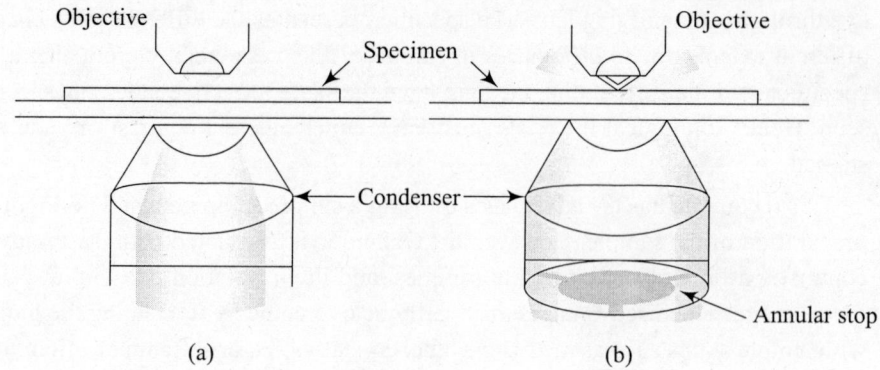

Fig. 5.3   Illumination in microscopy (a) bright-field and (b) dark-field

The dark-field effect can be brought about in the normal bright-field microscope by keeping the numerical aperture of the objective lower than that of the condenser lens. This is achieved by introducing a piece of opaque disk called *annular stop* or *stop* in front of the focal plane of the condenser to block out centre of the light beam and form a hollow cone of light (Fig. 5.4).

The diameter of the opaque disk should be large enough so that it can prevent any direct light entering the objective lens. In the absence of a specimen the light from the condenser misses the objective completely, giving a dark background. When the specimen containing reflective components or structures is placed in the path of illumination, light is scattered at all angles from the reflective structures and a part of the scattered light reaches the objective and the specimen appears bright.

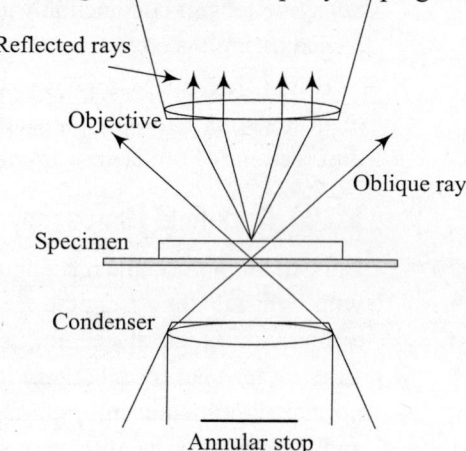

Fig. 5.4   Optical path in dark-field microscopy

Microscopes are usually provided with the required number of opaque disks that match objectives with different numerical apertures. Alternatively, a funnel-shaped cone called *funnel stop* may be inserted into the objective lens or an oil immersion-type objective lens with high numerical aperture with a built-in iris diaphragm can be used to reduce the numerical aperture of the objective lens. High magnification dark-field condensers require the specimen to be immersed in a proper medium. This is because the dark-field condenser produces an oblique angle light and if the angle is greater than the critical angle at any interface (critical angle at glass-to-air interface is 41° and for glass-to-water interface it is 61°) it will be totally reflected internally. Hence, it is best to immerse the dark-field condenser in the drop of oil placed on the slide even for the low power dark-field condenser.

Since dark-field imaging depends on scattered light from the sample the image formed includes that of the specimen as well as dust and other particles. Hence, it is necessary to take extra care in preparing the samples by cleaning the slides free from dust. Sample media, such as water, saline, or agar, need to be filtered to remove any contaminants. The sample has to be spread as a thin layer and prevent any overlapping layers. Samples containing highly refractive objects with empty space in between them are well suited for dark-field microscopic imaging. Biological fluids of plant or animal origin, cell cultures, microorganisms, thin sections of histological sections (either unstained or with silver staining of certain components, foods, colloids, fibres, crystals, etc.), samples prepared for autoradiography, gold-labelled specimens, etc., are suitable for dark-field microscopy. Samples containing dense objects which are too crowded and thick sections are not suitable for this imaging technique.

5.3.3 Phase-contrast Microscopy

Phase-contrast microscopy is a special illumination technique used in optical microscopy which is specifically useful for studying transparent and colourless specimens, particularly living cells without staining. The technique is based on transforming the phase shift in the light passing through a transparent and colourless specimen into amplitude to provide a contrast in the image of the specimen. Frit Zernike discovered that a destructive interference pattern caused by altering the optical path when light passes through transparent and colourless objects enhances the image of the object which led to the introduction of phase-contrast microscopy. He was awarded Nobel Prize in physics in 1953 for his contribution.

When light travels through a medium its interaction with the medium causes changes in wavelength, amplitude, and phase depending on the properties of the medium. The human eye is normally sensitive to changes in wavelength (change of colour) and changes in amplitude (changes in brightness or darkness, that is, contrast) but not changes in the phase caused by the passage of light through microscopic objects. Many transparent and colourless microscopic objects, for example, living cells, however, do not cause appreciable changes in colour or brightness but only cause a small change in the optical path (or phase) of light passing through them. Such objects are called *phase objects*. For example, the phase change of the colourless *silica valves of* diatoms which have a refractive index of about 1.5, when surrounded by a medium, such as air (refractive index of 1) or naphrax (refractive index of 1.7), is quite large and can be perceived by the human eye as differences in brightness.

Phase-contrast microscopy makes use of the phase differences of light caused by the passage of light through phase objects. The principle of phase-contrast is explained conveniently on the basis of diffraction theory. When light passes through a phase object consisting of regions or areas of different refractive indices (as exhibited by the different components of biological specimens) the optical path is retarded in phase by up to one-fourth of the wavelength but its amplitude remains unaltered. In phase-contrast microscopy the phase difference is increased to half wavelength by a transparent phase plate introduced into the compound microscope. A transparent specimen shines out in contrast to its surroundings as the phase difference is transformed into difference in amplitude of light manifested as bright and dark areas (or areas of contrast) which can be detected by the human eye.

A compound microscope may be used as a phase-contrast microscope by introducing two additional components, a *phase annulus* consisting of a clear ring on black field placed below the condenser and a phase plate located at the focal plane of the objective lens. Figure 5.5 shows the two components and the optical path in a phase-contrast microscope.

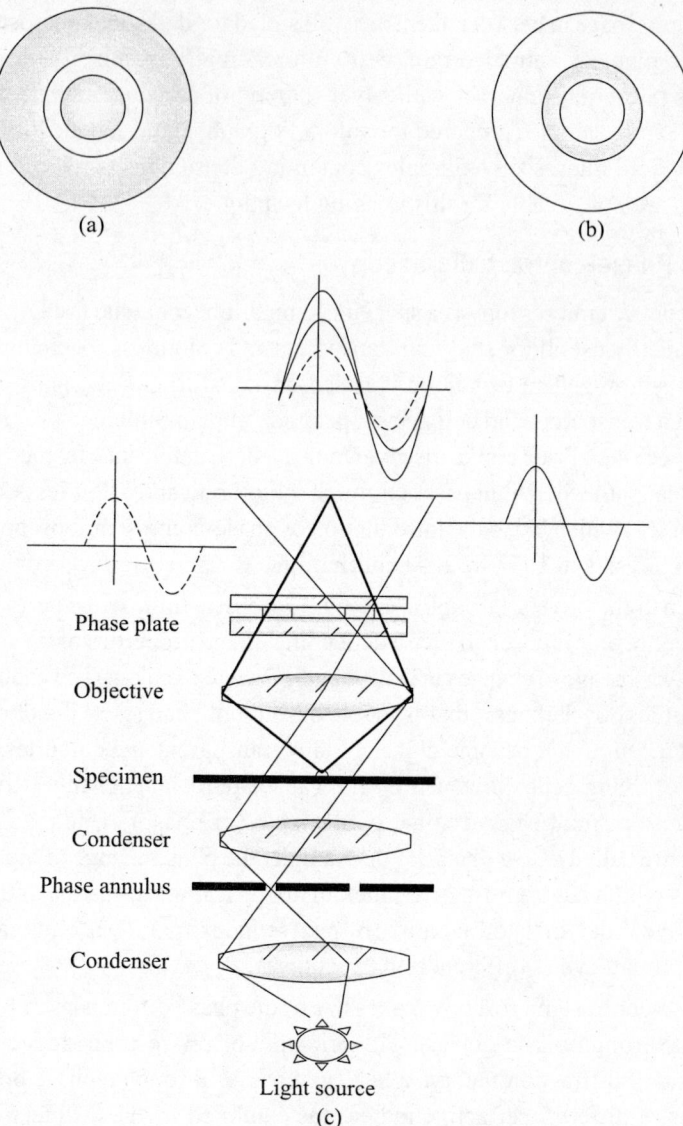

Fig. 5.5 Phase-contrast microscope: (a) phase annulus, (b) phase plate, and (c) optical path

The combination of these components creates two possible light paths: the specimen path and the background path. The specimen path light interacts with the specimen is refracted or diffracted away from the specimen and focused by the objective lens onto the image plane. The refracted wave is retarded in phase by one-fourth wavelength due to difference in the optical path between the specimen and the surrounding medium. The background path light of direct or unmodified background light does not interact with the specimen and is collected by the objective lens, passes through the phase plate ring and is retarded exactly by one-fourth of a wavelength and simultaneously attenuated. The phase shift in the background light is not detected by the eye and the image on the image plane appears as a normal bright background. The background light interacts with the specimen path light at the image plane creating an interference pattern. A constructive interference occurs at the image plane resulting in a *negative* or *bright phase contrast* in which cell components appear brighter due to the difference in the refractive index against a background of normal intensity derived from non-refracted light. The negative phase contrast has replaced the *positive* or *dark phase contrast* originally introduced by Zernike in which the cell components appear darker than the surrounding background in the present day phase-contrast microscopes.

Phase-contrast microscopy is highly useful in the study of processes that occur in a living cell (e.g., cell division) of transparent and colourless components and eliminates the necessity of dyeing to view the different cell components as dyeing invariably kills the organism. However, this technique is suitable only for phase objects and thin specimens which do not cause major changes in absorption. Another limitation is that phase-contrast causes a major reduction in resolution.

5.3.4 Fluorescence Microscope

Fluorescence microscope is based on the use of fluorescence or phosphorescence emission of the microscopic object itself as the source of light to view the object. The phenomenon of fluorescence emission occurs when certain materials on irradiation or exposure to light of specific wavelength emit radiation in the form of visible light. The material is said to undergo excitation in which the incident radiation on colliding with atoms in the material excite the electrons to higher energy levels. The excited atoms relax very quickly to a lower energy level or ground state energy level by emitting energy in the form of light. The material may be capable of fluorescing in its natural form—*auto fluorescence*—(e.g., certain minerals containing activators: scheelite, aragonite, ruby-corundum, etc., and biomolecules such as chlorophyll). Alternatively, the sample may fluoresce on treatment with fluorescing reagents called *fluorophores* or *fluorochromes* [e.g., flourescein isothiocyanate (FITC), auramine O, acridine yellow, green flourescent protein (GFP)] to label the specimen. The fluorescent emission is of lower energy (hence, of longer wavelength) as compared to the incident radiation. The fluorescence emitted from the specimen is passed through a fluorescent emission filter to the microscope eyepiece. In the microscope the fluorescence emitted by the different regions in the specimen appears bright and shining against a dark background.

The fluorescence microscope consists of a high intensity light source such as a mercury vapour lamp or a xenon arc lamp, the excitation filter, dichroic mirror or a beam splitter, and a fluorescence emission filter. The filters are chosen so as to match the excitation and emission characteristics of the fluorescent reagent chosen to label the specimen.

Epifluorescence microscope is the most commonly used instrument in biological sciences to identify cellular components with a high degree of specificity. It is a modified fluorescence microscope in which the excitation and observation of the fluorescence are from above the specimen. In conventional fluorescence microscopy the excitation light is passed first through the specimen and the transmitted light along with the fluorescent emission from the specimen reaches the objective. In epifluorescence microscopy the excitation light is passed through the objective onto the specimen and fluorescence emitted by the specimen is focused by the same objective onto the detector. The excitation light transmitted through the specimen is filtered by a filter placed in between the objective and the detector, thus, enhancing the signal-to-noise ratio. Fluorescence microscopes can reach optical resolution to less than 10 nm.

5.3.5 Confocal Microscopy

Confocal microscopy is an optical imaging technique with a shallow depth of field of the object, which enhances the resolution of images by eliminating the out-of-focus glare. The technique finds extensive applications in biological sciences for imaging fixed as well as living cells and tissues, optical sectioning of thick specimens, and for inspecting semiconductors.

In conventional fluorescence microscopy the entire specimen is flooded with light and the image is either viewed or projected on to a viewing screen. The secondary fluorescence emitted by the specimen affects the resolution of the regions of interest in the specimen, particularly in specimen of thickness greater than 2 μm.

The principle of confocal imaging technique (patented by Marvin Minsky in 1957) is based on eliminating the out-of-focus light or glare by using spatial filters in the form of pinholes (slits). The light from the source is passed through a pinhole onto the objective lens which focuses it at a desired focal plane in the fluorescent specimen. The light emerging from the specimen is focused again by a second objective lens through a second pinhole on a photodetector. The second pinhole has the same focus as the first pinhole (hence, confocal) and prevents the light from above or below the plane of focus in the specimen reaching the photodetector.

In practice confocal microscopy makes use of one or more beams of light from a light source (mostly a laser) to scan across the selected rectangular area of the specimen in a raster pattern (just as the needle of a record player scans the disk). The image is focused onto the photodetector through a pinhole in an optically conjugate plane in front of the photodetector to eliminate the out-of-focus light. The image formed is called an *optical section* to indicate the non-invasive method by which the image is collected using focused light rather than physical sectioning of the specimen. The image quality is better as only the light within the focal plane can be detected providing an improved axial and lateral resolution. Raster scanning produces two-dimensional and three-dimensional images of the specimen. The thickness of the focal plane is defined by the inverse of the square of the numerical aperture of the objective lens and also by the optical properties of the specimen and refractive index. Figure 5.6 shows the optical path in confocal microscopy.

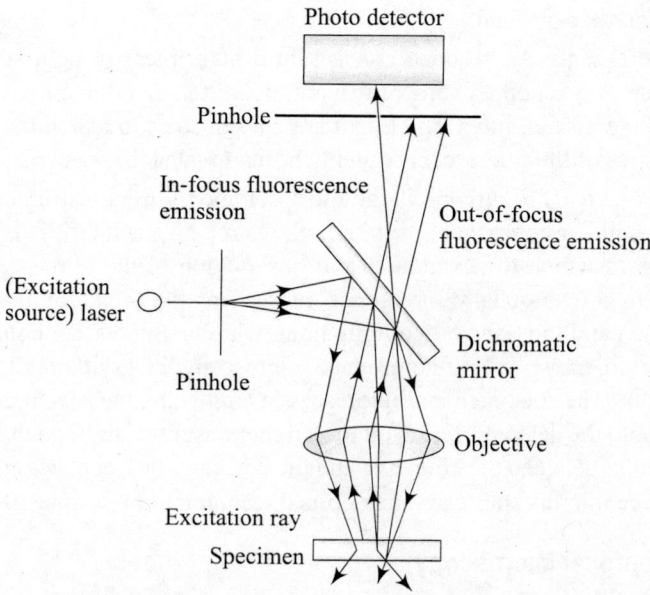

Fig. 5.6   Optical path in confocal microscopy

Three types of confocal microscopes are available commercially. These include (i) laser scanning confocal microscope (LSCM), (ii) spinning disk confocal microscope, and (iii) programmable array microscope. The LSCM is the most widely used instrument because of its better resolving power compared to fluorescence microscope. However, the resolving power of LSCM is less than that of the transmission electron microscope.

5.3.6 Polarizing Microscope

Polarizing microscope makes use of polarizing light as a contrast enhancing technique for viewing the birefringent specimens. When polarized light is passed through isotropic materials only negligible phase effects are brought about by the materials. However, when the polarized light is passed through anisotropic materials the plane of the polarized light is rotated. Polarizing microscope, thus, exploits the optical properties specific to anisotropy and reveals detailed information concerning the structure and optical properties of materials for identification and diagnostic purposes, for example, different phases in liquid crystals, minerals, lipid content in lipoproteins, birefringent uric acid needles in joint fluid from a patient with gout, etc.

Polarizing microscope is similar to the light microscope but with the condenser and ocular equipped with polarizing optics. The polarizing microscope uses a polarizing filter beneath the stage. This transmits all the light from the lamp through the specimen in the same plane. A second polarizing filter called the analyser is placed before the eyepiece so that it is out of phase with the substage polarizing filter. The analyser blocks all of the light causing a dark background unless the object on the slide is anisotropic. Birefringent objects rotate the light so that it passes through the analyser lens and the object appears light (white) against a dark background.

Polarizing microscope has a high degree of sensitivity which finds extensive use in investigating the optical properties of minerals, polymers, liquid crystals, biological specimens, and magnetic memory of materials.

5.3.7 Flow Cytometry

Flow cytometry is a technique for counting as well as studying the physical and chemical characteristics of microscopic particles suspended in a flowing stream of fluid. The technique is also useful for sorting (separating and isolating) the particles based on their characteristic features. The technique is used by biologists extensively to study the characteristics of cell suspensions.

The principle of flow cytometry involves passing a monochromatic beam of light, usually from a laser, through a hydrodynamically focused stream of fluid containing a suspension of microscopic particles. The characteristic properties of the particles are studied with the help of different photo and electronic detectors suitably placed and aiming at the point where the fluid stream passes through the light beam. The sample consisting of the particles, for example, a cell suspension is injected at the centre of a sheath flow of a fluid in the flow cell. The microscopic particles are allowed to pass in a single file at the centre of the stream by reducing the diameter of the flow cell at the exit end (Fig. 5.7). Each microscopic particle exiting the flow cell passes through the beam of light singly facilitating the counting and study of its physical and chemical characteristics. Each particle scatters the incident light and any fluorescent chemicals present in the particle naturally or attached to the particle, emit characteristic emission of fluorescence. The combination of scattered and fluorescent light is measured by detectors. Forward scatter (FSC) of light is detected by the detector placed in line with the beam of light and provides information on the cell volume. Side scatter (SSC) of light by detector placed in perpendicular direction to the beam of light is useful in the study of inner complexity of the cell such as the shape of the nucleus, amount and types of cytoplasmic granules, membrane roughness, and so on. In addition, the fluorescence detector placed perpendicular to the direction of light beam detects any fluorescence emission from the particles.

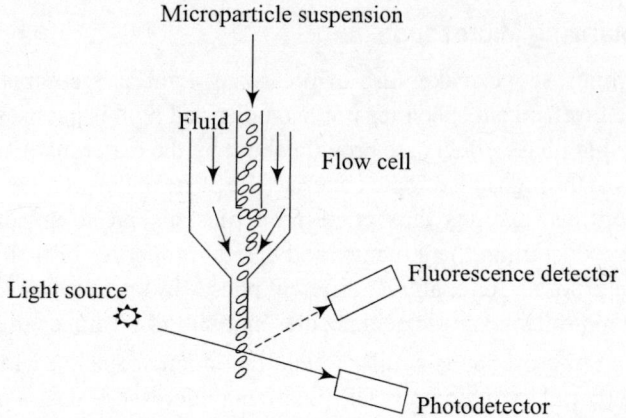

Fig. 5.7   A schematic representation of the principle of flow cytometry

Flow cytometer consists of the main components (i) a light source such as mercury vapour lamp, xenon lamp, or a laser providing a highly monochromatized beam of light (e.g., argon laser – 488 nm; helium–neon laser – 633 nm; diode lasers providing blue, green, red, violet radiation, dye lasers, argon, krypton lasers), (ii) a flow cell through which the fluid containing the suspended microscopic particles flows (the microscopic particles are aligned to pass through the light beam in single file), and (iii) detectors for measuring the scattered light and fluorescence emission. The light signals are converted to electrical signals, amplified, and fed to a computer for analysis of data. Some flow cytometers use only light scatter for measurement while others form images of cells from fluorescence, scattered light, as well as transmitted light.

Many modern flow cytometers have multiple lasers as light sources and fluorescence detectors. Such instruments are capable of analysing several thousand particles per second in real time. Instruments with more numbers of lasers and detectors facilitate multiple antibody labelling and even precisely identify the population by their phenotype. Some of the instruments can provide a digital image of the individual cell and fluorescent signal location within or on the surface of the cell. The experimental data generated by flow cytometers can be displayed in the form of one-dimensional histograms, two-dimensional dot plots, and three-dimensional images with the help of suitable software. The information provided by such instruments is of importance in diagnostic and clinical applications.

5.4   ELECTRON MICROSCOPE

Electron microscope uses a beam of electrons to illuminate the object and form an enlarged image. Electrons have much shorter de Broglie wavelengths compared to the wavelengths of visible light, and hence, a beam of electrons facilitates a much higher resolution. The resolving power of electron microscopes is much greater, about 2 million times as compared to 2000 times of light microscopes, and hence, details at atomic levels can be obtained.

The interaction between the primary beam of electrons and the object leads to a number of detectable signals as summarized in Fig. 5.8.

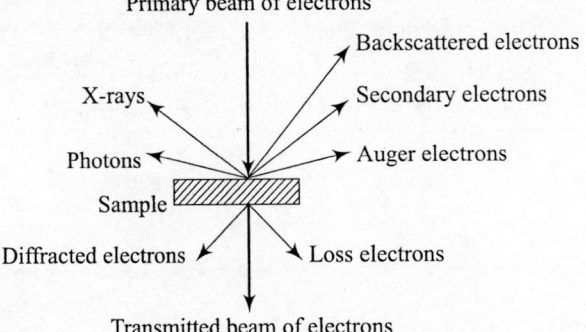

Fig. 5.8 Signals generated by the interaction of primary beam of electrons with a sample in electron microscopy

A fraction of the primary beam of electrons passes through the object depending on its thickness and density without suffering any energy loss. The transmitted electrons provide a two-dimensional image of the object. Electrons colliding with atoms of the object can be backscattered and backscattering is more effective with increasing mass of the atom, and hence, heavier atoms can be distinguished by a higher yield of backscattered electrons. Secondary electrons are emitted by the object as the primary electrons lose energy due to inelastic collisions. Auger electrons and X-rays (X-ray fluorescence) are emitted by the relaxation of core-ionized atoms. Electrons diffracted by particles in the object towards the primary beam provide the dark field image as well as crystallographic information of the object. Loss electrons are emitted as the primary beam of electrons lose energy due to vibrations in the object. The emission of a range of photons in the UV, visible, and IR regions also occurs mainly caused by the recombination of electron-hole pairs in the object, the phenomenon being called *cathodoluminscence*. The interaction of the primary beam of electrons with the sample, thus, provides a wealth of information on the morphology, crystallography, and chemical composition.

The major types of electron microscopes available at present include (i) transmission electron microscope, (ii) scanning electron microscope, and (iii) scanning tunnelling electron microscope.

5.4.1 Transmission Electron Microscope

Transmission electron microscope (TEM) is the original form of electron microscope introduced commercially in 1939. A beam of electrons in the wavelength range of 0.01 to 0.1 nm is produced by a high voltage electron gun consisting of a tungsten filament cathode which emits electrons on the application of a high voltage in the range of 40 to 400 kV. The beam of electrons is accelerated by an anode and focused by a set of electrostatic and electromagnetic lenses. The electron beam is partly transmitted through the object (a thin section of the sample), just as a beam of light passes through the object in optical microscope, and partly scattered. The information on the structure of the object contained in the emerging beam is magnified by an objective lens to create an image of the object. The image projected on a luminescent viewing screen is of several thousand times in magnification. A bright-field image or a dark-field image may be generated on the viewing screen by the use of a device called *selected area electron diffraction aperture*. When the transmitted beam consisting of electrons is not scattered by the sample a bright-field image is generated on the viewing screen. In contrast, when the transmitted beam consists of electrons which have undergone inelastic scattering with no deviation in their paths as well as with reflected electrons from various crystallographic planes of the sample a dark-field image is generated. The block diagram of TEM is shown in Fig. 5.9.

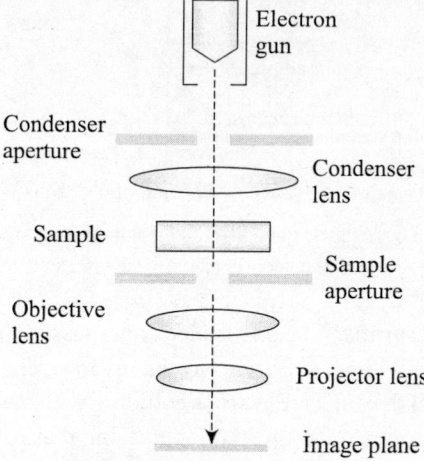

Fig. 5.9 Block diagram of TEM

The size, shape, and the relative arrangements of the particles of a sample can be imaged at atomic scale dimensions. TEM also provides crystallographic information in terms of the arrangement of the atoms in the sample as well as any atomic scale defects present in the sample. Electron microscope finds extensive use in research laboratories for the study of a variety of materials such as biological specimens, macromolecules, metals, crystalline materials, surface characteristics of materials, inspection, quality assurance, and failure analysis of semiconductor devices.

The modern high resolution TEM (HRTEM) makes use of software to correct spherical aberration normally seen in TEM, thereby producing images with high resolution and magnification of about 50 million times and can 'view' atoms. HRTEM has become an important tool for nano-technology research and development as it is capable of determining the positions of atoms within materials.

5.4.2 Scanning Electron Microscope

A scanning electron microscope (SEM) is basically a microscope generating an electron beam scanning forth and back over a sample. It provides information on (i) the topography or texture of the surface (appearance of the surface and detectable features), (ii) morphology, that is, the size, shape, and arrangement of particles lying on the surface, (iii) chemical composition of elements and compounds that are available on the surface and their relative ratios, and (iv) crystallographic information on the arrangement of atoms and their orientation (particularly in single crystals of size greater than 20 μm). SEM provides magnification up to 300,000x with the resolution approaching a few nanometers.

The basic principle in SEM involves the use of a primary beam of high-energy electrons (a few hundred eV to 100 keV) generated from a tungsten or lanthanum boride (LaB_6) electron gun. The electron gun is electrically heated to emit electrons which are focused by an anode. An accelerating voltage in the range 0.2 to 40 kV applied between the electron gun and the anode determines the energy and wavelength of the electron beam. The beam is focused by two or three electromagnetic condenser lenses to a fine spot and passed through deflecting scan coils to facilitate the movement of the spot back and forth in a raster pattern (similar to the needle of a record player moving over the surface of the disk) across the sample surface. The sample surface is scanned over a rectangular area horizontally as well as vertically by the electron beam focused over a very fine spot of about 0.5–5 nm. As the electrons strike and penetrate the surface a variety of interactions occur within a volume (called the interaction volume) extending to a depth of about 100 nm to about 5 µm from the surface. The interactions result in the generation of a variety signals such as back scattered electrons, diffracted backscattered electrons, secondary electrons, X-ray photons, and visible light (cathodoluminscence). Three types of signals are mostly used in SEM in the form of secondary electrons and backscattered electrons (BSE), diffracted backscattered electrons (EBSD), and X-ray photons. The primary beam of electrons on interacting with an atom on the sample surface undergoes either inelastic collision and scattering with atomic electrons or elastic collision and scattering with the atomic nucleus. In an inelastic collision the primary electron transfers part of its kinetic energy to the atomic electron facilitating its emission. If the emitted electron has energy less than 50 eV, it is called a *secondary electron*. Secondary electrons provide valuable information on the morphology and topography of the sample surface. When the primary electron undergoes elastic scattering its energy does not change and the elastically scattered electron is called the *backscattered electron*. Backscattered electrons are useful for providing contrasts in composition in multiphase samples (i.e., phase discrimination). The intensity of the BSE signal is strongly related to the atomic number (Z) of the specimen, and hence, BSE images can provide information about the distribution of different elements in the sample. Thus, both secondary electrons and backscattered electrons are used for imaging the surface. EBSD are used to determine crystal structures and orientations of minerals. X-rays are produced by inelastic collisions of the incident electrons with atomic electrons in the sample. As the excited electrons return to lower energy states, they yield characteristic X-rays for each element present in the sample surface, and hence, provide information on the chemical composition near the surface of the sample. SEM analysis is non-destructive as the X-rays generated do not lead to volume loss of the sample, and hence, it is possible to analyse the same materials repeatedly.

The samples should be solid with maximum size of about 10 cm horizontally and about 40 mm vertically, electrically conducting, and be stable in vacuum of the order of 10^{-6} torr. Any non-conducting surface may be coated with a thin layer of gold, platinum, or graphite to make the surface electrically conducting. The choice of material for conductive coatings depends on the data to be acquired. Carbon is the most desirable if elemental analysis is a priority, while metal coatings are most effective for high-resolution electron imaging applications. Biological samples are fixed in a fixative, such as glutaraldehyde or formalin, to preserve their structure and dehydrated to remove any water. Alternatively, an electrically insulating sample can be examined without a conductive coating in an instrument capable of *low vacuum* operation.

A block diagram of the scanning electron microscope is shown in Fig. 5.10.

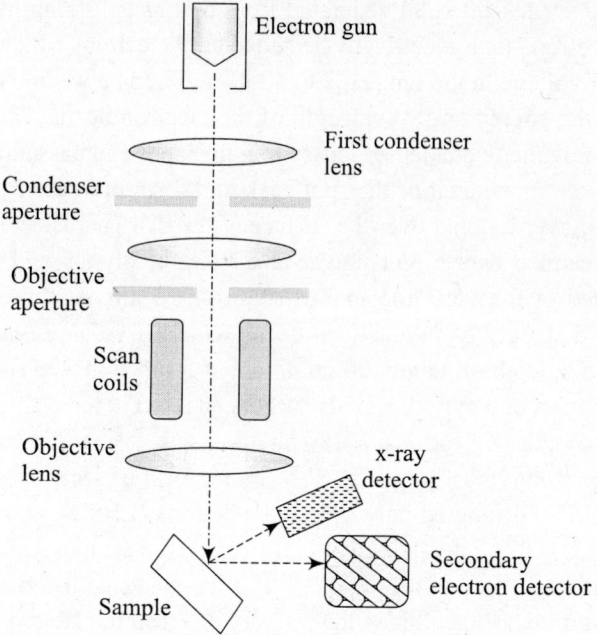

Fig. 5.10   Block diagram of SEM

The secondary electron detector is a scintillator–photomultiplier system. The low-energy (<50 eV) secondary electrons are collected by a collector grid and then accelerated towards a scintillator positively biased to about +2000 V. The accelerated electrons cause the scintillator to emit flashes of light (cathodoluminescence) which are guided to a photomultiplier tube. The amplified electrical signal output of the photomultiplier is displayed as a topographical image by a video display or stored as a digital image.

The backscattered electron detector is a scintillator or semiconductor device and detects high energy electrons that have been scattered backward providing information about the differences in atomic number of a sample. Elements of high atomic number appear brighter in the image as they backscatter electrons more strongly than lighter elements of low atomic number. Thus, BSE are used to detect contrast between areas with different chemical compositions. The beam current absorbed by the sample may also be detected to form images of the distribution of specimen current. The signals generated by the detectors are displayed as variations in brightness on a cathode ray tube (CRT) or digitally processed and displayed. Backscattered electrons can be used to form electron backscatter diffraction (EBSD) image for determining the crystallographic structure of samples.

SEM is a highly useful tool for probing the optical and electronic properties of semiconductor materials. The primary electron beam creates an electron-hole pair by promoting electrons from the valence band to the conduction band leaving behind holes. The recombination of the electron-hole pairs result in the emission of light in the form of cathodoluminescence. Cathodoluminescence is used in cathode ray tube monitors of computers and television sets. In the case of a *p–n* junction of a semiconductor sample an electron beam induced current (EBIC) flow is generated. Cathodoluminescence and EBIC are called *beam injection techniques*. These techniques find use for investigating the optoelectronic behaviour of semiconductor materials particularly the nanoscale features and defects.

The resolving power of SEM depends on the wavelength of the electrons and also the size of the interaction volume. The resolving power of SEM lies in the range of 1–20 nm much less compared to the resolution of TEM as the primary beam consists of longer wavelength electrons in SEM. However, SEM can image a relatively larger area of the sample and also image bulk materials whereas TEM can handle only thin films or foils. SEM has also an additional advantage in that it provides information on the nature and composition of the sample.

5.4.3 Scanning Transmission Electron Microscope

The scanning transmission electron microscope (STEM) incorporates the features of TEM and SEM as the electrons pass through the sample as in TEM and at the same time the optics focus the electron beam into a fine spot which scans over the sample in a raster fashion as in SEM. The microscope is useful for simultaneous acquisition of signals for techniques of energy dispersive X-ray (EDX) spectroscopy, electron energy loss spectroscopy (EELS), and annular dark-field imaging (ADF). The microscope in conjunction with a high angle detector provides images where the contrast is directly related to the atomic number unlike in electron microscopy where the phase-contrast image has to be interpreted by simulation. STEM is relatively more efficient than TEM for analysing biological samples as it allows contrast imaging without the necessity of staining the sample.

5.5 SCANNING PROBE MICROSCOPY

Scanning probe microscopy (SPM) is a relatively new technique for characterization of surfaces of materials at nanoscale. The SPM family includes the scanning tunnelling microscopy (STM) and atomic force microscopy (AFM). Both these surface imaging techniques provide topographic images of the surface in three dimensions at atomic scale resolution. The techniques can be used to study all solid surfaces—hard or soft, electrically conducting or non-conducting. The surfaces can be studied in vacuum, gas, or liquid.

5.5.1 Scanning Tunnelling Microscope

Scanning tunnelling microscope (STM) was developed by Gert Binnig and Heinrich Rohrer from the IBM Research Laboratory in Rüschlikon (Switzerland) in 1981 for which they were awarded Nobel Prize in physics in 1986. STM produces images of the surfaces of materials at extreme magnification so that individual atoms become visible. The surface should be electrically conducting. However, even electrically non-conducting surfaces (e.g., biological materials) can be studied by microscopes based on the principle of STM. In fact, about 20 different types of microscopes have been derived from the STM and these microscopes have been classified under a common name *scanning probe microscopes* (SPM).

The principle of STM is based on quantum tunnelling. STM has a sharp metallic tip of tungsten, platinum–iridium, or gold which functions as the probe to study the shape of the surface of a sample. The tip is mounted on a piezoelectric element. The piezoelectric element changes its length to a slight extent on the application of electrical voltage. A voltage in the range of a few millivolts to a few volts is applied between the tip and the surface of the sample. When the tip comes into contact with the surface, a current is generated and when the tip is not in contact with the surface the current is zero. During STM operation the tip does not come into contact with the surface and the distance between the tip and the surface of the sample is usually maintained in the range of 0.5–1.0 nm (2–4 atomic diameters) by using a feedback mechanism consisting of the piezoelectric element on which the tip is mounted. The feedback piezoelectric device varies in its length depending on the applied voltage and thereby regulates the distance between the tip and the surface of the sample (Fig. 5.11). With the tip being maintained at such short distances, electrons can jump through the vacuum from the tip to the surface or vice versa. This jumping is known *tunnelling*. The tunnelling current is quite low in the range of a few picoamperes (pA) to a few nanoamperes (nA). The magnitude of the tunnelling current depends on the distance between the last atom of the tip and the nearest atom on the surface of the sample. It decreases by 10^3 times when the distance increases by an atomic diameter, and hence, is a highly sensitive measure of the distance between the probe tip and the sample surface. The tip is moved over the sample surface in a raster fashion in the x–y plane by maintaining a constant height from the surface. Alternatively, it is moved in all the three directions by maintaining a constant tunnelling current. The tip movement is regulated by computer-controlled piezoelectric transducers in all the three directions. The tunnel current is extremely sensitive to vibrations, and hence, the STM has to be maintained free from vibrations and also free from eddy current effects.

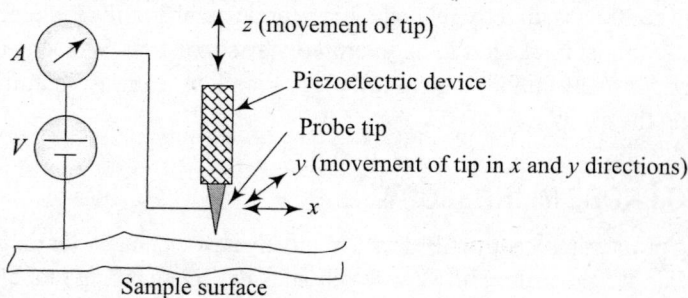

Fig. 5.11   Schematic diagram of the probe tip scanning the sample surface

5.5.2   Atomic Force Microscope

The atomic force microscope (AFM) or scanning force microscope (SFM) was invented in 1986 by Binnig, Quate, and Gerber as an offshoot of STM. The most important difference between STM and AFM is that the probe tip (or stylus) does not touch the sample surface in STM, while in AFM it does come into contact with the sample surface. The topographic information is gathered by 'feeling' the surface just like our fingers probe a material in dark for the brain to deduce an image of the material. As the tip gently touches the surface, it records the small attractive or repulsive forces of the order of a few pico-Newton exerted between the tip and surface and the data is processed by a computer to produce an image of the surface at atomic level. Over the years the AFM has undergone improvements to enable it to overcome the limitation of conventional optics and even measure properties such as local resistivity, temperature, elasticity and tribology. It is a very high-resolution scanning probe microscope capable of resolution in the range of fractions of a nanometer with more than 1000 times better resolution than optical diffraction limit. The AFM is one of the foremost tools for imaging, measuring, and manipulating matter at the nanoscale. Since AFM does not depend on the tunnelling current even electrically non-conducting samples (e.g., biological samples) can be studied.

The AFM consists of a thin flexible leaf spring or a microscale beam called *cantilever* made of silicon or silicon nitride (S_3N_4). The cantilever is usually V shaped or rectangular and about

0.1 mm long with a width of a hair and has a low spring constant. The tip is sharp and thin with a radius of 1–2 nm. When the tip is brought close to surface to be investigated various interactive forces, such as mechanical contact force, van der Waals forces, capillary forces, electrostatic interactions, magnetic forces, and chemical bonding, come into play between the tip and the sample surface. These forces deflect the cantilever according to Hooke's law. The deflection is measured by a laser spot reflected from the back surface of the cantilever into an array of photodiode detectors. In order to avoid any hard collision between the cantilever tip and the surface, a feedback mechanism is used to constantly monitor and adjust the distance between the tip and the sample surface. The topography of the sample surface is mapped by mounting the sample on a vertical piezoelectric scanner to monitor the changes in vertical position (z direction) and scanning the surface in the x- and y-directions using another piezo block.

The AFM can be operated in different operating or imaging modes depending on the application. The different operating modes may be classified into (i) contact or static mode and (ii) non-contact or dynamic mode. In the contact or static mode the tip is within a few Angstroms of the surface and the repulsive force between the tip and the sample surface is kept constant during scanning by maintaining a constant deflection. In the non-contact dynamic mode the distance between the tip and the surface is much larger in the range of 2–30 nm. The cantilever oscillates at or close to its fundamental resonance frequency above the adsorbed fluid layer on the surface of the sample with an amplitude of oscillation less than 10 nm. Any long-range force, for example, van der Waals forces, normally extends above the adsorbed fluid layer from 1 nm to 10 nm and during scanning the interaction between the tip and the sample surface modifies the amplitude, phase or the resonance frequency of the vibration and provides information about the sample's characteristics. In frequency modulation, the tip–sample surface interaction changes the vibration frequency and the changes in frequency are measured with very high sensitivity providing atomic scale resolution in ultra high vacuum conditions. In amplitude modulation the changes brought about in the oscillation amplitude or phase are used to discriminate between different types of materials on the surface and provide an image of the surface.

A dynamic mode called the *intermittent contact mode* or *tapping mode* (tapping AFM) provides better images. In this mode the cantilever is oscillated up and down near its resonance frequency at an amplitude of about 100–200 nm to facilitate modulation of the distance between the tip and the sample surface. Tapping mode is quite gentle in that the imaging process can be performed in a liquid cell filled with buffer solution to ensure that the biomolecules remain hydrated. Similarly, the conformation of single molecules, supported lipid bilayers, or adsorbed single polymer molecules can be studied under liquid medium.

AFM can be also used for *force spectroscopy* to determine the force–distance curves of forces operating at nanoscale with a resolution better than 0.1 nm. The different types of forces that can be measured include van der Waals forces, Casimir forces, dissolution forces in liquids, single molecular stretching, rupture forces, and so on. In force spectroscopy the tip is extended towards and retracted from the surface and the static deflection of the cantilever is monitored as a function of piezoelectric displacement.

The different advantages of AFM as compared to SEM include the following: (i) AFM provides better resolution and a three-dimensional surface profile is obtained compared to a two-dimensional projection in SEM, (ii) the instrument can be operated in ambient air and even in liquid environment without the requirement of vacuum, and hence, living organisms can be studied, and (iii) electrically non-conducting samples, such as biological samples and polymers can be studied and the samples do not require any pre-treatment.

AFM can view and provide an image of small area of the sample surface of the order of 150 μm^2, whereas the SEM can image a much larger area of the order of mm^2. The image of AFM is affected by incorrect choice of the probe tip as well as by the hysteresis of the piezoelectric material.

The scanning probe microscopes have to be isolated from all external mechanical vibrations and acoustic noise as such interferences can cause fluctuation in the distance between the probe tip and the sample surface.

SUMMARY

- Microscopy involves generating a larger or magnified image of the object by viewing the specimen under a microscope to study smaller details of the specimen.

- The resolving power of the microscope is its ability to distinguish two points apart from each other and depends on the wavelength of light used, the refractive index of the specimen, and the numerical aperture of the objective lens.

- Different imaging techniques find use in the study of specimens, particularly biological specimens, by microscopy. The simplest technique is the bright-field microscopy in which the incident light passes through the specimen and depending on the ability of the different parts of the specimen to refract light to different extents and the image formed is seen against the bright background light. Most biological materials require staining to bring about contrast.

- Dark-field imaging technique uses an oblique illumination surrounding the specimen so that the image of the specimen appears bright against a dark background.

- Phase-contrast microscopy involves bringing a phase shift in the light passing through a transparent and colourless specimen into amplitude to provide a contrast in the image of the specimen.

- Fluorescence microscopy is based on fluorescence emitted by the specimen itself on exposure to an intense radiation. The specimen may be a naturally fluorescing substance or fluorescence emission may be induced by treating the specimen with fluorescing reagents. The emitted fluorescence is passed through a filter to the microscope eyepiece. In the microscope the different regions in the specimen appear bright and shining against a dark background.

- Confocal microscopy is an optical imaging technique which enhances the resolution of the image by eliminating the out-of-focus glare. The technique is used for living cells and tissues, optical sectioning of thick specimens, and for inspecting semiconductors.

- Polarizing microscopy is a contrast-enhancing technique for viewing birefringent materials through polarizing light. The anisotropic optical properties of the specimen are useful for identification of the sample and for clinical diagnostic applications.

- Flow cytometry is useful for the study of microscopic particles, particularly biological cell suspensions, in a flowing stream of fluid and also for sorting the particles based on their characteristic features.

REVIEW QUESTIONS

1. What are the components of a compound light microscope? What are their functions?

2. Draw a neat sketch of the optical path of a compound light microscope.

3. Explain the terms 'resolving power' and 'theoretical resolution limit' with respect to microscopy.

4. List the different types of imaging techniques used in light microscopy.

5. Compare the bright-field and dark-field imaging techniques.

6. Explain the principle involved in phase-contrast microscopy with a neat sketch of the optical path.

7. What are phase objects?

8. Explain the principle of fluorescence microscopy and its applications.

9. Explain the principle of confocal microscopy.

10. What is a flow cytometer? How does it function? What are its applications?

11. Discuss the principles involved in the different electron microscopic techniques.

12. Give a block diagram of TEM and explain the functioning.

13. Describe the functional aspects of SEM with the help of a block diagram of the instrument.

14. Give an account of the scanning probe microscopy.

15. Discuss the principle and practice involved in (i) scanning tunnelling microscopy and (ii) atomic force microscopy.

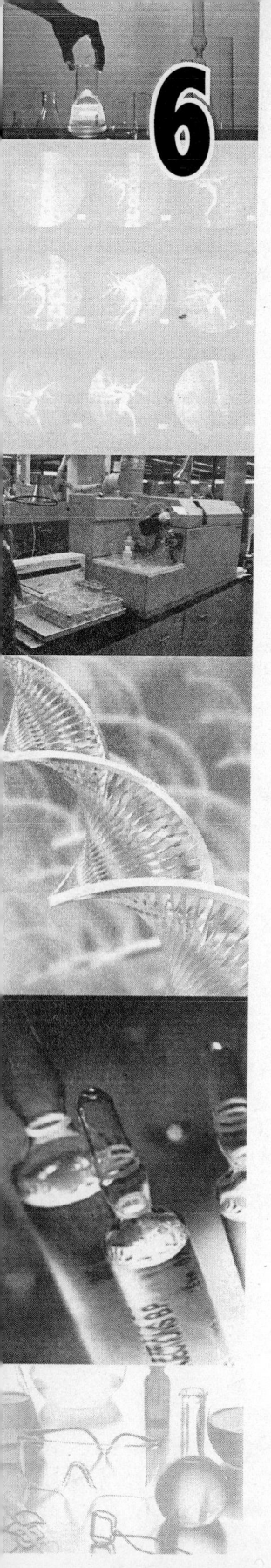

6 Spectroscopic Methods of Analysis

6.1 INTRODUCTION

Spectroscopic methods of analysis involve the study of interaction of electromagnetic radiation (energy) with matter made up of atoms and molecules. These methods provide information on the energy level structure of atoms and molecules and for understanding the molecular structure. These methods also find use in analytical aspects for qualitative and quantitative analyses. This chapter discusses the principles and practice involved in the spectroscopic techniques in different regions of electromagnetic spectrum mainly for analytical purposes.

6.2 ELECTROMAGNETIC RADIATION

Electromagnetic radiation simultaneously exhibits particle as well as wave nature. The dual nature of electromagnetic radiation has been used to explain the atomic and molecular processes of interaction of radiation with matter. The wave theory postulates that radiation emanating from a source consists of an electromagnetic field which varies periodically and in a direction perpendicular to the direction of propagation of radiation. The electromagnetic field is composed of electric and magnetic vectors oscillating in mutually perpendicular planes. The radiation may be described in terms of a sine wave based on the periodic variations of the electric and magnetic field components as shown in Fig. 6.1.

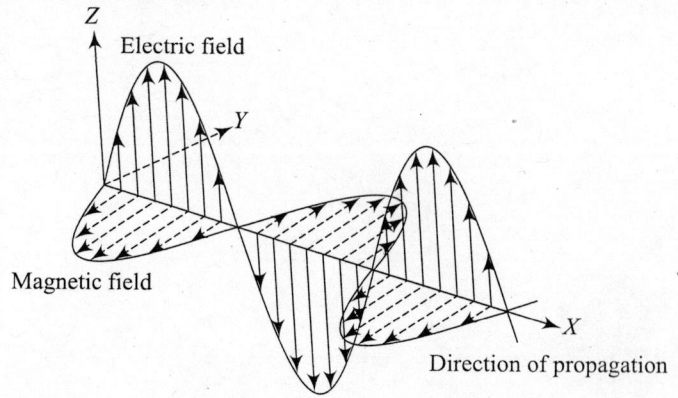

Fig. 6.1 Representation of electromagnetic radiation as a wave

The electromagnetic wave is characterized by three parameters, its frequency v, its length λ, and amplitude A. The *frequency* is defined as the number of oscillations or cycles per second expressed as Hertz (Hz) or simply as cycles per second. The *wavelength* is defined as the linear distance between any two equivalent points on successive cycles, the unit being metre (m). The more commonly used units for wavelength are micrometre (μm = 10^{-6} m), nanometre (nm = 10^{-9} m or millimicron, mμ), Angstrom (Å = 10^{-10} m), and picometre (pm = 10^{-12} m).

The wavelength and frequency are related by the expression $v = c/\lambda$, where c is the velocity of propagation of the electromagnetic wave in vacuum (3×10^8 ms^{-1}). The value of c is independent of frequency. Since the velocity of radiation is a constant, the wave is also characterized by a unit called wave number $\bar{v}$, which is the number of waves per cm and expressed in units of cm^{-1}. It is related to the wavelength and the frequency of radiation as

$$\frac{1}{\lambda} = \frac{v}{c} = \bar{v} \tag{6.1}$$

The unit of wavelength commonly expressed depends on the region of electromagnetic spectrum. For example, in the ultraviolet and visible regions the preferred unit is nm while in the infrared region it is usually wave number. The amplitude of a wave is defined as the maximum value of the vectors in a cycle and the intensity of the radiation is directly proportional to the square of the amplitude.

Electromagnetic radiation is a form of radiant energy and the energy of a unit of radiation is called the *photon* which is related to the frequency or wavelength as given by Einstein–Planck relationship

$$E = hv = h\frac{c}{\lambda} = hc\bar{v} \tag{6.2}$$

where E is energy in joules and h is Planck's constant (6.62×10^{-34} J-s (joule-second).

One mole of photons (Avogadro's number of photons) is called *einstein*. The energy of the radiation is directly proportional to the frequency of radiation while it is inversely proportional to its wavelength, and hence, decreases with increasing wavelength.

The different units of energy commonly used by chemists in addition to joule include eV (electron-volt), cal (calorie), erg, and cm^{-1} (inverse centimetre). The conversion factors of these different units are given in Table 6.1

Table 6.1 Conversion of different units of energy

Unit	joule	eV	cal	erg	cm^{-1}
1 joule	1	6.241×10^{18}	0.239	1.0×10^7	0.0836
1 eV	1.60×10^{-19}	1	23060	1.602×10^{-12}	8065.5
1	4.184	4.336×10^{-5}	1	6.948×10^{-17}	0.34975
1 erg	1.0×10^{-7}	6.241×10^{11}	1.439×10^{16}	1	5.034×10^{15}
1 cm^{-1}	11.96	1.240×10^{-4}	2.8591	1.986×10^{-16}	1

6.2.1 Electromagnetic Spectrum

The energy associated with electromagnetic radiation varies over a wide range of magnitude from millijoules to >10^9 joules based on the wavelength or frequency of radiation. Hence, it is convenient to identify the different ranges of energy in terms of wavelength or frequency regions. The different regions of electromagnetic spectrum include the gamma rays, x-rays, ultraviolet, visible, infrared, microwave, and radiofrequency waves (Fig. 6.2). The frequency and wavelength ranges of these regions overlap to some extent.

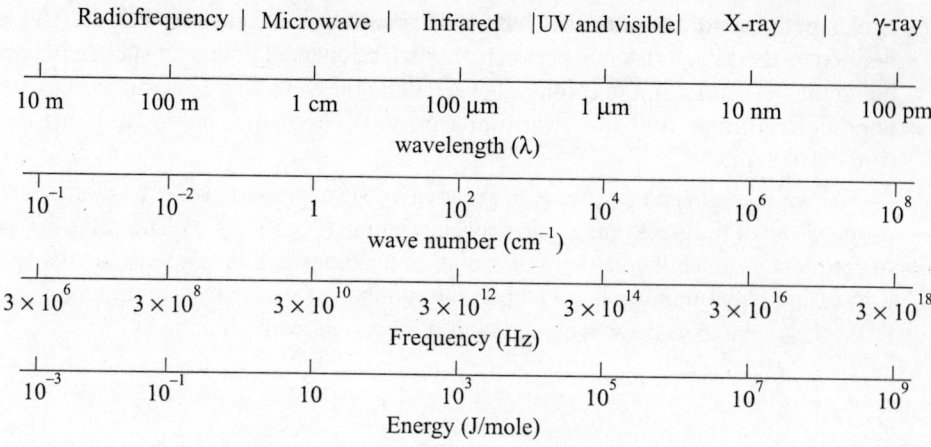

Fig. 6.2 Regions of electromagnetic spectrum

6.3 ENERGY LEVELS IN ATOMS

Before considering the interaction between radiation of the different regions of the electromagnetic spectrum with atoms and molecules it is necessary to have an overview of the energy structure of atoms and molecules. The total energy of an atom or a molecule is due to contributions from different sources. These include the nuclear energy, electronic energy due to interactions between electrons and nuclei, energy due to the spinning motions of the nucleus and the electron, vibrational energy due to the vibrational motion in molecules, the rotational energy in molecules, and the translational energy due to movement in space. All the forms of energy except the translational energy are quantized in the atoms and molecules.

The energy structure of atoms is simpler compared to that of molecules and consists of only the electronic energy levels as shown in Fig. 6.3. Atoms do not have vibrational or rotational energy levels. The electronic energy levels are usually designated as S_0 for the ground state or the lowest energy level and $S_1, S_2, S_3, \ldots$, etc., for the excited states having higher energies. The occupancy of the electronic energy levels in atoms (and molecules) under ambient conditions is determined by quantum theory and Maxwell–Boltzmann distribution

$$\frac{N_{\text{upper}}}{N_{\text{lower}}} = \exp\left(\frac{-\Delta E}{kT}\right) = e^{\left(\frac{-h\nu}{kT}\right)} \qquad (6.3)$$

where N_{upper} and N_{lower} are the populations (numbers) of electrons in the higher and lower energy levels, ΔE is the energy separation between the levels, k is Boltzmann constant (1.3806×10^{-23} J K^{-1}), and T is absolute temperature. Based on these rules the ground state will be the most occupied energy level, and hence, most of the atoms (and molecules) are said to be in their ground state under ambient conditions. The ground state of atoms is usually represented by the ground state electronic configuration, for example, the ground state electronic configurations are $1s^2 2s^1$ for lithium, $1s^2 2s^2 2p^6 3s^1$ for sodium, $1s^2 2s^2 2p^6 3s^2 3p^5$ for chlorine, and so on.

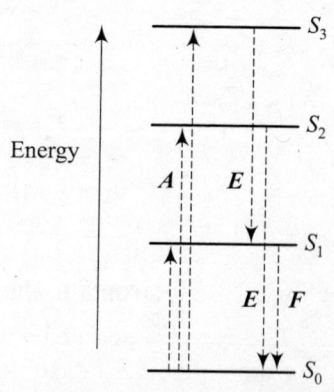

Fig. 6.3 Simplified energy level diagram of an atom and interaction with electromagnetic radiation resulting in absorption and emission of radiation

6.3.1 Interaction of Electromagnetic Radiation with Atoms

The interaction between electromagnetic radiation and atoms may involve the transfer of energy of the incident photons to the atoms which raises their energy from the ground state to an excited state. During excitation the electrons are promoted to higher energy levels from the ground state energy level. This process is called *absorption* represented as A and shown as vertical arrows in Fig. 6.3. The absorption of energy by atoms or molecules can occur only when the energy difference between any two levels is exactly matched by the energy (frequency) of the incident photons as given by the Bohr equation.

$$\Delta E = h \upsilon \tag{6.4}$$

where h is Planck's constant and υ is the frequency of incident radiation. The above equation represents a single spectral transition that is the transition of a ground state atom or molecule to the excited state. The absorption of radiant energy by atoms mostly occurs in ultraviolet and visible regions of the electromagnetic spectrum as the energy of separation between the various energy levels corresponds to these regions.

The absorption of energy by atoms results in the attenuation (decrease in the intensity) of the incident radiation at a particular frequency after it passes through the sample. Hence, the intensity of the transmitted light is relatively weaker compared to the intensity of incident radiation. *Absorption spectroscopy* of atoms (atomic absorption spectroscopy) involves the measurement of the intensity of transmitted light (I_t) by an appropriately placed suitable detector after a part of the incident light of intensity I_0 has been absorbed by the medium. The ratio $\log (I_0/I_t)$ is called the absorbance (A) or optical density (OD). The wavelengths at which absorption of electromagnetic radiation occurs are characteristic of the absorbing species while the intensity of absorption (that is A or OD) is proportional to the concentration of the absorbing species.

The atoms on absorbing energy go to the excited state. The excited atoms are stable only for a short time of the order of nanoseconds to microseconds and relax to the ground state by radiative (emitting radiation) or non-radiative pathways. In non-radiative pathways the energy is converted into translational energy and evolves as heat. The excited state atom may relax by a radiative pathway referred to as *fluorescence* (F) as shown in Fig. 6.3. Fluorescence emission occurs due to electronic transition from the first excited state S_1 to the ground state. Since the energy involved in the absorption and fluorescence emission processes in atom is the same, fluorescence emission occurs at the same frequency (or wavelength) of absorption and is referred to as *resonance fluorescence*.

The ground state atoms may also be excited by different excitation methods other than photo-excitation (excitation by the absorption of light energy). Atoms may be excited to higher energy levels by (i) thermal excitation provided by a flame or high temperature plasma or (ii) electrical excitation by a DC arc or a spark. The kinetic energy imparted to the atoms is utilized to excite electrons to higher energy levels from which they relax to the ground state by emitting radiation spontaneously at frequencies governed by Eq. (6.4). The phenomenon is called *emission* (E). Emission of radiation can occur from any of the higher excited states to other excited states of lower energy or ground state (e.g., S_3 to S_2 or S_1, S_2 to S_1, S_2 to S_0, etc.), as shown in Fig. 6.3 while fluorescence emission is always from the first excited state S_1 to the ground state S_0. Emission of radiation and fluorescence emission are usually observed in the ultraviolet and visible regions of electromagnetic spectrum.

Atomic emission spectroscopy and *atomic fluorescence spectroscopy* involve the measurement of the intensity of emitted radiation from the excited state atoms. As in absorption spectroscopy the wavelengths at which emission of radiation occur are characteristic of the individual atoms, and hence, useful in the qualitative analysis. The intensity of emitted radiation is proportional to the concentration of the excited state atoms.

The techniques of atomic absorption spectroscopy and atomic emission and atomic fluorescence spectroscopy are discussed in detail in Chapter 7.

6.4 ENERGY LEVELS IN MOLECULES

The energy structure of molecules is shown in Fig. 6.4. The electronic energy levels of the ground state S_0 and the excited states S_1, S_2, etc., consist of sub-energy levels called vibrational energy levels designated as $v = 0, 1, 2, 3$, etc. Each vibrational energy level in turn consists of sub-energy levels called the rotational energy levels designated $j = 0, 1, 2, 3$, etc. All the energy levels are quantized having discrete energies. The relative magnitudes of the energy levels is in the order

$$E_{electronic} > E_{vibrational} > E_{rotational} > E_{nuclear\ spin}$$

The differences between the nuclear spin levels are visualized only in the presence of an external magnetic field. Majority of the molecules normally exist in their ground state levels of rotational, vibrational and electronic energy levels. Molecules in addition to the electronic energy levels S_0, S_1, S_2, etc., (known as singlet energy levels) have also excited state levels called triplet energy levels usually designated as T_1, T_2, etc. The lowest triplet energy level is T_1 which has a slightly lower energy compared to the S_1 level as shown in Fig. 6.4.

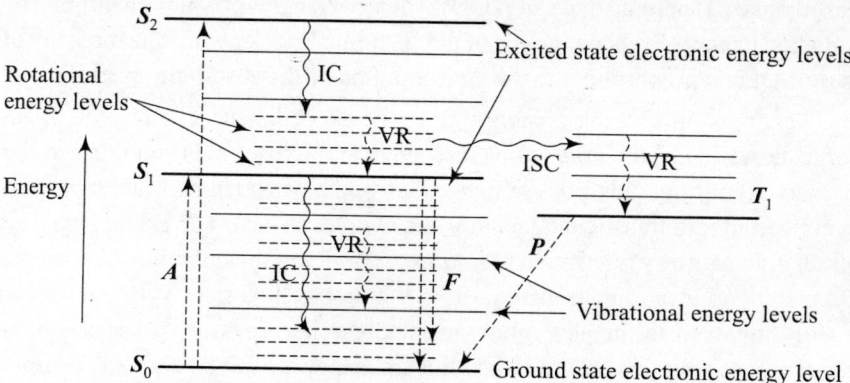

Fig. 6.4 Electronic, vibrational, and rotational energy levels in molecules

It is necessary to explain the difference between the singlet and the triplet electronic energy levels of molecules. The lowest electronic energy level or ground state of a molecule is said to be singlet state when the electrons occupying the level are spin paired. In contrast, the ground state of a free radical (contains an unpaired electron) is said to be a *doublet state* because the odd electron can have any of the two allowed spin states or orientations in the presence of a magnetic field. During excitation the energy absorbed by the molecule results in electronic transition or promotion of an electron from the lower to higher energy level. The excited state molecules can exist in a excited singlet or triplet state depending on the spin orientation and occupancy of the energy levels. In the excited singlet state the promoted electron retains its original spin of anti-parallel orientation to the electron in the ground state. However, in the triplet state the two electrons occupying the ground and the higher energy levels have parallel spins. The excited triplet state has a slightly lower energy compared to that of the excited singlet state. Figure 6.5 shows the arrangement of the spins of the electrons in the ground state, excited singlet, and triplet states.

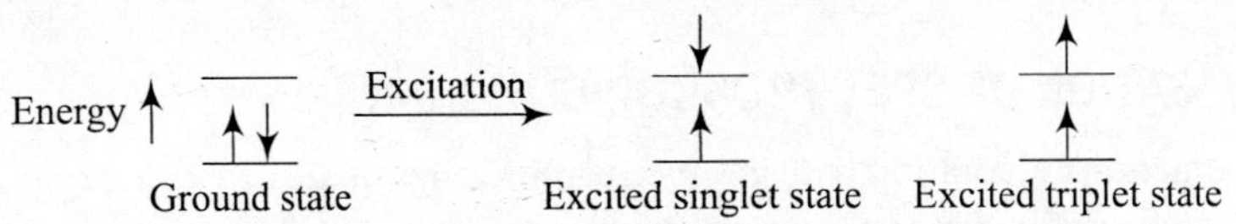

Fig. 6.5   Representation of spin orientations in different energy states of molecules

6.4.1   Interaction of Electromagnetic Radiation with Molecules

Molecules also interact with electromagnetic radiation by absorbing energy from the incident radiation. The *absorption* process (A) is shown in Fig. 6.4. The absorption process in molecules can occur in different regions of electromagnetic spectrum giving rise to *molecular absorption spectroscopy*. Molecular absorption spectroscopy, similar to absorption spectroscopy in atoms, involves the study of the wavelength of absorption and the attenuation of the intensity of incident radiation with the help of appropriately placed detectors. The wavelength of absorption is characteristic of molecules while the intensity of absorption is useful in quantitative analysis.

Molecules can absorb energy in the form of ultraviolet and visible radiation in which the valence shell electrons get promoted to higher energy levels. These electronic transitions give rise to *electronic absorption spectroscopy* more commonly referred to as *absorption spectroscopy*. Absorption of energy by molecules in the infrared region results in the vibrational transitions within the molecules, giving rise to *vibrational spectroscopy*. Molecules can also absorb in the microwave region undergoing rotational changes and such transitions give rise to *rotational spectroscopy*.

The energy involved in electronic transitions is about 400 kJ mole^{-1} while it is 20 kJ mole^{-1} for vibrational transitions, and about 0.4 kJ mole^{-1} for rotational transitions.

The excited state molecules can undergo relaxation to the ground state by emission of radiation in the form of *fluorescence* (F) or *phosphorescence* (P). Fluorescence emission always occurs from the first excited electronic energy state (S_1) to any one of the vibrational energy levels in the ground state electronic energy level. Fluorescence emission is quite intense as it is a spin-allowed transition. It occurs rapidly within 10^{-9}–10^{-7} s. Phosphorescence on the other hand involves the electronic transition from the triplet electronic energy state (T_1) to the ground state S_0. Phosphorescence is a spin-forbidden transition, and hence, weak in intensity. Because of the forbidden nature of transition phosphorescence emission occurs over a longer period of time of 10^{-3}–10 s.

Fluorescence spectroscopy or *fluorometry* and *phosphorescence spectroscopy* or *phosphorimetry* involve the study of fluorescence emission and phosphorescence emission of excited state molecules. In both the techniques the wavelength of emission mostly in the visible region is characteristic of the molecule while the intensity of emission is proportional to the concentration.

The different techniques of molecular absorption as well as fluorescence and phosphorescence spectroscopy are discussed in Chapter 8.

Figure 6.4 shows the different radiative processes of absorption and emission of radiation that occur in molecules as well as the non-radiative pathways of relaxation of the excited state molecules and is known as *Jablonski diagram*. The excited state molecule may relax to the ground state via non-radiative pathways, such as *vibrational relaxation* (VR) and *internal conversion* (IC), which occur very rapidly in 10^{-12} s. The process of non-radiative transfer from the singlet state to the lower energy triplet state is called *intersystem crossing* (ISC). The molecule in the vibrational excited state of the T_1 relaxes to the lowest vibrational level of T_1 by vibrational relaxation. The excited state molecules lose their excess energy in the form of heat released to the surroundings in these non-radiative pathways of relaxation.

6.5 CLASSIFICATION OF SPECTROSCOPIC TECHNIQUES

The interaction of radiation with atoms and molecules in the different regions of the electromagnetic spectrum leads to different levels of energy changes. The instruments called spectrometers used in the different regions are specific for the region of the electromagnetic spectrum. It is, thus, convenient to classify the spectral techniques and instruments depending on the region of electromagnetic spectrum on the basis of the types of changes brought about by the interacting radiation. Table 6.2 summarizes the changes in atoms and molecules that occur during interaction with radiation and the techniques/instruments used for studying such changes.

Table 6.2 Changes that occur in atoms and molecules during interaction with different regions of electromagnetic radiation and types of instruments

Regions of electromagnetic spectrum and wavelength range	Changes in atoms/molecules	Techniques and *instruments*
Absorption spectroscopic techniques		
γ-rays 0.005–1.4 Å	Nuclear configuration	γ-ray spectroscopy *γ-ray spectrometer*
X-rays 0.1–100 Å	Core electron distribution in atoms	X-ray absorption spectroscopy *X-ray spectrometer*
UV and Visible 190–780 nm	Valency shell electron transitions in atoms and molecules	Electronic absorption spectroscopy *UV–visible spectrometer* *Atomic absorption spectrometer*
Infrared 0.8–300 μm	Vibrational transitions in molecules	Vibrational spectroscopy *Infrared spectrometer* *Raman spectrometer*
Microwave 0.75–3.75 μm	Rotational characteristics of molecules	Rotational spectroscopy *Microwave spectrometer*
Radiofrequency 3 cm–10 m	Nuclear and electron spin in the presence of external magnetic field	Magnetic resonance spectroscopy *NMR spectrometer* and *ESR spectrometer*
Emission spectroscopic techniques		
γ-rays	Nuclear configuration	γ-ray spectroscopy *γ-ray spectrometer*
X-rays	Core electron distribution in atoms	X-ray emission and fluorescence *X-ray spectrometer*
UV and visible	Electronic transitions in atoms	Emission spectroscopy *Atomic emission spectrometer* and *Flame emission spectrometer*
UV and visible	Fluorescence emission in atoms and molecules	Fluorimetry *Spectrofluorimeter*
Visible	Phosphorescence emission in molecules	Phosphorimetry *Phosphorimeter*

The different spectroscopic techniques may also broadly be classified into (i) atomic spectroscopy and (ii) molecular spectroscopy involving the study of absorption as well as emission processes.

6.6 ABSORPTION AND EMISSION SPECTRA

The energy changes or transitions that occur during the interaction between electromagnetic radiation of different regions with atoms and molecules are measured by instruments called *spectrometers* in which a beam of radiation of specific frequency (or wavelength) is allowed to interact with a substance. Part of the incident radiation is absorbed by the substance and the rest of it is transmitted onto a detector. The intensity of the transmitted beam is much less compared to that of the incident beam. The detector output gives the amount of energy absorbed by the sample at that particular frequency. Scanning the amount of energy absorbed by the sample over a range of frequencies (or wavelengths) of the incident radiation provides an *absorption spectrum* in the form of a graphical plot of degree (intensity) of absorption of radiation as a function of energy (in terms of frequency or the inversely related wavelength) of the incident radiation.

The excited state atoms emit radiation mostly in the ultraviolet and visible regions and excited state molecules emit fluorescence or phosphorescence. The emitted radiation is analysed by the spectrometer and the graphical plot of intensity of emission as a function of frequency or wavelength is appropriately called *emission spectrum* in the case of atoms and *fluorescence spectrum* or *phosphorescence spectrum* in the case of molecules.

Ideally the absorption or emission spectrum of an atom or molecule should appear as a single line as shown in Fig. 6.6(a), because the energy levels in atoms and molecules are quantized and the Bohr equation [Eq. (6.4)] is valid only for a single frequency. However, the spectrum is usually obtained as a slightly broadened peak as shown in Fig. 6.6(b).

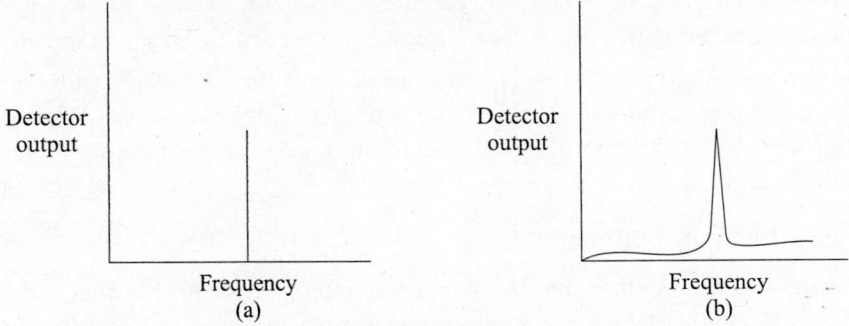

Fig. 6.6 The spectrum of an atom or a molecule undergoing a single transition (a) ideal and (b) actual appearance

6.6.1 Width of Spectral Lines

The width of a spectral line is governed by factors which include the (i) natural line width of an atomic or molecular transition and (ii) resolving power of the instrument.

The *natural line width* is the width inherent in the transition between the atomic and molecular energy levels which cannot be precisely determined. Thus, the natural line width is the minimum width of a spectral line and cannot be made sharper even with instruments of highest resolving capability. The natural line width depends on factors such as (i) Collision broadening, (ii) Doppler broadening, and (iii) Heisenberg uncertainty principle. These factors render the determination of energy levels in atoms and molecules difficult and contribute to a minimum natural line of width.

Collision broadening of the spectral lines occurs due to small changes in energies of atoms and molecules as they collide with each other during their random motion in the vapour and liquid phases. The extent of energy change in the liquid phase is relatively more compared to that in gas phase, and hence, spectra recorded in liquid phase are broader compared to spectra recorded in gas phase. The random movement of particles is less in solid phase, and hence, the spectra in the solid phase are expected to be sharper. However inter-particle interactions cause the spectral lines to be split into two or more components.

Doppler broadening of spectral line is due to the Doppler shift of transition frequencies. In gas and liquid phases the random movement of particles causes a shift in the absorption and emission frequencies to higher as well as lower frequencies (Doppler shift), and hence, the spectral line is broadened. Doppler effect is the predominant factor determining the natural line width in gas phase, while collision broadening is the important factor affecting line width in liquid phase spectra.

Heisenberg uncertainty principle enunciates that the energy levels even in an isolated atom or a molecule cannot be determined precisely. This is because for precise determination of the energy of a state its lifetime should approach infinity. If an atom or a molecule exists in an energy state over a limited period of δt seconds then the energy of that state will be uncertain to an extent of δE as given by the relationship $\delta t \times \delta E \sim h/2\pi \sim 10^{-34}$ J-s. The energy of the lowest energy level of an isolated system may be defined precisely because it will remain in that state infinitely ($\delta t = 0$ and $\delta E = 0$). The lifetime of the higher energy level (excited state) on the other hand is extremely short (less than 10^{-8} s), and hence, the uncertainty of δE will be about $10^{-34}/10^{-8}$ = 10^{-26} J. Thus, the transition between these two states will have an energy uncertainty of δE and the broadening will occur over a frequency range of δv because $\delta v = \delta E/h \sim h/2\pi$.

The resolving power of the instrument depends on the design aspects and the efficiency of the components of the optical system, and hence, the sharpness of the spectral line can be improved. However, even the instrument with highest resolving power cannot improve upon the natural line width.

6.6.2 Intensity of Spectral Lines

The number of lines that appear in an absorption or emission spectrum depend on the number of transitions that take place in an atom or molecule provided each individual transition is associated with unique energy. Hence, it is possible to deduce the energy level structure of atoms and molecules from spectral data.

The intensities of the individual spectral lines vary depending on different factors. The factors include (i) the magnitude of individual transition moments, that is, the probability of transitions occurring, (ii) the relative population of energy levels between which the transitions occur, and (iii) the concentration and path length of the sample.

The *transition probability* depends on the magnitude of the change in dipole moment associated with the transition. Spectral lines appear with greater intensity when the transition probability between the two energy levels is greater. The detailed calculation of absolute transition probabilities requires knowledge of the precise quantum mechanical wave functions of the two energy levels between which the transition occurs. A less rigorous approach commonly used involves the application of *selection rules*.

Selection rules determine whether a particular transition is allowed or forbidden. The spectral transition is considered as allowed when the transition probability is non-zero and the transition is said to be forbidden when the transition probability is zero.

The quantum mechanical selection rules applicable to electronic transitions for one-electron system (hydrogen atom) may be summarized as (i) Δn = anything, (ii) $\Delta l \pm 1$ only, and (iii) $\Delta j = 0$, ± 1 where n, l, and j refer to the principal quantum number, orbital angular momentum, and total momentum due to spin–orbit coupling respectively. In the case of many-electron atoms the relevant selection rules are (i) $\Delta S = 0$, (ii) $\Delta L = 0, \pm 1$ except $L = 0 \leftrightarrow L = 0$ (not allowed), (iii) $\Delta J = 0$ and ± 1 and except $J = 0 \leftrightarrow J = 0$ where S, L, and J represent the total spin angular momentum, total orbital angular momentum, and the total angular momentum respectively based on Russell–Saunders coupling scheme. In addition, according to Laporte selection rule based on symmetry considerations, transitions between states arising from the same configuration are forbidden and represented as $g \leftrightarrow g$ and $u \leftrightarrow u$, the symbols g and u referring to the *even* and *odd* with respect to the arithmetic sum of all orbital angular momenta of all the electrons. The transition $g \leftrightarrow u$ is allowed. In the case of electronic transitions in molecules the selection rules may be summarized as (i) $\Delta \Lambda = 0, \pm 1$, (ii) $\Delta S = 0$, (iii) $\Delta \Omega = 0, \pm 1$, (iv) $\Sigma^+ \leftrightarrow \Sigma^+, \Sigma^- \leftrightarrow \Sigma^-$ (allowed), $\Sigma^+ \leftrightarrow \Sigma^-$ (not allowed), and (v) $g \leftrightarrow u$ (allowed) whereas $g \leftrightarrow g$ and $u \leftrightarrow u$ are not allowed. The symbols Λ, S, and Ω refer to the total orbital momentum, total spin angular momentum, and the total angular momentum respectively. The selection rules (iv) and (v) are based on symmetry considerations the symbol Σ referring to the axial component of the total spin momentum along the inter-nuclear axis of the molecule.

In the case of molecules transitions between different vibrational states and rotational states also occur due to the interactions of the molecules with electromagnetic radiation in the infrared and microwave regions respectively. The selection rules for vibrational spectroscopy of molecules are stated as $\Delta v = \pm 1, \pm 2$, etc., and $\Delta J = \pm 1$. The selection rule for rotational spectroscopy is $\Delta J = \pm 1$.

Ideally forbidden transitions should not appear in the spectra but due to the breakdown of the selection rules attributed to various factors even the so-called forbidden transitions appear in the spectra of atoms and molecules. Allowed transitions in general give rise to spectral lines of greater intensity while the spectral lines of forbidden transitions are relatively weak in their intensity.

The *relative population of energy levels* between which transition occurs also affects the intensity of spectral lines. Maxwell–Boltzmann distribution law [Eq. (6.3)] is useful for the determination of the ratio of the atoms or molecules present in the lower and higher energy levels at a given absolute temperature T. The population of atoms or molecules in the lower energy state will be greater at equilibrium.

The third factor which affects the intensity of the spectral line includes the concentration of the species absorbing the radiation and the path length of the absorbing medium. This factor is quantitatively expressed in terms of Beer–Lambert law and is of importance in analytical applications.

6.7 ANALYTICAL APPLICATIONS OF SPECTROSCOPY

The analytical applications of spectroscopy are primarily based on Beer–Lambert law and include (i) qualitative analysis to identify the nature of the test sample, (ii) quantitative analysis of the desired analyte in a sample, and (iii) determination of percentage purity of a sample by comparison with a standard.

6.7.1 Beer–Lambert Law

When visible light from a natural or an artificial source falls on a coloured homogeneous medium contained in a glass container or cell, the colour of the medium is the complementary colour of the light that has been absorbed by the medium. During this interaction between the medium and the radiation, a portion of light is absorbed by the medium, a portion is reflected and the rest is transmitted.

The intensity of the incident light I_0 is obviously the sum of the intensities of absorbed (I_a), reflected (I_r), and transmitted (I_t) light.

$$I_0 = I_a + I_r + I_t \tag{6.5}$$

The magnitude of I_r is quite small, about 4% at air–glass interface and can be neglected by using a control (i.e., another glass cell of similar dimensions containing the colourless solvent). Hence, for practical purposes

$$I_0 = I_a + I_t \tag{6.6}$$

P. Bouguer (1729) and J.H. Lambert (1760) investigated the effect of the thickness of the absorbing medium on the intensity of light. Lambert's law states that when a monochromatic light passes through a transparent medium, the rate of decrease in the intensity with the thickness of the medium is proportional to the intensity of the incident light. The differential form of the law is given as

$$\frac{-dI_0}{dt} = kI_0 \tag{6.7}$$

where I_0 is the intensity of the incident light, t is the thickness of the medium (also called the optical path or path length), and k is the proportionality factor. The integrated form of the above equation is given as

$$\ln (I_0/I_t) = kt \tag{6.8}$$

or by changing from natural to common logarithm we get

$$I_t = I_0 \times 10^{-kt} \tag{6.9}$$

A. Beer (1852) investigated the effect of the concentration of the absorbing medium on the intensity of light and proposed the law that the intensity of transmitted beam of radiation passing through an absorbing medium decreases exponentially as the concentration of the absorbing substance increases arithmetically. Beer's law may be expressed as

$$I_t = I_0 \times 10^{-k'c} \tag{6.10}$$

where k' is another constant and c refers to the concentration of the absorbing substance.

Beer–Lambert law is the combined form of the above two expressions and is given as

$$I_t = I_0 \times 10^{-\varepsilon ct} \text{ or } \log (I_0/I_t) = \varepsilon ct \text{ or } A = \varepsilon ct \tag{6.11}$$

A diagrammatic representation of Beer–Lambert equation and the terms involved are shown in Fig. 6.7.

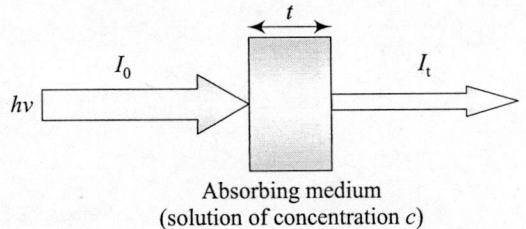

Fig. 6.7   Showing absorption of light and diminishing of intensity

The term A in Eq. (6.11) refers to *absorbance* or *optical density* (OD) given as $\log (I_0/I_t)$. Absorbance is a dimensionless quantity and does not have units. The concentration of the absorbing species c is expressed in $g\,L^{-1}$ and t in cm with the *absorption coefficient* or *absorptivity* ε having units of litre $g^{-1}\,cm^{-1}$. If the concentration is expressed in moles L^{-1} and t in cm, ε is called *molar absorption coefficient* or *molar absorptivity* the unit being liter $mole^{-1}\,cm^{-1}$.

The ratio I_t/I_0 is the *transmittance* T of the medium, that it, the fraction of incident light transmitted by the medium. It is usually expressed as percentage $T\,(\%T = I_t/I_0 \times 100)$. Absorbance A of the medium is related to transmittance as $A = \log (1/T)$ or $(-\log T)$.

6.7.2   Applications of Beer–Lambert law

Beer–Lambert law is useful for qualitative and quantitative analyses and also for determining the percentage purity of an analyte sample. The absorption spectrum of the sample including the number of peaks and shapes of the absorption peaks and the parameter molar absorption coefficient are useful in qualitative analysis. The analyte, may be identified by comparing its absorption spectrum with that of the standard. In addition, the molar absorption coefficient ε is of significance in qualitative analysis for the identification of the substance. The two factors which affect the magnitude of ε are (i) the wavelength of the incident beam of radiation and (ii) the chemical nature of the absorbing substance. The value of ε calculated at a specific wavelength (usually at λ_{max}, the wavelength of maximum absorbance) is a constant and depends on the chemical nature of the absorbing substance. The value of ε is calculated using Beer–Lambert equation from the measured absorbance value of a standard solution of the absorbing species. By comparing the ε values of the test sample with that of the standard or authentic substance, the percentage purity of the test sample may be determined.

Quantitative analysis is based on recording the absorbance at a specified wavelength, usually at λ_{max}, as a function of concentration for a series of standard solutions keeping the path length constant. Since absorbance is directly proportional to the concentration of the absorbing species the graphical plot of absorbance versus concentration of the absorbing species will be a straight line over a limited concentration range (Fig. 6.8). At higher concentrations the linear relationship does not hold. The linear plot is called the calibration graph or working graph. The absorbance of the test sample is measured under the same conditions as determined for the standards. The concentration of the test sample is read from the calibration graph by drawing a tie line from the y-axis to the straight line and dropping a perpendicular to the x-axis from the intercept as shown by the dotted lines in Fig. 6.8.

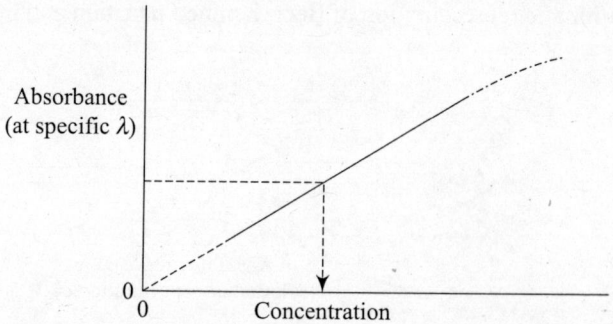

Fig. 6.8 Calibration graph showing conformity with Beer–Lambert law

6.7.2 Limitations of Beer–Lambert law

Beer–Lambert law is a limiting law and applicable only in dilute solutions where a linear relationship between the measured absorbance and concentration of the absorbing species is observed. It is necessary to determine from preliminary experiments the concentration range of the absorbing species in which linearity is exhibited. In addition to concentration other factors also affect the linearity. In general, three types of deviations from the law occur under experimental conditions limiting the application of the law. The deviations include (i) real deviations, (ii) chemical deviations, and (iii) instrumental deviations.

Real deviations from Beer–Lambert law occur at higher concentrations of the absorbing species. The absorptivity ε of the solution changes with concentration depending on the refractive index η of the solution according to the relationship

$$\varepsilon = \varepsilon_{\text{true}} \left[\eta/(\eta^2 + 2)^2 \right] \tag{6.12}$$

and since refractive index varies considerably at concentrations higher than 10^{-3} M, absorptivity also varies. In addition, refractive index also depends on the wavelength of the incident radiation and is a constant for a particular wavelength. Beer–Lambert law was derived on the assumption of an incident beam of monochromatic radiation and even the best monochromator system of an instrument provides only polychromatic beam of radiation spread over a few wavelengths.

Chemical deviations from Beer–Lambert law are observed due to the presence of more than one absorbing species in the solution. The measured absorbance in such cases is actually the sum of the absorbances of the individual species each having its own absorptivity value. For example, an aqueous solution of potassium dichromate containing dichromate ions $(Cr_2O_7^{2-})$ is orange in colour and shows absorption maxima at 350 nm and 450 nm in the visible region. Diluting this solution with water hydrolyses the dichromate ion to chromate ion $(CrO_4^{2-} - \lambda_{\text{max}} = 375$ nm) which is yellow in colour. Hence, the diluted solution contains two different absorbing species and results in non-conformity with Beer–Lambert law attributed to chemical deviation within absorbing medium. The presence of the two species may be represented by the equilibrium

$$Cr_2O_7^{2-} + H_2O \rightleftharpoons 2\,HCrO_4^- \rightleftharpoons 2CrO_4^{2-} + 2\,H^+$$

Conformity with Beer–Lambert law can be obtained by ensuring the presence of only one absorbing species by maintaining appropriate experimental conditions. For recording the absorbance of potassium dichromate solution at its absorption maximum of 450 nm a small amount of dilute sulphuric acid is added to ensure that the above equilibrium is shifted to the left and only dichromate ions remain in the solution. The absorbance of potassium chromate solution can be measured at 375 nm by adding a few mL of 0.05 M potassium hydroxide to ensure the presence of chromate ions only. Another example of chemical deviation involves the estimation of cobalt(II) in aqueous medium as the blue coloured $[CoCl_4]^{2-}$ species. The species is stable only in solution containing chloride ions at concentrations greater than 8 M. Hence, it is necessary to add sufficient quantity of concentrated hydrochloric acid to maintain the required concentration of chloride ions. Similar problems associated with chemical deviation can be eliminated in the case of weak acids and bases by maintaining suitable pH conditions to ensure the existence of only the protonated or the deprotonated species.

Instrumental deviations arise depending on the bandwidth of the instrument. The bandwidth of the instrument depends on the wavelength resolving capacity of the optical system consisting of the monochromator and slits and varies from instrument to instrument. More sophisticated instruments with better resolving power have narrow bandwidths and allow incident radiation of wavelengths within a narrow range. The absorptivity of the chemical species is wavelength dependent and any change in absorptivity over a narrow range of incident wavelengths may be negligible, particularly for absorbing species exhibiting a relatively broad absorption band. In such cases Beer–Lambert law holds good. The effect of instrumental deviation is more with instruments having wide or broad bandwidths and chemical systems exhibiting narrow absorption bands. Instrumental deviations can be minimized by using instruments with narrow bandwidths and proper selection of the wavelength for quantitative analysis. It is preferable to select a wavelength at or close to the λ_{max} and where a change in absorptivity is negligible for small changes in the incident beam wavelength. Figure 6.9 illustrates the effects of the wavelength selected for measuring the absorbance and the bandwidth of the instrument on the calibration plot. Measurement made at λ_1, at or close to the λ_{max}, with a narrow bandwidth gives a better conformity with Beer–Lambert law (curve I) compared to measurement at λ_2 or at λ_1 with a wide bandwidth (curve II).

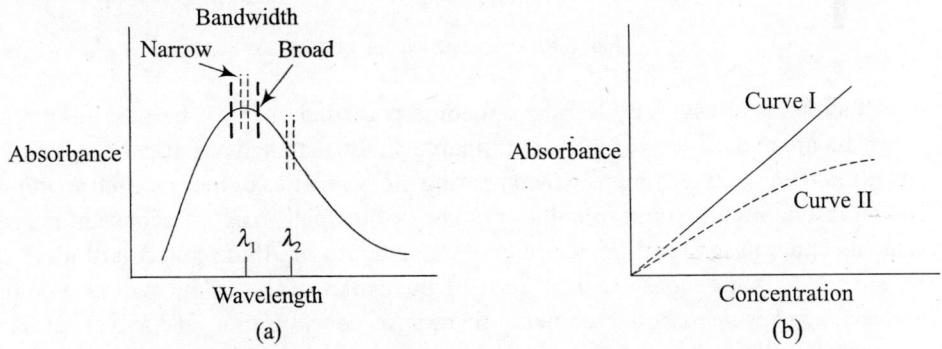

Fig. 6.9 Effect of selection of wavelength in the absorption spectrum and bandwidth (a) on the linearity of Beer–Lambert plot (b)

6.8 VISUAL COLORIMETRY

Visual colorimetry as the name implies is based on using white light or radiation in the visible region of electromagnetic spectrum. Visual colorimetry came into vogue long before the advent of instruments. The use of coloured compounds has been a well established practice in qualitative analysis of inorganic species and in the analysis of organic functional groups.

The basic principle of visible colorimetry for quantitative analysis consists of comparing the intensities of the colour produced by a standard solution of an analyte with that produced by a test sample solution of the analyte (whose concentration is to be determined) under specified conditions. Human eye is a very good 'detector' for comparing the intensities but cannot assign a numerical value for the same. In spectrophotometry the eye is replaced by a photodetector for measuring the intensity of colour.

Visual colorimetry is practiced by comparing the intensities of the colour of two solutions of the analyte, one a standard of known concentration and the second of the test sample, contained in cylindrical columns of identical dimensions. Two commonly used methods of visual colorimetry include (i) Nessler's method and (ii) Balancing column method using Dubosque colorimeter.

Nessler's method makes use of a pair of matched Nessler's tubes which are flat bottomed glass tubes of uniform cross-section with glass markings as shown in Fig. 6.10.

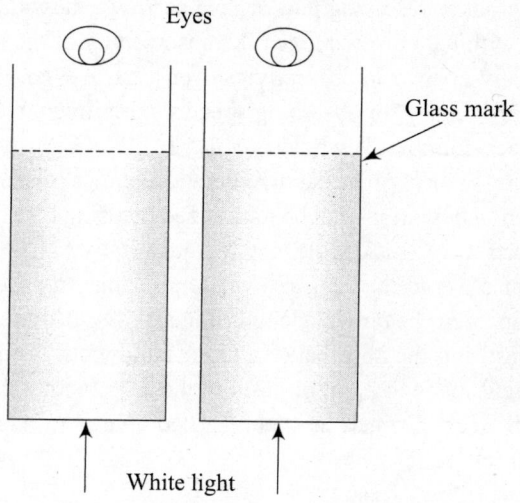

Fig. 6.10   Visual colorimetry by Nessler's method

Quantitative analysis by Nessler's method is carried out by filling the coloured solutions of the standard and the test sample of the analyte in the matched pair of tubes made up to the glass markings to a definite volume and comparing the intensities of the coloured solutions by looking down through the column of solutions. The optical path length is constant because the tubes contain columns of absorbing medium of the same length. If the concentrations of the absorbing species in both the tubes are the same, the intensities of the colour will also be the same. In a *standard series method* for the determination of concentration of the test sample, a series of standard solutions of the analyte are held in several matched Nessler's tubes and the intensities of the colour are compared with that of the test sample held in another matched Nessler's tube. Beer–Lambert law is valid because the intensities (absorbance) of the standard and test sample are the same under identical conditions of path length and concentration. It is necessary that the standards and the test sample are prepared by identical methods to ensure that the solutions contain only the same absorbing species.

Balancing column method makes use of Dubosque colorimeter (Fig. 6.11). The instrument consists of a light source providing visible light (e.g., tungsten filament lamp) and sample cups and plungers made of optical glass for holding the coloured solutions of the standard and the test sample. Light from the source passes through the solutions in the sample cups, through the plungers and the two beams of light are brought to a common axis by the optical system.

The concentration of the standard solution (C_S) and the optical path length (L_S) are kept constant by fixing the position of the plunger within the sample cup. The position of the plunger in the sample cup holding the test solution is varied till the intensity of the colour produced the test sample (I_T) is exactly equal to the intensity of the colour of the standard solution (I_S) as viewed in the eyepiece. Under this condition Beer–Lambert law is valid and $C_S L_S = C_T L_T$ where L_S and L_T are the optical path lengths of the absorbing species obtained from the scale readings. Hence, the concentration of the test solution (C_T) is calculated using the formula $C_T = C_S L_S / L_T$.

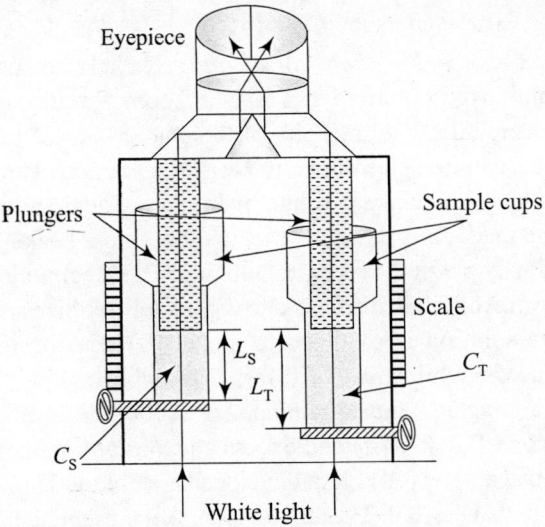

Fig. 6.11 Optical paths in Dubosque colorimeter

6.8.1 Quantitative Analysis

A few examples are discussed here to highlight the use of visual colorimetry.

Determination of iron content in ground water: The iron content in a water sample may be determined by visual colorimetry using a pair of matched Nessler's tubes. The principle is based on converting the iron(III) in a known amount of water sample to water soluble red-coloured complex by reacting with ammonium or potassium thiocyanate in the presence of dilute nitric acid.

$$Fe^{3+} + \text{excess SCN}^- \xrightarrow{\text{dil. HNO}_3} [Fe(SCN)_6]^{3-}$$

The intensity of the red colour of the iron–thiocyanate complex is directly proportional to the amount of iron(III) present in the sample. The intensity of the colour is matched with the intensity of iron(III) present in a standard solution in matched pair of Nessler's tubes. The experimental conditions are maintained to avoid chemical deviation by ensuring that all the iron is present only in the form of the complex by the addition of a sufficient excess of thiocyanate ion and by adding dilute nitric acid to oxidize any iron(II) to iron(III). Nitric acid also stabilizes the complex.

The procedure involves pipetting out quantitatively a known amount (say 5 mL) of sample water into one of the matched pair of Nessler's tubes and diluting with about 30 mL of distilled water. About 5 mL of dilute nitric acid is added to ensure that all the iron is present in the form of iron (III) followed by 5 mL of a 10% solution of ammonium or potassium thiocyanate. The reaction mixture is stirred well and made up to a final volume of 50 mL (up to the glass mark in the Nessler's tube). A known volume of a standard solution of iron(III) containing about 0.05–0.1 mg of iron(III) is taken in the second matched pair of Nessler's tube and treated in exactly the same manner as the test sample to develop the colour. The intensity of the colour of the two solutions is compared visually. The intensity of the colour of the standard solution is matched exactly with that of the test sample by appropriate use of the volume of standard iron(III) solution keeping all other conditions the same. When the intensities of the two solutions match visually the amount of the iron(III) present in the test sample is exactly the same as the amount of iron(III) in the standard solution. The total amount of iron present in the test sample is calculated and expressed usually in ppm.

Determination of nickel in cupro-nickel alloy: Nickel content of the alloy can be determined by forming a soluble red-coloured nickel–DMG complex after completely removing copper by precipitating as sulphide. The intensity of the colour is directly proportional to the amount of nickel, and hence, can be determined by Nessler's method. The procedure involves dissolving a weighed quantity of the alloy in dilute hydrochloric acid and making up to a known volume. An aliquot of the made up solution is pipetted out into a beaker and copper is precipitated out as cupric sulphide by passing hydrogen sulphide. The precipitate is filtered off and the filtrate is boiled with a few mL of dilute nitric acid to completely oxidize and expel any hydrogen sulphide. Nickel(II) in the solution is oxidized to nickel(III) with bromine water (nickel(II) forms a precipitate with DMG while nickel(III) forms a soluble complex) and the excess bromine water is destroyed by adding aqueous ammonia. An alcoholic solution of DMG (1%) is added to the colourless solution of nickel(III) and excess aqueous ammonia is added to render the solution distinctly ammoniacal. The red colour developed is stable and its intensity is directly proportional to the amount of nickel present. The amount of nickel present in the analyte solution is determined by comparing the intensity of the colour with that of a standard nickel–DMG solution prepared in a similar manner.

6.8.2 Instruments for Optical Spectrometry and Measurement of Absorbance

As mentioned earlier the human eye is capable of comparing the intensities of colours but cannot assign any numerical values. The intensity of the colour or the absorbance of the medium can be measured quantitatively by an instrument called *filter photometer*. In the instrument radiation from a tungsten lamp is allowed to pass through a colour filter to limit the radiation to a relatively narrow band of contiguous wavelengths. The narrow band of radiation passes through the absorbing medium and the transmitted light reaches a photodetector. The detector is used to quantify the intensity of the light by converting the light that falls on it into electrical energy in the form of a DC current which is then displayed by a micro ammeter in terms of per cent T or absorbance.

Filter photometers are useful only in the visible region of the electromagnetic spectrum. A more sophisticated instrument with higher sensitivity and accuracy is called *spectrometer* or *spectrophotometer* useful in the UV, visible, and IR regions. The spectrometers used for the different regions are different and are named appropriately. Electronic transitions within atoms and molecules take place in the ultraviolet and visible regions. The instruments called *ultraviolet-visible spectrophotometers* used in these regions are available commercially for ultraviolet and visible regions separately and also commonly covering both the regions.

Vibrational transitions within the molecule occur in the infrared region and the instruments used are known as *infrared spectrophotometers*. Rotational transitions within molecules occur in the microwave region, and hence, the instruments are called *microwave spectrometers*.

Spectrometry for analytical purposes (qualitative and quantitative analysis) is mostly confined to UV, visible, and IR regions. Spectrometry in the IR, microwave, and radiofrequency regions are mostly useful for the elucidation of molecular structure.

6.9 SPECTROMETERS AND THEIR COMPONENTS

The spectrometers used for UV, visible, and IR regions have certain common features and the functions of the components are similar. The basic components of spectrometers include (i) a radiation source to provide the desired frequency (or wavelength) range of radiation, (ii) dispersion devices or analysers to analyse the frequencies of absorbed or emitted radiation, (iii) sample holder, (iv) radiation detectors which convert radiant energy into electrical energy suitable for amplification, (v) signal or data processor, and (vi) display unit (or recorder). The phenomena of absorption, emission, fluorescence, phosphorescence, scattering, and chemiluminescence usually occur in the ultraviolet and visible regions. The arrangement of the basic components depends on the nature of measurement as shown schematically in Fig. 6.12. For measuring absorbance the radiation source, dispersion device, sample holder, and the detector are linearly arranged so that the transmitted radiation passes on to the detector. Fluorescence, phosphorescence and scattering are measured at right angle to the radiation source. In contrast a separate radiation source is not required for emission spectrometry and chemiluminescence spectrometry as the sample itself functions as the source of radiation.

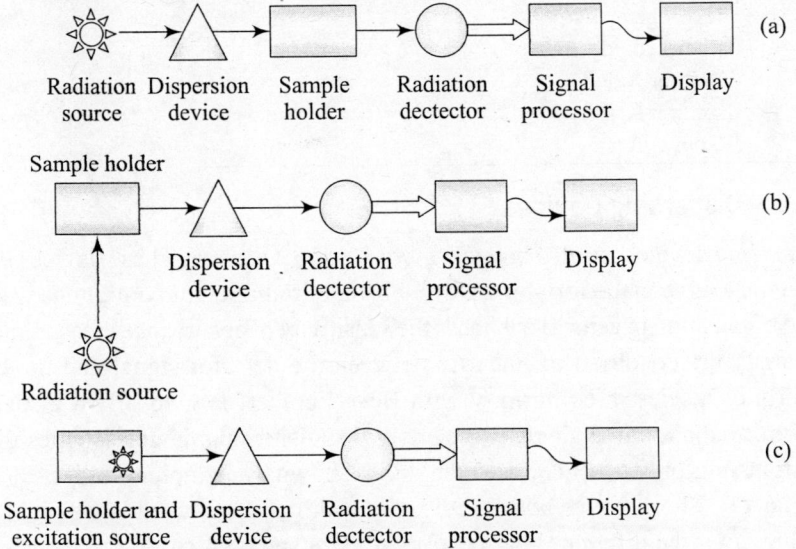

Fig. 6.12 Arrangement of components of spectrometers for measurement of (a) absorbance, (b) fluorescence, phosphorescence, and scattering, and (c) emission and chemiluminescence

6.9.1 Radiation Sources

The radiation provided by a source, such as a lamp, must be of sufficient power and stable for reasonable period of time for detection and measurement by the radiation detector. The different radiation sources used in ultraviolet, visible, and infrared regions include *continuum sources* which provide a radiation covering a range of wavelengths and *line sources* which emit a limited number of lines or bands of radiation covering a limited range of wavelengths. Table 6.3 lists the different radiation sources and the wavelength ranges in which they can be used.

Table 6.3 Radiation sources for UV, visible, and IR spectrophotometers

Radiation source	Spectral region	Wavelength range (nm)
Continuum sources		
Argon lamp	Vacuum UV	100–180
Xenon lamp	Vacuum UV, UV, visible, and near IR	160–800
Deuterium lamp	Vacuum UV and UV	180–360
Tungsten lamp	Visible and near IR	350–2800
Nernst glower ($ZrO_2 + Y_2O_3$)	UV, visible, near IR, and IR	300–20,000
Nichrome wire (Ni + Cr)	Near IR, IR, and far IR	800–30,000
Globar (SiC)	Near IR, IR, and far IR	1200–40,000
Line sources		
Hollow cathode lamps (of different elements)	UV and visible	280–700
Lasers	Vacuum UV, UV, visible, and near IR	180–3500
Examples:		
Ruby laser (Cr/Al_2O_3)		694.3
Nd:YAG		532
Nitrogen		337.1
Carbon dioxide		10,600 (10.6 µm)
Excimer XeF		351
Excimer KrF		248
Excimer ArF		193

6.9.2 Dispersing Devices

Dispersing devices are also known as *wavelength selectors*. These devices transmit the radiation from the source in the form of a narrow band of radiation of contiguous wavelengths. A narrow band of radiation in general enhances the sensitivity of absorbance measurements and is necessary to satisfy the condition of linearity between the detector signal and the concentration of the absorbing species in conformity with Beer–Lambert law. Ideally a dispersing device should transmit radiation of a single wavelength. No dispersing device satisfies the ideal requirement and transmits only a band of wavelengths as shown by a graphical plot of %T versus wavelength (Fig. 6.13). The *effective bandwidth* is the width of the peak at half-height of the peak or band as shown for the different filters in Figs 6.13(b) and 6.14(b).

Two types of wavelength selectors commonly used in the ultraviolet, visible, and IR regions include (i) filters and (ii) monochromators.

Filters are of two types: (i) absorption filters and (ii) interference filters. *Absorption filters* are made of coloured glass, or dye suspended in a thin film of gelatin sandwiched between glass plates. Absorption filters can be used only in the visible region and have effective bandwidths in the range of 30–250 nm as shown in Fig. 6.13(a).

Narrow bandwidth filters in general have low transmittance, as low as 10%, at their band peaks. *Cut-off filters* have almost 100% transmittance over a range of wavelengths but the transmittance decreases rapidly to almost zero beyond the cut-off wavelengths. Narrow bandwidths can be obtained by combining an absorption filter with a cut-off filter as shown in Fig. 6.13(b).

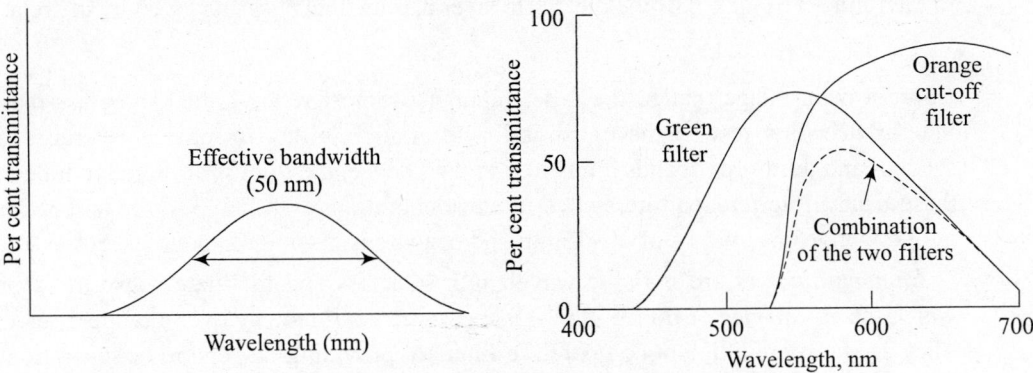

Fig. 6.13   Effective bandwidths of filters (a) Absorption filter and (b) combination of absorption and cut-off filters

Interference filters (also called Fabry–Perot filters) can be used in the UV, visible, and IR regions and are more efficient in transmitting narrower bandwidths. The interference filters consist of transparent dielectric material (e.g., calcium fluoride or magnesium fluoride) sandwiched between two highly reflecting but partially transmitting silver metal films, the entire assembly is again sandwiched between two glass plates (Fig. 6.14).

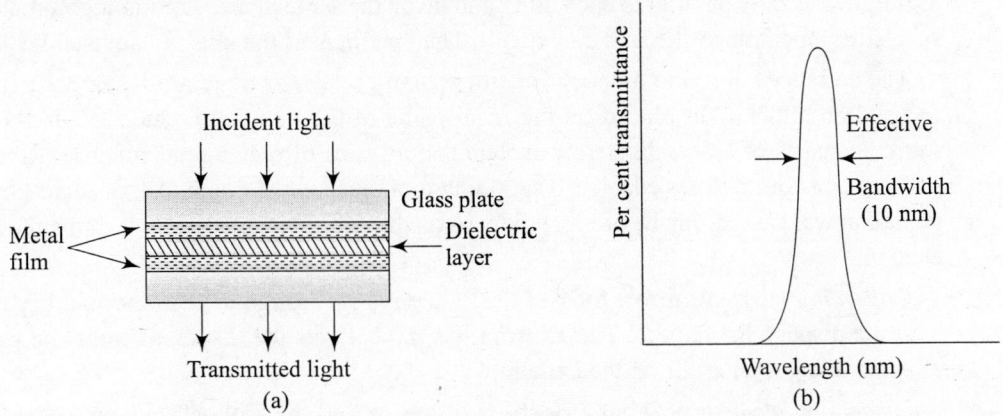

Fig. 6.14   Interference filter (a) cross section and (b) effective bandwidth

Interference filters work on the principle of optical interference effect depending on the thickness of the dielectric layer. When the incident light strikes the interference filter part of the light passes through the first metal film while the rest of the radiation is reflected. The light passing through the first metal film is similarly partially reflected back by the second metal film. If the reflected radiation is of proper wavelength it is partially reflected by the inner side of the first metal film and is in phase with the incoming light of the same wavelength. Thus, the particular wavelength is reinforced while the other wavelengths suffer destructive interference as they are out of phase with each other. For constructive interference and reinforcement of the light of certain wavelength to occur the thickness of the dielectric layer must be critically controlled to one half-wavelength or multiples of half-wavelengths of transmitted light.

The nominal wavelength λ at which full reinforcement occurs for an interference filter is related to the refractive index η of the dielectric layer and its thickness t as given by the relationship

$$\lambda = 2\,\eta\,t/n \tag{6.13}$$

where n is an integer called the order number or interference order. The glass plates of the interference filter restrict the transmission to a single order. In most interference filters the second- and third-order bands are narrower, and hence, the filter is arranged to transmit one of these bands. Interference filters cover a wavelength range of 253–380 nm and 380–1100 nm with effective bandwidths of 10–15 nm and peak transmission of about 25–50%.

Monochromators are variable wavelength selectors and facilitate *scanning*, a process in which the absorbance or transmittance is measured continuously over a range of wavelengths. They are far more efficient compared to filters in providing dispersion or separation between adjacent wavelengths.

A monochromator system consists of five basic components which include (i) an entrance slit to form a rectangular optical image, (ii) collimators (mirror or lens) to produce a parallel beam of radiation, (iii) dispersing element to separate the wavelengths, (iv) a focusing mirror or lens to reform the image of the entrance slit, and (v) an exit slit to isolate the desired band of wavelengths.

Slits are narrow openings between two metal jaws (slit jaws) through which radiation passes. Slit jaws are formed by careful machining of two pieces of metal to give sharp edges. The jaws are aligned exactly parallel to each other and lie on the same plane. The monochromator has two such slits, an entrance slit, and an exit slit. The openings of the slits are adjustable.

The entrance slit of the monochromator actually serves as a radiation source as its image is focused on the exit slit placed on the focal plane of the monochromator. When the radiation source consists of a few discrete wavelengths, a series of rectangular images of the entrance slit appear as bright lines each corresponding to a given wavelength. A line corresponding to a particular wavelength can be brought to focus on the exit slit by moving or rotating the dispersing element.

Collimator (lens or mirror) focuses the light passing through the entrance slit in parallel rays onto the dispersing element. The *focusing mirror* collects the dispersed light and reforms the image of the entrance slit on the exit slit.

Dispersing element is either a prism or a grating and the monochromator system is named accordingly as prism monochromator or grating monochromator.

Dispersion is defined as the spread of wavelengths in space or as the separation of a mixture of wavelengths into component wavelengths. Dispersion is achieved by a monochromator based on the phenomenon of (i) refraction using a prism or (ii) diffraction using a grating.

The linear dispersion D of a monochromator refers to the variation of wavelength $d\lambda$ as a function of the distance x along the focal plane of the monochromator.

$$D = \frac{dx}{d\lambda} \tag{6.14}$$

A more useful measure of dispersion is the *linear reciprocal dispersion*, D^{-1} defined as the spread of a range of wavelengths over a unit distance in the focal plane of a monochromator.

$$D^{-1} = \frac{d\lambda}{dx} \tag{6.15}$$

The dimensions of D^{-1} are nm/mm.

Prism monochromators are based on the phenomenon of refraction of light by the prism. The dispersive power of the prism depends on the variation of refractive index with the wavelength. A ray of light entering the prism at an angle of incidence i bends towards the normal of prism face and at the prism–air interface bends away from the normal. The resolving power of the prism is given by relationship

$$R = \frac{\lambda}{\Delta\lambda} = t\left(\frac{d\eta}{d\lambda}\right) \tag{6.16}$$

where t is the base length of the prism and $(d\eta/d\lambda)$ is the dispersive power of the prism material and η is the refractive index of the prism material. The dispersive power of the prism material depends on the wavelength of the incident light and increases from longer wavelengths to shorter wavelengths.

Prism monochromators are used for dispersing ultraviolet, visible, and infrared radiation. Prisms made from glass can be used in the visible region only as it is transparent to visible light and quartz prisms can be used in all the three regions. Two most common types of quartz prisms are the Cornu prism and the Littrow prism. *Cornu* prism is formed by cementing a right-handed 30-degree crystalline quartz prism to a left-handed quartz prism to form a 60-degree prism so that the optically active prisms do not cause any net polarization of the transmitted radiation. The *Littrow* prism is a 30-degree quartz prism with an aluminized rear surface which functions as a mirror. Refraction occurs twice at the same interface. Figure 6.15 shows the dispersion by the Cornu and Litrrow prisms.

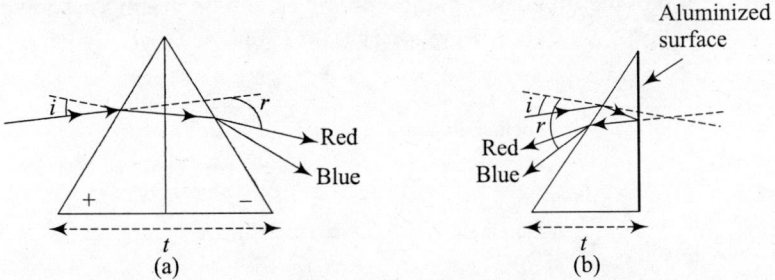

Fig. 6.15 Dispersion by (a) Cornu prism and (b) Littrow prism

The optical design of the prism monochromator system is such that the prism is illuminated through the entrance slit by a parallel light so that the light rays pass through a plane parallel to prism base. The rays pass through the prism symmetrically so that the incident and the emergent beams form equal angles to the faces of the prism. Such an optical design minimizes astigmatism of the prism. The image of the entrance slit is projected on to the exit slit as a series of images ranged next to each other, caused by light of shorter wavelengths being bent more than light of long wavelengths.

The arrangement of the components of a prism monochromator is shown in Fig. 6.16.

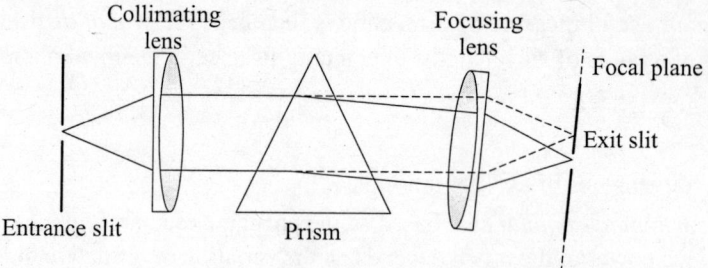

Fig. 6.16   Prism monochromator system

Grating monochromator makes use of the phenomenon of diffraction to isolate the desired wavelength. Diffraction gratings are mostly *reflection gratings*. A diffraction grating consists of a number of parallel and equally spaced grooves ruled on a hard, polished surface of glass or metal (Al) by a properly shaped diamond tool. The grating may contain 300–2000 grooves/mm depending on the wavelength region. For ultraviolet and visible regions 1200–1400 grooves/mm are most common. The grating for the infrared region contains 10–200 grooves/mm where about 100 grooves/mm being suitable for the IR range of 5–15 μm. *Replica gratings* made from a *master grating* are commonly used in most of the instruments. Master gratings are made by ruling with a diamond tool operated by a ruling engine provided with an interferometric control to make grooves of identical size, exactly parallel, and equally spaced over a length of 3–10 cm long surface. Replica gratings are made from the master grating by a replicating process in which a liquid resin is cast on the master grating preserving the optical accuracy of the master grating on the resin surface. The resin surface is then made reflective by coating with aluminum, gold, or platinum.

Echellette grating is the most common reflecting grating which is grooved or blazed such that it has relatively broad faces from which reflection occurs alternating with narrow unused faces. The diffraction of the incident radiation on the echellette grating is shown in Fig. 6.17.

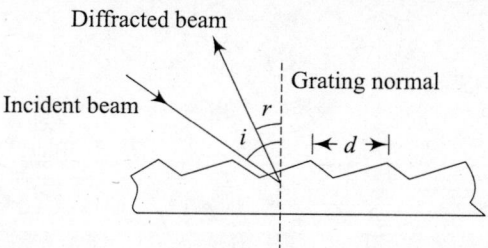

Fig. 6.17   Diffraction from an echellette grating

The incident beam of radiation strikes the broad face of the groove at an angle of incidence i relative to the grating normal and is reflected at an angle r with a maximum constructive interference or reinforcement when the wavelength of incident radiation satisfies the grating formula

$$n\lambda = d\,(\sin i + \sin r) \tag{6.17}$$

where n is an integer called the order of diffraction. When $n = 1$ for a diffracted beam of 600 nm, a second-order line of 300 nm, and a third-order line of 200 nm will appear at the same diffraction angle r. However, the first-order line is the most intense line and gratings are designed so that more than 90% of the incident radiation undergoes first-order diffraction. The higher order lines can be removed by filters. Echellette grating has the advantage of linear dispersion of radiation in contrast to the non-linear dispersion of a prism. The resolving power of a grating is related to the number of grooves or blazes N and the diffraction order n as

$$R = \lambda / \Delta\lambda = nN \qquad (6.18)$$

Concave grating is formed on a concave surface instead of a plane surface. It has the advantage in that the concave surface disperses and also focuses the radiation on the exit slit without auxiliary collimating and focusing mirrors or lenses. The design is simple, and provides a greater radiant energy throughput in the monochromator due to less number of optical surfaces besides being economical in space and cost.

Holographic grating is formed on a plane or concave glass surface by ruling with lasers instead of a diamond tool. The optical technique of ruling involves exposing a glass surface coated with a photoresist to a pair of identical lasers at suitable angles. The laser beams sensitize the photoresist to create interference fringes which on dissolving with a suitable solvent leaves a grooved structure on the photoresist. The photoresist is then coated with aluminum to get a reflecting grating. The spacing of the grooves can be altered by changing the angle between the two laser beams. Large-sized (~50 cm) holographic gratings containing about 6000 grooves/mm are commercially available. Holographic gratings have several advantages over mechanically formed gratings such as greater perfection in line shape and dimensions, ability to provide better spectra free from stray radiation and ghosts or double images, ease of fabricating replica gratings and relatively low cost.

The arrangement of the components in a grating monochromator system is shown in Fig. 6.18.

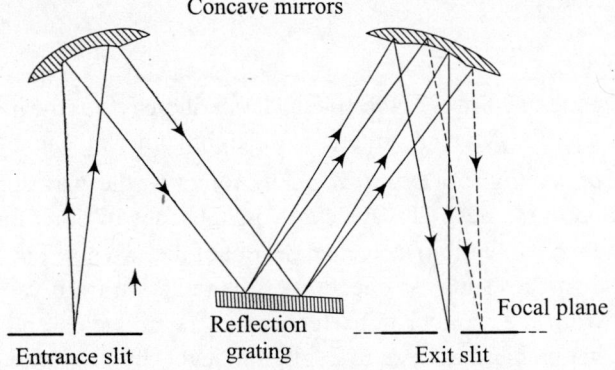

Concave mirrors

Reflection grating

Focal plane

Entrance slit

Exit slit

Fig. 6.18   Grating monochromator system

Echelle monochromator system contains two dispersing elements, a special type of grating called *echelle* grating and a low dispersion prism, arranged in series. The echelle grating differs from the echellette grating in several features: (i) it contains fewer grooves (~300 grooves/mm), (ii) the short side the blaze is used for diffraction, and (iii) the angle of refraction is much higher and approaches the angle of incidence ($r \approx i \approx \beta$). Hence, the grating equation becomes $n\lambda = 2d$ $\sin \beta$. In the echellette grating high dispersion is achieved by decreasing width d of the groove very small and increasing the focal length F. However an increase in focal length reduces the light-gathering power of the grating and the monochromator also is large and unwidely. In echelle grating high dispersion is achieved by increasing the angle β (to 63°) and also the order of diffraction n (to about 75). Echelle grating monochromators have higher linear dispersion, resolution, and light gathering power compared to echellette grating systems.

Performance characteristics of monochromators

The performance of a monochromator system involves three interrelated factors, namely, resolution, light-gathering power, and purity of light output. The resolution depends on dispersion and the perfection of image formation. Purity of light output depends on the amount of stray or scattered light mixed with the output light. Monochromators in general should have large dispersion and high resolving power.

The *resolving power* or *resolution R* of a monochromator is defined as its ability to separate adjacent images that have a slight difference in wavelength and is given as

$$R = \lambda/\Delta\lambda \tag{6.19}$$

where λ is the average wavelength of the two images and $\Delta\lambda$ is difference between the wavelengths. The resolution of the monochromator depends on the size and dispersing characteristics of the dispersing element (prism or grating), slit width, and the optical design of the monochromator system. In the case of recording instruments resolution also depends on the recording system at a given scan speed.

Light-gathering power of a monochromator depends on collimating lenses or mirrors of the monochromator system which collect the radiation from the entrance slit and focus onto the exit slit. In order to have an enhanced signal-to-noise ratio, it is necessary that sufficient light must reach the detector. The light-gathering power of the monochromator is expressed in terms of *f/number* or *speed* as the ratio of the focal length F of the collimating lens or mirror to its diameter d.

$$f/number = F/d \tag{6.20}$$

The light gathering power of the optical device increases as the inverse square of the *f/number*. Thus, the smaller the *f/number* greater is the light gathering capability.

Purity of output light or *spectral purity* refers to the wavelength purity of the radiation (i.e., separation of the desired wavelength and absence of other unwanted wavelengths due to scattering and stray radiation) as it emerges out of the exit slit. The unwanted radiation arises to due to several sources such as reflection of the radiation from various optical parts, scattering of radiation from mechanical imperfections of components and by dust particles, etc. The unwanted radiation is minimized to negligible levels by proper design and construction of the monochromator system, coating the interior surfaces with flat black paint, providing baffles at appropriate places to cut off spurious radiation, and preventing the dust particles from entering by proper sealing of the unit.

Bandwidth of monochromator

The wavelength of the radiation that exits the monochromator is variable over a considerable range. However, the wavelength range that emerges out of the exit slit for a given wavelength setting of the monochromator is called the *bandwidth* of the monochromator. The bandwidth is defined as the span of the monochromator settings in nm or in cm^{-1} required to move the image of the entrance slit across the exit slit. The *effective bandwidth* or the *spectral band-pass* is one half the bandwidth when the width of the entrance and exit slits are identical. The effective bandwidth is actually the range of wavelengths that exit the monochromator for a given wavelength setting. The effective bandwidth $\Delta\lambda_{eff}$ of a monochromator is ultimately controlled by the exit slit width w as given by the relationship

$$\Delta\lambda_{eff} = wD^{-1} \tag{6.21}$$

where D^{-1} is the linear reciprocal dispersion. Figure 6.19 shows the bandwidth and the effective bandwidth for a monochromator setting for radiation of a wavelength of λ_2 and its isolation from very close wavelengths of λ_1 and λ_3.

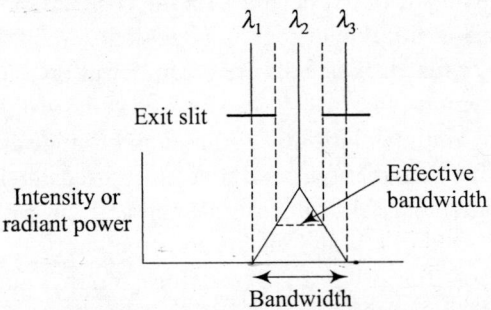

Fig. 6.19   Showing bandwidth and effective bandwidth of a monochromator

The effective bandwidth can be as small as 0.1 nm in the case of sophisticated high resolution instruments to more than 20 nm in less expensive instruments.

The effect of slit width on the effective bandwidth and the ability of the instrument to resolve adjacent peaks in the spectrum is shown in Fig. 6.20.

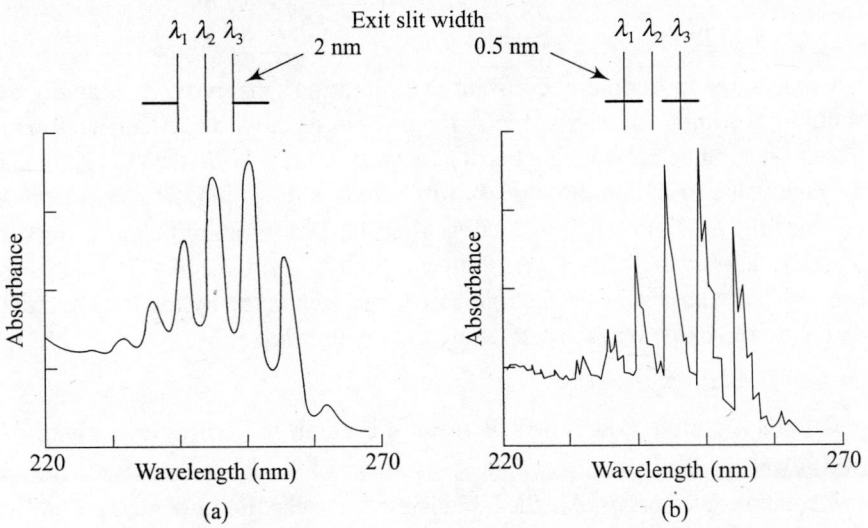

Fig. 6.20   Effect of slit width on spectral resolution of benzene vapour:
(a) slit width of 2 nm and (b) slit width of 0.5 nm

Monochromators have slits whose widths can be varied to control the effective bandwidth. Smaller or narrow slit widths are required for resolving narrow absorption or emission bands for getting finer details of the spectral bands (fine spectra) as seen in Fig. 6.20(b) which is useful for qualitative analysis and elucidation of structure of the analyte. However, narrow slit widths decrease the intensity or power of radiation reaching the detector, and hence, accurate determination of intensity for quantitative analysis becomes difficult.

6.9.3 Sample Holders

Sample holders called *cells* or *cuvettes* for holding liquid samples are usually made of optical glass for the visible region but cannot be used in ultraviolet region as glass absorbs UV radiation. Quartz or fused silica cells are useful for ultraviolet as well as visible regions. Plastic cells can be used in the visible region. Crystalline sodium chloride or potassium bromide are transparent to infrared radiation, and hence, are used in IR spectrometers. Figure 6.21 shows a sample cell of 1 cm path length commonly used in UV and visible spectrophotometers with a capacity of 2.5–3.0 mL. Semimicro, micro, and macro sample cells with path lengths of 0.1 cm to 10 cm are commercially available.

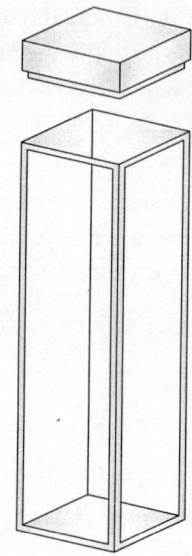

Fig. 6.21 A 1 cm path length sample cell or cuvette

6.9.4 Radiation Detectors

Radiation detectors are essentially transducers which convert radiant energy into electrical energy. An ideal transducer should generate an electrical signal S (mostly current) which is directly proportional to radiant power P as given by the relationship

$$S = kP \qquad (6.22)$$

where k is a proportionality constant or calibration sensitivity. In addition, an ideal transducer should have high sensitivity, high signal-to-noise ratio, and a constant response to radiation over a wide range of wavelengths. The *signal-to-noise ratio* (S/N) is defined as the ratio of the average value of the output signal to its standard deviation. Several methods are available to enhance the S/N ratio such as analog filtering, lock-in-amplification, smoothing, and Fourier transformation. Most transducers, however, exhibit a small constant response referred to as *dark current* in the absence of radiation. The signal generated by the transducer, thus, includes the dark current component k_d as given by the relationship

$$S = kP + k_d \qquad (6.23)$$

The dark current component is usually brought to zero by a compensating circuit in the instrument.

Radiation detectors used in the UV, visible, and near-IR regions are generally known as *photon transducers* (also called *photoelectric* or *quantum* detectors) as they generate an electrical signal in the form of photocurrent by absorbing the photons. In contrast, *thermal transducers* used in the IR region absorb photons and generate heat which is measured as temperature increase.

Photon transducers include a variety of devices such as photovoltaic cells, photoconductivity detectors, phototubes, photomultiplier tubes, silicon photodiode and photodiode array detectors, and charge transfer transducers.

Photovoltaic cell or *Barrier-layer cell* consists of a layer of semiconductor material, such as selenium deposited on copper or iron base plate, which functions as the positive electrode. A thin transparent film of silver or gold coated on the semiconductor layer functions as the collector electrode and is covered by a sheet of glass. The assembly is enclosed in a protective envelope made of plastic (Fig. 6.22).

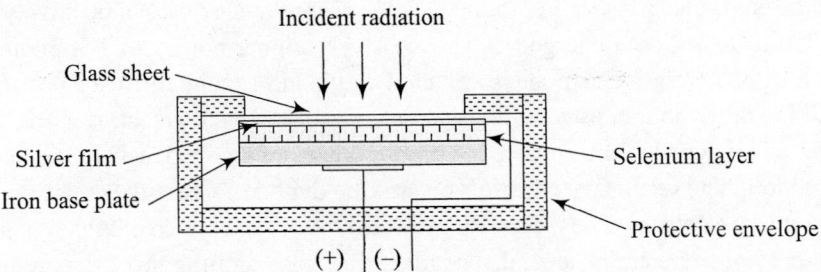

Fig. 6.22 Schematic diagram of a photovoltaic cell

When incident radiation passes through the transparent glass sheet and the thin metal film and strikes the selenium layer, photo-generated electrons, and holes are formed. The electrons migrate to the metal film and flow through the external circuit while the holes migrate to the base plate. The magnitude of the generated electric current, depends on the intensity (number of photons) of the incident radiation, is about 10–100 μA and can be measured by a microammeter.

Photovoltaic cells are simple in construction, rugged, and inexpensive, and hence, find use in portable instruments for reliable measurements. However, they suffer from the disadvantages such as lack sensitivity at low levels of illumination, inability to amplify the signal, and *fatigue* in which current output gradually decreases during continuous illumination.

Photoconductivity detector consists of a thin film of semiconductor material (e.g., lead sulphide, cadmium sulphide, cadmium selenide, mercury cadmium telluride, or indium antimonide) deposited on a glass or quartz plate which is housed in an evacuated sealed container. The valence shell electrons of the semiconductor material are promoted to higher energy level on absorption of radiation thereby increasing the electrical conductivity which is measured by an appropriate circuit. Lead sulphide and indium antimonide are sensitive to near-IR radiation in the range 12,500–3300 cm^{-1}. Mercury cadmium telluride is useful in the mid-IR and far-IR regions but requires cooling with liquid nitrogen to minimize thermal noise.

Phototube consists of a semi-cylindrical photocathode (cathode coated with a photoemissive material) and a wire anode sealed in evacuated transparent glass tube (Fig. 6.23). When the incident beam of radiation strikes the cathode the photoemissive material (e.g., made of K/Cs/Sb or red-sensitive Na/K/Cs/Sb or Ag/O/Cs) emits electrons, the number of emitted electrons being directly proportional to the intensity of the incident beam of radiation. The emitted electrons flow towards the wire anode maintained at a potential of about 90 V, generating a photocurrent. The magnitude of the photocurrent is relatively small

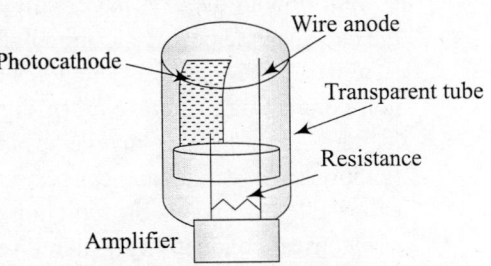

Fig. 6.23 Schematic diagram of a phototube

(~10%) compared to that generated in the photovoltaic cell but has the advantage of amplification by an electronic circuit.

Photomultiplier tube (PMT) is similar to the phototube but has the advantage of very high sensitivity due to built-in amplification of current, and hence, useful for measuring even radiation of low intensity. PMT also has a photoemissive cathode similar to that in phototube which emits electrons when irradiated with photons and a series of positively charged electrodes called *dynodes*, each charged at a successively higher potential. The photoelectron generated by the photoemissive cathode is attracted to dynode 1 maintained at a positive potential and on striking the dynode causes the emission of several secondary electrons. The secondary electrons emitted by dynode 1 are attracted to the dynode 2, maintained at a higher positive potential than dynode 1. The secondary electrons on striking dynode 2 cause the emission of more number of secondary electrons. Thus, each successive dynode maintained at higher positive potential than the previous one emits several secondary electrons and the process is repeated several times resulting in a cascade of secondary electrons for each incident photon. Thus, there is an inherent amplification (up to 10^6–10^7 times) in current output of PMT. The magnitude of the current is of several milliamps which is further amplified electronically and measured. Figure 6.24 shows the cross-section of the PMT and the arrangement of dynodes.

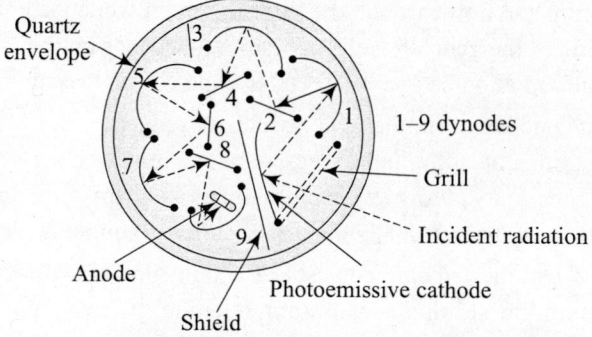

Fig. 6.24 Cross-section of PMT

Silicon photodiode transducer consists of a *p–n* junction made of a strip of *p*-type silicon in contact with an *n*-type silicon chip as shown in Fig. 6.25(a). The diode is said to be *forward biased* when the *p* region is connected to the positive terminal and the *n* region to the negative terminal of a DC source. The holes and electrons move towards the *p–n* junction where they annihilate each other. Electrons extracted by the positive terminal from the *p* region as more holes are formed migrate through the external circuit and reach the *n* region. Thus, there is a net current flow in the external circuit as indicated in Fig. 6.25(b). The diode is said to be *reverse biased* when the positive terminal of the DC source is connected to the *n* region and the negative terminal to the *p* region. The holes and electrons move away from the *p–n* junction forming a non-conducting *depletion layer* as shown in Fig. 6.23(c). The net flow of current is less by a factor of about 10^6–10^8 in reverse bias compared to that in forward bias. The diode in reverse bias functions as a rectifier and can serve as a photodetector because when the semiconductor diode is exposed to UV and visible radiation generates electron-hole pairs in the depletion layer resulting in electrical conductivity which directly proportional to the intensity of the incident radiation. The diode detector has a greater sensitivity than a single phototube but less than a PMT.

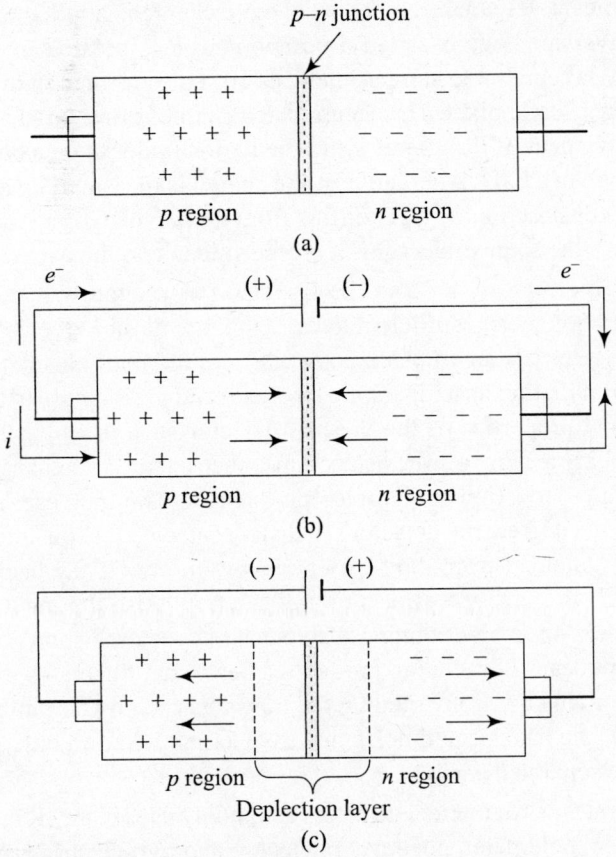

Fig. 6.25   (a) Representation of a silicon diode, (b) flow of electrons and current in forward bias, and (c) formation of depletion layer in reversed bias

Photodiode array detector consists of an array or a large number (1000 or more) silicon diodes (individual diode width is ~0.02 mm) formed on a single silicon chip and connected through an integrated circuit. The diodes monitor simultaneously all the wavelengths of radiation emerging from the monochromator and the signals are processed by a microprocessor. Thus, the diode array detector is a multichannel detector and the complete spectrum can be scanned almost instantaneously. However, the diode array detector is less sensitive compared to PMT and is of limited dynamic range and less *S/N* ratio.

Charge transfer device (CTD) is a solid state device with performance characteristics matching that of the PMT with the additional advantage of multichannel scanning, and hence, finds increasing use in spectroscopic instruments. Two types of charge transfer devices are available commercially, charge injection device (CID) and charge coupled device (CCD) based on the method of measurement of the output signal. A charge transfer device consists of a silicon chip of dimensions in the range of a few mm on which a large number of individual detectors called *pixels* are arranged in two-dimensional array of rows and columns. A pixel consists of two conductive electrodes overlying an insulating layer of silica which also separates the electrodes from a region of *n*-doped silicon. The assembly constitutes a metal oxide semiconductor capacitor capable of storing charges (holes) generated by radiation impinging on the doped silicon. The capacitor is negatively biased when a negative charge is applied to the electrodes of the pixel. A charge inversion region called a *potential well* is created under the electrodes which is energetically favourable to store as many as 10^6 holes formed in the n-type layer by absorption of photons in a single pixel. The accumulated charge is measured as the voltage change arising from the movement of the charge from the region under one electrode to the region under the other electrode in a CID. Alternatively, the charge is measured by a charge sensing amplifier in a CCD. The construction of CCD differs from that of the CID in that the pixel consists of three electrodes and the semiconductor is a p-type silicon and the capacitor is biased positively.

Thermal detectors are used in the IR region as photon transducers cannot be used as the energy of IR photons is insufficient to cause photoemission of electrons. The active element of a thermal detector is a small black body which is thermally insulated. When the black body is exposed to IR radiation its temperature increases and the increase in temperature causes a change in the physical properties of the thermal detector such as voltage induced in thermocouples and thermopiles, a change in resistance in a thermistor, an expansion of fluid in a Golay cell or an electric polarization in pyroelectric detector. The increase in temperature is just a few thousandths of a degree; the detector is usually protected by housing it in an evacuated envelope and carefully shielded from the surroundings to minimize the background radiation or noise. In addition to minimize the effects of any external noise the incident beam of radiation from the source is chopped by a rotating disk place between the source and the detector. Chopping of the incident beam of radiation produces a beam that fluctuates regularly from zero intensity to a maximum which is converted by the transducer to an alternating current signal which can be amplified. The background noise due to external radiation only generates a DC signal, and hence, can be separated easily.

The different types of thermal detectors commonly used in the IR region include thermocouples and thermopile, bolometer, pneumatic detector, and pyroelectric transducer.

Thermocouple is the simplest device fabricated by joining wires of two dissimilar metals or alloys (e.g., bismuth and antimony or copper and constantan) at the two ends (junctions). One of the junctions (called transducer junction) is blackened and exposed to IR radiation (hence, its temperature is higher) while the other junction (reference junction) is housed in the same chamber but shielded from IR radiation. A potential develops between the two junctions which also varies with the difference in the temperature of the two junctions. The electrical signal is chopped so that it can be amplified. *Thermopile* consists of several thermocouples connected in series so that their output signals add, and hence, a higher sensitivity is achieved. These transducers can detect temperature differences even as small as 10^{-6} K giving rise to potential difference of $6-8$ $\mu V/\mu W$. The response time is less than 100 ms.

Bolometer and *thermistor* are resistance thermometers. Bolometer is constructed from strips of platinum or nickel while thermistor consists of *therm*ally sensitive res*istors* made of semiconductor oxides of manganese, cobalt, or nickel. These oxides have a high temperature coefficient of resistance (~4%/ °C), that is, exhibit a relatively large change in electrical resistance as a function of temperature. The active element of the bolometer or the thermistor is small in size and blackened to absorb IR radiation while the reference element is optically shielded from IR radiation. The difference in the temperature between the active and reference elements gives rise to a difference in the electrical resistance which is measured by a bridge circuit.

Golay pneumatic cell is based on measuring the expansion of a gas when exposed to IR radiation. The cell consists of a small metal container with the wall facing the IR radiation source coated black while the opposite wall is a flexible silver diaphragm mirror. The container is filled with xenon. IR radiation falling on the blackened wall is absorbed and the heat is conducted to the gas which expands and deforms the flexible diaphragm mirror. Light from a lamp focusing on the diaphragm mirror is reflected onto a phototube detector. The distortion of the mirror leads to variation in the amount of light reaching the detector compared to when the mirror is at rest (in the absence of IR radiation) giving rise to an electrical signal which can be amplified. The Golay cell has a response time of about 20 ms and has sensitivity similar to that of a thermocouple. It is a superior detector for far-IR radiation.

Pyroelectric detector contains a single crystal wafer of a pyroelectric material like $LiTaO_3$, $LiNbO_3$, triglycine sulphate (TGS-$(NH_2CH_2COOH)_3 \cdot H_2SO_4$) either deuterated (DTGS) or with a fraction of glycine replaced with alanine, etc. Pyroelectric substances exhibit a strong temperature-dependent electric polarization in the presence of applied electric field which is retained even after the removal of the applied field. When the material is sandwiched between two electrodes, one of which is transparent to IR radiation it forms a temperature-dependent capacitor. On exposure to IR radiation the change in temperature alters the charge distribution within the crystal and generates a current flow in the external circuit connecting the two electrodes of the capacitor. The magnitude of the current is proportional to the surface area of the crystal and the rate of change of polarization with temperature. Pyroelectric materials loose their residual polarization at temperatures higher than Curie point. TGS with a Curie temperature of 47°C is the most commonly pyroelectric material used in IR spectrophotometers.

6.9.5 Signal Processors and Display Units

Signal processors are electronic devices which generally amplify the electric signal generated by the transducer. They may also alter the signal from DC to AC or reverse, filter unwanted components and perform mathematical operations on the signal such as integration, differentiation or conversion to a logarithm. They are coupled to display or readout devices, for example, digital meters, oscilloscopes, recorders, etc.

6.10 CONFIGURATIONS OF SPECTROMETERS

Two basic configurations of spectrometers are commercially available (i) single beam instruments and (ii) double beam instruments. In *single beam instruments* a single beam of radiation from the source passes through the dispersion device, sample (or in the reverse order first through the sample and then the analyser) and finally the transmitted radiation reaches the detector. The energy of the transmitted radiation is converted into electrical energy in the detector which is further amplified and passed on to the display unit.

Multichannel instruments make use of photodiode arrays and charge transfer devices in a single beam configuration and facilitate the scanning of the entire spectrum in the UV or visible region within a short time of about 1 s.

In *double beam instruments* the radiation from the source is split into two parallel beams which pass through the sample and reference or blank before finally reaching the detector. Thus, the radiation absorbed or emitted by the sample is automatically compared with the radiation absorbed or emitted by the reference thereby eliminating matrix effects and problems associated with instrumental noise and drift. These two configurations are further described in Chapters 7 and 8.

Multiplex instruments do not employ filters or monochromators for wavelength selection but obtain spectral information of all wavelengths simultaneously and analyse the information to determine the magnitude of the individual components through Fourier transformation.

6.11 FOURIER TRANSFORM SPECTROMETERS

Fourier transform (FT) spectroscopy was developed in the 1950s to record the weak infrared signals emanating from distant stars free from the environmental noise. The technique was incorporated into IR spectrophotometers much later to enhance the signal-to-noise ratio in conventional spectrophotometers (called dispersive instruments) using dispersing elements. FT spectrometry has several advantages over dispersive instruments which include (i) *Jaquinot advantage* which refers to the greater power of the radiation that reaches the detector as FT instruments require very few optical components and do not require slits which attenuate radiation, (ii) high resolving power and better wavelength reproducibility, (iii) time required for scanning the spectrum over a range of wavelengths is extremely short because all the wavelengths emitted by the radiation source reach the detector simultaneously in the absence of any monochromator device, and (iv) *Fellgett* or *multiplex advantage* which refers to the enormous increase in the signal-to-noise ratio due to multiple scanning and signal averaging. For example, the time required for recording an IR spectrum by a dispersive IR spectrophotometer over wavenumber range of 500–4000 cm^{-1} with a resolution of 5 cm^{-1} would be 350 s (~ 6 min) assuming that 0.5 s would be required for recording the transmittance for each frequency interval of 5 cm^{-1}. For obtaining a better spectrum the spectral resolution (slit width) should be less than 5 cm^{-1}. This leads to two disadvantages, namely, increase in the time required for recording the spectrum and more importantly the signal becomes weaker as the radiant power reaching the detector is cut down drastically due to narrower slit width. If S/N ratio is to be increased the number of scans should be increased so that signal averaging can be performed. In general the S/N ratio for the average of n measurements or scans increases with $\sqrt{n}$ (as given by the formula $S_x/N_x = \sqrt{n}$, where the subscripts indicate the averaged values). Thus, the number of scans required for doubling S/N ratio would be $n = 4$. The time required for scanning the wavenumbers four times would be 24 min since each scan requires about 6 min.

In FT spectrometer the time required for recording the spectrum over the same range of wavenumbers would be the same as required for a single frequency interval, that is, 0.5 s as all the wavenumbers will reach the detector simultaneously. Thus, in 24 min the FT instrument can scan 2880 times and S/N ratio would increase to 54.

The dispersive instrument records the spectral data in the form of *frequency domain spectra*, that is, variation of the radiant power as a function of the frequency of incident radiation or the inversely related wavenumbers. Hence, the instrument requires a monochromator system to isolate the frequencies and the detector generates a signal that corresponds to the radiant power of the isolated frequencies but not its periodic variation with respect to time.

The FTIR spectrometer is similar to the dispersive IR spectrometer but does not contain the dispersing device, namely, a grating or prism monochromator. Fourier transform basically involves generating a *time–domain spectrum,* which is variation of the radiant power as a function of time through a process called *modulation* or converting a high-frequency signal to a measurable frequency. It is essential to maintain a direct proportionality between the modulated and the original signal without distorting the time relationships in the latter. The process of modulation in the optical region is achieved by a device called *interferometer* developed in the late nineteenth century by Michelson for precise measurement of wavelengths of electromagnetic radiation and accurate measurement of distance.

The interferometer used in optical Fourier transform spectroscopy [Fig. 6.26(a)] consists of a radiation source from which the beam of radiation is collimated onto a beam splitter. The incident radiation is split approximately into two parallel beams A and B which are reflected by two separate mirrors, one of which is fixed while the other is movable.

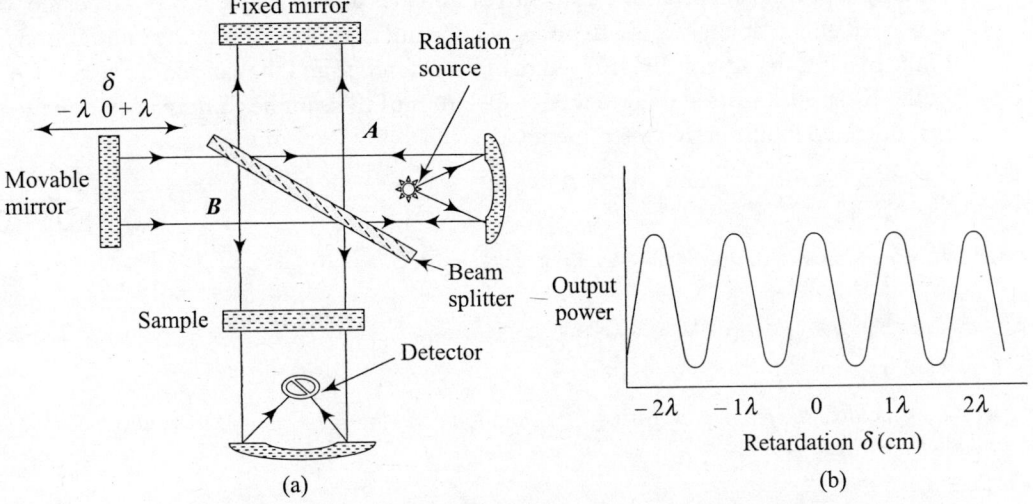

Fig. 6.26   Schematic diagram of (a) an interferometer for FT spectroscopy and (b) variation of intensity of radiation as a function of retardation

The beam A is reflected by the stationary mirror back to the beam splitter. Half of the beam A is transmitted to the sample and detector while the other half is transmitted towards the source. The beam B emerging from the beam splitter is similarly reflected by a movable mirror back to the beam splitter from which half of the beam is transmitted to the sample and detector while the other half is directed towards the source. Only the two halves of the beams A and B passing through the sample and detector are used for analytical purposes. When the two mirrors are at equal distance from the beam splitter the beams A and B are totally in phase with each other (constructive interference) and the power is at maximum. The power of radiation reaching the detector decreases when the movable mirror is moved horizontally and becomes zero when the movable mirror is at a distance which is one-quarter of a given wavelength of the incident radiation due to destructive interference. The interference pattern is due to the difference in the path lengths of the two beams of radiation termed as *retardation* δ. A plot of the output power of the detector as a function of the retardation is called an *interferogram* which appears as a cosine wave because the power is maximum when the path lengths of the two beams A and B are identical ($\delta = 0$) as shown in Fig. 6.26(b) for a single wavelength. Since the mirror is moved at a constant velocity the variation of the radiant power can be expressed as a function of time thereby generating a time–domain spectrum by the process of modulation.

Since the radiation source in the FTIR spectrophotometer emits polychromatic radiation, all the frequencies of radiation (usually in the range of 10^{12}–10^{15} Hz) reach the detector simultaneously. The variation of the output power of the detector as a function of retardation δ is accurately monitored for all the frequencies of the incident radiation simultaneously by a computer. The output power is large only at $\delta = 0$ since all the wavelengths are in phase, but when $\delta \neq 0$ total cancellation of the output power occurs as all the wavelengths are out of phase. The intense signal at $\delta = 0$ is known as the centre burst.

For recording the FTIR spectrum of a sample the first step involves recording the reference interferogram by scanning the power output over the entire frequency range of interest without the analyte sample. When the incident radiation is absorbed by the sample at a particular frequency the output power (intensity of the signal) diminishes at that frequency as the cosine wave is not cancelled out completely and can be seen in the interferogram. From the Fourier transformation of the two interferograms the IR spectra of the reference and the sample are computed. The ratio of the IR spectra is used to give the IR spectrum of the sample in frequency domain similar to that obtained in dispersive instruments.

SOLVED PROBLEMS

EXAMPLE 1   *Calculate the energy of the photon of wavelength 400 nm.*

SOLUTION

$$E = hc\ \bar{v} = 6.62 \times 10^{-34}\ \text{J-s} \times 3 \times 10^{10}\ \text{cm s}^{-1} \times 25000\ \text{cm}^{-1}$$
$$= 4.97 \times 10^{-19}\ \text{J}$$

EXAMPLE 2   *Calculate the molar absorptivity of a solution containing 5×10^{-5} M of a substance if %T in a 2-cm cell at 400 nm is 56%.*

SOLUTION

$$A = -\log T = -\log (56/100) = 0.2518$$
$$\varepsilon = A/ct = 0.2518/(5 \times 10^{-5}\ \text{mol L}^{-1} \times 2\ \text{cm}) = 2.518 \times 10^3\ \text{L mol}^{-1}\text{cm}^{-1}$$

EXAMPLE 3   *An analyte solution has molar absorptivity of 8.2×10^4 L mol^{-1}cm^{-1} at 470 nm. Calculate the absorbance and % T in (i) 1.0 and (ii) 2.0 cm cells if the concentration of the solution is 4.2×10^{-6} M.*

SOLUTION

(i) $A = \varepsilon ct = 8.2 \times 10^4\ \text{L mol}^{-1}\text{cm}^{-1} \times 4.2 \times 10^{-6}\ \text{mol L}^{-1} \times 1\ \text{cm}$

    $= 0.3444$

    $\text{Log } 1/T = A$

    $\%\ T = 45.25\%$

(ii) $A = 8.2 \times 10^4\ \text{L mol}^{-1}\text{cm}^{-1} \times 4.2 \times 10^{-6}\ \text{mol L}^{-1} \times 2\ \text{cm}$

    $= 0.689$

    $\%T = 20.47$

Absorbance scale is linear while %T scale is not linear.

EXAMPLE 4   *Calculate the molar absorptivity of a compound whose formula weight is 225 if its absorptivity is 427 L cm^{-1} g^{-1}.*

SOLUTION

Molar absorptivity = absorptivity × formula weight (or molecular weight)

$$= 9.61 \times 10^4$$

EXAMPLE 5   *A medicine tablet containing an active ingredient of formula weight 326 was dissolved in 500 mL and its absorbance was found to be 0.562 in 1.0 cm cell at its λ_{max}. Calculate the weight of the active ingredient in one tablet given its molar absorptivity is 1254 L mol^{-1} cm^{-1}.*

SOLUTION

Since $\quad A = \varepsilon c t$

$$c = 0.562 \times 0.5 \text{ L} / 1254 \text{ L mol}^{-1} \text{ cm}^{-1} \times 1 \text{ cm}$$
$$= 2.241 \times 10^{-4} \text{ mol}$$

Weight of the active ingredient in one tablet

$$= 2.241 \times 10^{-4} \text{ mol} \times 326 \text{ g mol}^{-1} = 0.073 \text{ g}$$

SUMMARY

- Electromagnetic radiation is a radiant form of energy and the unit of radiation is photon. The energy of the photon is directly proportional to the frequency and inversely to the wavelength of radiation. Electromagnetic spectrum covers radiation varying over a wide range of energies and for convenience subdivided into regions of radiofrequency, microwave, infrared, visible, ultraviolet, X-rays, and γ-rays with increasing order of frequencies.

- Spectroscopy involves the study of interaction of electromagnetic radiation with atoms and molecules giving rise to atomic spectroscopy and molecular spectroscopy, respectively.

- Atoms have a relatively simple energy level structure consisting of only electronic energy levels. Molecules, on the other hand, have a more complex structure of energy levels consisting of electronic energy levels, subdivided into vibrational energy levels and these in turn subdivided into rotational energy levels.

- Atoms and molecules with their quantized energy levels can absorb a part of incident radiation resulting in the formation of excited state atoms and molecules. The excited state atoms and molecules relax to their ground states by emitting the radiation. The absorption and emission of radiation must satisfy the Bohr condition $\Delta E = h \nu$.

- Based on the energy of interaction or the frequency of electromagnetic radiation involved in the interaction, spectroscopic techniques are classified into γ-ray spectroscopy, X-ray spectroscopy, ultraviolet and visible spectroscopy, infrared spectroscopy, microwave spectroscopy, magnetic resonance spectroscopy, atomic emission spectroscopy, molecular fluorescence spectroscopy, and molecular phosphorimetry.

- Absorption spectroscopy involves passing electromagnetic radiation of known intensity through a transparent medium so that a part of the incident radiation is absorbed by atoms and molecules and monitoring the resulting decrease in the intensity of the transmitted radiation.

- Emission spectroscopy involves producing excited state atoms and molecules by any of the different methods of excitation and analysing the frequency and intensity of emitted radiation.

- In accordance with the Bohr condition the width of the spectral line should be a sharp line. However, in practice the observed spectral line is actually a narrow peak consisting of several frequencies due to factors such as the natural line width for a given spectral transition and the resolving power of the instrument.

- The intensity of spectral lines is explained by quantum mechanical selection rules.

- Beer–Lambert law provides a quantitative relationship between the ratio of the intensities of incident and transmitted radiation to the nature of chemical species, its concentration and the path length through which radiation passes. Beer–Lambert law suffers from limitations such as real, chemical, and instrumental limitations.

- Visual colorimetry makes use of simple glassware in the form of Nessler's tubes or simple equipment such as Dubosque colorimeter for quantitative analysis of coloured species in accordance with Beer–Lambert law.

- The important components of dispersive spectro-meters used in UV, visible, and IR regions are (i) radiation source, (ii) dispersion devices, (iii) sample holder, (iv) radiation detectors, (v) signal or data processor, and (vi) display unit or recorder.

- Radiation sources commonly used include the tungsten filament lamp for the visible and the near-IR regions, deuterium lamp for UV region, and nichrome wire for the IR region.

- Dispersion devices include monochromators consisting of quartz prisms or replica gratings for UV and visible regions and gratings for IR region.
- Quartz sample holders are used in UV and visible region for solutions and liquid samples, whereas sample cells made of alkali metal halides (mostly potassium bromide) are used in IR spectrophotometers.
- Photomultiplier tube or silicon diode detector is used in UV and visible regions. Thermal detectors are used in IR spectrophotometers.
- Commercially available configurations of spectrophotometers include single- and double-beam spectrophotometers, multichannel spectrophotometers using diode array detectors, and multiplex instruments based on Fourier transformation.
- The FTIR spectrometer generates a time–domain spectrum in the form of the variation of the radiant power as a function of time through a process called *modulation* achieved by an interferometer. The interferogram obtained is transformed into the conventional frequency domain spectrum by a computer-aided process called *Fourier transformation*. FT spectrophotometers have the advantages of high-speed scanning and high signal-to-noise ratio thereby enhancing the sensitivity of the instrument.

REVIEW QUESTIONS

1. Identify the different regions of electromagnetic spectrum and the ranges of their wavelengths.
2. Write a note on the energy levels of an atom.
3. Give a detailed account of the type of interactions between electromagnetic radiation and atoms.
4. Explain the terms 'absorption', 'emission', and 'fluorescence' with reference to the interaction of radiation with atoms.
5. Draw a neat sketch of the energy levels in molecules and explain the different types of interaction of radiation with molecules.
6. Classify the different types of spectroscopic techniques.
7. How are absorption and emission spectra obtained?
8. Give an account of the factors affecting width of spectral lines.
9. What are the factors which influence the intensities of spectral lines?
10. State Beer–Lambert law. What is its importance?
11. Write a note on the different applications of Beer–Lambert law.
12. What are the different types of limitations of Beer–Lambert law?
13. What is visual colorimetry?
14. Discuss the principle and practice of visual colorimetry using Nessler's tubes with a suitable example.
15. Explain the principle of Dubosque colorimeter.

16. Give a block diagram of the absorption and fluorescence spectrometers and label the components.
17. List the radiation sources used in spectrometers for UV, visible, and IR regions.
18. Write a note on absorption and interference filters.
19. What are monochromators? How do they function?
20. Give a neat sketch of prism monochromator system and explain the functioning.
21. Explain the functioning of a grating monochromator system.
22. What are the performance characteristics of monochromators?
23. Write a note on the bandwidth of monochromators and the effect of slit width on the effective bandwidth of monochromators.
24. What are radiation transducers?
25. Describe the operating principle and working of (i) barrier cell, (ii) photo tube, (iii) photo multiplier tube, (iv) semiconductor detector, and (v) photo diode array detector.
26. Give a detailed account of the different types radiation detectors used in IR spectrometer.
27. Discuss in detail the principle and process involved in Fourier transformation of detector signal in spectroscopy.
28. List the advantages of FT spectroscopy.

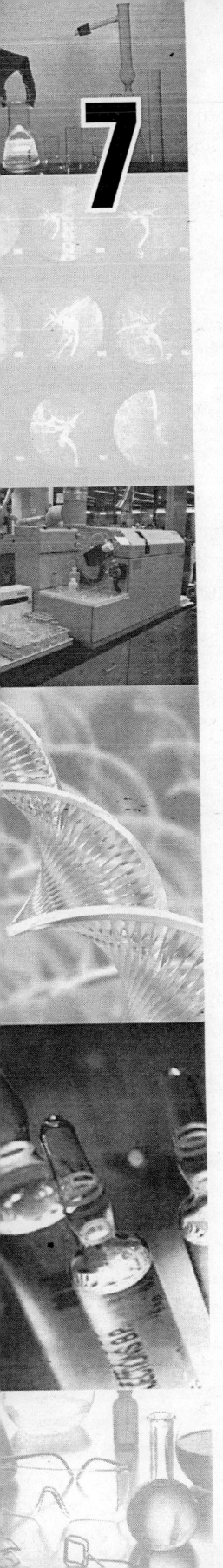

7 Atomic Spectroscopy

7.1 INTRODUCTION

Atomic spectroscopy, as mentioned in Chapter 6, involves the study of electronic transitions that take place by absorption as well as emission processes in atoms in the ultraviolet and visible regions of electromagnetic spectrum. Electronic transitions occur between quantized energy levels of atoms in accordance with the selection rules as stated in Chapter 6. Atomic spectral lines appear as sharp narrow lines. The wavelengths of absorption or emission are characteristic of individual elements and provide useful information for qualitative identification of the elements. Electronic transitions are associated with large changes in the dipole moments, and hence, the sensitivity of atomic spectrometric techniques is quite high, enabling the analysis of trace quantities of elements. The intensity of a given spectral line is proportional to the number of atoms absorbing or emitting radiation, and hence, useful in quantitative analysis.

7.2 CLASSIFICATION OF ATOMIC SPECTROMETRIC METHODS

The atomic spectroscopic techniques include the (i) absorption technique of *atomic absorption spectrometry*, (ii) emission techniques of *flame emission spectrometry* and *atomic emission spectrometry*, and (iii) *atomic fluorescence spectroscopy*. The methods are further classified depending on the type of atomization/excitation methods. However, all the methods require atomization as a first step.

7.3 ATOMIZATION

Atomic spectra are recorded with gaseous atoms or elementary ions which are obtained by a process known as *atomization*. The sample in the form of a solution or solid is converted into an atomic vapour by atomization. The gaseous atoms and ions then absorb or emit radiation giving rise to a spectral signal. The precision and accuracy of all atomic spectrometric methods depend critically on the atomization step. Atomization may be carried out by continuous or discrete atomizers.

Continuous atomizers make use of a flame, plasma, or electric spark for atomization process. The gaseous atoms and ions then absorb or emit radiation giving rise to a spectral signal which remains constant with time. Flame atomization is used for liquid samples in atomic absorption spectrometry and flame emission spectrometry. Plasma and electric spark atomization methods are used in atomic emission spectroscopy.

Discrete atomizers include electrothermal atomizer and electric arc atomization. The latter may be carried out as continuous or discrete atomization mode depending on the nature of the sample. In discrete atomization, a measured quantity of liquid or solid sample is introduced into the atomizer and converted into a vapour of atoms or ions by rapid electrical heating. The gaseous atoms and ions then absorb or emit radiation giving rise to a spectral signal which appears as a sharp peak. Electrothermal atomization is used for liquid and solid samples in atomic absorption spectrometry, whereas electric arc atomization method is used in atomic emission spectrometry.

7.4 ATOMIZATION METHODS

Atomization methods adopted in atomic absorption spectrometry as well as in atomic emission spectrometry include a variety of methods such as flame atomization, electrothermal atomization, glow discharge atomization, cold-vapour atomization, and hydride atomization.

7.4.1 Flame Atomization

Flame atomization is used in most atomic absorption, emission, and fluorescence spectrometric methods as it is a better method for introducing liquid samples. The primary step in flame atomization is known as *nebulization*. The process of nebulization converts the liquid sample to a mist of finely divided droplets or aerosol by a jet of compressed gas. A pneumatic *nebulizer* is most commonly used in flame atomization (Fig. 7.1).

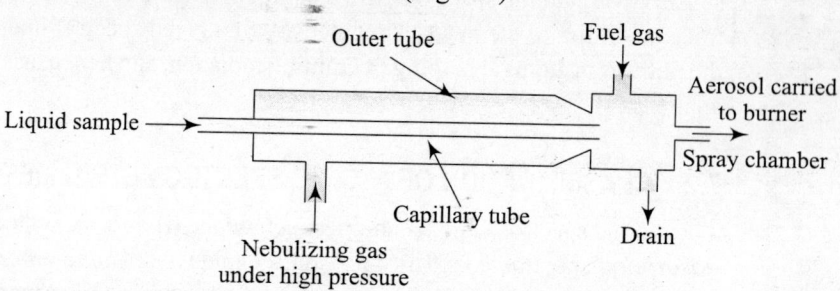

Fig. 7.1   Schematic diagram of a pneumatic nebulizer

It consists of concentric tubes, the liquid sample being sucked through the inner capillary tube while the nebulizing gas (mostly compressed air or an oxidant such as oxygen or nitrous oxide) flows through the outer tube and around the tip of the capillary, facilitating the *aspiration* of the liquid sample through the capillary. The high velocity flow of the nebulizing gas breaks down the liquid exiting the capillary into tiny droplets suspended in the gas (aerosol) which is then mixed with a fuel gas in the spray chamber and carried into the flame.

Atomization process is a multi-step process that takes place in the flame. The sequence of the steps includes (i) desolvation (removal of solvent) to produce a finely divided solid–gas aerosol, (ii) volatilization of the solid–gas aerosol to produce gaseous molecules, and (iii) dissociation of the molecules into atoms. Depending on the flame temperature, the gaseous atoms may undergo excitation to form excited atoms or ionization to form ions. Thus, atomization process yields a mixture of gaseous atoms and ions in their ground as well as excited states. The gaseous atoms and ions in the ground state can absorb radiation in the ultraviolet or visible region giving rise absorption spectra. On the other hand, excited state atoms and ions relax to the ground state giving rise to emission spectra.

Atomization process is highly dependent on the characteristics of the flame and as already mentioned the atomization process is a critical step in atomic spectrometric methods.

Flames can be produced by a combination of fuel and oxidant gases. The range of temperatures attained in a flame depends on the nature of the combination of gases as listed in Table 7.1. Most commonly used fuel–oxidant combinations include (i) liquefied petroleum gas/natural gas–air, (ii) hydrogen–air, (iii) and acetylene–air. With different fuel gases and air as oxidant, a temperature range of 1700–2400°C is sufficient for atomization of most samples. However, refractory samples require higher temperatures, and hence, oxygen or nitrous oxide is used as oxidant. In flame atomization, it is important that the flow velocity and the *burning velocity* of fuel–oxidant combination should match otherwise the flame propagates back into the burner at low flow rates or blows off the burner at high flow rates.

Table 7.1 Flames of different fuel gas–oxidant mixtures and their characteristics

Fuel	Oxidant	Temperature range (°C)	Burning velocity (cm s^{-1})
Natural gas/LPG	Air	1700–1900	40–43
Natural gas/LPG	Oxygen	2700–2800	370–390
Hydrogen	Air	2000–2100	300–440
Hydrogen	Oxygen	2500–2700	900–1400
Acetylene	Air	2100–2400	160–270
Acetylene	Oxygen	3000–3200	1100–2500
Acetylene	Nitrous oxide	2600–2800	280–300

The flame structure consists of a *primary combustion zone*, the *interconal* or *interzonal region*, and the *outerconal* or *secondary combustion zone* as shown in Fig. 7.2.

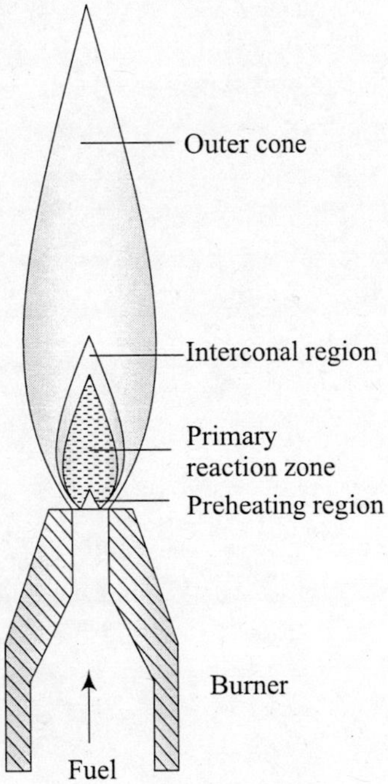

The primary combustion (reaction) zone appears as a blue luminescence due to the band spectra of C_2, CH, H_3O^+, HCO^+, and other radicals. The temperature in this region is relatively low compared to the maximum attainable temperature for a given combination of fuel–oxidant mixture. The zone does not attain thermal equilibrium, and hence, is not used in flame spectroscopy. The interconal region is somewhat narrow in hydrocarbon flames but reaches several centimetres height in fuel rich acetylene–oxygen or nitrous oxide flames. The maximum possible temperature for the given combination of fuel–oxidant mixture is attained and the region considerably rich in free atoms. The interconal region is the most widely used part of the flame for spectroscopic determinations. In the outer cone, the products of inner cone are converted to stable molecular oxides and dispersed to the surroundings.

Fig. 7.2 Flame structure

It is necessary to use different regions of the flame for different elements depending on their absorbance and emission profiles as a function of height of the flame measured from the tip of the burner. For example, magnesium exhibits a maximum absorbance at the middle of the flame, whereas silver shows maximum absorbance at the periphery of the flame in the outerconal region. Similarly, emission profiles also vary for different elements. Less sophisticated filter photometers facilitate absorbance or emission measurements over a large portion of the flame, and hence, control of flame position is less important. However, in sophisticated spectrometers the flame position must match with the entrance slit of the instrument.

Flame atomization is associated with certain disadvantages such as limited sensitivity as only about 10% of the sample reaches the flame as a fine aerosol with the rest of the sample being wasted in the nebulizer–spray chamber assembly. The aerosol of ground state atom has a very short residence time of about 10^{-3} s in the optical path of the spectrometer and for attaining a steady state atomization of at least 10 s the amount of sample required is very large (~ 10^4 times) compared to that is present in the optical path at any given moment. The attainable atom concentration in the flame is also limited due to the dilution effect of the relatively high flow rates of fuel–oxidant gas mixture and also due to expansion of the gas mixture due to combustion. The detectable limit of the analyte is affected because of the flickering nature of the flame and background spectrum (noise) of the flame.

The problems associated with the flame atomization have been mostly overcome with the proper design of the burner. Laminar flow burners provide a relatively quiet flame and a long path length containing the aerosol of atoms, thereby enhancing the sensitivity and reproducibility. A typical laminar flow burner with a concentric tube nebulizer, most commonly used for flame atomization in atomic absorption as well as in emission spectrometers, is shown in Fig. 7.3. The aerosol formed by the flowing oxidant mixes with the fuel and passes through a series of baffles to remove coarse droplets of the sample. Only the finest drops are allowed to pass into the burner and burned in a slotted burner which provides a flame of about 10 cm in length.

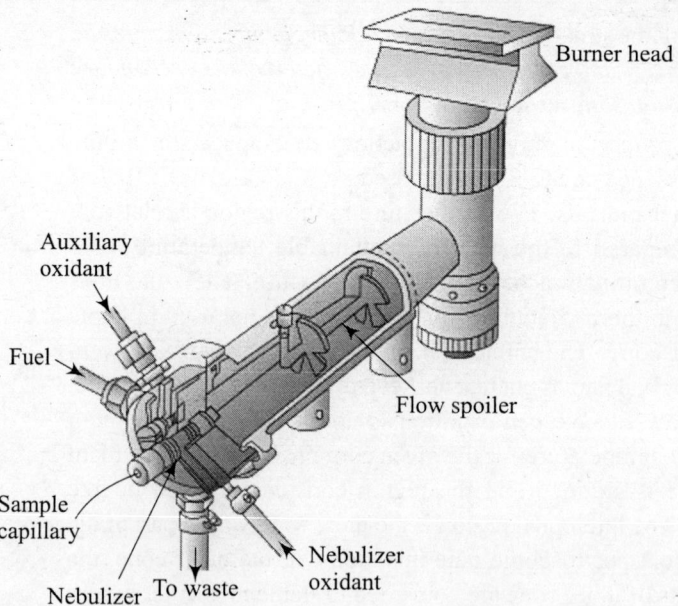

Fig. 7.3 A laminar flow slot burner of atomic absorption spectrometer (AAS) (Source: Principles of Instrumental Analysis – D.A. Skoog, F.J. Holler, and T.A. Nieman, Thomson Brooks/Cole 2005)

7.4.2 Electrothermal Atomization

Electrothermal atomization (also known as *flameless atomization*) is used for atomization in atomic absorption and atomic fluorescence measurements and also for vapourizing the sample before introduction into the plasma in inductively coupled plasma emission spectrometry. Electrothermal atomizers can handle both liquid and solid samples. Solids can be introduced as a finely powdered sample or as slurry of the powder prepared by ultrasonic agitation in aqueous medium. The sensitivity of electrothermal atomizers is quite high, particularly with volatile elements, such as zinc, cadmium, and magnesium, with detection limit being less than picogram level. The amount of sample required is also very small.

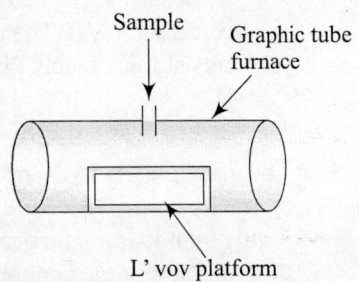

A typical electrothermal atomizer is a *graphite tube furnace* (Fig. 7.4). It consists of a cylindrical graphite tube of about 5–10 cm in length and about 0.3–0.5 cm in internal diameter. To prevent sample vapour diffusion into the porous graphite tube, it is coated with pyrolytic carbon to seal the pores. The tube is open at both the ends and has a central hole for sample introduction with the help of a micro pipette. The tube fits snugly into a pair of water-cooled electrical contacts at the two ends. The graphite tube is also provided with inlet and outlet for the flow of inert gas (argon). The axis of the tube furnace is aligned along with the optical path of the radiation source and the detector.

Fig. 7.4 Schematic diagram of a graphite tube furnace with L'vov platform

The graphite tube furnace is initially flushed with inert gas to prevent the formation of refractory oxides and oxidation of the graphite. The sample (a few μL) is introduced, evaporated, and ashed at a low temperature on the L'vov platform or shelf made of graphite and placed within the tube furnace. After ashing, a heavy current of several hundred amperes is supplied to raise the temperature of the furnace to about 2500 K within 1–2 min to atomize the sample.

7.4.3 Glow Discharge Atomization

Glow discharge atomization is a specialized technique applicable to electrically conducting samples. Alternatively, the solid-powdered sample may be mixed with finely ground graphite or copper and pressed into a pellet. Liquid samples may be deposited on graphite, aluminium, or copper cathode of the atomizer. The glow discharge atomizer consists of a cylindrical cell with a hole in the middle of the cell. The sample is pressed against the hole, sealing the cell. Argon gas is ionized within the cell by a current between an anode and the sample which acts as the cathode. The sample atoms are sputtered by the argon ions and the atomized vapour is drawn by vacuum to the axis of the cell for absorption measurement.

7.4.4 Cold-vapour Atomization

This method is especially suitable for atomization of mercury because of its high vapour pressure and volatility. Mercury in the sample is first converted to Hg^{2+} by treating the sample with an oxidizing mixture of nitric and sulphuric acids followed by reducing it to metallic mercury with $SnCl_2$. The mercury vapour is swept into a long quartz absorption tube by a stream of inert gas. The absorbance is measured at 253.7 nm in AAS.

7.4.5 Hydride Atomization

This method is suitable for elements, such as arsenic, antimony, tin, bismuth, selenium, tellurium, germanium, and lead, which form volatile hydrides. The hydride atomization enhances the detection limits of these elements by 10- to 100-fold. The sample containing any of these elements is treated by wet oxidation to destroy any organic component followed by reduction with an acidified aqueous 1% solution of sodium borohydride in a glass vessel. The volatile hydride formed is swept by argon gas into a hot silica atomization tube and decomposed to form atomic vapour of the element. Alternatively, the volatile hydride may be injected into a 'cool' hydrogen flame for decomposition and generate atomic vapour of the analyte element.

7.5 ATOMIC ABSORPTION SPECTROMETRY

Atomic absorption spectrometry makes use of a spectrophotometer called *atomic absorption spectrometer* (AAS). The sophisticated instrument is capable of analysing about 40–45 metallic elements at trace levels to micrograms.

7.5.1 Principle

The principle involves the conversion of analyte sample solution into a fine mist of ground state atoms by atomization and measuring the absorbance by exposing the ground state atoms to a highly monochromatic beam of light of specific wavelength. As the incident beam of light passes through the sample vapours, a part of it is absorbed and the intensity of the transmitted light (I_t) at a given frequency is less as compared to the intensity of the incident beam of light (I_0) and is related to the latter by the expression which is similar to Beer–Lambert law.

$$I_t = I_0 e^{-k'l} \tag{7.1}$$

where l is the thickness of the absorbing medium and k is determined by the concentration of atoms which can absorb the radiation at the given frequency. The magnitude of k' is given by the relationship

$$\int k'dv = (\pi e^2/mc)\, Nf \tag{7.2}$$

where e and m refer to the charge and mass, respectively, of the electron, N is the number of ground state atoms per cm^3 capable of absorbing radiation of frequency v and f is the *oscillator strength*, defined as the number of electrons per atom capable of being excited by the incident radiation. The *integrated absorption* is proportional to N, which approximates to the concentration of the analyte sample. Hence, the above equation can be simplified as

$$\log (I_0/I_t) = A = kC \tag{7.3}$$

where A is absorbance, C refers to the concentration of the absorbing species, and k is a constant. The relationship is useful in quantitative analysis.

The integrated absorption line width of the analyte sample vapour of ground state atoms at temperatures in the range of 2000–3000 K is about 10^{-2} nm which is quite narrow as compared to the absorption band of an analyte sample in solution. The absorption line width of an atomic absorption line depends on the following factors: (i) *natural width* which is about 10^{-5} nm, (ii) *Doppler broadening* due to thermal movement of atoms relative to the spectrometer, (iii) *collisional broadening* due to collisions between analyte atoms, (iv) pressure broadening due to collisions with atoms other than the analyte in the sample vapour, (v) *Stark broadening* resulting in disturbance in the atomic energy levels of the analyte caused by electric field, and (vi) *Zeeman broadening* due to magnetic fields which once again affect the atomic energy levels. In atomic absorption spectrometry using flame atomization, broadening is mostly due to Doppler effect and to a smaller extent due to collisional broadening.

Since the integrated absorption line is quite narrow it is necessary to provide a sharp line radiation source with a line width narrower than the absorption line width of the analyte sample.

7.5.2 Atomic Absorption Spectrometer

The block diagram and the main components of a *single beam AAS* are shown in Fig. 7.5. The main components include (i) radiation source, (ii) nebulizer, (iii) spray chamber, (iv) burner, (v) monochromator, (vi) photodetector, (vii) data processor, (viii) display unit, and (ix) deuterium lamp for background correction.

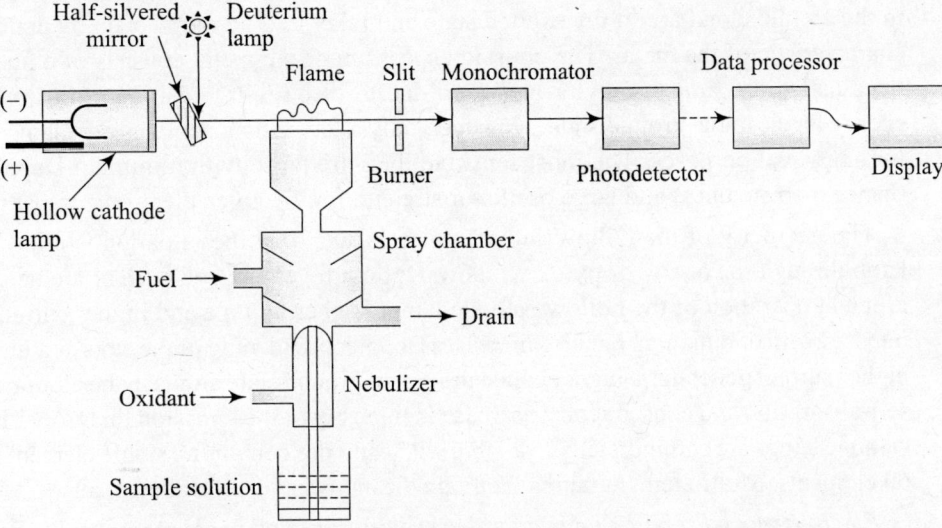

Fig. 7.5 Block diagram of a single beam AAS with deuterium background correction

The radiation source should be a sharp-line source capable of producing emission lines whose half-widths are much less as compared to that of absorption line of the analyte. Monochromators with high resolving power can at best provide only a bandwidth of about 1–10 nm which is much greater than the integrated absorption line width of an analyte vapour of the order of 10^{-2} nm. Hence, it is absolutely necessary to have sharp line sources in AAS to achieve greater sensitivity. Such sharp line sources include hollow cathode lamps and microwave-assisted electrode-less discharge tubes.

Hollow cathode lamp is the most commonly used source in AAS as it produces a highly monochromatized beam of radiation with characteristic wavelength of the analyte under investigation. The hollow cathode lamp is a cylindrical hard glass tube and contains a hollow cylindrical cathode made of the analyte metal or a coating of the metallic compound and an anode (Fig. 7.6).

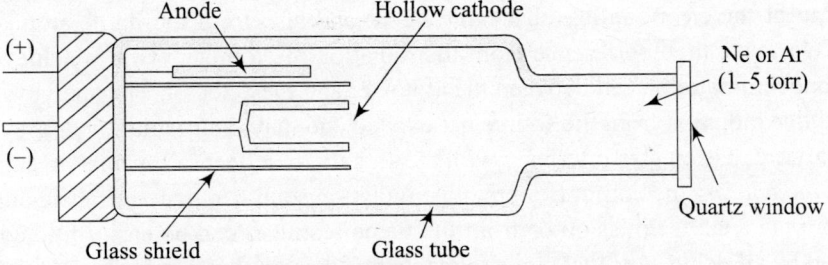

Fig. 7.6 Schematic diagram of a hollow cathode lamp

The glass tube is filled with neon or argon gas at low pressure (1–5 torr) and sealed. An applied voltage of approximately 300 V is applied across the electrodes to generate electrical current in the range of milliamps (5–50 mA). This current, in turn, generates a glow-discharge within the hollow cathode, which ionizes the inert gas. The gaseous cations have sufficient kinetic energy and bombard the cathode, vapourizing the metal atoms from the surface of the hollow cathode producing an atomic cloud. This process is called *sputtering*. Some of the metal atoms in the atomic cloud are in the excited state and relax to the ground state by emitting radiation characteristic of the metal. The emission spectrum of the metal consists of a number of sharp lines called *resonance lines* or *radiation* due to electronic transitions from different higher energy levels to the ground state. However, only a few resonance lines per element are suitable for analytical purposes. The most sensitive lines are those with minimum Doppler line width (hence, narrow lines) and large oscillator strength for the given electronic transition.

The geometry of the hollow cathode lamp is such that the radiation is allowed to exit the lamp through the quartz or pyrex window. Optimum voltage and current are necessary for the efficient operation of the hollow cathode lamp. Higher voltage and higher current generate an intense beam of radiation, but also increases Doppler broadening of the emission line. In addition, higher current generates a higher concentration of ground state atoms in the atomic cloud leading self-absorption and consequently decrease in intensity of the emission lines. A variety of hollow cathode lamps are commercially available with cathodes containing single element for more than 60 elements. Multi-element lamps facilitate the analysis of several elements.

Electrode-less discharge lamp may also be used as radiation source instead of hollow cathode lamps in AAS. The lamp is made of a sealed quartz tube containing an inert gas (neon or argon) at low pressure and a small quantity of metal or its salt. The quartz tube is exposed to an intense field of radiofrequency or microwave radiation to excite and accelerate the inert gas cations which in turn excite the metal atoms to emit characteristic radiation of prominent resonance lines. These lamps have the advantage in that the emitted radiation has higher intensity of about 100 times that derived from hollow cathode lamps. However, these lamps are not as reliable as hollow cathode lamps except in the case of arsenic, antimony, bismuth, selenium, and tellurium.

The other components, such as nebulizer, spray chamber, and burner, have been described earlier. Monochromator, photodetector, data processor, and display unit are similar to those used in UV–visible spectrophotometer (see Chapter 6).

To enhance the sensitivity and the signal-to-noise ratio, AAS are equipped with features such as source modulation and background correction.

Modulation of radiation source is necessary in atomic absorption spectrometry as the atomic vapour of the analyte is exposed to the radiation from the sharp-line source (hollow cathode lamp or the electrode-less discharge lamp) and also from the flame atomizer. It is necessary to eliminate the interference from the radiation of atomizer. This is achieved by the use of a monochromator placed between the atomizer and the detector. The analyte atoms on absorbing the line radiation from the source get excited and may emit radiation as they tend to relax. The analyte emission also interferes with the signal. The effect of the analyte emission is overcome by *modulating* the output from the line radiation source to generate an ac output. The dc signal from the analyte emission or from the flame atomizer can be easily filtered off electronically thereby allowing ac signal alone to reach the detector which is further amplified and displayed. Modulation is achieved by introducing a motor-driven partially mirrored chopper device between the hollow cathode lamp and the flame. The chopper rotates at a constant speed and allows the radiation to reach the analyte vapour half the time while reflecting it off during the other half. Thus, the intensity of the radiation from the lamp varies periodically from zero to maximum intensity and then to zero resulting in an ac current flow in the detector.

Background correction in single beam instruments aims at eliminating the spectral interferences due to broad band absorption of volatile molecular species by any of the following three techniques.

Continuous source correction method makes use of a deuterium continuum source as shown in Fig. 7.5. The half-silvered mirror facilitates the passage of light from both the hollow cathode lamp as well as the deuterium lamp to the atomic vapour at the same time. The light from the deuterium lamp is modulated 180° out of phase with the light from the sharp line source, and hence, the two signals can be distinguished electronically. The intensity of the light from the deuterium lamp will diminish to a large extent by the interfering broad band absorption but only to a negligible extent by the sharp line absorption. This variation of intensity in the light of the continuum source is used to compute a background correction for the analytical measurement.

Smith–Hieftje technique of background correction is based on self-source reversal or self-absorption of radiation emitted by the hollow cathode lamp operated at high currents. During high current operation of a hollow cathode lamp, a large concentration of the ground state atoms is generated which absorbs the radiation emitted by the excited species. The high current operation also broadens the emission band of the excited atoms. The emission band from the hollow cathode lamp has a minimum at its centre (due to self-absorption) which corresponds exactly to the absorption wavelength of the analyte atoms. The background correction is carried out by operating the hollow cathode lamp alternately at low and high currents. The total absorbance is obtained when the lamp is operated at low currents and the absorbance due to background is obtained at high current operation. The absorbance due to background is subtracted from the total absorbance by the data processor to provide the absorbance due to analyte only. The measurement cycle is repeated several times and thereby the signal-to-noise ratio is enhanced.

Background correction based on Zeeman Effect is a more sophisticated method. In this method, the atomic vapour is exposed to a strong magnetic field resulting in the splitting of energy levels of the atoms in the vapour. This leads to the formation of several absorption lines for each electronic transition, each line differing from one another by about 0.1 nm. The sum of the absorbance of these lines is exactly equal to that of the original line from which the split lines are formed. The splitting patterns vary depending on the type of electronic transition, the simplest splitting pattern being that of singlet transitions, which consists of a central or π line and two equally spaced satellite σ lines. The central line at the original wavelength is twice intense as compared to each of the σ line. The response of the two types of lines to polarized radiation is different, the π peak absorbing only when the plane polarized radiation is in a direction parallel to the external magnetic field while the σ peaks absorb only when the plane polarized radiation is perpendicular to the external magnetic field. AAS incorporating the Zeeman background correction is equipped with a rotating polarizer which separates the emission line from the hollow cathode lamp into two plane polarized component beams perpendicular to one another. The two components of the plane polarized light pass through the atomized sample vapour in a graphite tube furnace in alternate halves of a cycle. The magnetic field generated by the permanent magnet surrounding the graphite furnace splits the energy levels of the ground state atoms into three levels, consisting of a central π line and two equally spaced satellite σ

lines. The analyte absorbs the plane polarized light only when the beam of radiation parallel to the external magnetic field passes through the sample vapour but does not absorb the polarized beam component that is perpendicular to the magnetic field when it passes the sample vapour in second half of the cycle. Broad band absorption occurs due to molecular species, and scattering by matrix products occurs in both the half cycles. The data processor subtracts the absorbance during the perpendicular half cycle from the absorbance during the parallel half cycle thereby effecting background correction to the measured absorbance.

Double beam AAS employ a mirrored chopper to split the emitted beam of radiation from the hollow cathode lamp into two parallel beams. One beam (sample beam) passes through the sample vapour in the flame and while the other passes around the flame (reference beam). The two beams are recombined by a half-silvered mirror to a grating monochromator and a photomultiplier tube detector. The ratio between the sample and reference signals is amplified and fed to data processor.

7.5.3 Working of AAS

The analyte sample is brought into solution phase (preferably aqueous solution) by suitable treatment methods such as dissolving in hot mineral acids, oxidation with liquid reagents (e.g., nitric, sulphuric, or perchloric acids) or combustion in oxygen in a closed container. The solution is aspirated into the nebulizer or atomizer to produce a fine spray or aerosol of the solution. The aerosol of the analyte sample flows into the spray chamber where any large drops are condensed and drained off. Thus, only the finest aerosol reaches the flame. The flame is produced by burning the mixture of fuel gas (e.g., acetylene, hydrogen, or propane) and oxidant (air, nitrous oxide, or oxygen) in the burner. The flame temperature is in the range of 2000–3000°C which completely vapourizes the solvent, decomposes the metallic compounds, and produces a vapour consisting of neutral atoms as well as ions of the components of the sample.

A graphite furnace may be used instead of a flame to produce the vapour of ground state metal atoms. Flameless atomization has several advantages such as (i) elimination of interferences arising from the interaction between the sample components and the flame, (ii) greater sensitivity due to longer residence time of the atomic vapour in the light path and higher proportion of the analyte being converted into the vapour, (iii) capability to handle small quantities of sample, and (iv) ease of handling solid samples such as organic and inorganic substances, plant and animal tissues, and liquid samples such as organic solvents, petroleum products, and blood as they can be introduced into the furnace directly.

The emitted radiation from the hollow cathode lamp can be absorbed only by the vapour of the analyte metal atoms (in their ground state) even when the sample contains metal/non-metal atoms of elements other than the analyte metal. A fraction of the emitted radiation is absorbed by the analyte metal atom vapour and the decreased intensity of transmitted radiation is passed through the monochromator to filter-off any stray radiation. The intensity of the transmitted radiation is measured by a photodetector. The photodetector is usually a photomultiplier tube which converts light energy into electrical current and simultaneously amplifies enhancing the sensitivity of the instrument. The signal is processed by a data processor (computer) and displayed in the display unit. The decrease in radiant energy is directly proportional to the concentration of ground state metal atom vapour of the analyte in the flame in accordance with Beer–Lambert law.

The hollow cathode lamps capable of emitting the desired wavelength of incident radiation characteristic of individual elements or mixture of elements commercially available are fitted on to the instrument. The selected incident wavelengths for different metals include 328.1 nm for silver, 324.7 nm for copper, 240.7 nm for cobalt, 248.3 nm for iron, 357.9 nm for chromium, 232.0 nm for nickel, 217.0 nm for lead, 213.9 nm for zinc, etc., along with appropriate monochromator settings for analysis. The fuel–oxidant combination also varies with the nature of sample. Acetylene–air flame is used for the analysis of silver, copper, chromium, bismuth, cobalt, lead, antimony, zinc, nickel, etc., whereas acetylene–nitrous oxide flame is used for the analysis of vanadium, tungsten, titanium, silicon, molybdenum, etc. The calibration of the instrument is usually carried out with a blank (distilled water) and setting the display to 100% transmittance (0% absorbance). A series of standard solutions of the analyte metal is aspirated into the flame and the absorption determined as a function of the concentration to prepare a calibration graph. The calibration graph is a straight line plot of absorbance versus concentration of the analyte metal (Fig. 7.7). The analysis of the test sample is carried out by aspirating the solution into the flame under identical experimental conditions as carried out for the calibration and the absorbance is measured. The concentration of the analyte metal in the test sample is obtained graphically from the calibration chart.

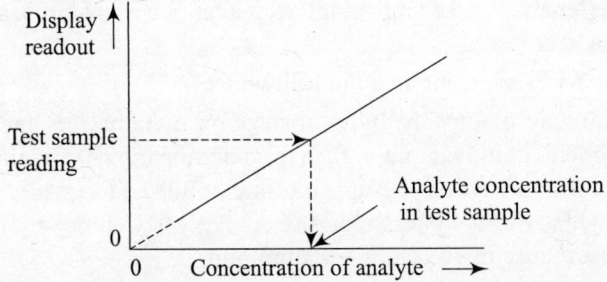

Fig. 7.7 Calibration chart for quantitative analysis in atomic spectroscopy

Modern AAS are mostly computer controlled and provided with background correction and double beam optics for recording. The built-in software can prepare the calibration chart and the data of the test sample is analysed and the concentration can be directly read from the display. About 60 elements can be analysed by AAS in a wide variety of samples, such as from metallurgical products, foodstuffs and beverages, body fluids, soils, petroleum products, plastics, environmental samples, etc., at concentration levels in the range of 0.1–100 ppb.

Standard addition method may be adopted for quantitative analysis by atomic absorption spectrometry. The method involves adding one or more increments of a standard solution to the analyte sample solution of the same size, diluting the solutions to a fixed total volume, aspirating the solution into the flame, and measuring its absorbance. For example, for determining the concentration C_X of the test sample, a minimum of three identical aliquots of the test sample of volume V_X are transferred to standard flasks of total volume V_T. To each of these flasks, different volumes (V_{S1}, V_{S2}, V_{S3}, etc.) of the standard solution of the analyte of known concentration C_S are added and made up to the total volume. Each of the made up solutions are aspirated into the flame and the absorbance is measured. A calibration plot of the measured absorbance values as a function of the volume or concentration of the standard solution added to the test sample gives a straight line. The concentration C_X of the test sample can be calculated from abscissa intercept obtained by back extrapolation of the straight line to the abscissa (Fig. 7.8) using the formula

$$C_X = b\, C_S/m\, V_X \qquad\qquad (7.4)$$

where m is the slope of the straight line plot and b is the intercept.

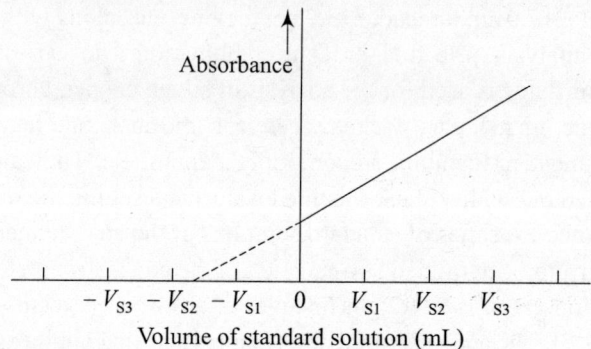

Fig. 7.8   Calibration plot for standard addition method

7.5.4   Interferences in Atomic Absorption Measurements

Spectral interferences and chemical interferences occur in atomic absorption measurements due to several reasons.

Spectral interferences include the following:

(i) Overlapping or closely lying absorption or emission bands of the analyte and other components of the sample, e.g., in the determination of aluminum at 3082.15 Å interference occurs from vanadium absorption line at 3082.11 Å as the two metals have absorption lines which differ by less than 0.1 Å. The interference may be overcome by measuring the absorbance at 3092.7 Å for aluminum.

(ii) Broad absorption bands of any combustion products and scattering of radiation by any particulate products of combustion, e.g., Ti, Zr, and W form refractory oxide particles in flame atomization of the sample and the bigger sized particles scatter the incident radiation.

(iii) Absorption and scattering characteristics of the matrix, e.g., the products of incomplete combustion of any organic matrix produces carbonaceous particles capable of scattering the incident radiation.

Another example is the interference caused by the broad absorption and emission bands (548–558 nm) of the combustion product CaOH formed during the analysis of barium at 553.6 nm in the presence of other alkaline earth metals. This interference is overcome by using nitrous oxide as oxidant instead of air. The higher temperature attained with nitrous oxide decomposes the CaOH and eliminates the interference.

Spectral interferences are relatively small in flame atomization, and in most cases, the interferences can be eliminated by operating at selected temperature and fuel-to-oxidant ratio. When the source of interference is known it is possible to use a *radiation buffer* method, in which a large excess of the interfering substance (radiation buffer) is added to both the sample and the standard and the absorbance measured. Due to the presence of a large excess of the radiation buffer the contribution of the interfering substance to the analyte signal of the sample will be negligible.

Chemical interferences occur more commonly in atomic absorption measurements and these can be minimized through suitable operating conditions. The chemical interferences are mainly due to (i) formation of low volatility compounds, (ii) dissociation reactions, and (iii) ionization.

Formation of compounds of low volatility with the analyte leads to reduced rate of atomization and thereby low results of the analyte. For example, sulphate or phosphate present in samples interferes in the determination of calcium and similarly aluminum is found to interfere in the determination of magnesium. This type of interference can be overcome by the use of higher temperatures or by the addition of *release agents* which react with the interfering substance preferentially and prevent its interaction with the analyte. For example, strontium or lanthanum functions as releasing agent preventing phosphate interaction with the analyte calcium. Addition of *preventive agents* results in the formation of stable but volatile species with the analyte and thereby prevents interference. For example EDTA eliminates the interference of aluminum, silicon, phosphate, and sulphate in the determination of calcium. Similarly, 8-hydroxyquinoline prevents interference of aluminum in the determination of calcium and magnesium.

Dissociation and association reactions occur more commonly at higher temperatures and some of these reactions are reversible. Dissociation reactions of metal oxides and hydroxides occur at different temperatures. For example, oxides of alkaline earth metals are quite stable at lower temperatures and cause interference in absorbance measurements due to more intense molecular bands. Another example is the interference from chlorine in the determination of sodium as it decreases the atomic sodium concentration and thereby lowers the line intensity.

Ionization in flames results in significant decrease in the concentration of ground state atoms particularly at higher flame temperatures and a decrease in the absorption (and emission) lines is observed as in the case of potassium, rubidium, and cesium. Hence, lower excitation temperatures are preferred for alkali metals.

7.6 ATOMIC EMISSION SPECTROSCOPY

In emission spectroscopy, excited state atoms of the analyte in the sample mixture are produced by different methods using flame, electrical arc or spark, and laser or high temperature plasma. The excited state atoms relax to the ground state by emitting radiation of characteristic wavelengths mostly in the ultraviolet and visible regions.

$$
\underset{\text{Ground state atom}}{M} \xrightarrow{\text{Excitation}} \underset{\text{Excited state}}{M^{*}} \xrightarrow[-h\nu]{\text{Relaxation}} \underset{\text{Ground state atom}}{M}
$$

The wavelengths at which emission of radiation occur are characteristic of the individual atoms, as in absorption spectroscopy, and hence, useful in the qualitative analysis. The intensity of emitted radiation I_E, is proportional to the concentration (C) of the excited state atoms as given by the relationship $I_E = k'C$, where k' is proportionality constant.

7.6.1 Excitation Methods

A variety of atomic emission spectroscopic techniques are in vogue, depending on the type of excitation method. The different excitation methods used in atomic emission spectroscopy to excite the ground state atoms include (i) thermal excitation by a flame, (ii) electrical excitation using arc or spark, (iii) laser excitation, and (iv) high temperature plasma excitation.

Thermal excitation by flame to produce a vapour of excited state atoms and ions is discussed in Section 7.4.1. Flames produced by the combination of different fuel–oxidant combinations yield temperatures up to a maximum of 3000°C. The other methods of excitation mentioned above provide more energetic atomization/excitation.

Electrical excitation by arc sources involves excitation of the sample in a gap between a pair of electrodes. The electrodes are made of graphite or metal and spaced few millimeters apart. The arc is ignited by a low-current spark initially so as to generate ions for electrical conduction in the space between the electrodes. Once the arc is struck, thermal ionization of sample maintains the current. Currents in the range of 1–30 A from a dc source voltage of 200 V or an ac source voltage of 2200–4400 V may be used for arc atomization/excitation. Solid, liquid, and gaseous samples may be used for atomization/excitation. Arc sources generate high temperatures in the range of 4000–10,000 K and a high degree of excitation of the sample atoms occurs. Emission lines characteristic of atoms appear from relatively cooler outer mantle of the arc while lines characteristic of ions originate from the high temperature core of the arc.

High voltage spark excitation of samples takes place by high voltage sparks produced intermittently by an ac line voltage that has been rectified and stepped up to 10–50 kV in a coil. The spark frequency and duration are controlled by a suitable solid state circuit. The average temperature of a spark source is relatively lower as compared to the arc source but the energy supplied in a small volume within the spark gap is quite high, and hence, excitation results in the formation of ions. Thus, emission line characteristics of ions are more pronounced in high voltage spark spectra than in arc spectra. Solid, liquid, and gaseous samples can be handled in spark sources as in arc sources.

Laser excitation of samples basically involves the use of a pulsed laser (e.g., ruby, Nd–glass, or Nd–YAG) to atomize and excite a small area on the surface of a sample. In *laser microprobe* technique, laser is used to vapourize the sample into a gap between a pair of graphite electrodes which serves as a spark excitation source.

High temperature plasma excitation is carried out by either a direct-current argon plasma source or an inductively coupled plasma torch assembly. This method of excitation is discussed in more detail in Section 7.8.

The most commonly used emission spectroscopic techniques include flame emission spectrometry and plasma emission spectroscopy.

7.7 FLAME EMISSION SPECTROMETRY

Flame emission spectrometry (FES) is also known as *flame atomic emission spectrometry* or *flame photometry*. As the name implies, the excitation source is a flame. The flame produced by burning the fuel gas in air is relatively a low energy source with a temperature range of 1700–2700°C. Only elements having low ionization energies, such as alkali metals (lithium, sodium, and potassium) and alkaline earth metals, calcium and magnesium, can be analysed by this technique. The principle is based on thermal excitation of the analyte atoms in the sample solution. The analyte solution of alkali or alkaline metal salts is aspirated into the nebulizer–spray chamber assembly to produce a fine spray or aerosol of analyte atoms for introduction into the flame. The metal atoms in the aerosol are thermally excited by the flame. The excited atoms emit radiation in the visible region as they relax to the ground state. The characteristic emission line for the particular element is isolated by a prism or grating monochromator and passed onto the photodetector such as a photomultiplier tube. The electrical current generated in the photodetector is amplified and fed to the data processor for display in a digital read-out. The intensity of the emitted radiation is directly proportional to the concentration of the analyte metal in the solution.

A block diagram of the flame emission spectrometer is shown in Fig. 7.9.

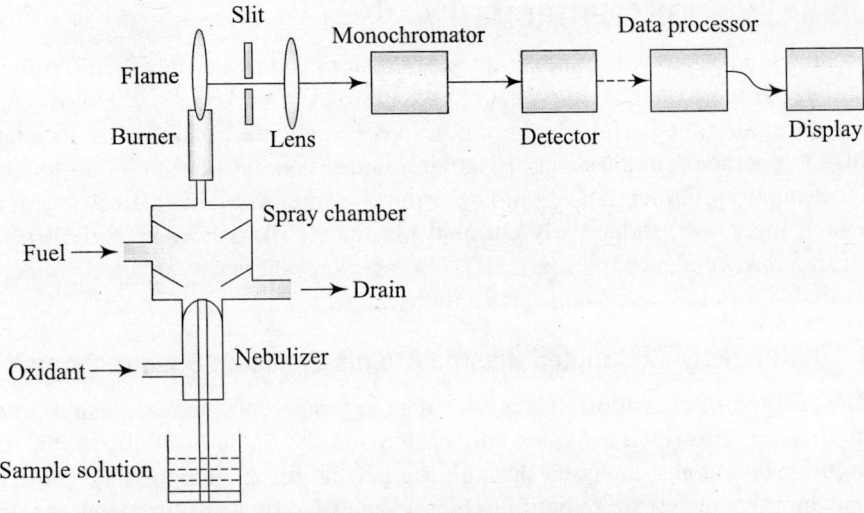

Fig. 7.9 Block diagram of a flame emission spectrometer

The instrument is calibrated by aspirating distilled water and setting the digital display reading to zero. A series of standard solutions are then aspirated and the corresponding display readings are recorded for preparing a calibration chart of detector signal versus concentration of the analyte as shown in Fig. 7.7. The test sample solution of the analyte, whose concentration is to be determined, is then aspirated into the flame, and the concentration of the analyte in the test sample is graphically determined from the calibration chart. Most of the present day instruments have the necessary software for preparing the calibration chart and displaying the concentration of the analyte in the test sample directly. Internal standard method is often used in analytical measurements. The method involves adding a fixed quantity of the chosen internal standard element to the test sample and standard solutions of the analyte. The intensity of the emission line is measured for the test sample and standard solutions and a calibration chart is prepared. The internal standard calibration chart is a plot of the ratio or the log ratio of the detector signal of the analyte to the detector signal for the internal standard as a function of log concentration of the analyte.

Flame emission spectroscopy is more sensitive with lower detection limits comparable to those of atomic absorption spectroscopy in the range of 0.1–10 ppb. Flame emission spectroscopy using different fuel gas–oxidant combinations can achieve higher temperatures, and elements other than alkali and alkaline earth metals also can be quantitatively analysed. Examples include the analysis of sodium, potassium, calcium, and magnesium in body fluids and biological samples, processed foods, soils and fertilizers and nutrients used in agriculture, etc. The wavelengths for the estimation of lithium, sodium, potassium, and calcium, respectively, are 670 nm, 590 nm, 766.5 nm, and 423 nm.

A less expensive, simple to operate and at the same time quite rugged instrument exclusively used for sodium, potassium, calcium, and magnesium makes use of a relatively low temperature LPG–air flame (<1500°C) and an optical filter (instead of a more sophisticated grating monochromator system and associated optics in flame emission spectroscopy). The low temperature excitation produces simple spectra consisting of fewer lines, and hence, narrow band pass filters are sufficient to isolate the desired emission lines of alkali and alkaline earth metals. The filter transmits only the characteristic radiation of the metal onto the detector by filtering off extraneous radiation.

7.8 PLASMA EMISSION SPECTROMETRY

Plasma is an electrically conducting gas of cations and free electrons in significant concentration and at the same time having a net charge close to zero. Argon is the most commonly used gas for generating the plasma as the argon ions formed are capable of absorbing energy to attain high temperatures of the order of 10,000 K and sustain the plasma for prolonged periods, almost indefinitely by further ionization. The argon plasma can be generated by any of the three methods which include (i) inductively coupled plasma (ICP), (ii) direct current plasma (DCP), and (iii) microwave-induced plasma (MIP). However, only the first two plasma sources are commercially available for use in emission spectrometry.

7.8.1 Inductively Coupled Plasma Atomic Emission Spectroscopy (ICP-AES)

The method of excitation makes use of high temperature argon plasma produced by an ICP torch assembly which provides temperatures in the range of 8000–10,000 K. The aerosol of ground state atoms is passed through the plasma for excitation. The excited state atoms and ions emit characteristic radiation which is analysed by a spectrometer consisting of a grating monochromator and photomultiplier detector system. The instrument is capable of analysing about 70 metals and non-metals in a wide variety of samples at ppm and ppb levels in a few minutes and finds applications in chemical, environmental, and biological sciences.

Inductively coupled plasma torch assembly consists of three concentric quartz tubes through which pure argon gas flows, the outermost quartz tube having a diameter of 2.5 cm (Fig. 7.10). Argon gas flows through the central tube at a flow rate of about 0.3–1.5 L min^{-1}, surrounded by the second and third concentric quartz tubes through which argon gas flows at a higher velocity (~15 L min^{-1}). The plasma is formed by igniting the flowing argon gas with a spark from a Tesla coil to attain an initial temperature of about 6000 K.

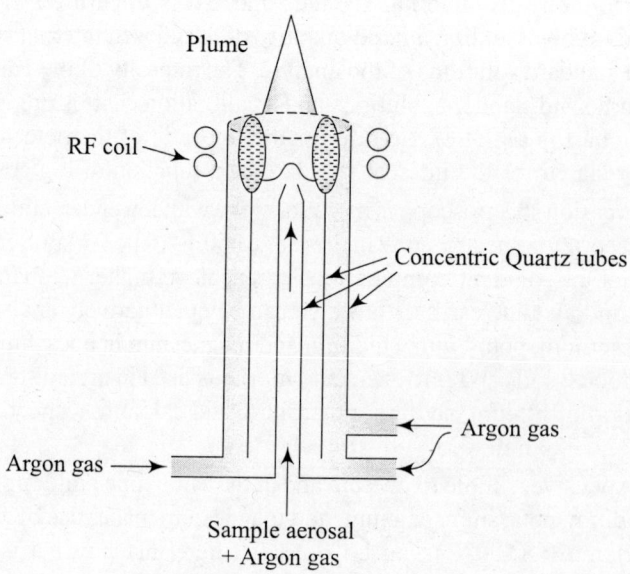

Fig. 7.10 ICP torch assembly

The plasma is sustained and raised to the operating temperature of 8000–10,000 K by induction heating. Induction heating is provided by the interaction of free electrons and ions in the plasma with radiofrequency generated by RF coils operating at about 2 kW and placed outside the outer quartz tube. The interaction causes the ions and electrons to flow in closed annular paths within the coil and ohmic heating occurs as a consequence of the resistance of the ions and electrons to this movement. The argon gas flowing tangentially through the outer quartz tubes at a higher velocity cools the centre tube and also lifts the plasma clear of the quartz tubes thereby protecting them from the high temperature.

The sample in the form of a solution is introduced into the nebulizer–spray chamber assembly by aspiration and the aerosol of ground state atoms formed flows through the central tube carried by a stream of argon into the ICP torch assembly. Liquid and solid samples, particularly small or micro samples, may also be introduced into the plasma by electrothermal vapourization using an electrically heated furnace or simply on a graphite rod. The vapour is carried by the argon stream into the plasma.

The sample aerosol passes through the high temperature region at the centre of the plasma and forms a plume containing the excited state atoms and ions of the sample elements free from any molecular associations.

A block diagram of the ICP-AES is shown in Fig. 7.11.

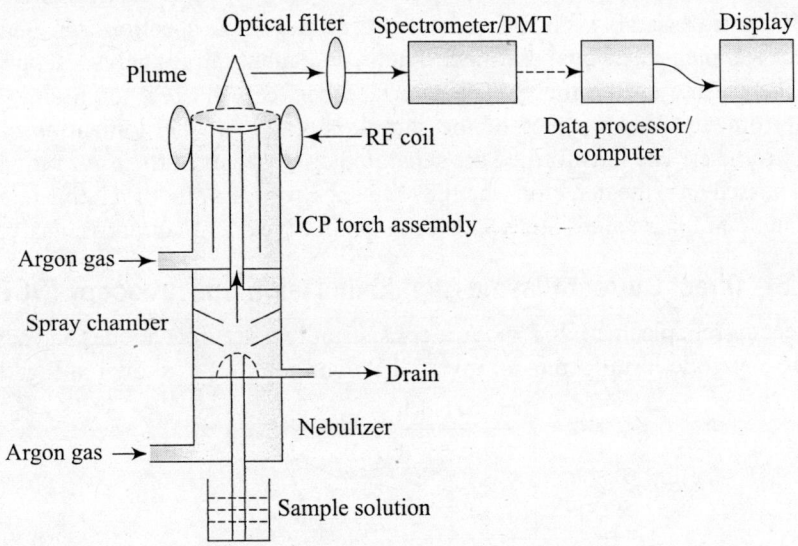

Fig. 7.11　Schematic diagram of ICP-AES

The excited state atoms and ions relax to their ground states emitting characteristic radiation. The radiation is analysed by a computer-controlled photomultiplier tube (spectrometer/PMT assembly). The instrument can be programmed to analyse the about 10–15 constituent elements either simultaneously or sequentially within a few minutes. The calibration procedure is similar to that discussed earlier for AAS and FES. Calibration plot in plasma emission spectroscopy is mostly the photocurrent of the detector plotted against the concentration of the analyte. Log–log plots may be used at higher concentration ranges. The straight line plots show bending towards the horizontal axis at high concentrations mainly due to self-absorption and due to incorrect background corrections. The method of internal standard is useful for quantitative analysis.

Figure 7.12 shows the emission spectrum of a sample solution containing different elements—zinc, molybdenum, antimony, and chromium.

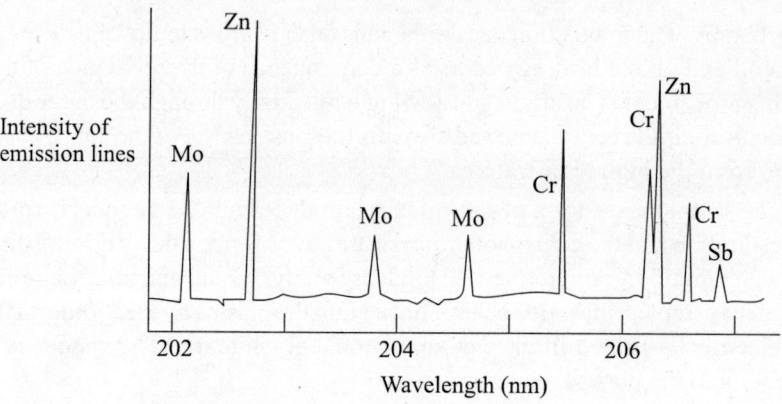

Fig. 7.12 ICP-atomic emission spectrum of a sample

7.8.2 Inductively Coupled Plasma–mass Spectrometry

Inductively coupled plasma–mass spectrometry (ICP-MS) is the hyphenated technique in which an ICP torch assembly is interfaced with quadrupole mass spectrometer. Qualitative identification of the sample is done by mass spectrometry and quantitative analysis at ppb level concentrations by the plasma spectrometry. The interface between the ICP torch and the quadrupole mass spectrometer allows a part of the sample stream into the ionization chamber of the mass spectrometer. The ions formed are separated on the basis of their m/z ratio by the mass analyser and passed onto the detector. The ICP-MS spectrum is a plot of relative intensities of the ions as a function of m/z values and yields both qualitative and quantitative analytical information.

7.8.3 Direct Current Plasma Atomic Emission Spectroscopy (DCP-AES)

Direct current plasma (DCP) source consists of two graphite anodes and a tungsten cathode, the three electrodes arranged in an inverted Y configuration as shown in Fig. 7.13.

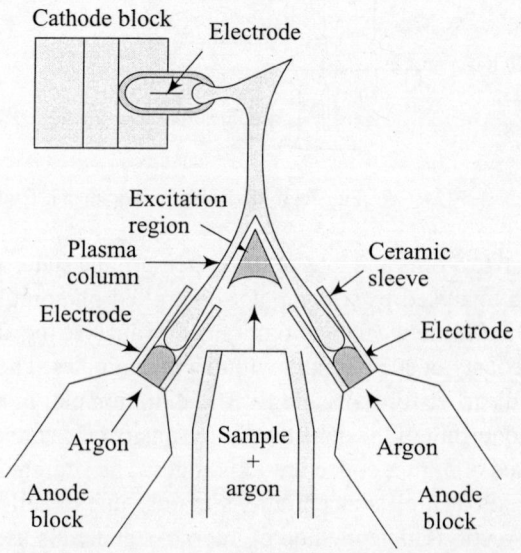

Fig. 7.13 Schematic diagram of a DCP jet (Source: Principles of Instrumental Analysis – D.A. Skoog, F.J. Holler and T.A.Nieman, Thomson Brooks/Cole 2005)

Argon gas flows at about 8 L min^{-1} through the anode blocs towards the cathode. The plasma ignition is initiated automatically by bringing the electrodes into contact momentarily. Ionization of argon occurs and a current of about 14 A develops which generates additional ions to sustain the plasma indefinitely. The plasma temperature is about 10,000 K while the excitation region at the junction of the argon stream is about 6000 K. The sample is nebulized and the aerosol formed is introduced into the excitation region to excite the atoms and generate emission spectra characteristic of the sample elements. The emission spectra are analysed by the spectrometer/photomultiplier assembly similar to that in ICP-AES instrument.

DCP source has advantages in that the emission spectra contain fewer lines as compared to those produced by ICP source and the lines are essentially due to atoms rather than ions. In addition, argon consumption is less and the auxiliary power supply is simpler and less expensive. Sensitivity and reproducibility of the two plasma sources are almost the same. However, DCP source requires the replacement of the graphite electrodes every few hours while ICP is maintenance-free.

7.8.4 General Features of Plasma Source Spectrometers

Commercially available plasma source spectrometers cover the entire range of ultraviolet and visible spectral regions from 170–800 nm. Some of the instruments have the extended capability to operate at even lower wavelengths of 150–170 nm in the vacuum ultraviolet region where phosphorus, sulphur, and carbon have emission lines. The instruments in general have high resolution capability (0.01 nm) and high S/N ratio. The plasma source has high stability and is not affected by environmental changes. The instruments are computer-controlled with facilities for easy background correction, acquisition, storage and processing of data as well as display. Detection limits of plasma spectrometers are quite low in the range of ppb levels.

The different types of plasma emission spectrometers include sequential spectrometers, simultaneous multichannel spectrometers, and FT spectrometers.

Sequential spectrometers are programmed to analyse several analyte elements present in a sample one by one sequentially up to about 20 elements. The instruments analyse the specified emission line of a particular element and then move to another emission line pausing in between for a few seconds to stabilize the background noise and enhance the S/N ratio. The instruments contain a grating monochromator system provided with a holographic grating containing 2400–3600 grooves/mm. The grating is rotated by a digitally controlled stepper motor so that different wavelengths are sequentially and precisely focused onto the exit slit. In order to scan a wide wavelength region, particularly in the case of complex spectra consisting of several lines, rapid scanning is achieved by *slew-scanning*. Slew-scanning makes use of a two-speed drive motor which rotates the grating or the detector and the slit quickly to a wavelength near the desired emission line and then scan the line of interest in a series of small steps of 0.001 nm.

Simultaneous multichannel spectrometers are capable of measuring the intensities of emission of lines of as many as 50–60 elements in the sample simultaneously. The detectors used are either a series (as many as 60) of photomultiplier tubes as in *polychromators* or two-dimensional charge injection devices or charge-coupled devices as in *array-based systems*.

7.9 ATOMIC FLUORESCENCE SPECTROSCOPY

Atomic fluorescence spectroscopy (AFS) is based on measuring the wavelength and intensity of the characteristic fluorescence emission of an atomic vapour of an analyte element following excitation by a primary radiation in the UV/visible region. The atomic vapour is generated by flame atomization as in atomic absorption spectrometry and the primary radiation source is either an element selective sharp line source such as a hollow cathode lamp or an electrode-less discharge lamp or a high-intensity broad band source such as a mercury discharge lamp. The atomic vapour of analyte element absorbs the primary radiation and the excited atoms relax to ground state by re-emitting radiation as fluorescence. The fluorescence emission occurs in all directions, and hence, may be monitored by placing the detector (e.g., photomultiplier tube detector) at angles other than in a direct line with the primary radiation source. The fluorescence emission is passed through a monochromator before reaching the detector in dispersive instruments while in non-dispersive instruments the use of sharp-line sources eliminates the requirement of monochromator. The intensity of the fluorescence emission is directly proportional to the concentration of the absorbing analyte element but is usually diminished due to *quenching* as the excited state atoms collide with other species in the flame. Nitrogen and hydrocarbon vapours have been found to enhance quenching. The problem due to quenching may be eliminated by avoiding flames incorporating the quenching vapours or the effect of quenching may be minimized by diluting with argon.

Atomic fluorescence spectroscopy is relatively a simpler and more sensitive technique compared to atomic absorption spectrometry and atomic emission spectroscopy but suffers from the disadvantages due to quenching and background noise arising from the scattering of radiation by particulate matter in the flame, particularly with refractory samples and in high temperature flames. AFS is particularly useful in analysis of volatile elements, for example, mercury and elements which form volatile hydrides (e.g., Hg, As, Sb, Se, Te, and Bi) or volatile organometallic compounds (e.g., Hg and Cd).

●   **SOLVED PROBLEMS**

EXAMPLE 1   *A calibration plot for chromium was prepared from the following data obtained by atomic absorption spectrometry.*

Cr content (mg L^{-1})	0	2	4	6	8	10
Absorbance	0.0	0.17	0.33	0.49	0.64	0.8

Calculate the concentration of chromium in the test sample with an absorbance of 0.42.

SOLUTION

From the calibration plot of absorbance vs. concentration of chromium the concentration of chromium in the test sample is determined as 5.2 mg L^{-1}.

EXAMPLE 2   *Five 10.0 mL aliquots of a test sample containing sodium were taken in five different 50-mL volumetric flasks and 0.0, 5.0, 10.0, 15.0, and 20.0 mL of a standard solution containing 12.5 ppm of sodium was added to each of the flask. The absorbance readings were found to be 0.24, 0.44, 0.61, 0.82, and 1.03 respectively. Calculate the concentration of sodium in the test sample.*

SOLUTION

A standard addition calibration plot of absorbance vs. volume of the standard solution added gives a straight line with a slope of 0.0364 and an intercept of 6.5 mL. Using Eq. (7.4) the concentration of sodium in the test sample is calculated as 8.24 ppm.

EXAMPLE 3   *A flame photometric analysis of lithium by internal standard method using sodium as the internal standard was carried out. The data is as follows.*

Li concentration (ppm)	2	4	6	8	10
Emission intensity reading-Li	15.2	26.8	38.6	45.2	54.6
Emission intensity reading-Na	8.5	8.7	8.7	8.5	8.6

A test sample was found to give an emission line of intensity for lithium as 36.8 and for sodium 8.5. Calculate the concentration of lithium in the test sample.

SOLUTION

The logarithm of the detector signal for (Li/Na) for the different concentrations of Li (log values) are calculated and used for preparing the calibration chart.

Log detector signal (Li/Na)	0.25	0.49	0.65	0.72	0.80
Log lithium concentration	0.30	0.60	0.78	0.90	1.0

For the ratio of the detector signal (Li/Na) of 0.64 the concentration of lithium in the test sample is found from the graph to be 5.8 ppm.

SUMMARY

- Atomization is an essential step in all the atomic spectroscopic techniques.
- Atomization involves the formation of a vapour of ground state atoms of the analyte. Atomization is carried out in atomic absorption spectrometry by different methods, the most common methods being flame atomization and electrothermal atomization.
- The ground state atomic vapour of the analyte absorbs a part of a highly monochromatic radiation emitted by a line source called *hollow cathode lamp*, specific to the analyte. The resulting decrease in the intensity of the transmitted radiation as monitored by a photomultiplier tube detector is directly proportional to the concentration of the analyte atoms in the vapour/sample.
- Atomic emission spectroscopy involves measuring the intensity of the characteristic radiation emitted by excited state atoms and ions by a photomultiplier tube detector. The first-step atomic emission spectroscopic techniques involve atomization of the analyte to produce a vapour of the atoms which also get excited to form excited state atoms and ions. Thus, both atomization and excitation processes occur together in emission spectroscopic techniques. In flame photometry, atomization and excitation processes are carried out by a flame, whereas in plasma emission spectroscopic techniques, high temperature, plasma is used for atomization and excitation.
- Atomic fluorescence spectroscopy involves measuring the intensity of the fluorescence emitted by excited state atoms of the analyte. Atomization to produce a vapour of the ground state atoms is carried out in a manner similar to that in atomic absorption spectrometry and the excitation of the ground state atoms is carried out by light emitted from sources such as mercury discharge lamp or a hollow cathode lamp. The excited state atoms relax to the ground state by emitting fluorescence.

REVIEW QUESTIONS

1. Classify the atomic spectroscopic methods. What are the principles involved?
2. What is atomization of a sample? What are the methods adopted?
3. Give a detailed discussion on the flame characteristics and flame atomization.
4. Write notes on electrothermal atomization, glow discharge atomization and hydride atomization.
5. Discuss the principle involved in atomic absorption spectrometry.
6. What are the sharp line sources used in AAS? What is the necessity for using such a source?
7. Give an account on the practice of AAS.
8. Draw a neat sketch of hollow cathode lamp and explain its functioning.
9. What is the necessity of background correction in AAS? How is it carried out?

10. What are the different spectral and chemical interferences encountered in atomic absorption measurements?

11. Draw a neat sketch of a single-beam AAS and describe the functions of the components.

12. How is quantitative analysis carried out by AAS?

13. Explain the basic principle involved in emission spectroscopy.

14. Write a note on the different excitation methods employed in emission spectroscopy.

15. Draw a neat sketch of the flame photometer and identify the components.

16. Describe a detailed procedure for the determination of the concentration of sodium in a sample solution by flame photometry.

17. What is plasma emission spectrometry? Describe the different plasma excitation sources.

18. Discuss the principle and practice of ICP-AES.

19. Give a brief description of atomic fluorescence spectroscopy.

8 Molecular Spectroscopy

8.1 INTRODUCTION

Molecular spectroscopy involves the study of the interaction of molecules with electromagnetic radiation of different regions. The fundamental aspects of this interaction and the classification of the different molecular spectroscopic techniques have already been discussed in detail in Chapter 6.

This chapter discusses the basic principles of molecular absorption spectroscopic techniques involving (i) electronic transitions which essentially occur in ultraviolet and visible regions, hence, known as *UV–visible spectroscopy* or *electronic absorption spectroscopy*, (ii) vibrational transitions in molecules that occur in the infrared region, giving rise to *infrared spectroscopy* or *vibrational spectroscopy*, (iii) rotational transitions in molecules which occur in the microwave region, called *microwave spectroscopy* or *rotational spectroscopy*, and (iv) scattering of radiation by molecules giving rise to *Raman spectroscopy*.

This chapter also discusses the molecular emission techniques of *fluorometry* and *phosphorimetry*.

8.2 UV–VISIBLE SPECTROSCOPY

The wavelength range of 200–350 nm of the electromagnetic spectrum is known as the ultraviolet region. The visible region of the spectrum extends between 350 nm and 800 nm. Molecules interact with radiation in the ultraviolet and visible region (wave number range 50,000–12,000 cm^{-1}) which results in transitions between electronic energy levels, particularly in the valence shell. Hence, UV–visible spectroscopy is also known as *electronic absorption spectroscopy*. The principles involved in the electronic transitions as well as instruments used for recording the spectra are common to both the ultraviolet and visible regions, and hence, it is convenient to discuss UV–visible spectroscopy together.

8.2.1 Electronic Spectra of Molecules

Electronic transitions in absorption as well as in emission of radiation involve interaction between the electric dipole of the molecule and the electric field component of the incident radiation. The electronic energy levels of molecules as shown in Chapter 6 consist of vibrational energy levels as sub-energy levels and rotational energy levels as sub-energy levels within vibrational states. Electronic transitions between two electronic states always are accompanied by vibrational transitions, referred to as *vibronic transitions*. These vibronic transitions in turn are accompanied by rotational transitions *(rovibronic transitions)*. Thus, an electronic transition accompanied by the vibronic and rovibronic transitions gives rise to bands in the spectrum called the *electronic band system*, consisting of the vibrational coarse structure and the rotational fine structure, usually observed in high-resolution spectra. In low-resolution spectra and the spectra of liquid phase samples the vibrational coarse structure may not be resolved and appears a single band.

Vibronic transitions are usually grouped into two types: (i) progressions and (ii) sequences as shown in Fig. 8.1 of a simple diatomic molecule.

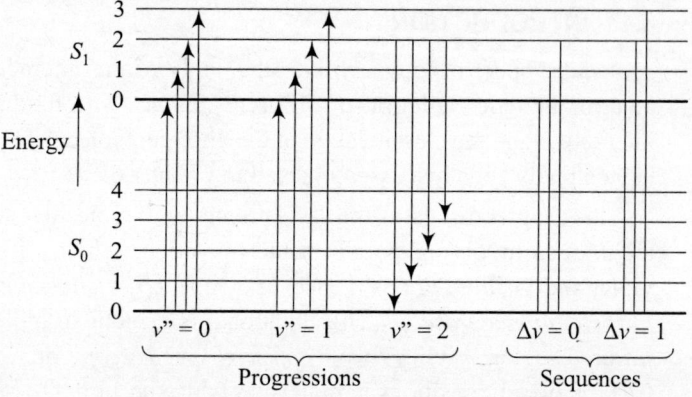

Fig. 8.1   Vibronic transitions in the electronic spectrum of diatomic molecule

Vibronic transitions from a common lower energy level (for absorption) or a common upper energy level (for fluorescence emission) are called *progressions*, for example, all transitions that occur from the $v'' = 0$ to higher energy levels for the progression series are designated as $v'' = 0$. Transitions that occur from $v' = 2$ to the lower energy level for the progression series are designated as $v' = 2$. A group of transitions having the same value for Δv (e.g., $\Delta v = 0$ or $\Delta v = 1$) are called *sequences* mostly observed in emission.

The intensities of the spectral lines due to electronic transitions depend on the selection rules as mentioned in Chapter 6. Although the selection rules do not restrict changes in the vibrational quantum number during an electronic transition, the vibrational lines in a progression are observed to have variations in their intensities. In some spectra the line due to $v'' = 0$ to $v' = 0$ transition is quite intense while in others the intensity is maximum at some other value of v'. In certain cases a continuum is observed. The variation in the intensities in the coarse structure during an electronic transition is explained on the basis of Franck–Condon principle.

8.2.2   Franck–Condon Principle

Franck explained the variation in the intensities of spectral lines in vibronic transitions qualitatively in 1925, before the development of Schrödinger equation, while Condon brought in quantum mechanical treatment to explain the same more quantitatively. Molecules undergo electronic transitions resulting in changes in their electronic structures. Since the energy changes involved in electronic transitions are quite large simultaneous changes in vibrational and rotational energies of molecules also occur. The energy of a simple diatomic molecule varies with the inter-nuclear distance diagrammatically shown in the form of the Morse potential energy curve with the position of one atom considered to fixed at $r = 0$ while the other atom oscillates between the limits of the curve (Fig. 8.2). The oscillating atom is most likely to be found at $r = r_{eq}$ at the centre of its vibrational motion for vibrational quantum number $v = 0$ and for $v = 1, 2, 3$, etc., the most probable positions steadily approach the extremities of the curve.

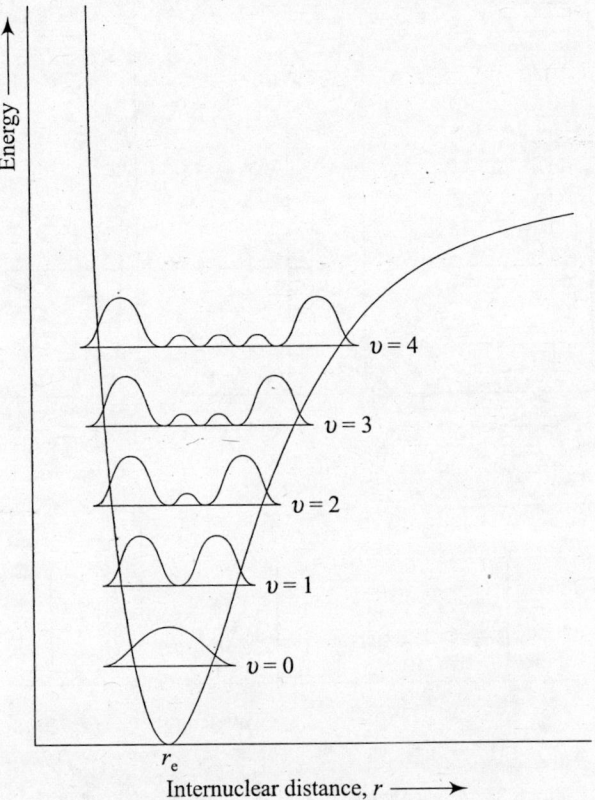

Fig. 8.2   Morse curve for a diatomic molecule

When the diatomic molecule undergoes a vibronic transition and goes to the excited state, the excited state of the molecule can be represented by a Morse curve similar to that of the ground state probably with small differences in equilibrium inter-nuclear distance and vibrational frequency.

Franck–Condon principle states that *electronic transition takes place so rapidly that a vibrating molecule does not change its inter-nuclear distance appreciably during the transition.* This is because electronic transitions occur much faster compared to vibrational transitions. Hence, the electronic transitions may be represented by vertical lines between the ground and the first excited state electronic levels. At room temperature almost all the molecules will be in the ground state electronic level and probably in the lowest vibrational energy level also. Four possibilities are shown in Fig. 8.3 arise considering the operation of Franck–Condon principle with regard to the equilibrium inter-nuclear distances. These include (i) the distances in the lower and upper state are equal, (ii) the distance in the upper state is slightly smaller than in the lower state, (iii) the distance in the upper state is slightly greater than that in the lower state, and (iv) the distance in the upper state is considerably greater.

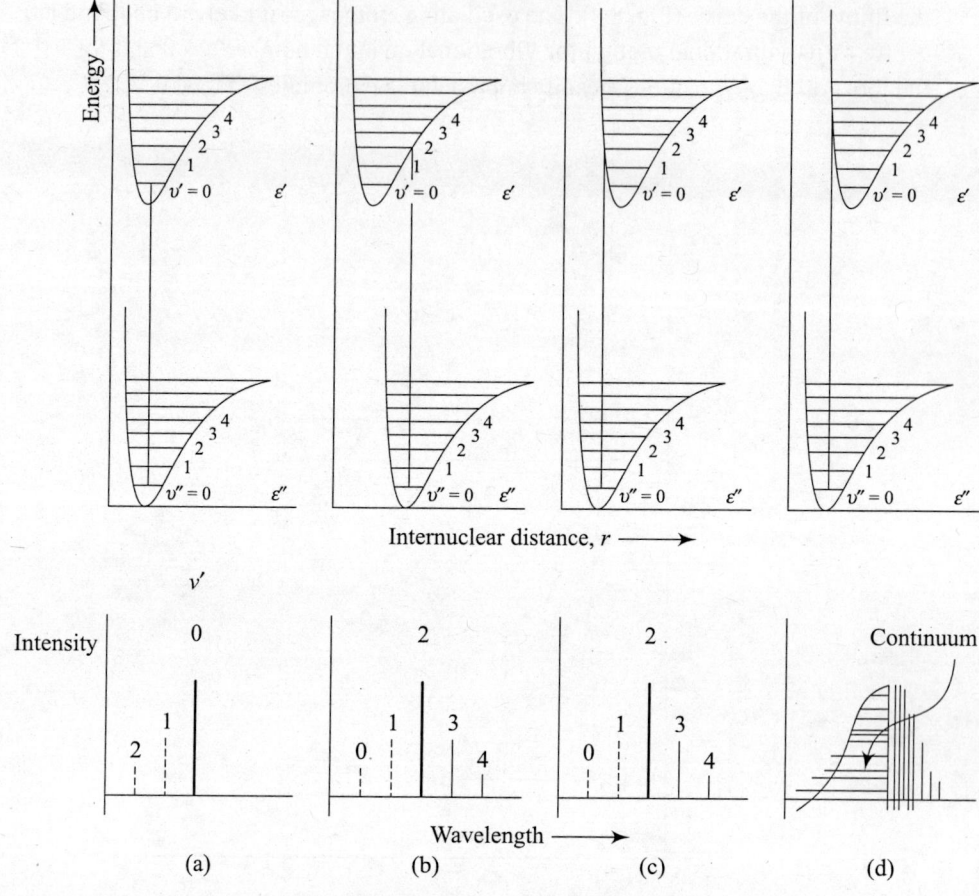

Fig. 8.3 Application of Franck–Condon principle when (a) $r_e' = r_e''$ (b) $r_e' < r_e'''$ (c) $r_e' > r_e''$, and (d) $r_e' \gg r_e''$ and the corresponding spectra

Based on the Franck–Condon principle the electronic transitions are represented by vertical lines in the diagram. The inter-nuclear distances do not change during the transition. The most probable transition in each of the four possibilities is shown by the continuous line while the other less probable transitions by dotted lines. Thus, for example, the most probable transition when $r_e' = r_e''$ is from $v'' = 0$ to $v' = 0$ transition. According to quantum mechanics the probability of finding the oscillating atom slightly away from r_e'' also exists to a lesser extent, and hence, transitions $v'' = 0$ to $v' = 1$ (1,0) and $v'' = 0$ to $v' = 2$ (2,0) are also allowed. However, the intensities of these transitions diminish rapidly as shown in Fig. 8.3(a).

When the inter-nuclear distance of the upper state is slightly smaller than that of the ground state Fig. 8.3(b) the most probable transition will be the $v'' = 0$ to $v' = 2$ and other less probable transitions include $v'' = 0$ to $v' = 1$, $v'' = 0$ to $v' = 0$ as well as $v'' = 0$ to $v' = 3$, $v'' = 0$ to $v' = 4$, etc. Figure 8.3(c) shows the transitions and the intensities of the spectral lines for a system when $r_e' > r_e''$. When the inter-nuclear distance is considerably greater compared to that of the lower state, transitions occur to vibrational levels higher in the Morse curve of the upper state and the transitions are not quantized and a continuum results as shown in Fig. 8.3(d).

The electronic transitions that occur in polyatomic molecules result in a more complex spectra. Gaseous molecules show *vibrational coarse structure* and *rotational fine structure*. However the fine structure is not observed in spectra of solutions due to collisions between the solute and solvent molecules resulting in overlapping of spectral lines into broad bands. The resulting overlapping bands coalesce to give one or more broad band envelopes each band characterized by the position of a *wavelength maximum* (λ_{max}). The intensity of the band corresponds to *molar absorptivity* (ε). For polyatomic organic molecules and metal complexes, the complete spectrum may contain several bands arising from a number of electronic transitions and their associated vibrational and rotational fine structures.

8.2.3 Electronic Transitions in Organic Molecules

The electronic transitions that occur in polyatomic organic molecules between valence shell electronic energy levels may be represented as shown in Fig. 8.4.

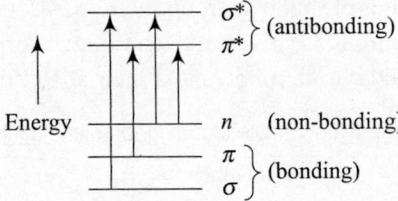

Fig. 8.4   Electronic transitions in polyatomic molecules

The bonding and non-bonding molecular orbitals in most of the organic molecules are filled and the antibonding orbitals are vacant. The various electronic transitions that can take place may be classified into (i) $\sigma-\sigma^*$, (ii) $n-\sigma^*$, (iii) $\pi-\pi^*$, and (iv) $n-\pi^*$. The relative energy changes involved in these transitions are in the increasing order

$$n-\pi^* < \pi-\pi^* \sim n-\sigma^* << \sigma-\sigma^*$$

The intensities of the $\pi-\pi^*$ and $\sigma-\sigma^*$ transitions are quite large (ε is large) while the other two transitions are considerably weak due to unfavourable selection rules.

The $\sigma-\sigma^*$ *transitions* between the bonding σ and the antibonding σ^* orbitals of organic molecules involve a large change in energy, and hence, take place in the far or vacuum ultraviolet region. These transitions occur in the range of 120–200 nm. Saturated hydrocarbons contain molecular orbitals of σ and σ^* type only and are transparent (do not absorb radiation) to near UV and visible radiation. Alkanes, for example, methane and ethane show λ_{max} at 122 nm and 135 nm respectively. Cycloalkanes, for example, cyclohexane absorbs at wavelengths shorter than 150 nm.

The $n-\sigma^*$ *transitions* are observed in molecules containing hetero atoms (N, O, S, or halogens) and involve excitation of an electron in unshared pair on the hetero atom to the antibonding σ^* orbital. Examples of substances which show $n-\sigma^*$ transitions include water (167 nm), methanol (183 nm), methyl chloride (173 nm), methyl iodide (258 nm), trimethylamine (227 nm), and 1-iodobutane (257 nm).

The $\pi-\pi^*$ *transitions* occur between the bonding π and antibonding π^* orbitals and are observed in the UV region for organic compounds containing double bonds involving hetero-atoms, for example, C=C, C=O, C=S, N=O, N=N, C≡N, etc. These groups called *chromophores* absorb intensely giving rise to strong absorption bands with very high molar absorptivity of the order of 5000–10,000 L mol^{-1} cm^{-1}. Thus, ethylene with >C=C< group shows a strong absorption band at 165 nm attributed to $\pi-\pi^*$ transition.

The $n-\pi^*$ *transitions* occur between the non-bonding orbitals and antibonding π^* orbitals. The absorption bands due to these transitions are less intense because the non-bonding orbitals are situated in a plane perpendicular to the π^* orbitals and the transition probability is strictly zero according to selection rules. However, these transitions are observed usually as weak absorption bands in the longest wavelength regions of UV and visible spectrum.

8.2.4 Factors Affecting Absorption Bands

Several factors affect the positions and the intensities of the absorption bands of organic compounds and these include: (i) presence of *chromophores*, (ii) presence of *auxochromes* as substituent groups in the chromophore, (iii) conjugation with other chromophores, and (iv) solvent effects.

Chromophore is a multiple bonded group such as C=C, C≡C, C=O, C=S, N=O, N=N, C≡N, etc. The presence of a chromophore gives rise to strong absorption of radiation. A few examples of isolated or unconjugated chromophores and their absorption maxima are listed in Table 8.1.

Table 8.1 Simple chromophores and their absorption maxima

Chromophore	Example	Type of transition	λ_{max} (nm)
>C=C<	$H_2C=CH_2$	$\pi-\pi^*$	165
−C≡C−	HC≡CH	$\pi-\pi^*$	173
>C=O	$(H_3C)_2C=O$	$\pi-\pi^*$	189
		$n-\pi^*$	279
−NO$_2$	H_3CNO_2	$\pi-\pi^*$	200
		$n-\pi^*$	275
Aromatic ring	C_6H_6	$\pi-\pi^*$	200, 255

Auxochrome, by definition, is a functional group that does not by itself absorb in the UV region but shifts the absorption peaks of chromophore groups to longer wavelengths. Mostly an auxochrome is a saturated functional group containing hetero atom(s) (e.g., −OH, −Cl, −OR, −NR$_2$, etc.). The presence of an auxochrome modifies the absorption characteristics of the chromophores, deepening the colour in most cases.

Bathochromic shift or *red shift* refers to the shift of the absorption maximum towards the longer wavelengths. Red shift occurs due to the presence of an auxochromes attached to a chromophore and also by a change in the solvent medium. For example, the absorption maximum of ethylene observed at 165 nm is shifted to longer wavelength of 185 nm in vinyl chloride ($H_2C=CHCl$). Similarly, alkyl-substituted ethylenes show a red shift compared to ethylene as in 1-butene (185 nm) and isobutene (188 nm). Red shift is also observed in carbonyl compounds having a double bond separated by two or more single bonds attributed to $n-\pi^*$ transitions occurring at much longer wavelengths (300–350 nm).

Red shift of absorption bands of $\pi-\pi^*$ transitions occurs in polar solvents. For example, phenol shows a red shift from 210 nm to 235 nm in polar solvents attributed to the phenolate anion.

Conjugated chromophores cause red shift of absorption bands and also enhance the intensity of the bands compared to isolated chromophores. The red shift is attributed to delocalization of π electrons and consequent lowering of the energy of the π^* orbital and give it less antibonding character. For example, the absorption maximum of ethylene shifts from 165 nm to longer wavelength of 217 nm in conjugated butadiene. Similarly, the π–π^* transition in acetone at 189 nm shifts to 219 nm in the α,β-unsaturated ketone $H_2C=CH-CO-CH_3$ because of conjugation of $C=O$ with $C=C$.

Red shift due to alkyl substitution in conjugated dienes is additive, and hence, it is possible to predict the position of absorption maximum in open chain dienes and six-membered ring compounds. Woodward put forward rules to predict the positions of absorption maximum in these compounds which were later modified by Fieser and Scott. These rules are known as *Woodward–Fieser rules* and are applicable to dienes and trienes. The rules are listed in Table. 8.2.

Table 8.2 Woodward–Fieser rules applicable for dienes and trienes

Particulars of substitution	Red shift of λ_{max} (nm)
Parent heteroannular	assigned base value (no shift) 214
Parent acyclic diene	assigned base value (no shift) 217
Parent homoannular diene	assigned base value (no shift) 253
Increment for	
(i) Each alkyl substituent or ring residue	5
(ii) Exocyclic double bond	5
(iii) double bonding extending by conjugation	30
(iv) polar groups as substituent −O−acyl	0
−O−alkyl	6
−Cl, −Br	5
−S−alkyl	30
−N(alkyl)$_2$	60

The shift due to a substituent group in the λ_{max} for any substituted diene can be calculated by adding the increment to the assigned base value of 214, 217, or 253 nm. For example, the calculated value of λ_{max} for the dimethyl-substituted open chain acyclic diene $H_2C=C(CH_3)$ $-C(CH_3)=CH_2$ is 227 nm (on the basis of assigned base value of 217 nm + 10 nm for the two alkyl substituents). The calculated value agrees well with the observed value of 226 nm. Woodward–Fieser rules are applicable to unsaturated compounds containing up to four double bonds.

Fieser and Kuhn formula may be used for calculating λ_{max} and ε_{max} of conjugated systems containing more than four double bonds as given below:

$$\lambda_{max} = 114 + 5M + n(48.0 - 1.7n) - 16.5R_{endo} - 10R_{exo} \tag{8.1}$$

where M refers to the number of alkyl substituents or ring residues in the conjugated molecule, n is the number of conjugated double bonds, R_{endo} is the number of rings with endocyclic double bonds, and R_{exo} is the number of rings with exocyclic double bonds.

The value of ε_{max} is given by the formula $n(1.74 \times 10^4)$, where n is the number of conjugated double bonds. The calculated λ_{max} and ε_{max} values for β-carotene containing 10 alkyl substituents, 11 conjugated double bonds and $R_{endo} = 2$, respectively are 453.3 nm and 19.4×10^4. The observed values λ_{max} and ε_{max} are 452 nm and 15.2×10^4 respectively.

Hypsochromic shift or *blue shift* refers to the shift of absorption maximum to shorter wavelengths. It is produced by the presence of auxochromes in compounds exhibiting absorption bands due to $n-\pi^*$ transitions. Polar solvents also cause a blue shift of absorption bands due to $n-\pi^*$ transitions. For example, aniline absorbs at 230 nm but in acid solutions the absorption maximum shifts to 203 nm. Similarly, the $n-\pi^*$ transition in acetone gives rise to an absorption band the maximum of which occurs at 279 nm in benzene while in water it is blue shifted to 264.5 nm.

Hyperchromic and *hypochromic effects* refer to the changes in the intensity of the absorption bands. Hyperchromic effect increases the intensity while hypochromic effect brings a decrease in the intensity of the absorption band. For example, phenol shows a bathochromic shift as well as a hyperchromic effect as the primary band at 210 nm with a molar absorptivity (ε) 6200 L mol^{-1} cm^{-1} shifts to 235 nm with a increased intensity (molar absorptivity 9400 L mol^{-1} cm^{-1}) for the phenolate anion. In contrast, benzoic acid shows hypsochromic shift and a hypochromic effect on becoming benzoate anion the band shifting from 230 nm (molar absorptivity 11,600 L mol^{-1} cm^{-1}) to 224 nm (molar absorptivity 8700 L mol^{-1} cm^{-1}).

8.2.5 Electronic Transitions in Inorganic Species

A number inorganic anions exhibit absorption peaks in the UV region attributed to $n-\pi^*$ transitions. Examples include nitrate (313 nm), nitrite (360 and 280 nm), carbonate (217 nm), and azido (230 nm).

Coordination compounds of transition metals containing organic and inorganic ligands are mostly coloured and absorb in the ultraviolet and visible regions of the electromagnetic spectrum. The absorption peaks are mostly broad and less intense. The electronic absorption spectra of coordination compounds are useful for structure analysis as well as for quantitative analysis. Three types of electronic transitions are observed in the spectra of transition metal compounds. These include (i) $d-d$ transitions within the transition metal ion of low intensity as they are Laporte forbidden, (ii) excitation within the organic ligand typically $\pi-\pi^*$ and $n-\pi^*$ transitions affected by the presence of the metal, and (iii) charge transfer transitions involving transfer of electron from the metal orbital to the ligand orbital (metal-to-ligand charge transfer or MLCT) or from the ligand orbital to the metal orbital (ligand-to-metal charge transfer or LMCT). The last two transitions give rise to intense bands, and hence, useful for trace analysis.

Lanthanide and actinide ions also absorb in the ultraviolet and visible regions exhibiting characteristic absorption peaks which are narrow and well-defined attributed to $f-f$ transition. The absorption peaks are unaffected by the type of ligand associated with the metal ion as the inner f orbitals are well screened from external influences.

8.2.6 UV–visible Spectrophotometer

Spectrophotometers for the ultraviolet region covering the wavelength range 190–350 nm and visible region (350–800 nm) as well as combined ultraviolet and visible regions are commercially available. The instruments may be of single beam or double beam recording-type spectrophotometers. The basic components of the instruments include (i) source of radiation, (ii) monochromator, (iii) sample cell, (iv) detector, and (v) display.

The schematic diagrams of the optical paths in a single beam and a double beam UV–visible spectrophotometer are shown in Figs. 8.5(a) and 8.5(b) respectively.

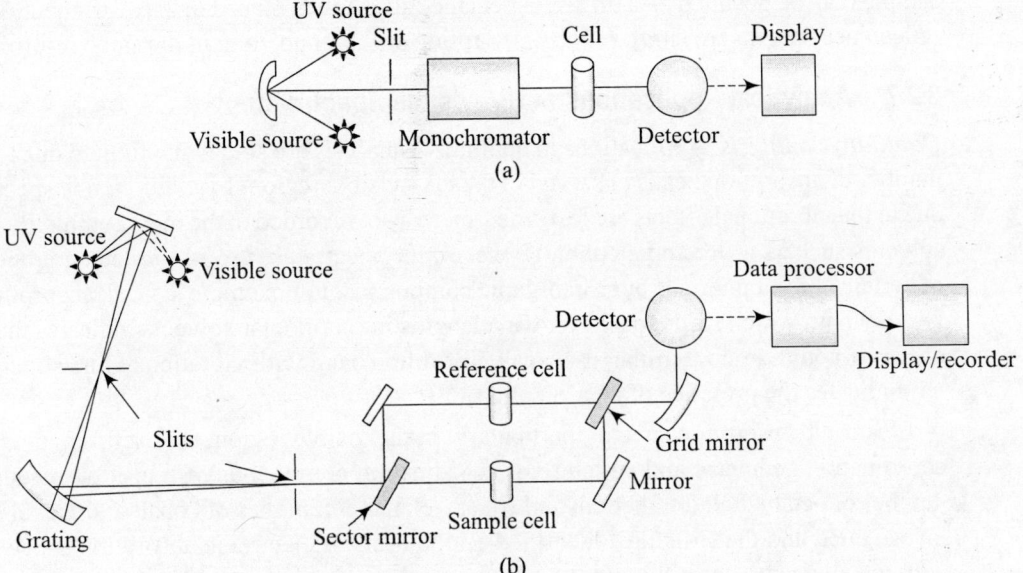

Fig. 8.5 (a) Schematic diagram of a single beam UV–visible spectrophotometer and
(b) Schematic diagram of a double beam UV–visible spectrophotometer

A detailed description of the various components of spectrophotometers has already given in Chapter 6. The components, such as the radiation sources, monochromators, sample cells, detector, and display unit, are common for both the single beam and double beam instruments. Double beam instruments are recording instruments and have additional components such as sector mirror or beam splitter and grid mirror or chopper.

The source of radiation for the visible region is usually a tungsten–halogen lamp and for the UV region a deuterium lamp is used. The UV–visible spectrophotometer contains both these sources covering the entire wavelength range of 190–800 nm. Monochromators disperse the polychromatic radiation from the source into narrow range of wavelengths. The dispersing element is either a quartz prism or mostly a grating which disperses the UV as well as visible radiation. Glass prisms can be used only for the visible region.

The monochromatic beam of radiation passes through the sample or solvent cell in a single beam instrument. The transmitted beam of attenuated intensity reaches the photodetector which converts the radiant energy to electrical energy which is displayed by a micro ammeter.

In the double beam instrument the monochromatic beam of radiation is split into two parallel beams by the sector mirror which pass through the sample and reference cells made of quartz (transparent to both UV and visible radiation) and reach the detector. The electrical signal generated by the detector is fed to a display or recorder unit.

Electronic absorption spectra are usually recorded for solutions. The spectrum of a sample (solid or liquid sample dissolved in a suitable solvent) is recorded by placing it in the sample cell and the solvent in the reference cell and scanning over the chosen wavelength region. The wavelength drive and chart drive of the recorder or the display unit are synchronized so that the detector signal converted into transmittance or absorbance units is recorded as a function of wavelength of the incident beam of radiation in the double beam instrument. Spectra of solid samples can be obtained by a diffuse reflectance attachment placed in the instrument. The diffuse reflectance spectra are similar to the absorption spectra and contain the same information.

8.2.7   Analytical Applications of UV–visible Spectroscopy

Qualitative analysis: Applications in qualitative analysis are somewhat limited due to the small number of absorption peaks observed in the UV–visible region. In addition, fine spectral details due to the vibrational effects are lost when spectra are recorded in the most commonly used polar solvents such as water and alcohol. However, the spectra are useful for detecting the presence of certain chromophoric groups in organic compounds. For example, a weak absorption band at 280–290 nm which shifts to shorter wavelengths in more polar solvents indicates the presence of the carbonyl group. Similarly, a weak absorption band with vibration at fine structure at 260 nm indicates the presence of an aromatic ring.

Electronic absorption spectra, particularly in the visible region, are useful in identifying the coordination geometry and structure of coordination compounds of transition metals. A large number of octahedral, tetrahedral, and square-planar complexes of cobalt and nickel have been investigated and the structures identified on the basis of electronic absorption spectra together with magnetic susceptibility measurements.

Quantitative analysis: Electronic absorption spectroscopy in the UV–visible region is highly useful technique for quantitative analysis essentially based on Beer–Lambert law. A detailed discussion on this aspect has already been given in Chapter 6 (Section 6.7).

Calibration chart method is the most widely used method in spectrophotometric analysis. In practice the absorption spectra of a solution of the analyte species in a suitable solvent is recorded to select the wavelength for quantitative analysis. A calibration chart is prepared by using a series of standards of the analyte and the absorbance is measured at the selected wavelength. The calibration plot will normally be linear over a narrow range of concentration, and hence, it is necessary to prepare standards within the concentration range of linearity. The absorbance of the test sample solution containing the analyte of unknown concentration is measured at the same wavelength under the same experimental conditions as carried out for the standards and the concentration is determined from the calibration chart as detailed in Section 6.7.

Standard addition method is helpful whenever interferences occur in the measurement of absorbance due to the presence of interfering species generally called *matrix effects*. For example, the absorbance of many metal complexes decreases in the presence of sulphate or phosphate as these anions tend to form colourless complexes with metal ions. The standard addition method involves the addition of one or more increments of a standard solution of the analyte to the test sample solutions of about the same size. Each solution is diluted to a final fixed volume. Absorbance is measured at a selected wavelength for the test sample solutions containing different amounts of the standard and also for the original test sample solution (without any added standard solution). A calibration plot of the measured absorbance values as a function of the volume of the standard solution added to the test sample gives a straight line with a slope $m = \varepsilon t C_S / V_t$ and an intercept $b = \varepsilon t V_S C_X / V_t$ to the *y*-axis (Fig. 7.8). The term ε is the absorptivity of the species, t is the path length of the absorbing medium, V_S and V_t refer to the volume of the standard solution and total volume respectively, and C_S and C_X are the concentrations of the standard solution and that of the test sample. The unknown concentration C_X is calculated using the formula

$$C_X = \frac{bC_S}{mV_X} \tag{8.2}$$

where V_X is the volume of the test sample taken.

8.2.8 Simultaneous Determinations

Simultaneous quantitative estimations of two analytes can be carried out even when their absorption spectra overlap to some extent as in the case of potassium dichromate and potassium permanganate (Fig. 8.6).

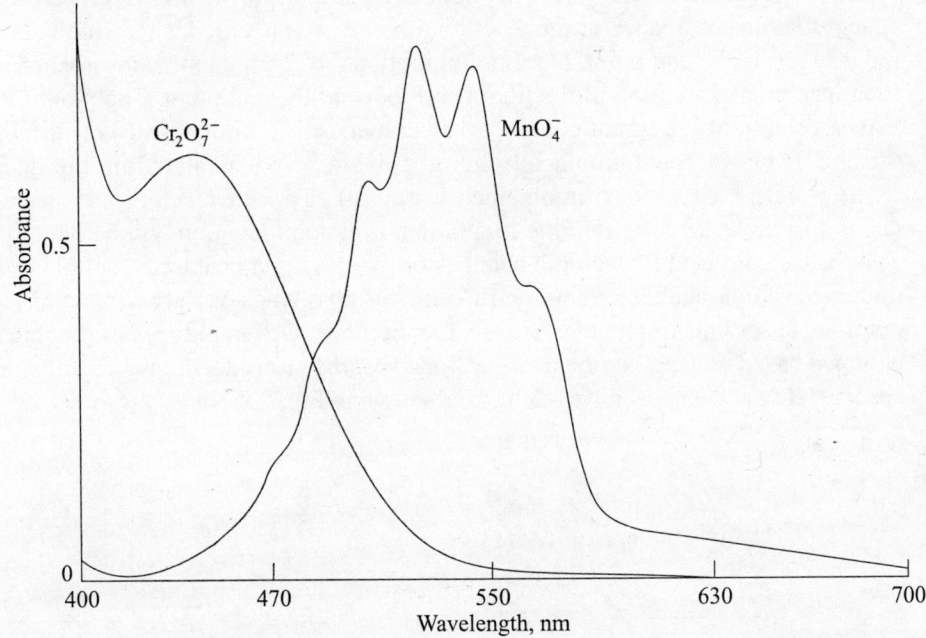

Fig. 8.6 Overlapping absorption spectra of potassium dichromate and potassium permanganate

The principle involved in simultaneous spectrophotometric determinations is based on Beer–Lambert law. The absorbance values are additive when the two components in a given mixture do not react with each other.

$$A_{\lambda 1} = A_{1(\lambda 1)} + A_{2(\lambda 1)} \tag{8.3a}$$
$$A_{\lambda 2} = A_{1(\lambda 2)} + A_{2(\lambda 2)} \tag{8.3b}$$

where $A_{\lambda 1}$ and $A_{\lambda 2}$ are the measured absorbance of the mixture of analytes at the two wavelengths λ_1 and λ_2 respectively and A_1 and A_2 are the absorbance values of the two analytes.

The wavelengths λ_1 and λ_1 are selected to coincide with the absorption maxima of the two analytes, analyte 1 absorbing strongly at λ_1 while analyte 2 absorbs strongly at λ_2. The Beer–Lambert equation for the mixture of analytes for a constant path length of the absorbing medium can be written as

$$A_{\lambda 1} = \varepsilon_{1(\lambda 1)} \, c_1 + \varepsilon_{2(\lambda 1)} \, c_2 \tag{8.4}$$
$$A_{\lambda 2} = \varepsilon_{1(\lambda 2)} \, c_1 + \varepsilon_{1(\lambda 2)} \, c_2 \tag{8.5}$$

The concentrations of the two analytes c_1 and c_2 can be obtained by solving the two simultaneous equations.

In practice the mixture of potassium dichromate and potassium permanganate is acidified with 1 M sulphuric acid to prevent hydrolysis of the anions. The two wavelengths for measuring the absorption are 440 nm and 545 nm for dichromate and permanganate ions respectively and the above equations can be solved simultaneously for determining the individual concentrations of the analytes.

8.2.9 Photometric Titrations

Photometric titrations also known as *spectrophotometric titrations* involve monitoring the changes in the absorbance of the reaction mixture at a specified wavelength during titration in order to locate the end point. During the titration the change in the absorbance is recorded for each incremental addition of the titrant well beyond the end point. The plot of the absorbance versus volume of the titrant consists of two intersecting straight lines of different slopes and the end point is located graphically by dropping a perpendicular from the intersection point to the x-axis. A curve may be obtained instead of sharp intersecting lines near the end point due to incomplete reaction or due to dilution. End point location will be better by plotting the absorbance corrected for volume changes or by using a concentrated solution of the titrant to minimize volume changes. Photometric titrations involving a variety of reactions may be carried out if the absorbing system obeys Beer–Lambert law. Different types of graphical plots can be obtained based on whether the analyte alone absorbs or product alone absorbs, or titrant alone absorbs, etc., at the chosen wavelength as shown in Fig. 8.7

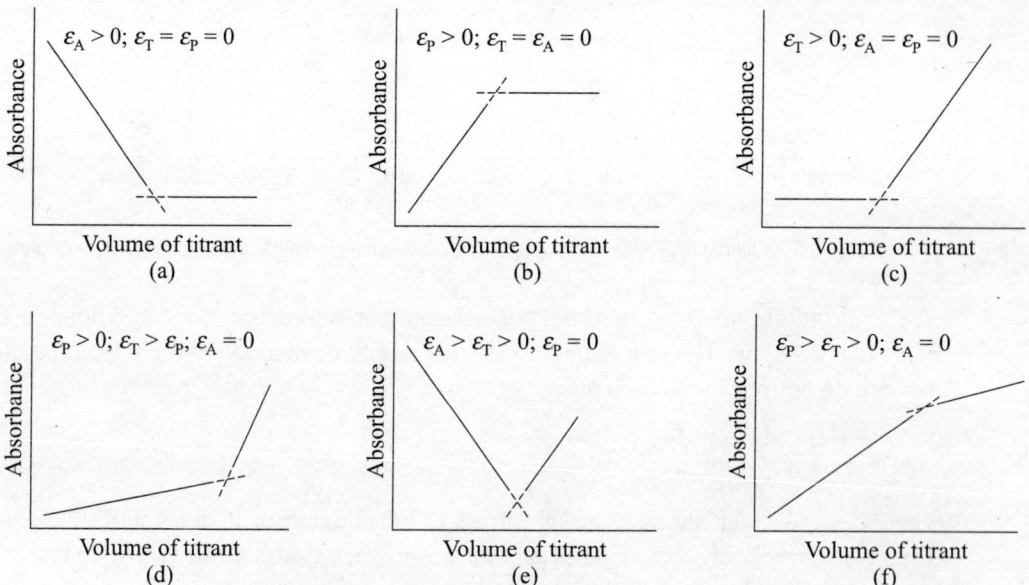

Fig. 8.7   Different types of photometric titration curves (ε_A, ε_P, and ε_T refer to the absorptivities of the analyte, product, and titrant respectively)

Examples of the titrations for some of the curves shown in Figs. 8.7(a)–(f) include curve a–iron(III)–salicylic acid complex vs. EDTA at 525 nm or p-toulidine vs. perchloric acid in butanol at 290 nm; curve b–copper(II) vs. EDTA at 745 nm; curve c–arsenic(III) vs. bromate-bromide mixture at 326 nm; and curve f–formation of coloured products of different compositions.

Even a mixture of metal ions can be quantitatively analysed by photometric titration without involving any preliminary separation step as in the case of the titration of a mixture of copper(II) and bismuth(III) with EDTA as shown in Fig. 8.8. Advantage is taken of the greater stability of Bi(III)–EDTA complex compared to that of Cu(II)–EDTA. At the chosen wavelength of 745 nm the cations and EDTA do not have any absorbance and the Bi–EDTA complex also does not

absorb. Hence, the absorbance remains close to zero till all the bismuth has reacted with EDTA. Copper reacts with EDTA only after all the bismuth has completely reacted and absorbance shows an increase as more of Cu–EDTA complex is formed. When all the copper has reacted absorbance levels off as the excess EDTA added does not absorb. Thus, two end points can be detected in a single titration curve.

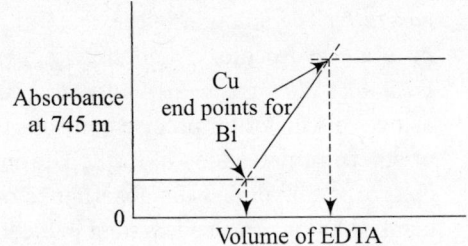

Fig. 8.8 Photometric titration curve for the titration a mixture of copper(II) and bismuth(III) vs EDTA

A similar photometric titration can be carried out with a mixture of arsenic(III) and antimony(III) with potassium bromate–bromide mixture at 326 nm. Arsenic(III) is oxidized first by the bromate–bromide mixture, the absorbance remaining unchanged. After all arsenic is oxidized antimony is oxidized, the absorbance decreasing to a minimum till the end point and increasing with the excess titrant.

Advantages of photometric titrations include (i) better precision and accuracy compared to direct absorbance measurement, (ii) very dilute solutions can be handled, (iii) greater choice of experimental conditions as it is sufficient that any one of species, namely, analyte, product, or titrant only need to absorb at the chosen wavelength, (iv) even slightly turbid solutions can be handled, and (v) relative changes in the absorbance alone are measured and not the absolute values and the end point can be easily detected by graphical means as long as straight line segments can be obtained and extrapolated.

8.2.10 Examples of Spectrophotometric Determinations

Iron content in a sample can be determined by complexing with 1,10-phenanthroline. Iron(II) forms an orange-red coloured complex $[Fe(C_{12}H_8N_2)_3]^{2+}$ while iron(III) forms a yellow coloured complex with the same reagent $[Fe(C_{12}H_8N_2)_3]^{3+}$. Both iron(II) and iron(III) can be determined from absorbance measurements of the complexes. The test sample containing both iron(II) and iron(III) is made slightly acidic with sulphuric acid and treated with 1,10-phenathroline reagent and buffered to pH 3.9 with potassium hydrogen phthalate. The absorbance at 396 nm gives the total amount while absorbance at 515 nm is due to the iron(II) complex alone.

The iron(II) complex is quite stable and the intensity of the orange-red colour is independent of pH in the range 2–9, and hence, is used for spectrophotometric determination. Any iron(III) that may be present is reduced to iron(II) with hydroxylammonium chloride. A calibration chart is prepared by using a series of standard solutions of ferrous ammonium sulphate, buffering the solutions with 0.2 M sodium acetate, reacting with 0.25% solution of 1,10-phenathroline monohydrate in water and measuring the absorbance at 515 nm. The test sample is treated in a similar manner to form the complex and its absorbance is measured under identical conditions. The calibration chart of measured absorbance as a function of iron concentration is linear and the concentration of iron in the test sample is determined graphically.

Lead content of a test sample is determined spectrophotometrically by the dithizone method. Lead forms a stable complex with dithizone which is extracted into chloroform at pH 9.5 and the absorbance measured at 510 nm. In practice a known volume of the lead solution is taken in a separating funnel and mixed with about 50 mL of ammonia–cyanide–sulphite solution (prepared by adding 0.15 g of sodium sulphite to a 100 mL solution containing 35 mL of concentrated ammonia and 3 mL of 10% KCN solution). *KCN is extremely poisonous and should be handled carefully.* The solution in the separating funnel is carefully adjusted to pH 9.5 by the addition of dilute hydrochloric acid slowly. *The pH should not go below 9.5 even temporarily as the poisonous HCN gas will be liberated.* It is safe to carry out the pH adjustment in a fume cupboard using a pH meter for precise control of pH. After the pH has been adjusted to 9.5 about 20 mL of chloroform is added to the separating funnel to extract the lead–dithizone complex into the chloroform layer and the absorbance of the chloroform layer is measured at 510 nm against a blank solution. The lead content in the test sample is determined graphically from the calibration chart prepared using standard solutions of lead nitrate.

Ammonia is determined by Nessler's method by reacting with Nessler's reagent. It forms an orange-brown colloidal product and absorbance measurement at 525 nm is made before the colloid flocculates. Nessler's reagent is potassium tetraiodomercurate(II) prepared by mixing potassium iodide (100 mL of 35% solution) with a 4% solution of mercuric chloride (~325 mL) till the red precipitate of HgI_2 initially formed is dissolved almost completely and only a slight red precipitate remains. The mixture is stirred with 250 mL of sodium hydroxide containing 120 g and finally made up to 1 L with distilled water. A small amount of mercuric chloride solution is added till a permanent turbidity is obtained. The reagent is allowed to stand for 24 hours and the decanted clear pale yellow coloured solution is stored in a brown bottle and used for analysis. The concentration of ammonia in a test sample is determined from the calibration chart prepared using standard solutions of ammonium chloride. The calibration chart will be linear only in very dilute solutions in the range of 0.01–1.0 mg of ammonia per mL. The reaction between ammonia and the reagent may be represented as

$$2\ K_2HgI_4 + 2\ NH_3 \rightarrow NH_2HgI_3 + 4\ KI + NH_4I$$

Nitrite in environmental samples is determined spectrophotometrically by measuring the absorbance of the dye formed by diazotizing sulphanilamide and coupling with *N*-(1-naphthyl)-ethylenediamine dihydrochloride. The nitrite sample (100 mL) is treated with 2 mL of sulphanilamide solution (prepared by dissolving 0.5 g of the amide in 100 mL of 20% hydrochloric acid). After about 5 minutes, 2 mL of *N*-(1-naphthyl)-ethylenediamine dihydrochloride solution (prepared by dissolving 0.3 g of the reagent in 100 mL of 1% hydrochloric acid) is added. The pH of the reaction mixture should be about 1.5 at this point. After about 10 minutes, absorbance of the dye solution is measured at 550 nm against a blank. The amount of nitrite in the test sample is determined from the calibration chart prepared by using standard nitrite solutions and treated in a similar manner.

Nitrate content in drinking water is determined by reacting with phenoldisulphonic acid to give a yellow coloured solution which has absorption maximum at 410 nm. Chloride and nitrite (at concentration >0.2 ppm) interfere in the determination. Chloride is precipitated out by adding silver sulphate solution (0.44%) and centrifuged. The clear centrifugate is used for nitrate determination. The sample water (100 mL) is adjusted to pH 7 by adding dilute sodium hydroxide solution and evaporated to dryness in a china dish. The dry residue is dissolved in 2 mL of phenoldisulphonic acid reagent (prepared by dissolving 2.5 g phenol in 15 mL of concentrated sulphuric acid and adding 7.5 mL of fuming sulphuric acid followed by heating for 2 hours on a hot water bath). The solution is diluted with about 20 mL distilled water and ammonia solution (~6 mL) is added till maximum colour is developed. If any turbidity is formed dissolve it by adding EDTA solution (prepared by forming a paste with 50 g of disodium salt of EDTA win 20 mL water and dissolving the paste in 60 mL of concentrated ammonia solution). The clear solution is transferred quantitatively to 50 mL volumetric flask and made up to the mark. Absorbance of the solution is read at 410 nm using a blank and the concentration of nitrate in the sample is determined from the calibration chart prepared by using standard nitrate solutions treated in a similar manner.

Protein concentration in solutions is determined by different methods, the most commonly used methods being (i) direct measurement of absorbance at 280 nm, (ii) Lowry method, and (iii) Bradford method.

Direct measurement of absorbance at 280 nm is based on the fact that most proteins absorb strongly ultraviolet light at 280 nm attributed to tyrosine and tryptophan residues in the protein. The measured absorbance can be taken as directly proportional to protein concentration in accordance with Beer–Lambert law. The analyte protein solution is taken in quartz cell or cuvette and the absorbance at 280 nm is directly read in the UV spectrophotometer against water or buffer solution taken in the reference cell. The protein concentration is determined from the calibration graph prepared using a series of standard solutions of bovine serum albumin (BSA) under identical conditions.

The method has the advantages of being simple, rapid, and non-destructive. The method is particularly useful in monitoring the concentration profile of proteins eluting out of chromatographic columns. Since the amounts of these amino acids vary in different proteins the method provides only an initial estimate of protein concentration. Further it suffers from the disadvantage of interference from nucleic acids commonly found in protein extracts which absorb strongly in the ultraviolet region with a λ_{max} at 260 nm. The interference from nucleic acids is overcome by using the empirical formula developed on the basis of measurements of absorbance at 280 and 260 nm of mixtures of pure protein and pure nucleic acid.

$$P = 1.55\,A_{280} - 0.76\,A_{260} \tag{8.6}$$

where P refers to protein concentration in mg/mL and the terms A_{280} and A_{260} refer to the absorbance measured at 280 nm and 260 nm respectively. The empirical formula can be used for protein solutions containing up to 20% nucleic acids.

Lowry method is quite sensitive and can detect protein content as low as 2 μg. The method involves the addition of alkaline copper sulphate solution. The Cu^{2+} ions form complexes with the peptide nitrogen and get reduced to Cu^+. On the addition of Folin–Ciocalteau reagent (a mixture of sodium tungstate, sodium molybdate, phosphoric acid, and hydrochloric acid) the Cu^+ ions get oxidized to Cu^{3+} and react with tyrosine, tryptophan, and cysteine residues of the protein to form imino peptide. The added Folin–Ciocalteau reagent is reduced to molybdenum–tungsten blue whose absorbance is measured at 727 nm or at 500 nm (when $A > 2$). The concentration of the protein in a test sample is usually determined graphically from the calibration chart prepared with standard solutions of BSA.

Bradford method is relatively widely used for protein estimation as it is more sensitive (detection limit being 0.2 μg) and rapid (~ 2–3 minutes) compared to Lowry method. The method involves the addition of the pale red solution of the dye Coomassie Blue G-250 which reacts with the arginine residues of the analyte protein to give blue colour. Absorbance is measured at 590 nm and the protein content in the test sample is determined from the calibration chart prepared using standard solutions of BSA.

8.3 INFRARED SPECTROPHOTOMETRY

Infrared spectrophotometry is also known as *vibrational spectroscopy* based on the nature of interaction of the infrared (IR) radiation with the vibrational modes of molecules. The technique is widely used for qualitative and quantitative analysis, checking purity of samples, and elucidation of molecular structure. Molecules undergo vibrational transitions on absorbing energy in the form of IR radiation. The transitions between vibrational energy levels of molecules give rise to the absorption bands characteristic of the molecule in the form of an IR absorption spectrum or vibrational spectrum.

8.3.1 Infrared Region

The frequency region of the electromagnetic spectrum between 12,800 cm^{-1} and 10 cm^{-1} is known as IR region. The IR region is usually sub-divided into three regions based on the nature of interactions of the regions of radiation with molecules as well as the requirements of instrumentation. The regions include (i) near-infrared (overtone region extending in the range 12,800–4000 cm^{-1}), mid-infrared (vibration–rotation region of 4000–200 cm^{-1}), and far-infrared (rotation region of 200–10 cm^{-1}). The near-infrared region is useful for quantitative analysis (e.g., mixtures of aromatic amines) while the far-infrared region is more useful for analysis of the structure of molecules. The mid-IR region commonly known as the conventional IR spectral region is used for identification of functional groups, quantitative analysis, and detecting impurities in samples.

The prerequisite for a molecule to interact with IR radiation is the presence of a permanent dipole which varies continuously in its magnitude. Molecules containing two atoms of the same element (*homonuclear diatomic molecules*), for example, H_2, N_2, O_2, Cl_2, etc., do not have a dipole as the electronegativity difference between the combining atoms is zero. Such molecules cannot absorb infrared radiation, and hence, do not exhibit vibrational spectra. *Heteronuclear diatomic molecules*, for example, HCl, CO, etc., have a permanent dipole as the difference in the electronegativities of the combining atoms is greater than zero. The magnitude of the dipole varies with time due to normal vibration of the molecule (molecular vibrations). Hence, heteronuclear diatomic molecules and polyatomic molecules which also have continuously varying dipole are capable of absorbing infrared radiation resulting in vibrational–rotational transitions, and hence, exhibit vibrational spectra.

8.3.2 Molecular Vibrations

The relative positions of the atoms in a molecule are not fixed rigidly but change or fluctuate due to different types of vibrations and all molecules exhibit molecular vibrations. A simple diatomic molecule, such as H_2 or N_2, has only one vibrational mode called the *stretching* mode. The stretching mode of vibration involves the stretching or elongation and compression resulting in a continuously varying interatomic distance (bond distance) along the bond axis between the combining atoms [vide Figs. 8.9(a) and (b)]. However, the stretching mode of vibration in such homonuclear diatomic molecules does not bring about an interaction with IR radiation as the molecules do not have a permanent dipole. In contrast, in a simple heteronuclear diatomic molecule, such as HCl or CO, the stretching vibration results in a continuously varying dipole, and hence, the molecule can absorb energy in the form of IR radiation. Since the molecule has only one vibrational mode it will absorb IR radiation only at one frequency called the *resonance frequency*, which is equal to the frequency of the vibrational mode of the molecule. The molecule on absorbing energy gets excited to a higher vibrational energy level and thereby exhibits one absorption band in the IR spectrum. HCl shows one IR absorption band at 2990 cm^{-1}.

$$
\begin{array}{ccc|ccc}
& & & \overset{\delta+}{} \quad \overset{\delta-}{} & \overset{\delta+}{} \quad \overset{\delta-}{} & \overset{\delta+}{} \quad \overset{\delta-}{} \\
H\text{——}H & H\text{—}H & H\text{-}H & H\text{——}Cl & H\text{——}Cl & H\text{-}Cl \\
\text{elongation} & \text{equilibrium} & \text{compression} & \text{elongation} & \text{equilibrium} & \text{compression} \\
& \text{(a)} & & & \text{(b)} &
\end{array}
$$

Fig. 8.9 Stretching vibrations in diatomic molecules (a) H_2 and (b) HCl

Molecular vibrations associated with a continuously varying dipole moment facilitate the interaction of the molecule with IR radiation leading to absorption of energy by the molecule. The molecule is said to undergo a vibrational transition as the energy absorbed is utilized to raise the energy of the molecule from a lower vibrational energy level (usually the ground state represented as $v = 0$) to a higher vibrational energy level (vibrational excited state).

Molecules containing more than two atoms (triatomic and polyatomic molecules) have more number of molecular vibrations depending on the number of atoms and shape of the molecule. A molecule consisting of N number of atoms has $3N$ degrees of freedom of which three are translational degrees of freedom and another three are rotational degrees of freedom along the three principle axes of inertia. Hence, the number of fundamental vibrations for a molecule is given by the formula $3N-6$ for a non-linear molecule. Linear molecules have only two rotational degrees of freedom as no energy change is involved in the rotation along the main axis. Hence, the number of fundamental vibrations for a linear molecule is $3N-5$ only.

A triatomic molecule such as CO_2 (linear) is associated with ($3N-5$) vibrations, that is, $3 \times 3 - 5 = 4$ fundamental vibrations. A bent (V shaped) triatomic molecule, such as H_2O, shows ($3N-6$), that is, $3 \times 3 - 6 = 3$ molecular vibrations. The number of fundamental modes of vibrations in polyatomic molecules conforms to ($3N-6$) formula.

The different molecular vibrations in triatomic and polyatomic molecules can be classified into two main types: stretching vibrations and bending or deformation vibrations.

Stretching vibrations are defined as those in which the atoms remain in the same bond axis but the distances between the atoms decreases or increases. The stretching vibrations can be of two types (i) *symmetric stretching* and (ii) *asymmetric stretching* as shown in Fig. 8.10(a).

Bending or deformation vibrations are those in which the positions of the atoms change relative to the original bond axis. Four different types of bending vibrations can be visualized and these include (i) scissoring, (ii) rocking, (iii) wagging, and (iv) twisting vibrations as shown in the Fig. 8.10(b).

The symmetric stretching vibration in CO_2 is not associated with a dipole as the two oxygen atoms (δ^- charges) are always at an equal distance from the carbon but in opposite directions cancelling out the dipole moment. Linear molecules cannot interact with IR radiation in their symmetric stretching vibration. Hence, the symmetric stretching vibration is IR inactive. In contrast, the asymmetric stretching vibration in a linear molecule is associated with a continuously varying permanent dipole, and hence, the molecule can absorb energy in the form of IR radiation giving rise to an IR absorption band.

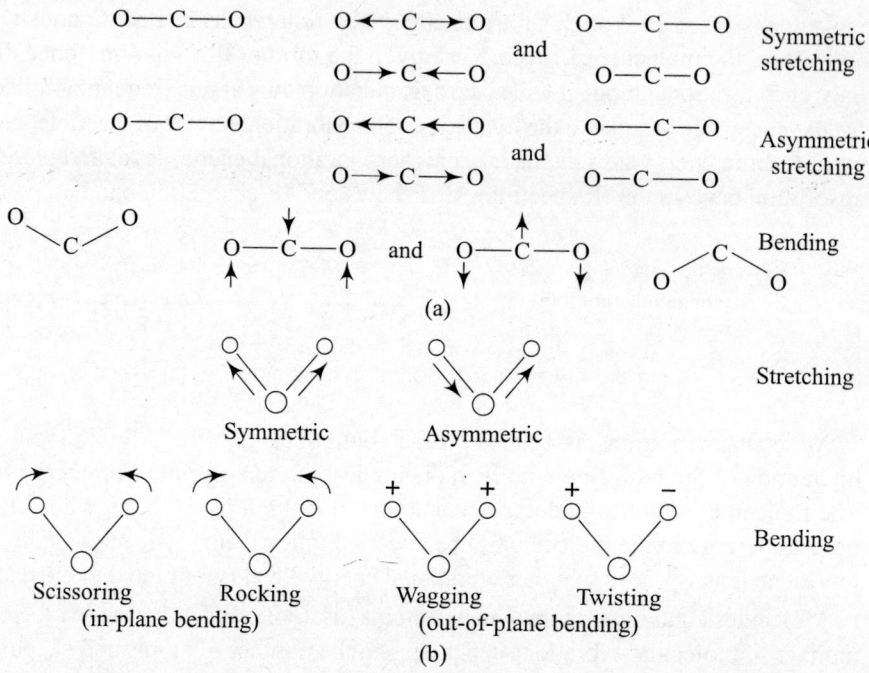

Fig. 8.10 Stretching and bending modes of vibrations in triatomic molecules
(a) CO_2 (linear) and (b) bent molecule (e.g., H_2O)

Linear molecules exhibit two bending motions as shown in Fig. 8.10 and both the molecular vibrations have the same frequency (energy), and hence, represent two degenerate (energetically equivalent) vibrational energy levels. The two bending molecular vibrations absorb at the same frequency in the IR region, and hence, a single IR absorption band is observed in the IR spectrum. Thus, carbon dioxide exhibits only two IR absorption bands at 2349 cm^{-1} for the asymmetric stretching and at 667 cm^{-1} for the symmetric bend.

A bent triatomic molecule, such as water, has three vibrational modes designated as symmetric stretching, asymmetric stretching, and symmetric bending (scissoring) modes all of which are associated with continuously varying dipole. Hence, the molecule can absorb IR radiation in all the three modes of vibration. Thus, the IR spectrum of H_2O shows three absorption bands at 3652, 3756, and 1596 cm^{-1} for the symmetric stretching, asymmetric stretching, and bending vibrations respectively.

8.3.3 Vibrational Frequencies and IR Absorption Bands

The vibrational frequency ω_{osc} of the normal mode of vibration in a diatomic molecule, and hence, the resonance frequency at which the molecule absorbs IR radiation may be calculated as given by the following equation.

$$\omega_{osc} = (1/2\pi) \sqrt{K/\mu} \text{ Hz} \tag{8.7a}$$

or

$$\omega_{osc} = (1/2\pi c) \sqrt{K/\mu} \text{ cm}^{-1} \tag{8.7b}$$

(Since the frequency in cm^{-1} = frequency in Hz divided by velocity of light)

where K is the force constant of the bond between the atoms and represents the strength of the bond, μ is the reduced mass of the atoms given by $(m_1 m_2 / (m_1 + m_2))$, and c is the velocity of light. The equation is valid on the assumption that a diatomic molecule can be represented as a classical mechanical model of a simple harmonic oscillator (Fig. 8.11). The vibrational frequencies increase with increasing bond strength and with decreasing mass of the vibrating atoms.

The vibrational energy levels in the molecule are quantized and the allowed vibrational energies for a simple harmonic oscillator based on Schrödinger equation is given as

$$E_v = (v + \tfrac{1}{2})\, h\, \omega_{osc} \text{ joules} \tag{8.8a}$$

or

$$E_v = (v + \tfrac{1}{2})\, \omega_{osc} \text{ cm}^{-1} \tag{8.8b}$$

(Since E_v in cm^{-1} = E_v in joules divided by hc)

where v is called the *vibrational quantum number*. The lowest vibration energy called the ground state vibrational energy of $E_{v0} = \tfrac{1}{2}\, \omega_{osc}$ is obtained for $v = 0$ in the above equation indicating that any molecule cannot have zero vibrational energy meaning that the atoms can never be at rest relative to each other. The quantity $\tfrac{1}{2}\, \omega_{osc}$ is called the *zero point energy* of the molecule.

The harmonic oscillator approximation is not realistic in that the bond between the atoms cannot break which is not true. A more realistic model is that of an anharmonic oscillator (represented by Morse potential curve as shown in Fig. 8.11) which takes into account the possibility of dissociation of the molecule by breaking the bond.

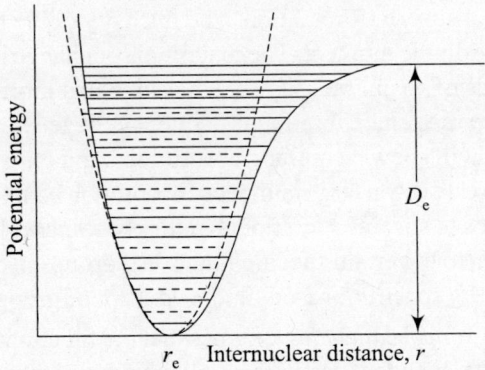

Fig. 8.11 Vibrational energy levels of a diatomic molecule showing a simple harmonic motion (dotted curve) and an anharmonic motion of Morse

Taking into consideration the quantum mechanical selection rules for the anharmonic oscillator of $\Delta v = \pm 1, \pm 2, \pm 3, \dots$ the observed frequency of a vibrational transition $\omega_{observed}$ representing the *fundamental absorption* is given by

$$\omega_{observed} = \omega_{osc}\, (1 - 2X_e) \text{ for the } v = 0 \rightarrow v = 1 \text{ transition; } \Delta v = \pm 1$$

The term ω_{osc} is actually the hypothetical equilibrium oscillation frequency of the anharmonic system and represents the vibrational frequency of infinitely small vibrations about the equilibrium point and X_e represents the anharmonicity constant.

Since vibrational transitions $\Delta v = \pm 2$ and $\Delta v = \pm 3$, etc., are also allowed the absorptions due to these transitions are called the *overtones*. The intensities of the overtone absorption bands are much weaker and it is sufficient to consider the first and second overtones only, the observed frequencies being given by

$$\omega_{observed} = 2\omega_{osc}(1 - 3X_e) \text{ for the } v = 0 \rightarrow v = 2 \text{ transition; } \Delta v = \pm 2$$

and

$$\omega_{observed} = 3\omega_{osc}(1 - 4X_e) \text{ for the } v = 0 \rightarrow v = 3 \text{ transition; } \Delta v = \pm 3$$

Since the value of X_e is quite low in the range of 0.01–0.2, the observed spectral transitions of fundamental, first overtone, and second overtone appear very close to ω_{osc}, $2\,\omega_{osc}$, and $3\,\omega_{osc}$ respectively.

The quantum mechanical selection rules also permit *combination bands* and *difference bands*. Combination bands arise due to the addition of two or more fundamental frequencies or overtones while difference bands arise due to the difference between two or more fundamental frequencies or overtones. The intensities of these bands are normally very weak but are often observed in the IR spectra.

8.3.4 Infrared Spectrum

In polyatomic molecules the number of vibrational modes increases with increasing number of atoms N. In addition to the number of fundamental absorption bands, overtones, combination, and difference bands also appear in the IR spectrum making the analysis of the spectrum complex. The IR spectrum may not show all the possible vibrational modes as absorption of IR radiation occurs only if a change in the dipole moment occurs during the vibration. In several vibrational modes the change in dipole moment may be nullified, and hence, such vibrations are IR inactive, that is, they will not absorb IR radiation. The intensity of the IR absorption band is proportional to the change in the dipole moment. The magnitude of change in dipole moment may be too small giving rise to bands of weak intensity. It is possible that more than one vibrational mode of the molecule may have the same vibrational frequency (degenerate modes) and all such degenerate vibrational modes will show a common absorption band only. The IR spectrum of a polyatomic molecule exhibits relatively a large number of absorption bands of varying intensity and assigning the individual bands to specific vibrational modes becomes difficult. However, many absorption bands can be assigned to certain specific functional groups in the molecules simplifying the task of interpreting the IR spectrum based on the concept of group frequencies.

The *concept of group frequencies* assumes that the functional groups in polyatomic molecules may be treated as independent oscillators and vibrational frequencies of most functional groups are independent of the rest of the structure of the molecule as a whole. The approximate frequency at which an organic functional group absorbs IR radiation can be calculated from the masses of the atoms and the force constants of the bonds between them. It is possible to assign a range of frequencies within which the absorption peak for the particular functional group appears in the IR spectrum. For example, the C=C double bond of the group $-CH=CH_2$ shows an IR absorption band at 1650–1652 cm^{-1} in the IR spectrum of propene ($H_3C-CH=CH_2$) as well as in 1-bromopropene ($BrH_2C-CH=CH_2$). Similarly, the $-CH_3$ group shows characteristic IR absorption bands quite independent of the nature of the molecule assigned to symmetric C–H stretching vibration at 2850–2890 cm^{-1}, asymmetric stretching vibration at 2940–2980 cm^{-1}, a symmetric deformation at 1375 cm^{-1}, and an asymmetric deformation at 1470 cm^{-1}.

Most of the group frequencies are observed extending between 3600 cm^{-1} and 1200 cm^{-1} in the mid-infrared spectrum (4000–600 cm^{-1}), and hence, this region in the spectrum is called the *group frequency region*. However, certain group frequencies do appear in the region beyond 1200 cm^{-1} (e.g., C–O–C stretch appears at about 1200 cm^{-1} and C–Cl stretching vibration appears at about 800 cm^{-1}.

The characteristic IR absorption frequencies of certain bonds and functional groups are highly useful in the identification of such groups in a wide variety of molecules. The characteristic frequencies of a few molecular groups are listed in the form of a simplified *correlation chart* in Table 8.3.

Table 8.3 Characteristic group frequencies of a few functional groups— A simplified correlation chart

Functional group	Nature of compound	Frequency (cm^{-1})
C–H	Alkanes	2850–2970; 1340–1470
	Alkenes	3010–3095; 675–995
	Alkynes	3300
	Aromatic rings	3010–3100; 690–900
O–H	Alcohols, phenols	3590–3650
	H-bonded alcohols and phenols	3200–3600
	Carboxylic acids	3500–3650
	H-bonded carboxylic acids	2500–2700
N–H	Amines, amides	3300–3500
S–H	Thiols	2580
C=C	Alkenes	1610–1680
C=C	Aromatic rings	1500–1600
C≡C	Alkynes	2100–2260
C–O	Alcohols, ethers, carboxylic acids, esters	1050–1300
C=O	Aldeyhydes, ketones, carboxylic acids, esters	1690–1760
C–N	Amines, amides	1180–1360
C=N	Imines and oximes	1640–1690
C≡N	Nitriles	2210–2280
C–F	fluoride	1050
C–Cl	chloride	725
C–Br	bromide	650
C–I	iodide	550

The region between 1200 cm^{-1} and 600 cm^{-1} in the mid-infrared spectrum is called the *finger print region*. In the finger print region, absorption bands due to most of the groups containing single bonds appear as the bond energies are about the same and strong interaction occurs between neighbouring bonds. The absorption bands, thus, arise depending on the various interactions and the overall skeletal structure of the molecule. It is difficult to interpret the spectral bands in this region due to the complex nature of interactions. However, the complex interactions give rise to absorption bands which are unique to a molecule, and hence, the finger print region is useful for final identification of the compound.

Infrared spectra of a few compounds are shown in Figs. 8.12(a)–(d). The IR spectrum of *n*-decane [Fig. 8.12(a)] shows absorption bands at ~2900 cm^{-1} assigned to sp^3 hybridized C–H stretching; ~1460 cm^{-1} assigned to CH$_2$ bending and asymmetric CH$_3$ bending; at 1380 cm^{-1} (symmetric −CH$_3$ bending) and at 720 cm^{-1} (rocking mode of the CH$_2$ groups).

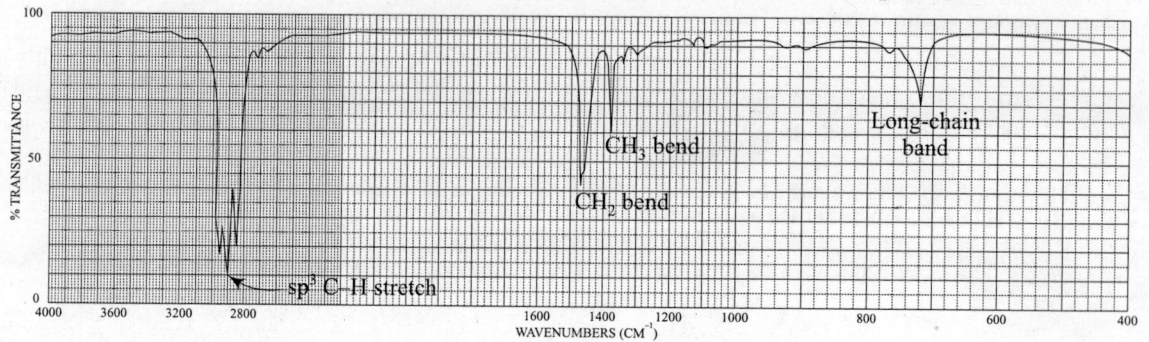

Fig. 8.12   (a) IR spectrum of n-decane

Figure 8.12(b) shows the IR spectrum of benzaldehyde. The absorption band at 1700 cm^{-1} is assigned to the C=O stretch of the carbonyl group conjugated to the aromatic ring. Conjugation of the carbonyl group with an aryl or an $\alpha\beta$ double bond shifts the normal C=O stretching band to a lower frequency at about 1690−1680 cm^{-1}. The IR absorption bands at 2750 cm^{-1} and at 2850 cm^{-1} are assigned to the C−H stretch of the aldehyde group and the sp^2 hybridized C−H stretch. These bands are important in distinguishing between aldehydes and ketones. The doublet seen in the region 2860−2750 cm^{-1} is due to Fermi resonance (coupling of the fundamental vibration with an overtone or combination band). The bands in the region 1600−1450 cm^{-1} are assigned to aromatic C=C stretch. The coupling of the aldehyde C−H stretching vibration to the first overtone of the C−H bending vibration results in IR bands appearing at 1400−1350 cm^{-1} region. The out-of-plane bending vibration of the mono-substituted group is seen in the region 800−650 cm^{-1}.

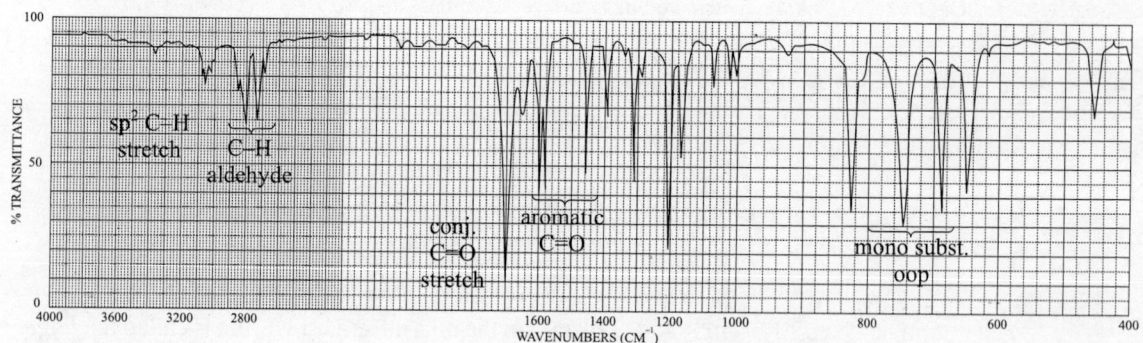

Fig. 8.12   (b) IR spectrum of benzaldehyde

The IR spectrum of butylamine is shown in Fig. 8.12(c). The N−H stretch of the primary amine is seen at 3500−3300 cm^{-1} as doublet. Secondary amines have one band while tertiary amines do not have N−H stretch. The N−H bending vibration in primary amines appears as a broad band at 1650−1640 cm^{-1} while secondary amines show an absorption band at 1500 cm^{-1}.

The out-of-plane bending vibration of N−H group is seen as a broad band centred around 700 cm^{-1}. The absorption band due to C−N stretch appears at 1100 cm^{-1}.

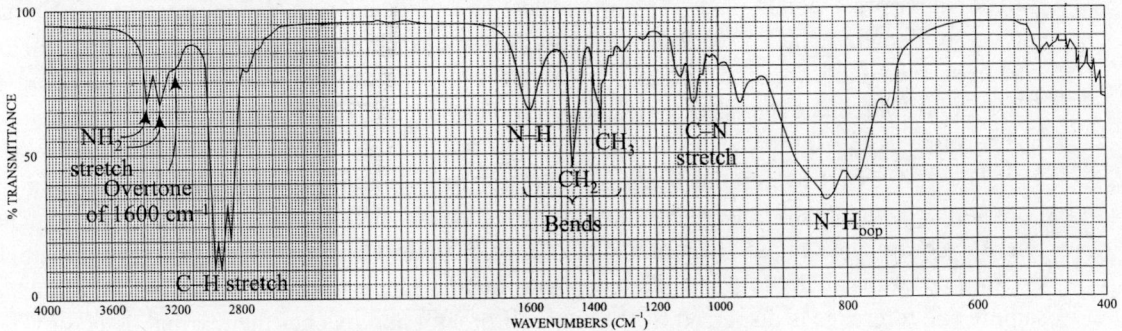

Fig. 8.12 (c) IR spectrum of butylamine

Figure 8.12(d) shows the IR spectrum of isobutryric acid. A very broad absorption band spread over $3400-2400$ cm^{-1} is assigned to hydrogen bonded O–H stretching which overlaps with the absorption due to C–H stretch. The absorption band due to C=O stretch is observed at 1710 cm^{-1} and that of the C–O stretch at 1200 cm^{-1}.

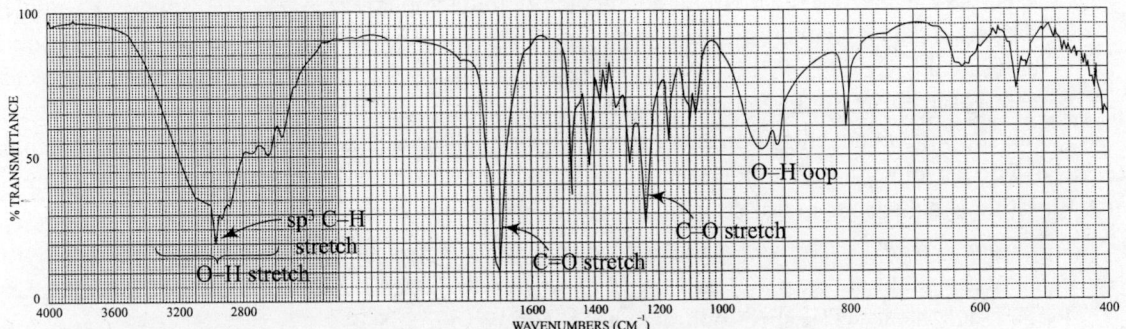

Fig. 8.12 (d) IR spectrum of isobutyric acid

8.3.5 IR Spectrophotometer

Most commonly used commercial IR spectrophotometers include dispersive instruments and FTIR instruments. In addition non-dispersive photometers find use in the analysis of volatile organic components in atmospheric gases while reflectance photometers are largely used for the analysis of solid samples.

Dispersive instruments are mostly double beam recording instruments using reflection grating monochromators for dispersing the incident radiation. A schematic diagram of the double beam IR spectrophotometer is shown in Fig. 8.13.

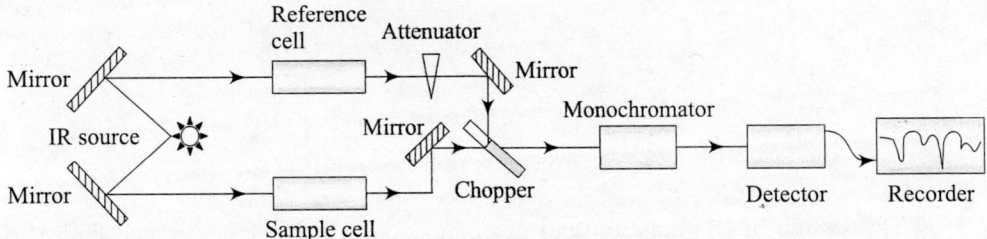

Fig. 8.13 Block diagram of a double beam IR spectrophotometer

The source of IR radiation is a (i) nichrome heating coil wound on a ceramic support, (ii) Nernst glower, a filament made of oxides of zirconium, yttrium, cerium, and thorium, or (iii) Globar, silicon carbide (carborundum) rod. Electrical heating of the source to temperatures in the range 1200–2000°C facilitates emission of IR radiation. The beam of polychromatic radiation generated by the source is split into two parallel beams for passing through the reference cell and the sample cell. The power of the reference beam is attenuated to match the power of the sample beam by imposing a device that removes continuously the variable fraction of the reference beam. The reference beam and sample beam pass on to a motor-driven chopper which reflects or transmits the sample beam alternately into the monochromator. The polychromatic radiation from the sample and reference is dispersed by the monochromator and reaches the thermal detector. The monochromators used in dispersive IR spectrophotometer are prisms made of alkali metal halides (mostly NaCl or KCl) or gratings. The optical signal is converted into an electrical signal in the detector (transducer), amplified, and displayed. Detectors used in IR instruments include selective or non-selective thermal detectors. Selective detectors are photon detectors, such as photocells or photoconductive cells or semiconductor devices (e.g., mercury cadmium telluride detector), which convert the incident radiation into electric current. Non-selective detectors are thermal detectors that convert the thermal radiant energy into temperature sensitive response. Examples of thermal detectors exhibiting different temperature sensitive responses include thermocouple (emf or voltage changes), thermistor and bolometer (changes in resistance), pyroelectric detector (changes in electric polarization), or pneumatic cell (changes in the pressure of an enclosed gas). The various components of different types of spectrophotometers have been described in Chapter 6. The wavelength selector and chart are driven by a synchronous motor simultaneously and a similar synchronous motor couples the pen drive and the attenuator.

Fourier transform infrared (FTIR) *spectrophotometer* has the advantages of greater sensitivity (greater signal to noise ratio) and speed compared to dispersive instruments. A block diagram of a single beam FTIR spectrophotometer is shown in Fig. 8.14.

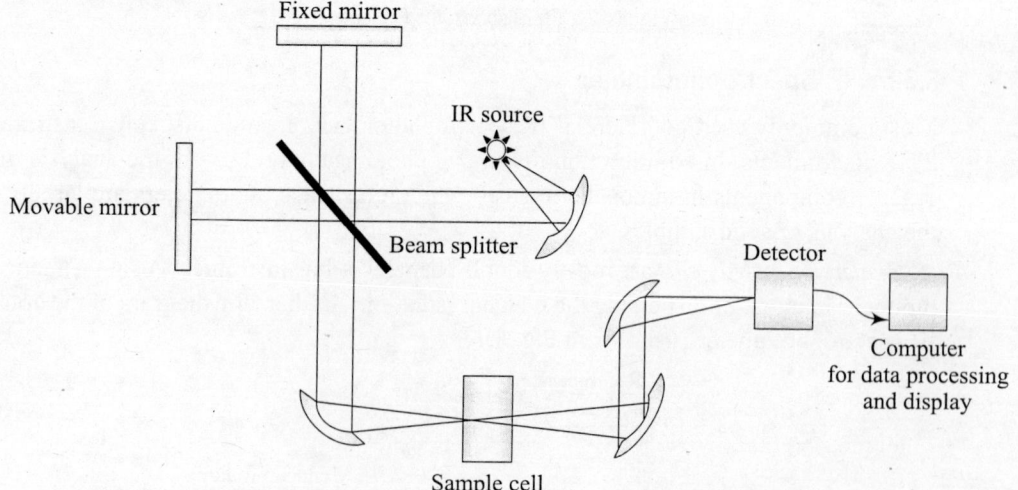

Fig. 8.14　Block diagram of a single beam FTIR spectrophotometer

The source of IR radiation and detector used in the dispersive and FT instruments are common. However in an FTIR spectrophotometer, a Michelson-type interferometer allows the polychromatic radiation from the source to impinge on the sample cell and enables all the absorption data to be collected simultaneously over the entire mid-IR region. In the interferometer the beam of radiation from the source is split into two beams of equal intensity by a beam splitter and passed on to fixed and moving mirrors. The two beams are reflected back and recombine at the beam splitter and then pass through the sample onto the detector. Continuous movement of the movable mirror forwards and backwards over a short distance generates a dynamic interference pattern due to the difference in the optical paths between the two halves of the recombined beam. The dynamic interference called the *interferogram* is monitored by the detector. The interferogram is recorded several thousand times by a microcomputer during one cycle of the movement of the mirror which takes about 0.1 second. The digitized interferogram is mathematically transformed by a point-by-point fast *Fourier transform* by the computer in less than 1 second into the conventional IR spectrum and displayed. Thus, FT instrument performs multiple scans within a short time giving the advantages of both enhanced sensitivity and speed of recording the spectrum.

8.3.6 Sample Preparation

Solid, liquid, and gaseous samples can be used for recording the IR spectrum. Solid samples can be handled in the form of KBr pellet, Nujol mull, solution, or film. (i) KBr pellet or wafer is prepared by mixing and grinding the sample (5–10 mg) with solid IR grade potassium bromide (40–100 mg) using an agate mortar and pestle followed by palletizing into a thin wafer in an evacuated die set by applying a pressure of about 3000–5000 kg cm^{-2}. The wafer is mounted onto the instrument using a sample holder for recording the spectrum. (ii) Powdered solid samples may be made into a paste or mull by mixing with a drop of Nujol (liquid paraffin) and smeared or spread on the KBr sample cell window for recording the spectrum. Nujol is mostly transparent to mid-IR region but shows absorption bands at 2857, 1449, and 1389 cm^{-1} assigned to C–H stretching and bending modes of vibration. (iii) Solids may be dissolved in suitable solvents (mostly carbon tetrachloride, chloroform, carbon disulphide or benzene) and the solution (about 0.1 mL) is placed in liquid sample cell for recording the spectrum. Carbon tetrachloride is transparent in the 4000–1333 cm^{-1} while carbon disulphide is transparent in 1333–650 cm^{-1}. (iv) Films of polymers and amorphous solids may be deposited on sample cell window by allowing the solvent to evaporate from a solution of the sample leaving behind a film of the sample on the window for recording the spectrum.

Liquid samples may be injected into liquid sample cells to record the *neat* IR spectrum. Alternatively a solution of the liquid sample dissolved in a suitable solvent may be placed in the sample cell for recording the *solution* spectrum. Gaseous samples are directly introduced into gas tight sample cells.

8.3.7 Applications

Qualitative analysis to identify organic compounds is the most important application of IR spectrometry. The process of identification involves two steps, the first step being the identification of functional groups in the *group frequency region* in the mid-IR spectrum. The second step involves a comparison of the spectrum of the test sample with that of the pure or authenticated compound over the group frequency region and in addition in the *fingerprint region*. The comparison of the spectra in the group frequency region confirms the presence of all the functional groups. Comparison in the fingerprint region reveals significant differences in the distribution, shapes, and intensities of absorption bands due to even small differences in the structure and constitution of the molecules of the test sample and the pure compound.

It must be emphasized that from a study of the IR spectrum and correlation chart, one can only guess the likely presence or absence of functional groups in the test sample. The study at best is only a preliminary step in the qualitative identification of the test sample and it is impossible to identify the compound unambiguously from IR spectrum alone.

Computer-based search is available in most of the modern IR spectrophotometers and the library contains the spectral data on thousands of compounds. The IR spectrum of the test sample is first coded according to the location of the strongest absorption peak followed by each additional strong band over a region of 200 cm^{-1} and finally the strong bands over regions of 100 cm^{-1} wide. The compounds in the library are coded in the same manner which facilitates narrowing down the search based on the appearance of the strongest band.

Quantitative analysis is rather difficult in IR spectroscopy because of the complexity of the spectra, narrowness of absorption bands and instrumental limitations due to the low intensity of IR sources, low sensitivity of the detectors, relatively wide bandwidths almost comparable to the width of absorption peaks, and difficulty in obtaining cells with identical transmission characteristics for the solvent and the solution of the test sample. All these limitations lead to deviations from Beer–Lambert law. However, quantitative analysis is carried out particularly when other analytical methods are not available or successful.

The most common method of quantitative analysis is called the *baseline method* in which the solvent transmittance is assumed to be constant or at least to change linearly between the shoulders of the chosen absorption peak. The method is based on Beer–Lambert law and the measured absorbance (($\log I_0/I_t$)) is taken as proportional to the concentration. The IR spectra of the analyte sample solution and the series of standards of known concentration are recorded in the region of interest. A baseline is drawn connecting the shoulders of the chosen absorption peak as shown in Fig. 8.15.

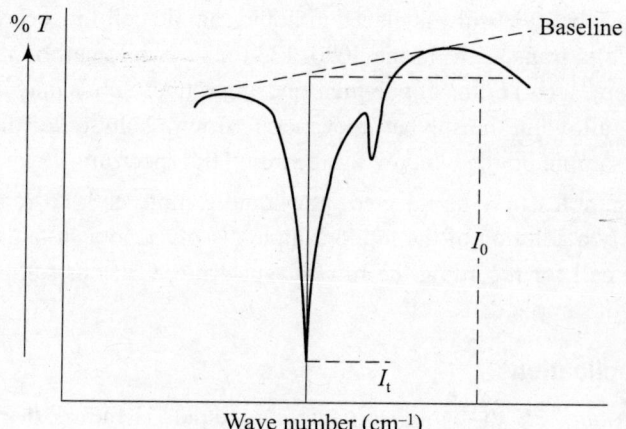

Fig. 8.15 Baseline method for quantitative analysis

The intensities of the incident radiation (I_0) and the transmitted radiation (I_t) are measured at the point where the perpendicular line from the peak maximum intersects the baseline. A standard calibration chart is prepared by plotting $\log(I_0/I_t)$ vs. concentration. The concentration of the analyte in the test sample can be obtained from the calibration chart.

Calibration graph or chart method can be used in quantitative analysis of additives in liquid products such as oils and solutions. A calibration graph is prepared by preparing standard solutions of the additive and measuring the absorbance at characteristic IR frequency of the additive. The absorbance of the analyte additive in the test sample is measured under identical conditions and the concentration of the additive is determined from the graph.

An *internal standard method* is adopted when the IR spectrum of the analyte sample is to be recorded in the form of KBr pellet. Potassium thiocyanate is used as internal standard. Finely ground and dried KSCN is mixed at a concentration of 0.2% by weight with dry KBr and stored over phosphorus pentoxide. A series of standards are prepared by mixing different weights of the analyte with the KBr–KSCN mixture, grinding, and pelleting. The amount of KBr–KSCN mixture and the pellet thickness should be maintained constant for all the standard pellets. The ratio of the intensity of the absorption band at 2125 cm^{-1} for KSCN to the intensity of the absorption band of the analyte at a chosen frequency is plotted against concentration of the analyte to prepare a calibration chart. The test sample is made into a pellet with KBr–KSCN mixture under identical conditions as for the standards and from the ratio of the intensities of KSCN band and the analyte band the concentration of the analyte in the test sample is determined graphically.

Other applications of IR spectroscopy include (i) determination of purity of samples, (ii) calculation of force constants, (iii) structural analysis of compounds, and (iv) a large number of studies involving inter-molecular and intra-molecular hydrogen bonding, cis–trans isomerization, progress of reactions, keto-enol tautomerism, progress of chromatographic separations, identification of reaction intermediates on catalyst surface, etc., to name a few.

8.3.8 Diffuse Reflectance Infrared Fourier Transform Spectrometry

Diffuse reflectance infrared Fourier transform spectrometry (DRIFTS) is highly useful in recording the IR spectra of powdered solids and liquids or gases adsorbed on the surface of a non-absorbing powdered solid substrate (e.g., alkali halide) with minimal sample preparation. A few micrograms of the sample are placed in a small cup for recording the spectrum. When the incident IR beam strikes the surface of a finely divided powdered sample, specular reflectance occurs from each randomly oriented plane of the surface and radiation is reflected in all directions. The intensity of the reflected radiation is roughly independent of the viewing angle. The relative reflective intensity of a sample powder is a function of the ratio R of the reflected intensity of the sample to that of a non-absorbing standard (e.g., finely powdered potassium chloride) and is related to the molar absorptivity ε, molar concentration C, and the scattering coefficient s of the sample as given by

$$R = \varepsilon C/s \tag{8.9}$$

The diffuse reflectance IR spectrum is a plot the relative reflectance intensity as a function of the wave number of the IR radiation. The reflectance spectrum is similar to the absorption spectrum of a sample with peaks appearing at the same frequencies in both the spectra. The relative heights of the peaks, however, differ considerably with minor peaks appearing larger in the reflectance spectrum.

A diffuse reflectance attachment to FTIR spectrometer consists of a specially designed curved or plain mirror system to collect the radiation diffusely reflected from the sample surface over a wide angle and focuses it onto the detector.

8.3.9 Attenuated Total Reflectance Spectroscopy

Attenuated total reflectance spectroscopy (ATR), also known as multiple internal reflection spectroscopy (MIR), is yet another method useful for the study of the surface of a solid. The method is based on the principle that when a beam of radiation passes from a denser to a less dense medium total reflection of incident radiation occurs at angles of incidence greater than a certain critical angle. During the reflection process the beam penetrates a small distance of the order of micrometers into the less dense medium before reflection occurs. The depth of penetration depends on the wavelength of the incident radiation, the angle of incidence at the interface between the two media and the refractive indices of the two materials. The penetrating radiation is called *evanescent radiation*. The absorption of the evanescent radiation by the less dense medium and attenuation of the beam at wavelengths of absorption bands is referred to as the attenuated total reflection.

The method involves placing the sample in contact with both the sides of a transparent crystalline material of high refractive index, for example, a flat thin mixed crystal of thallium bromide–thallium iodide or plates of germanium and zinc selenide as shown in Fig. 8.16.

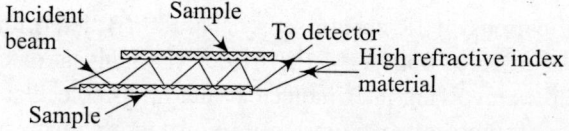

Fig. 8.16   Total internal reflection of radiation

The beam of radiation incident on the sample at proper angle (mostly 30°, 45°, or 60°) undergoes multiple reflections as shown in Fig. 8.16 before reaching the detector. Absorption and attenuation occur at each of the reflections, absorbance depending on the thickness of the sample but independent of the incident angle. The reflectance spectrum obtained is similar to the absorption spectrum with the same peaks but of differing intensities.

A wide variety of solid samples, such as polymers, fibres, rubbers, and fabrics can be handled with minimum preparation for recording the ATR spectra. Suspensions and pastes can be handled in a similar way. The ATR crystal can be dipped into liquid samples or water solutions provided the crystal is not soluble or affected chemically.

8.3.10 Near-infrared Spectroscopy

Near-infrared (NIR) region of electromagnetic spectrum extends between the visible and mid-infrared regions over the wavelength range of 770–2500 nm (13,000–4000 cm^{-1}). In the NIR spectrum the observed absorption peaks are due to the overtone and combination bands of the fundamental stretching vibrations of the C–H, N–H, and O–H bonds (which appear in the mid-IR region of 3000–1700 cm^{-1}). These bands are weak in their intensities with low molar absorptivities.

The NIR spectrophotometer is similar to the UV–visible spectrophotometer consisting of tungsten–halogen lamp, grating monochromators, and sample cells made of quartz but lead sulphide photodetectors. In fact many commercial UV–visible–NIR spectrophotometers having an extended range between 180 nm and 2500 nm using lead sulphide detector in addition to the photomultiplier tube are available. With a diffuse reflectance attachment it is possible to record the diffuse reflectance spectrum of the solid sample.

Absorption measurements are recorded usually in solutions using solvents such as carbon tetrachloride, carbon disulphide, and methylene chloride. Other solvents, such as acetonitrile, dioxane, benzene, heptanes, and dimethyl sulphoxide have a limited range of transparency.

NIR spectroscopy is more useful for quantitative analysis compared to mid-IR spectroscopy, particularly for functional groups containing hydrogen bonded to carbon, nitrogen, or oxygen. Examples of quantitative determinations include (i) water content in glycerol, organic films, fuming nitric acid, etc., (ii) O–H containing compounds, such as alcohols, phenols, organic acids, etc., based on the first overtone of the O–H stretching vibration at 7100 cm^{-1}, (iii) esters, carboxylic acids, ketones, etc., on the basis of the first overtone of the carbonyl stretching vibration in the region of 3300–3600 cm^{-1}, (iv) primary amines in the presence of secondary amines and tertiary amines based on the absorbance of combination band of N–H stretching vibration at 5000 cm^{-1}, and (v) primary and secondary amines together in the presence of tertiary amines from overlapping bands appearing in the 3300–10,000 cm^{-1} regions attributed to various N–H stretching vibrations and their overtones.

Near-infrared reflectance spectrometry has gained importance particularly as a quantitative technique for the analysis of finely powdered solids, particularly for the determination of moisture, protein, starch, oil, lipid, and cellulose content in a variety of agricultural products such as grains and cereals. The technique is simple, reliable, and fast. The diffuse reflectance spectrum is a plot of the ratio of the intensity of radiation reflected from the powdered sample to that of a standard reflector (barium sulphate) recorded as a function of the wavelength.

Moisture content of the sample is determined from the reflectance band at 1940 nm. Starch and protein in the sample give rise overlapping peaks at 2100 nm, and hence, their contents can be determined simultaneously from reflectance measurements at two wavelengths in the region.

8.3.11 Far-infrared Spectroscopy

Far-infrared region extends between the mid-IR and microwave regions with a frequency range between 600 cm^{-1} and about 80 cm^{-1}. Stretching vibrations of metal–ligand bonds (M–O, M–N, M–S, M–X (X=halide)) absorb in this region giving rise to absorption bands. Molecules containing only light atoms absorb in this region if they have skeletal bending modes of vibration involving more than two atoms other than hydrogen. For example, substituted benzene derivatives show characteristic absorption peaks, and hence, useful for qualitative identification of the sample. Absorption peaks due to rotational motion of gases (e.g., HCl, H_2O, O_3, etc.) are observed in the far-IR region. Far-infrared spectroscopy is useful in the study of inorganic solids for obtaining information on lattice energies of crystals and transition energies of semiconductors.

8.4 RAMAN SPECTROSCOPY

Raman spectroscopy is based on the scattering of the incident beam of light by molecules that can undergo a change in molecular polarizability (change in the shape of the molecule) during their vibrational transitions in conformity with quantum mechanical selection rules. The scattered light consists of incident frequencies as well as other frequencies. The shift in the frequencies of the scattered light occurs due to transfer of energy of the incident photon to the molecule and vice versa and depends on the chemical structure of the molecules. This phenomenon is known as *Raman scattering* or *Raman Effect* and the shift in frequencies is known as *Raman shift*. Raman shift is, thus, defined as the shift of frequency $\Delta v = v_0 \pm v_s$ where v_0 and v_s are the frequencies of incident and scattered radiation respectively, occurs usually in the mid-IR region, and hence, expressed in cm^{-1}.

The theory of Raman spectroscopy may be explained on the basis of the scattering of the incident beam of monochromatic radiation usually from a powerful laser source in the visible or infrared region by molecules of the sample through which the radiation passes. The wavelength of the incident or excitation radiation is normally away from the absorption peak of the sample. Most of the molecules in the sample exist in the ground state while a very few molecules exist in the first vibrational level of the ground electronic state. The molecules in the ground state interact with incident radiation and the acquired energy allows the molecules to attain any of the infinite number of values or states called *virtual states* between the ground state and the first electronic excited state as shown in Fig. 8.17. The absorption process is not quantized. The molecules in the virtual states relax to the ground state by releasing energy in the form of scattered radiation. Two types of scattered radiation are observed in the Raman spectrum of a sample. These include the Rayleigh scattering and Raman scattering, the latter consists of Stokes' and anti-Stokes' radiation.

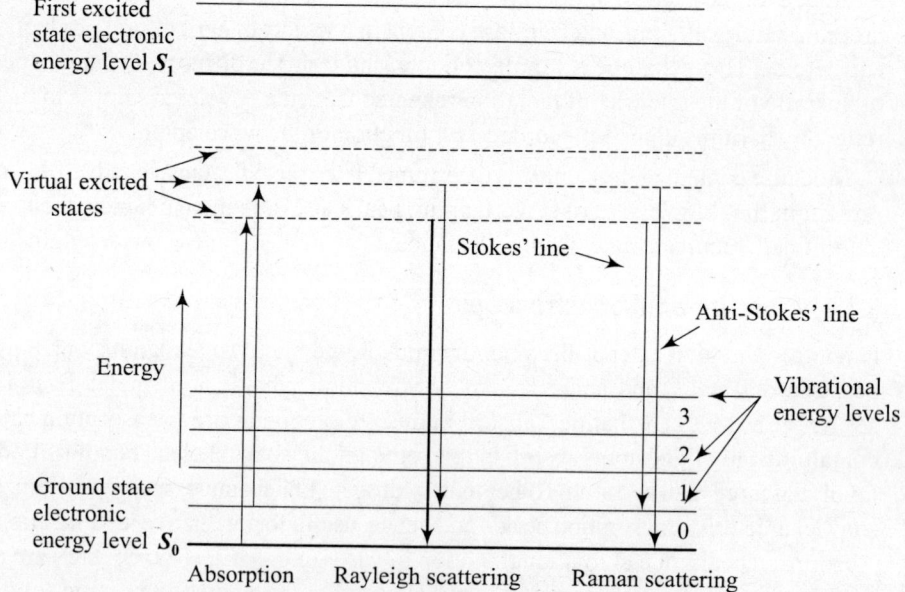

Fig. 8.17 Origin of Rayleigh and Raman scattering

Rayleigh scattering occurs due to elastic scattering of the photons of the incident light on colliding with molecules of the sample. The electric field produced by the polarized molecules oscillates at the same frequency of the incident radiation and as the incident radiation passes the molecules cease to oscillate without exchanging energy. Thus, the photons are scattered without energy exchange with molecules, and hence, the wavelength is exactly the same as that of the incident light. The intensity of Rayleigh scattered light is quite high as a majority of the molecules in the sample interact with the incident of beam light through elastic scattering.

A small number of molecules of the sample in the vibrational ground state, however, interact with the incident radiation exchanging energy as they undergo a change in polarizability during their normal vibrational modes and provide the basis for *Raman effect*. These molecules in the vibrational ground state absorb energy from the incident beam of frequency v_o and get excited to the virtual state. The excited state molecule may return to any of the vibrational energy levels higher than the ground state level by emitting radiation of lower frequency v_s compared to the frequency v_0 of the incident radiation. The emitted radiation with lower frequency than the incident frequency is known as the *Stokes' line*. The difference in the frequencies $(\Delta v = v_o - v_s)$ is equal to the natural vibrational frequency of the molecule in its ground state. Several such Stokes' lines corresponding to the different vibrations in the sample molecule will be observed in Raman spectrum.

A very small number of molecules which exist in the first vibration level of the ground electronic state absorb the incident radiation, attain any of the virtual states and relax to a vibrational ground state giving rise to Raman scattered light with frequencies greater than that of the incident light. Such lines are called *anti-Stokes' lines*. The intensity of the anti-Stokes' lines is much weaker as compared to the Stokes' lines and very much weaker as compared to the Rayleigh scattered lines. The intensity of anti-Stokes' lines will increase with increase in temperature due to an increase in the number of molecules in the first vibrational excited state.

Raman scattering, thus, consists of both the Stokes' and anti-Stokes' lines which appear along with Rayleigh scattered lines in the Raman spectrum. The interaction of the incident light from He–Ne laser source of frequency $v_o = 15{,}803$ cm^{-1} $(\lambda = 632.8$ nm$)$ with the vibrational energy levels of the molecule and the corresponding Raman spectrum are shown for the typical example of carbon tetrachloride in Fig. 8.18.

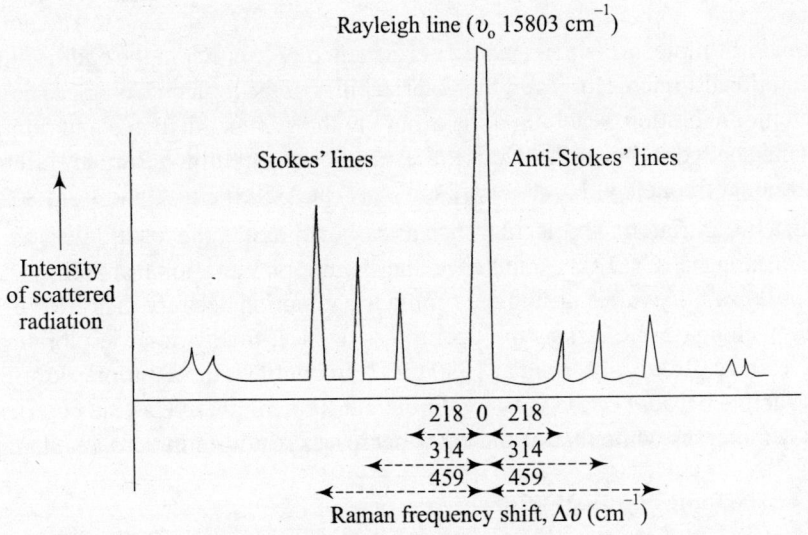

Fig. 8.18 Raman spectrum of CCl$_4$

The Stokes' and the anti-Stokes' lines appear on either side of the Rayleigh scattering line at $15{,}803$ cm^{-1}. The Stokes' lines appear on the lower frequency (longer wavelength) side at $15{,}585$ cm^{-1} (641.6 nm), $15{,}489$ cm^{-1} (645.6 nm), $15{,}344$ cm^{-1} (651.7 nm), $15{,}041$ cm^{-1} (664.8 nm), and $15{,}013$ cm^{-1}(666.1 nm). The anti-Stokes' lines appear on the higher frequency side at $16{,}021$ cm^{-1} (624.2 nm), $16{,}117$ cm^{-1} (620.5 nm), $16{,}262$ cm^{-1} (614.9 nm), $16{,}565$ cm^{-1}

(603.2 nm), and 16,593 cm^{-1} (602.7 nm). The intensities of the Stokes' lines are greater as compared to those of the anti-Stokes' lines.

8.4.1 Comparison of Raman and Infrared Spectra

The Raman shift between the incident frequency and the scattered frequencies are exactly the same for the Stokes' and the anti-Stokes' lines because the energy difference between the Rayleigh and Stokes' (or anti-Stokes') ΔE corresponds to the energy difference between the ground state and the first vibrational state. Hence, the Raman frequency shift and the infrared absorption peak frequency are identical. Thus, Raman spectrum and the IR spectrum of a substance are often similar. However, the intensities of the peaks in the two spectra are quite different.

An interesting feature is that certain peaks that appear in the Raman spectrum of a compound are absent in the IR spectrum and vice versa. The difference in the two spectra may be attributed to differences in the mechanistic aspects. The mechanisms that give rise to absorption bands in the IR and Raman spectra are dependent on the same vibrational modes of a bond. However, for IR absorption to occur the vibrational mode should be associated with a change in the dipole or charge distribution. On the other hand scattering, as in Raman effect, involves only a momentary distortion of the electrons distributed around a bond followed by re-emission of the radiation as the bond returns to the ground state. During the momentary distortion the molecule is temporarily polarized as it develops an induced dipole, which disappears as the molecule relaxes. Due to the difference in the mechanisms involved Raman activity may differ considerably from IR activity for a given vibrational mode. Thus, certain peaks that are IR inactive may be Raman active and vice versa. Hence, the two techniques are considered to be mutually complementary. For example, homonuclear diatomic molecules (e.g., H_2, N_2, F_2, etc.) do not absorb IR radiation because the molecules do not have a permanent dipole either in the equilibrium position or during a vibrational stretch. However, the polarizability of the molecule changes periodically, during the stretching vibration, reaching a maximum at the greatest distance and minimum at the shortest distance between the atoms. Hence, these molecules exhibit a Raman shift corresponding to the vibrational frequency, H_2 at 4395.2 cm^{-1}, N_2 at 2359.6 cm^{-1}, and F_2 at 802.1 cm^{-1}.

In a linear triatomic molecule, such as carbon dioxide, the net dipole is zero for the symmetric stretch along O−C−O axis, and hence, the symmetric vibrational stretch is IR inactive. However, the polarizability varies during the symmetric vibration because distortion of the bonds is greater as they elongate and less as they compress. Hence, the symmetric vibrational mode is Raman active and a peak is observed at 1330 cm^{-1}. In contrast, the asymmetric stretch of the molecule is IR active because of the varying dipole but Raman inactive as the polarizability of one of the bonds increases while that of the other decreases resulting in zero net change.

8.4.2 Raman Spectrometer

Raman spectrometer is similar in design to the ultraviolet and visible spectrophotometers and makes use similar optical components. The radiation source is mostly a laser to provide a narrow beam of highly monochromatic and coherent radiation which can be focused on to the sample. The most commonly used laser is helium–neon laser emitting monochromatic radiation at 632.8 nm. The laser beam is passed through the sample held in a narrow glass or quartz tube. The scattered light is collected by a lens and passed onto a grating monochromator and finally into a photomultiplier tube detector as shown in Fig. 8.19. The detector signal is amplified and displayed.

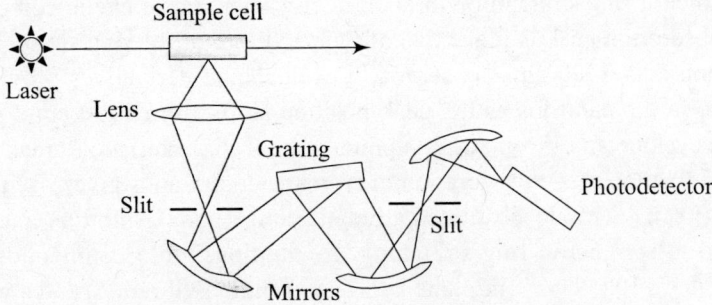

Fig. 8.19 A block diagram of Raman spectrometer

The sample cell is a glass or quartz tube. Liquid samples including aqueous solutions of samples can be used for recording the spectrum. Water is a weak scatterer of light, and hence, is an excellent solvent for Raman spectroscopic studies but not for IR spectroscopy. Spectra of solid and liquid samples also can be recorded by suitable methods of sample handling. Solid samples are ground to a fine powder and filled in a cavity for recording the Raman spectra.

A major limitation to Raman spectroscopy is the interference from fluorescence emission of the analyte or the matrix which gives rise to an intense background noise and it is extremely difficult to obtain a Raman spectrum of a fluorescent sample. However, Fourier transform Raman spectroscopy completely eliminates interference from the background signal due to fluorescence as it uses a Nd–YAG laser source emitting near-IR radiation at 1064 nm. This radiation is not energetic to cause electronic excitation and consequent fluorescence emission.

8.4.3 Applications of Raman Spectroscopy

Raman spectroscopy is useful in qualitative and quantitative analysis of inorganic, organic, and biological molecules and together with infrared spectroscopy for understanding the molecular structure.

The vibrations of metal–ligand bonds occur mostly in the $100–700 \text{ cm}^{-1}$ region, and, hence, difficult to assign from IR spectra. However, many of these vibrations are Raman active and Raman spectral data are useful for obtaining information on the composition, structure and stability of coordination compounds. Most of the important vibrations in organometallic compounds occur in low-frequency far-IR region and Raman spectroscopy is useful in the study of such compounds.

Raman lines are quite intense for symmetrical vibrations in relatively non-polar bonds with symmetrical charge distribution such as $>C=C<$, $-C\equiv C-$, $-C\equiv N-$, $-C=S$, $-C-S-$, $-S-S-$, $-N=N-$, and $-S-H-$ bonds. Thus, for example, the stretching vibrations of $-S-H$, $-C-S-$, and $S-S-$ are easily observed at 2500 cm^{-1}, 650 cm^{-1}, and 500 cm^{-1} respectively. The $-N-H$ and $-C-H$ stretching frequencies are easily observed in Raman spectra at about 3300 cm^{-1} while these bands may be obscured by the intense $-OH$ absorption in IR spectra.

Raman spectra of organic compounds provide more information as compared to IR spectra. For example, $>C=C<$ stretching vibration in olefins is not easily detected in IR spectra because of its weak intensity but in Raman spectra it appears as an intense peak at the same frequency of about 1600 cm^{-1}. In addition, the position of this peak provides more information as it is dependent on the nature of the substituent groups and their geometry.

Symmetric ring stretching vibration is characteristic of cyclic compounds and the position of the absorption peak is indicative of type and size of the ring size. Thus, the Raman peak in the region 700–1200 cm^{-1} observed in cycloalkane derivatives provides information on the ring size in the paraffins as the peak position shifts from 1190 cm^{-1} in cyclopropane to 700 cm^{-1} in cyclooctane. Aromatic compounds show characteristic Raman spectral lines. A strong ring deformation mode of vibration in aromatic compounds gives rise to an absorption peak at 1600 ± 30 cm^{-1}. Monosubstituted aromatic compounds exhibit an intense peak at 1000 cm^{-1} assigned to symmetric ring stretching. In addition, these compounds also show an intense peak at about 1025 cm^{-1} (in-plane hydrogen bending vibration) and a weak absorption peak at 615 cm^{-1} assigned to depolarized in-plane bending vibration. In di-substituted aromatic compounds Raman spectra show a line at 1037 cm^{-1} for ortho-substituted compounds and a weak peak at 640 cm^{-1} for the para-substituted compounds. Meta- and 1,3,5-trisubstituted compounds exhibit only one Raman line at 1000 cm^{-1}.

Quantitative analysis by Raman spectroscopy is much simpler as compared to IR spectroscopy and small sample sizes are sufficient. Further the laser sources can be focused precisely and instruments called *laser microprobes* have been used to determine analytes in smoke, fly ash, bacterial cells, and microscopic inclusions in minerals.

The intensity of Raman lines is directly proportional to the number of scattering molecules. However, it is difficult to measure the absolute value of intensity. Hence, the intensity of a Raman, line is measured in terms of arbitrarily chosen reference line, usually the line of CCl_4 at 459 cm^{-1} which is scanned before and after scanning the sample. The peak heights representing the scattering intensities are converted to *scattering coefficients* by dividing the peak height of the sample peak by the average of the peak heights of the dual traces of the CCl_4 peak. The concentration of the analyte in the test sample is determined from a comparison of the scattering coefficients of the standard and the test sample determined using the cells of same dimension.

Raman spectroscopy together with IR spectroscopy has been widely used for structure determination and the shapes of simple molecules. The *rule of mutual exclusion* states that if a molecule has a centre of symmetry then Raman active vibrations are infrared inactive and vice versa. If there is no centre of symmetry then some of the vibrations may be both Raman and IR active. The symmetry selectivity in Raman spectra gives rise to simpler spectra as compared to IR spectra and in addition the Raman lines are quite intense. Thus, for example, carbon dioxide with a centre of symmetry and the rule of mutual exclusion predicts that the IR inactive symmetric stretch will be Raman active and the IR active asymmetric stretch and symmetric bend will be Raman inactive. In the case of the SO_2 the structure has been shown to be bent (V-shaped) with no centre of symmetry as all the three possible vibrational modes of the molecule are both IR and Raman active showing peaks at 519 cm^{-1}, 1151 cm^{-1}, and 1361 cm^{-1}. Similarly, the structures of nitrate ion (NO_3^-) and chlorate ion (ClO_3^-) have been shown to be planar and pyramidal respectively from Raman and IR spectral data.

8.4.4 Resonance Raman Spectroscopy

The phenomenon of resonance Raman scattering refers to the intense Raman scattering that occurs with excitation wavelengths close to the electronic absorption band of an analyte that contains a chromophore. The resonance Raman spectra contain only few intense lines, and hence, the detection limits of analytes can be very low, about 10^{-8} as compared to normal Raman spectra.

In resonance Raman scattering the electron in the ground state of the molecule is promoted to an excited electronic state. The excited molecule relaxes to the ground state in about 10^{-14} s to a vibrational level of the electronic ground state. In contrast in fluorescence the electronic excitation occurs and excited molecule first undergoes a non-radiative vibrational relaxation to reach the lowest level of the first excited electronic state and then only relaxes to a vibrational level of the electronic ground state by fluorescence emission. The time scale for fluorescence emission is about $10^{-6} - 10^{-8}$ s. Thus, resonance Raman scattering occurs without a non-radiative vibrational relaxation to lowest level of the excited electronic state.

The intensity of resonance Raman lines increases as the excitation wavelength of the laser source approaches the electronic absorption peak of the sample. Hence, a tuneable laser source is used in resonance Raman spectroscopy to achieve maximum intensity. However, the sample may decompose due to exposure to intense laser radiation particularly when the electronic absorption peak is in the ultraviolet region. Hence, the sample is allowed to flow past the focused laser beam so that sample heating and decomposition are minimized.

Raman spectroscopy is useful for the study of single crystals for structure analysis, though less preferred as compared to X-ray diffraction studies.

Resonance Raman spectroscopy is useful to study biological molecules under physiological conditions in the presence of water. For example, the oxidation state and spin of the iron in haemoglobin and cytochrome-c have been determined from resonance Raman bands assigned to the vibrational modes of the tetrapyrrole chromophore.

8.5 MICROWAVE SPECTROMETRY

Microwave spectroscopy involves the study of the transitions between rotational energy levels of molecules, and hence, also known as rotational spectroscopy. Free molecular rotation is possible only in the gas phase. Molecules having permanent dipole moments, for example, HF, HCl, HBr, HI, CO, NO, H_2O vapour, NH_3, etc., absorb energy in the regions of radiofrequency and microwave and undergo rotational transitions to exhibit rotational spectra. On the other hand, gaseous molecules, for example, H_2, N_2, Cl_2, O_2, CO_2, etc., do not have permanent dipole moment, that is, they cannot absorb microwave radiation, and hence, do not show microwave spectra.

The rotational movement of molecules may be considered to consist of three rotational components about the three mutually perpendicular directions through the centre of gravity of the molecule or its principal axes of rotation. The three principal moments of inertia of a molecule, one about each axis of rotation, are designated as I_A, I_B, and I_C. The moment of inertia I of any molecule about any axis through the centre of gravity is given by the relationship

$$I = \sum_i m_i r_i^2 \qquad (8.10)$$

Based on the relative values of the principal moments of inertia (or shapes of molecules) molecules are classified into four main groups. The groups include:

(i) linear molecules (e.g., HCl, OCS, etc.) for which $I_A = 0$ and $I_B = I_C$,

(ii) symmetric top molecules (e.g., CH_3F) for which $I_A \neq 0$ and $I_B = I_C$,

(iii) spherical top molecules (e.g., CH_4) $I_A = I_B = I_C$, and

(iv) asymmetric top molecules (e.g., H_2O, $CH_2=CHCl$, etc.) for which $I_A \neq I_B \neq I_C$.

The rotational spectra of diatomic molecules are conveniently explained on the basis of a model called *rigid rotor*. The model assumes that the two atoms of the molecule having masses m_1 and m_2 are linked by a rigid rod of length r_0 (equilibrium distance between the atoms) which represents the bond distance. The moment of inertia I is given by

$$I = \mu r_0^2 \tag{8.11}$$

where μ is the reduced mass of the molecule. The reduced mass is calculated from the individual masses of the combining atoms by the relationship

$$\mu = \frac{m_1 m_2}{m_1 + m_2} \tag{8.12}$$

Based on quantum mechanics the rigid diatomic molecule can have allowed rotational energy levels as given by the expression

$$E_J = BJ(J + 1) \tag{8.13}$$

where J is the rotational quantum number with integer values (0, 1, 2, 3, …) and B is called the *rotational constant*. The value of the rotational constant is given by

$$B = \frac{h}{8\pi^2 cI} \tag{8.14}$$

where h is Planck's constant and c is the velocity of light. The rotational transitions for a rigid molecule is governed by selection rules, according to which only those transitions which satisfy the condition $\Delta J \pm 1$, that is rotational quantum number changes by unity only are allowed. For absorption spectra the selection rule will be $\Delta J = +1$, and hence, for the rotational transition $E_{J=0}$ to $E_{J=1} = 2B$ cm^{-1}, $E_{J=1}$ to $E_{J=2} = 2B$ cm^{-1}, and so on. Thus, for a rigid diatomic molecule the microwave spectrum consists of a series of lines equally spaced by $2B$ cm^{-1}. However, the rigid rotor model does not exactly describe the real molecule, and hence, the spectrum shows separations decreasing steadily between successive lines with increasing values of J.

The rotational spectrum of carbon monoxide in the far-IR region is shown in Fig. 8.20.

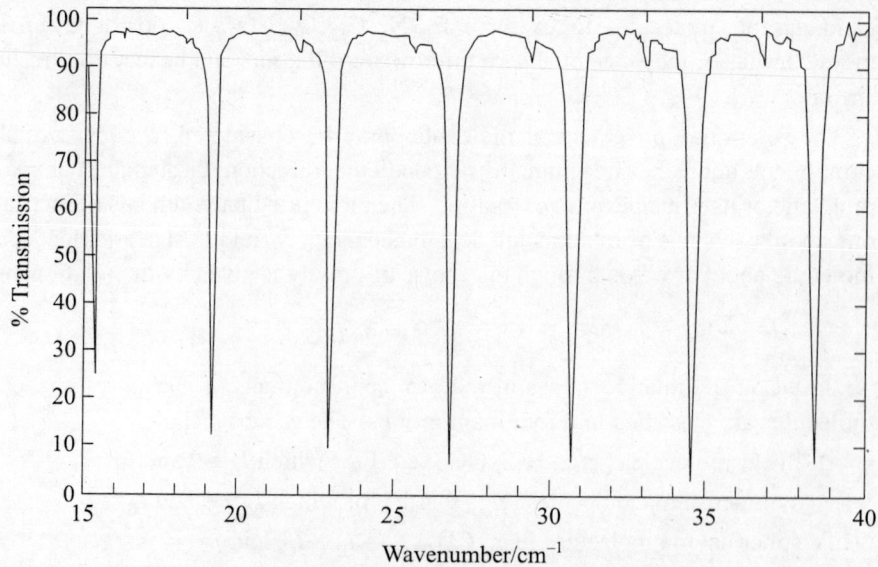

Fig. 8.20   Rotational spectrum of carbon monoxide

The absorption peaks correspond to the rotational transitions from $J = 3, 4, \ldots$ up to 9. From the total distance on the x-axis between the first peak to the seventh peak shown in Fig. 8.20 corresponds to six transitions and is equal to $12B$. The value of the rotational constant B is obtained as 1.929 cm^{-1}, from which the moment of inertia of the molecule is calculated. The reduced mass μ of $^{12}C^{16}O = 1.139 \times 10^{-26}$ kg. Substituting these values into $I = \mu r_0^2$, the bond distance in CO is calculated to be 1.13 Å.

Rotational spectra of symmetric top molecules, such as ammonia and chloroform, provide information on bond length and bond angles. Spherical top molecules in which all the three moments of inertia are identical do not have dipole moment because of their symmetry, and hence, do not exhibit rotational spectra. Majority of the molecules come under the classification of asymmetric top molecules and have different moments of inertia. These molecules have more complicated rotational energy levels and interpretation of the rotational spectra requires computational methods to achieve an agreement between the calculated and observed spectra. However, accurate data on bond lengths and bond angles have been derived in the case of small asymmetric top molecules, such as O_3 and H_2O, using computer-based methods.

8.5.1 Microwave Spectrometer

The instrument for recording the rotational spectra of gaseous molecules consists of the essential components of (i) source of radiation and monochromator covering the far-infrared and microwave regions, (ii) waveguides, (iii) sample holder, (iv) detector, and (v) data processor and display unit.

The source for microwave radiation is called the klystron, a special type of evacuated electron tube. The electrons generated within the klystron tube vibrate resulting in the emission of radiation in the microwave and radiofrequency region. The emitted frequencies cover a very narrow frequency range, and hence, the klystron tube acts as its own monochromator. The actual emission frequency can be varied electronically, and hence, it is possible to scan over a small frequency range using the klystron tube.

Waveguide is a hollow conductive metal tube of copper or silver, mostly of rectangular cross-section within which the radiation is confined. Waveguide is used to guide or direct the beam of radiation.

The sample holder is an evacuated waveguide and the sample gas is held at low pressures within the waveguide by thin mica windows. Even liquid and solid samples with appreciable vapour pressure generate sufficient amount of vapour for analysis by microwave spectroscopy.

The detector used in microwave spectrometer is a crystal diode detector. The signal is amplified and fed for display on an oscilloscope or to a chart recorder.

8.6 MOLECULAR FLUORESCENCE AND PHOSPHORESCENCE

Photoluminescence is the general term which refers to the re-emission of previously absorbed electromagnetic radiation. Atoms and molecules on absorbing electromagnetic radiation particularly in the ultraviolet and visible regions go to excited states. The excited state species are not stable with the singlet excited states having an average lifetime of the order only 10^{-7}–10^{-9} s while the triplet state has a relatively longer life time in the range of 10^{-4}–10 s. The excited state molecules lose their excess energy to return to their ground state. The process of relaxation can occur via non-radiative pathways such as *vibrational relaxation* (VR), *internal conversion* (IC), or *inter system crossing* (ISC). Alternatively the excited state molecules can relax to the ground state via radiative pathways of *fluorescence* (F) or *phosphorescence* (P). Excited state atoms can relax non-radiatively by internal conversion or radiatively by fluorescence only. The different pathways of relaxation of excited state molecules have already been discussed in Chapter 6 (Fig. 6.4, Section 6.4).

Photoluminescence is, thus, of two types (i) fluorescence and (ii) phosphorescence.

8.6.1 Molecular Fluorescence Spectroscopy

Fluorescence emission is observed when the excited state molecule in its lowest singlet excited state S_1 (and also the lowest vibration level of S_1) relaxes to the singlet ground state S_0. The electronic transition is spin allowed, and hence, fluorescence emission is quite intense and occurs over a period of 10^{-9}–10^{-7} s, of the same order of magnitude as the lifetime of the singlet excited state. Fluorescence emission also occurs from the lowest vibration level of S_1 to various higher vibrational energy levels of S_0 because the non-radiative relaxation processes of internal conversion and vibrational relaxation are quite fast as compared to fluorescence emission. Hence, the fluorescence emission spectrum consists of a band of closely spaced lines representing the transitions from the lowest vibrational level of S_1 to different vibrational levels of S_0. The fluorescence line originating from the lowest vibrational energy level of S_1 to the lowest vibrational level of S_0 corresponds to the energy difference between S_1 and S_0 and has the shortest wavelength. All other lines in the band appear at longer wavelengths corresponding to the smaller energy differences between the lowest vibrational level of S_1 and higher vibrational energy levels of S_0.

In organic molecules fluorescence emission is observed mostly when the molecule undergoes $\pi^* \rightarrow \pi$ and $\pi^* \rightarrow n$-type transitions. Transitions of $\sigma^* \rightarrow \sigma$ rarely give rise to fluorescence emission as the energy involved in such transitions is quite and may lead to deactivation of the excited states through dissociation.

The absorption spectrum, called the *excitation spectrum* in fluorescence studies, and the fluorescence emission spectrum of a molecule have a mirror–image relationship (Fig. 8.21) as the energy differences between the vibrational levels of ground and the excited states of a molecule are almost the same. In general, the longest wavelength band of the absorption spectrum overlaps with the shortest wavelength band of the fluorescence emission spectrum.

The mirror–image relationship may not be observed when the ground and the excited states of a molecule have different geometries or when different fluorescence bands originate from different parts of the same molecule.

In practice two types of spectra, namely, *excitation spectrum* and *emission spectrum* are usually recorded in fluorescence spectroscopic studies as they are characteristic of fluorescing molecules. The excitation spectrum is a plot of the

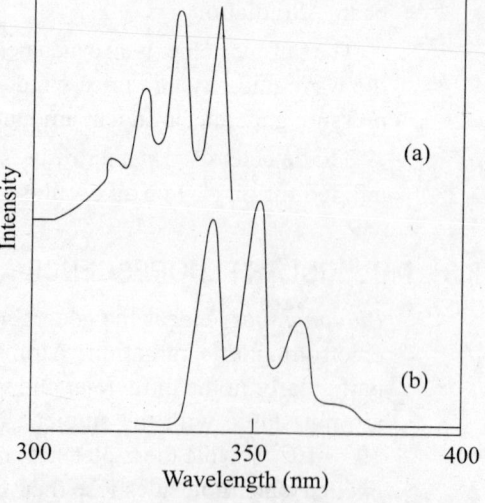

Fig. 8.21 (a) Excitation and (b) fluorescence emission spectra of anthracene

intensity of fluorescence emission as a function of exciting wavelength while emission spectrum is a plot of the intensity of fluorescence emission as a function of emitted wavelength.

8.6.2 Fluorescent Molecules

Theoretically all photo-excited molecules can give rise to fluorescence emission. However, most molecules do not fluoresce because the greater rate of the competing non-radiative pathways of de-excitation. The ratio of the number of molecules that fluoresce to the total number of photo-excited molecules called the *quantum yield* or *quantum efficiency* φ_F is a measure of fluorescence. It may also be defined as the ratio of number of photons emitted to the number of photons absorbed. The magnitude of φ_F depends on the relative rate constants of radiative and non-radiative relaxations

$$\varphi_F = k_F / (k_F + k_{NR}) \qquad (8.15)$$

where k_F and k_{NR} are the first-order rate constants for the fluorescence emission and non-radiative relaxation of the excited state molecule. Highly fluorescent molecules (e.g., fluorescein) have φ_F values approaching 1, while non-fluorescent molecules (e.g., nitrobenzene) have $\varphi_F = 0$.

8.6.3 Fluorescence and Molecular Structure

Compounds containing aromatic rings in general are highly fluorescent. Many un-substituted aromatic hydrocarbons fluoresce in solution, the quantum efficiency increasing with the number of rings and degree of condensation. Substitution in the aromatic ring affects the fluorescence characteristics of the molecules by shifting the wavelength of maximum absorption with the corresponding changes in fluorescence emission and altering the fluorescence efficiency. For example, the relative intensity of fluorescence emission of ethanol solution of benzene increases by the presence of electron donating substitute groups in the aromatic ring (e.g., toluene, propylbenzene, phenol, anisole, aniline, etc.) while it decreases due to the presence of electron withdrawing groups (e.g., fluorobenzene, chlorobenzene, iodobenzene, nitrobenzene, benzoic acid, etc.). Aliphatic and alicyclic carbonyl compounds fluoresce. In contrast, aromatic carbonyl compounds are mostly non-fluorescent but are strongly phosphorescent attributed to the involvement of $n-\pi$ transitions. The energy gaps between the singlet and triplet levels of the $n-\pi$ excited states are small. In addition, the life time of the triplet state is longer thereby favouring an increase in the population of the triplet state, which in turn leads to phosphorescence.

In general, molecules with rigid structures fluoresce with greater quantum efficiency as rigidity decreases the rates of non-radiative relaxation pathways of vibrational relaxation and intersystem crossing. Consequently the life time of the excited state molecules is greater facilitating fluorescence emission. For example, the quantum efficiency of biphenyl with a flexible structure is about 0.2 while that of fluorene which contains a rigid methylene bridge is five times greater.

Biphenyl   **Fluorene**

Compounds with highly conjugated double-bonded structures fluoresce. Thus, fluorescein and eosin with their rigid and conjugated structures exhibit strong fluorescence. However, phenolphthalein with its non-rigid structure is not fluorescent as its conjugate system is disrupted.

A large number of compounds have highly conjugated systems. Examples include aryl-substituted olefins (e.g., trans-stilbene), unsubstituted aromatic hydrocarbons (e.g., anthracene and pyrene), alkyl-substituted hydrocarbons (e.g., toluene, mesitylene), aromatic amines (e.g., 2-naphthylamine, aniline), amino acids (tyrosine and tryptophan), heterocyclic compounds (quinine), phenols, phenyl ethers (anisole), barbiturates, and aromatic acids (acetyl salicylic acid).

Organic chelating agents fluoresce with greater intensity on forming complexes with metal ions which is attributed to the formation of rigid chelate ring structures. For example, oxine (8-hydroxy quinoline) fluoresces with greater intensity as its zinc(II) complex. When fluorescing dyes are adsorbed onto a solid surface they fluoresce with greater intensity once again attributed to the structural rigidity achieved by immobilization of the molecule.

8.6.4 Factors Affecting Fluorescence Emission

Temperature, nature of solvent, pH of the medium, and the presence of diamagnetic or paramagnetic metal ions affect the fluorescence emission characteristics of organic compounds. Apart from these factors the intensity of fluorescence emission is directly related to the concentration of the fluorescing species, particularly at low concentrations.

Fluorescence emission of a compound is affected by temperature, generally decreasing with increasing temperature due to increased frequency of collisions between molecules resulting in enhanced rate of non-radiative relaxation.

The nature of solvent also affects the fluorescence characteristics of a compound. Fluorescence is enhanced in viscous solvents (and also in glass state) as internal vibrations are minimized. Solvents with lower viscosity decrease the fluorescence intensity. Solvents with substituent groups, such as Br, I, NO_2, or $-N=N-$, cause *quenching* of fluorescence emission because the strong magnetic fields surrounding the cores of the atoms in such groups favour spin decoupling and formation of triplet state. Hence, these solvents enhance phosphorescence. Solvents also cause a shift in the emission wavelengths, for example, the emission maximum of indole (excited at 285.0 nm) shifts from 297.0 nm in cyclohexane to 305.0 nm in benzene, 310 nm in dioxane, 330 nm in ethanol, and 350 nm in water.

Fluorescence emission is affected by the pH of the medium as a change in pH affects the charge on the chromophore. For example, phenol fluoresces at pH 7 but at pH 12 it does not fluoresce due to the formation of phenolate anion. In contrast, anisole remains fluorescent at both these pH values. Similarly, aniline is fluorescent at pH 7 as well as pH 12 in the visible region as resonance stabilizes the excited singlet state with more number of canonical structures facilitating fluorescence emission at longer wavelengths. However, resonance stabilization of the excited singlet state is relatively less in anilinium cation formed at pH 2 thereby making it non-fluorescent.

Transition metal ions with incompletely filled *d*-orbitals quench fluorescence but promote intersystem crossing and thereby promote phosphorescence. Coordination compounds of paramagnetic ions of copper(II) and nickel(II) with weakly fluorescent or non-fluorescent ligands exhibit phosphorescence but not fluorescence. In contrast, complexes of non-reducible and diamagnetic ions (e.g., zinc(II) or magnesium(II)) with the same ligands are fluorescent.

8.6.5 Analytical Aspects of Fluorescence Emission

The intensity of fluorescence emission I_F depends on the concentration of the fluorescing species which in turn depends on the intensity of the exciting radiation as given by the relationship

$$I_F = 2.3 \, K' \varepsilon bc I_o \tag{8.16}$$

where K' is a constant depending on the quantum efficiency of fluorescence, ε refers to molar absorptivity, b is the path length, c is the concentration (εbc = absorbance as per Beer–Lambert relationship), and I_0 is the intensity of the incident excitation radiation. The equation is applicable only to very dilute solutions in which less than 2% of the total excitation energy is absorbed, that is, where $\varepsilon bc < 0.05$. As per the above equation, I_F varies linearly with I_0. Hence, it is advantageous to use excitation radiation of high intensity to enhance the signal-to-noise ratio and thereby the sensitivity of measurement.

At a constant value of I_0 the equation can be simplified to

$$I_F = Kc \tag{8.17}$$

where K is a proportionality constant. At low concentrations, in the range 10^{-7}–10^{-5} M, I_F varies linearly with concentration but at higher concentrations absorbance becomes greater than 0.05 (%T is < 90) a non-linear variation is exhibited as shown Fig. 8.22.

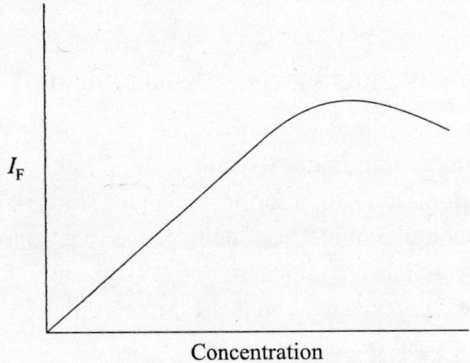

Fig. 8.22 Variation of intensity of fluorescence emission as a function of concentration.

Non-linearity is due to *inner-filter effect*, a combination of *primary absorption* and *secondary absorption*. Primary absorption refers to the strong absorption of incident radiation and fluorescence is no longer linearly dependent on concentration and secondary absorption refers to the phenomenon where other molecules absorb the emitted fluorescence resulting in its decrease with increasing concentration.

8.6.6 Fluorometers

A schematic diagram of a basic fluorometer (or fluorimeter) is shown in Fig. 8.23.

The excitation source is usually a xenon arc lamp which provides a continuum of radiation extending into the ultraviolet region. The monochromators used include filters in filter fluorometers and more efficient grating monochromators in spectrofluorometers which can scan over a wavelength range of 200–800 nm. Spectrofluorometers are capable of resolving Rayleigh and Raman scatter peaks. Sample cells are made of quartz, similar to cuvettes used in

UV–visible spectrophotometry, but with all four sides polished because fluorescence emission is monitored at 90° to the incident radiation. Photomultiplier tube (PMT) detector is the most common detector in fluorometers.

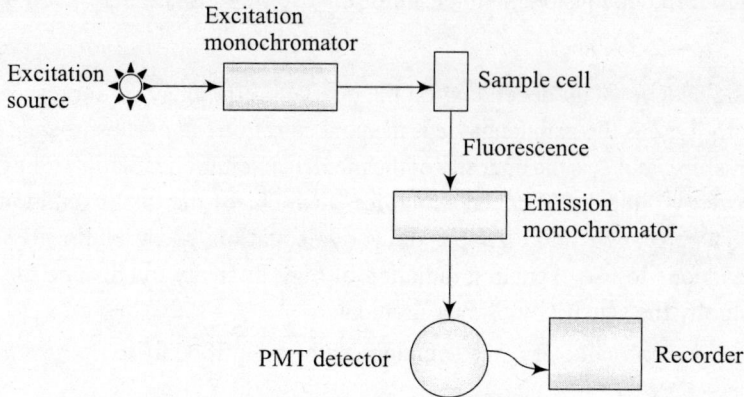

Fig. 8.23 Major components of a fluorometer

The detection limit of spectrofluorometers in 1 cm sample cells can be as low as 10^{-12} M, and, hence, fluorescence spectrometry is about 10^3 times more sensitive as compared to absorption spectrometry.

8.6.7 Applications of Fluorescence Measurements

Fluorescence spectroscopy finds wide applications for quantitative analyses of metal ions, drugs, and biochemicals, as an important detector in liquid chromatography, as a diagnostic tool in clinical tests, in environmental monitoring, and in the study of organic and inorganic compounds. In food and pharmaceutical industries fluorescence spectrometry finds applications in a variety of products. However, the technique is not a major tool for structure analysis of chemical compounds because fluorescence bands are broad, and hence, small structural differences in molecules cannot be identified.

Quantitative determination based on fluorescence measurements have been developed for a number of organic compounds. In the case of inorganic compounds quantitative methods are based on direct measurement of fluorescence or indirect method of measuring the decrease in fluorescence of the fluormetric reagent due to quenching by the analyte. Examples of flurometric reagents used in quantitative analyses of metal ions include oxine (for Al, Be, and other metal ions), benzoin (B, Zn, Ge, and Si), flavanol (Zr and Sn), and alizarin garnet R (Al) used in the direct measurement method. Fluoride can be determined by the indirect method by quenching the fluorescence of Al–alizarin garnet R complex.

In environmental monitoring and pollution control fluorescent tracers (e.g., rhodamine B dye) are useful for determining the time, place, and rate of discharge of effluents from industries into water bodies such as lakes, rivers, and streams and in ocean. The presence of carcinogenic aromatic hydrocarbons (e.g., 3,4-benzopyrene) in air samples and their concentrations can be determined by fluorescence measurements.

In clinical biochemistry fluorescence spectrometry is used for diagnostic purposes in the analysis of body fluids. Amino acids, particularly tyrosine and tryptophan, proteins, coenzymes, vitamins, nucleic acid, steroids, flavanoids, and alkaloids are analysed by fluorescence methods. Several medically important compounds, such as morphine, penicillin, adrenaline, lysergic acid diethylamide (LSD), are analysed by fluorometry.

8.6.8 Molecular Phosphorescence Spectroscopy

Phosphorescence is observed only after the excited state molecule relaxes to a slightly lower energy triplet state through the non-radiative pathway of intersystem crossing. The emission occurs at longer wavelengths as compared to fluorescence and also after a delay. The intensity is much weaker due to the spin-forbidden transition but the emission persists even after the removal of the excitation source and also over longer time duration from a few seconds to even minutes.

Phosphorimeter is similar to fluorimeter in construction and consists of a xenon excitation source, excitation monochromator, sample cell, emission monochromator, and detector. An additional component required in phosphorimeter is a rotating chopper placed in between the excitation monochromator and the sample cell. The rotating chopper facilitates periodic excitation of the sample so that the emitted phosphorescence can be measured after a suitable time delay without interference from fluorescence and any scattered radiation.

Phosphorescence measurements are performed at liquid nitrogen temperature to prevent collisional deactivation of the excited state molecules. The sample cell is a small Dewar made of fused silica and silvered in regions other than the optical path. The solvent is a mixture of diethylether, pentane, and ethanol in the volume ratio of 5:5:2 which solidifies to a clear transparent glass enclosing the analyte at liquid nitrogen temperature.

Phosphorescence methods are rarely used for analytical purposes as they are less precise and less accurate. In addition there is the inconvenience of cooling the substance to very low temperatures. The numbers of substances that exhibit phosphorescence are also small.

8.7 CHEMILUMINESCENCE

Chemiluminescence refers to the phenomenon of radiation emitted by an excited state species formed through a chemical reaction. Many organic compounds exhibit chemiluminescence which may be represented in a simplified manner as a reaction between two compounds A and B to give product(s), one of which exists in the excited state and relaxes through emission of radiation.

$$A + B \rightarrow C^* + D$$
$$\downarrow$$
$$C + h\nu$$

An important analytical application of chemiluminescence is that of the determination of atmospheric pollutants such as ozone, oxides of nitrogen, and sulphur compounds. Nitric oxide reacts with ozone to form excited state nitrogen peroxide which relaxes to the ground state through chemiluminescence in the wavelength range of 600–2800 nm.

$$NO + O_3 \rightarrow NO_2^* + O_2$$
$$NO_2^* \rightarrow NO_2 + h\nu$$

The method is useful for the determination of both NO and NO_2 in atmospheric gases in two stages. In the first stage the NO in the air sample is allowed to react with ozone from a generator and the intensity of chemiluminescence is measured with a photomultiplier tube to determine the concentration of NO alone. In the second stage the air sample containing both NO and NO_2 is passed through a thermal converter to convert all NO_2 to NO and the sample is then allowed to react with ozone. The chemiluminescence measured in the second stage gives the total value of NO + NO_2 in the air sample. The difference between the two determinations gives the amount of NO_2.

8.8 TURBIDIMETRY AND NEPHELOMETRY

The two techniques of turbidimetry and nephelometry are based on the phenomenon of scattering of light by particles suspended in a liquid medium. Depending on the size of the particles in the suspension, scattering of light can be differentiated into (i) Tyndall scattering, (ii) Rayleigh scattering, and (iii) Raman scattering. Tyndall scattering occurs when the particle sizes are equal or greater than the wavelength of the incident radiation. Rayleigh scattering and Raman scattering occur when the particle size is much smaller as compared to the wavelength of the incident radiation. Rayleigh scattering is essentially an elastic scattering as the scattering particles (molecules) do not absorb or lose energy, and hence, the incident and scattered light have the same wavelength. Raman scattering on the other hand consists of scattered radiation of both Stokes' lines of greater wavelengths and anti-Stokes' lines of shorter wavelengths as compared to the wavelength of the incident radiation.

Turbidity may be defined as the optical property of a suspension of particles which causes incident light to be scattered and absorbed instead of being transmitted in straight lines through the medium. The relationship between the intensity of the scattered light and the concentration of the scattering species is quite complex depending on parameters such as the size of the scattering particle, the wavelength of the incident light and its refractive index and the angle of observation relative to the incident light. However, an empirical linear relationship holds good between turbidity and concentration of the scattering species in a dilute suspension.

Turbidimetry involves measuring turbidity of an aqueous suspension by measuring the intensity of light transmitted through the suspension. In the turbidimeter light from a standard electric bulb is allowed to pass through a sample cell containing the analyte suspension. The intensity of the light transmitted through the sample is measured by a photodetector similar to that in absorption spectrometry.

Nephelometry involves measuring the turbidity of an aqueous suspension by measuring the intensity of scattered light by placing the light source and the photodetector at right angles to each other and expressed as nephelometric turbidity units (NTU).

Turbidity of the sample is compared with a standard solution of an insoluble polymer called *formazin* and expressed as formazin turbidity units (FTU). Formazin is prepared by condensation reaction between hydrazinium sulphate ($N_2H_6SO_4$) and hexmethylenetetramine. Accurately weighed (5 g) of hydrazinium sulphate ($N_2H_6SO_4$) and 50 g of hexamethylenetetramine are dissolved in one litre of distilled water and allowed to stand for about 48 hours. The formazin formed develops a white turbidity, the turbidity being assigned a value of 4000 NTU.

Applications of turbidity and nephelometry are mostly confined to checking of clarity of process water used in beverage, food processing, and pharmaceutical industries and measurement of haze or cloudiness in finished products.

● ● ● ● ● ● ● ● ● ● ● ● ● ● ● ● ● ● ●

EXAMPLE 1 *Iron content of a ground water sample was determined by the standard addition method. To six separate 10 mL aliquots of the test sample 0 mL, 2 mL, 4 mL, 6 mL, 8 mL, and 10 mL of standard ferric iron solution containing 8.2 mg L^{-1} were added and made up to a total volume of 50 mL in separate flasks. The measured absorbance values of the solutions after development of colour by the addition of potassium thiocyanate were found to be 0.22, 0.42, 0.61, 0.81, 1.0, and 1.18 respectively. Calculate the concentration of the iron in the ground water sample.*

SOLUTION

The formula for calculating the concentration of the analyte in the test sample by the standard addition method is given by Eq. (7.4) as

$$C_X = \frac{bC_S}{mV_X}$$

From the calibration graph obtained by plotting the given absorbance values (y-axis) as a function of volume of standard solution.

The value of slope $m = 0.0164$ and the intercept $b = 0.22$

$$C_X = (0.22 \times 8.2)/(0.0164 \times 10) = 11 \text{ mg L}^{-1}$$

EXAMPLE 2 *The absorptivity of two compounds A and B at their absorption maxima of 345 nm and 420 nm are 214 L g^{-1} cm^{-1} and 172 L g^{-1} cm^{-1} respectively. Compound A absorbs at 420 nm ($\varepsilon = 8.2$ L g^{-1} cm^{-1}) while compound does not absorb at 345 nm. Calculate the concentrations of the two compounds in a solution if the measured absorbance values in a 1 cm cell are 0.65 at 345 nm and 0.437 at 420 nm respectively.*

SOLUTION

Since only compound A absorbs at 345 nm, using Beer–Lambert equation

A (345 nm) = $\varepsilon_A c_A t$

Concentration of A = $0.65/(214 \text{ L g}^{-1} \text{ cm}^{-1} \times 1 \text{ cm})$

$\qquad\qquad = 0.003037 \text{ g L}^{-1} = 3.037 \text{ mg L}^{-1}$

A (420 nm) = $\varepsilon_A c_A t + \varepsilon_B c_B t$

$\quad 0.437 = (8.2 \text{ L g}^{-1} \text{ cm}^{-1} \times 0.003037 \text{ g L}^{-1} \times 1 \text{ cm}) + (172 \text{ L g}^{-1} \text{ cm}^{-1} \times c_B \times 1 \text{ cm})$

Concentration of B = $(0.437 - 0.025)/(172 \text{ L g}^{-1} \text{ cm}^{-1} \times 1 \text{ cm})$

$\qquad\qquad = 0.00239 \text{ g L}^{-1} = 2.39 \text{ mg L}^{-1}$

EXAMPLE 3 *A solution of pure compound A of concentration 0.008 M showed absorbance values of 0.536 and 0.035 at λ_1 and λ_2 respectively in a 1 cm cell. In the same cell a 0.0056 M solution of pure compound B had absorbance values of 0.080 and 0.256 at λ_1 and λ_2 respectively. Calculate the concentrations of the compounds A and B in a mixture of the two if the sample had an absorbance of 0.32 at λ_1 and 0.18 at λ_2 in the same 1 cm cell.*

SOLUTION

Using Beer–Lambert equation the value of the product εt for each compound at λ_1 and λ_2 are calculated.

At λ_1 for compound A $\varepsilon t = 67$ and for compound B $\varepsilon t = 6.25$

At λ_2 for compound A $\varepsilon t = 10$ and for compound B $\varepsilon t = 45.71$

Substituting the above values into simultaneous equations

At $\qquad \lambda_1$ 0.32 = 67 [A] + 6.25[B]

At $\qquad \lambda_2$ 0.18 = 10[A] + 45.714 [B]

Solving the equations for [A] and [B] gives the concentrations of compounds A and B in the test sample as [A] = 4.5×10^{-3} M and [B] = 2.95×10^{-3} M.

EXAMPLE 4 *The concentration of an analyte in the test sample was determined by standard addition method. To 3 mL of the sample solution taken in separate 25 mL flasks, 0 mL, 5 mL and 10 mL, of standard solutions of the analyte of 2.5×10^{-4} M were added and made up to 25 mL. The absorbance values recorded at a chosen wavelength were 0.18, 0.32, and 0.46, respectively. Calculate the concentration of the analyte in the test sample.*

SOLUTION

Using the equation given for the standard addition method in Chapter 2

$$\frac{C_X}{(C_X + C_S)} = \frac{R_X}{R_S} \text{ or on re-arranging we get } C_X = \frac{R_X C_S}{(R_S - R_X)}$$

where C_X and C_S refer to the concentrations of the analyte in the test sample and in the standard solution and R_X and R_S refer to the detector signal (absorbance) at the chosen wavelength for the test sample and the standard, the value of the unknown C_X can be calculated.

(i) For 5 mL of standard added $C_S = 2.5 \times 10^{-4} \text{ M} \times 5 \text{ mL}/25 \text{mL} = 5.0 \times 10^{-5} \text{ M}$

$R_S = 0.32; R_X = 0.18;$

$C_X = 0.18 \times 5 \times 10^{-5}/(0.32 - 0.18) = 6.43 \times 10^{-5} \text{ M}$

(ii) For 10 mL standard added $C_S = 2.5 \times 10^{-4} \text{ M} \times 10 \text{ mL}/25 \text{mL} = 5.0 \times 10^{-5} \text{ M}$

$R_S = 0.46; R_X = 0.18;$

$C_X = 0.18 \times 5 \times 10^{-5}/(0.46 - 0.18) = 6.43 \times 10^{-5} \text{ M}$

C_X average $= 6.43 \times 10^{-5} \text{ M}$

SUMMARY

- Electronic absorption spectra of molecules are observed in the ultraviolet and visible regions in the wavelength range of 200–800 nm (50,000–12,000 cm^{-1}) of the electromagnetic spectrum. Electronic transitions in molecules are accompanied by vibrational transitions and referred to as vibronic transitions.

- Franck–Condon principle explains the relative intensities of spectral lines of vibronic transitions of molecules.

- In organic molecules four types of electronic transitions that occur include (i) σ–σ^*, (ii) n–σ^* (iii) π–π^* and (iv) n–π^*. The relative energy changes in these transitions are in the order n–$\pi^* < \pi$–$\pi^* \sim n$–$\sigma^* \ll \sigma$–σ^*. The intensities of the π–π^* and σ–σ^* transitions are quite large (ε is large) while the other two transitions are considerably weak due to unfavourable selection rules.

- Presence of chromophores, auxochromes, and conjugation and solvent bring about changes in the absorption spectra of organic molecules. Red shift occurs due to the presence of auxochrome attachments to chromophores, polar solvents, and extended conjugation.

- The position of absorption maximum in open chain dienes and six-membered ring compounds can be predicted on the basis of Woodward–Fieser rules because red shift due to alkyl substitution is additive in dienes and trienes.

- Blue shift is attributed to the presence of auxochromes in compounds exhibiting absorption bands due to n–π^* transitions. Polar solvents also cause a blue shift.

- Coordination compounds exhibit three types of electronic transitions (i) Laporte forbidden d–d transitions within the transition metal ion of low intensity (ii) π–π^* and n–π^* transitions within the organic ligand affected by the presence of the metal, and (iii) charge transfer transitions involving transfer of electron from the metal orbital to the ligand orbital (MLCT) or from the ligand orbital to the metal orbital (LMCT).

- UV–visible spectrophotometers of single or double beam configuration are used to record electronic absorption spectra. Thee main components of the spectrophotometer include (i) source of radiation, (ii) monochromator, (iii) sample cell, (iv) detector, and (v) display.

- Qualitative and quantitative analyses are based on Beer–Lambert law. Two analytes in a sample mixture can be analysed simultaneously. Photometric titrations can be adopted for different types of analytes.

- Infrared spectroscopy involves the study of vibrational transitions in molecules in the frequency range 12,800 cm^{-1}–10 cm^{-1}. Mid-infrared region of 4000–200 cm^{-1} is most widely used to study the vibration–rotation transitions.

- Homonuclear diatomic molecules do not absorb IR radiation and hence do not exhibit IR spectra as they lack a permanent dipole. Non-linear triatomic and polyatomic molecules have 3N-6 normal modes of vibration whereas linear molecules have 3N-5 normal modes of vibration. IR radiation can be absorbed only by those vibrations associated with a continuously varying permanent dipole.

- Fundamental IR absorption occurs in accordance with the quantum mechanical selection rules for vibrational transitions satisfying the condition $\Delta v = \pm 1$. Vibrational transitions $\Delta v = \pm 2$ and $\Delta v = \pm 3$ are also allowed and such transitions give rise to much weaker overtone bands in IR spectra.

- Quantum mechanical selection rules also permit combination bands, which arise due to the addition of two or more fundamental frequencies and difference bands which arise due to the difference between two or more fundamental frequencies or overtones. The intensities of these bands are much weaker.

- The IR spectrum consists of the group frequency region (3600 cm^{-1}–1200 cm^{-1}) and fingerprint region (1200 cm^{-1}–600 cm^{-1}). The IR absorption bands in the group frequency region are useful in the identification of different functional groups in a wide variety of molecules. The IR absorption bands in the fingerprint region are unique to a molecule and hence the fingerprint region is useful for final identification of the compound.

- Commercially available IR spectrophotometers include dispersive instruments and FTIR instruments. In addition, non-dispersive photometers and reflectance photometers are also available.

- Dispersive IR spectrophotometers are double beam recording instruments using reflection grating monochromators for dispersing the incident radiation.

- FTIR spectrophotometers do not have the dispersing element but make use of a Michelson-type interferometer to allow the polychromatic radiation from the source to impinge on the sample cell. The absorption data is collected simultaneously by scanning over the entire mid-IR region several thousand times to generate an interferogram. The digitized interferogram is mathematically transformed by a point-by-point fast *Fourier transform* by the computer to generate the conventional frequency domain IR spectrum.

- The IR spectra of solid samples can be recorded in the form KBr wafers or Nujol mull or solution. The IR spectra of liquid samples can be recorded as neat spectra or in solution. The IR spectra of gaseous samples can be recorded using gas-tight IR sample cells.

- IR spectroscopy finds extensive use for qualitative analysis and identification of samples. Quantitative analysis is carried out by calibration method or internal standard method. In addition, IR spectroscopy is useful for determination of purity of samples, calculation of force constants, structural analysis of compounds, inter-molecular and intra-molecular hydrogen bonding, cis–trans isomerization, progress of reactions, keto-enol tautomerism, progress of chromatographic separations, identification of reaction intermediates on catalyst surface, etc.

- Diffuse reflectance infrared Fourier transform spectrometry (DRIFTS) is based on the specular reflectance of incident radiation from each randomly oriented plane of the solid surface, the intensity of the reflectance is related to the molar absorptivity, molar concentration, and the scattering coefficient of the sample. The graphical plot of the relative reflectance intensity as a function of the wave number of the IR radiation is similar to the IR absorption spectrum of the sample with peaks appearing at the same frequencies in both the spectra.

- Attenuated total reflectance spectroscopy (ATR), also known as multiple internal reflection spectroscopy (MIR), is based on multiple reflections of the incident beam of IR radiation at the surface of a solid sample to generate an attenuated reflectance spectrum similar to the absorption spectrum with the same peaks but of differing intensities.

- NIR spectra exhibit weak absorption peaks due to the overtone and combination bands of the fundamental stretching vibrations of the C–H, N–H and O–H bonds. The technique is useful for quantitative analysis of functional groups containing hydrogen bonded to carbon, nitrogen, or oxygen.

- Far-IR spectroscopy in the wave number region of 600 cm^{-1} to about 80 cm^{-1} exhibit absorption peaks due to stretching vibrations of metal–ligand bonds (M–O, M–N, M–S, M–X (X= halide)) and also due to rotational motion of gases (e.g., HCl, H2O, O3, etc.). Far-infrared spectroscopy is useful in the study of inorganic solids for obtaining information on lattice energies of crystals and transition energies of semiconductors.

- Raman spectroscopy is based on the scattering of the incident beam of light by molecules that undergo a change in molecular polarizability during their vibrational transitions. The scattered light consists of incident frequencies, due to elastic scattering called *Rayleigh scattering*, as well as other frequencies. The shift in the frequencies of the scattered light occurs due to transfer of energy of the incident photon to the molecule and vice versa and depends on the chemical structure of the molecules.

- Raman shift is defined as the shift of frequency $\Delta\upsilon = \upsilon_0 \pm \upsilon_S$ where υ_0 and υ_S are the frequencies of incident and scattered radiation respectively. The emitted radiation with lower frequency than the incident frequency is known as the *Stokes' line*. The difference in the frequencies ($\Delta\nu = \upsilon_0 - \upsilon_S$) is equal to the natural vibrational frequency of the molecule in its ground state.

- Raman scattered light with frequencies greater than that of the incident light called anti-Stokes' lines (of low intensity) are also exhibited in Raman spectra.

- The energy difference between the Rayleigh and Stokes' (or anti-Stokes') ΔE corresponds to the energy difference between the ground state and the first vibrational state. Hence, the Raman frequency shift and the infrared absorption peak frequency are identical. Thus, Raman spectrum and the IR spectrum of a substance are often similar. However, the intensities of the peaks in the two spectra are quite different.

- Raman spectroscopy and IR spectroscopy are considered to be mutually complementary because certain peaks that appear in the Raman spectrum of a compound are absent in the IR spectrum and vice versa.

- Raman spectroscopy together with IR spectroscopy has been widely used for structure determination and the shapes of simple molecules. The rule of mutual exclusion states that if a molecule has a centre of symmetry than Raman active vibrations are infrared inactive and vice versa. If there is no centre of symmetry then some of the vibrations may be both Raman and IR active.

- Microwave spectroscopy involves the study of the transitions between rotational energy levels of gas phase molecules. Only those molecules that have permanent dipole moments absorb energy in the regions of radiofrequency and microwave and undergo rotational transitions to exhibit rotational spectra.

- The rotational movement of molecules in the three mutually perpendicular directions gives rise to three principal moments of inertia. Based on the relative values of the principal moments of inertia (or shapes of molecules) molecules are classified into four main groups: (i) linear molecules, (ii) symmetric top molecules, (iii) spherical top molecules, and (iv) asymmetric top molecules.

- Microwave spectroscopy provides information on bond length and bond angles of molecules.

- Photoluminescence refers to the re-emission of previously absorbed electromagnetic radiation by molecules in the form of fluorescence and phosphorescence.

- Fluorescence emission involves the relaxation of the excited state molecule from the S_1 state to the ground or higher vibrational states of S_0 state. Fluorescence emission is a quantum mechanically spin allowed transition and hence quite intense and occurs over a period of $10^{-9} - 10^{-7}$ s.

- Fluorescence emission in organic molecules occurs mostly due to $\pi^* \to \pi$ and $\pi^* \to n$-type transitions. The absorption spectrum, also called the *excitation spectrum* and the *fluorescence emission spectrum* of a molecule have a mirror–image relationship.

- Molecules with rigid structures and highly conjugated double-bonded structures and metal complexes of organic chelating agents exhibit fluorescence with greater quantum efficiency.

- Factors affecting fluorescence emission include temperature, nature of solvent, pH of the medium, and the presence of diamagnetic or paramagnetic metal ions.

- The intensity of fluorescence emission is directly related to the concentration of the fluorescing species, particularly at low concentrations and hence can be used for quantitative analysis.
- Major applications of fluorescence spectroscopy include quantitative analyses of metal ions, drugs, and biochemicals, as an important detector in liquid chromatography, as a diagnostic tool in clinical tests, in environmental monitoring, and in the study of organic and inorganic compounds.
- Phosphorescence emission occurs from the T_1 state of the excited molecule to the S_0 state and hence is a spin forbidden transition and hence the intensity is much weaker. However, the emission persists even after the removal of the excitation source and also over longer time duration from a few seconds to even minutes.
- Phosphorescence measurements are performed at liquid nitrogen temperature to prevent collisional deactivation of the excited state molecules.
- Quantitative analyses based on phosphorescence measurements are less precise and less accurate. In addition, there is the inconvenience of cooling the substance to very low temperatures.
- Chemiluminescence refers to the phenomenon of radiation emitted by an excited state species formed through a chemical reaction.
- Chemiluminescence is used for quantitative determination of atmospheric pollutants such as ozone, oxides of nitrogen, and sulphur compounds.
- Turbidimetry and nephelometry are based on the phenomenon of scattering of light by particles suspended in a liquid medium.
- Turbidity of a suspension of particles causes incident light to be scattered and absorbed instead of being transmitted in straight lines through the medium. An empirical linear relationship holds good between turbidity and concentration of the scattering species in a dilute suspension.
- Turbidimetry involves measuring turbidity of an aqueous suspension by measuring the intensity of light transmitted through the suspension.
- Nephelometry involves measuring the turbidity of an aqueous suspension by measuring the intensity of scattered light by a photo detector placed at $90°$ to the light source. Turbidity is expressed as nephelometric turbidity units (NTU).
- Turbidity and nephelometry are used for checking the clarity of process water used in beverage, food processing and pharmaceutical industries, and measurement of haze or cloudiness in finished products.

REVIEW QUESTIONS

1. Classify the molecular absorption spectroscopic techniques.
2. Explain with a neat sketch the electronic transitions in molecules.
3. What is Franck–Condon principle? What is its importance?
4. Identify the different types of electronic transitions in organic molecules.
5. What are the various factors which affect the absorption spectral bands in organic compounds?
6. Explain the terms bathochromic shift and hypsochromic shift with suitable examples.
7. What are hypochromic and hyperchromic effects?
8. Give a brief note on the electronic spectra of inorganic species.
9. Give a neat sketch of the single beam UV–visible spectrophotometer and list the components and their functions.
10. How is a UV–visible spectrophotometer calibrated?
11. Write a note on the analytical applications of UV–visible spectrophotometry.
12. How is an analyte concentration determined by (i) calibration chart method and (ii) standard addition method?
13. Explain the principle involved in the simultaneous determination of the concentrations of two analytes in a given mixture.
14. What are photometric titrations? What are their advantages? Give a detailed account on the different types of photometric titration curves.
15. Discuss the principle and procedure involved in the determination of (i) iron, (ii) lead, (iii) nitrate, (iv) protein, and (v) nucleic acid.
16. Discuss in detail the principle involved in vibrational spectra of molecules.
17. Write a note on the different types of molecular vibrations in diatomic and polyatomic molecules.

18. How do molecular vibrations give rise to IR absorption bands? List the different types of IR absorption bands observed in the spectra of molecules.

19. Explain the terms 'group frequency' and 'finger print region' with reference to IR spectra. What is their significance?

20. Give a block diagram of a double beam dispersive IR spectrophotometer and describe the functions of the components.

21. Give a block diagram of FTIR spectrophotometer and explain the principle of Fourier transformation.

22. Give a note on sample preparation methods for recording the IR spectra.

23. Write a note on the applications of IR spectroscopy.

24. How is quantitative analysis carried out in IR spectroscopy?

25. Explain the principle in the DRIFTS, ATR, and NIR spectroscopic techniques.

26. Discuss the theory of Raman spectroscopy.

27. Explain the origin of Rayleigh and Raman scattering.

28. Give a block diagram of Raman spectrometer and identify the components.

29. Write a note on the applications of Raman spectrometry.

30. Justify the statement 'IR and Raman spectroscopic techniques are considered as complementary techniques'.

31. Explain resonance Raman spectroscopy.

32. Explain the principle involved in microwave spectrometry. How is it useful in the determination of bond distance in simple molecules?

33. What are photoluminescence methods?

34. Explain the phenomena of fluorescence and phosphorescence emission.

35. Explain the term quantum efficiency.

36. How is fluorescence emission affected by molecular structure?

37. Give a block diagram of a fluorometer and list the components.

38. Write a note on the analytical applications of fluorescence emission.

39. Write a note on phosphorimetry.

40. How is NO determined by luminescence method?

41. Write a note on nephelometry and its applications.

9 Magnetic Resonance Spectroscopy

9.1 INTRODUCTION

Magnetic resonance spectroscopy involves the study of the interaction between matter and electromagnetic radiation in the presence of an external magnetic field. The interaction involves absorption of radiant energy in the radiofrequency region of about 4–900 MHz of the electromagnetic spectrum. The absorption of energy is in accordance with Bohr's condition, $\Delta E = h\nu$, the energy between two quantized states of matter being quite small. Such small differences between quantized energy levels occur within the atom, and hence, the absorption process basically involves subatomic particles of nuclei and electrons. In order to facilitate the absorption of energy by nuclei and electrons the sample needs to be placed in an intense external magnetic field.

Magnetic resonance spectroscopic techniques include (i) nuclear magnetic resonance (NMR) spectroscopy involving the absorption of energy by nuclei and (ii) electron spin resonance (ESR) spectroscopy, also called electron paramagnetic resonance (EPR) spectroscopy, involving the absorption of energy by electrons. Both the techniques find extensive use in chemistry and biochemistry for the analytical purposes as well as elucidation of the molecular structure of chemical compounds.

9.2 NUCLEAR MAGNETIC RESONANCE SPECTROSCOPY

Nuclear magnetic resonance (NMR) spectroscopy involves the absorption of radiofrequency radiation by the nuclei of certain isotopes which spin on their own axes in the presence of a strong magnetic field. The absorption of energy results in nuclear transition in which the spinning nuclei change their orientation or alignment of spin to an opposite alignment. The energy required for bringing about the change in the alignment of a particular nucleus depends primarily on the magnetic field strength. Other factors such as the electronic configuration around the nucleus, the nature of the molecule, and intermolecular interactions also influence the absorption process and provide useful information on the nature and structure of the chemical species. NMR basically distinguishes between magnetically different atoms in a molecule and together with infrared spectroscopy (which provides information about the nature of functional groups) is often sufficient to elucidate the structure of the molecule.

9.2.1 Theory of Nuclear Magnetic Resonance

The nucleus within the atom is a positively charged region consisting of protons and neutrons. The quantum mechanical description is based on the assumption that nuclei of certain isotopes (e.g., ^1H, ^{13}C, ^{19}F, ^{27}Al, ^{29}Si, ^{31}P, etc.) have a property called *spin*.

The term refers to the rotation of the nucleus about an axis. The spin of the nucleus is given the symbol I designated as its *spin quantum number*. The nuclear spin quantum number I is a distinct physical constant for each nucleus. The total spin of the value of the nucleus depends on the number of protons and neutrons in the nucleus. The simplest nucleus, namely, hydrogen has only one proton and the spin of the hydrogen nucleus (^1H) is ½ while that of the heavier isotope of hydrogen, deuterium containing one proton and one neutron (^2H) can theoretically have a total spin of zero (proton and neutron having antiparallel spins) or one (both the particles having parallel spins). It has been observed that ^2H has a total spin of 1. On the other hand, ^4He with 2 protons and 2 neutrons has a total spin of zero. The total spin observed for different nuclei lead to the following rules:

(i) Nuclei with even numbers of protons and neutrons (hence, mass and charge are even) have zero spin or $I = 0$ (e.g., ^4He, ^{12}C, ^{16}O, etc.).

(ii) Nuclei with odd numbers of protons and neutrons (mass is even but charge is odd) have integral spin or $I = 1$ (e.g., ^2H, ^{14}N, etc.).

(iii) Nuclei with odd mass have half-integral spin (e.g., ^1H, ^{13}C, ^{15}N, ^{19}F, ^{31}P, etc.). NMR spectrum is obtained only for those nuclei with non-zero value for the spin quantum number

The *angular momentum* $\mathbf{I}$ associated with the nuclear spin is given by the relationship

$$\mathbf{I} = \sqrt{(I(I+1))}\,\frac{h}{2\pi} \tag{9.1}$$

The allowed spin components or states for the angular momentum range from $+I$ to $-I$ (I, $I - 1, \ldots -I+1, -I$). The value of I may be zero, half-integral, or integral multiples of $h/2\pi$. The maximum number of spin components or states values for the angular momentum for a nucleus is $2I + 1$ discrete quantized energy levels or spin states. These states are degenerate (have the same energy) in the absence of an external magnetic field.

The nucleus, because of its charge, generates a magnetic field as it spins on its own axis. In other words the spinning nucleus behaves as a tiny bar magnet along the spin axis. The *magnetic moment* or *magnetic dipole* μ is oriented along the axis of spin and is proportional to the angular momentum of the nucleus as

$$\mu = \gamma \mathbf{I} \tag{9.2}$$

The proportionality constant γ is called the *magnetogyric* (or gyromagnetic) *ratio* and is a constant for a given nucleus. For example, the value of γ is 2.68×10^8 radian $T^{-1}s^{-1}$ for ^1H while for ^{13}C it is 6.73×10^7 radian $T^{-1}s^{-1}$. The magnetic moment (also called magnetic field strength) is expressed in SI units of tesla (symbol T) the units being kg s^{-2} A^{-1} and 1 T = 10,000 Gauss.

9.2.2 Nuclear Energy Levels in an External Magnetic Field

In the presence of an external magnetic field, the angular momentum can assume only those orientations in space with its components in the direction of the field being an integral or half-integral number of angular momentum units as given by

$$\mathbf{I}_z = \frac{mh}{2\pi} \tag{9.3}$$

where $\mathbf{I}_z$ is the magnitude of the angular momentum component in the direction of the applied field (usually designated as z direction) and its value cannot exceed that of $\mathbf{I}$ as given by Eq. (9.1).

The symbol m represents the magnetic quantum number which can have a total of $(2I + 1)$ values varying as $m = 0, \pm \frac{1}{2}, \pm 1, \pm 3/2, \dots$. Thus, for nuclei with $I = \frac{1}{2}$ (e.g., ^{1}H, ^{13}C, ^{15}N, ^{19}F, and ^{31}P) the magnetic quantum number can have only two values, namely, $m = + \frac{1}{2}$ and $m = - \frac{1}{2}$. Hence, the magnetic moment μ_z can be given as

$$\mu_z = \frac{m\gamma h}{2\pi} = \pm \frac{1}{2}\left(\frac{\gamma h}{2\pi}\right) \tag{9.4}$$

Thus, when a nucleus with a spin quantum number $I = \frac{1}{2}$ is brought into an external magnetic field degeneracy is lifted and $2I + 1$ different energy levels result. For an applied magnetic field of strength B, its magnetic moment is oriented in any one of the two directions with respect to the external field as shown in Fig. 9.1.

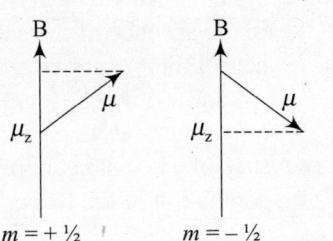

Fig. 9.1 Orientation of magnetic moments for a nucleus with $I = \pm \frac{1}{2}$

The potential energy of the nucleus in these two orientations is given by the relationship

$$E = -\mu_z B = -m\left(\frac{\gamma h}{2\pi}\right)B \tag{9.5}$$

Thus, the nuclei ^{1}H, ^{13}C, ^{15}N, ^{19}F, and ^{31}P with $I = \frac{1}{2}$ can have two energy levels (spin states) or two m values of $\pm\frac{1}{2}$ in the presence of an external magnetic field as shown in Fig. 9.2.

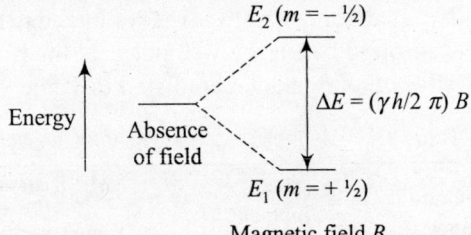

Fig. 9.2 Nuclear spin states for nuclei with $I = \frac{1}{2}$

The energies of the lower and the upper states respectively are given as

$$E_1 = -\frac{1}{2}\left(\frac{\gamma h}{2\pi}\right)B \quad \text{and} \quad E_2 = +\frac{1}{2}\left(\frac{\gamma h}{2\pi}\right)B \tag{9.6}$$

and the difference in the energy levels as

$$\Delta E = E_2 - E_1 = \left(+\frac{1}{2}\left(\frac{\gamma h}{2\pi}\right)B\right) - \left(-\frac{1}{2}\left(\frac{\gamma h}{2\pi}\right)B\right) = \left(\frac{\gamma h}{2\pi}\right)B \tag{9.7}$$

Since $\Delta E = h\nu$, the frequency of absorption for bringing about the nuclear transition can be arrived at as

$$\nu = \left(\frac{\gamma}{2\pi}\right)B \tag{9.8}$$

Thus, the absorption frequency for the nuclear transition is directly proportional to the applied magnetic field strength. Hence, greater the magnetic field strength, greater will be the separation between the energy levels and higher the absorption frequency.

9.2.3 Magnetic Resonance

The absorption frequency of a nuclear transition at a given magnetic field strength can be determined by exposing the nucleus in a sample to radiation of varying frequency, usually in the radiofrequency region. Alternatively the sample may be exposed to radiation of a fixed frequency and vary the applied magnetic field strength. When the conditions of Eq. (9.8) are satisfied the system is said to be in *resonance* and both upward and downward nuclear transitions will occur. In the absence of the external magnetic field the magnetic quantum states are degenerate, and hence, their energies are identical. The number of nuclei having the magnetic quantum states $+\frac{1}{2}$ and $-\frac{1}{2}$ will be identical. However, in the presence of the external magnetic field the nuclei orient themselves so that the number of nuclei having the lower the energy magnetic quantum state of $+\frac{1}{2}$ will be more as compared to the number of nuclei in the higher energy. The ratio of population of the two energy levels can be calculated by Boltzmann distribution law

$$\frac{N_u}{N_0} = e^{\left(-\frac{\Delta E}{kT}\right)} \tag{9.9a}$$

$$= e^{\left(-\frac{\gamma h B}{2\pi kT}\right)} \tag{9.9b}$$

where N_u and N_0 represent the number of nuclei in the higher energy and lower energy levels respectively, k is Boltzmann constant (1.38×10^{-23}) and T is absolute temperature. At room temperature (~ 300 K) the number of nuclei occupying the lower energy level N_0 will be slightly in excess (about 10–20 nuclei for one million) as compared to the population in the higher energy level. Hence, a net absorption of energy will occur at the resonance frequency. The magnetic resonance characteristics of a few nuclei are listed in Table 9.1.

Table 9.1 Magnetic resonance characteristics of a few nuclei

Nucleus	Spin quantum number I	Magnetic moment μ (A^2 m $\times 10^{27}$	NMR frequency MHz at		Relative sensitivity at constant B
			1.4092 T	2.349 T	
^{1}H	$\frac{1}{2}$	14.09	60.000	100.000	100
^{2}H	1	4.34	9.210	15.352	0.96
^{13}C	$\frac{1}{2}$	3.52	15.085	25.147	1.59
^{19}F	$\frac{1}{2}$	13.28	56.444	94.087	83.4
^{31}P	$\frac{1}{2}$	5.71	24.288	40.485	6.64

9.2.4 Classical Model of NMR Absorption

The process of absorption of energy by a nucleus may also be explained by classical model in which the nucleus is considered as a charged particle spinning on its axis and when placed in an external magnetic field aligns with or against the applied field. The nucleus spins on its axis and simultaneously moves in a circular path or *precesses* (*Larmor precession*) around the magnetic field in the *xy* plane. This circular motion or precessional motion is similar to that of a gyroscope in a gravitational field as shown in Fig. 9.3.

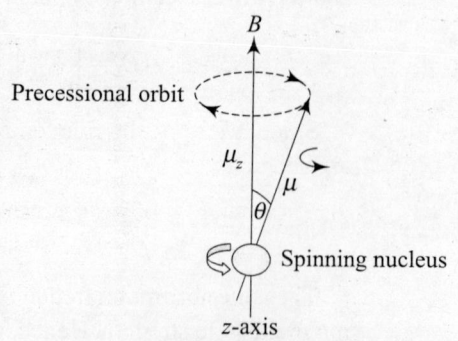

Fig. 9.3 Precessional motion of the spinning nucleus in a magnetic field

The magnetic field aligns with the spinning nucleus not exactly parallel or anti-parallel but such that the spin axis is inclined to the field and precesses around the field. The angular frequency of the precessional motion ω is given by

$$\omega = \gamma B \tag{9.10}$$

Dividing the angular frequency ω (in radians s^{-1}) by 2π we get the frequency of precession v, known as *Larmor frequency*.

$$v = \frac{\gamma B}{2\pi} \tag{9.11}$$

Equation (9.11) obtained in terms of the classical model is the same as Eq. (9.8) obtained by quantum mechanical model, and hence, the Larmor frequency and the absorption frequency are identical. When the applied magnetic field strength increases the nucleus precesses faster around the magnetic field.

The potential energy E of the precessing nucleus is given by

$$E = -\mu_z B = -\mu B \cos \theta \tag{9.12}$$

When energy is absorbed by the precessing nucleus in the form of radiation in the radiofrequency region, the angle of precession θ has to change which may be represented as the flipping (spin flip) of magnetic moment from a state in which it is oriented in the direction of the field to a state in which it is in the opposite direction as shown in Fig. 9.4.

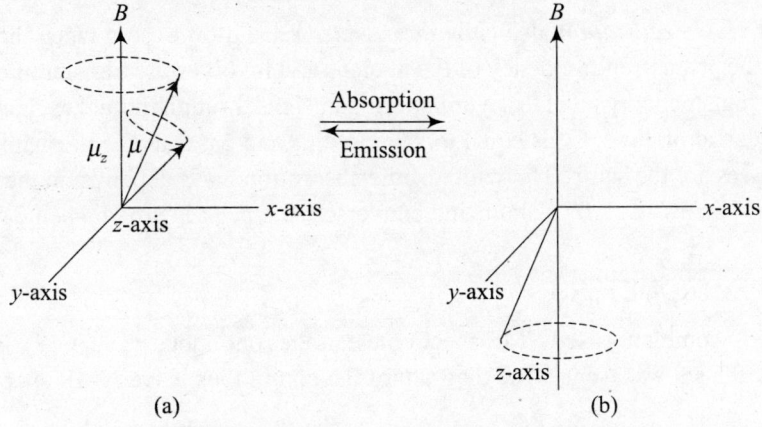

Fig. 9.4 (a) Precession of magnetic moment in the presence of B and RF and
(b) Spin flip of the nucleus at resonance condition

In order to bring about a flip of the precessing magnetic dipole, a magnetic force moving in a circular path in phase with the precessing dipole must be applied at right angles to the external magnetic field. The magnetic field component of radiation in the radiofrequency region consists of two superposed circularly varying fields of equal amplitude, one rotating clockwise and the other counterclockwise. Only the magnetic component of radiation that rotates in the precessional direction of the dipole can be absorbed. When the rotational frequency of the magnetic vector of the radiation and the precessional frequency of the nucleus are identical, resonance absorption of energy by the nucleus occurs and the magnetic dipole flips to the opposite direction. The radiofrequency radiation produced by an oscillator coil oriented along the perpendicular direction to the external magnetic field is capable of bringing about the spin flip of the nucleus.

9.2.5 Relaxation Processes

Nuclei in the ground state absorb energy and undergo nuclear transition to the higher energy level at the resonance frequency. The difference in the number of nuclei populating the lower and higher energy levels is quite small and the absorption signal will quickly attain a saturation level under experimental conditions. The intensity of the absorption signal will be constant at a given RF frequency only if the relaxation process by which nuclei of higher energy level comes back to the lower energy level is as rapid as the absorption process. The relaxation process also determines the width of the absorption band. Two types of relaxation processes have been identified, namely, (i) spin–lattice relaxation and (ii) spin–spin relaxation.

Spin–lattice relaxation is also called longitudinal relaxation in which the spin system of nuclei is considered to be embedded in fluctuating magnetic fields generated by random motions of neighbouring nuclei and electrons constituting a *lattice*. When the precession frequency of the embedded nucleus is equal to the fluctuation frequency of the magnetic field of the lattice, spin–lattice relaxation occurs in which the excess magnetic energy of the excited nucleus is transferred to the lattice as thermal energy. The relaxation is a first-order process with a rate constant called the *spin–lattice relaxation time*, T_1, which represents the time required to re-establish the Boltzmann distribution at a given magnetic field strength. The magnitude of T_1 decreases exponentially, the value being in the range of 0.01–100 s in the liquids while in viscous liquids and solids it is in the order of hours.

Spin–spin relaxation also called transverse relaxation occurs when the RF field is different from the precession frequency of the nucleus and involves the transfer of energy by the nucleus in the higher energy level to a neighbouring nucleus through mutual exchange of spin. The spin–spin relaxation time, T_2, is equal to T_1 in liquids and gases as the mechanisms of the relaxation processes are the same. The width of the absorption band depends on the magnitude of T_2, the peak is narrow when T_2 is long and conversely the peak is broad when T_2 is short.

9.3 NMR SPECTROMETERS

The most commonly used NMR spectrometer is the continuous wave (CW) or scanning instrument. Figure 9.5 shows a schematic diagram of the continuous wave NMR spectrometer.

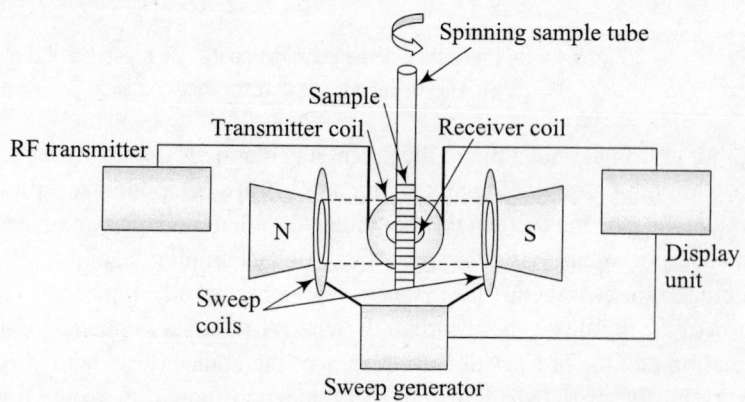

Fig. 9.5   Schematic diagram of continuous wave NMR spectrometer

The main components are (i) magnet which may be a permanent magnet or an electromagnet or superconducting solenoid providing a strong, stable, and homogeneous magnetic field of strengths in the range of $1-14.1$ T (tesla), (ii) sweep generator for varying the magnetic field over a small range by passing a variable direct current through coils that are coaxial with the direction of the main magnetic field, (iii) RF transmitter coil placed at right angles to the sweep coils to generate the exciting radiofrequency radiation in the range of $60-600$ MHz, (iv) sample tube, a long narrow (~5 mm outer diameter) glass/quartz tube, (v) receiver coil placed around the sample holder in the remaining orthogonal plane to function as a detector. A small current is generated in the receiver coil when the resonance condition is reached which is amplified and fed to a recorder, and (vi) display unit which may be a VDU screen or a chart recorder.

The NMR spectrum is most commonly recorded by operating the RF transmitter at a fixed frequency (so that the nuclei are forced to spin or precess in phase) and varying the magnetic field strength by a variable current passed through the sweep coils to enable the scanning. Alternatively, the magnetic field strength may be fixed and the spectrum scanned by varying the RF field. The sample tube is spun rapidly by means of an air turbine to improve the homogeneity of the field. The sample is usually dissolved in solvents such as carbon tetrachloride or deuterated solvents, such as chloroform ($CDCl_3$), benzene (C_6D_6), heavy water (D_2O), acetone ($CD_3.CO.CD_3$), etc. Liquids and gases may also be used directly for recording the spectra.

9.3.1 NMR Spectrum

NMR spectrum of a compound is a plot of relative abundance of the nucleus as a function of the magnetic field strength. All the protons in any molecule should absorb energy at the same value of the applied magnetic field strength B based on Eq. (9.8) or Eq. (9.11), and hence, should give a single absorption peak. However, the proton NMR spectrum (so also the ^{13}C NMR spectrum) of a compound shows more number of absorption peaks each corresponding to a particular nucleus or a group of nuclei and provides a wealth of information on chemical structure of the molecule. Further the absorption peaks may be split into several peaks of varying intensity (see Fig. 9.7). Thus, NMR spectroscopy provides useful information on structural aspects of compounds based on the position of absorption bands and any split in a given absorption band. In fact the usefulness of NMR spectroscopy to provide structural information depends entirely on the *chemical environment* surrounding the nucleus.

9.4 ENVIRONMENTAL EFFECTS

Two environmental effects are commonly encountered in all molecules and are responsible for the use of NMR spectroscopy as a tool for structural studies, particularly of organic compounds. The environmental effects include (i) chemical shift and (ii) spin–spin coupling or splitting.

9.4.1 Chemical Shift

The term *chemical shift* refers to the effect of the chemical environment around the nucleus which determines the exact position of absorption of radiofrequency radiation. The chemical shift is caused by the magnetic field generated by electrons orbiting around the nucleus. The hydrogen atom in a molecule is bonded to another nucleus by electrons. The applied magnetic field induces the electron in the hydrogen atom to circulate around the field direction generating a small localized magnetic field which opposes the applied magnetic field. Thus, the proton is subjected to a *diamagnetic shielding* by the electrons bonding it to another nucleus as shown in Fig. 9.6.

Consequently, the proton experiences a smaller effective applied magnetic field strength B_{eff} compared to the actual applied magnetic field B_0 given by

$$B_{eff} = B_0 (1 - \sigma)$$

(9.13)

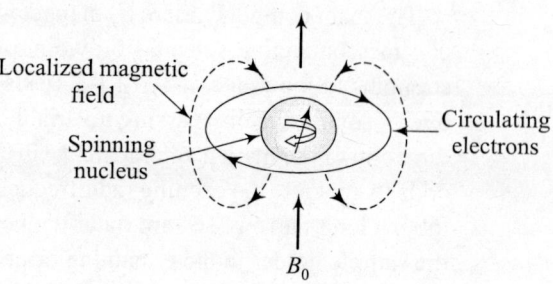

Fig. 9.6 Diamagnetic shielding of a nucleus

where σ is called the *shielding constant* or *screening constant*. For an isolated hydrogen nucleus the value of $\sigma = 0$. The extent of this diamagnetic shielding, and hence, the magnitude of the shielding constant depends on the strength of the opposing magnetic field generated by the electron density around the proton. This in turn depends on the nature of the nucleus, particularly its electronegativity, bonded to the proton. For example, a proton attached to the more electronegative oxygen in the –OH group experiences a greater magnetic field compared to the proton of a CH group containing the less electronegative carbon atom, as electron density is pulled away from the proton in the –OH group. Electron density about the hydrogen atom in the CH group is greater as compared to that in the –OH group, and hence, $\sigma_{C-H} > \sigma_{O-H}$. Consequently resonance condition is achieved with electromagnetic radiation of particular frequency at a smaller applied magnetic field for the –OH proton as compared to the CH proton.

The magnitude of the shielding constant depends on the state of hybridization of the carbon atom attached to the hydrogen atom. For example, the shielding constant is larger for the proton in methyl group (carbon atom is sp^3 hybridized) as compared to that for protons of methylene group (carbon atom is sp^2 hybridized). Thus, methyl protons achieve the resonance condition at a higher magnetic field strength for absorbing radiation of a given frequency than methylene protons. Thus, protons which are chemically non-equivalent due to the differences in their chemical environments in different functional groups are also magnetically non-equivalent, and hence, absorb energy at different applied magnetic field strengths.

Figure 9.7(a) shows the effect of the applied magnetic field on the energy levels of hydrogen atoms of the different functional groups in ethanol and the nuclear transitions between energy levels are shown by vertical arrows. Figure 9.7(b) shows the corresponding low resolution proton NMR spectrum of ethyl alcohol ($CH_3.CH_2.OH$). The three absorption bands arise due to the three chemically as well as magnetically different protons of the three functional groups.

All the three protons of the –CH_3 group are chemically as well as magnetically equivalent, and hence, give rise to a single absorption peak. The protons of this group are highly shielded by the electrons, and hence, absorb radiation only at relatively higher magnetic field strength.

The protons of the >CH_2 group also produce a single peak as they are also magnetically equivalent but absorb at lower applied magnetic field strength as they are less shielded compared to the protons of the methyl group. The electron density around the proton of the –OH group is much less, and hence, it absorbs at a lower magnetic field strength.

The intensity of the three peaks of the –CH_3, >CH_2, and the –OH groups obtained by integration of the peak areas and are in the ratio of 3:2:1 as shown in Fig. 9.7. The intensity of the peaks is, thus, related to the number of protons in each functional group.

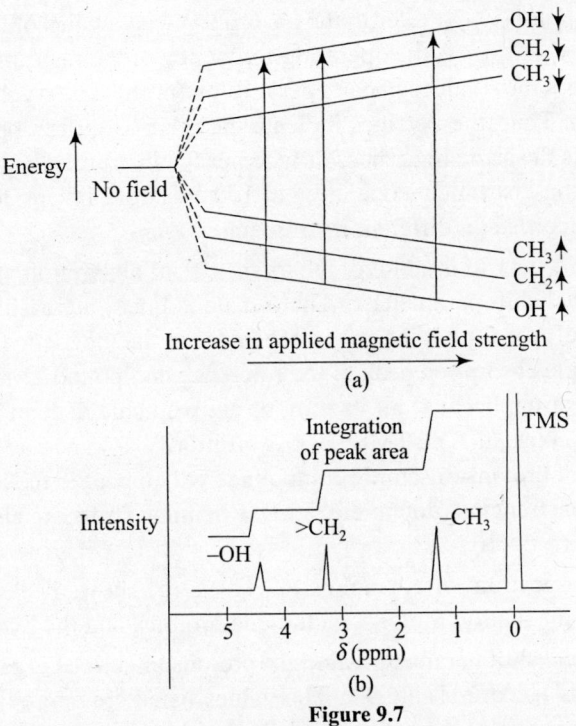

Figure 9.7

Fig. 9.7 (a) Variation of energy levels for the protons in ethyl alcohol as a function of increasing applied magnetic field. (b) Low-resolution proton NMR spectrum of ethyl alcohol

However, it should be emphasized here that the maximum shielding of the hydrogen nucleus occurs only in an isolated hydrogen atom for which $\sigma = 0$, and hence, theoretically it should absorb at the highest magnetic field strength. The methyl protons have $\sigma > 0$ (that is they are shielded to a lesser extent as compared to an isolated hydrogen nucleus), and hence, should absorb at a slightly lower magnetic field strength as compared to an isolated hydrogen atom. In other words, the magnetic field strength required to achieve resonance for the methyl protons shifts to lower values (down field shift) and this shift is referred to as the *chemical shift*. Since it is not possible to get an isolated hydrogen nucleus, a reference compound is required to calibrate the magnetic field strength so that the position of the resonance for different protons may be expressed in units independent of the applied magnetic field.

The reference compound is tetramethyl silane, TMS, $((CH_3)_4Si)$, which has 12 highly shielded protons which are chemically identical, and hence, magnetically identical. In addition the compound is chemically inert, soluble in most organic solvents, boils at 27°C, and hence, easily vapourizes off. All the 12 protons absorb energy at the same applied magnetic field strength giving rise to an intense and sharp signal. Since TMS in not soluble in water, sodium 2, 2'-dimethyl–2-silapentane–5-sulphonate (DSS) is used for aqueous samples. The reference compound is usually added to the sample before recoding the NMR spectrum.

The chemical shifts of a sample are usually expressed in a dimensionless unit called *chemical shift parameter* δ, obtained by using the formula

$$\delta = \frac{V_{(sample)} - V_{(TMS)}}{V_{(instrument)}} \times 10^6 \qquad (9.14)$$

where $v_{(sample)}$ and $v_{(TMS)}$ refer to the absorption frequencies of the sample and TMS expressed in Hertz and $v_{(instrument)}$ is the operating frequency of the instrument. The factor 10^6 is included so as to express the δ values in parts per million (ppm). The δ values for protons are in the range of 0–15 ppm. Thus, the δ values indicate the relative shift in ppm and for a given peak; the δ value remains the same regardless of whether the instrument is of low resolution at 60 MHz or better resolving instruments operating at 100 MHz or higher. Hence, it is possible to compare the spectra recorded on different instruments.

The NMR spectrum is a plot of the intensities of absorption peaks (peak areas) as a function of the chemical shift parameter δ. The δ scale is linear increasing from 0 on the right extreme towards the left corresponding with an increase in the magnetic field strength from left to right. Thus, the single absorption peak of the reference compound TMS will appear at the far right of the NMR spectrum because all its protons are well shielded, and hence, have a large value for the shielding constant. The TMS peak is arbitrarily assigned a value of 0.0 on the δ scale and all the peaks of protons of sample compounds will appear to the left of the TMS peak at smaller magnetic field strengths (downfield shift). Chemical shifts are also expressed in alternative tau (τ) scale, where $\tau = 10 - \delta$.

The NMR spectrum of ethanol as in Fig. 9.7(b) shows peaks at ~1.2 ppm, ~3.3 ppm, and ~4.2 ppm for the methyl protons, methylene protons, and the hydroxyl proton respectively.

The chemical shift parameter values of protons in various organic functional groups are well documented as listed in Table 9.2. The values listed are approximate as the exact values of δ depend on the nature of the solvent and also on the concentration of the solute due to hydrogen bonding and exchange effects.

Table 9.2 Approximate chemical shifts for protons in a few organic functional groups

Functional groups	Range of δ values (ppm)
TMS	0
$-CH_2-$, cyclopropane	0.3
CH_4	0.3
ROH monomer (dilute soln.)	0.5
CH_3-C- saturated	0.9–1.1
CH_2-C- saturated	1.2–1.35
RSH	1.1–1.5
RNH_2	1.1–1.5
$-C-H$ saturated	1.4–1.65
$CH_3-C=C<$	1.6–1.9
$CH_3-C=O$	2.1–2.6
CH_3-Ar	2.1–2.6
$H-C\equiv C-$ non-conjugated	2.45–2.65
$H-C\equiv C-$ conjugated	2.8–3.1
$ArNH_2$, $ArNHR$, and Ar_2NH	3.3–4.0
$CH_2=C<$ non-conjugated	4.6–5.0
$CH_2=C<$ acyclic non-conjugated	5.2–5.7

(Contd.)

(*Contd.*)

$CH_2 =C<$ cyclic non-conjugated	5.2–5.7
$CH_2 =C<$ conjugated	4.5–7.8
ArH benzenoid	6.5–8.0
ArH non-benzenoid	6.2–8.6
$ArNH_3^+$, $ArRNH_2^+$, and ArR_2NH^+	8.5–9.5
RCHO aliphatic, α, β-unsaturated	9.5–9.7
RCHO, aliphatic	9.7–9.9
ArCHO	9.7–10.0
ArOH	4.5–10.0
−COOH	8.9–12.2
−SO_3H	11.0–12.0

9.4.2 Diamagnetic Anisotropy and Chemical Shift

An anomalous feature of the data listed in Table 9.2 is that the proton peaks in certain compounds containing double bonds, triple bonds, and aromatic rings cannot be explained on the basis of simple diamagnetic shielding effects. For example, the acidic alkyne protons being bonded to the more electronegative triple-bonded carbon atoms should experience less diamagnetic shielding, and hence, should come to resonance at lower magnetic field strength (higher δ value) compared to the alkene protons which are bonded to relatively less electronegative double-bonded carbon atoms. However, the observed proton peaks are at ~2.7 ppm for the alkyne protons and ~4.5 ppm for the alkene protons, contrary to the expectation on the basis of diamagnetic shielding effect. The observed δ values for the protons in unsaturated compounds is explained on the basis of their anisotropic magnetic properties, which generates shielding and deshielding zones in space caused by the circulating π-electrons.

The symmetric distribution of π-electrons about the bond axis in acetylene allows the induced circulation of π-electrons when the molecular axis is parallel to the applied magnetic field. The field generated by the circulating electrons is similar to the flow of current in a wire loop and this secondary field acts against the applied magnetic field as it forms a shielding zone or core about the molecular axis. The protons are well shielded as they lie within the core as shown in Fig. 9.8(a). This shielding effect, thus, offsets the deshielding effect due to the acidity of the protons and electronic currents flowing in perpendicular directions to the bond. Hence, alkyne protons which are effectively shielded come to resonance only at a higher magnetic field strength. In contrast, in ethylene the nodal plane in the electron distribution of the double bond prevents the circulation of π-electrons about the bond axis. The π-electrons circulate in a plane along the bond axis as shown in Fig. 9.8(b). The protons experience lower shielding effect as they lie in the deshielded zone, and hence, come to resonance at higher δ values (lower magnetic field strength).

A similar diamagnetic anisotropic effect explains the observed higher δ values (~ 6–9 ppm) for aromatic protons. When the plane of the aromatic ring is perpendicular to the magnetic field the π-electrons circulate around the ring producing a *ring current*. The aromatic protons lie in the deshielded zones, and hence, come to resonance at a lower magnetic field strength. In contrast, when the plane of the aromatic ring is in other orientations the deshielding effect is absent or cancelled out. This diamagnetic anisotropic effect is clearly demonstrated in crystalline aromatic compounds, the magnetic susceptibilities varying with the orientation of the aromatic ring with respect to the applied magnetic field.

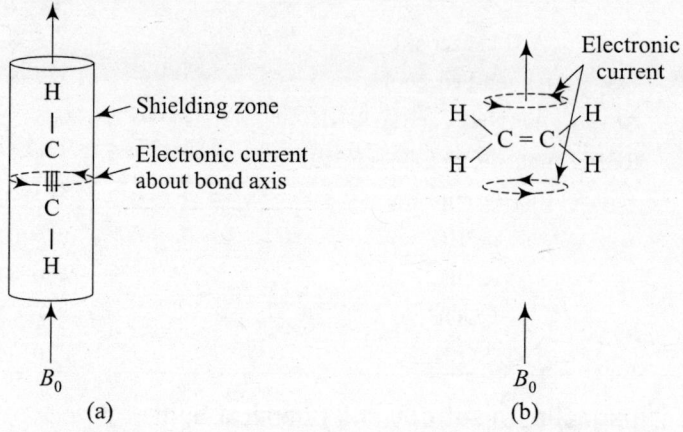

Fig. 9.8   Shielding of acetylene and deshielding of ethylene protons

In carbonyl compounds the aldehyde proton lies in the deshielded zone as the π-electron circulation occurs in a plane along the C=O bond axis similar to that in ethylene. This deshielding effect combines with the electronegative nature of the carbonyl group leading to a situation where the proton comes to resonance at much lower magnetic field strength, and hence, the observed δ value is large (~10 ppm).

The δ values of hydrogen-bonded protons are higher due to the deshielding effect as compared to the non-hydrogen-bonded protons. The effect of hydrogen bonding of chemical shifts is observed in alcohols, phenols, amines, and carboxylic acids. The chemical shift parameter values also depend on the nature of the solvent and the concentration of the substance.

9.4.3   Spin–spin Coupling

Spin–spin coupling or splitting is the second important environmental factor which affects the NMR absorption peaks. The coupling is due to the mutual interaction of the spins of adjacent nuclei which have two or more spin states or energy levels. Protons of the same carbon atom, however, do not split each other unless free rotation is hindered or an asymmetric centre is available adjacently. The coupling effect causes variations in the effective magnetic field strength experienced by the individual protons.

The splitting into multiplets is conveniently explained by considering the two protons designated H_A and H_B bonded to adjacent carbon atoms in a molecule as shown below:

$$
\begin{array}{cc}
H_A & H_X \\
| & | \\
-C & -C- \\
| & |
\end{array}
$$

Each of the hydrogen atoms has two allowed spin states which may be represented as $\uparrow$ and $\downarrow$, one spin aligned with (parallel) the magnetic field while the other one is aligned against the field. These two energy levels are equally populated in the sample. The net effective magnetic field experienced by H_A increases or decreases to a small extent depending on the spin states of the hydrogen atom H_X bonded to the adjacent carbon atom. This effect splits the absorption of H_A into two peaks (a doublet) with intensities in the ratio of 1:1. A similar effect causes the splitting of the absorption peak of H_X also into a doublet of equal intensity. The mutual effect of the spin–spin coupling on the NMR absorption peaks is shown in Fig. 9.9.

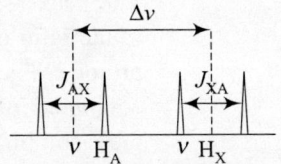

Fig. 9.9 First order AX spectrum ($\Delta v/J > 7$)

The components of each doublet are separated to the same extent and the spacing between the components is referred to as *coupling constant* J_{AX}, the units being Hz. The magnitude of the coupling constant is independent of the applied magnetic field. However, the chemical shift difference Δv between H_A and H_X varies in direct proportion to the strength of the applied magnetic field. The value of Δv is obtained by measuring the distance between the mid points of the doublets. The NMR spectrum consisting of the doublets for the two protons is called the *first-order spectrum* and the spectrum is classified as AX spectrum. The first-order spectrum due to spin–spin coupling of protons in a compound is observed only when the chemical shift difference Δv between the coupled groups is greater than the coupling constant at least to the extent $\Delta v/J > 7$.

When the ratio of chemical shift to coupling constant is smaller ($\Delta v/J \approx 2$) a *second-order splitting* occurs with the inner peaks of the multiplets gaining in intensity at the expense of the outer peaks, giving rise to a *second-order spectrum*. The conventional notation in NMR spectroscopy to use alphabets for labeling nuclei as A and B instead of A and X when the chemical shift difference Δv is small.

The chemical shift positions of vH_A and vH_X are displaced from the mid points of the respective doublets and the chemical shift difference Δv is measured between the centres of gravities of the multiplets, as shown in Fig. 9.10

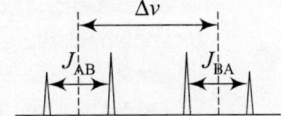

Fig. 9.10 Second order AB spectrum ($\Delta v/J \approx 2$)

When the chemical shift difference decreases further and reaches the limiting value of $\Delta v = 0$, the protons are equivalent, and hence, the inner peaks merge and the outer peaks disappear giving rise to a singlet (a single peak). The spectrum is then called A_2 spectrum.

9.4.4 Interpretation of First-order Spectra

The first-order spectra are relatively simple to interpret by considering the following aspects.

(i) The effect of spin–spin coupling in adjacent groups with two or more protons can be treated in a similar manner as in interpreting the AX spectrum.

For example, methylene groups contain two equivalent protons and two adjacent methylene groups give rise to A_2X_2 or A_2B_2 spectra depending on the magnitude of the ratio $\Delta v/J$. The first-order A_2X_2 spectrum consists of two triplets with individual peaks in a multiplet having intensities in the ratio 1:2:1.

Adjacent methyl and methylene groups give rise to A_3X_2 spectrum as seen in the high-resolution proton NMR spectrum of ethyl alcohol shown in Fig. 9.11.

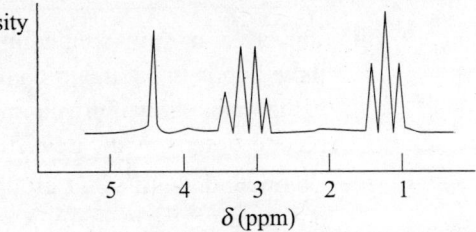

Fig. 9.11 High-resolution proton NMR spectrum of ethyl alcohol (A_3X_2 spectrum)

The peak assigned to methyl protons splits into three peaks (triplet) with intensities in the ratio of 1:2:1 due to the interaction between the spins of methyl and methylene protons. The two methylene protons have four possible combinations of spin states combined into two groups consisting of two combinations each. The magnetic effect of these combinations is transmitted to the methyl protons on the adjacent carbon atom. The effective field experienced by the methyl protons is either slightly decreased due to the opposing magnetic field effect resulting in an up-field shift or slightly increased due to parallel alignment of spins resulting in a downfield shift to bring them into resonance. Thus, the two magnetic effects split the absorption peak of methyl protons into three peaks of intensities of 1:2:1.

Similarly, the peak assigned to methylene protons is split into four peaks (quartet) of intensities in the ratio of 1:3:3:1 due to the effect of the adjacent methyl protons. The three methyl protons have eight possible combinations of spin states combined into four groups, two of them containing three combinations each producing equivalent magnetic effects. The magnetic effects transmitted to the methylene protons split the methylene peak into a quartet of intensities 1:3:3:1. The first-order spectrum of ethanol has a Δv value of about 140 Hz between the methyl and methylene multiplets and coupling constant J value for the peaks being 7 Hz.

(ii) In general the number of peaks in a multiplet is determined by the number (n) of magnetically equivalent protons on the neighbouring atoms and is given by the formula ($n + 1$). However, protons on the same carbon atom do not split each other as they are equivalent. The intensities of the individual peaks in a given multiplet depend on the statistical weights of the various possible combinations of the two spin states for the protons in the adjacent group. The intensity ratios of the peaks can be obtained in terms of Pascal's triangle or from the coefficients of binomial expansion of $(a + b)^n$. The simplified family tree is shown in Table 9.3

Table 9.3 Number of peaks in first-order multiplets and their relative intensities

No. of equivalent protons (n)	No. of peaks in the multiplet ($n + 1$)	Relative intensities of the peaks in the multiplet
0	1	1
1	2	1 1
2	3	1 2 1
3	4	1 3 3 1
4	5	1 4 6 4 1
5	6	1 5 10 10 5 1

(iii) The effect of spin–spin coupling is rarely observed for protons separated by more than three bonds in saturated compounds. In contrast, unsaturated compounds and aromatic compounds show more complex splitting patterns attributed to long-range coupling. The second-order spectra are difficult to interpret in such cases.

(iv) Compounds with vinyl groups and mono-substituted furans with three adjacent single protons give rise to an AMX spectrum consisting of a twelve line pattern of three groups of four peaks when the chemical shift differences are large. The AMX spectrum is observed because of the three unequal coupling constants J_{AM}, J_{MX}, and J_{AX}. When the chemical shift differences of the three protons in such compounds are small the peaks overlap giving rise to complex spectra due to distortion and second-order splitting.

(v) Protons attached to hetero atoms such as oxygen, nitrogen, or sulphur as in alcohols, phenols, carboxylic acids, amines, thiols, etc., are labile in solution, and hence, can be chemically exchanged with protons in the solvent. These exchangeable protons are essentially decoupled from neighbouring protons particularly when the rate of exchange is fast and give rise to sharp singlet peaks as in low-resolution NMR spectra. However, if the rate of exchange is slow the singlet peaks become broad and also give rise to multiplets at very slow exchange rates. The rate of exchange of OH protons is enhanced by trace quantities of acid or alkali added while treatment with D_2O removes the resonance completely. The exchange effect is observed in the case of the NMR spectrum of highly pure ethanol in the form of triplets for the OH protons and eight peaks for the methylene protons due to spin–spin coupling between the two sets of protons. The addition of a trace of an acid or base to pure ethanol effectively decouples, coalesces the peaks into singlets, and result in an NMR spectrum which is similar to the spectrum that is obtained at low resolution as shown in Fig. 9.7. The presence of exchangeable protons in a sample is easily detected by treating the sample with D_2O.

9.4.5 Simplification of Complex Spectra

NMR spectra of many compounds appear to be distorted showing overlapping absorption peaks due to similar chemical shifts of protons of different functional groups. Such complex spectra can be simplified into simpler and easily analysable patterns through the use of (i) instruments operating at higher frequencies, (ii) chemical shift reagents, (iii) double resonance, and (iv) spin tickling.

High-resolution NMR spectrometers operate at higher RF frequencies, for example at 100 MHz (2.35 T), 220 MHz (5.1 T), or 640 MHz (15 T) and give better resolution as compared to an instrument operating at 60 MHz because the chemical shift differences increase with increasing applied magnetic field strength. For example, the chemical shift differences in a 220-MHz spectrum are about twice that observed in a 100-MHz spectrum.

Chemical shift reagents offer a cheaper alternative to expensive high-resolution NMR spectrometers for obtaining well resolved spectra. The commonly used chemical shift reagents are the paramagnetic lanthanide complexes of fluorinated β-diketones which are coordinatively unsaturated. Examples include complexes such as tris-(dipivalomethanato)europium(III) (Eu(dpm)$_3$), tris-(6,6,7,7,8,8,8-heptafluoro–2,2-dimethyl–3,5-octanedionato)europium(III) (Eu(fod)$_3$), and tris-(trifuoromethylhydroxy–*d*-camphorato)ytterbium (Yb(TFC)$_3$) which induces downfield shifts. In contrast, Pr(dpm)$_3$ and Pr(TFC)$_3$ complexes induce up field shifts. Other examples of shift reagents include Ag(fod) and chromium acetylacetonato (Cr(acac)) complexes.

The shift reagent is added to a solution of the organic compound, dissolved in chloroform or carbon tetrachloride and the NMR spectrum is recorded. Increments of the reagent can be added till sufficient resolution is attained. Organic compounds with functional groups having lone pair of electrons on the hetero atom (e.g., alcohols, ethers, esters, aldehydes, ketones, amines, etc.) coordinate to the lanthanide metal and cause changes in the electron densities around the protons near the coordinating hetero atom. Theoretical studies have indicated that the induced shifts are due to contact and dipolar (pseudo contact) interactions between the shift reagent and the organic molecule.

Double resonance also known as *spin decoupling* is the opposite of spin–spin coupling and involves deliberate removal of spin–spin coupling between neighbouring nuclei. Spin decoupling is achieved when the sample is irradiated with a second, relatively strong alternating RF field v_2 perpendicular to the applied magnetic field in addition to the resonance RF frequency v_1. For example, it is possible to observe the AX spectrum of the coupled nuclei A and X operating at frequency v_1 corresponding to the resonance frequency of the nucleus A. The AX spectrum of nucleus X may also be observed by operating the instrument at v_2 corresponding to the resonance frequency of the nucleus X. Spin decoupling is achieved by applying simultaneously a second stronger RF frequency v_2, in addition to v_1. The spin and the magnetic moment of the nucleus X irradiated at the frequency v_2 is quantized in a direction perpendicular to that of its coupling partner A resulting in the decoupling of the spins of the AX couple. This in turn leads to the merging of the doublet of the nucleus A seen in Fig. 9.9 into a sharp single peak. The roles of the irradiated and observed nuclei (A and X respectively in the above example) may be reversed by setting the frequency v_1 to correspond to the resonance frequency of the nucleus X and v_2 to that of the nucleus A. The doublet of the nucleus X will also merge into a sharp single peak. In practice a frequency sweep technique is adopted for spin decoupling in which the second RF field is varied until the desired change occurs in the chosen multiplet.

The intensity of NMR lines is enhanced during double resonance which is attributed to the *nuclear overhauser effect* (NOE) (also known as *spin pumping*). The intensity of NMR signal depends on the small difference in the population between the excited and the ground energy levels and it is absolutely necessary that the relaxation processes must be efficient and rapid to restore the equilibrium conditions at which the population of ground state nuclei is higher. In the presence of a noise-decoupling radiation, a relatively more efficient non-radiative relaxation process occurs through a dipolar spin–spin interaction between the excited nucleus and an adjacent magnetic nucleus within the same molecule. This interaction occurs through space and is distinct from the spin–spin coupling transmitted through bonds resulting in the splitting of energy levels. The non-radiative relaxation restores the equilibrium conditions more rapidly, and hence, the intensity of the NMR signal is enhanced up to about 50%.

Spin tickling technique is similar to double resonance but involves the application of a weak secondary radiofrequency so that only partial decoupling occurs or in some cases splitting of the peaks resulting in extra multiplets occurs. The technique is useful for the determination of fundamental NMR constants, particularly the sign of coupling constants.

9.5 NUCLEAR MAGNETIC RESONANCE SPECTROSCOPY OF NUCLEI OTHER THAN HYDROGEN

Nuclei with non-zero spin quantum number can give rise to NMR signals and in fact more than 200 such isotopes are known to have non-zero magnetic moments. However, the most widely studied nucleus with spin quantum number of ½ is that of ^{13}C. Other nuclei with spin values of ½ include ^{19}F, ^{31}P, and ^{15}N which find use in study of organic compounds and biologically important molecules.

9.6 CARBON-13 NMR SPECTROSCOPY

The principles involved in ^{13}C NMR spectroscopy are the same as that in proton NMR spectroscopy as both the nuclei have half-integral spins. However, ^{13}C NMR spectroscopy is about 6000 times less sensitive as compared to the proton NMR due to two factors: (i) the natural abundance of the isotope ^{13}C is only 1.1% and (ii) the magnetogyric ratio is also much less (about one fourth) as compared to that of hydrogen.

The characteristic features of ^{13}C NMR spectra are (i) the same sample can be used to obtain the NMR spectra of ^{13}C as well as the proton; (ii) the chemical shift of a nucleus depends on the chemical environment or more precisely on the electron density surrounding the nucleus and the electronegativity of the atoms attached to it as in proton NMR spectra; (iii) the chemical shifts can be expressed in terms of ppm in δ scale using TMS as the standard with a δ value of zero. The chemical shift range is much wider spreading over from 0 to 250 ppm (Table 9.4) as compared to that in proton NMR spectra with a range of 0–15 ppm; (iv) the effect of spin–spin coupling between neighbouring ^{13}C nuclei is not observed in practice though coupling can occur. This is because of extremely small per cent of natural abundance of the ^{13}C isotope and one can discount the presence of more than one ^{13}C nucleus in each molecule of a non-enriched sample containing less than 100 carbon atoms per molecule; and (v) spin–spin coupling between neighbouring ^{13}C and ^{1}H nuclei occurs and is observed in ^{13}C NMR spectra.

Table 9.4 Chemical shift ranges of ^{13}C resonances

Functional groups	Ranges of ^{13}C resonances (ppm from TMS)
Saturated carbons	0–100
primary (RCH_3)	0–25
secondary (R_2CH_2)	15–45
tertiary (R_3CH)	20–50
quaternary (R_4C)	25–40
Alkenes	75–175
$R_2C=CH_2$	100–110
$RCH=CH_2$	110–120
$RCH=CHR$	115–140
Acetylenic carbons ($-C\equiv C-$)	25–90
Acids ($-COO$)	160–190
Aldehydes ($-CHO$)	190–210
Ketones ($>C=O$)	180–220
Esters ($RCOOR$)	150–190
Nitriles ($RC\equiv N$)	110–130
Benzenoid carbons	80–170
Azomethine ($R_2C=NR$)	150–170

Advantages of carbon-13 NMR spectra over proton NMR spectra include (i) direct information on the carbon skeleton or backbone of the molecules is useful in structural analysis; (ii) carbon-13 spectra are much simpler as compared to proton NMR spectra and facilitate easy interpretation because of wide separation between shifts of chemically different nuclei with no overlapping of absorption peaks because of negligible spin–spin coupling between neighbouring ^{13}C nuclei and availability of proton decoupling techniques (noise decoupling) for eliminating the spin–spin coupling between ^{13}C and neighbouring protons. However, it should be mentioned here that coupling between ^{13}C and neighbouring protons is sometimes helpful in assigning absorption peaks. For example, the presence of groups such as (C–H), (>C=C–H), and (–C≡C–H) can be easily identified because they exhibit doublets with widely differing values for the coupling constant J_{CH} of 125 Hz, 170 Hz, and 250 Hz respectively. The coupling pattern due to spin–spin coupling is similar to that in proton NMR spectra with the C–H, CH_2, and CH_3 groups exhibiting respectively a doublet, triplet, and quartet. However, spin–spin coupling between ^{13}C and neighbouring protons becomes a hindrance also because the peak may not be observed easily as its intensity decreases significantly.

Double resonance technique also known as *noise decoupling* eliminates the coupling between ^{13}C nuclei and protons and thereby provides a simpler spectrum as the multiplets coalesce to give single lines. Noise decoupling involves irradiating the sample at ^1H frequency (e.g., 100 MHz) over a wide frequency range (white noise irradiation) covering all the proton resonances in the sample while observing ^{13}C signals. The intensity of the noise decoupled ^{13}C spectral lines also increases due to *nuclear Overhauser effect* (NOE). Figure 9.12 shows the noise decoupled spectrum of 2-butanol and the assignments of the four absorption bands.

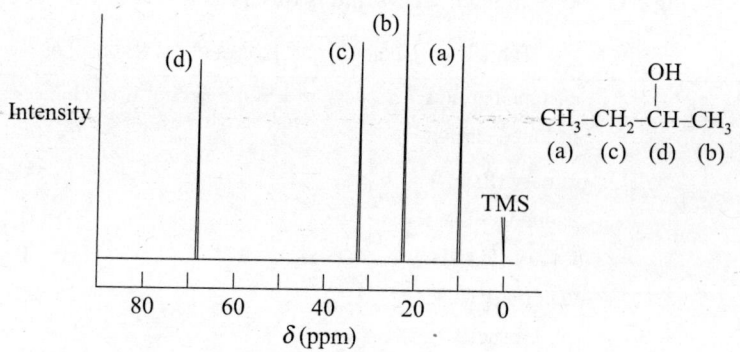

Fig. 9.12   Noise decoupled ^{13}C spectrum of 2-butanol

9.7   APPLICATIONS OF NMR SPECTROSCOPY

Qualitative identification of the functional groups and elucidation of structures of pure organic, organometallic, and biomolecules are the major applications of proton NMR spectroscopy.

Quantitative analysis is based on the direct proportionality between the integrated peak areas and the number of protons in the molecule. The technique is useful for quantitative analysis of functional groups of organic compounds. However, quantitative analysis is rarely done by NMR spectroscopy due to the possibility of overlap of peaks, particularly in complex molecules and the prohibitive cost of the instrument.

Magnetic resonance imaging (MRI) based on proton NMR spectroscopy has an important medical application for generating a three-dimensional picture of the detailed internal structure of whole or parts of the patient's body. The technique is relatively safe as compared to computed tomography (CT) scan and X-ray imaging techniques as it does not use any ionizing radiation.

9.8   FOURIER TRANSFORM NMR SPECTROSCOPY

Fourier transform NMR spectroscopy (FT NMR spectroscopy) involves the excitation of the sample with a series of high power RF pulses of wide frequency range in the presence of an applied magnetic field of about 2.5 T. Each RF pulse lasts several microseconds in duration during which all the nuclei of a particular isotope are excited as they experience radiation of their particular resonance frequency. The duration of the RF pulse and its frequency spread are inversely related, shorter pulse giving a wider range of frequencies and vice versa. The excited nuclei relax to establish the equilibrium population of the spin states by emitting radiation in the form of a signal which decays rapidly and finally lost in the noise. The relaxation process is about 4–5 times the spin–lattice relaxation time T_1 and the time over which the signal is collected is called the acquisition time T_{acq}. The exponentially decaying signal is known as *free induction decay* (FID) signal which is a time domain signal of NMR data which can be converted to the conventional frequency domain signal obtained in CW NMR spectrum by computer-aided Fourier transformation. The FIDs can be collected several times within a short span of a few seconds by repeating the pulse sequence several times to enhance the signal-to-noise (S/N) ratio.

FT NMR spectroscopy has several advantages over CW NMR technique in that the sample requirement for recording proton NMR spectra can be as low as a few micrograms and nuclei other than protons can be studied because of its greater sensitivity. In addition the development of multiple pulse techniques giving rise to *one-dimensional* (1-D) and *two-dimensional* (2-D) NMR spectra facilitate the identification of chemically different groups of nuclei and correlation between spectra of different elements in the same compound. In addition the technique is useful for elucidation of structures of complex biomolecules, for example, proteins.

The 1-D NMR spectrum obtained by the pulsed Fourier transform technique is similar to the conventional two-dimensional representation of the absorption versus frequency data. In 2-D Fourier transform technique the NMR signal is recorded as a function of two time variables, t_1 and t_2, and the resulting data is Fourier transformed twice to yield a spectrum as a function of two frequency variables. The 2-D spectrum is actually a three-dimensional topographical map or the *contour plot* in which the contour lines show the intensities of the absorption peaks as a function of two frequency axes F_1 and F_2 at right angles to each other. In the 2-D technique the sample is subjected to a pulse sequence consisting of a single RF pulse to initiate the *preparation period* followed by one or more pulses at controlled time intervals during *evolution and mixing period* t_1. After the mixing period the signal is recorded as a function of the second time variable t_2. The FID signal generated is acquired at the end of the *detection and acquisition period* t_2. The sequence of events is called the pulse sequence and is repeated several times for varying increments of t_1 and recording the FID signal as a function of t_2 for each value of t_1.

9.9 MAGIC ANGLE SPINNING NMR SPECTROSCOPY

In the condensed phase the nuclear spin in the sample experiences three main interactions, namely, dipolar, chemical shift anisotropy, and quadrupolar interactions which lead to broad NMR lines. In liquid samples these interactions average out due to time-averaged molecular motion. While recording the NMR spectra of liquid samples the sample is spun at about 20 Hz (1000–1500 rpm) so that the sample experiences homogeneity in the external magnetic field strength and the NMR lines are quite sharp. However, in solids the dipolar, chemical shift anisotropy, and quadrupolar interactions are time-dependent. In addition the molecules are oriented in different directions within the solid and experience a non-homogenous magnetic field resulting in broadening of the NMR lines. Fortunately, the broadening of NMR lines of solid samples can be avoided and sharp signal obtained by *magic angle spinning*.

Magic angle spinning NMR spectroscopy (MAS NMR) involves spinning the sample at an angle called the *magic angle* θ_m (54.74°) with respect to the direction of the magnetic field.

The nuclear dipole–dipole interaction between the magnetic moments of the nuclei averages to zero when the solid sample is spun at the magic angle. The chemical shift anisotropy due to nucleus–electron interaction also averages to a non-zero value while the quadrupolar interaction is partially averaged to a residual value. The net effect of magic angle spinning results in NMR spectra consisting of narrower NMR lines giving rise to isotropic value useful for the determination of structure of solids and spinning sidebands which occur at multiples of spinning speed useful for determining the chemical shift anisotropy of the nuclei.

In practice, it is not possible to orient every molecule in the solid at the magic angle, and hence, while recording the NMR spectra of solid samples and samples which cannot be dissolved, for example, proteins (dissolution may alter or even destroy the structure), the sample is spun rapidly by air turbines at a frequency of $1-70$ kHz (600×10^3 to 1500×10^3 rpm) so that each molecule behaves magnetically as though it is the only molecule oriented at the magic angle. The magic angle spinning gives rise to NMR signals similar to that obtained from liquid samples.

9.10 ELECTRON SPIN RESONANCE SPECTROSCOPY

Electron spin resonance (ESR) spectroscopy also known as *electron paramagnetic resonance* (EPR) spectroscopy involves absorption of energy in the form of microwave radiation by unpaired electrons in the presence of an external magnetic field. Diamagnetic species such as molecules with even number of electrons or paired electrons do not exhibit ESR spectra as the number of electrons in the two spin sates is equal and the magnetic effects of the spinning electrons get cancelled. In contrast, paramagnetic species with one or more unpaired electrons generate an induced magnetic field which aligns parallel to the applied field and reinforces it. Electron has a spin quantum number $s = \frac{1}{2}$ and the half-integral value can have $(2\,s + 1)$, that is, two energy levels or spin states with values of $+\frac{1}{2}$ and $-\frac{1}{2}$. The two spin states remain degenerate in the absence of an external magnetic field. However, in the presence of an applied magnetic field the degeneracy is lifted and the spin states resolve into higher and lower energy levels. The difference in energy of the two spin states ΔE is greater than for nuclear spin. The spinning electron in the lower energy state can absorb energy which results in the change of the spin state giving rise to ESR spectrum when the frequency of the radiation satisfies the resonance condition as given by

$$\Delta E = h\upsilon = g\beta H_0 \tag{9.15}$$

where g is a dimensionless constant called the Lande splitting factor (also called the g-factor or spectroscopic splitting factor) which varies with the electron environment, β is a constant called Bohr magneton (9.274×10^{-21} erg gauss^{-1} or 9.274×10^{-24} JT^{-1}) and H_0 is the strength of the applied magnetic field. Usually the microwave frequency is held constant and the strength of the applied magnetic field varied to the particular value of H_0 satisfying the resonance condition. The principle and practice involved in ESR spectroscopy are, thus, similar to that in NMR spectroscopy.

9.10.1 ESR Spectrometer

The main components of ESR spectrometer include (i) an electronic oscillator called *klystron*, as a source of microwave radiation, most commonly operating in the X-band (wavelength ~3 cm or frequency of 9500 MHz), (ii) a homogeneous and steady applied magnetic field ranging between 50 and 5500 G provided by an electromagnet, (iii) hollow rectangular copper or brass tubes (2.2×10 cm) the inner walls coated with silver or gold called *wave guides* through which the microwave radiation is transmitted, (iv) sample cavity where the magnetic vector is at a maximum, (v) sample tube to hold solid or liquid samples, usually made of quartz of ~ 3 mm outer diameter or glass capillary or rectangular cells particularly when the sample is dissolved in polar solvents with appreciable dielectric constant, (vi) crystal detector to determine the microwave power absorbed by the sample, (vii) modulating coils connected to an oscillator to superimpose a variable amplitude sinusoidal modulation on the varying magnetic field, (viii) phase sensitive detector, and (ix) recorder. Sample tube is not rotated while recording the ESR spectrum.

The detection limit of X-band spectrometers is about 10^{-7}–10^{-9} M which is the minimum sample concentration in aqueous solutions to get a reasonably good spectrum, but for quantitative analysis and studies on structural aspects the sample concentration should be at least 10^{-6} M. Concentrations $>10^{-4}$ M lead to spin–spin broadening of the ESR peaks. ESR spectrometer operating at K band at frequencies $>25,000$ MHz has a greater sensitivity, about 20 times that of X-band spectrometer. Dissolved oxygen should be removed for high-resolution studies because oxygen is paramagnetic and contributes to line broadening.

Figure 9.13 shows a block diagram of the components of ESR spectrometer.

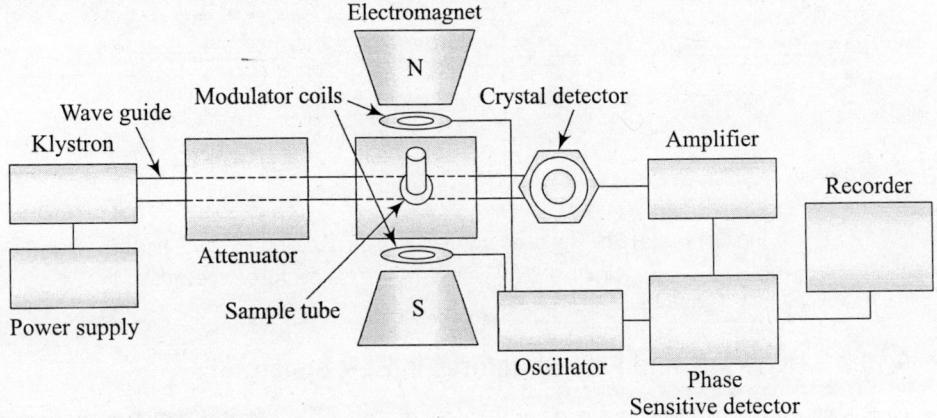

Fig. 9.13   Block diagram of the major components of ESR spectrometer

9.10.2   ESR Spectrum

The splitting of energy levels of the unpaired electron in the presence of the external magnetic field and the ESR absorption spectrum as a plot of RF power absorbed versus magnetic field in accordance with the resonance condition are shown in Figs 9.14(a) and (b) respectively. Usually the ESR spectrum is recorded as the first derivative [Fig. 9.14(c)] since a phase-sensitive crystal detector is used for the microwave power absorbed by the sample. The peak-to-peak line width is given as ΔH in the derivative spectrum.

The value of the Lande splitting factor g for a free electron is precisely 2.0023 and is almost a constant for all the free radicals and some ionic crystals because the free electron is completely delocalized in the molecular orbital encompassing the entire molecule and behaves similar to a free electron in space. The g value is slightly higher for free radicals containing oxygen or nitrogen whereas it is lower in free radicals containing halogen. Many ionic crystals also give a value for the g-factor close to that of the free electron either because of the ion contributing the free electron exists in an S state, and hence, $L = 0$ (e.g., Fe^{3+} with five unpaired electrons giving rise to the term symbol ^6S) or because Russell–Saunders coupling breaks down and the electron spin precesses independently about the applied magnetic field.

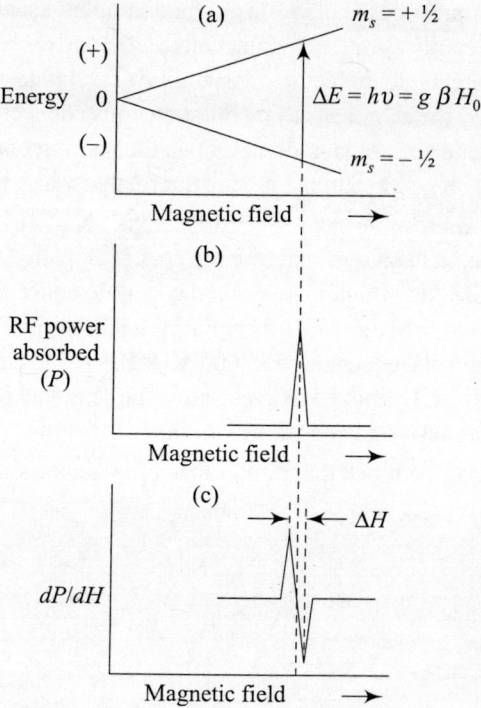

Fig. 9.14 (a) Splitting of energy levels of an unpaired electron, (b) ESR absorption spectrum, and (c) first derivative of absorption spectrum

9.10.3 Hyperfine and Fine Structures in ESR Spectra

ESR spectra exhibit two types of multiplets due to (i) hyperfine structure or hyperfine splitting and (ii) fine structure.

Hyperfine splitting is relatively a smaller effect and arises due to interaction of the magnetic moment of the unpaired electron with a neighbouring nucleus with a non-zero spin. This electron–nucleus coupling is similar to the spin–spin coupling in NMR spectra. The simplest example of hyperfine splitting is observed in the ESR spectrum of hydrogen atom. The nucleus of the hydrogen atom with a spin quantum number $I = \frac{1}{2}$ interacts with the electron spin and the ESR spectrum consists of two peaks (a doublet). The splitting of energy levels in hydrogen atom and the corresponding ESR spectrum are shown in Figs 9.15(a) and (b) respectively.

The two peaks are separated by 506.8 gauss (~0.05 T) with the resonant magnetic field centred between the peaks so that the Lande splitting factor $g = 2.00232$. The parameter A_0 is the isotropic electron spin–nuclear spin hyperfine coupling constant, a measure of the *hyperfine coupling energy*.

The pattern of hyperfine splitting of an ESR peak is given by the formula $(2nI + 1)$ where n is the number of equivalent nuclei interacting the unpaired electron and I is the spin quantum number of the nucleus. Thus, when an unpaired electron interacts with a single proton the resulting ESR spectrum of H atom splits into $(2 \times 1 \times \frac{1}{2} + 1 = 2)$ peaks as shown in Fig. 9.15. For a compound containing a single $^{63}Cu(II)$ ion ($I = 3/2$) four lines are exhibited due to hyperfine splitting $(2 \times 1 \times 3/2 + 1 = 4)$. Methyl radical ($\cdot CH_3$) with three equivalent hydrogen nuclei exhibits four lines. The natural abundance of ^{13}C isotope ($I = \frac{1}{2}$) is quite low (1.08%) and the satellite peaks due to the interaction of the unpaired electron with ^{13}C atoms are not resolved. Hence, in hydrocarbon radicals the unpaired electron can be considered to interact only with n number of equivalent hydrogen atoms giving rise to $(n + 1)$ lines due to hyperfine splitting. The number of lines in the ESR spectrum is helpful in identifying the radical. If an organic radical has two or more sets of equivalent atoms, the total number lines in ESR spectrum is given by the formula

$$(2n_1I_1 + 1)(2n_2I_2 + 1) \cdots (2n_jI_j + 1) \quad (9.16)$$

The intensity pattern of the ESR spectral lines can be predicted on the basis of 'family tree' method. Table 9.5 lists the relative intensities of ESR lines for organic radicals containing n number of equivalent nuclei each with $I = \frac{1}{2}$.

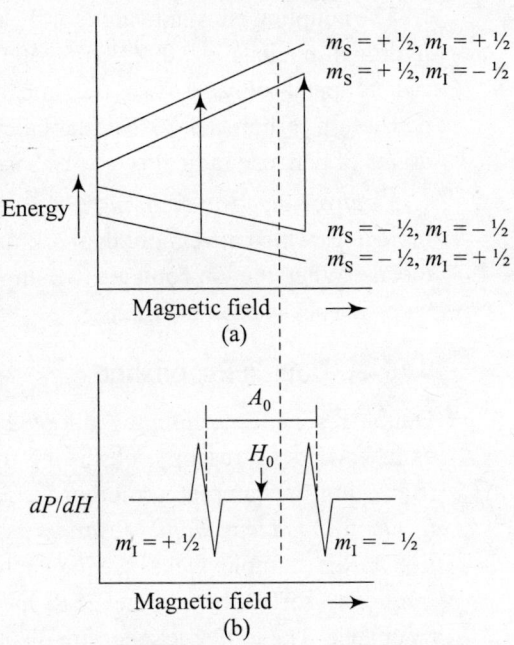

Fig. 9.15 (a) Splitting of energy levels and (b) ESR spectrum of hydrogen atom due to hyperfine splitting

Table 9.5 Relative intensities of ESR lines due to hyperfine splitting

No. of equivalent nuclei each with $I = \frac{1}{2}$	Total number of lines $(n + 1)$	Relative intensities of ESR lines
1	2	1:1
2	3	1:2:1
3	4	1:3:3:1
4	5	1:4:6:4:1
5	6	1:5:10:10:5:1

The hyperfine coupling energy or coupling constant A_0 for most organic radicals with one unpaired spin lies in the range 10^{-4}–10^{-3} T, much smaller as compared to that of hydrogen atom (~ 0.05 T). This is because the unpaired electron in the organic molecule spends only a part of its time at any one nucleus or bond as it can easily move over the different nuclei, whereas in the hydrogen atom the unpaired electron is confined to the hydrogen atom. In other words, the electron density is completely localized on the hydrogen atom. The coupling constant is therefore the related to the electron density near a nucleus (time the unpaired electron spends in the s orbital on the nucleus), the greater the magnitude of A_0 greater is the probability of finding the electron at the nucleus. The relationship between the electron density ρ and A_0 is expressed as

$$A_0 = \rho R \quad (9.17)$$

where R is intrinsic coupling for unit density. For hydrogen atom, $\rho = 1$, and hence, $R = 0.05$ T.

The coupling constant value for the methyl radical is 2.3 mT and the electron density on each hydrogen nucleus is 0.0023/0.05 = 0.046 or 4.6%. Thus, the unpaired electron of the methyl radical spends only < 14% of the time on the three hydrogen nuclei or in other words the electron density on carbon is 86%. Similar calculations show that the electron density on the six carbon atoms of benzene radical is ~ 95% based on the value of $A_0 = 0.38$ mT.

Electron–electron coupling giving rise to *fine structure* in the ESR spectrum occurs only in molecules which have more than one unpaired electron, for example, complex ions of transition metals. When the ion contains two unpaired electrons the ESR spectrum shows a fine structure with two peaks or fine structure lines.

9.10.4 Double Resonance

Double resonance technique is adopted to improve the resolution of ESR spectrum and involves, as in NMR spectroscopy, observing the ESR spectrum at one frequency and simultaneously irradiating the sample at another frequency. Two double resonance techniques are in vogue (i) *electron nuclear double resonance* (ENDOR) and (ii) *electron double resonance* (ELDOR). ENDOR is a simpler technique in which the sample is irradiated simultaneously with microwave frequency for observing the ESR spectrum and with radiofrequency for achieving nuclear resonance. The ENDOR spectrum displays the ESR signal height as a function of the sweeping radiofrequency. The ELDOR technique involves irradiating the sample with two microwave frequencies, one for observing the ESR spectrum while the other is swept through to give an ESR signal height as a function of difference of the two microwave frequencies.

9.10.5 Applications of ESR Spectroscopy

Analytical applications of ESR spectroscopy include qualitative analysis of species containing unpaired electrons and detection and determination of paramagnetic metal ions such as copper(II), iron(III), chromium(III), manganese(II), titanium(III), and vanadium(IV) at ppb levels. However, quantitative analysis based on integration of the intensities of ESR lines is relatively less accurate and time consuming.

Another interesting application of ESR technique is the determination of active surface area of catalysts. The method is based on the fact that the total area of the ESR signal is proportional to the number of unpaired electrons in a sample. The method involves adsorbing a paramagnetic magnetic substance such as oxygen on the surface of the catalyst and determining the total area of the ESR signal. The number of unpaired electrons on the surface (hence, number of adsorbed oxygen molecules) is obtained by comparing with the total area of the ESR signal obtained for diphenylpicrylhydrazyl (DPPH) radical, which has 1.53×10^{21} unpaired electrons per gram.

ESR spectroscopy is useful to study the disease-state of the biological systems by spin *labeling* technique. The technique involves binding paramagnetic molecules (spin labels) such as nitroxide derivatives of iodoacetamide which bind specifically to methionine, lysine, and arginine residues of amino acids or nitroxide derivatives N-ethylmaleimide to –SH groups or incorporating nitroxide derivatives of strearic acid or cholesterol or phospholipids into biological systems and recording their ESR signals. Spin labeling technique is useful to get information on the dynamic or static nature of the biological system including structure, conformational changes, phase transitions, chemical reactions, polarity, and fluidity.

EXAMPLE 1   *Deduce the structural formulae of two compounds (a) C_2H_4O and (b) C_3H_8O if the low-resolution proton NMR spectrum of compound (i) shows two peaks of areas in the ratio 1:3. Identify the compound while that of the compound (ii) shows three peaks of ratios 2:3:3.*

SOLUTION

Based on the available information the compound (i) has two chemically and magnetically non-equivalent protons in the ratio of 1:3. The three equivalent protons can be from the CH_3 group while the single proton can be from OH group and the compound is identified as CH_3CHO.

On a similar basis the compound (ii) consists of three sets of non-equivalent protons in the ratio 2:3:3. The compound is identified as $CH_3OCH_2CH_3$ which has three types of protons, two methyl and one methylene groups, with one of the methyl groups being different as it is adjacent to the electronegative oxygen atom.

EXAMPLE 2   *How many non-equivalent protons are available in the following molecules?*

(i) toluene, (ii) o-xylene, (iii) m-xylene, (iv) nitrobenzene, (v) o-bromotoulene, and (vi) m-chlorotoulene.

SOLUTION

(a) 3 sets   (b) 2 sets   (c) 3 sets   (d) 3 sets   (e) 4 sets   (f) 4 sets

EXAMPLE 3   *Assign the peaks with chemical shift parameter values for the protons at 1.0, 1.5, 3.8, and 7.3 ppm observed in the low-resolution NMR spectrum of propyl benzoate.*

SOLUTION

$$\underset{\text{(d)}}{C_6H_5}-\overset{\overset{\displaystyle O}{\|}}{C}-O-\underset{\text{(c)}}{CH_2}\underset{\text{(b)}}{CH_2}\underset{\text{(a)}}{CH_3}$$

The peaks observed at chemical shift parameters 1.0, 1.5, 3.8, and 7.3 ppm respectively in the NMR spectrum of propyl benzoate are assigned to magnetically non-equivalent protons designated as (a), (b), (c), and (d).

- Magnetic resonance spectroscopy involves the study of the interaction between spinning nuclei (NMR spectroscopy) and electrons (ESR spectroscopy) with electromagnetic radiation in the microwave or radiofrequency region in the presence of an external magnetic field.

- Nuclei with non-zero spin quantum number (e.g., 1H and 13C) exhibit splitting of nuclear energy levels in the presence of external magnetic field and undergo nuclear transitions by absorbing energy from the incident electromagnetic radiation. The absorption of energy occurs when the resonance condition occurs between the frequency of the spinning nuclei and the frequency of the electromagnetic radiation giving rise to NMR

- absorption signal and an absorption spectrum in the form of a peak.

- The use of NMR spectroscopy in the study of chemical species entirely depends on two environmental effects, namely chemical shift and spin–spin coupling.

- The proton is subjected to a diamagnetic shielding by the electrons bonding it to another nucleus. Hence, the proton experiences a smaller effective applied magnetic field strength as compared to the actual applied magnetic field. Consequently, resonance condition is achieved with electromagnetic radiation of particular frequency at smaller applied magnetic field as compared to the reference compound TMS which has chemically identical and magnetically identical 12 highly shielded protons.

- The NMR spectrum is a plot of the intensities of absorption peaks as a function of the chemical shift parameter δ. The δ scale is linear increasing from 0 on the right extreme towards the left corresponding with an increase in the magnetic field strength from left to right.

- The single absorption peak of TMS is arbitrarily assigned a value of 0.0 on the δ scale and appears at the far right of the NMR spectrum. The peaks of protons of sample compounds will appear to the left of the TMS peak at smaller magnetic field strengths (down field shift).

- Chemical shift in proton NMR spectroscopy occurs due to the differences in the diamagnetic shielding effect experienced by the protons bound to different nuclei in different functional groups.

- Chemical shift facilitates the distinction between chemically and magnetically equivalent and non-equivalent protons. The chemically and magnetically non-equivalent protons absorb radiation at different frequencies giving rise to NMR absorption signals which can be assigned to protons in different functional groups.

- The chemical shift parameter values of protons in various organic functional groups are well documented in literature.

- The proton peaks in certain compounds containing double bonds, triple bonds, and aromatic rings appear in the NMR spectrum at different chemical shift parameter values from the expected values on the basis of diamagnetic shielding effects. This is attributed to anisotropic magnetic properties exhibited in such compounds due to the circulating π-electrons which generate shielding and deshielding zones in space.

- Spin–spin coupling is the second important environmental factor which affects the NMR absorption peaks. The coupling is due to the mutual interaction of the spins of adjacent nuclei which have two or more spin states. The coupling effect causes variations in the effective magnetic field strength experienced by the individual protons.

- The first-order spectrum due to spin–spin coupling of protons in a compound containing two protons bonded to two adjacent carbon atoms is called the AX spectrum and consists of two doublet peaks of equal intensity for the two protons.

- The chemical shift difference Δv is obtained by measuring the distance between the midpoints of the doublets. The components of each doublet are separated to the same extent and the spacing between the components is referred to as coupling constant J_{AX}.

- The first-order spectrum due to spin–spin coupling of protons in a compound is observed only when the chemical shift difference Δv between the coupled groups is greater than the coupling constant at least to the extent $\Delta v/J > 7$.

- A second-order spectrum, called AB spectrum is observed when the ratio of chemical shift to coupling constant $(\Delta v/J \approx 2)$ is smaller. The inner peaks of the multiplets gain in intensity at the expense of the outer peaks in the second-order spectrum.

- The effect of spin–spin coupling in adjacent groups with two or more protons gives rise to the first-order spectra classified as A_2X_2 or A_2B_2 spectra for two adjacent methylene groups. The A_2X_2 spectrum consists of two triplets with individual peaks in a multiplet having intensities in the ratio 1:2:1.

- Molecules with adjacent methyl and methylene groups (e.g., ethyl alcohol) give rise to A_3X_2 spectrum.

- The number of peaks in a multiplet is determined by the number (n) of magnetically equivalent protons on the neighbouring atoms and is given by the formula ($n + 1$).

- The ratio of the intensities of the individual peaks in a given multiplet can be obtained in terms of Pascal's triangle or from the coefficients of binomial expansion of $(a + b)^n$.

- NMR spectra of many compounds consist of peaks due to similar chemical shifts of protons of different functional groups and hence are difficult to analyse. Such complex spectra can be simplified into simpler and easily analysable patterns through the use of high-resolution instruments operating at higher frequencies; chemical shift reagents; double resonance technique, or spin tickling.

- Carbon-13 NMR spectroscopy is similar to proton NMR spectroscopy as both the nuclei have half-integral spins. However, ^{13}C NMR spectroscopy is much less sensitive as compared to the proton NMR due to two factors, namely lower abundance (1.1%) of the 13C isotope and smaller magnitude of the magnetogyric ratio.

- The chemical shift in ^{13}C NMR spectrum depends on the electron density surrounding the nucleus and the electronegativity of the atoms attached to it as in proton NMR spectra. The chemical shifts can be expressed in terms of ppm in δ scale using TMS as the standard with a δ value of zero and are spread over a range of 0–250 ppm.

- The effect of spin–spin coupling between neighbouring ^{13}C nuclei is relatively weak and is not observed. However, spin–spin coupling between neighbouring ^{13}C and ^1H nuclei occurs and is observed in ^{13}C NMR spectra.

- Carbon-13 NMR spectra are much simpler and easy to interpret as compared to proton NMR spectra and also provide direct information on the carbon skeleton or backbone of the molecules is useful in structural analysis.

- Spin–spin coupling between ^{13}C and neighbouring protons causes some difficulties in interpretation and the same is overcome by noise decoupling which eliminates the coupling between ^{13}C nuclei and protons.

- Fourier transform NMR spectroscopy involves the excitation of the sample with a series of high power RF pulses of wide frequency range in the presence of an applied magnetic field of about 2.5 T.

- FT NMR spectroscopy has the advantage of obtaining one-dimensional (1-D) and two-dimensional (2-D) NMR spectra for the identification of chemically different groups of nuclei and correlation between spectra of different elements in the same compound and for elucidation of structures of complex biomolecules.

- Magic angle spinning NMR spectroscopy is useful for recording the NMR spectra of solid samples. The sample is spun at an angle called the magic angle θ_m (54.74°) with respect to the direction of the magnetic field which gives rise to narrower NMR lines facilitating the determination of structure of solids.

- Electron spin resonance spectroscopy involves absorption of microwave radiation by unpaired electrons in the presence of an external magnetic field. The unpaired electrons generate an induced magnetic field which aligns parallel to the applied field and reinforces it. The degeneracy of the spin states is lifted and the electron in the lower energy state absorbs microwave radiation giving rise to ESR spectrum when the frequency of the radiation matches the precessional frequency of the electron.

- The ESR absorption spectrum is a plot of RF power absorbed versus the applied magnetic field and is usually recorded as the first derivative.

- The ESR spectra exhibit multiplets due to hyperfine structure or hyperfine splitting and fine structure. Hyperfine splitting arises due to interaction of the magnetic moment of the unpaired electron with a neighbouring nucleus with a non-zero spin. Thus, the ESR spectrum of hydrogen atom shows two peaks, the separation being a measure of the hyperfine coupling energy.

- The pattern of hyperfine splitting of an ESR peak is given by the formula $(2nI + 1)$, where 'n' is the number of equivalent nuclei interacting the unpaired electron and I is the spin quantum number of the nucleus.

- In hydrocarbon radicals the unpaired electron can be considered to interact only with n number of equivalent hydrogen atoms giving rise to $(n + 1)$ lines due to hyperfine splitting. The number of lines in the ESR spectrum is helpful in identifying the radical.

- The intensity pattern of the ESR spectral lines can be predicted on the basis of 'family tree' method similar to that in NMR spectroscopy.

- The fine structure in the ESR spectrum is due to electron–electron coupling and occurs only in molecules which have more than one unpaired electron, e.g., transition metal ions.

REVIEW QUESTIONS

1. Discuss the theory of nuclear magnetic resonance.
2. Explain the concept of Larmor precession and its use in understanding the nuclear transition.
3. Write a note on relaxation processes in NMR spectroscopy.
4. What are the major components of CW NMR spectrometers?

5. Explain the term chemical shift and its significance with a suitable example.

6. What is the effect of diamagnetic anisotropy on chemical shift?

7. Explain the term spin–spin coupling with reference to AX- and AB-type NMR spectra.

8. What are the different methods adopted to simplify complex spectra?

9. Write notes on (i) chemical shift reagents and (ii) double resonance.

10. Discuss the principle involved in ESR spectroscopy.

11. Write a note on hyperfine structure of ESR spectrum.

12. List the applications of ESR spectroscopy.

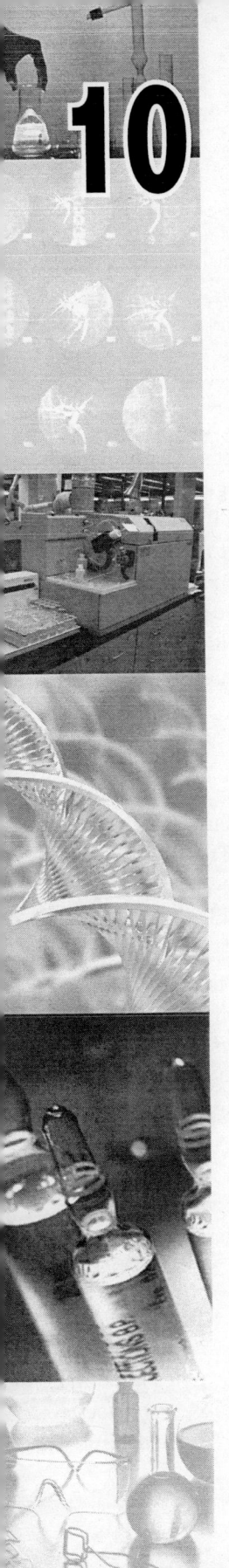

10 Mass Spectrometry

10.1 INTRODUCTION

Mass spectrometry is a widely used technique for characterizing vapour phase molecules on the basis of their ionization and fragmentation patterns. The molecules ionize and also undergo fragmentation when bombarded with a stream of ionizing particles. The technique is not a spectrometric or spectroscopic method as it does not involve interaction of molecules with electromagnetic radiation. However, the name is used as the data is obtained in a spectral form of relative abundance of fragments of different masses. The instrument called the *mass spectrometer* separates and detects the ions formed from the molecules according to their mass using electrical and magnetic focusing elements. The technique has the twin advantages of high sensitivity and specificity over other analytical techniques, and hence, finds extensive use in analysing organic and inorganic substances as well as biological molecules including biopolymers. The high sensitivity is essentially due to the mass analyser which functions as a filter, the instrument requiring only a few nanograms of the solid, liquid, or gaseous sample for analysis. The high selectivity is also due to the characteristic fragmentation pattern of the sample molecule which provides useful information on molecular weight and molecular structure.

10.2 PRINCIPLE

The sample is converted into a rapidly moving stream of positive ions by ionizing the vapour phase sample. Different ionization methods are employed such as electron impact ionization (EI), chemical ionization (CI), field ionization (FI), electrospray ionization (ESI), fast atom bombardment (FAB), and matrix-assisted laser desorption and ionization (MALDI).

The kinetic energy of the fragment ions of mass m and charge z, obtained by ionization and fragmentation of the sample molecule, accelerated through a potential gradient V is given as

$$\text{Kinetic Energy} = \frac{1}{2}mv^2 = zeV \tag{10.1}$$

where e is the electronic charge (1.6×10^{-19} C) and v is the velocity of the fragment ions. The accelerated fragment ions are then subjected to an applied magnetic field at right angles to the direction of their motion. The fragmented ions are deflected to a curved path traversing a 90° sector (or 60° or 180°) of a circumference. The centrifugal force mv^2/r, where r is the radius of the sector, due to the kinetic energy is balanced by the centripetal force Bzv, where B is magnetic field strength of the magnetic field. Hence,

$$\frac{mv^2}{r} = Bzev \tag{10.2}$$

Rearranging the above equation gives

$$v^2 = \left(\frac{Bzer}{m}\right)^2 \tag{10.3}$$

Substituting Eq. (10.3) into Eq. (10.1) and rearranging gives

$$\frac{m}{z} = \frac{B^2 r^2 e}{2V} \tag{10.4}$$

As per Eq. (10.4) for ions carrying a single positive charge the mass is proportional to the square of the radius of the sector. At a specific accelerating voltage V, the fragment ions of a given mass only will pass along the mass analyser tube and exit through the slit to impinge on the collector plate of the detector. The fragment ions of a particular mass (or more specifically m/z) that reach the collector plate generate an ion current which is then amplified and fed to a recorder. The accelerating voltage is continuously varied so that fragment ions of different masses get focused on to the collector plate. The mass spectrum is obtained as a plot of the relative intensities of the ion currents generated by fragment ions of different masses as a function of m/z.

The capability of a mass spectrometer to differentiate between masses of species is expressed in terms of *resolution R*, which is defined as

$$R = \frac{m}{\Delta m} \tag{10.5}$$

where Δm is the mass difference between two adjacent peaks in the mass spectrum and m is the mass of the first peak (sometimes the mean mass of the two peaks is used). The two peaks are said to be separated if the height of the valley between the peaks is less than 10% of the peak height. A mass spectrometer is said to have a resolution of 1000 if it is capable of giving well resolved peaks of 1000 and 1001.

The resolution required for a mass spectrometer depends on the nature of the sample. If the species whose molecular masses defer by a unit mass (e.g., NH_3 and CH_4 having molecular weights 17 and 16 respectively) can be easily distinguished by mass spectrometer having low resolution of about 50. On the other hand, a mass spectrometer of high resolution of a few thousands is required to identify species, such as C_2H_4, N_2, and CO, all having a nominal mass of 28, but the exact masses being 28.0313, 28.0061, and 27.9949 respectively.

Most instruments have a resolving power of 1000. However, the study of macromolecules requires instruments of higher resolving power of 20,000 or more. Mass spectrometers with higher resolving power usually employ double-focusing technique in which separate electrostatic and magnetic focusing systems are incorporated in the analyser tube.

10.3 MASS SPECTROMETER

The main components of a mass spectrometer include (i) sample inlet, (ii) ionization source and acceleration chamber, (iii) mass analyser to separate the ions according m/z ratios, (iv) detector to generate a current signal as the separated ions impinge on it, and (v) recording and display system to display the mass spectrum. The instrument is maintained at low pressure by an efficient vacuum pumping system capable of reaching 10^{-7} torr. Figure 10.1 shows the schematic diagram of the basic components of a single-focusing 90 degree mass spectrometer.

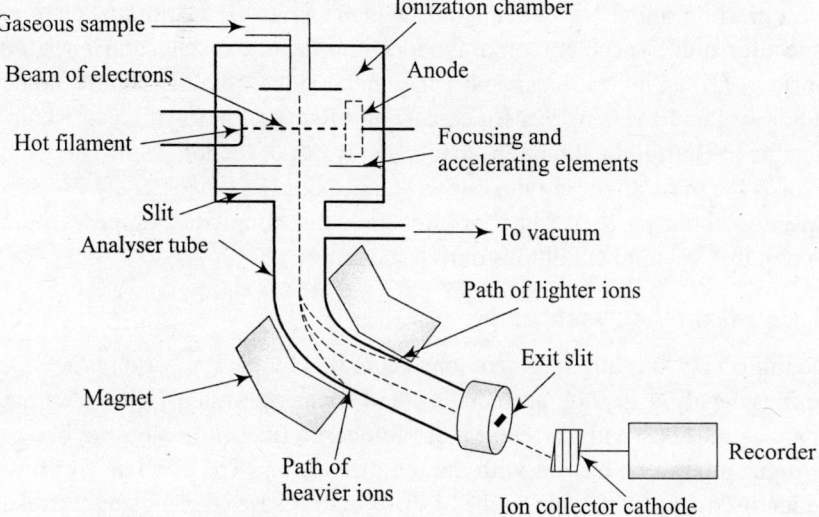

Fig. 10.1 Schematic diagram of the single-focusing mass spectrometer with electron impact ionization source

10.3.1 Sample Inlet

The sample inlet system facilitates the introduction of a small amount of sample of the order micromole or less into the ionization chamber with minimal loss of vacuum. Mass spectrometers are usually provided with three different types of sample inlets, namely, batch inlet, direct probe inlet, and chromatographic inlet.

Batch inlet facilitates the introduction of the sample that has been volatilized externally, into the evacuated ionization chamber. Gaseous and volatile liquid samples can be introduced into the batch inlet connected to a small manifold coupled to a pressure gauge. The sample is injected by a hypodermic syringe and let into a large volume reservoir to allow the expansion of the volatile sample and let into the ionization chamber through a pin hole as a steady stream of sample molecules. The sample pressure in the ionization chamber will be about 10^{-4}–10^{-5} torr.

Direct probe inlet facilitates the introduction of non-volatile liquid and solid samples. The sample is held on the surface of a glass or aluminum capillary tube or wire of the probe and volatilized by electrical heating. The volatilized sample is introduced directly into the ionization chamber through a vacuum lock. The inlet is positioned within a few millimeters of the ionization source and the slit leading to the analyser tube. The low pressure within the ionization chamber builds up sufficient concentration of non-volatile samples, for example, carbohydrates, steroids, low molecular weight polymers, organometallic species, etc. Since the sample is introduced very close to the ionization source, mass spectra of even thermally unstable compounds can be obtained before decomposition occurs.

Chromatographic inlet is required when the mass spectrometer is coupled as a detector to a gas chromatograph or a HPLC system. The inlet links the chromatographic column to the mass spectral detector so that the sample components that elute out of the chromatographic column are subjected to mass spectral analysis.

10.3.2 Ionization Source and Acceleration Chamber

The ionization source has two functions which include (i) the production of ions from the sample molecules and (ii) acceleration of the ions into the mass analyser. Hence, the ionization source consists of (i) an ionization chamber to generate ions from the sample molecules and (ii) an ion withdrawal and focusing system consisting of several pairs of electrostatic focusing elements and slits to control the direction, shape, and width of the ion beam The ion beam is accelerated towards the mass analyser with only a small spread of kinetic energies and usually generates a current of about 10^{-10} A. The ionization source is coupled to a high vacuum pump to maintain the required vacuum conditions during operation.

10.3.3 Mass Analyser

The molecular ions and fragment ions generated by the ionization process are accelerated into the mass analyser by the accelerating and focusing system. The mass analyser or separator separates the ions with different mass-to-charge (m/z) ratios on the basis of its capability to distinguish between masses with minute differences. The efficiency of the mass spectrometer to identify different species on the basis of their masses depends on the resolving power of the mass analyser. Different types of mass analysers are used in mass spectrometers and the mass spectrometer itself is named based on the type of mass analyser used.

Magnetic sector analyser used in a single-focusing mass spectrometer is a permanent magnet or an electromagnet through which the ions are accelerated. The beam of ions is directed to travel in a circular path through 60, 90, or 180 degrees governed by the relationship expressed by Eq. (10.4). Single-focusing instruments have a maximum resolution of about 2000.

A *double-focusing* spectrometer makes use of a carefully selected combination of electrostatic and magnetic fields to improve the resolution of the instrument (Fig. 10.2).

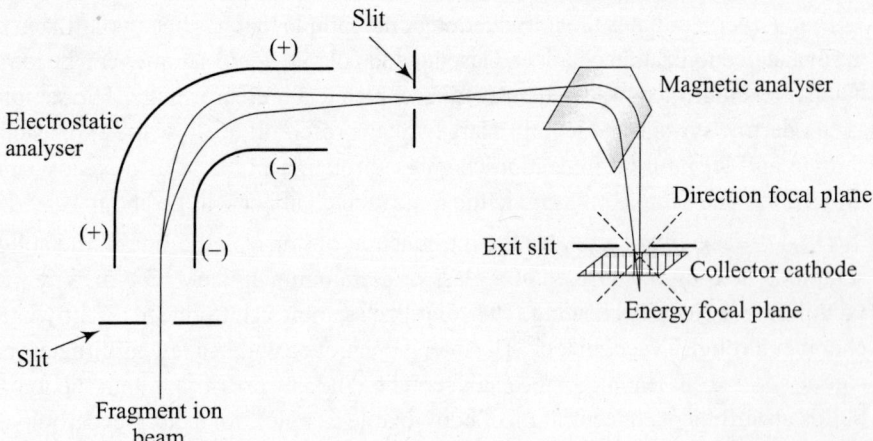

Fig. 10.2 Schematic diagram of a double-focusing mass analyser with Nier–Johnson geometry

The ion beam is passed first through an electrostatic analyser consisting of two smooth curved metallic plates across which a dc potential is applied. The dc potential is regulated to limit the kinetic energy of the ions reaching the magnetic analyser to a narrow range so that ions having different kinetic energies other than the selected ranges do not reach the magnetic analyser. The magnetic analyser separates the ions by directional focusing so that only ions of one particular m/z value reach the detector. The ion collector cathode is placed exactly at the intersection point of the focal planes of the energy focusing and direction focusing. Double-focusing instruments are capable of resolution of the order 10^5.

10.3.4   Detector

Electron multiplier is the most commonly used detector in mass spectrometer. It is similar in design and operation to the photomultiplier detector used in UV–visible spectrophotometers. It consists of a cathode and a series of dynodes (up to 20) with Cu/Be surfaces, each successive dynode held at increasing potential. The ion beam separated by the mass analyser is focused on to the cathode which generates electrons which are accelerated and strike the first dynode. The dynode emits secondary electrons which are once again accelerated and strike the second dynode producing more number of secondary electrons. The successive dynodes continue to amplify the current by a factor of 10^5–10^8.

Scintillation counter employs a dynode which on impact of the fragment ions of the sample emits electrons. These electrons strike a phosphorous-coated screen causing the emission of light photons, the latter being detected by a photomultiplier tube.

Faraday cup detector consists of a collector electrode placed within a cage so that the particles striking or leaving the electrode are reflected away from the entrance of the cage and the reflected ions and ejected secondary electrons do not escape. The charge generated on the electrode and the cage by the striking positive ions is neutralized by the flow of electrons from the ground through a large resistor. The resulting potential drop across the resistor is amplified through a high-impedance amplifier. The Faraday cup detector has the advantages of an inexpensive simple mechanical/electrical design and operation and a response which is independent of the nature of fragment ion, its energy, and mass. However, the detector suffers from the disadvantages of low sensitivity and low scan speed as compared to electron multiplier.

10.3.5   Recording System

The recorder and display system consists of VDU screen or chart recorder. Most mass spectrometers are interfaced to a dedicated computer to facilitate the operation of the instrument, processing and display of data, and to scan and compare the data with the library data.

10.4   IONIZATION METHODS

A variety of ionization methods are made use of in a mass spectrometer depending on the nature of the sample to be analysed. The relatively 'hard' ionization methods include the most commonly used (i) electron impact ionization and (ii) chemical ionization. Soft ionization methods have come into vogue for the analysis of solids and high molecular weight substances mostly of biological importance. These methods generate mostly parent molecular ion and only a few fragments and thereby avoid the problems associated with thermal stability and non-volatile nature of macromolecular analytes. The soft ionization sources include (i) field ionization, (ii) field desorption, (iii) fast atom bombardment, (iv) electrospray ionization, and (v) matrix-assisted laser desorption and ionization.

10.4.1 Electron Impact Ionization

Electron impact ionization (EI) is a hard ionization method and involves the bombardment of the sample molecules with a high energy beam of electrons within the ionization chamber maintained at about 200°C and a pressure of about 5×10^{-3} torr. The electron beam is generated by a rhenium, thoriated iridium, or carbonized tungsten filament cathode (electron gun) connected to a power source (Fig.10.1). The power source is capable of providing a variable electric field in the range of 6–100 V. The standard operating voltage is usually 70 V because the mass spectrum obtained is nearly independent of the electric field and is reproducible for quantitative analysis. The kinetic energy of the beam of electrons generated at 70 V is given by the relationship

$$KE = eVN \qquad (10.6)$$

where e is the charge on the single electron (1.6×10^{-19} C), V is the applied voltage, and N is the Avogadro number (6.023×10^{23} electrons/mol). The kinetic energy of the electron beam is $\sim 6.7 \times 10^3$ kJ/mol and is more than the average energy of a chemical bond (200–600 kJ/mol), and hence, ionization/fragmentation of the sample molecule can be easily achieved.

The path of the electron beam is at right angles to the path of the sample molecules and as the two beams intersect ionization of sample molecules occurs. The primary ionization product is mostly the single positively charged ions M^+ formed by the removal of one electron from the sample molecule M. However, the electron beam is of sufficient energy not only to ionize but also cause a relatively high degree of characteristic fragmentation of sample molecules and thereby generate positive ions of varying mass. The mass spectrometer is usually calibrated using the electron impact ionization of perflouoroalkanes as mass markers as other methods provide very few ions. However, many molecules undergo excessive fragmentation because of the high energy of the electron beam and do not give the characteristic molecular ion. Hence, when fragmentation is not desirable the energy of the electron beam is reduced by maintaining the operating voltage in the range of 6–14 V.

Electron impact ionization method has the advantages of generating high ion current and extensive fragmentation of the sample molecules giving rise to a large number of peaks in the mass spectrum thereby facilitating unambiguous identification of the analyte. However, excessive fragmentation is also disadvantageous in that the molecular ion peak disappears and the sample molecular weight cannot be determined. Another limitation of this method is that the sample needs to be volatilized which may lead to thermal degradation of the sample. This problem is overcome by using the direct probe inlet.

10.4.2 Spark Ionization

Spark ionization method is suitable for ionizing non-volatile inorganic samples and residues from ashed organic samples for the determination of elemental composition. The method involves generating a radiofrequency (RF) spark by applying about 30 kV between two electrodes, one of which is made from the sample itself. The sample is vapourized and ionized and the ions are accelerated into the mass analyser.

10.4.3 Chemical Ionization

Chemical ionization (CI) refers to the process of ionizing the gaseous sample molecules by colliding with ions of a reagent gas. Chemical ionization can be either positive chemical ionization (PCI) or negative chemical ionization (NCI). Almost all neutral molecules form positive ions through reactions with the reagent gas. The most commonly used reagent gas in PCI is methane. The reagent gas molecules are ionized by electron impact ionization using 200–500 V to generate ions, such as CH_4^+, CH_3^+, and CH_2^+, in the first step. These ions react with additional methane molecules to generate secondary ions such as CH_5^+ and $C_2H_5^+$. In the second step, the primary and secondary ions in turn ionize the sample molecules XH through charge, proton, or hydride transfer.

$$CH_4^+ + XH \rightarrow CH_4 + XH^+ \text{ (charge transfer)}$$
$$C_2H_5^+ + XH \rightarrow C_2H_4 + XH_2^+ \text{ (proton transfer)}$$
$$C_2H_5^+ + XH \rightarrow C_2H_6 + X^+ \text{ (hydride transfer)}$$

Proton transfer generates $(M + 1)^+$ ions with one mass unit greater than that of the sample molecule, whereas hydride transfer generates $(M - 1)^+$ ions with one mass unit less than that of the sample molecule. In addition sometimes $(M + 29)^+$ ions are also generated due to the transfer of $C_2H_5^+$ ion to the analyte molecule.

Reagent gases other than methane also find use in chemical ionization, particularly for achieving better control of the fragmentation pattern of sample molecules. For determining molecular weight of a compound, isobutane is useful as the reagent gas as it generates a low energy $t\text{-}C_4H_9^+$ ion which in turn ionizes the sample molecule with minimum fragmentation. Ammonia is yet another reagent gas and generates the weak protonating agent NH_4^+ ion to yield sample ions, such as $(M + H)^+$ and $(M + NH_4^+)$, useful for characterizing polyhydroxy compounds such as carbohydrates. Deuterium oxide is used as the reagent gas for the detection of hydrogen. All alcohols give ions $(M - 17)^+$ when nitric oxide is used as the reagent gas. Primary and secondary alcohols give additionally ions $(M - 1)^+$ and $(M + 30 - 2)^+$. Nitrogen, helium, and argon as reagent gases produce fragmentation pattern of sample molecules identical to electron impact ionization but with higher sensitivity. Helium has the advantage of being used as a carrier gas in gas chromatography and in GC-MS systems the gas facilitates charge transfer reactions with sample molecules.

In NCI the sample molecule must be capable of forming negative ions or acquire a negative charge by electron capture. Since only analytes containing electronegative elements, like halogens or acidic groups are capable of generating negative ions NCI is selective towards such analytes. Hence, NCI is mostly used for the analysis of environmentally hazardous polychlorinated biphenyls (PCBs), chlorine containing pesticides, and fire retardants. Oxygen or hydrogen is used as reagent gas in NCI.

Atmospheric pressure chemical ionization (APCI) is often used in mass spectrometers in conjunction with HPLC to ionize the analyte by a corona discharge at atmospheric pressure. The mobile phase containing the analyte is converted into an aerosol spray with high flow rates of nitrogen and subjected to corona discharge to generate ions of the analyte.

Chemical ionization is relatively a softer ionization method as compared to electron impact ionization with a better control of the extent of fragmentation thereby providing a simpler mass spectrum. However, CI does not lead to C—C bond cleavage, and hence, the mass spectrum provides little skeletal information of the analyte sample. CI has also an advantage over electron impact ionization in that the limited extent of fragmentation increases the sensitivity by about hundred times for the molecular ion. Sensitivity is further enhanced because of the high cross section for chemical ionization process and also the long residence time of the order of 10^{-3}–10^{-5} s.

10.4.4 Field Ionization

Field ionization (FI) method involves subjecting the sample molecule to a large electric field of the order of 10^7–10^8 V cm^{-1} generated between two closely spaced electrodes. The anode consists of a number of specially formed carbon dendrite or whisker emitters (diameters less than 1 μm) deposited on a thin needle or wire of tungsten of a few micrometers in diameter. The anode is positioned immediately behind the exit slit of the ionization chamber. The exit slit itself functions as the cathode. High voltage of the order of 10–20 kV is applied between the electrodes to generate the electrostatic force. Under high vacuum conditions of 10^{-4} torr the force removes an electron from the sample molecule without imparting too much energy to the molecule to induce fragmentation. The electron is extracted from the molecule by the electric field concentrated at the tips of the carbon dendrites through quantum mechanical tunneling and field-induced surface reactions to form the molecular ion under experimental conditions. Sensitivity of this method is an order of less in magnitude as compared to that of electron impact ionization with a maximum current of about 10^{-11} A.

10.4.5 Field Desorption

Field desorption is similar to field ionization in that it makes use of an emitter wire electrode. However, the electrode is mounted on a probe that can be removed from the ionization chamber, coated with a solution of the analyte and reinserted into the chamber. Ionization of the sample is carried out by the application of a high potential as in field ionization method. The sample may require heating which is provided by passing electric current through the wire.

Field desorption method facilitates direct formation of gaseous ions without involving volatilization and subsequent ionization, and hence, the mass spectra are quite simple often containing only the molecular ion or the protonated molecular ion. The method is useful for the study of surface characteristics of the deposited analyte, high molecular weight substance, and even delicate biochemicals.

10.4.6 Fast Atom/Ion Bombardment

Fast atom/ion bombardment (FAB) method was introduced in 1981 for ionization of macromolecules. A schematic diagram of FAB ionization is shown in Fig. 10.3.

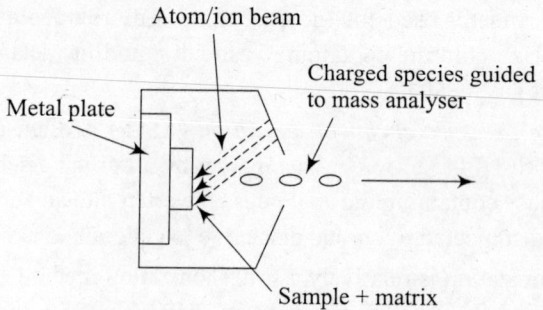

Fig. 10.3 FAB ionization of sample molecules

The method makes use of a high energy (about 7–8 keV) beam of xenon or argon atoms, cesium (Cs^+) ions, or glycerol–NH_4^+ clusters to sputter the sample–matrix mixture supported on a metal plate. The sample is usually dissolved in a non-volatile matrix substance, such as glycerol or 3-nitrobenzyl alcohol, coated onto the metallic plate and inserted into the ionization chamber. Sputtering the sample with a beam of atoms or ions transfers most of the energy to the solvent molecules, and hence, excessive fragmentation of the sample molecules is avoided. The solvent and sample molecules vapourize and if they are not already charged, they get charged by colliding with the surrounding gas phase ions. The charged molecules are then guided electrostatically to the mass analyser. The method is highly sensitive, and hence, requires a total volume of sample and matrix of the order a few microlitre only.

Fast atom bombardment of samples generates mainly molecular ions and some fragment ions. The method is suitable for the determination of molecular weights, high molecular weight substances, and for getting structural information of organic, biochemical as well as thermally labile compounds.

10.4.7   Electrospray Ionization

Electrospray ionization (ESI) method is a soft ionization method capable of producing molecular ions of biomacromolecules from aqueous solutions, and hence, highly suitable for protein and DNA analyses. The liquid sample is introduced into the ionization chamber at a flow rate of $5-20$ μL min^{-1}, sometimes about hundred to thousand times less in the range of $10-100$ nL min^{-1}, through a stainless steel needle. The stainless steel needle is maintained at about $4-5$ kV relative to the surrounding cylindrical cathode, both of which are housed in the ionization chamber as shown in Fig. 10.4.

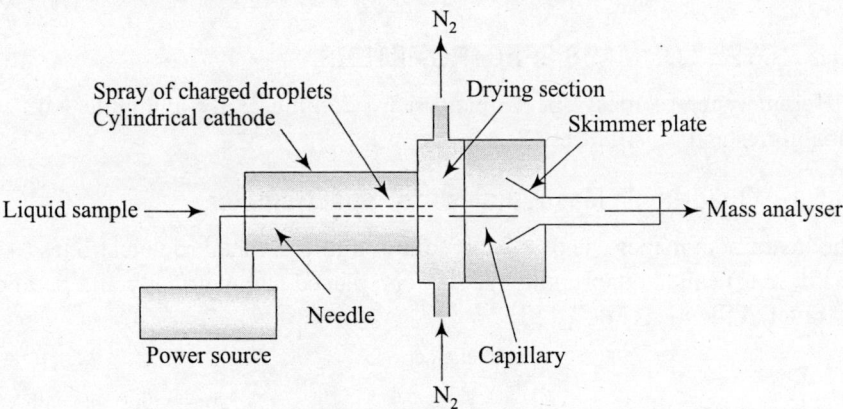

Fig. 10.4   Electrospray ionization

The sample liquid is ejected out of the needle in the form a fine spray of charged droplets by the electrostatic force at the tip of the needle. The charged droplets migrate in the field toward the glass capillary inlet tube which introduces the droplets in the drying section or first pumping stage. The droplets are dried by nitrogen gas flowing at about 100 mL s^{-1}. As the solvent evaporates, the droplets become smaller in size and consequently the surface charge increases. At a critical limit of size-to-surface charge density called the Rayleigh limit, the electrostatic repulsion becomes equal to the surface tension of the droplet resulting in the explosion of the droplet into smaller droplets. This process of breaking the droplets further into smaller sized droplets ultimately results in the ionization of the sample molecules. The ionized molecules, mostly molecular ions and multiple charged species are then propelled into the mass analyser through the skimmer.

10.4.8 Matrix-assisted Laser Desorption/Ionization

Matrix-assisted laser desorption/ionization (MALDI) method involves the vapourization and ionization of the sample by using a laser beam. The ionization chamber is similar in arrangement to that of FAB ionization chamber shown in Fig.10.3, with the modification of a laser source replacing the ion beam source. The sample is mixed with the matrix compound, such as 2,5-dihydroxybenzoic acid, nicotinic acid, sinapinic acid, or 2-cyanocarboxylic acid, in the molar ratio of 1:1000 and evaporated onto a metal plate. The metal plate is introduced into the ionization chamber and the sample–matrix is exposed to short pulses of laser radiation from a nitrogen laser (337 nm) or suitable ultraviolet or infrared lasers. The temperature of the matrix compound increases rapidly as it absorbs the laser radiation strongly and desorbs and ionizes the sample. The sample ions are then guided into the mass analyser.

The combination of matrix-assisted laser desorption/ionization and time of flight mass spectrometer (MALDI-TOF MS) is used in analysing biological samples. MALDI generates the short bursts of ions from the sample and after a lag of about $10-500$ ns a strong electrostatic field is switched on to impart a fixed kinetic energy to the ions. The ions are then allowed to pass through a field-free region in the flight tube of the TOF. Ions with low m/z values travel faster and arrive at the detector first. The MALDI-TOF MS has the advantage of an unlimited mass range but its resolution is low, less than 500. Higher resolution MALDI-TOF MS instruments make use of an electrostatic mirror called *reflectron* to correct the distribution in the initial kinetic energy so that the ions do not enter the flight tube precisely at the same time.

10.5 OTHER TYPES OF MASS SPECTROMETERS

Different types of mass spectrometers are commercially available with varying resolving capabilities and sensitivities.

10.5.1 Quadrupole Mass Analyser or Spectrometer

The instrument makes use of a set of four cylindrical metal rod electrodes (6 mm diameter and 15 cm long) called quadrupole mass filters, placed symmetrically and parallel to the direction of travel of the ions (Fig. 10.5)

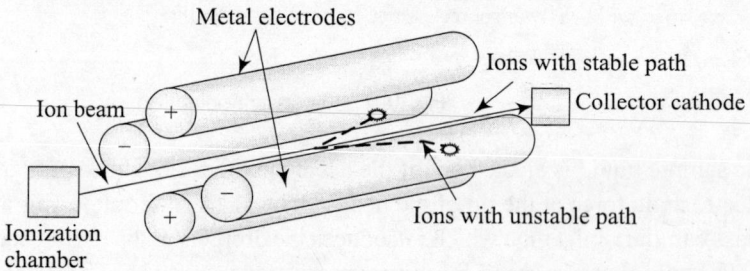

Fig. 10.5 Schematic diagram of a quadrupole mass spectrometer

The fragment ions are accelerated by a potential of 5–15 V and allowed to travel through the space between the metal electrodes. The diagonally opposite pair of rods are connected electrically to a positive terminal of a variable dc source and the other pair to the negative terminal. In addition an oscillating radiofrequency ac potential is applied across each pair of electrodes. Only fragment ions of one m/z value can pass through the quadrupole analyser following a stable path to reach the detector, and thus, be detected for a given radiofrequency potential and frequency. Fragment ions of other *m/z* values are forced to follow an unstable path leading to collision with the electrodes. By careful progressing alteration of the dc potential and RF field the fragment ions of different *m/z* values can be analysed. The quadrupole analyser does not discriminate between the polarities of the ions, and hence, even negative ions obtained by chemical ionization can be analysed.

Quadrupole mass spectrometers are quite compact and rugged besides being less expensive, and hence, are common as bench top mass spectrometers. They have the advantage of fast scanning in less than about 100 ms and particularly useful as mass spectral detectors for real-time scanning of chromatographic peaks in GC-MS and HPLC-MS.

10.5.2 Time of Flight Mass Spectrometer

Time of flight (TOF) mass spectrometer allows the separation of fragment ions in field-free conditions. The sample fragment ions are produced by periodic bombardment of samples with short duration pulses of electrons, secondary ions, or laser generated photons. The fragment ions are accelerated through a field-free drift tube of about one meter length. As all the ions entering the drift tube have the same kinetic energies they drift within the tube at velocities inversely related to their masses as governed by Eq. (10.1). Thus, the lighter ions reach the detector earlier than the heavier ones; the time flight being in the range of 1–30 μs. Figure 10.6 shows the schematic diagram of a time of flight mass spectrometer.

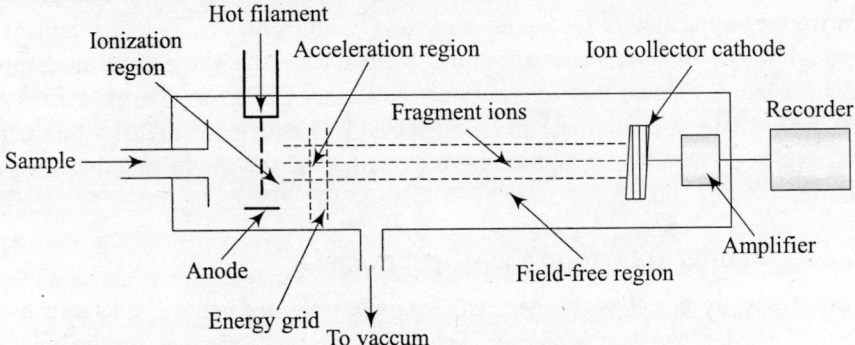

Fig. 10.6 Schematic diagram of a time of flight mass spectrometer

The transit time *t* of ions through a distance of *L* cm is of the order microseconds and is given by the relationship

$$t = L\sqrt{\frac{m}{z}\frac{1}{2V}} \quad \text{or} \quad \frac{m}{z} = \frac{2Vt^2}{L^2} \tag{10.7}$$

Time of flight mass spectrometer has several advantages over other types of mass spectrometers such as simple design, ruggedness, ease of accessibility of ion source, and virtually unlimited mass range. The instrument is highly useful for kinetic studies of fast reactions as the complete mass spectrum can be generated within about 50 μs. However, the instrument has limited resolution and sensitivity.

10.5.3 Ion Trap Analyser (Spectrometer)

Ion trap analyser consists of an ion trap in which gaseous ions can be formed and confined by electric and/or magnetic fields for extended periods. A simple ion trap device developed originally as a detector for GC/MS consists of a doughnut-shaped ring electrode and a pair of end-cap electrodes (Fig.10.7). A variable radiofrequency voltage is applied to the ring electrode while the end cap electrodes are grounded.

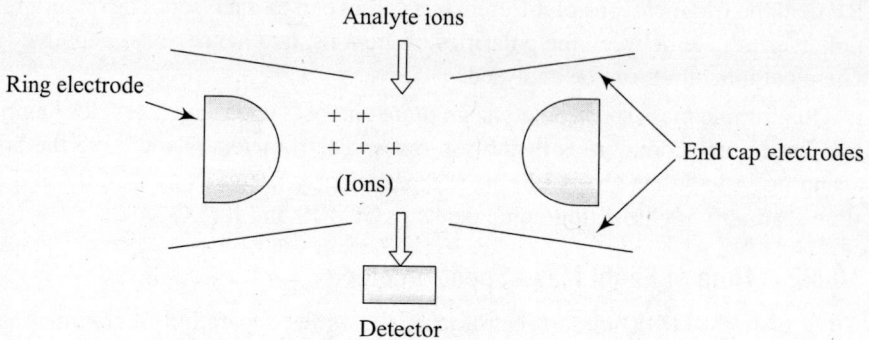

Fig. 10.7   Cross-section of ion trap mass analyser

The ions formed with appropriate *m/z* values are confined to circulate in a stable orbit within cavity surrounded by the ring electrode. With an increase in radiofrequency voltage the orbits of heavier ions get stabilized. In contrast, the orbits of the lighter ions get destabilized and the ions collide with the wall of the ring electrode. A burst of analyte ions formed by electron impact or chemical ionization are introduced through a grid in the upper end cap. The ions are trapped by scanning the radiofrequency voltage and finally leave the ring electrode cavity through the opening in the lower end cap electrode. The ejection of the trapped ions from the cavity is controlled so that scanning on the basis of mass-to-charge ratio is possible. Ion trap spectrometer is quite compact and rugged besides being less expensive compared to quadrupole mass spectrometer. The spectrometer is capable of resolving ions with a mass difference of one unit in the 500–1000 mass range.

10.5.4   Fourier Transform Mass Spectrometer

FT mass spectrometer has a higher signal-to-noise ratio, and hence, higher sensitivity and resolution, besides greater speed of scanning. The FT instrument operates on the basis of *ion cyclotron resonance* (ICR) phenomenon. Gaseous ions, when subjected to an external magnetic field, move in a well-defined circular orbit with an angular frequency called *cyclotron frequency*, ω_c, in a plane perpendicular to the direction of the magnetic field. When a fluctuating ac electric field is applied the ion trapped in a circular orbit in the magnetic field absorbs energy from the electric field at a frequency which matches the cyclotron frequency. The absorbed energy causes an increase in the velocity of ion, and hence, its radius of the circular path without affecting the cyclotron frequency. When the ac field is terminated the radius of the path becomes constant. Ions with different *m/z* ratios have different cyclotron frequencies and only those ions with a particular *m/z* ratio, and hence, a particular cyclotron frequency will absorb energy from the ac field with a frequency matching with the cyclotron frequency. Other ions with different *m/z* ratios, and hence, different cyclotron frequencies will be unaffected. This phenomenon is known as *ion cyclotron resonance*.

The FT mass spectrometer contains a *trapped ion analyser cell* (Fig.10.8) in which ions are formed from gaseous sample molecules by a pulse of an electron beam which can be switched on or off periodically.

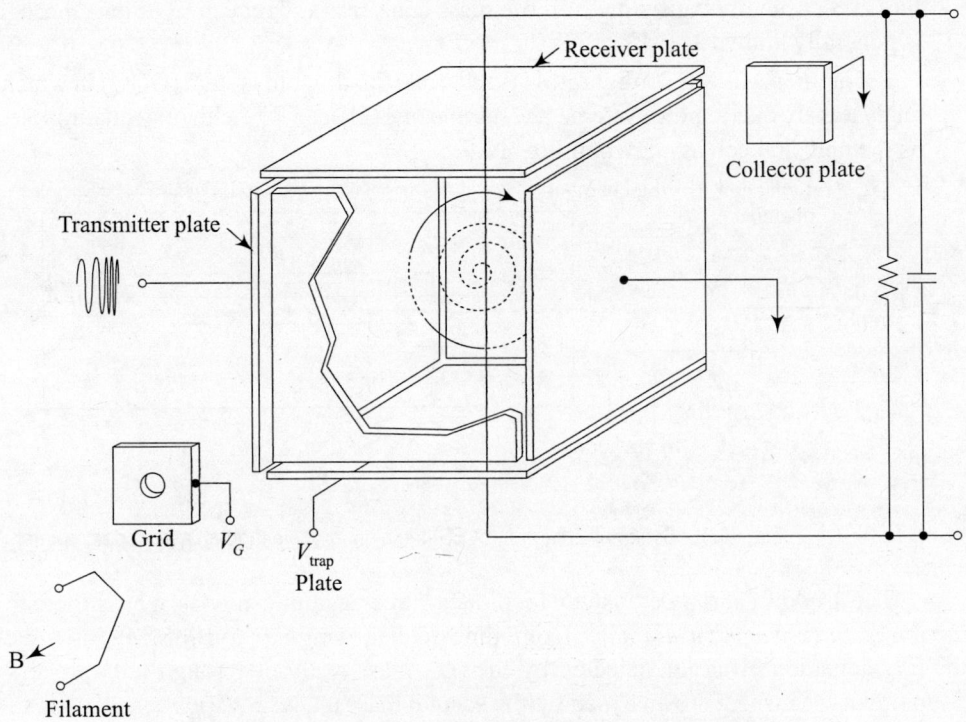

Fig. 10.8 Trapped ion analyser cell of FT mass spectrometer
(Source: D.A.Skoog, F.J.Holler and T.A.Nieman. Principles of
Instrumental Analysis, Thomson Brooks/Cole, 2005)

The ions are confined to the cell by a 1–5 V potential applied to the trap plate and accelerated by a radiofrequency field applied to the transmitter plate. The trapped ions with a particular cyclotron frequency absorb energy at the resonant frequency from the radiofrequency field thereby increasing the radius of their circular path. The coherent circular motion of the ions with the same m/z ratio generates an *image current* (a capacitor current) which can be observed after the termination of the frequency sweep signal. The image current in turn induces an ac current in an external circuit, the frequency of which is characteristic of the m/z value of the ion and the magnitude is a measure of the concentration of the ions. The decay of the image current occurring over a period of a few tenths of a second to several seconds is amplified, digitized, and stored in memory and then transformed to yield a frequency domain signal which in turn is converted to a mass-domain signal by a computer.

10.6 TANDEM MASS SPECTROMETRY

Tandem mass spectrometry or MS/MS makes use of two or more mass analysers in tandem to allow a secondary fragmentation and characterize the selected ion(s) generated by the first analyser. The secondary fragmentation process commonly referred to *collision-induced dissociation* (CID) is usually carried out by collisions with inert gas (Ar) molecules contained in a collision cell, positioned in between the first and the second mass analysers. The mass analysers may be a combination of magnetic, electrostatic, quadrupole, or time of flight analysers. Tandem mass spectrometry consisting of more mass analysers and recycling of ions through the system is generally known as $(MS)^n$.

A common MS/MS configuration is the triple quadrupole system (QQQ) in which all the three analysers are quadrupole systems, the middle one (Q_2), an RF field, only quadrupole, functioning as the collision cell as shown in Fig. 10.9.

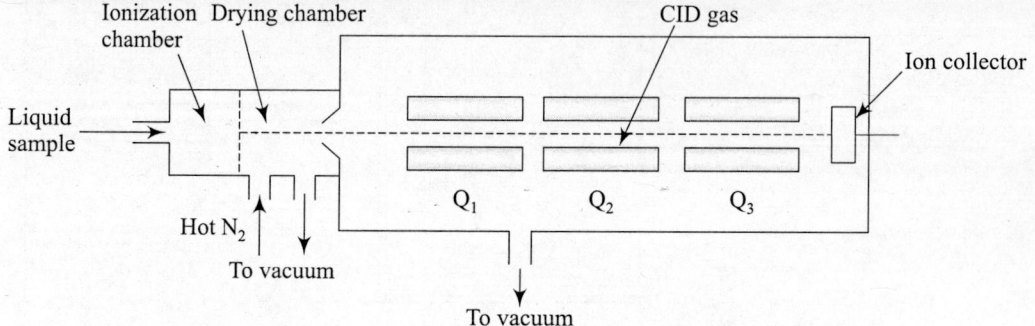

Fig. 10.9 Schematic diagram of ESI-triple quadrupole tandem mass spectrometer

The triple quadrupole system facilitates three scanning modes which include (i) *product ion scan* or *fragment ion scan* (sometimes called *daughter* ion MS/MS) involves secondary fragmentation of an ion selected by the first mass analyser through CID process to generate product ions which are analysed by the second mass analyser, (ii) *precursor ion scan* or *parent ion scan* involves selecting a product ion by the second mass analyser and a scan of the precursor ion which gave rise to the product ion is scanned by the first analyser, and (iii) *neutral loss scan* involves selecting a neutral fragment and scanning by both the mass analysers simultaneously after offsetting the second mass analyser by a value equal to the mass of the neutral fragment. For example, if water molecule or CH_3 group (neutral fragment) is lost from an ion during CID process between the two mass analysers the second mass analyser is set to lag behind the first mass analyser by 18 or 15 atomic mass units (amu). The scan reveals the precursor ion which gives rise to the product ion with a loss of 18 or 15 mass units.

Another interesting example is that of the alkyl phenols ($HOC_6H_4CH_2R$) present in solvent-refined coal. All alkyl phenols generate a fragment ion $HOC_6H_4CH_2^+$ with $m/z = 107$ independent of the nature of R in the first ionization. Thus, the mass spectrum of 1,4-*t*-butyl phenol (MW = 150) shows a base peak at $m/z = 135$ and a prominent peak at $m/z = 107$. The fragment ion with peak at $m/z = 107$ can be formed by the loss of C_3H_7 (43 amu) from the parent ion in a single step or by the loss of CH_3 (15 amu) followed by a neutral loss of CO or C_2H_4 (28 amu). A precursor ion scan for the fragment of m/z-135 showed the abundant ion to be of $m/z = 107$ confirming that the initial loss from 1,4-*t*-butyl phenol was due to the loss of CH_3 and not C_3H_7. A similar analysis with the precursor ion of $m/z = 137$ indicated that the neutral loss was due to C_2H_4 and not due to CO.

Tandem mass spectrometry is, thus, useful in the determination of exact configuration of complex molecules and in the hyphenated technique of HPLC-MS.

10.7 INTERPRETATION OF MASS SPECTRUM

The mass spectrum is a graphical presentation of relative abundance of each fragment ion expressed as a percentage of the most abundant one against the mass-to-charge (m/z) ratio in the form of a bar chart or line diagram. The general features of the mass spectrum of a sample molecule include the following. The tallest peak due to most abundant fragment ion is called the *base peak*. The most abundant ion formed during ionization of the sample molecule gives rise to the base peak, which is taken as having a relative abundance of 100% and all other peaks in the mass spectrum are usually reported as percentages of the abundance of the base peak. The process of ionization may result in the removal of a single electron from a molecule to yield an ion (a radical cation), and hence, the ion is called the *parent* or *molecular ion*, M^+. The molecular ion is the heaviest ion assuming that only one electron is lost to form M^+ during ionization. The value of m/z at which the peak due to the molecular ion appears in the mass spectrum gives the actual molecular weight of the sample molecule. Thus, if the peak due to the molecular ion can be identified, the molecular weight of the sample can be determined.

All atoms have heavier isotopes that occur in characteristic natural abundance and this natural abundance of the isotopes of the constituent atoms is reflected in the molecular weights of the molecules. For example, hydrogen contains 0.02% of the heavier isotope, ^2H, whereas carbon contains 1.08% of the isotope ^{13}C. Peaks due to ions having heavier isotopes of atoms also appear in the mass spectra of sample molecules, the intensities of the peaks being proportional to the natural abundances of the isotopes. Hence, besides the peak due to the molecular ion M^+, peaks due to M + 1, and M + 2 also appear in the mass spectrum.

The sample molecule on ionization undergoes fragmentation as the ionizing beam of particles has sufficiently high energy to break some of the bonds in the molecule forming a series of molecular fragments with positive charges. These fragment ions are also accelerated and passed through the mass analyser and the ion current is recorded. Thus, in addition to the molecular ion peak fragment, ion peaks also appear at different m/z values in the mass spectrum of a sample. The molecular ion formed initially may not be stable and may disintegrate to form fragment ions within a time span less than 10^{-6} s. Often one of the fragment ions is the most abundant ion produced in the mass spectrum, and hence, gives rise to the base peak. The fragmentation or cracking patterns of a pure sample molecule is unique and characteristic of the sample, and hence, useful for qualitative identification of the sample and also provides structural information.

Molecular and fragment ions formed in the ionization chamber having lifetimes of about 10^{-6} s or more are accelerated in the ionization chamber. However, some of these ions may disintegrate into fragments while passing through the analyser tube. The disintegration process is usually a one-step process in which the original ion forms a daughter ion with the loss of a neutral fragment. Such fragment ions have low energy as compared to other ions and follow abnormal flight paths on their way to the detector. These ions give rise to broad peaks at non-integral m/z values in the mass spectrum depending on their mass as well as the masses of the original ions from which these fragment ions were formed. The peaks are called *metastable ion peaks* and the position of such peaks in the mass spectrum is related to the mass of the original ion as

$$m^* = (m_2)^2/m_1 \tag{10.8}$$

where m^* is the apparent mass of the metastable ion and m_2 refers to the mass of the new fragment ion formed by the disintegration of the original ion of mass m_1. The presence of metastable ion peaks in the mass spectrum is useful to prove a proposed fragmentation pattern of a sample molecule and in structure analysis.

The interpretation of the mass spectrum is explained taking the typical example of *n*-octane (vide Fig. 10.10).

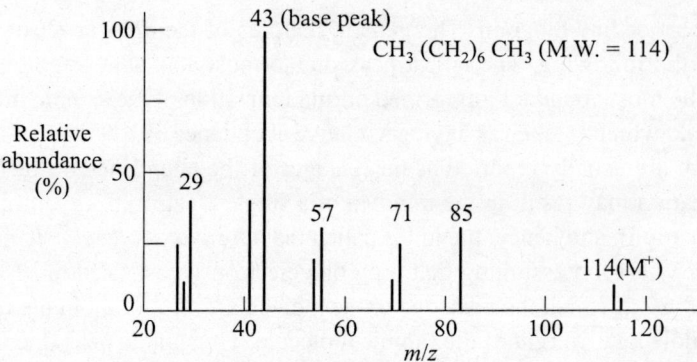

Fig. 10.10 Mass spectrum of *n*-octane

The tallest peak (base peak) in the mass spectrum of *n*-octane appears at *m/z* value of 43 a peak due to a fragment ion. The molecular ion peak appears at *m/z* value of 114 for *n*-octane. The M+1 peak for octane can be seen at *m/z* value of 115, with much reduced intensity. The peak in the mass spectrum of *n*-octane at *m/z* value of 85 may be explained due to the breaking off an ethyl group (mass 29) from the parent molecule. The other peaks at *m/z* values of 71, 57, 43, and 29 may be explained on the basis of fragmentation pattern involving the breaking off the methylene groups (CH_2) to gives fragments of masses with difference of 14 atomic mass units.

The mass spectrum of branched chain iso-alkanes, for example, isooctane, shows a base peak at *m/z* = 57 attributed to the two main fragment ions formed by the breaking of the bond shown by the dotted line. The intensity of the molecular ion peak is too weak to be observed (Fig. 10.11). This feature is common to all molecules with highly branched carbon skeleton. The peak at *m/z* = 99 may be assigned to the loss of CH_3 group from the molecular ion. The other peaks at *m/z* values of 56, 43, 41, 29, and 27 arise due to loss of CH_3 and CH_2 groups from fragment ions with mass values of 99 and 57.

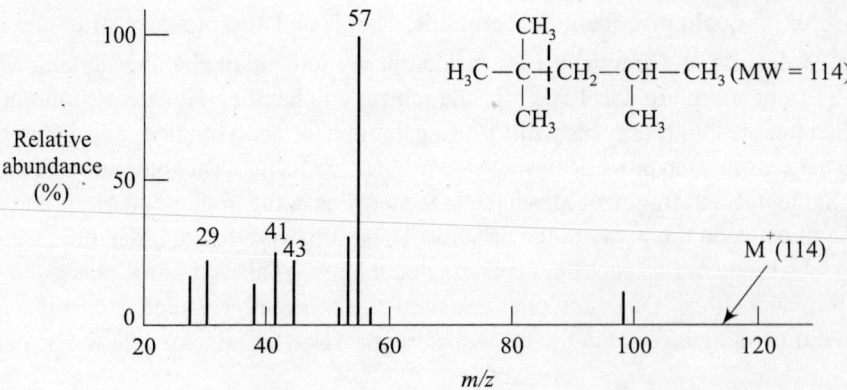

Fig. 10.11 Mass spectrum of isooctane

10.8 APPLICATIONS

The various applications mass spectrometry may be grouped into (i) determination of molecular weight, (ii) determination of molecular formula, (iii) structural analysis, and (iv) identification of the sample compound.

10.8.1 Molecular Weight Determination

Precise determination of the molecular weight of a low molecular weight (<500) sample molecule is an important application of mass spectrometry and primarily depends on the ability to determine the *relative molecular mass* (RMM) with a high degree of accuracy and recognize the molecular ion. For sample molecules that can be ionized to give a molecular ion (M^+), a protonated molecular ion $(M+1)^+$ or a deprotonated molecular ion $(M-1)^+$ by any one of the ionization methods, mass spectrometer is a unique and an unsurpassed tool for the determination of molecular weight, provided the ions have a life time of at least 10^{-5} s, and reach the detector without undergoing fragmentation. The position of the peak on the *x*-axis indicates the accurate molecular weight of the sample molecule.

The determination of the molecular weight, however, is not so simple because of the following factors.

(i) The mass of an ion as detected in a mass spectrometer is its true mass and NOT its mass obtained based on chemical atomic weights. The chemical scale of atomic weights is based on the weighted averages of the weights of all isotopes of a given element. All atoms have heavier isotopes that occur in characteristic natural abundance and this natural abundance of the isotopes of the constituent atoms is reflected in the molecular weights of the molecules. For example, hydrogen contains 0.02% of the heavier isotope, ^2H, whereas carbon contains 1.08% of the isotope ^{13}C. Peaks due to fragment ions having heavier isotopes of atoms also appear in the mass spectra of sample molecules, the intensities of the peaks being proportional to the natural abundances of the isotopes. Hence, besides the peak due to the molecular ion M^+, peaks due to $(M+1)^+$, and $(M+2)^+$ also appear in the mass spectrum. However, the intensities of the isotope peaks that occur at higher masses are usually much lower as compared to that of the molecular ion peak.

(ii) The molecular ion peak may be totally absent or its relative intensity may be negligibly small as compared to other peaks, particularly when the sample molecule is subjected to electron impact ionization. Further EI ionization results in fragmentation giving rise to a complex mass spectrum with peaks appearing at different *m/z* ratios. Identification of the molecular ion peak becomes difficult when the molecular ion is less stable and undergoes fragmentation easily. Molecular ions of highly branched molecules undergo fragmentation easily and the molecular ion peak will not be observable in the mass spectra.

(iii) Chlorine and bromine containing molecules pose additional problems in identifying the molecular ion peak because of the higher relative abundances of their isotopes. Chlorine has two isotopes of masses 35 and 37 with relative abundances being 75.77% and 49.5% respectively. Similarly, the isotopes of bromine with masses 79 and 81 have relative abundances of 50.5% and 49.5% respectively.

(iv) The molecular ions formed from alcohols lose water (18 amu) and similarly acetates lose acetic acid (60 amu) as neutral fragments before acceleration, and hence, it may be necessary to add the respective numbers to the heaviest fragments observed in the mass spectra of alcohols and acetates.

(v) Oxygen and nitrogen containing compounds form stable oxonium and ammonium ions respectively and the mass spectra of such compounds recorded at sample pressures higher than 0.5 mm of Hg may show peaks with one mass unit higher than the mass of the molecular ion. This is due to ion–molecule collision resulting in the transfer of a hydrogen atom to the ion, which is then accelerated in the mass analyser.

The molecular ion peak can be identified by considering the following: (i) the molecular ion peak will appear at the highest m/z value apart from the isotopic peaks which appear at even higher m/z values. However, the intensities of the isotopic peaks are much lower as compared to that of the molecular ion peak, (ii) the molecular ion will have an odd number of electrons. This is because the molecule on ionization by the loss of one electron generates a radical-cation with a charge of one unit making it an ion with odd number of electrons, (iii) the molecular ion must be capable of forming important fragment ions, particularly fragments with relatively high mass, by the loss of neutral fragments. For example, the mass spectra n-alkanes shows strong fragment ion peaks at M-14, M-28, M-42, etc., due to the loss of CH_2 groups. Similarly, the mass spectra of aliphatic ketones show a series of peaks due to fragment ions at M-15, M-29, M-43, etc. Aromatic hydrocarbons containing side chain attached to the benzene ring undergo fragmentation cleaving the side chain forming a benzyl cation which spontaneously rearranges to tropylium cation ($C_7H_7^+$) which shows a strong peak at $m/z = 91$, (iv) the observed abundance of the molecular ion should satisfy the requirements of the assumed molecular structure. The lifetimes of the molecular ions vary in decreasing order as shown for various organic substances.

Aromatic compounds > conjugated alkenes > alicyclic compounds > organic sulphides > straight chain hydrocarbons > mercaptans > ketones > amines > esters > ethers > carboxylic acids > branched hydrocarbons > alcohols

Nitrogen rule is useful for verifying whether a given peak corresponds to the molecular ion. According to this rule a compound with an even number of nitrogen atoms or no nitrogen atoms generates a molecular ion of an even mass value while a molecule containing an odd number of nitrogen atoms generates a molecular ion with an odd mass. The nitrogen rule is based on the fact that nitrogen has an odd-numbered valence and extra hydrogen is included in the molecule giving molecule as an odd mass.

Confirmation that a particular peak corresponds to that of a molecular ion can be obtained by varying the energy of ionizing electron beam in the electron impact ionization. Since fragmentation of a sample molecule is lower at lower energy of the ionizing beam the intensity of the molecular ion peak increases with decreasing applied potential while the intensities of the peaks of the fragment ions decreases. Chemical ionization and field ionization methods are preferred as they generate molecular ion or $(M + 1)^+$ ion peaks with higher intensity.

Molecular weight of biomacromolecules, such as proteins, can be determined by mass spectrometry. ESI, FAB, and MALDI methods of ionization are useful. Biomacromolecules undergo protonation and deprotonation reactions during ionization. Molecules which undergo such reactions at more number of sites yield larger families of peaks for each fragment. The peaks of different fragments may overlap at different m/z values complicating the mass spectrum. Hence, it is necessary to avoid more fragmentation by the use of soft ionization methods.

Electrospray ionization yields multiply charged ions without fragmentation. Two advantages of this method of ionization facilitate (i) the detection of large molecules even with a mass spectrometer of low mass range and (ii) estimation of molecular weights of large molecules. For example, when a molecule of about 10,000 Da forms MH^+ ion it would be detected as m/z ~10,001 and when the same molecule forms multiply charged MH_9^{9+} ion it can be detected at m/z ~1001. The second advantage is that the formation of multiply charged ions results in more number of peaks at different m/z values which in turn allows the estimation of molecular weight with greater accuracy by multiple peak average technique with the help of computer-based algorithms. The algorithms are based on the assumption that if two adjacent peaks from the same molecule differ only by one proton the relevant equations relating the m/z values of the two peaks and the molecular weight of the biomolecule may be written as

$$m_1 = (M + nH)/n \qquad (10.9)$$

and

$$m_2 = (M + (n + 1)H)/(n + 1) \qquad (10.10)$$

where m_1 and m_2 refer to the m/z values of the two contiguous peaks respectively, M is the molecular weight of the biomolecule to be evaluated, n is the number of charges, and H is the mass of proton and $m_1 > m_2$. The molecular weight M of the biomolecule is then calculated by solving the following equations.

$$n = (m_2 - H)/(m_1 - m_2) \qquad (10.11)$$
$$M = n (m_1 - H) \qquad (10.12)$$

Computer-based algorithms make use of all identical peaks to calculate more precise values of the molecular weight. Multiply charged molecular ions facilitate precise calculation of molecular weights of biomacromolecules. The charge states of molecular ions are determined from the isotopic peaks by a high-resolution mass spectrometer. The molecular ion peaks are formed by the coalescence of different isotopic peaks. For example, electrospray ionization of egg-white lysozyme gives rise to molecular ion peaks at m/z values of 1302, 1432, 1592, and 1791 corresponding to the charge states of $(M + 12H)^{12+}$, $(M + 10H)^{10+}$, $(M + 9H)^{9+}$, and $(M + 8H)^{8+}$ respectively as shown in Fig. 10.12(a) and the calculated molecular by computer-based algorithm is shown in Fig. 10.12(b).

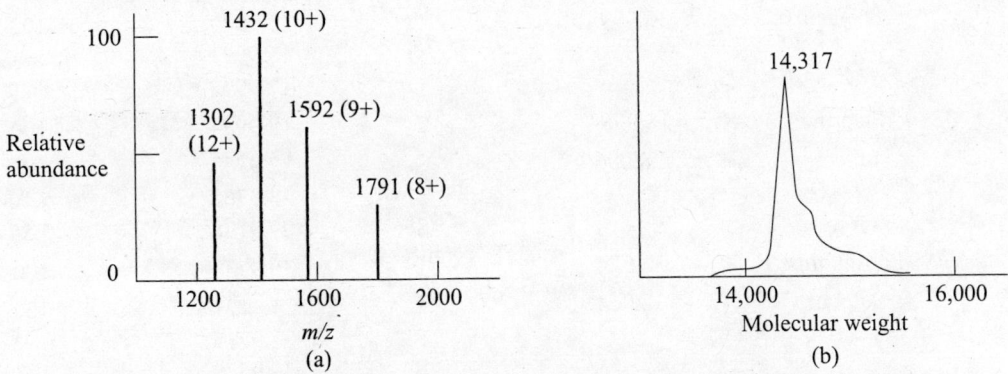

Fig. 10.12 Determination of molecular weight of egg-white lysozyme: (a) mass spectral lines and (b) molecular weight calculated by computer-based algorithm

For proteins the isotope peaks are produced mainly by the ^{12}C and ^{13}C isotopes with a mass difference of ~1 mass unit. If the molecular ion peak has a charge state of +1 the difference in the mass between the isotopic peaks will be 1 Da and for a charge state of +2 the difference in the mass between the isotopic peaks will only be 0.5 Da. Hence, greater resolution is necessary to observe isotope peaks of molecular ions with greater than +2 charge. Most of the modern instruments are capable of showing distinct isotope peaks up to triply charged molecular ions.

10.8.2 Determination of Molecular Formula

The molecular formula of a compound can be determined by (i) determining the molecular weight exactly or (ii) evaluating the intensities of isotope peaks $(M + 1)^+$ and $(M + 2)^+$ as compared to the molecular ion peak in the mass spectrum of the compound.

The exact determination of molecular weight of compound requires a high-resolution mass spectrometer capable of detecting mass differences of the order of a few thousandths of a mass unit. For example, for a molecular weight of 60 the molecular formula could be C_3H_8O, $C_2H_4O_2$, CH_4N_2O, or $C_2H_8N_2$. The exact masses of these compounds are 60.05754, 60.02112, 60.03242, and 60.06884 respectively. If the measured mass of the molecular ion is 60.0332 ± 0.005 then it is possible to assign the molecular formula of CH_4N_2O.

Molecular formula can be conveniently determined on the basis of the intensities of the isotope peaks observed in the mass spectrum of the sample even from a low-resolution instrument that can only discriminate between ions differing in mass by whole numbers. As already mentioned isotope peaks appear in the mass spectrum of a compound due to the natural abundance of heavier isotopes of elements. The natural abundance of the isotopes of common elements is listed in Table 10.1.

Table 10.1 Precise masses and natural abundance of stable isotopes of some common elements

Element	Atomic weight	Isotopes	Mass	Relative abundance
Hydrogen	1.00797	^1H	1.00783	100
		^2H	2.01410	0.016
Carbon	12.01115	^{12}C	12.0000	100
		^{13}C	13.00336	1.08
Nitrogen	14.0067	^{14}N	14.0031	100
		^{15}N	15.0001	0.38
Oxygen	15.9994	^{16}O	15.9949	100
		^{17}O	16.9991	0.04
		^{18}O	17.9992	0.20
Fluorine	18.9984	^{19}F	18.9984	100
Silicon	28.086	^{28}Si	27.9769	100
		^{29}Si	28.9765	5.10
		^{30}Si	29.9738	3.35
Phosphorus	30.974	^{31}P	30.9738	100
Sulphur	32.064	^{32}S	31.9721	100
		^{33}S	32.9715	0.78
		^{34}S	33.9679	4.40
Chlorine	35.453	^{35}Cl	34.9689	100
		^{37}Cl	36.9659	32.5
Bromine	79.909	^{79}Br	78.9183	100
		^{81}Br	80.9163	98.0
Iodine	126.904	^{127}I	126.9045	100

The intensity of the $(M + 1)^+$ peak can be calculated using the formula

% intensity of $(M + 1)^+$ peak = $(M + 1)/M$

$$= [(1.08 \times nC) + (0.016 \times nH) + (0.38 \times nN) + 0.04 \times nO)] \tag{10.13}$$

where nC, nH, nN, and nO refer to the number of carbon atoms, hydrogen atoms, nitrogen, and oxygen atoms that may be present in the sample molecule. Similarly, the approximate intensity of the $(M + 2)^+$ ion peak can be calculated using the formula

% intensity of $(M + 2)^+$ peak = (M + 2)/M

$$\approx [(1.08 \times nC)^2/200 + (0.016 \times nH)^2/200 + (0.2 \times nO)] \quad (10.14)$$

In practice the contributions due to hydrogen isotopes are rarely included.

Thus, for example, a sample of molecular mass 120 may correspond to the molecular formula of $C_5H_5N_4$ (purine-120.044), C_8H_8O (acetophenone-120.058), or C_9H_{12} (ethyltoluene -120.096). The molecular ion (M^+) peak will be observed at 120 for all the three compounds but the intensities of the $(M + 1)^+$ isotopic peaks will differ. Purine with five carbon atoms, five hydrogen atoms, and four nitrogen atoms will give rise to an $(M + 1)^+$ isotopic peak $[(5 \times 1.08 = 5.4) + (5 \times 0.016 = 0.08) + (4 \times 0.38)] = 7.00\%$ in intensity as compared to that of the (M^+) peak based on the natural of abundances of ^{13}C, ^1H, and ^{15}N isotopes respectively. In contrast, the intensity of the $(M + 1)^+$ isotopic peak of acetophenone will have an intensity of $[(8 \times 1.08 = 8.64) + (8 \times 0.016 = 0.128) + (1 \times 0.04)] = 8.808\%$ as compared to the intensity of the M^+ ion peak, based on the natural of abundances of ^{13}C, ^1H, and ^{17}O isotopes respectively. Ethyltoluene will give rise to an $(M + 1)^+$ isotopic peak of intensity $[(9 \times 1.08 = 9.72) + (12 \times 0.016 = 0.192)] = 9.912\%$ as compared to the M^+ ion peak. Thus, it is possible to assign the molecular formula on the basis of the relative intensities of the isotope peaks.

In the case of bigger molecules, the number of possible combinations that give rise to $(M + 1)^+$ and $(M + 2)^+$ ion peaks increase. However, for a particular combination of atoms, the relative intensities of the isotope peaks as compared to that of the molecular ion peak are unique. Hence, the isotope ratio method is highly useful for establishing the molecular formula of a compound.

The relative intensities of the $(M + 1)^+$ ion and $(M + 2)^+$ ion peaks for all possible molecular formulae up to molecular weight of 500 for various compounds containing carbon, hydrogen, oxygen, and nitrogen have been worked out and published in the form of tables by Beynon. Table 10.2 lists the possible empirical formulae for a molecular mass of 100.

Table 10.2 Partial list of possible molecular formulae for a molecular mass of 100 and the relative intensities of $(M + 1)^+$ ion and $(M + 2)^+$ ion peaks. (J.H.Beynon and A.E.Williams "Mass and abundance tables for use in mass spectrometry", Elsevier, 1963)

Empirical formulae	Relative intensities of peaks	
	$(M + 1)^+$	$(M + 2)^+$
$C_2H_2N_3O_2$	3.42	0.45
$C_2H_4N_4O$	3.79	0.26
$C_3H_2NO_3$	3.77	0.65
$C_3H_4N_2O_2$	4.15	0.47
$C_3H_6N_3O$	4.52	0.28
$C_3H_8N_4$	4.90	0.10
$C_4H_4O_3$	4.50	0.68
$C_4H_6NO_2$	4.88	0.50
$C_4H_8N_2O$	5.25	0.31
$C_4H_{10}N_3$	5.63	0.13
$C_5H_8O_2$	5.61	0.53
$C_5H_{10}NO$	5.98	0.35
$C_5H_{12}N_2$	6.36	0.17

(Contd)

(Contd)

Empirical formulae	Relative intensities of peaks	
	$(M + 1)^+$	$(M + 2)^+$
$C_6H_{12}O$	6.72	0.39
$C_6H_{14}N$	7.09	0.22
C_7H_2N	7.98	0.28
C_7H_{16}	7.82	0.26
C_8H_4	8.71	0.33

In the case of molecules containing atoms of elements other than carbon, hydrogen, oxygen, and nitrogen, it is necessary to first identify the presence of other elements and then calculate the expected intensities for the $(M + 1)^+$ ion and $(M + 2)^+$ ion peaks.

Molecules containing chlorine or bromine give rise to intense $(M + 2)^+$ ion peaks as the heavier isotopes of these elements differing by two mass units from the lighter isotopes have greater natural abundance. The natural abundance of ^{37}Cl is 32.45% as compared to the lighter isotope ^{35}Cl while the natural abundance of ^{81}Br is 98.0% that of ^{79}Br. When two chlorine or bromine atoms are present in a molecule a distinct $(M + 4)^+$ ion is also observed in addition to the intense $(M + 2)^+$ ion peak. Sulphur containing compounds also show a distinct $(M + 2)^+$ ion peak. It is possible to determine the number of S, Cl, or Br atoms in the molecule, for example, a compound containing two chlorine atoms gives rise to an $(M + 2)^+$ peak which is about 65% of the intensity of M^+ ion peak. The relative intensities of the $(M + 2)^+$, $(M + 4)$, and $(M + 6)^+$ ion peaks for chlorine and bromine containing compounds are listed in Table 10.3

Table 10.3 Relative intensities of isotope peaks for combination of chlorine and bromine (J.H.Beynon and A.E.Williams "Mass and abundance tables for use in mass spectrometry", Elsevier, 1963)

Halogen	Relative intensities of peaks (%)			
	M^+	$(M + 2)^+$	$(M + 4)$	$(M + 6)^+$
Br	100	97.7	---	---
Br_2	100	195.0	95.5	---
Br_3	100	293.0	286.0	93.4
Cl	100	32.6	---	---
Cl_2	100	65.3	10.6	---
Cl_3	100	99.8	31.9	3.47
Cl_4	100	131.0	63.9	14.0
Cl_5	100	163.0	106.0	34.7
Cl_6	100	196.0	161.0	69.4
BrCl	100	130.0	31.9	---
Br_2Cl	100	228.0	159.0	31.2
Cl_2Br	100	163.0	74.4	10.4

The determination of molecular formula also provides information on the number of unsaturated sites in the molecule which in turn can lead to inferring a possible structure. The unsaturated sites in a molecule include rings, double and triple bonds. Triple bonds are counted as two double bonds. The formula for calculating the total number N of unsaturated sites is

$$N = [nC + 1 - (nH - nN)/2 - (nX/2)]$$

(10.15)

where nC, nH, nN, and nX represent the number of carbon, hydrogen, nitrogen, and halogen atoms in the molecule. Thus, for an empirical formula of $C_7H_7NO_2$, the number of unsaturated sites is 5 and the possible structures include $NO_2C_6H_4CH_3$ and $NH_2C_6H_4COOH$.

10.8.3 Structural Information

Information on the structure of a molecule can be obtained from mass spectral data by a systematic study of fragmentation pattern. Based on the concept of relative strengths of chemical bonds and the relative stabilities of fragments formed it is possible to arrive at the structural formula of a compound. In general it is sufficient to identify and assign the peaks of greater intensity as it may not be possible to assign for all the peaks observed in the mass spectrum of a compound. A few general rules that have been evolved on the fragmentation pattern of various organic compounds include:

(i) The relative intensity of the parent ion peak usually decreases with increasing molecular weight in a homologous series. It also decreases with increasing degree of chain branching and the parent ion peak may not be observable in branched chain compounds of higher molecular weight.

(ii) Cyclic compounds including aromatics and molecules containing double bonds give rise to intense parent ion peaks. For example benzene shows a strong molecular ion peak at $m/z = 78$ without much fragmentation, as fragmentation of the benzene ring requires high energy.

(iii) Cleavage of C–C bond occurs more easily at branches in the aliphatic chain with the largest substituent at the branch being eliminated, for example, in alkanes secondary or tertiary carbocations are formed as they are relatively more stable than primary ions, for example, isopropyl cation is formed from iso-butane while tert-butyl carbocation is formed from iso-octane. In straight chain paraffins cleavage of successive cleavage of C–C bonds results in the loss of successive CH_2 groups and clusters of peaks differing in mass by 14 amu are observed. In most cases stable hydrocarbon fragments containing three or four carbon atoms are formed and the corresponding peaks are quite intense.

Alkyl-substituted aromatic hydrocarbons undergo cleavage of the side chain to form a benzyl cation which spontaneously rearranges to a tropylium cation which gives a prominent peak at $m/z = 91$.

Benzyl carbocation   Tropylium ion

(iv) The presence of a hetero atom (N, O, or S) in the molecule promotes the cleavage of C–C single bond adjacent to the hetero atom. Thus, for example, in primary alcohols cleavage of the C–C bond adjacent to oxygen is quite common giving rise to a strong peak at $m/z = 31$ attributed to the CH_2OH^+ ion. The C–C bond cleavage leads to the loss of the largest alkyl group at the carbon atom adjacent to the oxygen, for example, 2-butanol loses an ethyl group to form CH_3CHOH^+ to give a strong peak $m/z = 45$ while 2-methyl-2-propanol (t-butanol) loses a methyl group to form $(CH_3)_2COH^+$ ion to give an intense peak at $m/z = 59$. Alcohols give rise to a very weak molecular ion peak, not observed at all in the case of tertiary alcohols, as the molecular ion loses water to give rise to an intense peak at $m/z = (M - 18)$.

(v) Quite intense peaks may be observed in the mass spectrum at different m/z values not related to the possible fragments obtained from the original molecule. Such peaks may be attributed to the formation of more stable fragments due to (i) rearrangement of the fragments originally formed, (ii) loss of neutral fragments, or (iii) migration of hydrogen atoms in molecules containing hetero atoms.

10.8.4 Identification of the Sample Compound

Identification of the sample compound is based on the fundamental assumption that the fragmentation pattern is unique for a compound, and hence, mass spectrum of the compound is also unique. However, this assumption is not valid in the case of stereo-isomers (geometrical and optical isomers) and also for closely related compounds. The observed mass spectrum of a compound will be unique and the probability of different compounds giving rise to almost same mass spectra becomes less particularly when the spectrum contains more number of peaks. Electron impact ionization yields more fragment ions, and hence, more peaks in the mass spectrum making it the preferred method of ionization for spectral comparison. Thus, the sample compound can be identified by comparing and matching the mass spectral data available in most modern mass spectrometers equipped with computerized library search facilities. In addition, hyphenated techniques of mass spectrometer coupled to chromatograph facilitate the sample identification.

It must be mentioned here that the peak intensities of the fragment ions depend strongly on the experimental conditions, such as the energy of the electron beam, the location of the sample with respect to the beam, sample pressure, and temperature as well the geometry of the mass spectrometer. Hence, it is possible that different instruments may give rise to differences in the peak intensities. Considering these aspects the identity of the sample compound may be confirmed by comparing the spectral data with those obtained from an authentic or standard compound under the same experimental conditions.

Protein identification is of crucial importance in biological research. Mass spectral determination of the protein molecular mass provides adequate information to confirm the identity of a protein. The method is particularly useful for characterization of recombinant proteins and also to investigate the modification of molecular mass of a protein by post-translational modifications.

A protein may be identified either by (i) MS fingerprint method or (ii) using peptide sequence tags. In the first method the purified protein is digested enzymatically to produce a mixture of peptides. The peptide mixture is analysed and the mass spectral pattern called MS fingerprint is recorded and used to search the protein database available in the internet. The origin of the protein and a rough estimate of its molecular weight are required as input information to narrow down the search. The net-based search involves the digestion of all the available proteins in the database with a chosen enzyme to match the theoretical and experimental masses. The search yields a list of best matches along with a confidence parameter.

The second method based on peptide sequence requires prior information of the protein or gene sequence and involves the use of sequence tags to identify the peptides generated by enzyme digestion. The basis of identification is that in a MS/MS peptide spectrum some consecutive amino acids can be unambiguously identified. Since the enzymatic reaction is highly selective, the peptide sequence, and hence, the identification of protein becomes possible even if the peptide contains just two amino acid residues. A protein can be unambiguously identified if its origin is from an organism whose gene sequence is completely known (e.g., *E. coli* or *B. subtilis*).

10.8.5 Applications in the Study of Proteins and Nucleic Acids

Mass spectrometry has been used for the determination of the sequence of peptides and proteins. A typical procedure for protein sequencing involves the following steps: (i) digestion of the protein by trypsin (or other enzymes) to yield tryptic peptides and (ii) electro-spray ionization of the tryptic peptides followed by analysis using tandem mass spectrometry. In tandem mass spectrometry the fragments are separated according to their *m/z* ratios by the first quadrupole, the separated fragments are subject to CID process, and the resulting fragments analysed by the third quadrupole. The mass spectral data obtained may be compared with peptide database to determine the sequence.

DNA and RNA are difficult to ionize and most of the mass spectral analyses have been successful for short oligonucleotides with less than 30 nucleotides. Mass spectrometry has been used mostly as a sequence confirmation method for sequence information obtained by PCR amplification.

● ● ● ● ● ● ● ● ● ● ● ● ● ● ● ● ● ● **SOLVED PROBLEMS**

EXAMPLE 1   *Calculate the kinetic energy of an ion with a mass of 88 and a charge of +1 when it is accelerated at 100 V in an electron impact ionization chamber.*

SOLUTION

According to Eq. (10.1)

$$KE = zeV = 1 \times 1.6 \times 10^{-19} \, C \times 100 \, V = 1.6 \times 10^{-17} \, J$$

EXAMPLE 2   *Calculate the resolution of the mass spectrometer required for the separation of (a) CH_4^+ and NH_3^+ (masses being 16 and 17 respectively) and (b) N_2^+ and CO^+ (masses being 28.0061 and 27.9949 respectively)*

SOLUTION

According to Eq. (10.5), $R = m/\Delta m$

  (a) $R = 16.5/1 = 16.5$

  (b) $R = 28.0005/0.0112 = 2500$

EXAMPLE 3   *Predict the probable composition of a sample molecule which gave a molecular ion peak at m/z=156 and an $(M + 1)^+$ peak of an intensity of 1.1% of the parent ion peak.*

SOLUTION

The presence of an $(M + 1)^+$ peak of an intensity only 1.1% of the parent ion peak indicates the presence of only carbon and hydrogen apart from a monoisotopic element other than N, O, S, or Cl. The probable molecular formula is C_2H_5I.

EXAMPLE 4   *Predict the empirical formula of a compound containing only C, H, and O if its exact molecular weight was determined to be 94.0417.*

SOLUTION

Taking the data from Table 10.1 the formula can be worked out as follows:

$$^{16}O - 15.9949 \times 1 = 15.9949$$
$$^{12}C - 12.0000 \times 6 = 72.0000$$
$$^{1}H - 1.0085 \times 6 = 6.0468$$
$$\text{Total} = 94.0417$$

The formula of the compound is C_6H_6O, possibly C_6H_5OH.

EXAMPLE 5 *Deduce the formulae of three compounds A, B, and C from the following mass spectral data*

Compound A		Compound B		Compound C	
m/z	Peak intensity (%)	*m/z*	Peak intensity (%)	*m/z*	Peak intensity (%)
151	100 (M$^+$)	168	100 (M$^+$)	49	100
152	10.4	169	7.4	51	30
153	32.1			84	60 (M$^+$)
154	2.9			86	40
				88	8

SOLUTION

Compound A: The molecular ion peak is at an odd *m/z* value, hence, nitrogen rule indicates the presence of one N atom. Absence of (M + 4)$^+$ and (M + 6)$^+$ peaks indicate the presence of only one Cl atom. Since the base peak is that of the parent ion, the compound is aromatic. The residual mass of the compound after removing one N and one Cl atoms is 102. Dividing the residual mass by atomic mass of C, the presence of 8 C atoms, and 6 H atoms are indicated. Thus, the possible formula of the compound is C_8H_6ClN.

Compound B: The molecular ion peak at *m/z* 168 is the base peak indicating the compound to be aromatic. The presence of (M + 1)$^+$ alone indicates the presence of N and O and even molecular weight the presence of 2 N atoms. The residual mass is 76 indicating the presence of 6 C and 4 H atoms. Considering the formula to be $C_6H_4N_2O_4$ the intensity of the (M + 1)$^+$ peak will be (six ^{13}C + two ^2H + two ^{15}N + four ^{17}O) 7.44 % as compared to that of M$^+$ ion. The formula is in agreement with the ratios of the intensities of the M$^+$ and the (M + 1)$^+$ ions.

Compound C: The presence of intense (M + 2)$^+$ peak and also (M + 4)$^+$ indicates the presence of two Cl atoms. The base peak at *m/z* = 49 and the peak at *m/z* = 51 (^{37}Cl) may be attributed to the loss of one Cl atom. The possible formula is CH_2Cl_2 (MW = 84.9).

● **SUMMARY**

- Mass spectrometry is an essential technique in the identification and understanding of the structure of chemical species. The sample is vapourized and the sample molecules undergo ionization and fragmentation to yield positively charged ions and get separated depending on their mass-to-charge ratio (*m/z*).

- The mass spectrum is a plot of the relative intensities of the fragment ions of different masses as a function of *m/z*. The fragmentation pattern is characteristic of the structure of the sample molecules and hence useful in the identification and structural analysis of the sample molecules.

- The mass spectrometer has the main components of (i) sample inlet, (ii) ionization source and accelerator chamber, (iii) mass analyser to separate the ions according to *m/z* ratios, (iv) detector, and (v) recording and display system. The instrument is maintained at low pressure of about 10^{-7} torr.

- Different ionization methods are employed to ionize the sample molecules including electron impact ionization (EI), chemical ionization (CI), field ionization (FI), electrospray ionization (ESI), fast atom bombardment (FAB), and matrix-assisted laser desorption and ionization (MALDI).

- Quadrupole mass spectrometer makes use of a set of four cylindrical metal rod electrodes or quadrupole mass filters, placed symmetrically and parallel to the direction of travel of the ions. The fragment ions are accelerated by a potential of 5–15 V and allowed to travel through the space between the metal electrodes under an oscillating radiofrequency ac potential applied across each pair of electrodes. Only fragment ions of one *m/z* value can pass through the quadrupole analyser following a stable path to reach the detector while fragment ions of other *m/z* values follow an unstable path leading to collision with the electrodes.

- Time of flight mass spectrometer allows the separation of fragment ions in field-free conditions through a field-free drift tube of about 1 m length. The ions get separated depending on their masses.

- Ion trap mass analyser consists of an ion trap in which gaseous ions can be formed and confined by electric and/or magnetic fields for extended periods. The trapped ions leave the trap in a controlled manner so that scanning on the basis of mass-to-charge ratio is possible

- FT mass spectrometer contains a trapped ion analyser cell within which the gaseous ions are subjected to an external magnetic field and move in a well-defined circular orbit with an angular frequency called cyclotron frequency in a plane perpendicular to the direction of the magnetic field. Ions with different m/z ratios have different cyclotron frequencies. The trapped ions of different m/z ratios get separated by absorbing energy from an applied electric field at a frequency which matches the cyclotron frequency of a particular m/z ratio.

- Tandem mass spectrometry makes use of two or more mass analysers in tandem to allow a secondary fragmentation and characterize the selected ion(s) generated by the first analyser.

- The mass spectrum is a graphical presentation of relative abundance of each fragment ion expressed as a percentage of the most abundant one against the mass-to-charge (m/z) ratio in the form of a bar chart or line diagram.

- The tallest peak due to most abundant fragment ion is called the base peak. The molecular ion is the heaviest ion formed by the loss of only one electron. The value of m/z at which the peak due to the molecular ion appears in the mass spectrum gives the actual molecular weight of the sample molecule.

- Metastable ion peaks appear in the mass spectrum as broad peaks at non-integral m/z values. The presence of metastable ion peaks in the mass spectrum is useful to prove a proposed fragmentation pattern of a sample molecule and in structure analysis.

- Mass spectrometry is useful for the determination of molecular weight, determination of molecular formula, structural analysis, and identification of the sample compound.

REVIEW QUESTIONS

1. Explain the basic principle involved in mass spectrometry.

2. Draw a neat sketch of the single-focusing mass spectrometer and describe the functions of the components.

3. What is electron ionization? How is it carried out?

4. Give a detailed account of the different methods of ionization.

5. What is double focusing? How is it achieved in mass spectrometry?

6. How does a quadrupole mass analyser function?

7. Discuss the principle involved in time of flight mass spectrometry.

8. What is tandem mass spectrometry? What is its use?

9. How is mass spectrum useful for the determination of the molecular weight of a compound?

10. Write a note on the determination of molecular weight of a protein from mass spectral data.

11. Give an account of the application of mass spectra in the determination of molecular formula of compounds.

12. How is a protein identified by mass spectral analysis?

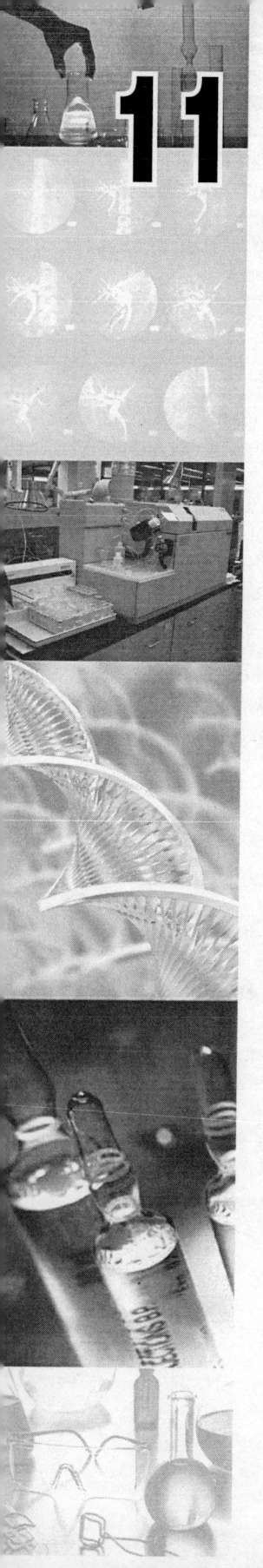

11 X-ray Methods

11.1 INTRODUCTION

X-rays are short wavelength electromagnetic radiation in the range between 10^{-5} Å and 100 Å, the most commonly used wavelength region being 0.1 Å to about 25 Å. Similar to the electromagnetic radiation of the more familiar ultraviolet, visible, and infrared regions, X-rays are useful in chemical analysis of samples based on their interaction with individual elements in a sample. The interaction of X-rays with elements includes different phenomena such as absorption, emission, fluorescence, scattering, and diffraction.

X-rays are mostly obtained by three different methods, namely (i) bombarding a target metal with a beam of high energy electrons, (ii) exposing a target substance to a primary beam of X-rays (obtained by the first method) to generate a secondary beam of X-ray fluorescence, and (iii) using a radioactive source which emits X-rays by decay process. A less common method of production of X-rays is from synchrotron radiation.

The different methods produce X-rays in the form of continuous radiation called *white radiation* or *Bremsstrahlung* (meaning radiation that arises from retardation of the particles) and discontinuous radiation of specific wavelengths.

11.2 X-RAY SPECTROSCOPIC INSTRUMENTS

Instruments using X-rays for different analytical methods such as absorption or emission spectroscopy and diffraction have common components such as (i) X-ray source, either a generator or a radioisotope, (ii) filters, (iii) monochromator, collimator and goniometer assembly, and (iv) detector. Hence, it is convenient to discuss the general features of these components and the configuration of the instrument before introducing the different analytical methods.

11.2.1 Production of X-rays by Electron Bombardment

The method involves bombarding a target metal anode with a high-energy beam of electrons emitted by a heated cathode (electron emitter) in an X-ray tube. The X-ray tube (Fig. 11.1) is an evacuated glass or ceramic tube into which a cathode and the water-cooled target metal anode are sealed.

The electrons produced by the hot cathode made of tungsten filament are accelerated through a high voltage field of the order of 10 kV to about 100 kV towards the anode. The high-energy electron beam on hitting the target anode is brought to rest quickly. As the kinetic energy of the incident electrons is lost due to their collisions with the atoms constituting the target metal, X-rays are generated which pass through the window.

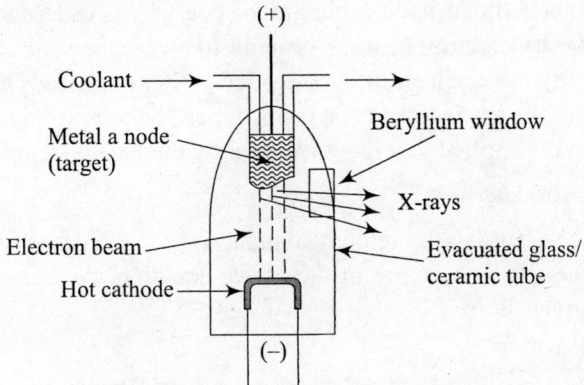

Fig. 11.1 Schematic diagram of an X-ray tube

Two types of X-rays are generated in the X-ray tube constituting the X-ray spectrum: (i) a continuous spectrum of X-rays (white radiation) and (ii) a discontinuous or line spectrum of X-rays superimposed on the continuous spectrum, which depend only the nature of target anode and independent of the applied voltage.

The *continuous spectrum* of X-rays is characterized by a broad maximum in intensity and a definite short wavelength limit, λ_o, depending only on the applied voltage to the X-ray tube and is independent of the target anode (Fig. 11.2).

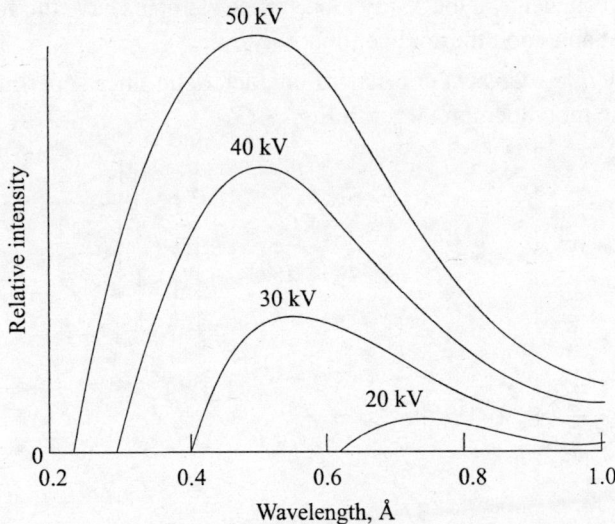

Fig. 11.2 X-ray continuum from a target metal (tungsten) at different voltages

Thus, the λ_o for the X-rays produced by bombarding different target metals will be identical for a given applied voltage as shown in Fig. 11.2. The deceleration of the electron of the primary electron beam due to its collision with the target metal generates an X-ray photon whose energy is equal to the difference in the kinetic energies of the primary electron before and after its collision. The electrons in the primary electron beam are decelerated over a series of collisions with the target, the loss in the kinetic energy differing from collision to collision. Hence, X-ray photons of differing energies are generated resulting in a continuous spectrum of X-rays. The

X-ray photons generated can have a maximum energy (v_0 in terms of frequency) when the electron of the primary electron beam is brought to a complete stop (zero kinetic energy) and all the kinetic energy of the electron is converted to X-ray photon in a single collision. Since the kinetic energy of the primary electron beam depends on the voltage applied, the relationship between the v_0 and the applied voltage V is given by the Duane–Hunt law as

$$hv_0 = Ve \quad \text{or} \quad hc/\lambda_0 = Ve \tag{11.1}$$

where h is Plank's constant, c is the velocity of light, and e is the charge on the electron. Substituting the numerical values for h, c, and e in the above equation, the cut-off limit or the minimum wavelength λ_0 is given as

$$\lambda_0 = 12,398/V \tag{11.2}$$

where λ_0 is in Angstroms and V in volts. Equation (11.2) provides a way to determine the value of Planck's constant accurately. As expressed by the equation, an increase in the applied voltage to the X-ray tube increases the kinetic energy of the primary electron beam which in turn generates X-ray photons of shorter wavelengths. The intensity I of the X-ray photons in the continuum is related to the current i (mA), voltage (V), and the atomic number Z of the target element as

$$I = k\,i\,Z\,V^2 \tag{11.3}$$

where k is a proportionality constant. The intensity of the continuum increases with increasing applied voltage as shown in Fig. 11.2 and also with increasing atomic number of the target element. Hence, usually higher atomic number elements, such as tungsten or platinum, are often used as target elements in the X-ray tube. These elements have the additional advantages of high melting point and good thermal conductivity.

The *X-ray line spectrum* consists of characteristic lines superimposed on the continuum as shown for the molybdenum target in Fig. 11.3.

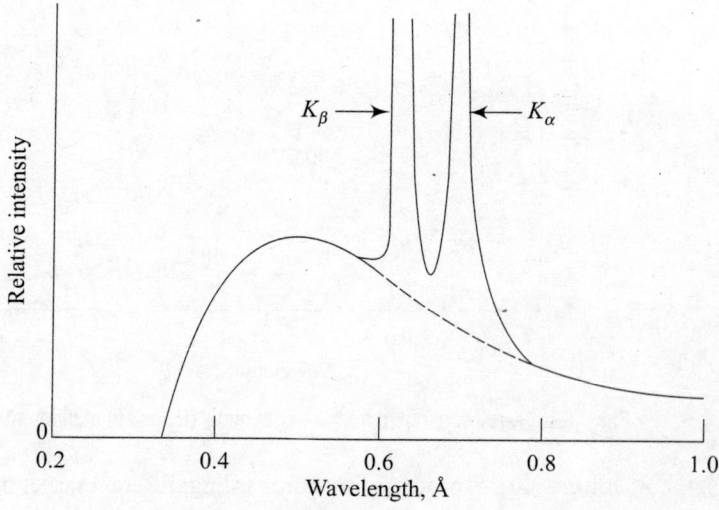

Fig. 11.3 Continuum and line spectrum of Mo target at 35 kV

The line spectrum depends on the nature of the target metal and is characteristic of the target element as given by Mosley's law

$$v = k\,(Z - \sigma)^2 \tag{11.4}$$

where σ is a constant whose value depends on the particular series (for K_α series $\sigma = 1$ and for L_α series $\sigma = 7.4$). The characteristic line spectrum is independent of the applied voltage to the X-ray tube. The minimum voltage required for observing the characteristic line spectrum depends on the atomic number of the target element. Thus, for example, a minimum of 20 kV is required for exhibiting the characteristic K lines in the case of molybdenum ($Z = 42$) while a minimum of 70 kV is required in the case of tungsten ($Z = 74$).

The X-ray line spectra are exhibited due to the electronic transitions that occur within the innermost atomic energy levels of the target element. When an atom of the target element is bombarded by a primary beam of high energy particles, such as electrons or protons or X-ray photons, an electron is ejected from one of the inner shells of the target atom. The vacancy caused is immediately filled by an electron from a higher energy shell, creating a vacancy in that shell. This vacancy in turn is filled by an electron from a still higher energy shell. Each electronic transition results in the emission of an X-ray photon of specific frequency or energy depending on the difference in the binding energies of the two electrons involved. Thus, for example, when an electron in K shell, closest to the nucleus (principal quantum number $n = 1$) of the target atom is ejected by the primary beam of electrons, the vacancy created is filled by an electron from the next higher energy L shell ($n = 2$) or still higher energy shells of M ($n = 3$), N ($n = 4$), etc. The electronic transitions from the higher energy shells to the K shell give rise to the characteristic K series of X-ray photons as shown in Fig. 11.4. Similarly, electronic transitions to the L shell from higher energy shells are called L series and so on.

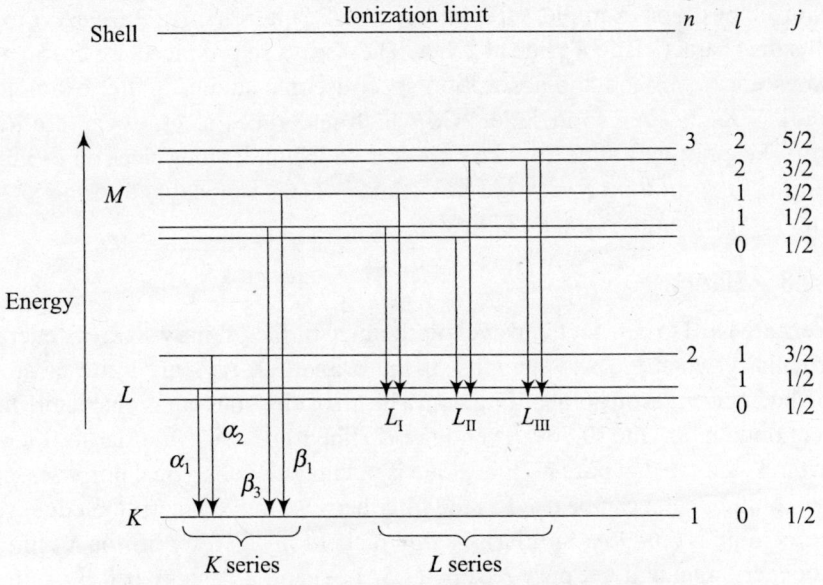

(The letters α and β are used to designate the transitions to lower quantum number when $\Delta n = -1$ and -2 respectively.)

Fig. 11.4 Common X-ray transitions giving rise to line spectra in atoms

Based on quantum mechanical selection rules only certain electronic transitions are allowed. The X-ray spectral lines are designated as $K\alpha_1$, $K\alpha_2$, $K\beta_1$, $K\beta_2$, $L\alpha_1$, $L\alpha_2$, etc. The alphabets K and L indicate that the lines arise due to the removal of an electron from K shell and L shell respectively and the lines are called K series lines and L series lines. Each line in a series is designated by a Greek alphabet α, β, etc. to indicate the sub-shell of the outer electron involved in the transition while the numerical subscript indicates the relative strength or intensity of each line in a given series. The intensity of the line $K\alpha_1$ is greater than $K\alpha_2$. Since the energy difference between the K and the L shells is quite large as compared to the energy difference between the L and the M shells in an atom, the K series lines appear at shortest wavelengths. The energy difference between the α_1 and α_2 and similarly between β_1 and β_2 is quite small, and hence, separate lines can be observed only in high-resolution X-ray spectrometers.

X-ray line spectra are much simple as compared to the line spectra in the ultraviolet and visible regions as they consist of a limited number of lines. In addition the intensity and the wavelengths of the emitted X-rays are independent of the chemical environment and also the physical state of the element, particularly for elements of higher atomic numbers. Thus, X-ray line spectra are highly useful in qualitative identification of the element.

11.2.2 X-rays from Radioactive Sources

Decay of radioisotopes often results in the emission of gamma rays and X-rays. Decay processes involving α and β particle emission results in the formation of an excited state nucleus which releases gamma rays as it relaxes to the ground state. Electron capture or K capture involving the capture of a K electron by the nucleus and the resulting formation of an element of lower atomic number leads to electronic transitions to the vacated orbital and consequent X-ray line spectral emission of the newly formed element. Monochromatic X-rays can be obtained from artificial radioisotopes. For example, ^{55}Fe undergoes K capture ($t_{1/2}$ = 2.6 years) to form ^{54}Mn which emits the characteristic K_α line at 2.1 Å. The X-rays so produced are useful as a source in X-ray fluorescence emission and absorption spectroscopic studies. Other examples of radioisotopes useful as X-ray sources include ^{57}Co which undergoes electron capture [($t_{1/2}$ = 270 days) to form ^{56}Fe emitting X-rays (6.4 keV), ^{109}Cd which undergoes electron capture ($t_{1/2}$ = 1.3 years) to form ^{108}Ag emitting X-rays (22 keV), and ^{125}I ($t_{1/2}$ = 60 days) forming ^{124}Te through electron capture, which emits X-rays (27 keV)].

11.2.3 Filters

Filters are used to restrict the wavelength range of the primary X-rays generated from a source, particularly when the wavelengths of the primary X-rays are quite close. For example, the molybdenum target on bombarding with a beam of electrons emits characteristic X-rays consisting of continuum, K_α line (0.709 Å) and the K_β line (0.632 Å). In order to filter off the continuum and the K_β line so that pure K_α line alone is obtained for analytical purposes, a thin foil (0.01 cm) of zirconium metal can be used as the filter because the K absorption edge (see Section 11. 4.1) of zirconium is 0.689 Å. Similarly, a thin nickel foil can filter off the K_β line (1.392 Å) emitted by copper target and get pure K_α line (1.541 Å) and a vanadium foil can filter off the K_β line (2.085 Å) to get pure K_α line (2.290 Å) from Cr target.

11.2.4 Monochromator, Collimator, and Goniometer Assembly

Monochromator used in X-ray spectroscopic instruments consists of a pair of collimators and a dispersing element essentially a single crystal (called diffracting or analysing crystal) mounted on *goniometer* or rotating table. Collimators are closely spaced parallel plates or bundle of tubes of < 0.5 mm in diameter and allow only parallel beams of X-rays to pass through. The analysing crystal may have a flat or curved surface, the latter has greater efficiency as it diffracts and also focuses the divergent beam onto the detector. The analysing crystal diffracts the incident beam of X-rays in accordance with the Bragg equation

$$n\lambda = 2\,d\sin\theta \tag{11.5}$$

where λ is the wavelength of X-ray beam, d is the inter-planar spacing within the crystal, θ is the angle of diffraction and n is the order of diffraction (usually first order). X-rays in the wavelength region 0.1 Å to about 10 Å are used in analytical applications and the ability of the crystal to diffract and monochromatize the incident X-rays depends on the inter-planar spacing. Different crystals are used depending on their lattice spacing and therefore their ability to disperse X-rays over a given wavelength region. Topaz with a lattice spacing of 1.356 Å is useful for dispersing X-rays in the wavelength range of 0.24–2.67 Å. Other crystals include LiF (d = 2.014 Å) for dispersing 0.35–3.97 Å, Aluminum (d = 2.338 Å) for wavelength range of 0.326–4.52 Å, NaCl (d = 2.820 Å) for wavelength range of 0.49–5.55 Å, Calcium fluoride (d = 3.16 Å) for wavelength range of 0.44–6.11 Å), Ethylenediamine d-tartrate, EDDT (d = 4.404 Å) for wavelength range of 0.77–8.67 Å, and Ammonium dihydrogen phosphate, ADP (d = 5.325 Å) for wavelength range of 0.93–10.50 Å. ADP has a greater wavelength range but suffers from lower dispersion.

Monochromators effectively function between angles (2θ) 10° and 160°. At angles of $2\theta < 10°$, a large amount of polychromatic radiation is scattered by the crystal while at $2\theta > 160°$ measurements cannot be made as the detector will receive bulk of the source radiation directly.

Goniometer (Greek *gonia* = angle; *metron* = measure) facilitates the rotation of the crystal mounted on it and also measures the angles between the crystal face and the collimated beams of radiation precisely.

11.2.5 Detectors

Detectors used in analytical X-ray instruments are common to detection and measurement of radioactivity from natural and artificial isotopes. The detectors include proportional counter, Geiger–Muller tube, scintillation detector, or photographic emulsion. The signal generated by these detectors is subjected to pulse height analysis involving electronic circuits for amplification and elimination of background noise. The basic principles involved and the constructional aspects of these detectors along with pulse height analysers are discussed in detail in Chapter 18 on Radiochemical methods of analysis.

11.3 CLASSIFICATION OF X-RAY METHODS

The different X-ray methods include (i) *X-ray absorption spectroscopy* based on the different absorption characteristics of substances, (ii) *X-ray emission spectroscopy*, (iii) *electron probe microanalysis* based on the characteristic emission of X-rays by an excited element in a sample, (iv) *X-ray fluorescence spectroscopy* involves the irradiation of the sample with primary X-rays from a suitable source and the elements in the sample emitting characteristic fluorescence or secondary X-ray emission, and (v) *X-ray diffraction* involves the identification of crystal phase and study of crystal structures. The different methods, thus, provide valuable information on analytical aspects of qualitative identification and quantitative estimation and also on the composition and structure of substances in bulk as well as at the surface.

11.4 X-RAY ABSORPTION SPECTROSCOPY

As the name implies the technique involves exposing a sample to a beam of monochromatic X-ray and determining the extent of attenuation of the intensity of the incident beam as it passes through the sample to a detector. The experimental methodology is similar to any other absorption spectroscopy encountered in UV–visible or IR regions. The process of absorption of X-rays by a target atom depends on the quantized energy levels of the atom.

11.4.1 Absorption of X-rays

Absorption of X-rays occurs when a beam of X-rays is passed through a thin layer of a sample and the intensity of the incident beam decreases giving rise to an absorption spectrum consisting of well-defined peaks as shown in Fig. 11.5. The absorption peak wavelengths are characteristic of the element and are independent of the chemical and environmental state of the element.

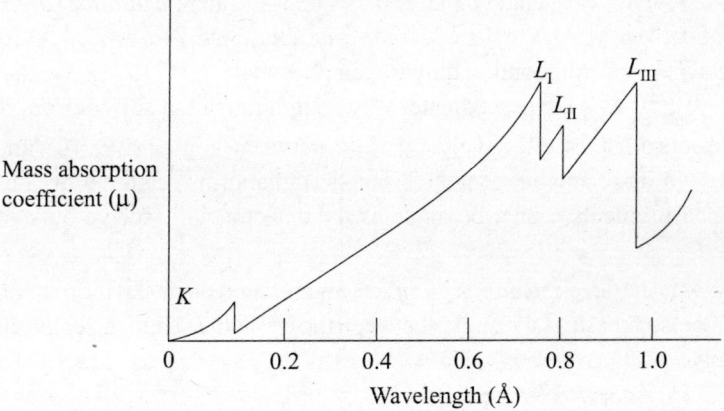

Fig. 11.5   X-ray absorption spectrum of lead

When a quantum of the incident beam of X-rays is absorbed by an atom its energy is transferred to the atom resulting in the ejection an electron, called the *photoelectron*, from the innermost energy level and the formation of an excited ion. The energy of the incident beam is partitioned between the kinetic energy of ejected photoelectron and the potential energy of the excited ion. The probability of absorption of the X-ray photon is highest when the energy of the incident photon is exactly equal to the energy required to remove the electron just to the periphery of the atom (the energy is less than that required for complete ionization as shown in Fig. 11.4) and the kinetic energy of the photoelectron approaches zero. For example, the energy required to eject the K electron from an atom of lead corresponds to a wavelength of 0.14 Å of the incident X-rays as seen in the absorption spectrum of lead shown in Fig. 11.5. At wavelengths shorter than 0.14 Å, the probability of absorption gradually increases with increasing wavelength till 0.14 Å and a smooth curve is observed. The kinetic energy of the ejected photoelectron increases continuously. As the wavelength increases slightly beyond the critical value the energy of the X-rays is insufficient to remove the K electron, and hence, the absorption shows an abrupt decrease. Thus, there is a sharp discontinuity or *absorption edge* in the absorption spectrum at 0.14 Å. Similar absorption edges are also seen at longer wavelengths in the absorption spectrum at wavelengths corresponding for the removal of an electron from the next higher energy level

L. Atoms have three sets of *L* levels differing slightly in their energies as shown in Fig. 11.4, and hence, three peaks or absorption edges are seen in Fig. 11.5.

The absorption of X-rays by a target atom provides useful information for the qualitative and quantitative analysis. The absorption edges are characteristic of an element because each element has its own characteristic set of *K*, *L*, etc. absorption edges. The wavelengths corresponding to the absorption edges, thus, can be used to identify the element present in a sample. The absorption of a part of the incident beam of X-rays as it passes through the sample leads to the diminution of the intensity of the transmitted beam that reaches the detector. The attenuation of the intensity can be related to the amount of the target element in the sample by an exponential expression similar to the Beer–Lambert equation as

$$I_t = I_o e^{(-\mu\rho d)} \tag{11.6}$$

where I_o is the intensity of the incident beam of monochromatic X-rays and I_t is the intensity of the transmitted beam after passing a distance of *d* cm through the sample and density ρ (g cm^{-3}) and μ is the *mass absorption coefficient*, the units being cm^2 g^{-1}. The mass absorption coefficient depends on the atomic number of the element and the wavelength of the incident radiation only. It is independent of the physical and chemical states of the sample. Thus, for example, the mass absorption coefficient value for bromine has the same value in gaseous HBr and the solid sodium bromide. In a compound or mixture containing different elements the total mass absorption coefficient of the sample is the sum of the mass absorption coefficients of the constituent elements μ_1, μ_2, μ_3, etc. of weight fractions W_1, W_2, W_3, etc. respectively and is given as

$$\mu_{total} = \mu_1 W_1 + \mu_2 W_2 + \mu_3 W_3 + \cdots \tag{11.7}$$

11.4.2 X-ray Absorption Spectrometer

A commonly used X-ray absorption spectrometer is a non-dispersive X-ray absorptiometer consisting of an X-ray tube with tungsten target operating at 15–45 kV, a motor driven chopper to interrupt one-half of the X-ray beam alternately, a variable thickness aluminum attenuator to vary the intensity of the reference beam, reference and sample cells for handling solid, liquid and gases, and a phosphor-coated photomultiplier tube detector. The polychromatic radiation generated by the X-ray tube generator is used as such.

In practice the attenuator is adjusted so that the absorption of the incident beams of X-rays on reference and sample cells are balanced. The intensities of the incident and transmitted beams are measured by the detector for use in qualitative and quantitative analyses of the sample.

11.4.3 Applications of X-ray Absorption Spectrometry

X-ray absorption spectrometry finds analytical applications particularly in process control when the sample composition remains constant except for one analyte element of interest of variable amount. Examples include determination of the amount of lead added in the form of tetraethyl lead to gasoline, sulphur content in crude petroleum, chlorine content in plastics, lead or barium content in special glass, etc. If the sample composition is not known the measurement of the intensities of two monochromatic X-ray lines one on either side of the characteristic absorption edge of the analyte element is useful. In multi-element samples quantitative analysis is based on the additive absorption of different elements as given by Eq. (11.7). Mass absorption coefficients of different elements at different wavelengths are available in literature.

Microradiography is based on the different absorbing powers of incident X-ray beam by different elements in a sample. The technique is useful in determining the gross structure of small specimens. The X-ray beam passes through the sample and finally impinges on a photographic film placed behind the sample. Positions of elements that strongly absorb the X-rays appear light and positions of elements that do not absorb X-rays appear light in the photographic film. Even biological samples can be studied by impregnating the sample with a high molecular weight substance.

11.5 X-RAY FLUORESCENCE SPECTROSCOPY

When a sample is irradiated with a beam of short-wavelength X-rays, the atoms in the sample get excited and emit characteristic fluorescent X-rays. The wavelengths corresponding to the emission are useful in the identification of the elements present in the sample while the intensities of emission lines are useful in quantitative analysis.

11.5.1 Fluorescence Emission of X-rays

Fluorescence emission of X-rays occurs when a sample is exposed to an intense beam of primary X-rays. A part of the incident beam of primary X-rays is absorbed by the atoms of a target element in the sample resulting in the ejection of K or L electrons from the atoms. The excited state atoms formed may relax to the ground state either by a non-radiative process or radiative process. The non-radiative process is called *Auger effect* in which valence shell electrons are ejected from the excited state atom. Alternatively the excited atoms of the target element relax to the ground state by losing their excess energy in the form secondary X-ray photons as electrons from the higher orbitals drop into the K or L level. The phenomenon of emission of secondary X-rays is called *X-ray fluorescence* (XRF). Figure 11.6 shows a schematic representation of X-ray fluorescence emission process.

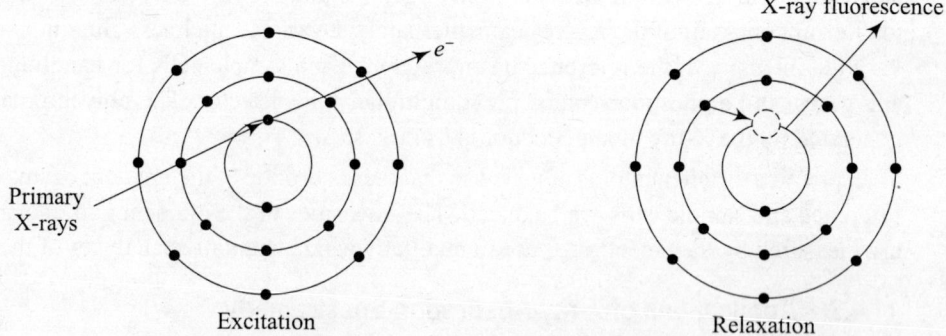

Fig. 11.6 Schematic representation of X-fluorescence emission process

Only a proportion of the excited state atoms emit the X-ray fluorescence. The *fluorescence yield factor* φ is given as the ratio of the intensity of the fluorescent emission I_F to the intensity of the primary radiation absorbed I_A.

$$\varphi = I_F/I_A \tag{11.8}$$

The magnitude of φ is 0.5 for heavy elements but is small (~ 0.01) for lighter elements with atomic number less than 15.

The intensity of the X-ray fluorescence emission I_F depends on the (i) concentration or weight fraction of the analyte element in the sample, (ii) mass absorption coefficients of elements other than the analyte, generally referred to as matrix elements, for the primary (incident) X-rays, and (iii) mass absorption coefficient of the matrix elements for the fluorescence emitted by the analyte. The magnitude of I_F is related to the intensity of the incident or primary X-rays I_0 as given by the expression

$$I_F = I_0 \, (\varphi \mu_X W_X)/(\mu + \mu_f) \tag{11.9}$$

where φ is fluorescence yield factor, μ_X is the mass absorption coefficient of the analyte element and W_X its weight fraction in the sample, μ is the combined mass absorption coefficients of the matrix elements for primary X-rays, and μ_f is the combined mass absorption coefficients for the fluorescence emitted by the analyte element.

X-ray fluorescence spectroscopy involves the analysis of the emitted X-ray fluorescence from the target and is an important analytical technique of greater importance as compared to X-ray absorption spectroscopy.

11.5.2　X-ray Fluorescence Spectrometer

The instrument for detecting and measuring X-ray fluorescence consists of the major components of (i) X-ray tube to generate primary X-rays, (ii) sample holder, (iii) collimators, (iv) analyser or diffracting crystal mounted on the goniometer, and (v) detector (vide Fig. 11.7).

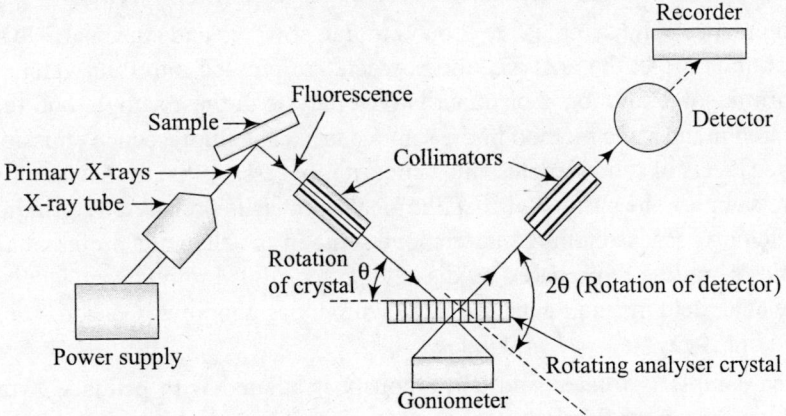

Fig. 11.7　Schematic diagram of an X-ray fluorescence spectrometer

The X-ray tube generates primary X-rays as a continuous or characteristic radiation of the primary target by the application of high voltage as governed by Eq. (11.2). The wavelength of the primary radiation should be shorter than the absorption edge of the spectral lines desired. For qualitative analyses of multi-element samples, the X-ray tube is operated at the highest possible voltage so that largest possible number of elements in the sample will be excited and can be identified through their characteristic fluorescence emission. However, the applied voltage can be much less than the highest possible voltage for selective excitation of certain elements or when only longer wavelength emission lines from the sample are to be analysed.

The primary X-ray source may be a radioisotope (e.g., ^{55}Fe, ^{241}Am, or ^{238}Pu) instead of an X-ray tube. Though the range of energies available in radioisotopes is limited, they have the advantage of independence from electrical power. Radioisotopes as source of primary X-rays are suitable for portable X-ray fluorescence instruments as such instruments are relatively simple in design and smaller in size.

X-ray fluorescence emission can also be obtained by bombarding the sample with (i) electrons as in *scanning electron microscope* (SEM), *electron probe* (EP) and in *transmission electron microscope* (TEM); (ii) monochromatic X-rays emitted by certain radioisotopes (e.g., ^{125}I); or (iii) protons generated by ion accelerators as in *particle induced X-ray emission* (PIXE).

The sample holder is an aluminum cylinder or a plastic container to hold acidic or alkaline solutions. The sample held in the sample holder is usually rotated to improve the uniformity of exposure to primary X-rays. A thin film of Mylar is used to support the specimen and an aluminum mask is used to restrict the exposure to a rectangular area of about 18 mm × 27 mm.

A portion of the fluorescence emitted by the elements in the sample is collimated onto the entrance slit of the goniometer and focused onto the plane surface of the diffracting crystal, a flat single crystal plate of size 2.5 cm × 7.5 cm. The incident radiation is reflected satisfying the Bragg condition and the reflected radiation passes through the exit slit or collimator of the goniometer to the detector. The detector converts the energy of the X-rays into electrical impulses or counts. The entrance slit, the diffracting crystal, and the exit slit of the goniometer assembly are placed on the focal circle so as to satisfy Bragg's condition throughout the rotational movement of the goniometer. The detector is rotated at twice the angular rate of the analysing crystal by the goniometer. The goniometer chamber is evacuated or purged with helium gas to reduce the loss in intensity of fluorescence emission due to absorption by air.

In practice solid samples are ground to fine powders and since particle size and particle size distribution affect the analysis, the powders are pressed into thin wafers or fused with borax to form a solid solution. For quantitative analysis the test sample and the standards must be prepared in the same method in the same matrix. The fluorescence emission is mostly from the surface layers of solid samples, and hence, it is necessary that the surface is representative of the entire sample. The surface of metallurgical samples is prepared by grinding and polishing but not etching because etching may remove some of the elements preferentially from the surface. Liquids samples are prepared by dissolving the solids preferably in solvents, such as water or nitric acid, held in small cups provided with mylar windows. Gaseous samples require special cells with X-ray transparent windows.

The sample is rotated and continuously irradiated with primary X-rays while gradually increasing the angle θ between the surface of the diffracting crystal and the incident fluorescence beam while the detector rotates through an angle of 2θ. The incident fluorescence beam is reflected at certain angles and the detector measures the counts. The analytical data is displayed in the form of a graphical plot consisting of peaks of varying intensity as a function of goniometer reading (2θ). The goniometer readings can be converted into wavelengths of the emission peaks from the known value of the inter-planar spacing of the diffracting crystal (using Bragg equation).

Qualitative identification of elements present in the sample is based on the characteristic fluorescence emission lines appearing at specific wavelengths.

Quantitative analysis of elements is based on the measured intensity of fluorescence emission at characteristic wavelength of the analyte. As expressed by Eq. (11.9) the intensity depends on the analyte concentration as well as the matrix. Several methods are in vogue for reducing the matrix effects, such as (i) comparison of the intensities of the test sample and standards of similar composition, (ii) dilution of standards and the test sample with a common solvent so the weight fraction becomes small enough to make the intensity–concentration relationship linear, (iii) use of an internal standard in which an element of one atomic number higher or lower than the analyte element is added, and (iv) standard addition method in which the same element as the analyte is added as the standard and the intensity of the test sample is measured before and after the addition.

11.5.3   Applications of X-ray Fluorescence Spectroscopy

XRF spectroscopy finds applications in the qualitative and quantitative analyses of elements in a wide variety of samples such as ores, minerals, metals and alloys, metallurgical samples, films, coatings, ceramics, plastics, etc. The method is non-destructive, and hence, the samples can be used for further analyses.

11.6   X-RAY EMISSION AND ELECTRON PROBE MICROANALYSIS

X-ray emission and electron probe microanalysis are called direct X-ray methods and involve the exciting atoms of a sample with a beam of high energy electrons. The excited atoms relax to the ground state by emitting characteristic X-ray emission lines. Hafnium was discovered by X-ray emission method. However, this method has a major disadvantage in that the sample must be coated or smeared on the target anode of the X-ray tube. This requires a demountable target anode and re-evacuation of the X-ray tube every time a new sample is introduced. Other disadvantages include heating of the sample by the electron beam may cause volatilization, melting, or chemical reaction. Hence, X-ray emission spectroscopy is virtually practiced only as electron probe microanalysis for the study of solid surfaces.

Electron probe microanalysis (EPMA) is a direct X-ray method in which the solid sample is used as a target for bombarding with a beam of electrons and the primary X-rays generated are analysed. A very small area (~1 μm in diameter) of the solid surface is exposed to a collimated beam of electrons of 10–50 keV energy. The electron beam penetrates to a depth of only 1–2 μm into the sample, and hence, the method is essentially a surface analytical method. The emitted primary X-rays have intensities higher than that of fluorescent emission, and hence, detection limit is quite low of the order of 10^{-14} g in the small area of the sample.

A schematic diagram of an electron probe microanalyser is shown in Fig. 11.8. The instrument consists of (i) an electron gun for producing a stable electron beam of 1 μm in diameter, (ii) light optical system to view the microscopic area under study, (iii) a movable sample stage for moving the sample accurately under the electron beam, and (iv) a curved crystal X-ray spectrometer to analyse the characteristic wavelengths of X-rays generated from the sample. The technique is also known as *wavelength dispersive X-ray analysis* (WDX).

Electron microprobe analysis is also done by scanning electron microscope (SEM) in which the sample image is obtained from backscattered or transmitted electrons and elemental composition is determined by *energy dispersive X-ray analysis* (EDX analysis or EDAX). In EDAX the sample is bombarded with a beam of electrons inside the SEM. The incident electrons knock off some of the inner shell electrons of the atoms of the sample. The vacancy created in the inner shell of an atom is filled by a higher energy electron from an outer shell with a simultaneous release of excess energy in the form of an X-ray. The kinetic energy of the X-ray photon is measured by solid state lithium-drifted detector. The energy of the X-ray photon is characteristic of the element, and hence, useful in identifying the element. The EDX spectrum is a plot of the intensity of X-rays emitted as a function of their energies. The peak positions are useful in qualitative analysis while the peak heights are used in quantitative analysis.

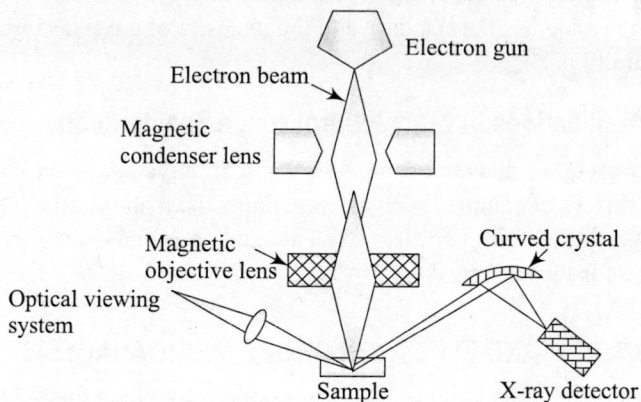

Fig. 11.8 Schematic diagram of an electron probe microanalyser

11.7 X-RAY DIFFRACTION METHODS

X-ray diffraction technique is used for the determination of crystal structures and also to identify the phases and determine the phase purity of polycrystalline materials and rarely for quantitative analyses. The basic principle involved is based on the fact that each and every atom in a crystalline solid scatters the incident X-rays in all directions. The scattering power of an atom depends on its number of electrons (atomic number), the greater the atomic number greater is its scattering power. Since a crystal is made of a regular and orderly arrangement of atoms in a repetitive manner diffraction of the incident X-ray beam through constructive interference becomes possible when the Bragg condition given by Eq. (11.5) is satisfied. The angle of diffraction of a monochromatic incident beam of X-rays depends on the size and shape of basic repetitive unit of the crystal called the *unit cell*. The intensity of the diffracted beam depends on the type of atoms and their relative positions in the crystal. Constructive interference occurs when diffraction of the incident beam occurs from atoms positioned exactly on the crystal planes resulting in maximum intensity of the diffracted beam. In contrast, the intensity of the diffracted beam is negligible due to destructive interference when diffraction occurs from atoms located halfway between the planes. Other atoms located at various other positions give rise to either constructive or destructive interference, and hence, the intensity of the diffracted beam varies. The combination of the angle of diffraction of the incident X-ray beam and the intensity of the diffracted beam, called the *diffraction pattern*, is unique for a given crystal because no two crystals have identical unit cells or types of atoms and their positions. Thus, the diffraction pattern serves as a fingerprint of a crystalline compound and highly useful for identification of the compound.

Two diffraction methods find extensive use in the study of crystalline substances by X-ray diffraction. The *rotating crystal X-ray diffraction method* is exclusively used for crystal structure determination by diffracting a monochromatic beam of X-rays on a well-grown single crystal of the sample. The diffraction pattern provides information on the types and positions of the atoms within the unit cell facilitating the determination of the structure. The *powder X-ray diffraction method* is used to identify the phases present in a powder sample which is essentially polycrystalline. The method is useful for substances which cannot be obtained as single crystals.

● ● ● ● ● ● ● ● ● ● ● ● ● ● ● ● ● ● **SOLVED PROBLEMS**

EXAMPLE 1 *Calculate the minimum applied voltage required to obtain the characteristic K_α X-ray emission lines from copper, molybdenum, and chromium targets.*

SOLUTION

From Eq. (11.2)

$$V = 12,398/\lambda_o$$

Substituting the values of λ_o for copper as 1.541 Å, for molybdenum 0.709 Å, and chromium as 2.290 Å the applied voltages are 8045.4 V for copper 17481 V for molybdenum and 5414 V for chromium respectively.

EXAMPLE 2 *Calculate the inter-planar distance in a diffracting crystal if the K_α line of chromium is 2.287 Å is observed on diffraction by the crystal at $2\theta = 56.24°$.*

SOLUTION

Using Eq. (11.5)

$$d = 2.287 \text{ Å}/2 \sin 28.12° = 2.246 \text{ Å}$$

EXAMPLE 3 *Calculate the X-ray fluorescence emission wavelength which appears as a strong peak at $2\theta = 46.82°$ on being diffracted by the analyser crystal of having a d spacing of 2.24 Å.*

SOLUTION

Using Eq. (11.5)

$$\lambda = 2 \times 2.24 \text{ Å} \times \sin 23.41° = 1.779 \text{ Å}$$

EXAMPLE 4 *The lead content of a leaded petrol sample of density 0.75 g cm^{-3} and an average molecular weight of 114 corresponding to C-8 alkane was determined by X-ray absorption spectroscopy using an incident X-ray beam of Cu K_α line and a sample cell of 0.5 cm thick. Calculate the per cent of lead in the leaded petrol if the 82% of incident radiation was absorbed by the sample given the mass absorption coefficients for the Cu K_α line for lead, carbon, and hydrogen are 230, 4.52, and 0.48 cm^2 g^{-1} respectively.*

SOLUTION

Weight fraction of an element in a compound W_X

= atomic weight/molecular weight of compound

Weight fractions of C and H in C-8 alkane (C_8H_{18})

$$W_C = 8 \times 12.01/114 = 0.842; \quad W_H = 18 \times 1.008/114 = 0.159$$

Mass absorption coefficients for petrol corresponding to C-8 alkane

$$\mu_{(C\text{-}8 \text{ alkane})} = [(4.52 \times 0.842) + (0.48 \times 0.159)]$$
$$= 3.88 \text{ cm}^2 \text{ g}^{-1}$$

Re-arranging Eq. (11.6), $I_t = I_o \exp^{(-\mu\rho d)}$

$$\log (I_o/I_t) = \mu\rho d$$
$$\log (100/18) = (\mu_{(sample)} \times 0.75 \text{ g cm}^{-3} \times 0.25 \text{cm})/2.303$$
$$\mu_{(sample)} = 9.152 \text{ cm}^2 \text{ g}^{-1}$$

Weight fraction of lead in petrol = $W \mu_{(lead)} + (1 - W) \mu_{(C8 \text{ alkane})}$

$$9.152 \text{ cm}^2 \text{ g}^{-1} = W \times 230 + (1 - W) \times 3.88$$
$$W = 0.0233$$

The amount of lead in petrol = 2.33%

SUMMARY

- X-rays are obtained most commonly by exposing a target metal to a beam of high energy electrons in an X-ray generator. Alternatively, X-rays are obtained by exposing the target to primary X-rays or from a radioactive source.

- X-ray emission accompanies electronic transitions that occur from a higher electronic energy level to the lower energy K or L shell of an excited atom.

- X-ray spectroscopic methods involve absorption of incident beam of X-rays as in X-ray absorption spectroscopy, emission of characteristic X-rays as in emission spectroscopy and electron microprobe analysis, and emission of X-ray fluorescence and X-ray diffraction.

- In X-ray absorption spectroscopy the wavelength at the absorption edge is characteristic of an element useful for qualitative identification, whereas the decrease in intensity of the incident X-rays is proportional to the amount of the element in the sample.

- X-ray fluorescence spectroscopy is a highly sensitive analytical technique in which the sample on exposure to a primary beam of X-rays emits characteristic fluorescence. The wavelength of fluorescence emission is useful in qualitative analysis, whereas the intensity of emission depends on the amount of the element in the sample.

- Electron probe microanalysis involves bombarding the sample with high energy electrons and analysing the characteristic X-rays emitted by the sample. The method is suitable to study the surface composition.

- X-ray diffraction is primarily used to determine the internal structure of crystalline solids.

REVIEW QUESTIONS

1. List the main components of X-ray spectroscopic instruments.

2. How are X-rays produced in a generator?

3. Give a detailed account of the characteristic features of the X-ray spectrum of a target element.

4. Distinguish between X-ray continuous and line spectra. How are characteristic X-ray emission lines generated?

5. Describe the functions of filters and monochromators in X-ray spectroscopic instruments.

6. What are the different types of detectors used for detection of X-rays?

7. Give a detailed account of the functional aspects of scintillation counter, GM tube detector, and proportional counter.

8. Explain the term absorption edge in the X-ray spectrum of an element. What is its significance?

9. Discuss the principle involved in X-ray absorption spectroscopy.

10. Distinguish between X-ray emission and X-ray fluorescence spectroscopic techniques.

11. Discuss the principle and practice of electron probe microanalysis.

12. Give a detailed account of XRF.

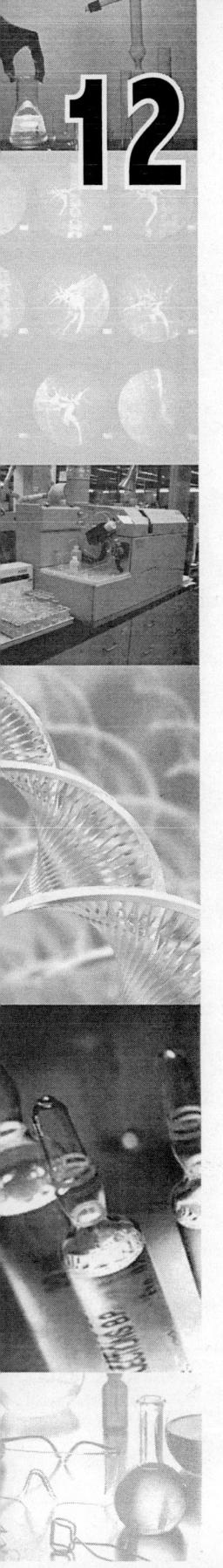

12 Separation Methods

12.1 AN OVERVIEW OF SEPARATION METHODS

Analysis of substances, whether by qualitative or quantitative methods, requires the analyte to be of 100% purity. However, in practical situations, the analyte is always a part of a complex substance or mixture and it is absolutely essential to isolate and purify the analyte for further identification and quantitative estimation. Hence, separation methods are inherently involved in any analytical procedure. Similarly in manufacturing industries, purification of raw materials, intermediate products or finished products means separation and removal of impurities, contaminants or by-products as the critical processing step. Thus, separation methods are of significance both for analytical and industrial production purposes.

The separation methods may broadly be classified into two main groups: (i) *bulk or large-scale separations* and (ii) *analytical separations*.

Bulk separations or *large-scale separations* handle quantities in the range of milligram to grams in analytical and research laboratories and kilogram quantities in process industries. These techniques include solvent extraction, ion exchange, membrane separation techniques, crystallization, precipitation, lyophilization, and solid–liquid separation techniques of filtration and centrifugation. It must be emphasized here that though the various methods have been classified under large-scale methods, they are still amenable and used for small- or laboratory-scale operations for both preparative and analytical applications. Of these solvent extraction and ion exchange separations are most widely used both in analytical laboratories and in industrial-scale operations, and hence, discussed in more detail in this chapter. The other separation techniques such as membrane separation, crystallization, precipitation, and lyophilization are mostly preparative separations and an overview of principles of these techniques have been included in this chapter. Centrifugation and ultracentrifugation are dealt with in a separate chapter (Chapter 15) with applications oriented towards biotechnology.

Analytical separations as the name implies are used for analytical purposes often involving amounts of substances in micrograms or even less. The separated components of the mixture are used for qualitative identification and quantitative estimation only and the samples are not collected. Most of the methods under this group are chromatographic methods making use of sophisticated instruments such gas chromatograph (GC), high-performance liquid chromatograph (HPLC), and high-performance thin layer chromatograph (HPTLC). Electrophoresis and related techniques are important analytical separation techniques mostly used by biochemists and biotechnologists. Most of these methods are based on selective and sometimes specific interactions between the analyte molecules and tailor made separating media.

These instrumental methods based on chromatography have been included in Chapter 13 and those on electrophoresis and related techniques in Chapter 14.

The separation methods may be based on (i) differences in the diffusion/transport characteristics of the individual components of the mixtures through a separating medium as in filtration and membrane separation methods, (ii) differences in solubility of the components in two immiscible solvents in contact with each other as in solvent extraction, (iii) specific chemical interaction of the desired component (or the contaminant) to form a solid phase involving precipitation, (iv) crystallization from supersaturated solutions, and (v) density differences between solid and liquid components exploited in centrifugation. Precipitation and crystallization result in the formation of a solid which needs to be separated from the mother liquor by solid–liquid separation techniques such as centrifugation and filtration.

12.2 SOLVENT EXTRACTION

The general technique of solvent extraction involves the extraction of analyte(s) into a liquid phase by the use of a suitable solvent. When the analyte(s) is extracted from a solid phase the technique is appropriately called *solid–liquid extraction*. On the other hand, if the analyte is present in a liquid phase from which it is extracted into another immiscible liquid phase the technique is known as *liquid–liquid extraction* or simply *solvent extraction*. When the extracting solvent is a supercritical fluid the extraction is called *supercritical fluid extraction*.

Solid–liquid extraction is practiced in industry for extraction of edible oils such as sunflower seed oil, sesame seed oil, peanut oil, etc., from oil seed meal, and essential oils such as jasmine oil, turmeric oil, etc., from plant sources, in the extraction of beverages such as coffee, tea, etc. In the analytical laboratory, a Soxhlet apparatus is used to extract and determine the soluble solid/liquid component of the analyte sample.

Solid–liquid extraction is based on the solubility of the desired component of the solid in a chosen solvent. The extracting solvent is contacted with finely divided solid to solubilize the desired component. An important parameter influencing the extraction process apart from the nature of the solvent itself is temperature, higher temperature generally enhancing the efficiency of extraction.

Solvent extraction or liquid–liquid extraction in widely used both for industrial and for analytical applications. In the industry the technique finds use for bulk separation of antibiotics and organic acids while in the analytical laboratory it is used for rapid and clean separation of organic and inorganic analytes. Liquid–liquid extraction involves the partitioning or distribution of the sample analyte (or solute) between two immiscible liquid phases. A simple apparatus called the separating funnel is used for extraction in the laboratory. One of the two liquid phases is mostly aqueous (pure water or buffered solutions) while the extracting liquid phase is an organic solvent such as aliphatic hydrocarbons (e.g., *n*-hexane, isooctane, etc.), aromatic hydrocarbons (e.g., toluene, xylenes), chlorinated hydrocarbons (e.g., dichloromethane, chloroform), and water immiscible alcohols, esters, ketones, and ethers.

12.2.1 Principles of Liquid–Liquid Extraction

An important requirement for extracting a substance into a non-polar organic phase is that the substance should exist or form a neutral species. The absence of electrostatic interactions between a neutral solute and water lowers the solubility of the substance in the aqueous solvent.

Similarly, extraction into the organic phase is enhanced if the substance is not hydrated or when the coordinated water is easily displaced by bulky hydrophobic organic molecules.

The principle of solvent extraction is based on partition or distribution of the analyte between two liquid phases in contact with each other. At equilibrium the ratio of the activities a_O (activity in the organic phase) and a_A (activity in the aqueous phase) of the analyte will be a thermodynamic equilibrium constant as given by

$$K = \frac{a_O}{a_A} \tag{12.1}$$

where K is called the *partition* or *distribution coefficient*. Since the analyte may be present in each phase in more than one form such as dissociated, associated, or complex species, a more useful relationship involves the *distribution ratio D*. According to *Nernst partition* or *distribution law*, the distribution ratio is the ratio of the concentrations of all the species of the analyte in the two phases given as

$$D = \frac{C_O}{C_A} \tag{12.2}$$

where C_O and C_A represent the total concentrations of all the species of the analyte in the organic and aqueous phases respectively. The value of D depends on the nature of the solute, solvents used, and temperature but independent of the volume ratio of the two solvents. However, the fraction of solute extracted depends on the volume ratio of the two solvents. If W_A g of the solute dissolved in V_A mL of aqueous medium is extracted n times with V_O mL portions of organic solvent then the amount of solute remaining in the aqueous medium W_n g may be calculated by the formula

$$W_n = W_A \left[\left(\frac{V_A}{(D V_O + V_A)} \right) \right]^n \tag{12.3}$$

The amount of the solute extracted after n times in the organic solvent will be $(W_A - W_n)$ g.

It is always advantageous to extract with small portions of organic solvent more number of times as compared to extraction with a single large volume portion.

The percentage extraction (E) is given by

$$E = 100 \frac{D}{\left[D + \left(\frac{V_A}{V_O} \right) \right]} \tag{12.4}$$

When $V_A = V_O$

$$D = (V_A/V_O) E/(100 - E) \text{ or } D = E/(100 - E) \tag{12.5}$$

12.2.2 Selectivity of Extraction

When two (or more) solutes are present in the aqueous medium both the solutes will be extracted into the extracting solvent. The ability to separate the two solutes depends on the relative magnitudes of the two distribution ratios. The *separation factor* or *coefficient* β is given as

$$\beta = \frac{[X_o]/[X_A]}{[Y_o]/[Y_A]} = \frac{D_X}{D_Y} \tag{12.6}$$

where X_o and X_A are the analytical concentrations of the solute X in the organic and aqueous phases respectively. Similarly, Y_o and Y_A are the concentrations of the solute Y in the organic and aqueous phases and D_X and D_Y represent the distribution ratios of the two solutes X and Y respectively. A value of $\beta = 1$ for a given pair of solutes in a chosen organic (extracting) solvent will not result in the separation of the two solutes as both will get extracted into the organic solvent to the same extent. The best separation of the solutes will be possible only when the values of D_X and D_Y are widely different. For example, if D_X is very large as compared to D_Y the solute X will be extracted into the organic solvent leaving the solute Y in the aqueous phase in a single extraction itself. Hence, knowledge of the distribution coefficients of the desired analyte in different aqueous–organic solvent systems is helpful for separating the desired analyte from contaminants.

The separation of two solutes by extraction can be achieved to some extent by adjusting the ratio of the volumes of the organic and aqueous phases to an optimum value as given by Bush–Densen equation

$$\frac{V_o}{V_A} = \frac{1}{(D_X D_Y)^{1/2}}$$ (12.7)

12.2.3 Parameters Affecting the Extraction Process

The parameters that affect the extraction process include (i) partition coefficient, (ii) the pH of the aqueous phase, (iii) association, (v) ion-pair formation, and (v) complex formation. Knowledge of the magnitude of the partition coefficient for a given solute in a given extracting solvent–water system is essential for selecting the extracting solvent by *physical extraction* of the solute.

The pH of the aqueous phase is quite important in the extraction of weak organic acids and bases and by controlling the pH it is possible to enhance the efficiency of extraction which is generally known as *dissociative extraction*.

Association of the solute molecules to form dimers, ion pair formation, and formation of complexes by reagents dissolved in the extracting solvent to enhance the efficiency of extraction is known as *reactive extraction* or *ion pair extraction*.

12.2.4 Extraction Methods

Based on the various factors which affect the extraction process the extraction methods may be classified into (i) physical extraction, (ii) dissociative extraction, and (iii) reactive extraction.

Physical extraction involves preferential dissolution of the desired analyte in a chosen organic solvent. The selection of the organic solvent for analytical extractions depends on the distribution coefficient. In large-scale extractions other factors, such as density, viscosity, recoverability, cost and interfacial tension between the aqueous and organic phases, also have to be taken into account. For analytical separations it is theoretically possible to recover maximum amount of the analyte based on the knowledge of distribution coefficients in different aqueous–organic solvent systems. However, the availability of organic solvent satisfying the requirement of high distribution coefficient value for the analyte is rather limited.

Dissociative extraction involves the modification of the physical properties of the analyte to enhance the partition coefficient in the chosen aqueous–organic solvent system and is particularly applicable to solute molecules which undergo dissociation such as organic weak acids or bases.

The partition coefficient values of such solutes are easily enhanced by adjusting the pH of the aqueous phase. For example, in the extraction of organic acids adjustment of pH of the aqueous phase to a value below the pK_a (negative logarithm of the dissociation constant of the acid) ensures the existence of the acid in the undissociated protonated form and facilitates its extraction into the organic phase. The neutral solute molecules are easily solvated by organic solvents through weak hydrophobic interactions.

Thus, for a solute which is a weak acid the distribution ratio D may be expressed as

$$D = \frac{K_i}{\left(1 + \frac{K_a}{[H^+]}\right)} \text{ or } \log_{10}\left[\left(\frac{K_i}{D}\right) - 1\right] = \text{pH} - pK_a \tag{12.8}$$

where the intrinsic partition coefficient $K_i = [HA]_O/[HA]_A$ and the acid dissociation constant $K_a = [H^+][A^-]/[HA]$. The magnitude of K_i is independent of K_a or pH but the value of K is dependent. In strongly acidic solutions $[H^+] \gg K_a$ and therefore $D \sim K_i$, that is, all of the acid exists as unionized HA.

Similarly for solutes, which are weak bases the partition coefficient K may be given as

$$\log_{10}\left[\frac{K_i}{D} - 1\right] = pK_b - \text{pH} \tag{12.9}$$

The changes in partition coefficients due to changes in pH may be used to isolate as well as purify a desired solute. The selectivity of separation, β, between two weak acid solutes A and B is given by

$$\beta = \frac{K_i(A)}{K_i(B)} \frac{1 + \frac{K_a(B)}{H^+}}{1 + \frac{K_a(A)}{H^+}} \tag{12.10}$$

Extraction of penicillin from the aqueous fermentation liquor into butyl acetate is feasible by dissociative extraction. The pH of the aqueous phase is adjusted to 2.0–2.5 by the addition of sulphuric acid so that undissociated form of the antibiotic is the predominant species since its pK_a is 2.75. However, at such a low pH the antibiotic loses its activity over a period of time, the half-life for deactivation being only about 20–25 minutes and extraction process is hindered because of the formation of emulsion. These problems are overcome by enhancing the rate of extraction through the use of centrifugal contactors which give a short contact time and facilitate rapid phase separation. The penicillin in the organic phase is stabilized and recovered by re-extracting (stripping) into the aqueous phase by controlling the pH of the aqueous buffer at 6.0.

The separation of penicillin F from penicillin K is also achieved by dissociative extraction in amyl acetate–water system as the two forms have different pK_a values of 3.51 and 2.77 respectively.

Reactive extraction also called *ion pair extraction* involves the formation of an organic solvent soluble complex of the desired solute. A typical example of reactive extraction involves the use of amine extractants, such as long chain aliphatic amines (e.g., tri-octylamine and di-octylamine), for the recovery of organic acids such as citric acid and antibiotics such as penicillin from fermentation broths.

Reactive extraction of solutes may be due to (i) association as in the case of carboxylic acids, (ii) ion pair formation, and (iii) complex formation. The last two are specifically applicable to extraction of metals.

Association of the molecules of the substance in the organic phase increases the magnitude of the distribution ratio D as in the case of carboxylic acids which form dimers in organic solvents of low polarity such as carbon tetrachloride or benzene. The formation of a dimer may be represented by the equilibrium

$$2 \text{ RCOOH} \rightleftarrows (\text{RCOOH})_2 \; ; K_{\text{dimer}} = \frac{[(\text{RCOOH})_2]_O}{[\text{RCOOH}]_O^2} \tag{12.11}$$

where K_{dimer} is the constant for dimerization in the organic phase. The distribution ratio D for the organic solvent–aqueous system is given by

$$D = [\text{RCOOH}]_O + 2\,[(\text{RCOOH})_2]_O/[\text{RCOOH}]_A \tag{12.12}$$

Substituting for $[(\text{RCOOH})_2]_O$ in the above Eq. (12.12) and rearranging gives

$$D = [\text{RCOOH}]_O + 2\,K_{\text{dimer}}\,[\text{RCOOH}]_O^2/[\text{RCOOH}]_A \tag{12.13}$$

or

$$D = K\,(1 + 2\,K_{\text{dimer}}\,[\text{RCOOH}]_O) \tag{12.14}$$

where K is the partition coefficient as given by Eq. (12.1).

When the substance shows greater tendency to dimerize particularly at low pH, the values of K_{dimer} and D will become large and the efficiency of extraction increases.

Ion pair formation involves the extraction of bulky cations and anions as pairs or aggregates. The bulky pairs do not have primary hydration shell, and hence, cause disruption of the hydrogen bonded water structure exhibiting the tendency to dissolve in the organic phase. Greater the size of the ion pair greater is the extent of disruption of the water structure and also the tendency of the ion-pair to dissolve in the organic phase. For example, the bulky cation tetraphenyl arsonium $[(C_6H_5)_4As]^+$ forms ion pairs with bulky anions, such as permanganate $[MnO_4]^-$ and perrhenate $[ReO_4]^-$ and facilitates their extraction into chloroform.

$$[(C_6H_5)_4As]^+Cl^- + [MO_4]^- \rightarrow (C_6H_5)_4As^+MO_4^- + Cl^-$$
$$(M = \text{Mn or Re}) \qquad \text{ion pair}$$

Similarly, tetrabutyl ammonium $[(C_4H_9)_4N]^+$ ion forms ion pairs with anionic complexes of $ZnCl_4^{2-}$, $GaCl_4^-$, and $Co(CN)_6^{3-}$ enhancing their solubility in organic solvents.

Long chain alkyl ammonium salts (e.g., tricaprylmethyl ammonium chloride and tri-iso-octylamine hydrochloride, TIOA) are liquids. These salts are referred to as *liquid anion exchangers* as they are capable of forming extractable ion pairs with anionic metal complexes.

Ion pair formation involving these bulky cations, however, is not specific as they form ion pair with any large anion. Polyvalent ions have greater hydration energy, and hence, their extraction into organic phase is not that easy.

Complex formation is an important factor which enables extraction of inorganic species (e.g., metal ions and anions) into an organic phase. The methodology involves the formation of (i) neutral chelate complexes with ligands containing hydrophobic groups or (ii) ion-association complexes. In either case the changes in hydration sphere of the metal or anion enhance the solubility of the species in organic solvents.

Neutral chelate complex formation involves the reaction of the hydrated metal ion with the anion of the chelating ligand which may be represented as

$$M(H_2O)_x^{n+} + nR^- \rightleftarrows MR_n + xH_2O; \quad K_f = [MR_n]_A / ([M(H_2O)_x^{n+}]_A [R^-]_A^n) \quad (12.15)$$

where K_f refers to the formation constant of the chelate complex. The metal chelate complex distributes itself between the organic and aqueous phases according to Nernst law which may be given by the relationship

$$D = [MR_n]_O / [MR_n]_A + [M(H_2O)_x^{n+}]_A \quad (12.16)$$

Based on certain assumptions that the metal exists only as the chelate complex MR_n, that the organic reagent HR and the chelate complex MR_n exist only as undissociated molecules in the organic phase and the concentration of hydrated metal ion is negligible the above Eq. (12.16) can be written as

$$D = K^* [HR^n]_O [H^+]_A^{-n} \quad (\text{where } K^* = K_f K K_a^n / K_R^n) \quad (12.17)$$

where K_f is the formation constant of the chelate complex; $K = [MR_n]_O / [MR_n]_A$; K_a is the dissociation constant of the chelating ligand $[H^+]_A [R^-]_A / [HR]_A$; and $K_R = [HR]_O / [HR]_A$.

According to Eq. (12.17) the extraction of the metal chelate complex depends only the pH and the concentration of the reagent in the organic phase and is independent of the initial concentration of the metal. Under experimental conditions the concentration of the reagent in the organic phase is usually in large excess, hence, all the metal exists only as the chelate complex and the equation simplifies to

$$D = K^{**} [H^+]_A^{-n} \text{ or } \log D = \log K^{**} = n \, pH \quad (\text{where } K^{**} = K_f K K_a^n [HR]_O / K_R^n) \quad (12.18)$$

As per the equation the value of D depends only on the pH of the aqueous phase.

The distribution ratio D and percentage extracted E are related as per Eq. (12.5) when the volumes of the aqueous and the organic phases are equal. Hence,

$$\log D = \log E - \log (100 - E) = \log K^{**} + n \, pH \quad (12.19)$$

The above equation defines that the extraction of a metal by a chelating ligand depends only on the pH of the aqueous medium. When E is plotted against pH a sigmoid curve is obtained for a monovalent metal forming 1:1 chelate complex as shown in Fig. 12.1. The position of the curve along the pH axis depends only on the magnitude of K^{**}. Greater the value of K^{**} either due to a greater stability of the metal chelate or due to greater acidity of the chelating ligand, the lower will be the pH range in which the metal will be extracted. The slope of the sigmoid curve depends on the value of n as shown in Fig. 12.1 for metal chelate complexes where $n = 1, 2,$ and 3 respectively formed by monovalent, divalent, and trivalent metal ions with the monoanionic chelating ligand R^-.

The pH at which 50% of the metal is extracted ($E = 50\%$) is defined as $pH_{1/2}$ and given as

$$pH_{1/2} = - 1/n \, (\log K^{**}) \quad (12.20)$$

When two metal ions are to be extracted by the same chelating ligand, the separation factor β depends on the distribution ratios of the two metal ions as given by Eq. (12.6) ($\beta = D_X / D_Y$). When the metal ions have the same valence and form chelate complexes with the same value of n, the separation factor is given by the relationship

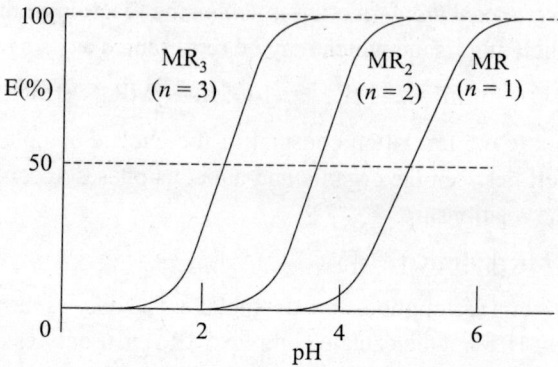

Fig. 12.1 Extraction of neutral metal chelate complexes of mono, di, and trivalent metal ions forming 1:1, 1:2, and 1:3 complexes respectively

$$\log \beta = n\Delta pH_{1/2} \tag{12.21}$$

As per Eq. (12.21) if the $pH_{1/2}$ values of the two metal chelate complexes are sufficiently apart, the metals can be separated efficiently by single extraction by proper control of the pH of extraction. For metal chelate complexes of 1:1 type for complete separation by single extraction the $\log \beta$ should be equal to 5 or more. Hence, the value of $\Delta pH_{1/2}$ should also be 5. For metal chelate complexes of 1:2 and 1:3 type $\Delta pH_{1/2}$ should be 2.5 and 1.7 respectively to achieve quantitative separation of the two metal ions in a single extraction.

The $pH_{1/2}$ values for extracting two metal ions by a chelating reagent may be altered by masking or sequestration. Masking is carried out by the addition of another chelating agent to the extracting system. For example, EDTA can be used as the masking agent for the separation of copper from mercury in a given aqueous solution by extraction with dithizone in carbon tetrachloride at pH 2. The masking agent forms a stable water-soluble complex with copper alone at pH 2 thereby facilitating the extraction of mercury alone into the organic phase.

Typical examples of chelating ligands for extracting metals include:

(i) 8-hydroxyquinoine (oxine) and its derivatives for extracting about 50 metals in total and particularly Al, Mg, Sr, V, and W,

(ii) acetylacetone for extracting about 50 metals in total and especially alkali metals, Be, Sn, Cr, Mn, and Mo,

(iii) dithizone (diphenyldithiocarbazone) for extracting Pb, Hg, Cu, Pd, Pt, Ag, Bi, and Zn,

(iv) dimethylglyoxime for extracting Ni and Pd,

(v) cupferron (ammonium salt of N-nitrosophenyl hydroxylamine) for extraction Fe(III), Ga, Sb(III), Ti(IV), Sn(IV), Hf, Zr, V(V), U(VI), and Mo(VI), and

(vi) 1-nitroso-2-naphtol for extracting Co(III), Cu, and Ni.

Ion-association complexes of metals are of three types formed by (i) non-chelating ligands, (ii) neutral chelating ligands, and (iii) oxonium systems and solvents. The first type of ion-association complexes are also called ion pairs formed by bulky hydrophobic anionic ligands as exemplified by tetraphenyl arsonium and tetrabutyl ammonium cations and liquid ion exchangers. The second type of ion-association complexes includes complexes of neutral chelating ligands such as o-phenanthroline, EDTA and oxine, and acidic alkyl phosphoric acid esters called *liquid cation exchangers*. These chelating ligands form cationic or anionic complexes with metals and require counter anions or cations to form ion-association complexes for extraction into organic solvents. The third type of ion-association complexes include oxygen containing solvents, such as alcohols, ethers, esters, and ketones, which directly participate in complex formation with metals. Typical examples include diethyl ether, methyl isobutyl ketone, and isoamyl acetate which form oxonium cations with protons under strongly acidic conditions. The oxonium cations form ion-association complexes with anionic metal complexes facilitating their extraction into the organic phase. For example, $FeCl_4^-$ is extracted into diethyl ether from 7 M hydrochloric acid solution as the solvent forms a stable ion-association complex. Another example is tributyl phosphate (TBP) containing the semipolar phosphoryl group ($-P^+-O^-$) of good coordinating ability which has been exploited for the extraction of uranyl nitrate to separate it from nuclear fission products. Macrocyclic ligands, such as Crown ethers containing 9–60 atoms, including 3–20 oxygen atoms in a ring, form stable ion-association complexes with metal ions particularly alkali metal ions. Typical examples of macrocyclic ligands include 18-crown-6, dibenzo-18-crown-6, and cryptand-222.

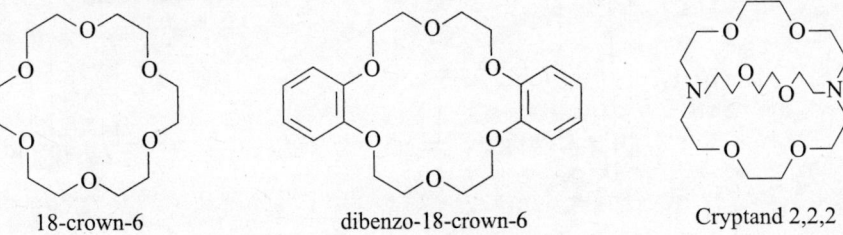

18-crown-6   dibenzo-18-crown-6   Cryptand 2,2,2

12.2.5 Modes of Extraction

Liquid–liquid extraction may be carried out in (i) batch mode, (ii) continuous mode, and (iii) discontinuous counter-current distribution mode.

The *batch mode* of extraction is the simplest and may be carried out using a separating funnel in the analytical laboratory. The two phases are mixed in the separating funnel by shaking until equilibrium is reached and then the separating funnel is allowed to stand to separate the two phases. It is advantageous to use a solvent of higher density (e.g., chloroform or carbon tetrachloride) to extract the solute from the aqueous phase so the aqueous phase can be left in the separating funnel for multiple extractions.

The *continuous mode* of extraction is particularly useful when the distribution ratio is small and multiple extractions are necessary. The operating method involves distilling the organic solvent from a reservoir, condensing the vapours, allowing the liquid to contact the aqueous phase, and finally returning to the reservoir for recycling. Organic solvent of lower as well as higher density as compared to that of the aqueous phase may be used. Figure 12.2 shows the apparatus required for the continuous mode extraction.

The *discontinuous counter-current distribution mode*, also known as Craig extraction, or *fractional extraction with a stationary phase* or *counter-current extraction* has been developed for separating substances with similar distribution ratios. The extraction is carried out in specially designed apparatus consisting of a large number (50 or more) of identical interlocking units made of glass and mounted on a metal frame. Figure 12.3 shows a pair of tubes in the extraction unit. The frame is rocked and tilted mechanically to mix and separate the phases during each extraction step.

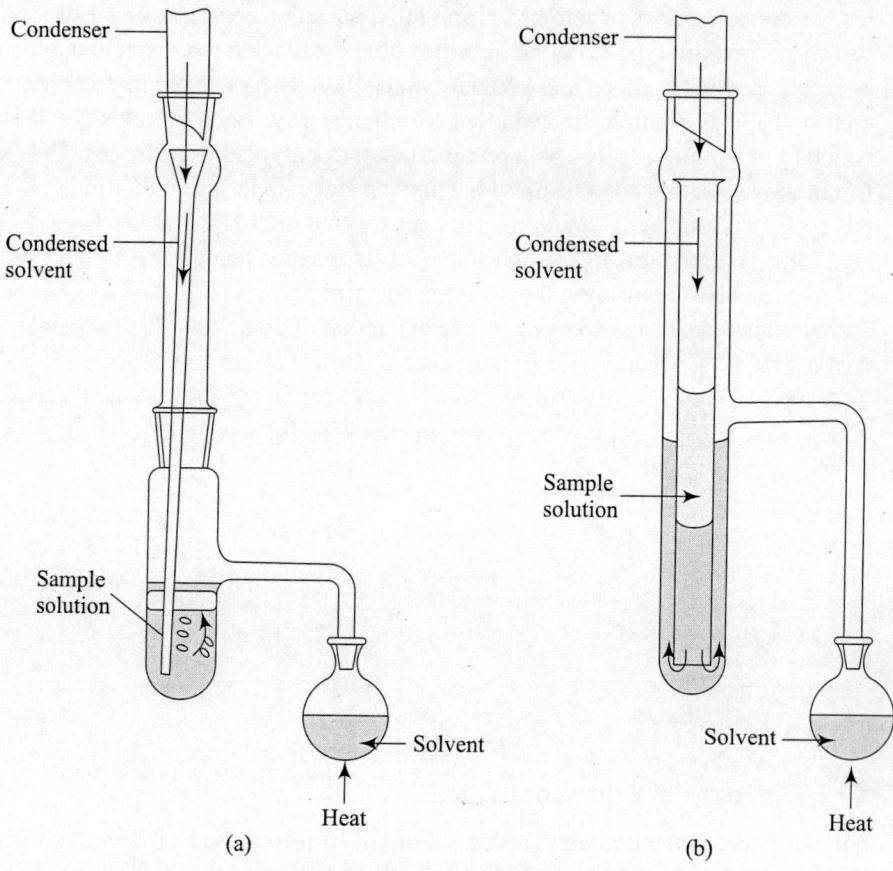

Fig. 12.2 Apparatus for continuous mode extraction using (a) lower density solvent and (b) higher density solvent

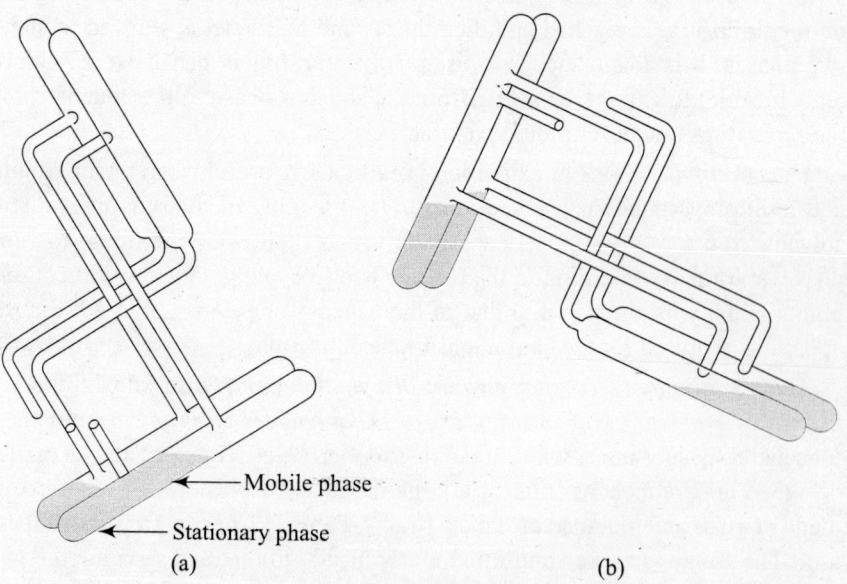

Fig. 12.3 Craig extraction unit (a) position during extraction (b) position during transfer

The first tube contains the solute dissolved in the lighter phase and the pure denser phase of equal volumes. The first tube is gently rocked till the solute concentrations in the two phases reach equilibrium by keeping the tube in position (a) in Fig. 12.3. After equilibrium is attained the tubes are brought to position (b) and the lighter phase is transferred from the first tube (hence, called mobile phase) to the second tube. The transfer of the mobile phase is completed by returning the tubes to position (a). Fresh solute-free lighter phase is added to the first tube. The denser extracting solvent is called the stationary phase as it remains in the same extraction unit throughout extraction process. The lighter phase moves from the first tube to the second tube and so on with fresh lighter phase being always added to the first tube till the initial portion reaches the final tube and all the tubes contain portions of both the phases. The sequence of operations is repeated several times.

The distribution of the solute after a series of transfers may be expressed in terms of the binomial expansion of $(x + y)^n$ where x and y represent the fractions of the solute in the mobile and the stationary phases and n is the number of extractions. The values of x and y depend on the distribution ratio D for the given solute and on the proportions of the two phases used. The solute moves through the extracting tubes at a rate proportional to the value of D as shown in Fig. 12.4. With increasing number of extractions the solute is spread over more number of tubes as shown in Fig. 12.5 though separation of the components in a mixture improves.

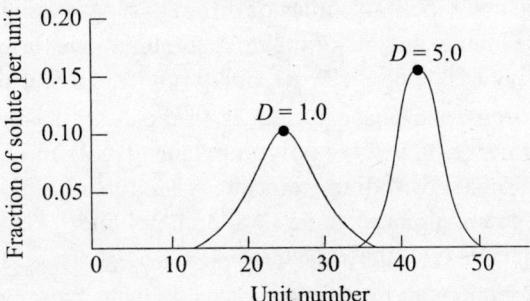

Fig. 12.4   Effect of the magnitude of D on solute distribution after 50 extractions

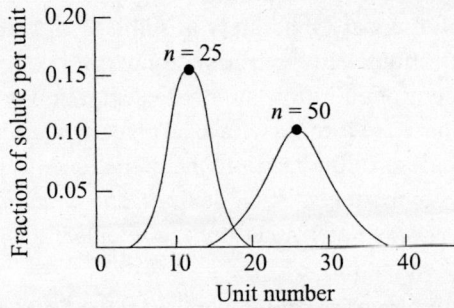

Fig. 12.5   Effect of number of extractions on solute distribution ($D = 1$)

Craig extraction finds use in the separation of amino acids, fatty acids, and polypeptides as in most cases the distribution ratios are very close, the difference being less than 0.1. However, the method can never result in 100% separation because of the Gaussian nature of the distribution of the solute and in addition suffers from the disadvantages of requiring large volumes of solvents and time-consuming procedure.

12.3 AQUEOUS TWO-PHASE EXTRACTION

Liquid–liquid extraction involving organic solvent–aqueous systems cannot be used for extracting biomolecules, such as proteins, enzymes, and nucleic acids, as these substances undergo denaturation and lose their biological activity in organic solvents. Hence, liquid–liquid extraction of biomolecules is carried out using solvents rich in aqueous phase. Liquid–liquid extraction involving two mutually insoluble solvent systems each consisting of a high concentration of water is called *aqueous two-phase extraction* or *aqueous biphase extraction*. Both the solvents contain mixtures of hydrophilic polymers or a hydrophilic polymer and salt dissolved in water. The water content of the solvent systems may be as high as 80–85%, but the two phases are immiscible. As the water content is high the phases do not denature proteins and other biopolymers.

12.3.1 Aqueous Two-phase Systems

Aqueous two-phase systems are obtained by mixing aqueous solutions of hydrophilic polymers such as polyethylene glycol and dextran. Polyethylene glycol (PEG) is a linear chain polymer with a high density of lone pairs of electrons, while dextran is globular and does not have any tendency for dipole formation. The mutual insolubility of these two hydrophilic polymers may be attributed to the molecular form of each polymer, which results in mutual repulsion and separation into two distinct phases. Each polymer tends to attract molecules of similar shape, size, and polarity and repel molecules of different type leading to partitioning of solutes in the separate phases. Other examples of polymer combinations forming aqueous two-phase systems include PEG-polyvinyl alcohol (PVA), polypropylene glycol-dextran, PVA-dextran, etc.

A similar aqueous two-phase system is formed with a mixture of aqueous solutions of a hydrophilic polymer (e.g., PEG or polypropylene glycol) and a low molecular weight salt such as potassium phosphate or sodium chloride. A mixture of aqueous solutions of polyelectrolyte (e.g., sodium dextran sulphate) and a non-ionic polymer + salt (e.g., PEG + sodium chloride) also forms an aqueous two-phase system.

The miscibility range and the formation of aqueous two-phase systems between two phase forming polymer–polymer or polymer–solute combinations may conveniently be discussed with the help of phase diagrams. Figure 12.6 shows the phase diagram of polyethylene glycol–dextran system. Both the polymers are separately miscible with water in all proportions and with each other at low concentrations. However, as the polymer concentrations increase, the miscibility depends on the concentration ratios and phase separation occurs. A PEG-rich upper phase and dextran-rich lower phase are formed with each phase containing more than 80% water. The curve separating the two regions of biphase and the homogeneous phase is called the binodal curve. A mixture of the three components, PEG, dextran, and water within the biphase region, separates into two distinct phases as determined by the intersections of the tie-line passing through the mixture point with the binodal.

For example, a mixture of the three components of certain composition 95% water + 2.5% PEG + 2.5% dextran, represented by point A in the phase diagram lies above the binodal curve, and hence, exists as a homogeneous phase. On the other hand, a mixture represented by point B with a composition of 85 % water + 5% PEG + 10% dextran, is below the binodal curve and is not stable as a homogeneous phase. It will separate into two phases, one phase with 90% water + 10% PEG (represented by point C in the phase diagram) and a second phase with 80% water + 19% dextran + 1% PEG (represented by point D in the phase diagram). The relative volumes of the two phases are given (by inverse liver arm rule) as $V_C/V_D = CB/BD$, where V_C and V_D are

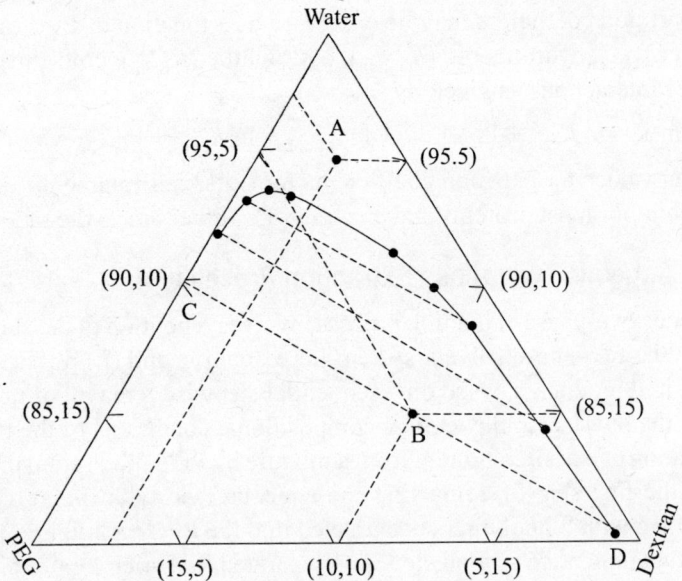

Fig. 12.6 Phase diagram of PEG–dextran aqueous two-phase system

the volumes of the phases with compositions represented by points C and D respectively and CB and CD are the lengths of the tie-line segments connecting point C with B and point B with D respectively. All mixtures represented by points on the same tie-line give rise to identical phase systems but with different volume ratios.

A variety of factors, such as molecular weights of the polymers, viscosity, density, interfacial tension, and temperature, affect the properties of aqueous biphasic systems. The concentrations of the polymers necessary to split a homogeneous solution into two phases depend on the molecular weights of the polymers, high molecular weight polymers requiring relatively low concentrations for phase separation. The viscosities of the two phases are important in bringing about a good contact between the phases for mass transfer and phase separation. The viscosities of the phases are affected by the presence of various other constituents in the phases. Difference in density between the two phases is preferable for better contact and phase separation. However, the density ratio of the two phases is mostly close to unity in most cases. A low interfacial tension is preferable for effective dispersion and creation of high interfacial area. Temperature changes affect the viscosity and phase equilibrium. An increase in temperature decreases the miscibility and enhances phase separation in PEG–potassium sulphate–water system. In contrast, the miscibility of PEG–dextran system increases with increase in temperature.

12.3.2 Theoretical Principles of Aqueous Two-phase Extractions

The basic principle in aqueous biphase extraction, involves the differential partitioning of the proteins in a given mixture to extract and separate/purify a desired protein.

The distribution of biopolymers between two water-rich liquid phases may be due to a variety of forces such as charge interaction, hydrogen bonding, and van der Waals interactions between solute molecules and the polymer molecules of the liquid phases. These interactions are influenced by various parameters such as molecular weights of the polymers, type, concentration of the salt, pH, and temperature. The partitioning coefficient, K, is expressed as the ratio of the concentration of the desired protein in the lighter upper phase to that in the denser lower phase.

The partition coefficient may be expressed as a function of five important factors involving electrical (K_{elec}), hydrophobic (K_{hphob}), hydrophilic (K_{hphil}), conformational (K_{conf}), and ligand (K_{lig}) type interactions as given by

$$\ln K = \ln K_{elec} + \ln K_{hphob} + \ln K_{hphil} + \ln K_{conf} + \ln K_{lig} \qquad (12.22)$$

Typical values for partition coefficients for cells, cell fragments, and DNA are in the range of 100 while for most proteins and enzymes the values are in the range of 10.

12.3.3 Aqueous Two-phase Extraction Process

The aqueous two-phase extraction process involves repetition of the steps (i) mixing the aqueous feed with the two-phase solvent system for extraction and (ii) phase separation. The recovery of the desired product is based on a sequential stepwise removal of products and impurities by adjusting the physical and chemical compositional conditions of the two phases. For example, the separation of a desired protein from a mixture by PEG–potassium phosphate biphasic system involves the first step of contacting the aqueous feed with the solvent system. The desired protein along with biopolymers is extracted into the PEG-rich lighter phase which is separated by centrifugation in the second step. The separated PEG-rich lighter phase is once again mixed with an aqueous solution of the salt at controlled pH and ionic strength to remove the unwanted biopolymers and proteins into the denser salt phase, leaving the desired protein in the lighter phase. The desired protein may be recovered from the separated phase by ultrafiltration or by chromatography.

The extraction process may be carried out in batch or continuous mode as in conventional solvent extraction. Large preparatory scale extractions of proteins, enzymes, and nucleic acids are carried out using a variety of centrifugal separators taking into consideration the viscosity and low interfacial tension of chosen aqueous two-phase solvent systems.

Applications of aqueous two-phase extraction are mostly in biotechnology area for large-scale separation and purification of proteins and enzymes, recovery of valuable proteins, enzymes from waste streams, removal of unwanted cell debris, and cell wall fragments from fermentation broths.

12.4   REVERSED MICELLAR EXTRACTION

Reversed micellar extraction is yet another method for the separation and purification of biopolymers. Surfactants form nanometer scale aggregates called *reversed* or *inverted micelle* spontaneously in organic solvents. In the reversed micelle the polar groups of the surfactant are buried within the core while the hydrophobic tail groups of the surfactant extend into the surrounding organic solvent, which is not miscible with water. The reversed micelles also contain significant quantities of water and hydrophilic solutes and exhibit unique solubilizing characteristics, which are made use for extracting proteins from an aqueous solution.

The solubility of proteins in the reversed micelle is due to electrostatic interactions between the charged proteins and the surfactant polar head groups. Partitioning and extraction of proteins from the bulk aqueous phase into the reversed micellar phase is influenced by factors such as pH, ionic strength, and the type of salt as well as the nature of organic solvents and surfactants. Any factor, which enhances the electrostatic interaction between the protein and the polar groups of the surfactant, would facilitate the extraction of the protein. For example, at pH < pI (isoelectric point) a protein will have a net positive charge, and hence, its solubility is greater in an anionic surfactant. When the pH > pI, the solubility of the protein in the reversed micellar phase will be less or negligible as the protein would have a net negative charge. The converse is also true in that the solubility of the protein would be more at pH > isoelectric pH in a reversed micelle formed by a cationic surfactant.

Higher ionic strength due to higher salt concentrations screens the electrostatic interactions and thereby decreases protein solubility in reversed micelle. Higher ionic strength may also lead to salting-out of the protein from the micellar phase.

Increase in surfactant concentration increases the number of reversed micelles, and hence, increases the solubility of proteins. Reversed micelles formed by ionic surfactants have been found to be too small to accommodate large proteins while non-ionic surfactants form large-sized micelles capable of dissolving a wide range of proteins.

Affinity partitioning of proteins is possible in reversed micellar extraction. The surfactant with a polar head group involving a specific substrate or inhibitor is chosen to extract the target enzyme selectively.

12.5 SUPERCRITICAL FLUID EXTRACTION

Supercritical fluid extraction as the name implies makes of use of a supercritical fluid as the extracting solvent. The method is exclusively useful for extracting highly labile or unstable components such as food flavours and aroma chemicals.

A substance exists as a supercritical fluid at temperatures and pressures above but close to the critical temperature and critical pressure (close to its critical point). The critical temperature of a substance is the temperature above which a distinct liquid phase cannot exist regardless of the pressure. The vapour pressure of the substance at its critical temperature is called the critical pressure. A variety of substances, such as carbon dioxide, nitrous oxide, sulphur dioxide, ammonia, ethane, propane, butane, pentane, and ethylene, are known to exist as supercritical fluids.

In general the properties, such as density, viscosity, and diffusion coefficients of a supercritical fluid lie in between those of the corresponding liquid and gaseous phases. Because of their low viscosity and greater diffusivity as compared to liquids, supercritical fluids diffuse more rapidly into a sample and even penetrate solid samples. The solvent strength of a supercritical fluid can be controlled easily by increasing the pressure on the fluid. Increase in pressure increases the density of the fluid and thereby its solvent strength. Near the critical point small changes in pressure create large changes in the density of the supercritical fluid. The heat capacity of the supercritical fluid is several times greater than that of the normal liquid. Supercritical fluids in general have greater advantages over liquids as extracting solvents.

The principle of supercritical fluid extraction is based on the solvent power (the ability of the solvent to extract a particular component in a mixture) of the supercritical fluid. The solvent power is highly dependent on the density of the supercritical fluid, which in turn depends on the pressure and temperature. In general the solvent power of a supercritical fluid increases with increasing density at constant temperature and at constant density it increases with increasing temperature. In addition the solvent power also depends on the similarity of the physical and chemical properties of the solute and the solvent, close similarity results in high solubility.

Supercritical fluid carbon dioxide finds wide use as the extracting solvent because of its advantages of being non-toxic and non-hazardous nature. As the solvent power of supercritical fluid carbon dioxide increases the percentage solubility of a solute increases and the range of extractable solutes also increases. Thus, at high solvent power the selectivity is low. In contrast at low solvent power the supercritical fluid carbon dioxide exhibits high selectivity for dissolving solutes. Thus, supercritical fluid carbon dioxide offers the possibility to tailor the extracting conditions to optimize recovery and purification of the desired solute components from a mixture.

The solubility of organic compounds in supercritical fluid carbon dioxide depends on the polarity and the molecular weight of the compounds. Non-polar substances, such as aliphatic hydrocarbons and monoterpenes, are completely soluble. Weakly polar compounds, such as triglycerides, are soluble to some extent, the solubility decreases with increase in molecular weight. Polar compounds, such as carboxylic acids and water, are soluble to a negligible extent. The solubility of a solute in supercritical carbon dioxide may be enhanced by the addition of a cosolvent or entrainer. The addition of a cosolvent modifies the thermodynamic affinity between the solute and the extractant favourably.

Supercritical fluid extraction using carbon dioxide as solvent has several advantages such as (i) a high degree of control over selective extraction of the desired component from a complex mixture, (ii) the solvent has advantageous thermophysical properties useful for extraction such as greater heat transfer efficiency, low viscosity for efficient mass transfer, and low heat of vapourization for efficient separation of the product from the solvent, (iii) higher recovery rates as compared to conventional solvent extraction, (iv) absence of contamination of the final products as carbon dioxide volatilizes off, (v) the solvent is non-toxic for bioproducts, (vi) a variety of samples such as solids, semisolids, and liquids of different chemical nature can be used in the extraction process, and (vii) ecofriendly as well as economical.

Applications of supercritical fluid extraction are mostly in food and biotechnology industries. Typical examples include extraction of bitter flavour from hops, decaffeination of coffee, extraction of food flavours, β-carotene from plants, edible oils and oils used in perfumes, flavours from different plant sources, monoglycerides from vegetable oils, etc.

12.6 SOLID PHASE EXTRACTION

Solid phase extraction (SPE) as the name implies makes use of a solid as the extracting phase. It has replaced solvent extraction to a large extent in the analytical laboratory for extracting and pre-concentrating the analyte(s) prior to chromatographic analysis. The technique is widely used for organic compounds present in very low concentration in aqueous solutions, for example, drugs, narcotics, amino acids and their metabolites in urine and body fluids, pesticides, fungicides from fruits and fruit juices, and pollutants such as polychlorinated and polyaromatic hydrocarbons in drinking water, surface waters, and industrial effluent samples. The solid phase extractant powder is packed in the barrel of a disposable plastic syringe and conditioned by passing a small volume of organic solvent (e.g., alcohol) through the packed bed. The organic analyte is adsorbed (extracted) on to the SPE bed by passing the aqueous sample solution containing trace quantities of the organic compound by suction or pumping. The organic analyte retained on the SPE bed is recovered by washing the bed with a small volume of organic solvent and subjected to further analysis.

Solid phase extraction adsorbents for extracting a variety of organic and inorganic solutes from aqueous solutions are commercially available in the form of disposable cartridges, pipette tips, or disks. The cartridges contain 100–500 mg of SPE adsorbents packed in plastic syringe barrels of 1–5 mL capacity. SPE adsorbent filled pipette tips are used in automated liquid handling systems.

SPE disks or plates with small wells containing SPE adsorbents are used for rapid analysis of drugs by liquid chromatography–mass spectrometry (LC-MS) hyphenated technique.

12.6.1 Solid Phase Micro Extraction

Solid phase micro extraction (SPME) is a solvent-free extraction technique gaining importance in analytical laboratory for concentrating analytes in trace quantities for chromatographic analysis. The SPE adsorbent is coated to a thickness of $10-100$ μm on the outer surface of a silica capillary fibre of about 1 cm in length. The fibre is attached to a syringe needle and covered by a stainless steel protective sheath. In practice the fibre is exposed to analyte samples in gaseous or liquid matrices for specific time of a few minutes at fixed temperature to adsorb the analyte. The fibre is then inserted into the injection port of gas chromatograph and the analyte is thermally desorbed onto the column for normal chromatographic analysis.

12.7 ION-EXCHANGE SEPARATION

Ion-exchange separation is based on reversible competitive exchange of ions of the same sign (e.g., positive or negative charged ions) between a water insoluble solid matrix called the *ion exchanger* and the surrounding liquid medium. Hence, the method is limited to samples containing ionized or partially ionized solutes. For example, when a sample solution containing cations, anions, and neutral solute components represented as M^+, M^{2+}, M^-, M^{2-}, and M^0, is allowed to come into contact with the ion exchanger containing exchangeable cations (usually H^+) held to the negatively charged matrix, ion exchange of ions of the same sign occurs between the solid and solution phases. Consequently the cations M^+ and M^{2+} are bound to the solid matrix with the simultaneous release of H^+ ions into the surrounding liquid phase. The anionic and the neutral species remain in solution as no exchange is possible due to electrostatic repulsion between the negatively charged matrix and anion. Similarly no exchange occurs due to the absence of any electrostatic interaction between the matrix and the neutral species. The ion exchange process is schematically shown in Fig. 12.7.

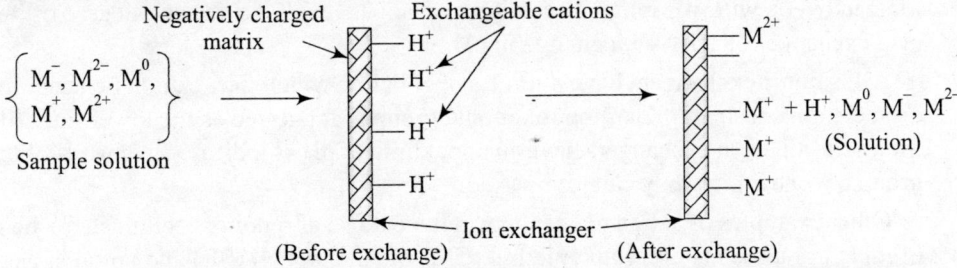

Fig. 12.7 Schematic diagram of ion-exchange process in a cation exchanger

Cations with greater positive charges (M^{2+}) are exchanged and bound to the matrix strongly as compared to those with less positive charges (M^+). Since in the above example, the matrix is negatively charged and cations undergo exchange between the solid matrix and the surrounding solution the ion exchanger is called *cation exchanger*. In contrast, the *anion exchanger* consists of exchangeable anions (usuallly OH^- or Cl^-) bound to the positively charged matrix.

12.7.1 Ion Exchangers

Ion exchangers are of natural as well as of synthetic origin. Natural exchangers include aluminosilicate minerals such as clays and zeolites. For most analytical works synthetic ion exchangers based on organic polymers are used. Commercially available synthetic ion-exchange resins are in the form of water insoluble, rigid porous structured spherical beads made from long chain organic polymers with controlled cross-linking. The most commonly used polymer resin is that of polystyrene cross-linked with varying amounts of divinyl benzene in the range of 2–20%. The beads contain functional groups responsible for the ion-exchange properties of the resins. The functional groups are either acidic groups such as sulphonic or carboxyl capable of exchanging H^+ ions for other cations (hence, called cation exchangers) or basic groups such as quaternary ammonium groups ($N^+ R_3$), primary, secondary, or tertiary amino groups capable of exchanging anions (hence, called anion exchangers) in water. The structures of the cation- and anion-exchange resins may be represented as

Cation-exchange resin   **Anion-exchange resin**

Sulphonic and quaternary amino groups are used to form *strong ion exchangers* while other functional groups such as carboxyl or primary, secondary, or tertiary amines form *weak ion exchangers*. The terms strong and weak are relative and refer only to the extent of variation of ionization with pH and not to the strength of binding. Strong ion exchangers are completely ionized over a wide pH range. Weak ion exchangers exhibit a varying degree of ionization and also exchange capacity with change in pH.

All cationic exchangers have a limiting pH below which they cannot be used. In general the dissociation constant, pK, of the functional group is suggested as the lower limit of pH for use. Similarly, all anion exchangers have an upper limit of pH as indicated by the pK of the functional group beyond which they cannot be used.

Other examples of cation exchangers include those of modified cellulose by the introduction of ionic groups such as carboxymethyl (CM) or sulphoethyl (SE) and anion exchangers based on cellulose include diethylaminoethyl (DEAE) or triethylaminoethyl (TEAE). Inorganic ion exchangers include those of hydrated zirconium oxide, ammonium molybdophosphate, and zirconium phosphate.

12.7.2　Ion-Exchange Equilibrium

The ion-exchange process is reversible in that the ions of like charge are exchanged between the solid matrix and the surrounding solution and may be represented generally as

$$A_{matrix} + B_{soln.} \rightleftarrows B_{matrix} + A_{soln.} \tag{12.23}$$

The equilibrium constant K_{IE} is called the *selectivity coefficient* for the ion-exchange process and is given as

$$K_{IE} = [B_{matrix}][A_{soln.}]/[A_{matrix}][B_{soln.}] \qquad (12.24)$$

where the concentrations of the ions are represented by terms in the brackets. The magnitude of K_{IE} indicates the relative affinities of the ions for the matrix. In the above example when $K_{IE} > 1$, B ions have greater affinity for the matrix. The values of the selectivity coefficient have to be determined experimentally for a given set of exchangeable ions and the chosen matrix.

The ion-exchange equilibrium for a given matrix is influenced by factors such as (i) the charge on the ion, (ii) size of the hydrated ion, and (iii) concentration of the exchanging ions in solution. The nature of the matrix, the nature of the functional groups attached to the matrix, and the extent of cross-linking in a polymer matrix also influence the ion-exchange equilibrium.

For a strong cation-exchange polymer matrix the extent of ion exchange increases with increasing charge on the exchanging cations in solution at low concentrations (<0.1 M) and ambient temperature in the given order

$$Th^{4+} > Al^{3+} > Ca^{2+} > Na^{+}$$

A similar order is observed for a strong anion-exchange polymer matrix with the extent of ion exchange increasing with increasing charge on the exchanging anions in solution.

The extent of ion exchange on a cation exchanger increases with decreasing size of the hydrated cation in the case of monovalent cations

$$Li^{+} < H^{+} < Na^{+} < K^{+} < Rb^{+} < Cs^{+}$$
$$\xrightarrow{\hspace{5cm}}$$

Decreasing size of hydrated radius

In the case of divalent cations the ionic size and the extent of dissociation of the metal salt together determine the extent of exchange on a cation exchanger.

$$Cd^{2+} < Be^{2+} < Mn^{2+} < Mg^{2+} = Zn^{2+} < Cu^{2+} = Ni^{2+} < Co^{2+} < Ca^{2+} < Sr^{2+} < Pb^{2+} < Ba^{2+}$$

The extent of ion exchange on an anion exchanger also depends on the size of the hydrated exchanging anions in the solution, the order being similar to that of cations.

The concentration of the exchanging ions in solution also plays a role in determining the extent of ion exchange. Higher concentrations of exchanging cations of lower charge in solution favour exchange of cations of higher charge on the ion exchanger. In contrast, dilution of exchanging cations of higher charge in solution favours exchange of cations of lower charge on the matrix.

The nature of the matrix and the nature of the functional groups attached to resin matrix and the extent of cross-linking also affect the selectivity in the ion-exchange process. The functional groups may be strongly acidic (e.g., sulphonic acid groups) or weakly acidic (e.g., carboxyl groups) in a cation exchanger and the ion exchange is facilitated depending on the degree of dissociation of the acidic groups. The strongly acidic groups facilitate ion exchange over the entire pH range while a weak cation exchanger can function only in pH above 7 as the weak carboxyl groups have pKs in the range of 3.5–4.0. Similarly, strongly basic functional groups (quaternary ammonium groups) facilitate ion exchange in a strong anionic exchanger over the entire pH range but weakly dissociating functional groups (DEAE) of weak anion exchanger require pH values less than 8. The degree of cross-linking also influences the selectivity of ion exchange process, selectivity increasing with increasing degree of cross-linking.

12.7.3 Capacity of Ion Exchangers

The capacity of an ion-exchange resin is a quantitative measure of its ability to take exchangeable counter ions. The *total ion-exchange capacity* depends on the number of charged functional groups per gram of dry ion exchanger or per mL of swollen gel. It is usually determined by titration with a strong acid or base. Ion exchangers have a high degree of ionic capacity in the range of 100 to 500 µM/mL of bed. The *available capacity* of the ion-exchange resin packed in a column is the amount of the charged solute that can be bound to gel under specified experimental conditions. The available capacity is determined by batch method in which a series of solutions of different concentration of a given solute (e.g., protein) are added to known quantity of ion-exchange resin taken in test tubes, equilibrated at suitable pH, and ionic strength for binding to occur. After equilibration the supernatants are assayed for the solute concentration and the available capacity of the ion-exchange resin is calculated. The *dynamic capacity* of the ion-exchange resin is its available capacity under chromatographic conditions particularly for large molecules such as proteins. It is determined by saturating the given volume of the ion-exchange resin in a column (usually 1 mL) with the protein solution at a specific flow rate, washing the column to remove excess of protein, and then eluting the bound protein with appropriate buffer. The total amount of the bound protein eluted out gives the dynamic capacity of the ion-exchange resin at the specified flow rate. The dynamic capacity of ion-exchange resin is usually in the range of about 20% of the total ion-exchange capacity.

The available and dynamic capacities depend critically on the properties of the ion exchanger matrix, particularly the degree of cross-linking. Greater the degree of cross-linking smaller is the pore size in the ion-exchange resin and smaller is its capacity. Experimental conditions such as pH, ionic strength, and temperature, also influence the capacity of an ion-exchange resin. The dynamic capacity usually decreases with increasing flow rate of the mobile phase in a chromatographic column.

12.7.4 Regeneration of Ion Exchangers

The cation-exchange resin after exhaustion is regenerated by equilibrating with a dilute solution of an acid to its original H form and washing finally with water while an anion-exchange resin is regenerated by equilibrating with sodium hydroxide or sodium chloride followed by washing with water.

12.8 FILTRATION

Filtration is defined as the separation of solid from the mother liquor by passing the slurry through a septum called the filter medium. The slurry is allowed to pass through the filter medium by applying a higher pressure on the upstream side or vacuum on the downstream side. The slurry may be pumped in perpendicular direction to the filter medium as in normal flow (dead end filtration) or in parallel direction as in cross-flow filtration, usually adopted in membrane separations (Figs 12.8 and 12.9).

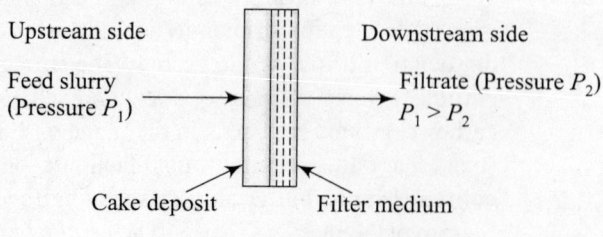

Fig. 12.8 Normal flow filtration

Batch filtration

Normal flow filtration is adopted at laboratory scale operations for analytical purposes. The filter medium is either a filter paper or a porous-sintered glass. The filter medium allows the fluid (filtrate) to pass through and retains the solids in the form of a cake. The thickness of the cake increases gradually during filtration as the solids accumulate on the upstream side. Filtration is usually a batch mode operation in the laboratory. Normal flow filtration in batch mode is also practiced in industrial scale separations using equipment, such as filter presses and pressure leaf filters, in which the filtering medium is mostly canvas cloth or fine wire mesh.

As the fluid passes first through the cake and then the filter medium in normal flow filtration it experiences a pressure drop ($\Delta p = P_1 - P_2$). The pressure drop is the sum of the pressure drop across the filter cake (Δp_c) and across the filter medium (Δp_m). The rate of filtration is given by the equation.

$$dV/dt = 1/(KV + B) \tag{12.25}$$

where V is the volume of feed slurry, $K = [(\eta \, C \, \alpha)/(A^2 \, \Delta p)]$, and $B = [(\eta r_m)/(A \, \Delta p)]$. In Eq. (12.25) A is the area of the filter medium, C is the concentration of solids in the feed, η is the viscosity of the filtrate, α is the specific resistance of the cake, r_m is the resistance of the filter medium, and Δp pressure drop of the filtrate. In the case of biomass which forms compressible cakes the above equation is modified by substituting $(\Delta p)^{1-s}$ (where s is the cake compressibility with values ranging between 0.1 and 0.8), for Δp in the above equation.

Continuous filtration is practiced, particularly in biochemical industries, for filtering biomass, and other large volume feed using a rotary drum filter.

12.9 MEMBRANE SEPARATION TECHNIQUES

Membrane separation techniques are based on a variety of physical principles and are used for the separation of a wide range of miscible components most commonly in bioprocess industries. The different membrane separation techniques have a common feature, namely, a thin semipermeable membrane acting as a barrier between two fluid phases. The components of the phase on the upstream side are separated by their selective movement across the membrane, the membrane allowing some components to pass through while retaining others.

Membrane separation techniques are broadly classified into three groups depending on the nature of the driving force responsible for mass transfer across the membrane. The driving force is hydrostatic pressure in the first group which includes three techniques (i) microfiltration (MF), (ii) ultrafiltration (UF), and (iii) reverse osmosis (RO). Dialysis, belonging to the second group of membrane separation techniques, makes use of concentration difference between different streams on the two sides of the membrane. Electrodialysis belongs to the third category wherein the driving force is an applied electric field. The membranes used in the different techniques have varying pore sizes–from non-porous nature in RO to small-sized pores in UF to relatively bigger pores in MF. The characteristic features of the different membrane separation techniques are summarized in Table 12.1.

12.9.1 Theory of Membrane Separation

The transport of the solute molecules across the membrane is conveniently explained on the basis of two theoretical models of mass transport: (i) capillary flow model which is based on filtering or sieving mechanism and (ii) solution–diffusion model based on Fick's law diffusion. It is possible that both the mechanisms are valid in mass transport across the membrane.

Table 12.1 Characteristic features of membrane separation techniques

Technique	Driving force	Characteristic features		
		Membrane pore size	Separation mechanism	Applications
MF	Applied pressure $0.1-1$ kg/cm^2	$0.02-10$ μm	Sieving	Separation of coarse organic/inorganic particulates, bacterial cells, yeasts and fungi of size 10^2-10^4 nm
UF	Applied pressure $2-10$ kg/cm^2	$0.001-0.02$ μm	Sieving	Separation of biopolymers (proteins, enzymes, polysaccharides–MW 10^4-10^6), virus, and colloids (2–1000 nm)
RO	Applied pressure $10-100$ kg/cm^2	Non-porous	Solution diffusion	Separation of inorganic salts and simple organic acids, sugars, antibiotics with MW 10−1000 and size 0.1−1.2 nm
Dialysis	Concentration difference	$1-5$ nm	Sieving and diffusion	Separation of biopolymers, colloids from unwanted low molecular weight solutes
Electro-dialysis	Electrical potential	MW < 200	Ion migration	Separation of ions

In the *capillary flow model* the membrane is considered to be microporous with a pore size structure described as 'loose'. The membrane is permeable to particles less than 10 Å while retaining bigger particles. A filtering or sieving process occurs as the feed solution flows by convective flow. Smaller solute molecules flow through the pores along with the solvent while larger molecules exit as part of the retentate as they cannot pass through the pores. The filtering mechanism depends on the molecular size of the solute molecules and is not affected by chemical nature or the structure of the membrane. The capillary flow model is applicable to microfiltration and ultrafiltration as transport across the membrane in these techniques may be considered as an extension of filtration operation with the membrane functioning as a sieve.

The *solution diffusion model* is more relevant in reverse osmosis. The membrane is virtually non-porous or 'tight' in that even small molecules of about 10 Å or less are retained. The model postulates that molecular species dissolve in the membrane material and are transported across the membrane by diffusion process governed by Fick's law of diffusion. The driving force is the concentration gradient set up in the membrane by the applied pressure difference. The model is capable of explaining the selectivity exhibited by reverse osmosis process in allowing certain molecules while retaining others on the basis of solubility and the rate of diffusion. The chemical nature and membrane structure seem to have a role in deciding the transport of solute molecules across the membrane.

12.9.2 Retention Coefficient

Retention coefficient also called rejection coefficient is a measure of the separating ability of a membrane in pressure driven membrane separation methods of MF, UF, and RO. The true or theoretical retention coefficient R is related to the concentrations of the solute in the permeate (C_p) and at the membrane surface C_m as given by

$$R = \frac{C_m - C_p}{C_m} \tag{12.26}$$

However, the observed retention coefficient, R' is less than the true retention coefficient and is related to the concentrations of the solute in the bulk phase, C_b, and permeate C_p as given by Eq. (12.27).

$$R' = \frac{C_b - C_p}{C_b} \qquad (12.27)$$

The lower value of the observed retention coefficient as compared to that of true retention coefficient is attributed to concentration polarization at the membrane surface as the ratio $C_m/C_b > 1$. Concentration polarization affects separation efficiency of the membrane. Due to concentration polarization solute leakage through the membrane increases and the observed retention coefficient decreases, particularly for RO process.

12.9.3 Factors Affecting Membrane Separation

The selectivity and permeation rate of the membrane, and hence, the overall performance of membrane separation are affected by two factors, namely, (i) concentration polarization and (ii) fouling of the membrane.

Concentration polarization is a short-term effect which affects the performance characteristics of the membrane. Concentration polarization sets in on the upstream side of the membrane due to accumulation of chemical species which are not allowed to pass through the pores of the membrane. In addition, concentration polarization also sets in due to different rates of transport of various species. The net result is that the osmotic pressure increases thereby decreasing the permeation rate. The phenomenon is important particularly in ultrafiltration.

Fouling of the membrane occurs over a period of time and this long-term effect is caused by several factors which include (i) slime formation, (ii) growth of microorganisms, (iii) deposition of macromolecules and colloids on the membrane surface particularly in UF, and (iv) physical compaction of the membrane due to high pressure operation. Fouling is an irreversible phenomenon. Hence, the membrane has to be replaced. Fouling of the membrane can be inhibited by proper selection of the membrane material. For example, a hydrophilic membrane surface is more resistant to fouling by proteins. Fouling can also be minimized by adopting appropriate steps such as (i) pretreatment of the feed by pH control or desalting, (ii) frequent cleaning of the membrane with chemicals, and (iii) back flushing with permeate.

The problems associated with concentration polarization and fouling can be minimized by cross-flow filtration instead of the conventional dead-end filtration. In the conventional filtration the feed flows in perpendicular direction to the filter medium resulting in accumulation of solute particles and concentration polarization on the upstream side of the membrane. In cross flow filtration (Fig. 12.9) the feed is pumped tangentially across the filter medium thereby avoiding setting-in of concentration polarization. Cross-flow filtration has other advantages also in that the membrane separation characteristics and the permeation rate are not affected as a function of filtration time.

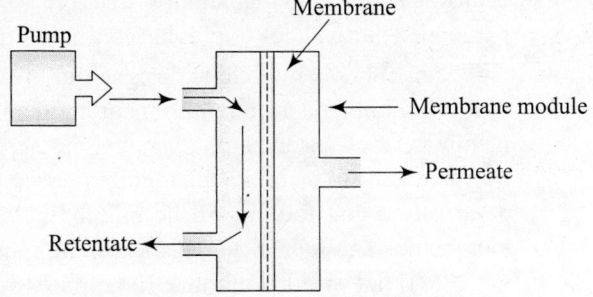

Fig. 12.9 Cross-flow filtration

12.9.4 Membranes and Their Characteristics

The membranes used in pressure driven processes should satisfy the requirements such as (i) high selectivity and good separation efficiency, (ii) high permeation rate, (iii) mechanical strength to withstand the high pressure operation without altering the pore dimensions, (iv) consistent performance throughout lifetime, (v) resistance to corrosion and microbial attack, (vi) ease of fabrication in different modules, and (vii) in addition the membrane material should be of low cost and readily available. The requirements for RO membrane are relatively more stringent in that (i) the membrane should be able to discriminate between the solvent molecules and low molecular weight solutes and ions and (ii) since the osmotic pressure in aqueous salt solutions is quite large the membrane should be capable to withstand high pressures in the range of 50–60 atmospheres.

The membranes used in RO and UF are of 0.1–0.2 mm in thickness and consist of two layers (i) a thin (0.5–10 μm) top layer of membrane material with microporous structure which is responsible for the separation process and (ii) a thick support layer (50–125 μm) of macroporous material to provide the required strength to the membrane. The membrane is supported on a rigid, porous backing structure.

The membrane material is chosen depending on the technique and the solutes to be separated. Membranes commercially available are based on cellulose acetate, cellulose phthalate, polyamides, polyacrylonitrile, polyethylene, and polytetrafluoroethylene. Cellulose acetate membrane is mostly used to remove salts, such as sodium chloride, sodium bromide, calcium chloride, and sodium sulphate, during desalination of brackish water by RO. Cellulose acetate membrane can be used only for aqueous solutions and at temperatures below 60°C.

Membranes made from polypropylene are hydrophobic, and hence, have to be primed with alcohol–water mixtures before using for aqueous solutions. Membranes with most monodisperse pores and of very low porosity are made from non-porous films of mica or polycarbonate and irradiating with α radiation. The radiation tracks in the film are then etched away with hydrofluoric acid. Composite membranes consisting of several layers of metallic oxides (usually zirconia, ZrO_2) on a calcined carbon support have advantages such as capability to withstand high stress during operation, high shock resistance, suitability to undergo steam sterilization, and capability to handle hot process fluids, high concentrations, and high viscosity of the feed.

12.9.5 Equipment for Membrane Separation

The main components of the different membrane separation methods are the same as shown in the block diagram (Fig. 12.10). These include (i) high pressure pump (piston, diaphragm, or centrifugal type) and regulators and (ii) membrane module. The feed is usually filtered to remove suspended particulates to prevent fouling of the membrane.

The membrane module is designed to provide (i) a high membrane surface area to volume ratio to minimize space requirement and capital cost, (ii) structural support to allow the thin membrane to withstand high operating pressures, (iii) a low pressure drop on the upstream side of the membrane to maintain the required driving force, (iv) turbulence to minimize concentration polarization and fouling of the membrane, and (v) easy back flushing and replacement of membrane. The four different membrane module configurations used in commercial practice include (i) flat sheet membrane, (ii) spiral-wound membrane, (iii) tubular, and (iv) hollow fibre modules.

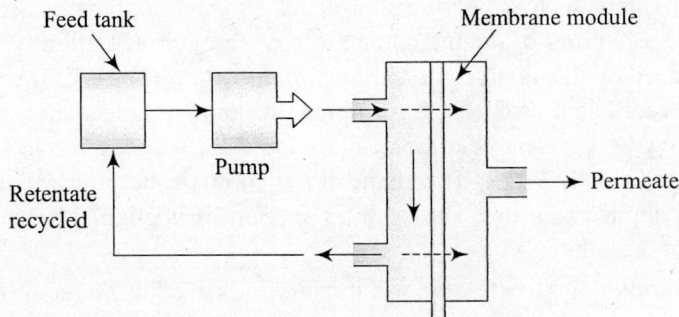

Fig. 12.10 Block diagram of membrane separation unit

Flat sheet membrane modules consist of several flat sheets of membrane stacked as a multilayer sandwich in a plate-and-frame filter-press type arrangement. The module is capable of withstanding high pressures of 30–40 kg/cm^2. However, it has a low surface area to volume ratio. *Spiral-wound membrane module* consists of two layers of a sheet of membrane wound spirally to give a high surface-to-volume ratio. The feed is passed axially from the outer side of the double layer and the permeate flows to a pipe located at the centre of the spiral. *Tubular membrane module* also called the *shell and tube module* consists of several tubes of membranes connected to a common header and packed into a perforated outer shell. The feed enters the lumen of the tubes, the permeate passes through the wall while the retentate passes out at the other end of the tubes. This module is useful for high viscosity feed or feed containing particulate matter. *Hollow fibre membrane module* consists of hollow fibres of membrane packed into a shell-and-tube arrangement to provide a high surface area per unit volume.

12.9.6 Membrane Separation Methods

The different pressure driven membrane separation methods include microfiltration, ultrafiltration, and reverse osmosis. Dialysis depends on concentration gradient while electrodialysis involves electrokinetic mobilities of ions.

Microfiltration (MF) is similar to conventional filtration with a filter membrane made of synthetic polymer, such as polypropylene or polytetrafluoroethylene (PTFE) or inorganic alumina or zirconia, to filter suspended solids in the size range of 0.1–10 µm. Dissolved solids including macromolecules and colloids pass through the membrane. The solvent and the dissolved solutes pass through the membrane primarily due to convective flow through well defined pores due to applied pressure in the range of 1–2 bar. Microfiltration is mostly used for harvesting microbial cells from fermentation broths and also for separating blood cells and plasma from whole blood. Two major disadvantages of this technique as compared to conventional filtration are (i) the frequent fouling of the membrane due to deposition of solid particles on the membrane surface which necessitates periodical purging and (ii) only a concentrated suspension containing the desired solids (e.g., microbial cells or blood cells) is obtained which requires conventional filtration to get dry cake. As the concentration of the suspension increases during filtration the viscosity also increases causing pumping problems.

Ultrafiltration (UF) is also a pressured driven separation method operating at pressures in the range of 2–10 bar. The membrane made from synthetic polymers, such as polysulphone, is microporous with the pore size being smaller than the pore size of microfiltration membrane. Dissolved solutes of low molecular weight pass through the membrane while high molecular weight products, such as polymers, proteins, and colloidal materials, are not permitted by the controlled size pores of the membrane to pass through. Ultrafiltration membranes are usually characterized by the *nominal molecular weight cut-off* (NMWCO) based on tests in which the rejection coefficient profile of a membrane to known molecular weight solutes is evaluated. The NMWCO represents the molecular weight for which the rejection coefficient is a fixed percentage (usually 90%). The liquid flows through the pores of the membrane due to the moderate applied pressure. The osmotic pressure is negligible because of the high molecular weights of the solutes.

Ultrafiltration finds extensive use in dairy industry for the recovery of whey proteins from the by-product stream of whey or serum obtained by precipitating off casein during cheese manufacture. It is also used in fruit processing industry to concentrate and clarify fruit juices and in pharmaceutical industry for concentrating a variety of products from cell-free fermentation broths.

Reverse osmosis (RO) is also known as *hyperfiltration*. The RO membrane is virtually non-porous with pore sizes in the range of 0.0001−0.001 µm. Hence, the membrane is permeable to water only but not dissolved salts of even low molecular weight. The nomenclature is based on the fact that the direction of normal osmotic flow of a solvent across a semipermeable membrane is reversed due to an applied pressure which is greater than the osmotic pressure of the liquid. In osmosis the solvent (usually water) diffuses through the membrane from the solvent side (or solution of lower concentration) to the solution (of higher concentration) side. The driving force is the osmotic pressure difference between the two sides. In reverse osmosis, an applied pressure greater than the osmotic pressure (π) in the range of 20−100 bar is imposed on the membrane reversing the flow of the solvent from the solution side to the pure solvent side. Thus, the solution-side becomes more concentrated as the solvent is removed. Figure 12.11 shows the schematic representation of osmosis and reverse osmosis.

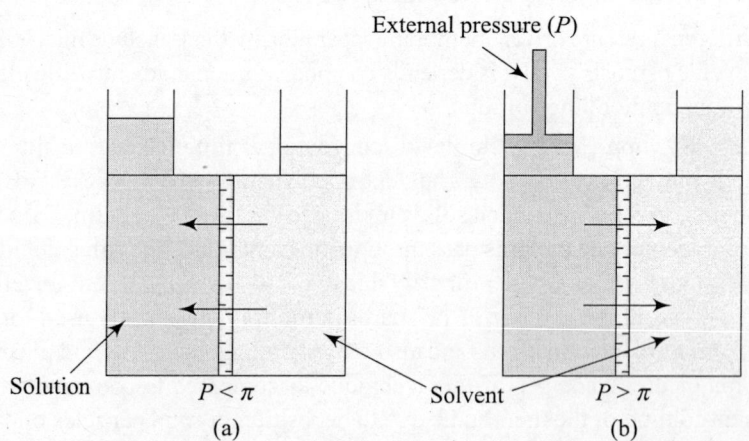

Fig. 12.11   Showing the direction of flow of solvent in (a) osmosis
($P < \pi$) and (b) reverse osmosis ($P > \pi$)

RO is useful for the separation of low molecular weight products, such as salts (as in desalination of seawater), sugars, or organic acids, from aqueous solutions. Its use has been extended to food and dairy industries to concentrate fruit juices, vegetable juices, and milk and treatment of wastewater.

Dialysis is less known to chemists than biologists. Dialysis involves the separation of solutes by diffusion across the semipermeable membrane which separates two aqueous phases containing different concentration of inorganic or organic solutes. The membrane is made of cellulose acetate or synthetic polymer with pore sizes of 1−5 nm and solutes diffuse through the pores on the basis of molecular size and molecular conformation. The sole driving force is concentration difference across the membrane. The solutes diffuse from the more concentrated to the less concentrated phase (or pure solvent) containing the same solvent till the concentrations in both the phases become equal. The concentration difference between the two phases should be sufficiently large and the membrane should be thin to reduce the diffusion path. A schematic representation of dialysis is shown in Fig. 12.12.

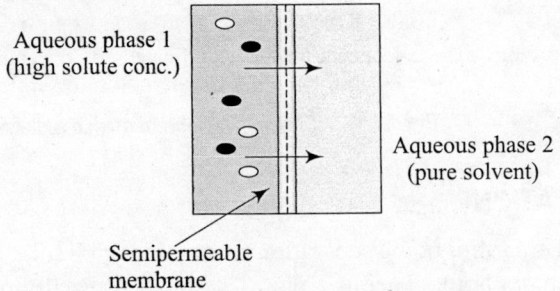

Aqueous phase 1 (high solute conc.)

Aqueous phase 2 (pure solvent)

Semipermeable membrane

Fig. 12.12   Schematic representation of dialysis

The diffusion coefficients of solutes are inversely proportional to the square root of molecular weight with small molecules diffusing faster than larger ones. The solute flux across the membrane J_s is related to the resistance r offered by the membrane, and concentration difference, ΔC between the two phases as given by the equation

$$J_s = r\,\Delta C \tag{12.28}$$

Dialysis is a major method used for purification of blood and is known as *haemodialysis* or artificial kidney. The patient suffering from renal failure is hooked to the haemodialysis unit and blood drawn from the patient passes through the machine as one the phases while water containing solutes, such as potassium salts, is passed through the unit as the second phase. The aqueous phase contains dissolved salts so that it has the same osmotic pressure as blood to minimize transfer of water. The waste products from the blood, such as urea, uric acid, creatinine, phosphates, and excess of chloride, diffuse into water thereby purifying the blood. Dialysis has potential applications in the separation and concentration of inorganic colloids, separation of alcohol from beer, etc.

Electrodialysis makes use of ion selective cationic and anionic ion-exchange membranes to separate ions and was developed for the desalination of brackish water to yield potable water. The electrodialysis unit consists of a tank fitted with a series of cationic and anionic exchange membranes placed alternatively so as to form a stack of compartments. A cathode and an anode are placed at the two opposite ends of the tank. The ion-exchange membranes are selective in that the cationic membrane allows only cations to pass through while the anionic membrane allows only anions to pass through. Electrokinetic transport of ions occurs due to the applied electric field and depending on the ionic mobilities within the membranes selective separation of ions occurs. The alternate compartments contain solutions of lower concentration (or desalinated water) and higher concentrations of ions. Figure 12.13 shows a schematic diagram of an electrodialysis unit.

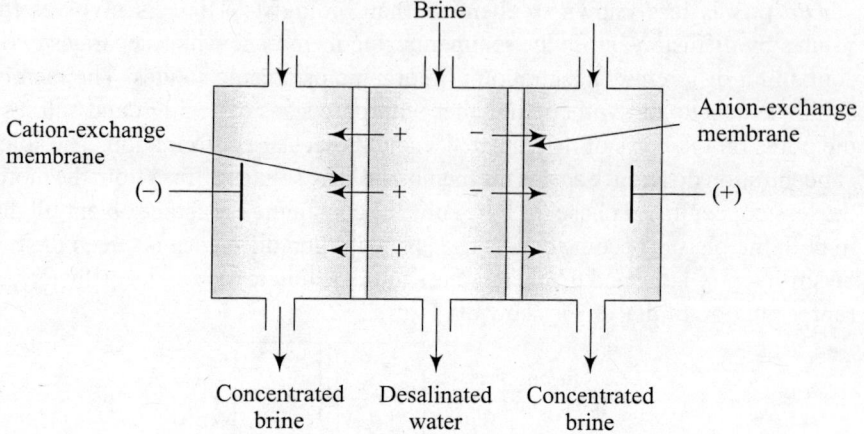

Fig. 12.13 Schematic diagram of electrodialysis unit

12.10 CRYSTALLIZATION

Crystallization is a solid–liquid separation process in which solid particles of the dissolved solute separate out from a homogeneous phase. Crystals are usually of exceptional purity free from even closely related impurities, and hence, crystallization is considered as a method of final purification of the desired product. Crystals are obtained in specified size and shape facilitating their separation by filtration or centrifugation and drying. A crystalline product has always a better appearance and consumer acceptance. Crystallization is adopted mostly in industrial practice, well known examples being common salt crystallizing out from brine and sugar from sugarcane juice. Crystallization requires that the solution of the desired solute must be in supersaturated condition, a thermodynamically unstable state. The degree of supersaturation of a solution is measured in terms of the supersaturation coefficient S, which is the ratio of the concentration of solute (C_t) in a solvent at a given temperature to that of the concentration of solute in a saturated solution (C_o) at the same temperature.

$$S = C_t/C_o \qquad (12.29)$$

The solution is said to be supersaturated when $S > 1$.

Crystallization may be initiated either by cooling or by evaporation so that the solution is concentrated to a supersaturated solution. Crystallization is considered to occur involving two basic steps (i) nucleation and (ii) crystal growth. The driving force for both these steps is supersaturation. The supersaturated state may be considered to consist of three zones, a metastable zone, an intermediate zone, and a labile zone. In the labile zone, nuclei are formed spontaneously from a clear solution while in the intermediate zone, new nuclei and crystals are formed. In addition growth of crystals also occurs. In the metastable zone only crystal growth occurs by the deposition of the dissolved solute in excess of the equilibrium concentration on existing crystals. No new crystals or nuclei are formed in the metastable zone of supersaturation.

Crystallization is initiated by primary nucleation also called homogeneous nucleation in which the solute molecules come together to form clusters or by heterogeneous nucleation involving the addition of solute crystals (seed) to form additional crystal nuclei (secondary nucleation). Heterogeneous nucleation may also be brought about by external agents such as dust, gas bubbles, mechanical shock, or ultrasonic shock.

Crystal growth occurs by the dissolution of smaller crystals and subsequent deposition of the dissolved solute on bigger crystals. The rate of growth of a crystal depends both on the transport of material to the surface of the crystal and on the mechanism of surface deposition. Stirring the solution during crystallization helps in the transport of material to the surface. However, the rate of growth of crystals is diffusion controlled. The presence of impurities usually reduces the rate of crystal growth.

The rate of growth of a crystal face is defined as the distance moved per unit time in a direction perpendicular to the face. The growth of the crystal takes place only at the outer face in a layer-by-layer process with the solute molecules diffusing through the solution to reach the face and getting integrated into the space lattice at the crystal surface. The diffusion and interfacial (surface reaction) steps determine the overall rate of crystal growth. The growth is measured on the basis of increase in length ΔL, in mm, in linear dimension of one crystal and the growth rate is expressed in terms of $(\Delta L/\Delta t)$.

12.11 PRECIPITATION

Precipitation involves the addition of a specific or a selective reagent to decrease the solubility of a desired solute so that it precipitates out from the solution so that centrifugation or filtration can be used to separate the solid. Precipitation of inorganic substances mostly used for analytical purposes has been discussed in Chapter 3 under gravimetry. Precipitation is used in laboratory scale and also in bioprocess industries particularly to precipitate proteins from fermentation broths. A desired protein may be precipitated out by (i) isoelectric precipitation–controlling the pH of the medium as the solubility of a protein is minimum at its isoelectric point, (ii) salting out technique–using ammonium sulphate to increase the ionic strength of the medium and thereby decrease the solubility of the protein, and (iii) organic solvent-mediated precipitation–by adding acetone or isopropanol to decrease the dielectric constant of the medium. Sometimes unwanted proteins may also be precipitated by selective denaturation of such proteins by control of pH, temperature, or addition of organic solvents.

12.12 LYOPHILIZATION

Lyophilization is yet another solid–liquid separation technique useful for thermally labile substances such as biopolymers, cells, tissues, food products, etc. It is also known as *freeze drying* or *sublimation drying* as it involves freezing the material by exposure to cold air to convert the solvent (mostly water) to solid ice followed by sublimation of the ice at low pressures from the frozen state to produce a dried product. The drying process actually consists of sublimation which is the primary drying process and desorption as secondary drying process. The water content of the product is reduced to about 3% by weight of the freeze-dried product.

The equipment for freeze drying consists of a drying chamber, condensers, cooling system, and a vacuum system. The equipment is usually operated under sterile conditions particularly for lyophilization of injectible pharmaceutical products.

Lyophilization involves a series of distinct, integrated steps which include product preparation, freezing, primary drying, and secondary drying. The product mostly in the form of solution (or a suspension of biomass) is frozen in the first step under controlled conditions depending on the nature of the product. The freezing rate is critical since it affects the structural integrity of the freeze-dried product. Rapid cooling and freezing may result in the formation of very small ice crystals that lead to higher water vapour resistance and necessitates longer drying time. On the other hand, slow freezing will result in larger ice crystals affecting the product structure and making it coarser. Drying time is also enhanced. Hence, an optimum rate of freezing is to be selected depending on the nature of the product. Once the product is frozen the second step of drying is initiated by evacuating the chamber to low pressures and simultaneously heating.

During the second step of primary drying the drying and heating rates are to be critically controlled as they are crucial for the success of lyophilization process. The pressure within the chamber and the shelf temperature to which the product is to be heated during the primary drying stage depends on the nature of the product, its composition and eutectic or collapse temperature. The eutectic temperature is the highest temperature at which the product exists as a solid and during primary drying the temperature must be maintained below the eutectic temperature. The supply of heat to the product and the loss of heat from the product due to sublimation must be such that the product reaches a temperature which is lower than the shelf temperature. As water within the product is freezing supercooling may set in due to depression in freezing point. Supercooling can cause structural changes in the product and alter its physical appearance such as the formation of a skin which hinders the escape of water vapour. All the sublimed water must be removed by an efficient condenser otherwise the water may migrate back to the product hindering the drying process. Excessive heating may cause the melting of the product, and hence, should be avoided. When the primary drying is successful the final freeze-dried product will be stable with a long shelf life and can also be reconstituted properly. The increase in the temperature of the product to the temperature of shelf life indicates the end of the primary drying by sublimation. The control of pressure within the chamber is also critical for efficient lyophilization process. Extremely low pressures also result in inhibiting the primary drying process due to inefficient heat transfer. An inert gas (nitrogen) atmosphere within the chamber can enhance the heat transfer and improve the drying process thereby reducing overall cost.

After primary drying the residual moisture on the product surface is reduced by secondary drying. The temperature is slowly increased and simultaneously decreasing the partial pressure of water vapour within the chamber. The bound water desorbs from the product until the water content decreases to desired levels. The remaining residual water may no longer support chemical reactions or biological growth thereby promoting shelf life of the product. The nature of the product determines the final residual water content and thereby the duration of secondary drying. Excessive heat or duration of drying may cause the product to shrink or even char.

Lyophilization has several advantages over conventional drying such as (i) absence of crystallization or precipitation which usually accompany conventional drying, (ii) preservation of chemical and biological potency and homogeneity of the final product, (iii) ease of dispensing the product during final packaging, (iv) the freeze-dried product has higher stability, longer shelf life and broader temperature tolerance, (v) the product is stable against chemical degradation and can be easily reconstituted, and (vi) the freeze-dried product is free from contamination, and hence, lyophilization is a convenient preprocessing method for samples which cannot be analysed immediately. Industrially freeze-dried food products are of highest quality with no loss of flavour and aroma. The main disadvantages of lyophilization include high capital and energy cost and long process time.

Lyophilization is particularly useful for chemically and biologically unstable or labile products and is the only method available for providing long shelf life for biologically active products such as vaccines and high value proteins.

SOLVED PROBLEMS

EXAMPLE 1 *Calculate the weight of organic substance remaining in aqueous solution if 0.2 g of the substance is extracted from 100 mL of an aqueous solution once with 50 mL of an organic solvent. The distribution ratio D for the organic solvent–aqueous system is 50. Calculate the weight of organic substance remaining in aqueous solution if the extraction is carried out 3 times using 10 mL lots.*

SOLUTION

Substituting the given values in Eq. (12.3)

$$W_n = 0.2 \, [(100/(50 \times 50 + 100)]^1$$
$$= 0.2 \, [100/2600] = 7.69 \times 10^{-3} \text{ g} = 3.85\% \text{ remains in aqueous solution}$$

For extracting 3 times with 10 mL lots

$$W_n = 0.2 \, [(100/(50 \times 10 + 100)]^3$$
$$= 0.2 \, [100/600]^3 = 9.26 \times 10^{-4} \text{ g} = 0.46\% \text{ remains in aqueous solution}$$

Thus, it is clear that extracting several times with smaller volumes of organic solvent is more efficient than extracting once with a large volume.

EXAMPLE 2 *Calculate the efficiency of extraction in per cent for a given extraction of a metal chelate from 75 mL of aqueous solution with 25 mL of organic solvent if the value of D for the given system is 26.*

SOLUTION

Substituting the given values into Eq. (12.4)

$$E = 100 \times 26/[26 + (75/25)] = 89.66\%$$

EXAMPLE 3 *Calculate the volume of an organic solvent required to remove 99% of an organic substance with a single extraction from an aqueous solution of 60 mL containing 5 mg of the substance. The value of the distribution ratio D is 3.2 for organic solvent/water system.*

SOLUTION

Substituting the given values in Eq. (12.3)

$$W_n = W_A \, [(V_A/(D \, V_o + V_A)]^n$$
$$1 = 100[60/(3.2 \, V_o + 60)]^1$$
$$V_o = 1856 \text{ mL}$$

Volume of organic solvent required for single extraction = 1856 mL

EXAMPLE 4 *Calculate the number of extractions required to remove 99% of the substance in problem 3 with 50-mL the same solvent.*

SOLUTION

$$W_n = W_A \, [(V_A/(D \, V_o + V_A)]^n \qquad (12.3)$$
$$1 = 100[60/(3.2 \times 50 + 60)]^n$$
$$1/100 = (60/220)^n$$
$$\log 0.01 = n \log 0.27$$
$$n = 3.51 \text{ rounded off to } 4$$

Number of extractions involving 50 mL of organic solvent required = 4

Hence, multiple extractions involving small volumes are more efficient.

EXAMPLE 5 *Calculate the value of the distribution ratio of a substance between an organic solvent and water if 97.5% of the substance is extracted from 100 mL of aqueous solution by 100 mL of the organic solvent.*

SOLUTION

Substituting the values of $E = 97.5\%$; $V_A = 100$ mL, and $V_0 = 100$ mL into Eq. (12.5)

$$D = 97.5/(100 - 97.5) = 39$$

- Separation methods are broadly classified into (i) bulk- or large-scale separation techniques which include solvent extraction, ion exchange, membrane separation techniques, crystallization, precipitation, lyophilization, filtration and centrifugation and (ii) analytical separation techniques which include mostly chromatographic methods such as gas chromatography, HPLC, HPTLC, and electrophoresis.

- Solvent extraction is based on the selective distribution of the desired solute into one of the two immiscible phases brought into contact with each other.

- Selectivity is achieved by proper selection of the extracting solvent and the pH of the aqueous medium.

- Reactive extraction of solutes can be achieved by the use of certain additives to bring about (i) association as in the case of carboxylic acids, (ii) ion pair formation, and (iii) complex formation. The latter two are specifically applicable to extraction of metals.

- Liquid–liquid extraction may be carried out in (i) batch mode, (ii) continuous mode, and (iii) discontinuous counter-current distribution mode.

- Aqueous two-phase extraction involves two mutually insoluble solvent systems, each consisting of high concentration of water. Both the solvents contain mixtures of hydrophilic polymers or a hydrophilic polymer and salt dissolved in water and as the water content is high the phases do not denature proteins and other biopolymers.

- Proteins and other biopolymers in a given mixture can be separated by aqueous two-phase extraction based on the differential partitioning of proteins into the two solvent systems.

- Reversed micellar extraction involves the formation of reversed micelle by dissolving surfactants in organic solvents. The reversed micelles contain significant quantities of water and hydrophilic solutes and exhibit unique solubilizing characteristics towards proteins and hence the technique is useful in the separation and purification of proteins.

- Partitioning and extraction of proteins from the bulk aqueous phase into the reversed micellar phase is influenced by factors such as pH, ionic strength, and the type of salt as well as the nature of organic solvents and surfactants.

- Supercritical fluid extraction uses a supercritical fluid as the extracting solvent. The principle of supercritical fluid extraction is based on the solvent power of the supercritical fluid. The solvent power depends on the density of the supercritical fluid, which in turn depends on the pressure and temperature. In general, the solvent power of a supercritical fluid increases with increasing density at constant temperature and at constant density it increases with increasing temperature.

- Supercritical fluid carbon dioxide finds wide use as the extracting solvent because of its advantages of being non-toxic and non-hazardous in nature. In addition, the extracting conditions can be easily optimized through parameters such as temperature, pressure, and use of a co-solvent for the recovery and purification of the desired solute components from a mixture.

- The solubility of organic compounds in supercritical fluid carbon dioxide depends on the polarity and the molecular weight of the compounds. Non-polar substances, aliphatic hydrocarbons, and monoterpenes are completely soluble whereas polar substances are not soluble. The solubility of weakly polar substances decreases with increasing molecular weight of the compound.

- The method is exclusively useful for extracting highly labile or unstable components such as food flavours and aroma chemicals.

- Solid phase extraction involves the use of a solid as the extracting phase. The technique is useful for pre-concentrating organic analytes present in low concentration in aqueous solvents. The organic analyte in the aqueous phase is extracted by contacting with a packed bed of the solid and recovered by washing with small quantity of a suitable organic solvent for subjecting it to further analysis by chromatographic or spectroscopic methods.

- Ion-exchange separation is based on reversible competitive exchange of ions of the same sign (e.g., positive- or negative-charged ions) between a water-insoluble solid matrix called the ion exchanger and the surrounding liquid medium. The method is limited to samples containing ionized or partially ionized solutes.

- Cation and anion exchangers of natural as well as of synthetic origin are commercially available with strong or weak ion-exchange capabilities.

- Ion-exchange capability of an ion exchanger is influenced by factors such as the charge on the ion, the size of the hydrated ion, and concentration of the exchanging ions in solution. In addition the nature of the ion exchanger matrix, the nature of the functional groups attached to the matrix, and the extent of cross-linking in a polymer matrix also influence the ion-exchange equilibrium.

- Filtration is a separation method for separating the solid from the mother liquor. The separating is effected by passing the slurry through a septum called the filter medium.

- In normal flow or dead end filtration the slurry is pumped in perpendicular direction to the filter medium, whereas in cross-flow filtration it is pumped in parallel direction.

- Filtration is carried out in batch mode as well as in continuous flow mode.

- Membrane separation techniques include microfiltration, ultrafiltration, and reverse osmosis based on hydrostatic pressure as the driving force. Dialysis makes use of the concentration difference between different streams whereas electrodialysis uses applied electric field as the driving force to bring about separation of a variety of solutes in a slurry.

- Microfiltration makes use of a filter membrane made of polypropylene or polytetrafluoroethylene or alumina or zirconia to filter suspended solids in the size range of 0.1 to 10 microns by applying a pressure of 1–2 bar.

- Microfiltration is mostly used for harvesting microbial cells from fermentation broths and also for separating blood cells and plasma from whole blood.

- Ultrafiltration operating at pressures in the range of 2–10 bar makes use of a membrane made from synthetic polymers such as polysulphone. The membrane is microporous with the pore size being smaller than the pore size of microfiltration membrane. The technique separates low molecular weight solutes from high molecular weight solutes.

- Ultrafiltration is used for the recovery of whey proteins from whey in dairy industry, to concentrate and clarify fruit juices in fruit processing industry and concentrating a variety of products from cell-free fermentation broths in pharmaceutical industry.

- Reverse osmosis membrane is virtually non-porous with pore sizes in the range of 0.0001–0.001 μm. The membrane is permeable to water only but not dissolved salts of even low molecular weight. In reverse osmosis, an applied pressure greater than the osmotic pressure in the range of 20–100 bar is imposed on the membrane reversing the flow of the solvent from the solution side to the pure solvent side.

- Reverse osmosis is mainly used for desalination of seawater and for separating low molecular weight compounds such as sugars or organic acids from aqueous solutions. Its use has been extended to food and dairy industries to concentrate fruit juices, vegetable juices and milk, and treatment of wastewater.

- The separating ability of a membrane in pressure driven membrane separation methods of MF, UF, and RO is expressed in terms of retention coefficient or rejection coefficient.

- Membrane separation processes are affected by the two factors, concentration polarization, and fouling of the membrane.

- Dialysis is used for purification of blood and is known as haemodialysis. Dialysis has potential applications in the separation and concentration of inorganic colloids, separation of alcohol from beer, etc.

- Electrodialysis makes use of ion selective cationic and anionic ion-exchange membranes to desalinate brackish water to yield potable water.

- Crystallization involves separation of dissolved solutes from the mother liquor. The technique yields solids of high purity and hence crystallization is considered as a method of final purification of the desired product.

- Precipitation involves the conversion of a soluble substance into an insoluble one by the addition of a specific or selective reagent so that centrifugation or filtration can be used to separate the solid.

- Lyophilization involves the isolation of thermally labile solids from the mother liquor by freezing the solvent water to solid ice followed by sublimation of the ice at low pressures to yield a dried product.

REVIEW QUESTIONS

1. Give an overview of separation methods.
2. Explain the principles of solvent extraction.
3. Explain the term separation factor.
4. Write a note on the different extraction processes.
5. Give a detailed account on reactive extraction of metal complexes.
6. What is dissociative extraction? Give examples.
7. Write a note on the various factors which affect extraction process.
8. What are ion-association complexes? What is their significance?
9. Give an account of the different modes of extraction.
10. What are aqueous biphasic systems? Give examples. How are they useful?
11. Give an account of the theoretical principles involved in the aqueous two-phase extraction of an enzyme.
12. How is reverse micellar extraction of a biopolymer carried out?
13. What is a supercritical fluid? What are its characteristics?
14. Discuss the principles of supercritical fluid extraction.
15. What is solid phase extraction? What is its use?
16. Classify the membrane separation methods and explain the principles involved.
17. What is cross-flow filtration? What is its advantage?
18. What is reverse osmosis? How is brackish water desalinated by reverse osmosis?
19. Write a note on electrodialysis.
20. Give a neat sketch of membrane separation unit and describe its working.
21. What are the different configurations of membrane separation modules?
22. Discuss the theoretical and practical aspects of crystallization.
23. What is lyophilization? How is it carried out? What are its advantages?

13 Chromatographic Separations

13.1 INTRODUCTION

Chromatographic methods include a group of closely related separation techniques based on the differences in reversible physicochemical interactions between the different components of a mixture and the separating medium. Chromatographic methods are mostly used for analytical purposes in qualitative and quantitative analysis of sample mixtures of microgram to milligram quantities. Chromatographic methods also find use in research laboratories and bioprocess industries for isolating milligram to gram quantities of materials.

Mikhail Tswett, a Russian botanist, invented the technique to separate and identify the pigments of tree leaves and coined the term *chromatography* in 1906 (Greek chroma–colour and graphein–writing). He used a simple experimental set-up consisting of a glass column packed with chalk powder (calcium carbonate) as adsorbent to separate the pigments from an ether extract of green leaves. He applied a small volume of the extract at the head of the column and irrigated the column with a solvent. The adsorbed components of the mixture separated into five component pigments as coloured bands which were easily observed on the white background of the adsorbent. He extruded the adsorbent from the column and extracted the individual coloured bands with alcohol in separate containers. He was able to identify the component pigments as chlorophyll-β (pale yellow), chlorophyll-α (bluish green), viola xanthene (yellow), xanthophyl (yellow), and carotene (yellow-orange).

The experimental set-up and methodology remain the same even at present with suitable addition of instrumental components such as a pump to deliver the solvent at the head of the column and a detector attached at the exit end of the column with associated electronics as most of the present day instruments are microprocessor/computer controlled.

Chromatography may be defined as the selective distribution of the components of a mixture between two immiscible phases in intimate contact with each other. One of these phases is called the *stationary phase* and consists of a bed of solid particles while the other phase is called the *mobile phase* (*eluent* or *carrier gas* depending on whether it is a liquid or gas). The mobile phase percolates through the bed of stationary phase particles. The sample is usually dissolved in the mobile phase.

13.2 CLASSIFICATION OF CHROMATOGRAPHIC METHODS

Chromatographic methods may be classified in three different ways. The first classification is based on the physical configuration of the stationary phase and the manner in which the mobile phase is brought into contact with the stationary phase.

On this basis chromatographic methods may be classified broadly into two types: (i) column chromatography and (ii) planar chromatography. In column chromatography the stationary phase particles of uniform size are packed in a glass, plastic, or metal column. The mobile phase may be a liquid or a gas or a supercritical fluid which flows through the column at a constant flow rate. Examples of column chromatographic techniques include gas chromatography (GC), high-performance liquid chromatography (HPLC), and supercritical fluid chromatography (SCFC). In contrast, in planar chromatography the stationary phase particles are coated on a flat plate of glass, plastic, or metal foil as in thin layer chromatography (TLC) and high-performance thin layer chromatography (HPTLC). Planar chromatography also includes paper chromatography (PC) in which the stationary phase consists of water molecules held in the interstices of a filter paper. The mobile phase moves through the stationary phase by capillary suction or by gravity in planar chromatography.

A second classification is based on the nature of mobile phase. Accordingly chromatographic methods may be classified into (i) liquid chromatography, (ii) gas chromatography, and (iii) supercritical fluid chromatography. Liquid chromatography employs a liquid as the mobile phase. An important and sophisticated liquid chromatographic technique is HPLC. The mobile phase in gas chromatography is also called the carrier gas and gases such as nitrogen, helium, argon, and hydrogen are usually used as the carrier gas depending on the type of detector used in GC. Supercritical fluid chromatography makes use of a supercritical fluid (e.g., supercritical fluid carbon dioxide) as the mobile phase.

A third more fundamental classification is based on the mechanism of separation. The five different mechanisms that often find extensive use in chromatographic separations include (i) adsorption, (ii) partition, (iii) ion exchange, (iv) size exclusion, and (v) affinity. HPLC instrument is amenable to introduce all the five mechanisms by appropriate selection of the operating mode, namely selection of the stationary phase column in conjunction with the mobile phase, and hence, is a versatile technique for the separation of a wide variety of substances. In contrast, GC makes use of only partition or adsorption as the separation mechanism. A summary of the classification is listed in Table 13.1.

Table 13.1 Classification of chromatographic methods

Mobile phase	Stationary phase	Mechanism of separation	Technique
1. Liquid chromatography			
Liquid	Solid	Adsorption	LSC, TLC, HPTLC
	Liquid coated on inert solid support	Partition	PC, LLC, normal phase HPLC
	Surface-modified solid stationary phase	Hydrophobic interaction and affinity	HIC, RPC, RP- HPLC; AC
	Ion-exchange resin	Ion exchange	IEC
	Liquid in the pores of polymeric beads	Size exclusion	SEC
2. Gas chromatography			
Gas	Solid	Adsorption	GSC
	Adsorbed liquid or organic-bonded phase	Partition	GC
3. Super critical fluid chromatography			
Supercritical fluid	Organic-bonded phase	Partition	SCFC

AC–affinity chromatography; GC–gas chromatography; GF–gel filtration; GPC–gel permeation chromatography; GSC–gas solid chromatography; HIC–hydrophobic interaction chromatography; HPLC–high-performance liquid chromatography; HPTLC–high-performance thin layer chromatography; IEC–ion-exchange chromatography; LLC–liquid–liquid chromatography; LSC–liquid–solid chromatography, PC–paper chromatography, RPC–reversed phase chromatography; SCFC–supercritical fluid chromatography; TLC–thin layer chromatography.

13.3 COLUMN CHROMATOGRAPHY

Column chromatography consists of a cylindrical column packed with uniform-size particles of the solid stationary phase. The packing density of the particles is uniform and the mobile phase liquid flows through the bed of stationary particles at a constant flow rate.

A simple experimental set-up of column chromatography is shown in Fig. 13.1. The main components are (1) solvent or eluent reservoir, (2) pump (e.g., peristaltic pump), (3) sample inlet, (4) column containing the packed stationary phase, (5) a suitable detector (e.g., spectral or fluorescence detector), (6) a data processor with display or recorder provision, and (7) fraction collector. The eluent is delivered by the pump, from the reservoir to the head of the column, at a constant flow rate.

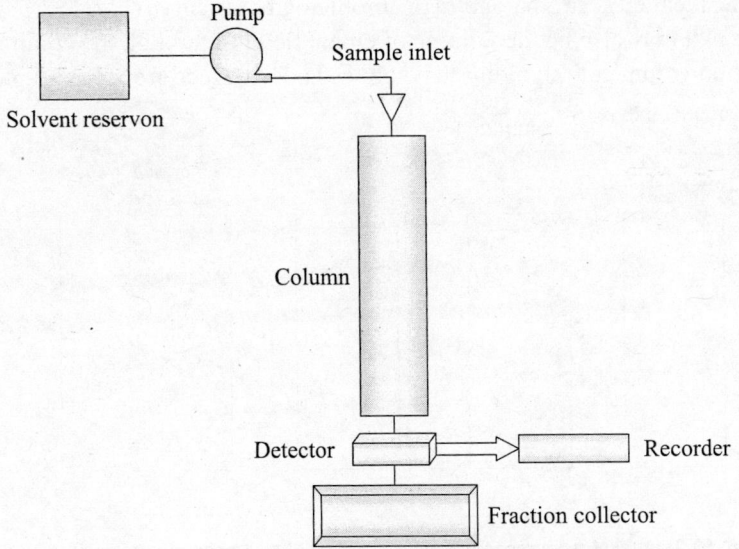

Fig. 13.1 Experimental set-up for column chromatography

13.3.1 Principle of Separation in Column Chromatography

Chromatography for analytical purposes is carried out by *elution analysis*. The primary aim of the technique is to separate the components of the sample mixture and thereby facilitate qualitative and quantitative analyses.

Elution analysis involves the application of the sample mixture at the head of the stationary phase column followed by pumping the mobile phase at a constant flow rate. The separation of a two component mixture of A and B may be described to explain the separation process. The column containing the chosen stationary phase is equilibrated with the mobile phase. A small volume (or pulse) of the sample mixture (about 3–5% of column volume) dissolved in the mobile phase or a compatible solvent is introduced at the head of the column through the sample inlet (e.g., a dropper pipette or capillary tube). Almost immediately the components of the sample solution adsorb onto the stationary phase. Once the adsorption of the sample is complete, the mobile phase is introduced continuously, at a constant flow rate, at the head of the column. The flow of the mobile phase results in desorption of the solute components by dissolution and their migration down the column along with the mobile phase. The solute components are exposed to fresh surfaces of the stationary phase leading once again to their adsorption on the stationary phase resulting in the distribution of the components of the mixture between the two phases. The mechanism of interaction between the component molecules and the stationary phase particles may be any one of the five mechanisms mentioned earlier. A series of such distributions between the mobile and the stationary phases occurs as the solute components move down the column leading to different migration rates of the solutes A and B. The differences in the migration rates of the two solutes in the mixture depend on the fraction of time each solute remains in the mobile phase. The different migration rates of the solutes lead to their separation into separate bands. The individual bands elute out of the column at different times. Figure 13.2 shows a schematic representation of separation in a column. The concentrations of the solutes are monitored with a suitable detector placed at the exist end of the column. The graphical plot of the detector signal, which is proportional to the amount of the solute components on the y-axis shown as a function of time, is called a chromatogram or chromatographic curve (Fig. 13.3). Instead of time the x-axis of the plot may also be the volume of eluent flowing through the column. The retention time or retention volume of the individual solutes is determined from the graphical plot.

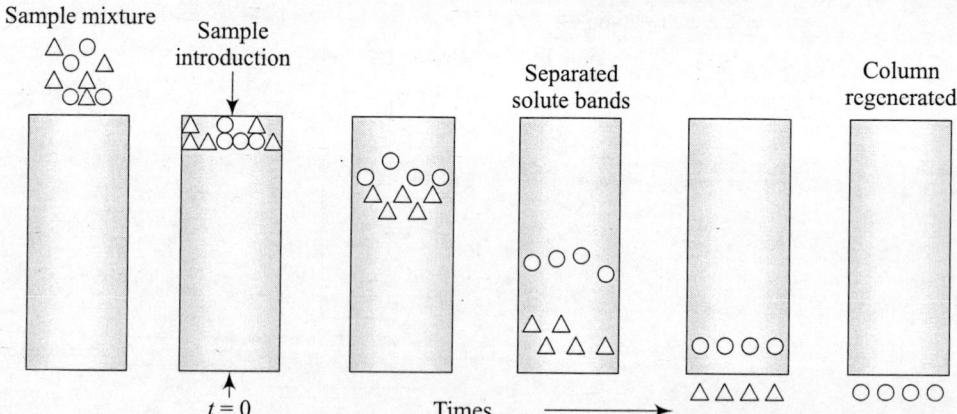

Fig. 13.2 A sketch representing the separation of a two-component mixture by column chromatography

The chromatogram shows peaks corresponding to the elution of the different solute components of the mixture, indicated as A and B. The small peak close to $t = 0$ is due to any totally unretained component of the mixture. The detector signal is proportional to the amount of the individual solute and is calculated from the area of the peak for the specific solute. The terms t_{R1} and t_{R2} refer to the retention times of the individual components of the mixture. The term t_m refers to the time taken by the mobile phase molecules to pass through the column. Retention times of individual solutes are useful in qualitative analysis. The chromatogram is also useful for quantitative analysis as the peak height or peak area is proportional to the amount of solute eluting out of the column, and hence, to the amount of the solute in the sample mixture.

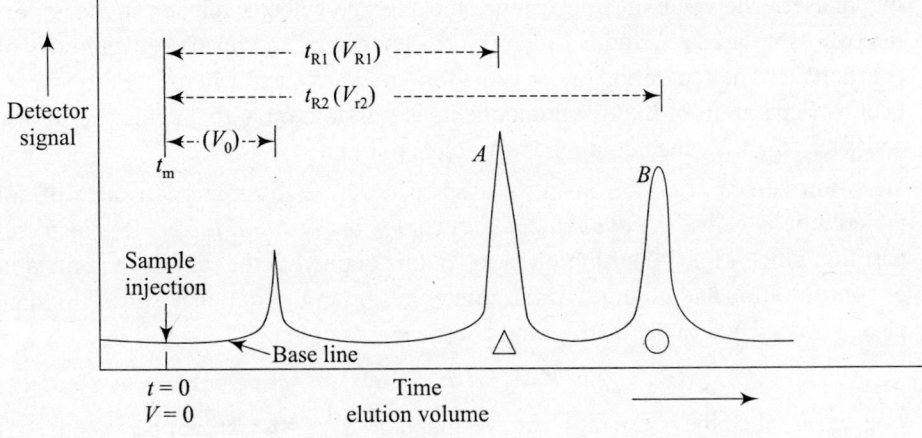

Fig. 13.3 Chromatogram of the two-component mixture

13.4 CHROMATOGRAPHIC PARAMETERS

Chromatographic parameters are of significance in the practice and evaluation of the performance of chromatographic separation. The various parameters discussed below are applicable to all types of chromatographic techniques.

13.4.1 Retention Time

Retention time of a solute is defined as the time taken by the solute to reach the detector from the moment of its injection into the column. It is determined by measuring the distance between the sample injection point to the apex of the peak from the chromatogram as shown in Fig. 13.3. It is expressed in time scale.

The retention time of a solute depends on factors such as (i) nature of the stationary phase, (ii) composition of the mobile phase, (iii) column length (and diameter to a minor extent), and (iv) mobile phase flow rate. The retention time of a solute increases with increasing length of the column and with decreasing flow rate of the mobile phase (i.e., solute spends more time in the column before eluting out). The retention time of a solute is independent of the quantity of the sample pulse introduced into the column and depends only on the nature of the sample. Hence, it is useful in qualitative identification of the solute. The solute may be identified by comparing its retention time with that of the authentic solute (chromatographic standard) determined under identical experimental conditions.

Retention time of the solute is related to the *partition ratio* or *distribution ratio* (also called *partition coefficient* or *distribution coefficient*) K, which describes the way in which a compound or solute distributes itself between the stationary and the mobile phases as

$$t_R = t_m \left(1 + K \left(\frac{V_s}{V_m} \right) \right)$$

$$t_R = t_m \left(1 + K \left(V_s / V_m \right) \right) \tag{13.1}$$

where $K = C_S / C_M$ (ratio of the concentration of the solute in the stationary phase to that in the mobile phase). The terms V_s and V_m represent the volumes of the stationary phase and the mobile phase respectively. If the value of $K = 1$ the solute is equally distributed between the two phases. The value of K determines the average velocity of each solute zone. Separation of the two components A and B of a mixture depends on the relative magnitudes of the partition coefficients. The separation factor is the ratio of K_A/K_B and when the ratio is one no separation occurs. Separation of the two components is possible on a given column only when $K_A \neq K_B$ and when $K_B > K_A$ then component A elutes out first.

A normalized retention quantity used to describe the migration rates of solutes through the column is called the *column capacity factor* or *retention factor*. It is also called the solute partition ratio or mass distribution ratio, k'. It is defined as the ratio of the total amount (product of concentration and volume) of the solute present in the stationary phase to that in the mobile phase. It is related to the distribution coefficient as given by

$$k' = (C_s V_s)/(C_m V_m) \text{ or } k' = (K/\beta) \tag{13.2}$$

where β is called the phase ratio and is equal to (V_m/V_s –the ratio of the volume of mobile phase to that of the stationary phase). The capacity factor is a measure of the time spent by a solute in the stationary phase relative to the time spent in the mobile phase. The value of k' can be calculated from the retention time of the solute as given by

$$t_R = t_m (1 + k') \quad \text{or} \quad k' = \frac{t_R - t_m}{t_m} \tag{13.3}$$

The value of k' should be between 1 and 5 in practice. If $k' = 0$ then the solute is not retained on the column.

In planar chromatographic techniques of paper chromatography and thin layer chromatography the *retardation factor* or *retention ratio*, R_f, is defined as the mobility of the solute relative to the mobility of the solvent or mobile phase and is related to the capacity factor k' as

$$k' = (1 - R_f)/R_f \tag{13.4}$$

13.4.2 Retention Volume

Retention volume (V_R) is defined as the volume of the mobile phase (in mL) required to elute a solute from the point of its introduction at the head of the column to its exit from the column. It is obtained from the chromatographic plot of the detector signal vs. volume of eluent. It may also be calculated by multiplying the retention time of the solute with the flow rate, F (mL per unit time), of the mobile phase ($V_R = t_R \times F$). The retention volume is related to the partition coefficient as

$$V_R = V_0 (1 + K(V_s/V_m)) \quad \text{or} \quad V_R = V_0 + K V_s \tag{13.5}$$

where V_0 is called the void volume or dead volume of the column and is equal to the product ($t_m \times F$). It is a measure of the total volume of the mobile phase contained within the column. The capacity factor k' can also be calculated from the retention volume.

$$k' = V_R - V_0/V_0 \tag{13.6}$$

13.4.3 Relative Retention

Relative retention of a chromatographic system describes the differential migration rates of two solutes and indicates its ability to separate the peaks of two solutes. It is also called the *separation factor* or *selectivity factor* and given the symbol α. It is calculated from the relationship

$$\alpha = K_2/K_1 \quad \text{or} \quad k'_2/k'_1 \quad \text{or} \quad t_{R2}'/t_{R1}' \text{ provided } t_{R2}' > t_{R1}' \tag{13.7}$$

where t_{R2}' and t_{R1}' refer to the *corrected retention times* calculated from the relationship $(t_R' = t_R - t_m)$. A proper choice of the two phases is necessary to achieve the relative retention of a pair of solutes to be at least 1.05 or greater up to a maximum of 2. Values of $\alpha > 2$ increase the time required for completing the chromatographic run.

13.4.4 Column Efficiency

The efficiency of the separation effected by the column is related to the width of the chromatographic peak. Ideally the chromatographic peak is symmetrical with a Gaussian shape.

A Gaussian peak is described by the equation, the peak widths at the inflection point at half height and at the base of the peak being related in terms of the standard deviation as shown in Fig. 13.4.

$$y = y_0\, e^{-x^2/2\sigma^2} \tag{13.8}$$

where y is the height at any point on the curve and y_0 is the height at the peak maximum, x is the distance between the point on the curve and the ordinate passing through the peak maximum, and σ is the standard deviation. The square of the standard deviation is variance.

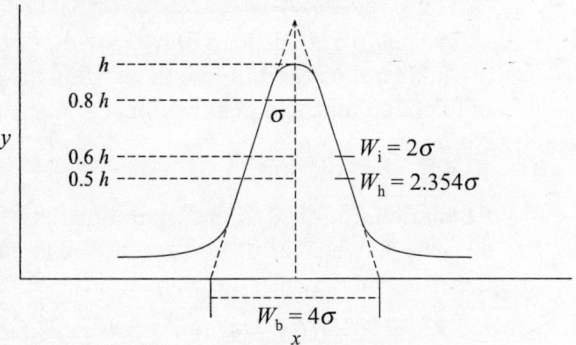

Fig. 13.4 An ideal chromatographic peak

The Gaussian peak is symmetrical with respect to the ordinate, and hence, the peak width w at any point is $= 2x$. The inflection points $(x = \sigma)$ of the peak occur at a height of $0.607\, y_0$ and the tangents to the peak drawn from the inflection points intercept the baseline at $x = 2\sigma$.

Hence, the peak widths at the inflection points W_i, at half height W_h, and at the base, W_b are related to standard deviation as

$$W_i = 2\sigma\,;\ W_h = 2\sigma\sqrt{(2\ln 2)} = 2.354\ \sigma\,;\ W_b = 2W_i = 4\sigma \tag{13.9}$$

The efficiency of a chromatographic column is conveniently described in terms of two related parameters, namely, the *number of theoretical plates*, n and the *plate height*, h. The concept of theoretical plate is useful to characterize the performance of the chromatographic column. All symmetrical peaks in a chromatogram indicate approximately the same plate number which is a measure of the total efficiency of the column. The number of theoretical plates is calculated from retention time and peak widths using the relationships

$$n = 16\,(t_R/W_b)^2 \tag{13.10a}$$

or

$$n = 5.545\,(t_R/W_h)^2 \tag{13.10b}$$

where W_b and W_h refer respectively to the peak widths at the base and half height.

The efficiency of a column is also expressed in dimensionless quantity as the *number of effective theoretical plates* N using the corrected retention time t_R' which is more representative of the actual retention of the solute. Equation (13.10a) may be rewritten as

$$N = 16\,(t_R'/W_b)^2 \tag{13.11}$$

The parameters N and n are related as

$$N = n\,(k'/(1 + k')^2 \tag{13.12}$$

As the value of k' increases, the term $(1 + k')$ becomes approximately equal to k', the value N approaches the value of n.

The plate height h is defined as the length of the column occupied by one theoretical plate (also called *height equivalent to a theoretical plate*–HETP). It is related to the length of the column (bed length), L and variance σ^2 as

$$h = \sigma^2/L \qquad (13.13a)$$

and

$$h = L/n \qquad (13.13b)$$

The *height equivalent to an effective theoretical plate* (HEETP) or $H = L/N$.

The relative efficiencies of two columns of same or different lengths may be compared in terms of number of plates/unit length by using the same compound as reference for the determination of retention time and peak width.

13.4.5 Resolution

Resolution R_s, refers to the separation between two chromatographic peaks. It is defined as the distance between the peak maxima as compared with the average base widths of the two peaks. It is calculated by using the equation

$$R_s = 2\,[(t_{R2} - t_{R1})/(W_{b1} + W_{b2})] \qquad (13.14)$$

with the retention time and peak width being measured in the same units to give a dimensionless value to resolution.

Three factors that affect chromatographic resolution include (i) the column selectivity or separation factor (α), (ii) the retention factor or capacity factor (k'), and (iii) the number of theoretical plates (n) as given by the equation

$$R_s = \tfrac{1}{4}\,[(\alpha - 1/\alpha)\,(k'/1 + k')\,(N)^{1/2}] \qquad (13.15)$$

The combined effect of the three terms on the appearance of chromatogram is shown in Fig. 13.5.

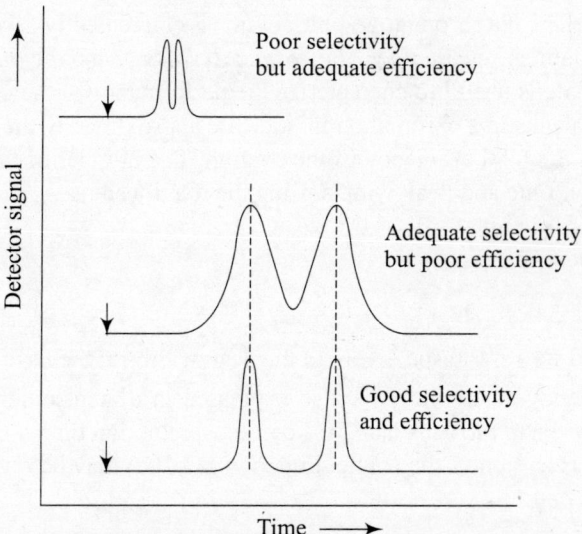

Fig. 13.5   The combined effect of the three terms of Eq. (13.15) on the separation of chromatographic peaks

The value of R_s should be ≥1.5 for achieving complete baseline separation of a pair of peaks. At this or greater resolution the purity of the peak is considered to be 100%. However, a completely resolved peak need not necessarily indicate a pure substance. A well-resolved peak may simply represent the presence of more than one of solute component, which have not been resolved under the selected experimental conditions.

13.4.6 Peak Asymmetry

Chromatographic peaks are usually asymmetric as shown in Fig. 13.6. Asymmetric peaks are observed at lower concentrations of the solute mainly due to higher k' values or due to poorly packed columns and sample injection problems. Peak asymmetry quantified by the degree of asymmetry or skewness, the asymmetry factor A_s, being defined as the ratio of the back half width b to the front half width a of the peak both measured at 10% peak height.

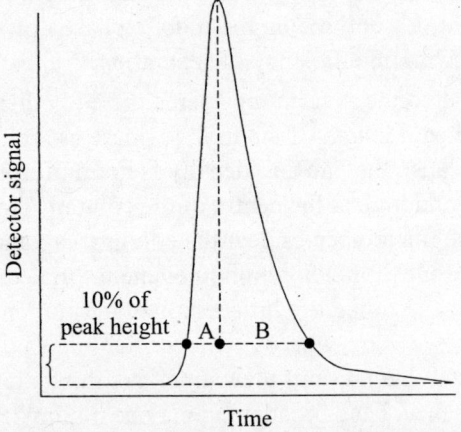

Fig. 13.6 An asymmetric chromatographic peak

Asymmetry of a chromatographic peak may also be due to tailing or fronting (Figs 13.7a and b). The tailing factor T_f is given as $(a + b)/2a$. For a Guassian peak, $a = b$ and both A_s and T_f are exactly equal to 1. For most of the packed columns the asymmetry factor ranges between 0.90 (fronting) and 1.10 (tailing) and are considered acceptable.

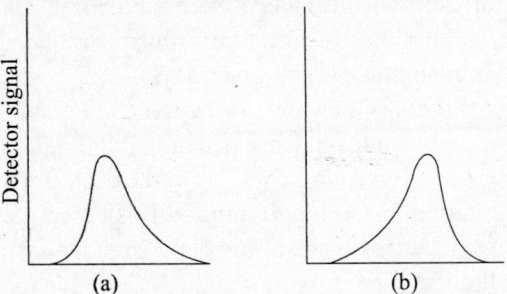

Fig. 13.7 Asymmetry in chromatographic peaks due to (a) tailing and (b) fronting

13.4.7 Broadening of Chromatographic Peaks

The separation efficiency of a chromatographic column generally decreases due the broadening of peaks (or band broadening) during the elution process. The plate height of a chromatographic column is a function of thermodynamic and kinetic processes that take place within the column, and hence, both thermodynamic and kinetic factors affect the separation efficiency of the column. Thermodynamic aspects such as partition coefficient cannot explain the broadening of chromatographic peaks. The broadening of elution bands has been explained by the kinetic theory developed by van Deemter and others. The kinetic factors that contribute to the band broadening in a chromatographic column as identified by the theory include the (i) inhomogeneity in the flow rate of the mobile phase, (ii) longitudinal diffusion of the solute molecules, and (iii) resistance to mass transfer between the two phases.

The relationship between plate height and the factors responsible for band broadening in liquid–liquid chromatography is given by van Deemter equation as

$$h = A + B/u + (C_{\text{stationary}} + C_{\text{mobile}})\, u \qquad\qquad (13.16)$$

The equation represents the variation of the plate height with the linear velocity of the mobile phase u and is useful for optimizing the mobile phase flow rate for a given column to achieve h_{min}, and hence, maximum efficiency of separation.

The first term called the A term represents the eddy diffusion due to the inhomogeneity of flow velocities and path length of the mobile phase as well as the solute molecules within the column. This is because the packing density is not uniform within the column; it is low close to the column wall and high at the centre of the column. This results in higher flow rates of the mobile phase and solute molecules near the column wall and smaller flow rates at the centre of the column. As a result of this flow inhomogeneity, molecules of the same solute elute out at slightly different times leading to a broadening of the elution band. The magnitude of A depends on the relationship $A = \lambda\, d_p$, where λ is a constant depending on the packing uniformity and column geometry and d_p represents the diameter of the stationary phase particle. The magnitude of the A term, and hence, band broadening may be minimized by decreasing the mean diameter of the stationary phase particles. In addition the packing density should be uniform and a constant flow rate of the mobile phase should be ensured. In the case of capillary columns used in gas chromatography there is no stationary phase particle packing, and hence, the first term is eliminated from the van Deemter equation.

The second term called the B term in the van Deemter equation relates to the effect of longitudinal or axial diffusion of the solute molecules as they pass through the column. The solute molecules diffuse from the concentrated centre of the band to the more dilute regions both forward and backward from the band centre within the mobile phase resulting in a broadening of the elution band. The magnitude of B is given as $B = 2\gamma D_m$, where γ is called the obstructive factor or tortuosity factor and D_m is the solute diffusion coefficient in the mobile phase. Longitudinal diffusion is minimized by uniform dense packing of the stationary phase particles in packed bed columns. The value of γ is equal to 1 for most of the coated capillary columns and 0.6 for packed bed columns. In addition to the longitudinal diffusion solutes with high diffusion coefficient values also exhibit axial diffusion resulting in the broadening of the elution bands particularly at low mobile phase flow rates because more time is available for diffusion. Band spreading is thus inversely proportional to mobile phase velocity. The magnitude of B, and hence, band broadening may be minimized by optimizing the flow rates of the mobile phase for a given column.

The resistance to mass transfer at the solute–stationary phase interface is represented by the $C_{\text{stationary}}$ term in the van Deemter equation. The resistance to mass transfer is proportional to t^2/D_s, where t is the effective thickness of the stationary phase and D_s is the diffusion coefficient of the solute in the stationary phase. Solutes having low D_s values exhibit slower rate of mass transfer leading to peak broadening. The rate of mass transfer can be improved by decreasing the effective thickness of the stationary phase and by choosing non-viscous liquids as the active stationary phase.

The C_{mobile} term represents the radial mass transfer resistance. It is proportional to the square of the particle diameter of the stationary phase packing and inversely proportional to D_m.

The effects of the individual terms of the van Deemter equation and the effect of the composite term on the plate height in shown in Fig. 13.8. The optimum flow rate range of the mobile phase is selected based on the h_{min} as shown in the same figure.

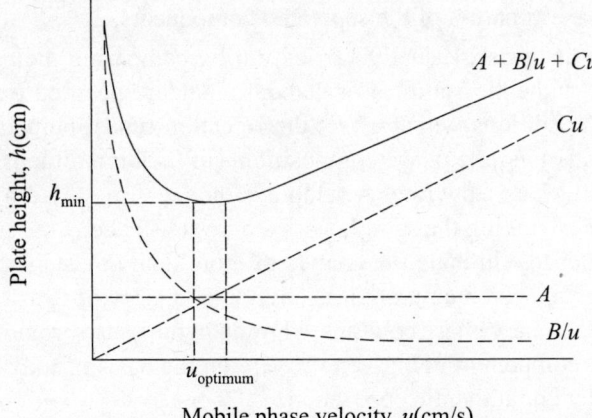

Fig. 13.8 van Deemter plot—plate height as a function of velocity of the mobile phase

Other factors generally called as *extra column factors* also contribute to band broadening. These include the void volume or dead volume of the instrument, the injector design, sample volume, and the flow properties of the detector, its volume, and response time. The term *void volume* refers to the volume of mobile phase held in the tubing connecting the different components of the instrument, the space between stationary phase particles within the column, unused space in the injector, and detector. The extra column band broadening may be minimized by using (i) short and narrow diameter connecting tubes, ferrules, and nuts of low void volume for connecting the different components of the instrument, (ii) direct on-column injection, (iii) a small sample volume as compared to the column volume, (iv) a properly designed injector, and (v) a detector with a small cell volume.

13.4.8 Optimization of Column Performance

The optimum conditions for a clean separation of the components of a mixture can be arrived at by considering Eq. (13.15). The three fundamental parameters, namely, α, k', and N can be varied independently to achieve maximum resolution. The parameters α and k' can be conveniently varied by changing the composition of the mobile phase in liquid chromatography through gradient elution or solvent programming and by increasing the temperature through temperature programming technique in gas chromatography. They can also be varied by changing the stationary phase. The parameter N can be varied by changing the length of the column, N increases with increasing length of the column as practiced in gas chromatography or by varying the particle size of the stationary phase and achieving maximum packing density within the column as practiced in HPLC. In addition the plate height h may also be minimized by varying the viscosity of the mobile phase to increase the diffusion coefficient of the solute in the mobile phase besides decreasing the particle size and increasing the packing density. The optimum flow rate of the mobile phase for a given stationary phase column is arrived at by van Deemter plot.

13.4.9 Applications of Chromatography

All chromatographic techniques are essentially separation techniques. Since separation is achieved even for closely related chemical species, chromatographic techniques are also useful in qualitative analysis for the identification of the different components of a sample mixture and for quantitative estimation of the separated components.

Qualitative analysis is usually carried out by comparing their retention times or retention volumes with authentic samples or standards under specified experimental conditions. The experimental conditions which affect the retention times include (i) nature of the stationary phase column, (ii) mobile phase composition, (iii) column dimensions, particularly the length, and (iv) mobile phase flow rate. A sample component may also be identified by the increased peak obtained by spiking the sample with a pure substance, whose presence in suspected in the sample. In order to eliminate the chance of more than one substance having almost the same retention time on a given column, it is advisable to carry out gas chromatographic analysis on two different stationary phase columns. Chromatograms also provide the evidence of the absence of a particular component in a given mixture on the basis of the comparison of retention time with the standard or authentic compound.

Alternatively, the eluting sample components may be trapped and subjected to further analysis or directly sent to a mass spectrometer or infrared spectrophotometer for identification. Gas chromatograph coupled to either mass spectrometer (GC-MS) or an infrared spectrophotometer (GC-IR) and similarly HPLC coupled to a mass spectrometer (LC-MS) are called coupled or hyphenated techniques.

Quantitative analysis is based on measuring the peak height or for more accurate determinations the integrated area of the peak assigned to the particular compound on the basis of retention time. Most of modern chromatographs have electronic integrators for precise determination of the peak areas taking into account any drift in the baseline and provide information of peak areas along with the retention times in the display. However, for accurate quantitative estimations, methods such as (i) calibration chart, (ii) internal standardization, (iii) internal normalization, or (iv) standard addition are employed.

The simplest and straightforward method is to use a calibration chart or graph prepared by recording the chromatograms for a series of standard solutions that approximate of the composition of the test sample. The graphical plot of peak height or peak area versus the concentrations of the standard in the series of standard solutions will yield a straight line passing through the origin. The concentration of the analyte in the test sample mixture is determined from the measured peak height or area in the chromatogram using the calibration chart.

In the internal standard method, an accurately measured amount of a standard is added to the sample and the mixture is chromatographed. The chosen internal standard should have a retention time close to that of the analyte component in the sample but well separated and its concentration also should be comparable to that of the analyte. The ratio of the peak areas of the standard and that of the analyte in the sample whose quantity is to be determined is calculated.

Since the peak ratio is the response ratio of the instrument for the internal standard and the analyte it is a reliable function of the amount of the analyte component and is independent of experimental conditions and size of the sample. The calculated ratio is used to compute the amount of the analyte component from previously prepared calibration chart. This method has the advantage of the highest precision for quantitative analysis in chromatography. In addition the detector response need not be known.

Internal normalization method is adopted when the components of the sample are similar in chemical composition and the response of the instrument is similar facilitating the detection of all the components. The percentage of a particular analyte A in the sample is given by the formula

$$\text{Percentage } A = (\text{peak area of } A/\text{total area of all components}) \times 100 \tag{13.17}$$

The formula assumes that the detector response is the same for each component of the sample mixture. When the detector response is not the same for all the components in the given mixture as in gas chromatographic separations, it must be determined using a set of standards and the peak areas have to be calculated using correction factors setting the response of one component equal to unity.

Standard addition method is adopted when the sample component in a given mixture is also available separately in 100% purity or as a standard. The sample is chromatographed before and after the addition of an accurately measured amount of the standard. The amount of the sample component present in the mixture is determined from the ratio of the peaks in the two chromatograms. This method is useful particularly in the analysis of complex mixtures for which a suitable internal standard may not be readily available.

13.5 LIQUID CHROMATOGRAPHY

Liquid chromatography (LC), also called liquid–liquid chromatography (LLC), makes use of a liquid as the mobile phase in conjunction with a variety of stationary phases held in a cylindrical column. A large number of chromatographic separations are carried out in liquid phase as compared to gas chromatography or supercritical fluid chromatography. All the five mechanisms mentioned earlier (see Section 13.2), namely, adsorption, partition, ion-exchange, size exclusion, and affinity, are exploited in liquid chromatography and discussed in more detail in Sections 13.6–13.10. Liquid chromatography may be practiced in the laboratory using a simple set-up described earlier (Fig. 13.1) or using the sophisticated instrument called high-performance liquid chromatograph (HPLC) (see Section 13.11).

13.5.1 Practice of Liquid Chromatography

The stationary phases commonly used vary widely depending on the mechanism of separation to be exploited. Adsorption chromatography makes use of packed columns of inorganic solids such as silica, alumina, and calcium carbonate as the stationary phase for separation of mixtures of components by selective or differential adsorption. In all other chromatographic techniques the stationary phase is a *bonded phase*, that is, chemically modified support particles. The active stationary phase is supported or immobilized as a thin film on an inert solid support mostly by covalent coupling. The support particles used include silica, alumina, porous glass beads, cellulose, polyacrylamide, agarose, dextran, etc. The support matrix material has to satisfy certain pre-requisites. The support should be water-insoluble, physically rigid to withstand the operational stress and at the same time porous and permeable to the mobile phase and solutes. The support should be chemically stable but at the same time amenable for chemical derivatization of the surface to bind the solute components through different mechanisms. It should have minimal non-specific adsorption characteristics. The material should be reusable and preferably of low cost.

The stationary phase particles of uniform particle size are packed in a cylindrical column made of glass, plastic or stainless steel by dry or wet packing methods. Dry packing involves pouring the solid particles into the column with gentle tapping to pack the column with uniform density. A better method is the wet method of packing which involves using slurry of the stationary phase particles in water and pouring the slurry into the column and simultaneously allowing the liquid to drain from the bottom end. The wet method is easy and also a better packed column of uniform packing density is obtained. The dimensions of the laboratory columns are usually 5–25 mm of inner diameter and a length of 30–75 cm. After packing the column the bed of stationary phase particles is washed with 4–5 column volumes of suitable solvent or equilibrating solvent prior to use.

The mobile phase, also called the eluent for LC may be pure solvent, binary, ternary, and quaternary mixtures of solvents, buffers, and salt solutions. The chosen mobile phase should be compatible with the chosen stationary phase in that it should be chemically inert towards the stationary phase and also to the sample components. The sample should be readily soluble in the mobile phase.

The sample is introduced by injecting a small amount (pulse) of sample with the help of a capillary or hypodermic syringe at the head of the column or a special arrangement called loop or adaptor. Solid as well as liquid samples are dissolved preferably in the mobile phase or some solvent compatible with the mobile and stationary phases.

The components of the sample mixture are eluted out of the column by different elution techniques, such as isocratic elution, step, or continuous gradient elution, involving a change of pH, ionic strength, concentration, and polarity of the eluent.

Isocratic elution is the simplest elution technique used for chromatographic development in which the eluent composition is maintained constant throughout the experiment. In isocratic elution the retention time of a component in a mixture depends on the partition coefficient of the components. Isocratic elution is practiced only when it is known that the sample components have retention times which give rise to well-resolved peaks. However, in the separation of most sample mixtures containing components which differ widely in their chemical compositions and nature, resolution is not accomplished by isocratic elution and this problem is referred to as the general elution problem. The general elution problem is conveniently overcome by gradient elution.

In *gradient elution* the composition of the mobile phase is changed during the chromatographic development either continuously or step-wise so as to change the partition coefficient values of each component with time. The gradient is formed by mixing two or more solvents differing in their polarities either incrementally (step-wise) or continuously (e.g., water–methanol mixture with varying composition as a function time). When aqueous buffers or salt solutions are used as the mobile phase the pH or ionic strength of the mobile phase may be varied to achieve gradient elution. The gradient elution technique is useful for samples with a wide range of capacity factor values, which are not easily separated under isocratic conditions. The most frequently used gradients are binary solvent systems with a linear, concave, or convex increase in the per cent volume fraction of the stronger or more polar solvent.

13.6 ADSORPTION CHROMATOGRAPHY

Adsorption chromatography also known as liquid–solid chromatography (LSC) involves adsorption as the mechanism of separation. The solute components of the sample mixture adsorb onto the stationary phase particles of silica or alumina. Other examples of stationary phases include calcium carbonate, talc, magnesium carbonate, charcoal, sugar, starch, and cellulose. The stationary phase particles are mostly spherical or sometimes irregular shaped, the average particle size varying between 3 and 10 μm. The stationary phase is highly polar and the order of retention of organic solutes is olefins < aromatic hydrocarbons < halides < ethers < nitro compounds < esters, carbonyl compounds < alcohols < amines < amides < carboxylic acids. The separation of the solutes is effected by altering the composition of the mobile phase by mixing solvents of varying polarity or eluent strength. The eluting power of the mobile phase is determined by its overall polarity and in general eluting power increases with increasing solvent polarity. Polar solvents are considered as strong solvents. Table 13.2 gives a short list of the solvents used for adsorption chromatography in the order of their eluting power, generally known as *eluotropic series*.

Table 13.2   A short list of eluotropic series of solvents

Solvent	Viscosity (C_p at 25°C)	Refractive index (at 25°C)	UV cut off wavelength (nm)	Solvent polarity (p') (partition)	Solvent polarity (ε^0) (adsorption)
n-Hexane	0.30	1.372	190	0.1	0.01
Cyclohexane	0.90	1.423	200	− 0.2	0.04
Carbon tetrachloride	0.90	1.457	265	1.6	0.18
Toluene	0.55	1.494	285	2.4	0.29
Benzene	0.60	1.498	280	2.7	0.32
Methylene chloride	0.41	1.421	233	3.1	0.42
n-Propanol	1.9	1.385	240	4.0	0.82
Tetrahydrofuran	0.46	1.405	212	4.0	0.57
Ethyl acetate	0.43	1.370	256	4.4	0.58
iso-Propanol	1.9	1.384	205	3.9	0.82
Chloroform	0.53	1.443	245	4.1	0.40
Acetone	0.30	1.356	330	5.1	0.56
Ethanol	1.08	1.359	210	4.3	0.88
Acetonitrile	0.34	1.341	190	5.8	0.65
Methanol	0.54	1.326	205	5.1	0.95
Water	0.89	1.333		10.2	

The mobile phase composition required for a good separation is arrived at by trial and error. In most cases better separations are achieved with solvents of low polarity.

Adsorption chromatography is useful mostly for the separation of mixtures of non-polar from moderately polar low molecular weight organic compounds (e.g., polyaromatics, fats, oils, etc.)

and particularly for the separation of mixtures of isomers (e.g., positional isomers, geometric isomers).

13.7 PARTITION CHROMATOGRAPHY

Partition chromatography is classified into two types based on the relative polarities of the stationary and mobile phase: (i) *normal phase chromatography* and (ii) *reversed phase chromatography*. Both the techniques are based on the distribution or partitioning of the solute components between the stationary phase and the mobile phase depending on the relative solubilities of the solute components in the two phases.

13.7.1 Normal Phase Chromatography

In normal phase chromatography the stationary phase consists of a thin layer of organic liquid (e.g., polyethylene glycol, β-, β-oxydipropionitrile, etc.,) coated on particles of an inert support material such as silica. The stationary phase is highly polar, and hence, the mobile phase is a non-polar solvent such as *n*-hexane or isopropyl ether. The least polar component of a sample mixture is the most soluble one in the non-polar mobile phase, and hence, elutes out first in normal phase chromatography. The eluting power of the mobile phase can be modified by mixing the solvents according to their eluting power (Table 13.2).

13.7.2 Reversed Phase Chromatography

Reversed phase chromatography is one of the widely used liquid chromatographic techniques for the separation of a variety of organic compounds. The term *reversed phase chromatography* (RPC) was coined to describe the use of a non-polar stationary phase in conjunction with a polar mobile phase in contrast to the normal phase chromatography. The polarity of the stationary phase is reversed from highly polar (used in normal phase chromatography) to non-polar by covalent binding of non-polar hydrocarbon groups, such as *n*-alkyl chains of octadecyl, octyl, or butyl groups and phenyl groups, on the surface of silica support particles. The covalent binding is carried out by treating the surface silanol groups with appropriate silanizing agents. The nature and density of the non-polar groups at the surface affect the retention, the loading capacity, and selectivity of the column. The most commonly used hydrocarbon bonded stationary phase is the octadecylsilane (ODS), in which a hydrocarbon chain of 18 carbon atoms is covalently bound to the silica particles. The stationary phase known as ODS or C-18 or RP-18 is highly non-polar. Similar relatively more polar octyl or C-8 (RP-8) and C-2 stationary phases are also used in RPC. Other commercially available silica-based polar bonded stationary phases include cyanopropyl (nitrile), diol, and aminopropyl.

The mobile phase in RPC is polar. Most commonly used mobile phase with RP-18 stationary phase is water–organic solvent mixture containing a buffer component and an organic modifier such as methanol, isopropanol, acetonitrile, or tetrahydrofuran. In addition an ion pairing agent may be added to affect the selectivity. The surface tension of the mobile phase is important in determining the magnitude of retention. Solute retention decreases with increasing concentration of the organic modifier as it affects the gross properties of the mobile phase such as surface tension, viscosity, dielectric constant, etc. The organic modifier also influences the efficiency of separation by its effect on diffusion rates of the solutes in the sample. Table 13.2 provides the list of the solvents (*eluotropic series*) applicable for the reversed phase chromatographic separations. The most commonly used mobile phase for C-18 column is methanol–water mixture, for C-8 column it is acetonitrile–water mixture, and for the C-2 column it is 1,4-dioxane–water mixture.

The sample mixture is dissolved in the initial mobile phase which is usually aqueous solvent and introduced into the column. The column is then washed with the initial mobile phase to remove any unbound and occluded solute components. Non-polar organic molecules are strongly retained on RPC columns when neat water (solvent with the highest surface tension) is the initial mobile phase. Elution of the adsorbed solute components is carried out by adjusting the polarity of the mobile phase composition so that the bound components desorb sequentially. The polarity of the mobile phase is usually decreased by gradually increasing the percentage of the organic modifier (increasing hydrophobicity) from an initial level of 0% to a final maximum content of 100%. The solute components elute out in decreasing order of polarity according to their individual hydrophobicities. The retention factor thus decreases with decreasing surface tension of the mobile phase. After elution of all the solute components any residual components that are still retained are washed out by using a mobile phase of 100% of organic modifier. The column is then re-equilibrated with the initial mobile phase. The RPC stationary phase gel may be washed finally with and stored in 20% ethanol or pure methanol.

In many applications of RP column *ion suppression* technique is employed to separate polar substances, such as organic acids and bases, by controlling the pH of the mobile phase with a suitable buffer (e.g., phosphate buffer). The ionization of the acids or bases is suppressed by the chosen pH in the range 2–8 facilitating the separation of undissociated molecules by RP column. Alternatively *reversed phase ion pair partition chromatography* (RP-IPC) may be used for the separation of polar substances on RP columns. The method involves the use of a mobile phase containing a pairing ion so that the ionic solute component of the mixture forms an ion-pair facilitating its separation on RP columns. For example, the mobile phase of methanol–water mixture containing a long chain alkyl sulphonate facilitates the formation of ion-pairs in the separation of mixtures of vitamins.

RP-18 column is suitable for the separation of a wide range of moderately polar organic mixtures and typical examples include carboxylic acids, amino acids, drugs, and pharmaceuticals. It finds application in biochemistry/biotechnology for purification and analysis of proteins, separation of diastereomeric peptides, membrane proteins, peptide sequencing, and structural studies, determination of protein purity, etc. RP-8 column is more suitable for the separation of more polar mixtures such as peptides, metabolites in body fluids, pesticides, herbicides, etc. RP-2 column is used for the separation of more polar substances and multiply charged species, for example, dyes, organic bases, etc.

13.7.3 Hydrophobic Interaction Chromatography

Hydrophobic interaction chromatography (HIC) is similar to reversed phase chromatography and is specifically applicable to separation and purification of proteins.

In both RPC and HIC the interaction between the solute components of a mixture and the stationary phase involves hydrophobic interaction. The stationary phase consists of covalently bound hydrophobic ligands, the density of ligands being much higher in RPC as compared to that in HIC. Though hydrophobic interaction is the main cause for separation, the mechanism at the molecular level is different for the two techniques. In RPC the stationary phase may be considered as a continuous hydrophobic phase and the interaction between the stationary phase and the solute molecules is relatively strong. Hence, only low-molecular weight organic molecules can be separated on RPC columns. The eluting conditions are relatively harsh with organic solvents being used as mobile phase. Hence, high-molecular weight biopolymers denature when applied to RPC columns. In contrast in an HIC column where the hydrophobic character of the stationary phase is much weaker as compared to an RPC column, retention factor increases with increasing salt concentration. Surface tension increases with increasing salt concentration in aqueous solutions and aids retention.

HIC stationary phase gels based on agarose, dextran, and cellulose with covalently bound phenyl, octyl, or butyl groups are commercially available for the separation of proteins and other biopolymers. HIC gels for use in HPLC are silica or organic polymer resin-based gels of uniformly small size (~ 1.5 μm) so that they can withstand high pressure operation. The HIC gels are packed into columns and equilibrated with of $0.01-0.05$ M buffers containing ammonium sulphate in the concentration range of $0.75-2.0$ M or sodium chloride in the range of $1-4$ M.

The mobile phase is an aqueous buffer with varying salt concentration. The sample mixture is usually dissolved in a suitable aqueous buffer containing salt and applied to the HIC column. The column is washed with the mobile phase to remove any physically occluded solute components. Since the retention factor decreases with decreasing salt concentration of the mobile phase the salt concentration is gradually decreased during elution of the adsorbed solute components.

HIC finds extensive use in biotechnology for the separation of mixtures of peptides and proteins as these solutes undergo denaturation when applied to RP columns.

13.8 ION-EXCHANGE CHROMATOGRAPHY

Ion-exchange chromatography (IEC) was developed in 1940s for the separation of ions of lanthanide and actinide series for nuclear applications. It is used extensively for the separation and purification of charged species because of its simplicity in operation. IEC is also a versatile technique with high resolving power and high capacity. The basic principle involves the use of ion-exchange resin as the stationary phase with capability to bind competitively and reversibly oppositely charged species. For example, when a sample mixture containing ions as well as neutral species is applied to a column containing cationic (negatively charged) exchange resin the solute components with more positive charges (M^{2+}) are adsorbed (exchanged) strongly as compared to those with less positive charges (M^+). Components with no net charge (M^0) or a net negative charge (A^-) pass through the column without getting adsorbed. The ion-exchange column is washed to remove any physically occluded solute components. The adsorbed positively charged species may be eluted out with a suitable buffer sequentially.

The ion-exchange reactions that occur in the resin bed and subsequent elution may be represented as follows:

$$RSO_3^-H^+ + M^{n+}, M^0, A^- \rightarrow (RSO_3^-)_n M^{n+} + M^0 + A^- + n\,H^+ \text{ (exchange)}$$
$$\quad\; \text{resin} \quad\;\; \text{sample solution} \qquad\quad \text{resin} \qquad\quad\;\; \text{solution}$$

$$(RSO_3^-)_n M^{n+} + H^+X^- \rightarrow RSO_3^-H^+ + M^{n+} + X^- \text{ (elution)}$$
$$\quad\;\; \text{resin} \qquad\quad \text{eluent} \qquad \text{resin} \qquad\quad \text{solution}$$

The sequence of steps, namely, adsorption of the sample components leading to exchange, washing, and the elution of adsorbed ions giving rise to a chromatogram are shown in Figs 13.9(a) and (b).

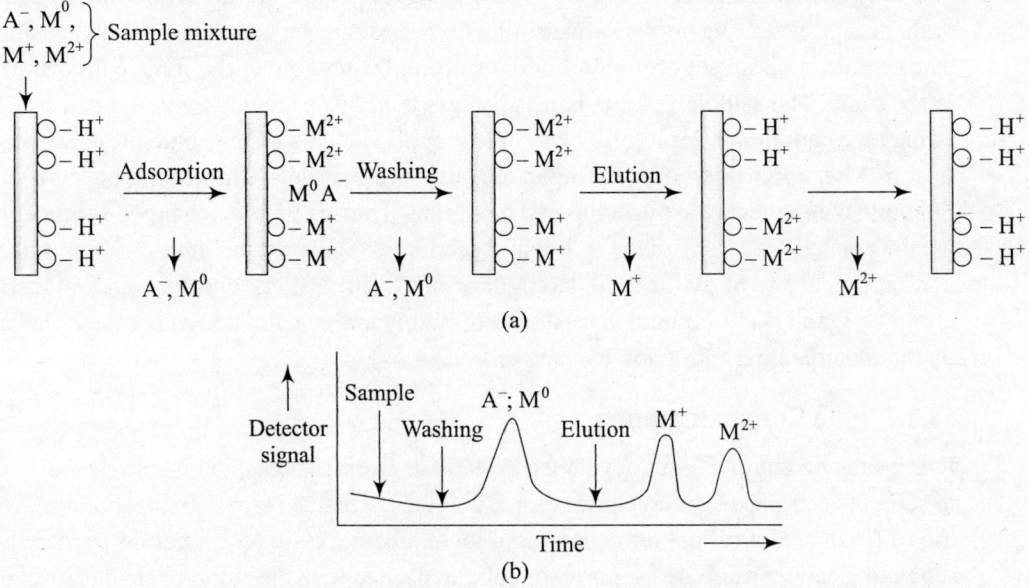

Fig. 13.9 (a) Schematic representation of principle of IEC and (b) the corresponding chromatogram

IEC can be performed using a simple set-up consisting of a peristaltic pump to deliver the mobile phase buffer, column packed with the chosen ion-exchange resin, and a UV/visible spectral or conductance detector.

The stationary phase can be either a cation- or an anion-exchange resin. Commercially available ion-exchange resins are in the form of water-insoluble porous-structured beads made from organic polymers, such as polystyrene cross-linked with divinyl benzene, and contain functional groups responsible for the ion-exchange properties of the resins. The functional groups are either acidic groups or basic groups. Cation exchangers have acidic functional groups, such as sulphonic ($-SO_3^-H^+$) or carboxyl ($-COO^-H^+$), with a net negative charge on the matrix and positively charged exchangeable counter ions (H^+) and are capable of exchanging H^+ ions for other cations (hence, called cation exchangers). Anion exchangers have basic functional groups, such as substituted amino (quaternary amino) groups ($\equiv N^+OH^-$) or carboxymethyl groups, with a net positive charge on the matrix. These resins are capable of exchanging OH^- ions for anions (hence, called anion exchangers). Sulphonic and quaternary amino groups are used to form strong ion exchangers while other functional groups, such as carboxyl and carboxymethyl, are used in weak ion exchangers. The terms *strong* and *weak* are relative and refer only to the extent of variation of ionization with pH and not to the strength of binding. Strong ion exchangers are completely ionized over a wide pH range. Weak ion exchangers exhibit a varying degree of ionization and also exchange capacity with change in pH.

The mobile phase is usually an aqueous buffer containing a non-buffering salt such sodium chloride to maintain ionic strength. The pH and ionic strength are the two important parameters influencing the property of the mobile phase.

In practice the steps involved in IEC include (i) equilibration of the column, (ii) preparation of sample and its application, (iii) washing to remove unadsorbed species, (iv) elution of the adsorbed/exchanged solute components, and (v) column regeneration. The column is equilibrated with the chosen starting buffer to ensure that the ion exchanger contains the exchangeable cations. The sample is dissolved preferably in the starting buffer so that the ionic composition of both is the same. The sample volume is usually less than 5% of the bed volume and free from any suspended particles or turbidity. The sample is applied with a syringe onto the column or through a loop. After application of the sample, the column is washed with starting buffer till no free or unbound components are available in the column. The adsorbed/exchanged solute components are then eluted out by any of the following techniques (i) isocratic elution, (ii) step-wise gradient elution by change of pH or ionic strength or both, (iii) affinity elution, and (iv) displacement chromatography. The column is washed thoroughly and regenerated as per the procedure given by the manufacturer of the ion exchanger.

13.8.1 Ion Chromatography

Ion chromatography (IC) was developed in 1970s as an extension of ion-exchange chromatography to separate cationic or anionic mixtures using ion-exchange resins. The technique involves the use of (i) an analytical column containing ion-exchange resin (e.g, a cation-exchange resin for separating cations) and (ii) a suppressor column packed with the anion-exchange resin to remove the ionic species other than the desired analyte ions in the eluting mobile phase. The sample cationic mixture is applied at the head of the analytical (cation-exchange resin) column and the exchanged cations are eluted with hydrochloric acid. The reactions may be represented as

$$RSO_3^-H^+ + M^{n+} \rightarrow (RSO_3^-)_n M^{n+} + n\,H^+ \text{ (exchange)}$$
$$\text{resin} \quad \text{sample solution} \quad \text{resin} \quad \text{solution}$$

$$(RSO_3^-)_n M^{n+} + HCl \rightarrow RSO_3^-H^+ + M^{n+} + Cl^- + \text{excess HCl (elution)}$$
$$\text{resin} \quad \text{eluent} \quad \text{resin} \quad \text{solution} \quad \text{eluent}$$

The solution exiting from the cation-exchange resin column passes through the suppressor column containing anionic resin in hydroxide form. The suppressor column converts the H^+ of the solution to molecular water without affecting the analyte cations.

$$R^+OH^- + M^{n+} + H^+ + Cl^- \rightarrow R^+Cl^- + M^{n+} + H_2O \text{ (suppression of } H^+)$$
$$\text{resin} \quad \text{solution} \quad \text{resin} \quad \text{solution}$$

The analyte cations are not retained in the suppressor column and can be detected and quantitatively determined by conductivity detector placed at the exit end of the suppressor column.

Similarly for the separation of anions the analytical column contains anion-exchange resin and the suppressor column contains a cation-exchange resin in the acid form. The mobile phase is usually sodium bicarbonate or carbonate. The reaction in the suppressor column involves the suppression of carbonate ions exiting the analytical column to form un-dissociated carbonic acid.

$$R^-H^+ + X^- + Na^+ + CO_3^{2-} \rightarrow R^-Na^+ + X^- + H_2CO_3 \text{ (suppression of } CO_3^{2-})$$
$$\text{resin} \quad \text{solution} \quad \text{resin} \quad \text{solution}$$

The analyte anions are not retained in the suppressor column and can be analysed by conductivity detector.

13.9 SIZE-EXCLUSION CHROMATOGRAPHY

Separations based only on molecular size and shapes of the constituents of a mixture are classified as size-exclusion chromatography (SEC). SEC may be broadly classified into two types: (i) gel filtration or gel filtration chromatography (GF) carried out using an aqueous mobile phase for the separation of biopolymers such as proteins, enzymes, polysaccharides, nucleic acids, etc., and (ii) gel permeation chromatography (GPC) for the separation of synthetic polymers in conjunction with organic solvents as mobile phase.

The principle involves the separation of molecules of different sizes brought about by the differences in the time spent by the solute molecules in the liquid mobile phase entrapped within the pores of the stationary phase particles. The stationary phase used in SEC consists of small ($\sim 10 \, \mu m$ size) porous particles or beads (also called gel) made of silica or cross-linked polymers. The particles have a network of uniform-size pores. Gels with different pore sizes in the range of 50 Å to about 10^6 Å are commercially available. The pore dimensions are such that the smaller analyte molecules depending on their hydrodynamic volumes (that is their size and shape) can diffuse into the pores along with the solvent molecules and are retained for longer time. Bigger sized molecules cannot enter the pores and are said to be excluded from the pores.

Thus the accessible volume of solvent is very much less for molecules totally excluded from the gel than for small molecules which are free to penetrate the gel as shown in the Fig. 13.10.

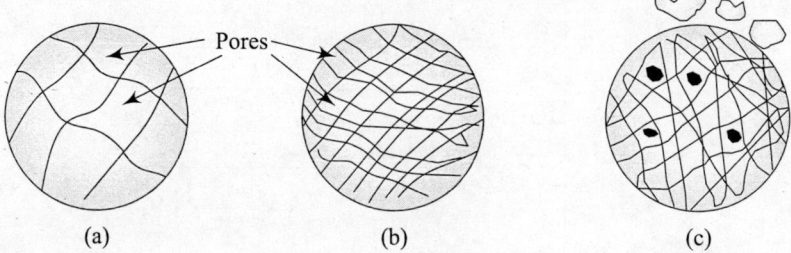

(a)　　　　　　　(b)　　　　　　　(c)

Fig. 13.10　Pore size variation in beads: (a) large pores, (b) small pores, and (c) exclusion of large-size molecules (⬭) from pores and inclusion of small molecules (⬤)

The time spent by molecules within the pores is related to the fraction of the pores that are accessible to the solute. When a mixture of small and large molecules of the solute is introduced into a column packed with stationary phase particles of controlled pore size and eluted with a suitable solvent, small molecules diffuse into the pores of the gel and follow a longer path than the larger molecules which are completely excluded from the stationary phase particles. This leads to differential migration rates within the column, the larger molecules moving down the column faster compared to smaller molecules. Eventually complete separation occurs, and the larger molecules elute out of the column first. The smallest molecule in the sample mixture elutes out of the column last. Molecules of intermediate sizes, whose diffusion into the pores of stationary phase particles depends on their individual sizes, elute out of the column at different time intervals between the time required to elute out the largest and the time required to elute out the smallest molecule, and hence, can be separated on a column of given gel.

Figure 13.11 shows the separation of a sample mixture of a polymer (or a mixture of biopolymers such as proteins) containing molecules having four different molecular weights (M_1, M_2, M_3, and M_4) on a column packed with stationary phase gel with a specified molecular weight cut off.

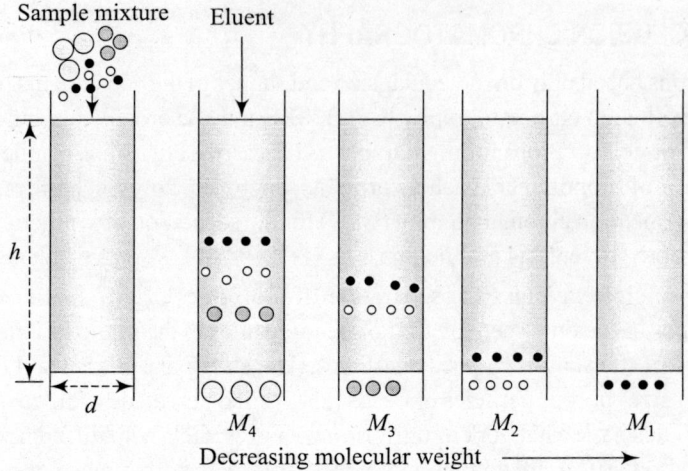

Fig. 13.11 Elution of molecules of different molecular weights of a polymer by SEC (Molecules of M_4 type are totally excluded, and hence, exit the column first.)

The corresponding gel permeation chromatogram and the distribution curve are shown in Figs 13.12(a) and (b) respectively. The distribution curve also known as selectivity curve is a plot of K_d versus the log molecular weight.

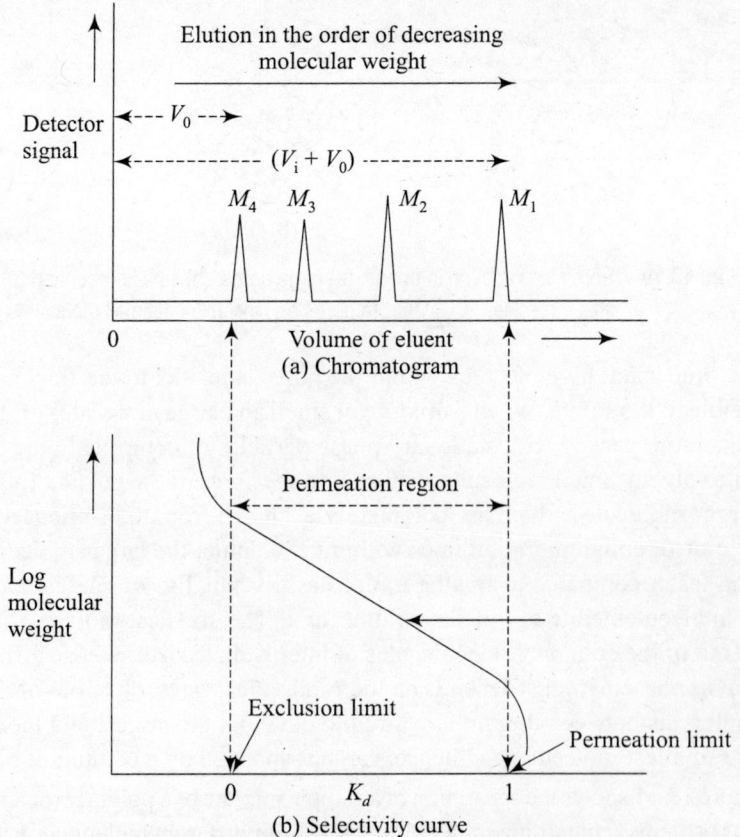

Fig. 13.12 (a) Size-exclusion chromatogram of a polymer sample and (b) selectivity curve

The solute molecules of molecular weight M_4 are totally excluded from the pores of the gel and elute out of the column along with the mobile phase first as shown in the chromatogram in Fig. 13.12. All solute molecules with greater molecular weights than M_4 also will not be retarded in the column, and hence, will elute out of the column together without getting separated.

Solute molecules of intermediate sizes (e.g., M_2 and M_3 in Fig. 13.11) elute out of the column at eluent volumes between the two limits V_o and $(V_i + V_o)$. Only solute molecules which have molecular sizes within the permeation limit of a given gel can be separated and will elute out of the column at different elution volumes.

Solute molecules which are small enough to have free access to the pores within the stationary phase particles elute out of the column at an eluent volume V_e which is the sum of inner volume (also called the pore volume) V_i and outer volume or void volume V_o. Thus for small molecules (e.g., M_1 in Fig. 13.11) $V_e = V_i + V_o$ corresponds to permeation limit. The inner volume can be calculated using the formula $V_i = WS$, where S is the grams of solvent (water) taken up by 1 g of xerogel (gel with no solvent) and W is the dry weight of the gel. The permeation limit refers to the molecular weight below which all solute molecules have complete access to the pores of the gel. All solute molecules having molecular weights below the permeation limit will elute out the column together giving rise to a single chromatographic peak and cannot be separated.

In general the elution volume V_e of a solute is given by the equation.

$$V_e = V_o + K_d \cdot V_i \quad \text{or} \quad K_d = (V_e - V_o)/V_i \tag{13.18}$$

where K_d is the distribution coefficient (also called selectivity coefficient). The distribution coefficient indicates the fraction of the inner volume accessible to a particular solute molecule and is independent of the geometry of the column and cannot be altered by a change in mobile phase composition. The value of $K_d = 0$ for large molecules which are excluded from the gel, and hence, $V_e = V_o$. For very small molecules $K_d = 1$ and such molecules diffuse freely into the pores of stationary phase particles. In such cases $V_e = V_o + V_i$. For molecules of intermediate size K_d will vary between 0 and 1. A value of $K_d > 1$ is indicative of adsorption or some other specific interaction of the solute on the stationary phase and it is usual to avoid such a situation in SEC.

A graphical plot of log molecular weight versus elution volume (or log molecular weight vs. K_d) gives a sigmoid selectivity curve as shown in Fig. 13.12(b). A linear relationship can be seen in the centre portion of the selectivity curve which may be expressed as

$$K_d = a - b \log M \tag{13.19}$$

The slope of the selectivity curve depends on the width of the pore size distribution of the stationary phase gel and the value of the intercept is a function of the mean pore size. The sigmoid shape of curve reduces the practical working range of the stationary phase gel to about $0.1 < K_d < 0.9$.

HPLC instrument is used for size-exclusion chromatography (GPC) particularly for synthetic polymers in conjunction with a refractive index detector. The GPC column is usually placed in a thermostatic oven for precise control of the temperature. A pure or mixture of organic solvents capable of dissolving synthetic polymers are used as the mobile phase.

Gel filtration exclusively used for the separation of biomolecules using aqueous buffers is mostly operated at atmospheric pressure using laboratory packed columns and peristaltic pump for mobile phase delivery. UV/visible spectral or fluorescence detectors are commonly used.

The practice of SEC involves the following steps: (i) selection of stationary phase gel, packing into the column, and equilibration, (ii) sample application, (iii) elution and sample collection, and (iv) column regeneration.

The stationary phase particles (gels) made from silica and polymers, such as polystyrene cross-linked with divinyl benzene and polyacrylamides, are mostly used in GPC as they are compatible with organic solvents. Stationary phase particles made from biopolymers, such as polysaccharides of dextran and agarose, are also commercially available for use in GF in conjunction with aqueous media for the separation of peptides, proteins, enzymes, nuclei acids, hormones, polysaccharides, etc.

Table 13.3 lists a few examples of stationary phase gels for use in GPC and GF.

Table 13.3   Stationary phase gels for SEC and their exclusion limits

1. Gel permeation		2. Gel filtration	
Stationary phase	Molecular weight exclusion limit	Stationary phase	Molecular weight exclusion limit
Silica	$2 \times 10^3 - 5 \times 10^4$	Cross-linked dextran (Sephadex)	
	$3 \times 10^3 - 1 \times 10^5$	G-10	up to 700
	$5 \times 10^3 - 5 \times 10^5$	G-15	up to 1500
	$5 \times 10^5 - 2 \times 10^6$	G-25	$1 \times 10^3 - 5 \times 10^3$
Polystyrene-divinyl benzene	up to 700	G-50	$1500 - 3 \times 10^4$
	$1 \times 10^3 - 2 \times 10^5$	G-75	$3 \times 10^3 - 7 \times 10^4$
	$1 \times 10^4 - 2 \times 10^5$	G-100	$4 \times 10^3 - 15 \times 10^5$
	$1 \times 10^5 - 2 \times 10^6$	G-150	$5 \times 10^3 - 4 \times 10^6$
		G-200	$5 \times 10^3 - 8 \times 10^6$
		Cross-linked beaded agarose (Sepharose)	
		CL-4B	$6 \times 10^4 - 2 \times 10^7$
		CL-6B	$1 \times 10^4 - 4 \times 10^6$

The sample is applied by injecting through a loop in the GPC mode or simply at the head of the column in GF. The sample volume to be applied on a given column of specific volume should be less than the difference in the elution volumes of two solutes giving adjacent peaks. Thus for a two-component mixture the sample volume, V_s, that can be applied to a column is given by equation.

$$V_s = V_{e1} - V_{e2} \tag{13.20}$$

The solute components are eluted by isocratic or gradient elution techniques.

SEC finds applications in the study of both biological and synthetic macromolecules. One of the important applications of size-exclusion chromatography is the determination of molecular weights.

Gel filtration finds applications exclusively in the study of biopolymers. For the determination of molecular weights of biopolymers calibration of the gel filtration column is carried out with chosen molecular weight markers or standards and the elution volumes of the different standards are noted. The sample mixture of biomolecules such as proteins and enzymes is chromatographed under identical conditions and the molecular weights of individual biomolecules are determined from their respective elution volumes using the linear graphical plot of elution volume versus log M as shown in Fig. 13.13.

Other applications of gel filtration include fractionation of proteins, separation of monomers, dimers, and higher aggregates, determination of equilibrium constants for reactions involving binding of low molecular weight substances such as drugs to proteins, study of the competition of two biomolecules for the same site, estimation of reaction rates, termination of reactions between macromolecules with low molecular weight substances and desalting of biopolymers. Desalting is an important preliminary step for the removal of small molecules prior to the use of ion-exchange chromatography, hydrophobic interaction chromatography, and affinity chromatography. Desalting is useful for the removal of phenol from preparations of nucleic acids, removal of unincorporated nucleotides during DNA sequencing, and removal of free low molecular weight labels from solutions of labelled proteins.

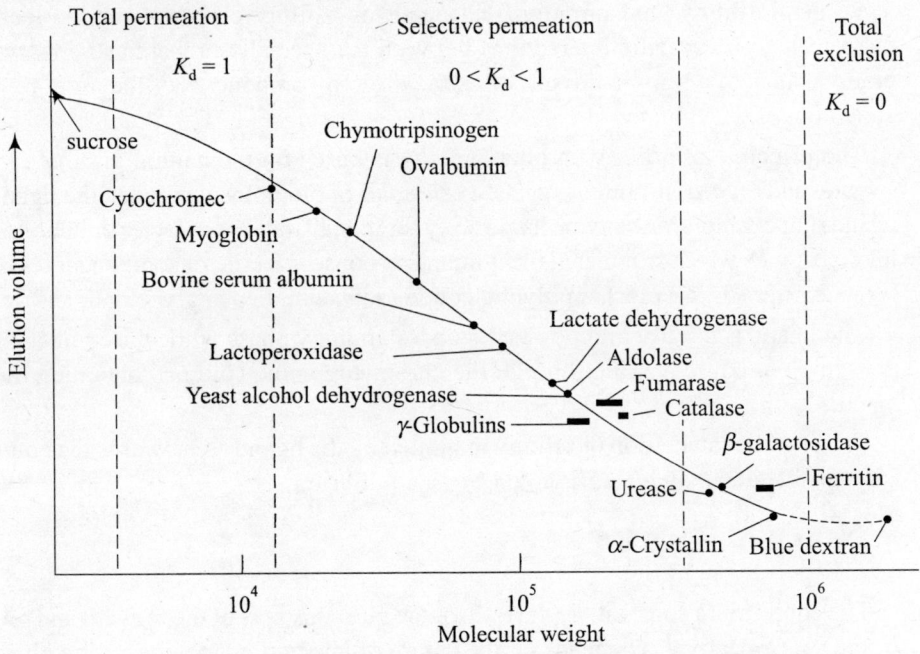

Fig. 13.13   Gel filtration of proteins of different molecular weights

Gel permeation chromatography is useful for the study of synthetic polymers, for the determination of molecular weights and distribution of molecular weights. It is difficult to correlate the elution volume with molecular weights of polymers of different chemical compositions as the differences in the structure and solvent–polymer interactions lead to different hydrodynamic volumes for equivalent molecular weight substances. In the case of linear polymers the hydrodynamic radius of the molecule is proportional to the logarithm of the product of molecular weight and intrinsic viscosity η, as $(\ln M [\eta])$. In the practice of GPC the concept of universal calibration is adopted in order to avoid the necessity of calibrating the column with narrow molecular weight dispersed fractions of each polymer of interest. The concept is based on the Mark–Houwink equation which relates the intrinsic viscosity with the molecular weight of the polymer as

$$[\eta] = KM^a \tag{13.21}$$

where K and a are constants whose values are available for a number of polymer–solvent systems in the literature.

Another application of GPC includes the determination of the molecular weight distribution of the polymer. The parameters of interest, such as weight average molecular weight M_w, number average molecular weight M_n, and the dispersity M_w/M_n, are obtained from the chromatogram.

13.10 AFFINITY CHROMATOGRAPHY

Affinity chromatography is a versatile and high resolution technique used in the separation and purification of biomolecules, such as peptides, proteins, enzymes, hormones, nucleic acids, sugars, etc. The technique includes a group of closely related techniques, such as bioaffinity, dye–ligand affinity, and immobilized metal ion affinity chromatographic techniques. In all these techniques specific interactions between the specially prepared stationary phase and the biomolecules of a sample mixture bind the desired components of the mixture which are then eluted out of the column.

The principle of affinity chromatography is based on the natural affinity and formation of specific and reversible complexes between a pair of biomolecules called the ligand and counter-ligand. For example, an enzyme has affinity for its substrate (or cofactor or inhibitor), an antibody for the antigen which stimulated the immune response, a lectin for some sugar residue, a hormone for its carrier (or receptor), a polynucleotide to its complementary strand, and so on.

The ligand is usually immobilized onto a stationary phase while the counter-ligand is part of the sample mixture passing through the chromatographic column containing the immobilized ligand.

The specific interaction or affinity exhibited by the ligand (A) towards the counter-ligand (B) is reversible and may be represented by an equilibrium as

$$A + B \underset{k_{-1}}{\overset{k_1}{\rightleftharpoons}} AB \tag{13.22}$$

The equilibrium constant, K (the ratio of the rate constants of the forward and reverse reactions k_1 and k_{-1} respectively) depends on the nature of interaction and interacting biomolecules. The K values for enzyme–substrate interaction lie in the range of 10^4–10^6 while for antigen–antibody or hormone–receptor interactions K values are in the range of 10^6–10^{12}. In general, if the K value is too low affinity adsorption of the counter-ligand does not occur while if the K value is very high the strong affinity may lead to irreversible adsorption. In either case, separation or purification of the counter-ligand may be difficult.

In practice the various steps include (i) preparation of the stationary phase gel, packing, and equilibration of the column, (ii) preparation of sample and its application, (iii) washing the column to remove the unbound species, (iv) elution of the bound solute components, and (v) column regeneration. In the fist step the ligand is usually attached covalently (immobilized) to a water-insoluble polymer stationary phase matrix or gel (e.g., agarose, dextran, cellulose, polyamide or porous glass) to form a tailor-made chromatographic stationary phase suitable to adsorb specifically the desired component (counter-ligand) from a mixture. The stationary phase gel is packed into a column and equilibrated with the chosen mobile phase. The sample mixture is applied to the adsorbent bed facilitating the adsorption of the desired component. All other constituents of the mixture do not show any affinity towards the immobilized ligand and pass through the column unrestrained. The column is washed with a suitable buffer to remove any non-specifically adsorbed components of the sample leaving behind the desired counter-ligand bound to the stationary phase matrix. The adsorbed component may then be eluted out in pure form by using a suitable eluent or buffer. The column is regenerated by washing with a suitable buffer (usually the starting buffer). A typical separation procedure is conceptually shown in Fig. 13.14. The chromatogram consists of two peaks, a first peak due to all the non-retained components of the sample and the second peak due to the elution of the desired component.

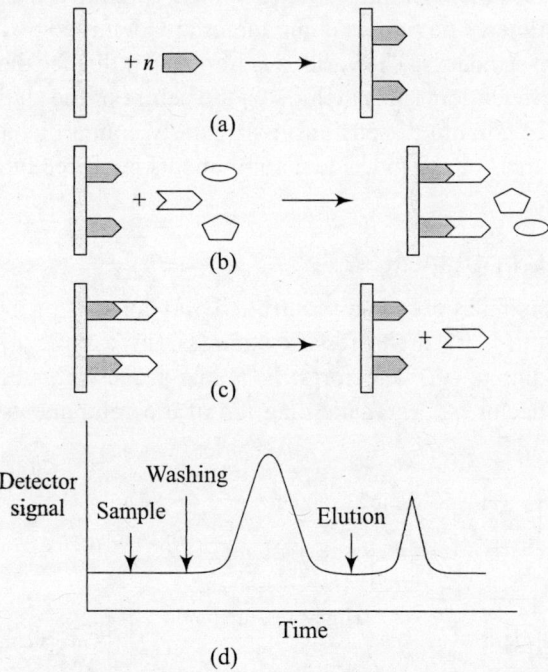

Fig. 13.14 Affinity chromatographic separation of biomolecules : (a) binding of ligand to the matrix; (b) application of sample mixture containing counter ligand; (c) recovery of counter ligand by elution after washing the column; and (d) affinity chromatogram

Applications of affinity chromatography are numerous. Typical examples include purification, recovery or concentration of biomolecules such as proteins, enzymes, monoclonal antibodies, allergens, lectins, sugars, etc., affinity scavenging for selective removal of unwanted contaminants, such as detergents and endotoxins from bioproducts such as proteins; and clinical diagnosis for highly precise and rapid biochemical analysis.

13.11 HIGH-PERFORMANCE LIQUID CHROMATOGRAPHY

High-performance liquid chromatography (HPLC) is also known as high-pressure liquid chromatography. The technique involves the use of the versatile and sophisticated instrument called high-performance liquid chromatograph. The instrument is useful for the separation of a wide variety of liquid samples and solid samples dissolved in a suitable solvent. Gaseous samples cannot he handled.

13.11.1 Principle

All the major liquid chromatographic separation mechanisms, such as adsorption, partition, ion exchange, size exclusion, and affinity can be exploited in HPLC simply by the selection of an appropriate stationary phase column. The instrument is useful for both qualitative and quantitative analyses. The samples dissolved in the liquid mobile phase are introduced into the HPLC instrument through the sample injection port and pass through the stationary phase column at a controlled flow rate. The components of the sample mixture get distributed between the stationary phase and the mobile phase. Based on the mechanism of separation, the components of the mixture pass through the stationary phase column at different migration rates and elute out at different times. The sample components are identified by comparing their retention times with those of the standards. The various factors that influence the separation of the components, and hence, their retention times include (i) the nature of the stationary phase, (ii) mobile phase composition, (iii) column dimensions, particularly column length, and (iv) mobile phase flow rate. The peak areas of the individual components are directly proportional to the amounts of the components.

13.11.2 HPLC Instrument

The instrument is microprocessor-controlled and consists of the main components (i) solvent or eluent reservoir(s), (ii) high-pressure pump(s), (iii) sample injection port, (iv) guard column, (v) analytical column, (vi) detector, (vii) a data processor, (viii) display unit or recorder, and (ix) fraction collector. A schematic diagram of the components of a binary HPLC instrument is shown Fig. 13.15.

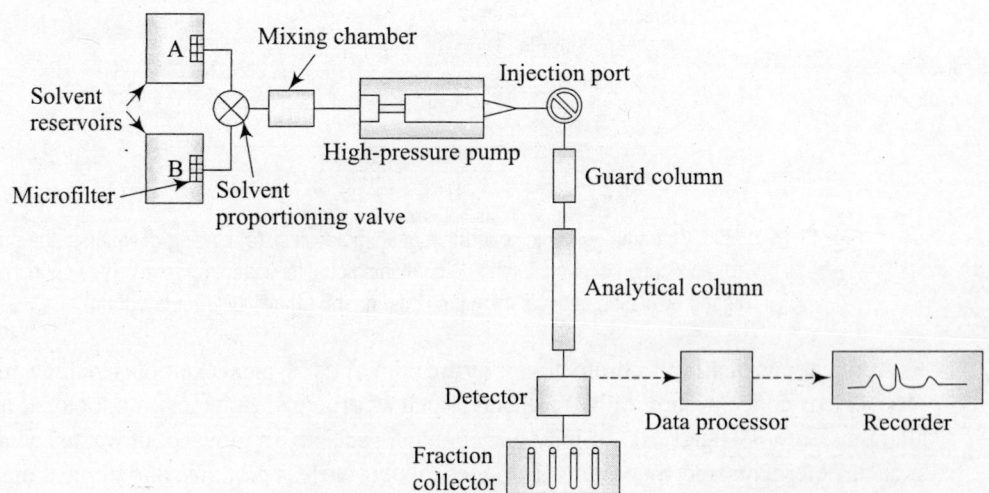

Fig. 13.15 A schematic diagram of the components of a binary HPLC system

The mobile phase delivery system consists of solvent reservoir(s) (one reservoir for isocratic elution and two reservoirs for gradient elution), solvent degassing unit, microfilter, microprocessor-controlled high-pressure pump, and a mixing chamber. The components are inter-connected through SS tubing (1/16th inch outer diameter). The mobile phase solvent is purified, degassed, and stored in the reservoir, such as a glass or plastic bottle. The solvent is degassed to remove any dissolved gas which otherwise would damage the column packing and degrade the pump and detector performance. In addition any dissolved oxygen in the mobile phase causes oxidative degradation of the sample and affects the sensitivity and operating stability of UV, refractive index, fluorescence, and electrochemical detectors. Degassing is carried out by applying vacuum above the solvent, by sparging with nitrogen or helium or better by ultrasonic treatment. The mobile phase solvent is filtered through a microfilter (2 μm pore size) to remove any suspended and particulate impurities before it is pumped from the reservoir. A flexible PTFE hose connects the reservoir to the high-pressure pump.

The high-pressure pump is one of the important components of the instrument as its performance has a profound influence on the overall performance of the instrument. A good HPLC pump should satisfy certain criteria such as (i) capability to function at pressures up to 500 atmospheres, (ii) deliver the mobile phase in an accurate, precise, and pulse-free manner at flow rates in the range 0.02–10 mL/min with minimum drift, and (iii) should have minimum response time of the order of microseconds for changing the flow rates. Two types of pumping systems are commonly used in HPLC, constant pressure pumps such as gas displacement, and pneumatic amplifier pumps and constant flow pumps such as reciprocating and syringe pumps. A single pump and a single solvent reservoir are sufficient for isocratic elution technique. Gradient elution, however, requires two or three independently controlled high-pressure pumps each connected to a separate solvent reservoir. Alternatively a single pump may be used in conjunction with a solvent proportioning valve (Fig. 13.15) for gradient elution. A mixture of two, three, or four solvents (binary, ternary, quaternary) may be used for gradient elution. The solvents for gradient elution are blended in the mixing chamber. The high-pressure pump has the facility for flow programming wherein the flow rate of the mobile phase is changed during chromatographic development.

The sample injection port consists of a valve (called a valve injector) for introducing a range of sample volumes into the column as a sharp pulse or plug without adversely affecting the column even when the instrument is operating at a high pressure. The sample of about 10–20 μL (sometimes as high as 100 μL) is preferably dissolved in the mobile phase solvent and injected with a micro syringe through a sample loop while the mobile phase flows through the sample injection valve at high pressure directly to the column (eluent bypass). Then the sample injection port valve is turned to allow the mobile phase to sweep the injected sample in the sample loop on to column at the same pressure. Sample loops of varying sizes in the range of 20–200 μL are provided with the instrument.

The stationary phase column for analytical purposes is a stainless steel tube (25 cm long and 3–4 mm inner diameter) containing the packed bed of stationary phase particles. A pre-column or guard column is usually included between the sample injector and the main column to protect the main column from damage.

The stationary phase for adsorption chromatography and normal phase partition chromatography consists of small rigid particles mostly of silica or alumina of uniform particle size. The stationary phase for reversed phase partition chromatography consists of surface-modified silica particles while for size-exclusion chromatographic mode of HPLC it consists of rigid, porous polymeric beads. Ion-exchange chromatography uses ion-exchange resins based on synthetic or natural polymers. For affinity and hydrophobic interaction chromatographic techniques the stationary phase consists of specific ligands coupled or immobilized on beads of agarose or other suitable supports.

The stationary phase particles are packed in the column uniformly at a very high density. The high-density packing results in small plate heights and a large number of theoretical plates per unit length of the column thereby giving rise to high separation efficiency. However, the high packing density also increases the resistance to the flow of the mobile phase through the bed of particles, necessitating the use of high-pressure pump.

Two types of detectors, namely, *solvent property detectors* and *solute property detectors*, are used in HPLC. A bulk or solvent property detector detects changes in the property of the mobile phase such as refractive index or conductance. The refractive index (RI) detector and conductance meter are typical examples of solvent property detectors.

An important solvent property detector is the refractive index detector. The most common RI detector is the *deflection refractometer* (Fig. 13.16) used in liquid chromatography particularly for the analysis of synthetic polymers by gel permeation chromatography.

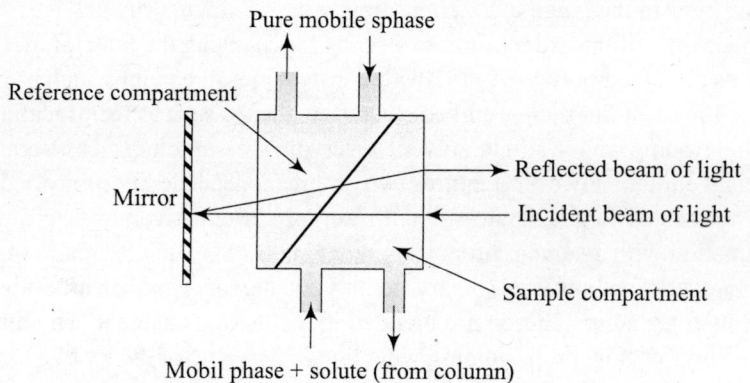

Fig. 13.16 Schematic diagram of a deflection refractometer

The detector is based on the deflection of a beam of light from a light source as it passes through a flow cell consisting of two compartments of the detector before it reaches the photodetector. One of the compartments (reference compartment) contains the reference solvent (pure mobile phase) whose refractive index remains constant. The effluent from the chromatographic column passes through the sample compartment. As long as the pure mobile phase passes through both the compartments no deflection of light beam occurs and a base line is obtained. When the solute elutes out of the column into the detector the composition of the mobile phase in the sample compartment is different (as also its refractive index) and the altered refractive index causes the light beam to be deflected. The magnitude of deflection is proportional to the amount of solute in the mobile phase giving rise to an on-line chromatographic peak.

Conductance meter is particularly useful as a chromatographic detector to detect the presence of solute ions in a sample mixture when the conductivity of the mobile phase is quite low. The detector is used in ion chromatography.

A solute property detector such as UV spectral detector, UV–visible spectrophotometer, fluorescence detector, or electrochemical detector detects the property of the solute. The detector signal is proportional to the amount of the solute passing through the detector. The signal is amplified and fed to data processor system for visual display and printout of the chromatogram.

The *spectral detectors* monitor the absorbance of the solute species at specific wavelengths as they pass through the detector. The fixed wavelength UV spectral detector of wavelengths 220 or 254 nm and visible spectral detectors of wavelengths 436 and 546 nm are quite common. UV–visible spectrophotometers have variable wavelength over a wavelength range of 190–800 nm (Fig. 13.17). These spectral detectors show linear response to concentration of the absorbing species conforming to Beer–Lambert law and employ flow-through cells of low volume of about 10 μm or less. The fixed wavelength spectral detectors have greater sensitivity than spectrophotometers but the latter are useful because of the possibility of selecting a more suitable wavelength. In addition the spectrophotometers are mostly microprocessor-controlled with built-in software to record the absorption spectrum of a selected species.

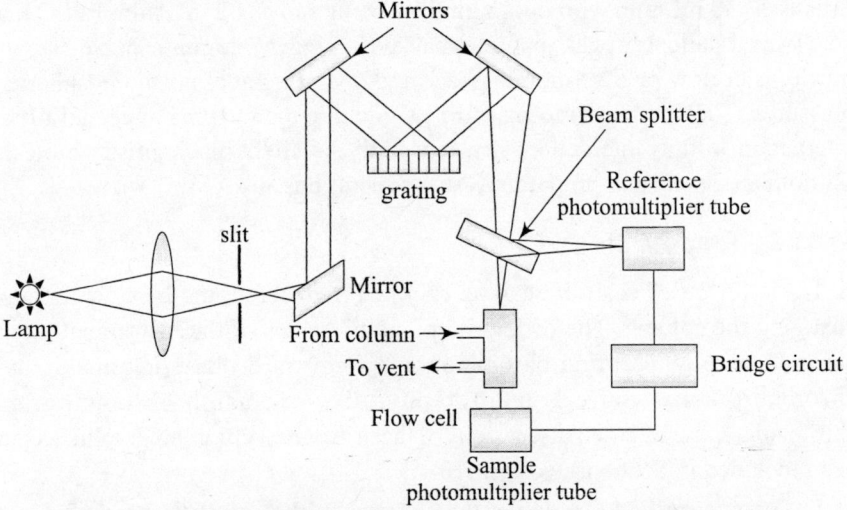

Fig. 13.17 Schematic diagram of a variable wavelength spectrophotometer detector

Diode array spectrophotometers have an array of photo detectors and are capable of rapid scanning but are less sensitive and expensive compared to dispersive spectrophotometers. The detector provides spectral information which can be processed with graphics software to get 3D chromatogram of time/absorbance/wavelength.

Fluorescence detectors are highly selective and also highly sensitive. The basic principle is based on exciting the solute species with an intense beam of light from a xenon lamp at specific wavelength (excitation wavelength) and monitoring the fluorescence emission at specific wavelength (emission wavelength) as the solute species passes through the flow cell of the detector. Figure 13.18 shows a schematic diagram of fluorescence detector for HPLC.

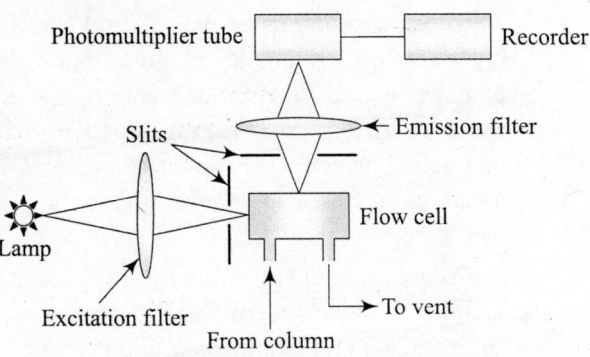

Fig. 13.18 Schematic diagram of a fluorescence detector

Amperometric detectors are electrochemical detectors based on the principle of polarography and are suitable for detecting solute species capable of undergoing oxidation or reduction at an electrode. The solute species is subjected to a constant applied voltage at the surface of gold, platinum, or glassy carbon micro-electrode in the detector and the current generated is measured.

The fraction collector in most of the sophisticated HPLC systems is a programmable unit synchronized with detector signal to collect the components of the sample mixture separately.

Fast protein liquid chromatography (FPLC) is especially suitable for the separation and purification of proteins and other biomolecules. The FPLC instrument is similar to HPLC in construction, the components, and operation. High-resolution HPLC makes use of columns containing stationary phase particles of 5–40 μm size, rigidly packed at very high packing density, and capable of withstanding high pressure operation up to even 400 bar. However, the columns have limited sample loading capacity and maximum flow rate of the mobile phase is usually 5–10 mL/min with optimum flow being only 1–2 mL/min. FPLC makes use of similar small-sized stationary phase particles packed in bigger columns capable of higher sample loading, much faster flow rates of mobile phase and operating at almost atmospheric or slightly greater pressures. The operating modes of FPLC include the ion exchange, gel filtration, hydrophobic interaction, affinity interaction, etc., especially suited for biochemistry/biotechnology laboratory for both analytical and preparative-scale operations.

13.11.3   Practice of HPLC

HPLC is versatile because it can be operated in different modes of chromatography simply by changing the column. The different operating modes of the instrument include the adsorption chromatography, partition chromatography, reversed phase chromatography, ion-exchange chromatography, ion chromatography, size-exclusion chromatography, and affinity chromatography. Detailed discussions of these different chromatographic techniques have already been included in Sections 13.6–13.10.

The practice of HPLC involves the injection of the sample (~20 μL) into the sample injection port. The chromatogram is developed under specified conditions of composition and flow rate of the mobile phase. The retention times and peak areas of the components of the mixture are determined from the on-line chromatogram. The retention times are useful in qualitative identification of the components and the peak areas (or heights) are proportional to the amounts of the components in the sample, and hence, useful in quantitative analysis.

Chiral chromatography makes use of chiral stationary phases (CSPs) to separate optically active isomers or enantiomers. Chiral chromatographic stationary phases are mostly based on silica supports. Examples include ionically or covalently bonded chiral amino acids, chiral peptides (e.g., bovine serum albumin, α_1-acid glycoproteins, and chiral cyclodextrins). These chiral phases form complexes with analyte molecules through hydrogen bonding, $\pi-\pi$ interactions or dipolar interactions. Parameters such as pH, steric repulsion, ionic strength, temperature, and nature of solvent also help in differential retention of a particular enantiomer on the column facilitating its separation.

Optimization of HPLC operating conditions is essential for successful separation of a given sample mixture. Modern HPLC instruments are mostly interfaced with dedicated computers. Optimization of operating conditions particularly with respect to mobile phase composition is achieved based on statistical design of experiments and mathematical analysis of resolution parameters. Gradient elution is used mostly as the first step for unknown samples to select the mobile phase composition for isocratic elution. Computer-based automated methods to optimize mobile phase composition are helpful to achieve adequate chromatographic resolution of the components of a sample mixture in the shortest possible time. An automated method commonly used is a simultaneous technique in which a predetermined number of chromatograms are recorded and each chromatogram is evaluated based on a single numerical value called *chromatographic response function* (CRF) or *chromatographic optimization function* (COF). The CRF or COF represents the quality of the chromatogram as a function of resolution for each pair of adjacent peaks in the sample and the total elution time. The individual CRF values are then used to draw a contour map to identify the optimum mobile phase composition.

A typical example of the simultaneous technique is called the *simplex lattice design* based on *solvent selectivity triangle* developed for a reversed phase chromatographic separation. The method involves mixing three solvents (e.g., methanol, acetonitrile, and tetrahydrofuran) of differing chromatographic selectivity with a fourth solvent, for example, water as solvent strength adjuster, to provide three *isoeluotropic* (equal solvent strength) mobile phases. The three mobile phases form the apices of a solvent selectivity triangle. A minimum of four other mobile phases are prepared by mixing the three initially formed mobile phases in different proportions. The compositions of these four mobile phases lie either along the sides of the triangle or within the triangle. The sample mixture is then chromatographed with each of the seven (or more) mobile phases and the resolution-based COF values are computed for each chromatogram. A resolution contour map is obtained by superimposition of the individual contour maps to give an *overlapping resolution map* (ORM). The ORM provides the information on optimum mobile phase composition for the separation of the sample mixture for a specified resolution.

Several other computer-based mathematical methods are also available for determining the optimum conditions of operating the HPLC for a given separation.

13.11.4 Applications of HPLC

HPLC can be used for qualitative analysis for the identification of solutes of a give sample on the basis of comparison of retention times of the sample components with chromatographic standards under specified conditions of nature of stationary phase, column dimensions, mobile phase composition, and its flow rate.

Quantitative analysis is based on measuring the peak heights or preferably the peak areas which are directly proportional to the amount of the sample components. The sample mixture is usually introduced through sample loops of 50–200 μL capacity provided with the instrument and subjected to isocratic elution because of its reproducibility. Quantitative analysis is performed using previously prepared calibration charts of individual components.

Table 13.4 summarizes the selection of suitable columns for liquid chromatography in general and applications of the different techniques.

Table 13.4 Applications of HPLC

Mode of HPLC	Stationary phase	Applications
For compounds with molecular weights <2000		
Adsorption (LSC)	Microparticulate silica	Non-polar to moderately polar organic mixtures (e.g., fats, oils, aromatics, and mixtures of isomers)
Partition (LLC) or BPC		

(Contd.)

(*Contd.*)

Mode of HPLC	Stationary phase	Applications
For compounds with molecular weights <2000		
Normal phase	Diol bonded phase	Polar water-soluble compounds (e.g.,food and beverage additives)
	Nitrile bonded phase	Non-polar compounds
	Aminoalkyl bonded phase	Polar water-soluble organic compounds (e.g., carbohydrates)
Reversed phase and IPPC	ODS (C-18) bonded phase	Moderately polar compounds–most widely used for a variety of compounds
	Octyl (C-8) bonded phase	More polar compounds
	Short chain (C-2) bonded phase	Highly polar compounds
IEC	Cation- and anion-exchange resins	Ionic and ionizable compounds
CC	Amino acids bonded to aminopropyl, chiral peptides, cyclodextrins	Mixtures of enantiomers, particularly drugs, and pharmaceuticals.
For compounds with molecular weights >2000		
SEC-GPC	Silica or polymers with controlled pore size	Mixtures of polymers
SEC-GF	Agarose, dextran polymers with controlled pore size	Mixtures of biopolymers

BPC–bonded phase chromatography; CC–chiral chromatography; LC–liquid–liquid chromatography; IPPC–ion pair partition chromatography; IEC–ion-echange chromatography; SEC-GPC–size-exclusion chromatography–gel permeation chromatographpy; SEC-GF–size-exclusion chromatography–gel filtration.

13.11.5 HPLC-Mass Spectrometry (HPLC-MS)

HPLC coupled to mass spectrometer is a versatile and sophisticated instrument which provides structural information of the separated components. A variety of interfaces have been developed to couple HPLC with mass spectrometer to facilitate the introduction of separated organic sample components exiting the reverse phase column along with the aqueous mobile phase directly into the mass spectrometer.

Particle beam interface nebulizes the mobile phase with helium gas producing an aerosol from which the sample components are evaporated at near ambient conditions of temperature and pressure. The mixture of mobile phase vapour containing the separated components and helium gas is accelerated into a low-pressure two-stage separating chamber wherein the vapour-mix expands supersonically. The solvent vapour and helium are pumped out. The heavier sample molecules because of their greater momentum pass through skimmer plates along a narrow transfer tube into the ionization chamber of the mass spectrometer. The sample molecules are subjected to electron impact ionization and the ions produced are accelerated through analyser tube to separate the fragments on the basis of their m/z ratios.

Thermospray interface facilitates the transfer of the mobile phase (mostly ammonium acetate buffer) containing the separated sample components through a hot tube maintained at 350–400°C into an evacuated chamber. The mixture expands supersonically in the evacuated chamber into a mist of electrically charged droplets. The mobile phase liquid evaporates and the liberated charged solid particles are subjected to chemical ionization process with ammonia gas to form positively charged species such as MH^+ and MNH_4^+ which are then skimmed off into mass spectrometer.

Atmospheric pressure chemical ionization (APCI) interface nebulizes the column effluents by nitrogen gas. The mobile phase and the sample components exiting the column are nebulized with nitrogen gas and the aerosol formed is heated to about 120°C in a desolvation chamber. The mobile phase molecules are ionized to produce reactant ions by electrons generated by a corona discharge. The reactant ions of the mobile phase molecules collide with sample component molecules to form MH^+ or MH^- ions by chemical ionization. The analyte ions are then skimmed off into the mass spectrometer. This method is highly efficient as the chemical ionization occurs at high collision frequency occurring at atmospheric pressure. The method has the advantages of minimum thermal degradation of the sample due to rapid desolvation and vapourization process and high sensitivity as even trace quantity of analyte in the range of 50–100 picogram can be detected.

Electrospray interface also operates at atmospheric pressure and highly useful for generating ions from biological samples.

13.12 SUPERCRITICAL FLUID CHROMATOGRAPHY

Supercritical fluid chromatography (SCFC) combines the advantageous features of gas chromatography and HPLC by using a supercritical fluid as the mobile phase solvent. Mixtures of compounds (e.g, non-volatile compounds and thermally labile compounds) which cannot be handled by gas chromatography and compounds which do not have any specific functional group(s), and hence, cannot be detected by detectors of HPLC can be conveniently handled by SCFC.

13.12.1 Supercritical Fluid Solvents and Their Properties

Supercritical fluid phase of a substance exists at temperatures and pressures above but close to its critical temperature and critical pressure (close to its critical point). The critical temperature of a substance is the temperature above which a distinct liquid phase cannot exist regardless of the pressure. The vapour pressure of the substance at its critical temperature is called the critical pressure. Critical pressure may also be defined as the pressure required for liquefying a gas at its critical temperature. Supercritical fluids have physical properties such as density, viscosity, and diffusivity and other properties that are intermediate between those of the substance in its gaseous and liquid state. For example, the densities of supercritical fluids lie in the range 0.2 to 0.5 g/cm^3 in between the density values of a normal gas at STP $(0.6–2.0) \times 10^{-3}$ g/cm^3) and a normal liquid (0.6–2.0 g/cm^3). The relatively low viscosity and greater diffusivity of supercritical fluids compared to liquids allow supercritical fluids to diffuse more rapidly into a sample and even penetrate solid samples. In addition, the solvent power or strength (the ability of the solvent to extract a particular component in a mixture) of a supercritical fluid can be controlled easily by increasing the pressure on the fluid. Increase in pressure increases the density of the fluid and thereby its solvent strength. In general, small changes in pressure near the critical point result in large changes in the density of a supercritical fluid. These properties of supercritical fluids are advantageously used in SCFC as well as in supercritical fluid extraction.

13.12.2 SCFC Instrument

The instrument is similar to HPLC because the operating pressures and temperatures for most supercritical fluids are similar to that in HPLC. The major components of the instrument include (i) mobile phase delivery system consisting of a reservoir and syringe pump, (ii) sample injection valve, (iii) stationary phase column housed in a thermostat for precise control of temperature of the column, (iv) a device called restrictor placed at the exit end of the column to regulate the back-pressure in the column, and (v) detector and data processor with display/recorder unit. The block diagram of the instrument is shown in Fig. 13.19.

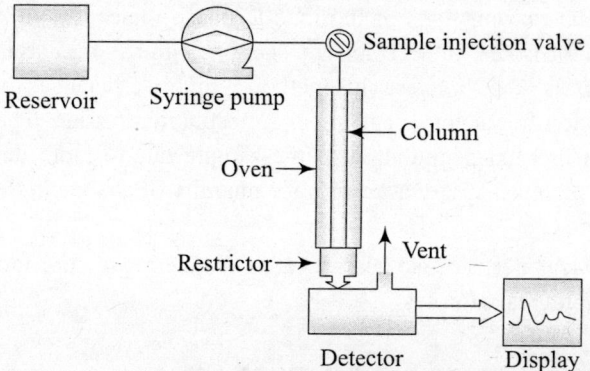

Fig. 13.19 Block diagram of supercritical fluid chromatograph

The most commonly used supercritical fluid in SCFC is carbon dioxide because of its non-toxic and non-hazardous nature. Carbon dioxide has additional advantages such as its dissolving capability of a wide variety of non-polar organic substances, transparency to ultraviolet light, absence of odour or colour, and low cost. Organic modifiers such as methanol (~1%) may be added to carbon dioxide for analysis of polar organic substances. The mobile phase transports the solute molecules similar to the carrier gas in gas chromatography and also interacts with the solute molecules influencing the selectivity factor because of its solvent power. The solvent power of a supercritical fluid is highly dependent on its density, which in turn depends on the pressure and temperature. As pressure increases the density of the mobile phase and consequently its solvent power increases resulting in decreasing the elution time required for solutes. The selectivity factor can thus be varied in supercritical fluid chromatography unlike in gas chromatography.

The stationary phases used in SCFC are of two types: (i) packed columns similar to those used in HPLC containing stationary phase support particles of 3–10 μm in diameter coated with liquid stationary phase for partition chromatography and (ii) wall coated (with different types of coatings) open tubular-fused silica columns 10–20 m long and internal diameters of 0.05–0.1 mm.

The restrictor maintains the pressure within the column at the required level and at the same time facilitates the conversion of the mobile phase from the supercritical fluid to a gas phase before it reaches the detector.

The most commonly used detector in SCFC is flame ionization detector, similar to the detector used in GC. Other detectors such as ultraviolet and infrared spectral, fluorescence emission, flame photometric, and mass spectral detectors commonly used in HPLC also find use.

Supercritical fluid chromatography finds use in the qualitative and quantitative analyses of foodstuffs, drugs, polyaromatic hydrocarbons, pesticides, surfactants, oligomers, polymers, fuels, explosives, and propellants most of which cannot be handled either by GC or HPLC.

13.13 GAS CHROMATOGRAPHY

Gas chromatography (GC) involves the separation of the components of a mixture in vapour or gas phase by interacting with stationary phase placed in a column. The technique is also called *gas liquid chromatography* (GLC), *vapour phase chromatography*, or *vapour–liquid partition chromatography*.

13.13.1 Principle

Gas chromatography can be used for separating thermally stable volatile liquids (with boiling points less than 350°C) and gaseous mixtures. Solids and high boiling liquid samples cannot be directly introduced into the instrument. However, such samples can be chemically modified into volatile derivatives for introduction into the instrument. The liquid samples in microgram quantities are vapourized immediately after their introduction into the instrument, mixed with the mobile phase or carrier gas stream and passed through the stationary phase column at a constant flow rate. The stationary phase is housed in an oven maintained at a predetermined temperature which is usually higher than the boiling temperature of the liquids so as to prevent condensation of the samples within the column. The stationary phase consists of uniform-size particles of an inert solid coated with a stationary phase liquid. The separation mechanism is mostly partition of the components of the sample mixture between the mobile phase gas and the stationary phase. The partitioning of the components is brought about due to the differences in their boiling points and relative affinities for the stationary phase and the mobile phase. The components get separated due to their differential migration through the column. The various factors that influence the separation of the components include (i) the nature of the stationary phase, (ii) temperature of the column, (iii) length of the column, and (iv) flow rate of the carrier gas.

Gas solid chromatography (GSC) is a similar technique where the stationary phase column contains solid particles packed through which the vapour of the sample mixture is carried by the mobile phase gas. The separation mechanism is adsorption. In all other respects the technique is similar to that of GC.

13.13.2 GC Instrument

The main components of the gas chromatograph are shown in Fig. 13.20. The main components include (i) regulated mobile phase (carrier gas) supply system, (ii) injection port for introducing liquid or gas samples, (iii) thermal compartment or column oven, (iv) stationary phase column, (v) detector, and (vi) data processor and recorder.

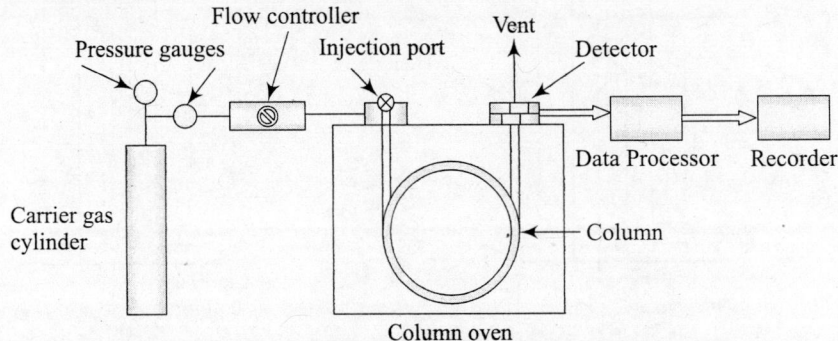

Fig. 13.20 Schematic diagram of the components of a gas chromatograph

The *regulated mobile phase supply system* consists of carrier gas cylinder(s), pressure gauges, and flow controller. The carrier gas flows from the cylinder regulated by a mass flow controller which maintains a constant predetermined flow rate. Carrier gases used in GC are selected depending on the type of detector available on the instrument and also on the type of column (packed or capillary columns) and to some extent on the nature of the sample. Commonly used carrier gases include nitrogen, hydrogen, and helium. Hydrogen or helium is used as carrier gas in conjunction with thermal conductivity detector (TCD) because of the high thermal conductivity of these gases. Helium or hydrogen is the preferred gas for capillary columns because of greater chromatographic efficiency with increasing flow rates. Nitrogen, helium, or argon is used as the carrier gas when the instrument is fitted with a flame ionization detector (FID). Flame ionization detector requires support gases in addition to the carrier gas. The support gases include hydrogen and air. When the instrument has an electron capture detector (ECD) nitrogen is used as the carrier gas.

The *sample injection port* facilitates the introduction of the sample into the carrier gas stream as a narrow band or pulse. The injection port for packed columns called *flash vapourizer* [Fig. 13.21 (a)] is provided with a self-sealing silicone rubber septum through which the sample is injected with the help of a micro syringe or an auto sampler. The injection port is usually maintained at a predetermined temperature which is greater than the boiling point of the sample by at least 15–20°C to ensure complete instantaneous volatilization of the sample.

Volatile liquid or gaseous sample is introduced into the injection port and the sample vapour gets mixed with the carrier gas in the injection port itself well before it enters the column. The optimum volume of the liquid sample for most columns is usually about 0.2–2 μL and about 20–50 μL in the case of gases. Non-volatile compounds and solids may be analysed by pyrolysis gas chromatography wherein the sample is pyrolysed (thermally decomposed) to yield volatile products before they enter the column. Alternatively, the non-volatile sample may be derivatized into volatile substance before injection into GC.

A *split injector* is used for introducing samples into a capillary column as the sample capacity of capillary columns is much smaller. The split injector [Fig. 13.21(b)] is provided with a needle valve which allows only about 2–3% of the injected sample volume into the column while venting the rest of the sample into the atmosphere. Split injection is associated with the disadvantage of *discrimination effect*, in that higher amount of lower boiling components of a mixture of

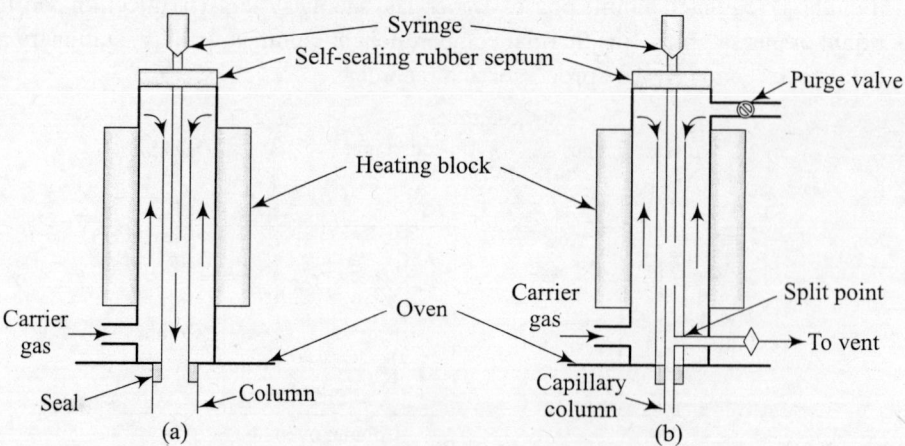

Fig. 13.21   Schematic diagram of sample injection systems (a) flash vapourizer and (b) split injector

components with widely differing boiling points will enter the column than the high boiling components. Split injection is not suitable when higher sensitivity is required.

Splitless injection of samples avoids these problems. In one of the common methods of splitless injection, suitable for sample components having high boiling points, the injected sample of several µL is collected in a cold trap at the top end of the column. The cold trap is maintained at a temperature of about 100°C below the boiling point of the most volatile component of the sample. The cold trap is then heated to boil off the components sequentially. Alternatively the sample may be mixed with a high boiling solvents, for example, octane (boiling point 126°C) and injected into the sample inlet. The solvent condenses as a thick layer at the top of the column. This layer retains and concentrates the components which are then volatilized by raising the temperature of the column.

On-column injection is yet another method which allows very small amounts of liquid samples to be placed directly at the cooled top of the column. The sample is injected by a syringe fitted with a fine quartz needle into a specially designed septumless valve. Air cooled to about 20°C below the boiling point of the solvent containing the sample is simultaneously circulated through the valve to facilitate the condensation of the solvent at the top of the column. Hot air is subsequently circulated through the valve to volatilize the sample in situ in the column. The method is useful for thermally sensitive samples as the risk of thermal decomposition is minimized. The method has the advantage eliminating discrimination effect.

Automatic injectors are usually controlled by computers and have the advantages of handling a large number of samples to be analysed under different operating conditions and a high degree of precision and reproducibility for sample introduction.

The *column oven* or *thermal compartment* is an air oven which houses the column. The oven can be programmed to maintain a desired temperature up to a maximum of about 400°C. In *isothermal operation* the oven temperature is maintained at a selected value with an accuracy of ±0.1°C. In contrast, in *temperature programming* operation the temperature is increased in a programmed manner from an initial value to a higher value at a predetermined heating rate. The heating rates can be varied in the range 1°C/min to 40°C/min. Linear as well as non-linear temperature programming is available in modern instruments.

The *stationary phase column* is a coil made of stainless steel, fused silica, or glass tube. The column is placed inside the thermal compartment. Columns may be packed columns or capillary columns. The column contains the actual stationary phase that is responsible for the separation process. A large number of stationary phases are commercially available for use in GC. Polar liquids such as high molecular weight polyesters, polyethylene glycol, amines, ethers, etc., and non-polar liquids, such as hydrocarbon and silicone oils are used as stationary phases. Packed columns made of glass or stainless steel tubing are mostly 2–10 m long. Usually the outer diameters of the columns are 1/8″ or 1/4″ and contain porous granular solid supports coated with a thin film of the stationary phase liquid. The solid supports commonly used include silica, alumina or finely divided fire brick of uniform particle size of the order of 75–150 µm depending on the diameter of the tube. Solid stationary phases such as silica or alumina are used in gas–solid chromatography. Capillary columns also known as open tubular columns are made of high purity fused silica (quartz) of 5–50 m length and have their inner walls coated with a thin film (0.1–5 µm thick) of the liquid stationary phase. Commercially available general purpose columns used for the separation of widely differing samples include carbowax 20 M and SE30 for the separation of alcohols, amines, ethers, and aromatic compounds. Dedicated columns containing stationary phases of dinonyl phthalate are used for the separation of esters and alcohols, diethylene glycol succinate for the separation of fatty acid esters, squalane for the separation of high molecular weight saturated hydrocarbons, apiezon L for the separation of high boiling hydrocarbons, etc. Other columns such as Porapak series and porous layer open tubular (PLOT) columns containing molecular sieves for the separation of permanent gases (e.g., helium, nitrogen, oxygen, carbon monoxide, carbon dioxide, methane, etc.) are also commercially available.

Detectors most commonly used in GC include the thermal conductivity detector (TCD), flame ionization detector (FID), electron capture detector (ECD), and flame photometric detector (FPD).

Thermal conductivity detector is based on monitoring the difference in the thermal conductivity (ability to dissipate the heat of a hot body) between the pure carrier gas passing through the reference cell of the TCD and the carrier gas mixed with vapours of the sample components (usually of low thermal conductivity) eluting out of the column through the sample cell of the TCD. The detector consists of two electrically heated filaments of a metal placed in the separate cells or compartments, a reference cell, and a sample cell, in the heating block (Fig. 13.22). Pure carrier gas flows through the reference cell. The sample cell is connected to the exit end of the column and carrier gas exiting the column (sometimes pure and sometimes mixed with the eluted solute components) flows through the sample cell. As long as pure carrier gas flows through both the reference and sample cells the temperature, and hence, the resistance of the two filaments remain the same.

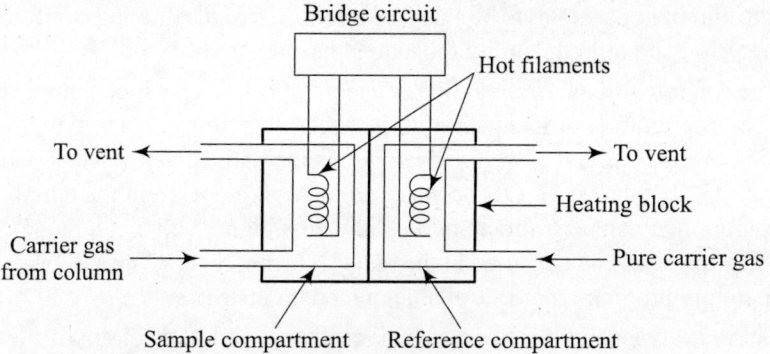

Fig. 13.22   Schematic diagram of thermal conductivity detector

A change in thermal conductivity of the carrier gas flowing through the sample cell occurs when the sample components elute out of the column. This results in an imbalance in the resistances of the two filaments and generates an electrical signal in the bridge circuit connecting the two filaments. The electrical signal is registered as a deflection in the base line and appears as a peak in the on-line chromatogram. The apex position of the peak indicates the retention time (or retention volume) of the sample component and the area under the peak is directly proportional to the amount of the component in the injected sample, and hence, useful for quantitative analysis.

Flame ionization detector requires two support gases, namely, hydrogen and air in addition to the carrier gas (usually nitrogen). In the FID, hydrogen is burnt in the presence of air to produce a flame through which the carrier gas and sample components pass through. As long as the pure carrier gas passes through the detector only a residual ion current is generated. When the sample component is eluted out of the column by the carrier gas and reaches the detector it ionizes in the flame and generates an ion current proportional to the amount of the component in the sample. FID is more sensitive as compared to TCD and has the widest linear range for any detector. However, the detector does not respond to air, inorganic gases, and formic acid. A schematic diagram of the FID is shown in Fig. 13.23.

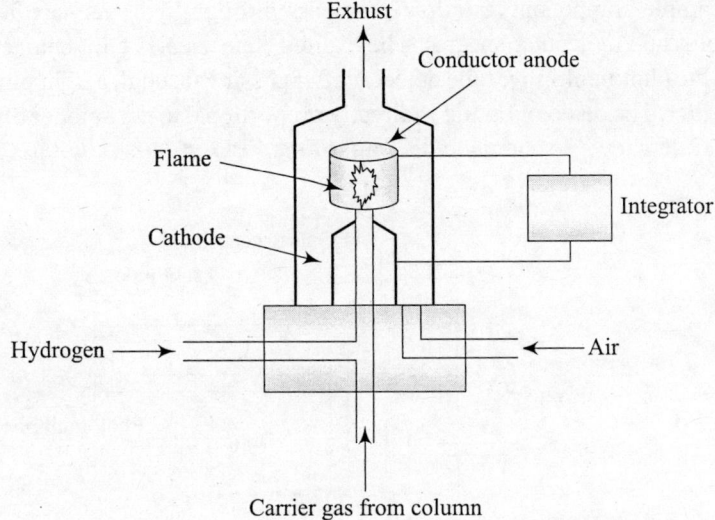

Fig. 13.23 Schematic diagram of flame ionization detector

Electron capture detector contains a tritium (^{3}H) isotope adsorbed on titanium or scandium or ^{63}Ni foil which emits a steady stream of electrons (β rays) and ionizes the nitrogen carrier gas passing through the detector (a cylindrical tube which functions as the cathode) forming slow electrons. The slow electrons are attracted to a wire anode maintained at a potential difference of about 50 V producing a steady current. When the carrier gas exiting from the column contains sample components some of the slow electrons are captured thereby reducing the current. The extent of decrease depends on the amount of the sample component. Figure 13.24 shows a schematic diagram of the ECD.

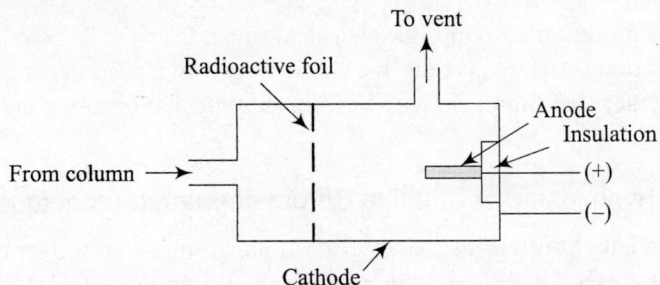

Fig. 13.24 Schematic diagram of electron capture detector

ECD is highly sensitive to halogen and sulphur containing compounds, anhydrides, peroxides, nitrites, nitrates, conjugated carbonyl, and organometallic compounds. It is totally insensitive to hydrocarbons, alcohols, ketones, and amines. The linear range of the detector is relatively low and the detector is also sensitive to changes in temperature. In addition, the carrier gas for ECD must be extremely pure with oxygen, air, or water levels at less than 10 ppm. Halogenated solvents should not be used for sample preparation as any residual traces can deactivate the detector. ECD is used specifically for the analysis of pesticides, drugs, and biologically active compounds.

Flame photometric detector is a flame emission photometer in which the carrier gas containing eluted sample components is allowed to pass through hydrogen–air flame to produce excited state atoms and molecular species. The excited state species emit characteristic radiation which reaches the photomultiplier tube detector after passing through a light pipe and narrow band pass optical filter. The detector output is directly proportional to the amount of the sample component. Figure 13.25 shows a schematic diagram of the FPD.

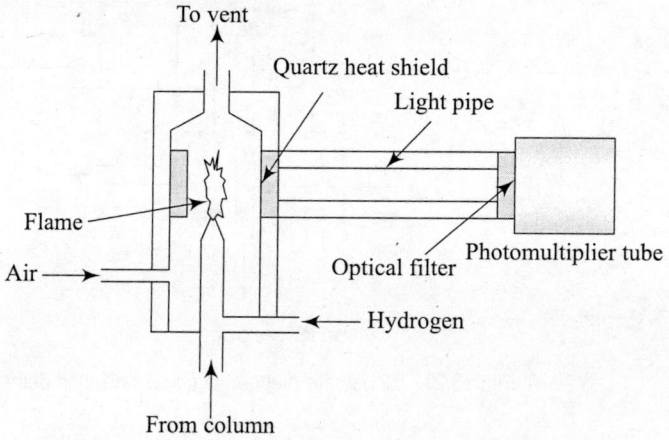

Fig. 13.25 Schematic diagram of flame photometric detector

The detector is particularly useful for detecting sulphur and phosphorus containing compounds as these elements form S_2 with characteristic emission at 394 nm and HPO species at 526 nm respectively. The detector has a high sensitivity but has a relatively limited linear range for phosphorus while for sulphur the sensitivity of the detector is proportional to the square of sulphur concentration. It also varies with the nature of the sulphur compound and is linear over a still smaller range.

The data processor amplifies the electrical signal from the detector and transfers the signal to display unit or recorder to give on-line chromatograms. Data processors are mostly controlled by microprocessors or computers with built-in analytical software to analyse the data and display the same.

13.13.3 Hyphenated or Coupled Chromatographic Techniques

Hyphenated techniques use detectors such as mass spectrometer (GC-MS), infrared spectrophotometer (GC-IR), and atomic emission spectrometer (GC-AES) in conjunction with GC. The first two hyphenated techniques are more common. In the hyphenated technique the separation of sample components is carried out by the gas chromatograph and the separated components are analysed by the mass spectrometer or IR spectrophotometer for identification of the component.

In GC-MS, interfacing the two instruments is achieved by the use of precisely aligned sample enrichment and introduction devices called *separators* because the GC operates at atmospheric pressure while the MS operates at very low pressures in the range of 10^{-6}–10^{-8} torr. A jet-orifice separator is used in conjunction with packed columns.

The separator consists of an evacuated glass tube into which the carrier gas flowing at 30–50 mL/min from the exit end of the column of the gas chromatograph is let in through a fine bore jet. Another similar smaller jet, placed within a short distance of about 1 mm from the carrier gas jet, carries the separated/enriched sample molecules into the mass spectrometer as shown in Fig. 13.26. As the carrier gas containing the sample component passes through the separator, major amount of the lighter carrier gas (helium) is removed by the vacuum while the heavier sample molecules because of their greater momentum cross the gap between the jets and move on to the mass spectrometer.

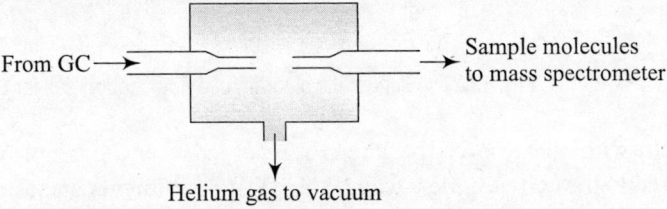

From GC →   Sample molecules to mass spectrometer

Helium gas to vacuum

Fig. 13.26   Schematic diagram of a jet-orifice separator

A separator is not required if the GC has a capillary column as the volumetric flow of the carrier gas is much smaller and the carrier gas is easily pumped out as it enters the mass spectrometer, leaving the sample molecules to be introduced directly into the ionization chamber of the mass spectrometer. Hence, the capillary column is directly inserted into the mass spectrometer.

In GC-MS separations, the solvent in the sample mixture gives rise to a large peak and minor components of the sample may elute out on the tails of the solvent or major component peaks. In such cases the solvent and the major components cause the swamping of the mass spectrometer. This may be avoided by diverting the solvent and any other major component (if their analysis is not required) through the use of a *solvent dumping valve*. The valve is placed between the exit end of the column and the separator and by a carefully timed operation the solvent and unwanted components may be vented out to the atmosphere.

Quantitative analysis by GC-MS can be carried out measuring the *total ion current* (TIC) which is the sum of the currents generated by all the fragment ions of a particular sample component and is directly proportional to the amount of the component introduced into the ionization chamber. Alternatively a stream-splitter is incorporated at the exit end of the column to allow the simultaneous detection of the eluting sample component by FID as well as the mass spectrometer. *Selected ion monitoring* (SIM) involves monitoring the ion current of a selected fragment ion with an *m/z* value which is a characteristic of the particular sample component. Selected ion monitoring is more sensitive than TIC, and hence, finds use in trace analysis.

In GC-MS separations, fast scanning of the mass range of interest is necessary particularly of partially resolved GC peaks because the sample components may elute out of the column in the space of a few seconds. *Peak slicing* of partially resolved GC peaks several times by rapid scanning facilitates the identification of the individual sample components even when they are not well separated. However, peak slicing is possible only when the dead volume of the interface is smaller than the peak volume of the component.

In GC-IR, since both the instruments operate at atmospheric pressure, the carrier gas from the exit end of the column is directly let via a heated transfer line into a heated IR gas flow-through cell. The vapour phase IR spectrum of the sample is recorded usually by FTIR spectrophotometer. The flow-through gas cell (Fig. 13.27) is a glass tube about 50 cm long and 0.2 cm internal diameter. The inner wall of the tube is coated with gold to maximize the transmission of IR radiation. The ends the tube are sealed with KBr or CsI windows transparent to IR radiation.

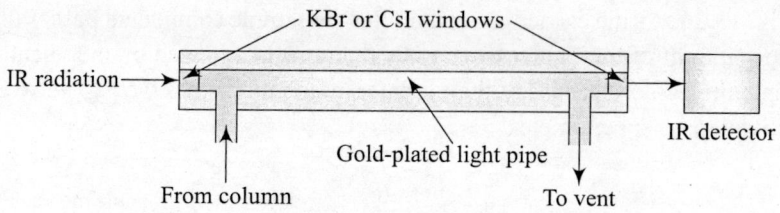

Fig. 13.27 Schematic diagram of a flow-though cell for GC-IR

IR spectra with well-resolved peaks can be recorded when the sample volume is appreciably more as compared to the cell volume because the FTIR instrument can scan the partially resolved GC peaks several times by peak slicing. On the other hand, if the sample volume is smaller as compared to the cell volume cross-contamination from adjacent peaks may occur.

13.13.4 Practice of GC

The stationary phase column is selected depending on the nature of the sample mixture to be analysed and fitted into the thermal compartment of the instrument. The carrier gas is allowed to flow into the instrument at controlled flow rates in the range of 10–100 mL/min (optimum flow rates being 20–30 mL/min) at a pressure head of about 3–4 kg/cm^2. The optimum flow rate of the carrier gas for a given column may be determined with the help of van Deemter plot. The temperature of the injection port is set usually at a value about 10°C greater than the boiling point of the sample. The detector temperature is set as per the instruction of the manufacturer, the value being about 200°C for TCD as well as for FID. Elution of the sample components can be carried out either by *isothermal chromatography* or *temperature programming*.

In *isothermal chromatography* the temperature of the thermal compartment (and the column) is set at the desired value usually 10–15°C above the boiling point of the sample and controlled with an accuracy of ±0.1°C. When the pure carrier gas is flowing through the instrument the on-line chromatogram will show a stable base line running almost parallel to the *x*-axis indicating that the instrument is ready for introducing the sample mixture.

The sample mixture of about 1–2 µL (for liquids) or 20 µL (for gases) is injected into the injection port. The on-line chromatogram is recorded from the moment the sample is introduced. Most modern instruments show the on-line chromatogram on a video display and with built-in software to record the retention times of the individual components of the sample mixture along with the heights (as wells areas) of the individual chromatographic peaks and store the same. Under specified chromatographic conditions of the nature of the stationary phase, column length, the temperature of the column, and the flow rate of the carrier gas the retention time of a sample component is useful in qualitative analysis. The peak areas (or heights) are proportional to the amounts of the components in the sample, and hence, useful in quantitative analysis.

In *temperature programming* the temperature of the thermal compartment is programmed to increase from a chosen initial value to a chosen final value at a specified heating rate during chromatographic run. For example, the initial value of the column may be set at 50°C and the final value at 250°C and the heating rate as 10°C/min. The temperature programming run starts the moment the sample is injected into the sample injection port. The temperature of the thermal compartment (and hence, that of the column) increases linearly with time from 50°C at 10°C/min and reaches, say, 100°C at the end of 5 min and the final value of 250°C at the end of 20 min. Thus the total elution time is 20 min during which all the sample components would have eluted out. After completing the scan up to 250°C the temperature of the thermal compartment is automatically brought back to the initial value of 50°C. Temperature programming is highly useful for separating components of a mixture whose boiling points differ widely or when the sample mixture contains components of different chemical nature. The separation of a mixture of hydrocarbons differing widely in their boiling temperatures by isothermal and temperature programmed GC are shown in Figs. 13.28(a), (b), and (c) respectively.

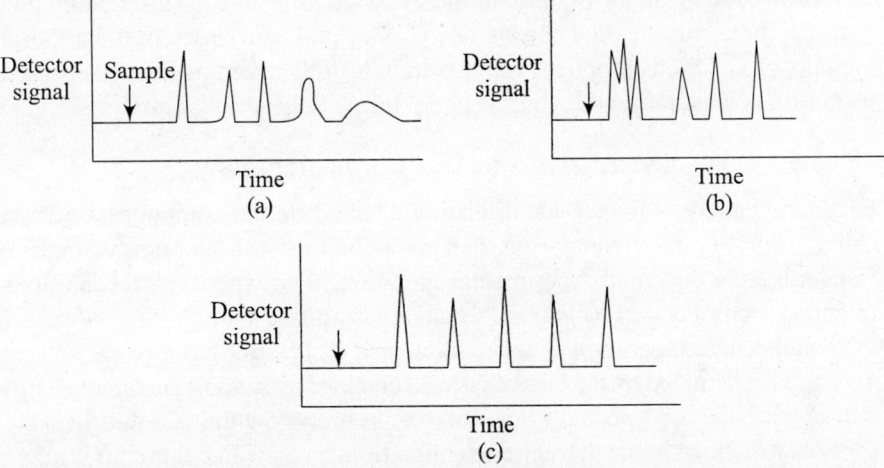

Fig. 13.28 Gas chromatographic separation of a hypothetical mixture of components of widely differing boiling points under; (a) isothermal GC at low column temperature; (b) isothermal GC at high column temperature; and (c) temperature programming

It can be seen that in isothermal GC if the temperature of the column is set at a lower value the higher boiling components of the mixture do not elute out properly giving rise to tailing of peaks. In contrast, if the temperature of the column is set at a higher value the separation of the lower boiling components is not proper as they elute out almost immediately after injection and sometimes undergo pyrolysis giving rise to multiple peaks. Temperature programming on the other hand separates the components with good resolution.

Derivatization involves converting non-volatile polar compounds and thermally sensitive compounds into stable volatile derivatives suitable for gas chromatographic analysis. Compounds having polar functional groups such as hydroxyl, carboxyl, and amino groups (e.g., carbohydrates, fatty acids, phenols, amino acids, etc.) are derivatized with suitable chemical reagents to produce less polar and volatile methyl, trimethysilyl, or trifluoroacetyl derivatives for subjecting them to gas chromatographic analysis.

Pyrolysis involves thermally decomposing non-volatile and thermally sensitive compounds under controlled conditions to produce characteristic low molecular weight products which are then separated and identified by gas chromatograph. Pyrolysis GC is particularly useful for analysing polymers, plastics, paints, and even biological substances.

Head space analysis involves the analysis of vapours in the head space above the sample in a partially filled and sealed container or vapours obtained by warming the volatile sample. The analysis requires an automated head space sample injection system consisting of thermostatically heated compartment in which the samples can be equilibrated and from which the head space vapours are let into the carrier gas stream for injection into the gas chromatograph. Head space analysis is particularly useful for the analysis of perfumes, flavours, and solvents in partially filled sealed containers, samples containing a mixture of volatile and non-volatile components such as residual monomers in polymers, alcohol in blood samples, etc.

Thermal desorption involves the adsorption of vapours by an adsorbent held in a closed tube in the first step and in the second step the adsorbed vapours are thermally desorbed by rapidly heating the adsorbent and swept by the carrier gas into the gas chromatograph. The method concentrates the analyte vapours in the first step prior to gas chromatographic analysis and is particularly useful for the analysis of traces of pollutant gases in the atmosphere and solvent vapours or volatile component vapours in industrial environments. Thermal desorption is also useful for concentrating the volatile components in head space analysis.

13.13.5   Qualitative Analysis by Gas Chromatography

Qualitative analysis for the identification of the different components of a sample mixture is usually carried out by comparing their retention times or retention volumes with authentic or standards under specified experimental conditions. The experimental conditions which affect the retention times include (i) nature of the stationary phase column, (ii) its dimensions, particularly the length, (iii) temperature of the column, and (iv) carrier gas flow rate. A sample component may also be identified by the increased peak obtained by spiking the sample with a pure substance, whose presence is suspected in the sample. In order to eliminate the chance of more than one substance having almost the same retention time on a given column, it is advisable to carry out gas chromatographic analysis on two different stationary phase columns.

Alternatively the eluting sample components may be trapped and subjected to further analysis or directly sent to a mass spectrometer or infrared spectrophotometer for identification.

Kovats retention index (RI) was proposed by Kovats in the 1950s to convert retention times of compounds in GC into system-independent constants. It is applicable only to organic compounds. The retention times of sample components determined by GC depend on various factors such as the nature of the stationary phase, the film thickness of the stationary phase, length and diameter of the column, carrier gas flow rate and pressure, void time, and temperature. In contrast, the derived Kovats retention indices are quite independent of these parameters, and hence, allow comparison of retention times measured under different experimental conditions.

The concept was proposed on the basis of a linear relationship between logarithm of the corrected retention time (t_R') and the number of carbon atoms for a homologous series of compounds. The indices are based on *n*-alkanes as standards because they cover a wide range of boiling points, have low polarity, chemically stable and non-toxic, get readily separated on almost all stationary phase columns following the order of their vapour pressure, easily available, and relatively cheap. The Kovats retention index for each *n*-alkane is defined as 100 times the number of carbon atoms in the compound at all temperatures and is independent of the stationary phase. Thus the values of RI for the *n*-alkanes pentane, hexane, heptane, and octane are 500, 600, 700, and 800 respectively. The plot of log t_R' against RI for *n*-alkane series is linear as shown in Fig. 13.29.

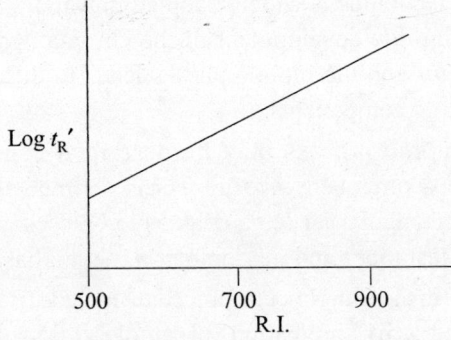

Fig. 13.29 Linear relationship between log t_R' and RI for *n*-alkanes

The RI value of a substance having a retention time in between an adjacent pair of *n*-alkanes can be determined from the graphical plot or calculated by linear interpolation. The determined RI value is useful for the possible identification of the sample compound from published data or from previously collected data on a given column stored in a computer database. Kovats retention indices are useful for comparing stationary phases with regard to their retention capabilities for a set of compounds. Based on the RI values it is possible to select a stationary phase for the separation of a group of compounds and also for identifying similar stationary phases from a large number of liquid stationary phases quoted in literature.

13.13.6 Quantitative Analysis by Gas Chromatography

Quantitative analysis is based on measuring the peak height or for more accurate determination measuring the integrated area of the peak assigned to the particular compound on the basis of retention time. Most of modern gas chromatographs have electronic integrators for precise determination of the peak areas taking into account any drift in the baseline and provide information of peak areas along with the retention times in the display. However, for accurate determinations methods such as (i) internal standardization, (ii) internal normalization, or (iii) standard addition are employed as discussed in Section 13.4.9 because the response of the GC detectors is not uniform for all compounds.

13.14 PLANAR CHROMATOGRAPHIC TECHNIQUES

The name *planar chromatographic techniques* implies that the stationary phase is suitably placed on a planar chromatographically inert support over which the mobile phase (eluent) moves aided by capillary action or sometimes by gravity. The different techniques include paper chromatography (PC), thin layer chromatography (TLC), and high-performance thin layer chromatography (HPTLC). The techniques are mostly used for analytical purposes. However, TLC can be used for preparative work for isolating/purifying products up to a few hundred milligrams. The practical aspects of planar chromatographic techniques are similar.

13.14.1 Paper Chromatography

In paper chromatography a filter paper strip functions as the stationary phase support. The water molecules bound to the cellulose molecules of the filter paper function as the actual stationary phase. The mobile phase is a homogenous aqueous solution or an aqueous–organic solvent mixture. Ascending technique is widely adopted in which the mobile phase moves upwards by capillary action during the development of the chromatogram. The partitioning of the solute between the stationary and the mobile phases leads to differential migration and thereby the separation of the solute components.

The chromatographic process may be described conveniently by the separation and identification of a two-component mixture. The experimental set-up is quite simple and consists of a cylindrical or rectangular jar (e.g., beaker) to be used as the development chamber, spray bottles, cover glass, test tubes and micropipettes, or capillaries.

The steps involved in the experimental procedure include (i) equilibration of the chromatographic development chamber, (ii) application of sample and standard(s), (iii) development of the chromatogram, (iv) detection or location of the solute spots, and (v) calculation of retention ratio R_f and sample identification.

At the start of the experiment the chosen eluent is placed in the chromatographic development chamber provided with a cover glass. A petri dish or beaker placed inside the jar functions as the eluent reservoir (Fig. 13.30). The atmosphere within the development chamber is allowed to get equilibrated with the mobile phase vapour. Equilibration of the chamber requires about 10–20 min.

A filter paper strip (Whatman No.1) of about 25 cm long and 10 cm wide is used as the stationary phase. A small amount of the sample mixture and the pure components A and B (standards) are spotted with the help of a capillary or micropipette on the start line or origin line marked (with a pencil) located about 2–3 cm from the bottom end of the filter paper strip. The sample spots are dried using a current of air from a hair drier or dried in an air oven. The sample spots should be compact with diameters less than 5 mm.

The chromatogram is developed by placing the filter paper strip vertically inside the development chamber and held in position with the help of paper clips to avoid direct contact between the sample spots and the eluent as shown in Fig. 13.30. The chromatogram develops as the mobile phase moves up the paper by capillary action. The solute components of the sample and the standards also move up the paper depending on their partition coefficient values between the stationary and mobile phases. Differential distribution with consequent differential migration of the solute components of the mixture results in the separation of samples. The filter paper strip is removed from the chamber once the solvent front has moved up wetting almost to 80–90% of the strip. The solvent front is marked with a pencil and the strip is dried.

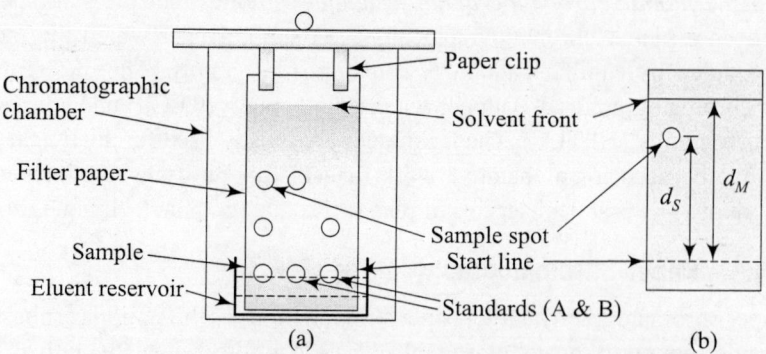

Fig. 13.30   (a) Paper chromatographic assembly and (b) measurement of R_f

The solute spots are located by spraying the dried filter paper strip with a suitable locating agent to develop a colour. For example, a metal ion such as copper can be located by the formation of brown coloured spot obtained by spraying with potassium ferrocyanide while nickel can be located by the formation of a red coloured spot by spraying with DMG solution. Similarly, amino acids can be located by using ninhydrin as the locating agent. Alternatively the dried filter paper may be exposed to UV light to visualize the emitted fluorescence. The linear distances of the individual solute spots (d_S) and that of the solvent front (d_M) from the start line are measured using a scale. It must be noted that the solute spots are not perfectly symmetric and the measurement of d_S is usually based on the position of maximum intensity of the solute spot. The retardation factor R_f for a given solute is calculated as

$$R_f = d_S/d_M \qquad (13.23)$$

The R_f values are related to the partition coefficient K. The values of R_f vary between 0 ($K = \infty$) and 1 ($K = 0$).

The retention parameters t_R and t_M which correspond to the times required for the solute and the mobile phase respectively to travel a fixed distance d_S on the planar stationary phase may be related to the distances d_S and d_M as follows. The time t_M required by the mobile phase to travel the distance d_S depends on its linear velocity u and is given as

$$t_M = d_S/u \qquad (13.24)$$

The time taken by the solute to reach the same distance, namely, d_S, is however, equal to the time taken by the mobile phase to reach the distance d_M, and hence,

$$t_R = d_M/u \qquad (13.25)$$

Substituting these equations into the relationship $k' = t_R - t_M/t_M$ gives

$$k' = d_M - d_R/d_R \qquad (13.26)$$

Equation (13.25) may be rewritten relating the R_f and k' as

$$k' = \frac{(1 - d_R/d_M)}{(d_R/d_M)} = (1 - R_f)/R_f \qquad (13.27) \text{ [see Eq. (13.4)]}$$

The solute components of the sample mixture are identified by comparing the R_f values with those of standards spotted in the same filter paper strip. The magnitude of R_f values depend on the composition of the mobile phase, and hence, selectivity and separation efficiency can be altered by altering the eluent composition.

Paper chromatography is useful in qualitative analysis. Quantitative estimations may be carried out by spotting exactly known amount of the sample mixture, cutting out the spots of the separated components after chromatographic development and extracting quantitatively the individual spots into a suitable solvent for estimation by spectrophotometry, gas chromatography, or HPLC. Alternatively the solute spots may be sprayed with a suitable locating agent to develop a stable colour, the intensity of which is proportional to the amount of the solute, and measuring the intensity with the help of a densitometer. A wide variety of inorganic (e.g., metal ions and anions), organic, and biochemical compounds can be analysed by PC by appropriate selection of mobile phases with variations in polarity and pH.

13.14.2 Thin Layer Chromatography

The stationary phase support in thin layer chromatography (TLC) is usually a flat surface of glass, plastic, or metal plate (mostly Al). The active stationary phase is fine particles (~20 μm) of an adsorbent such as alumina or silica gel coated on the support plate as a thin layer of about 250 μm thickness. The mobile phase moves over the stationary phase wetting the surface. The separation of solute components of a mixture is effected due to the different migration rates of individual solutes depending on their relative affinities towards the adsorbent and the mobile phase. The mechanism of separation involves partition and/or adsorption depending on the amount of water immobilized during drying of the TLC plate during its preparation. The separation of solutes is expressed in terms of R_f values as in paper chromatography.

The experimental setup and procedure are similar to that in paper chromatography. The steps involved in the experimental procedure include (i) preparation of TLC plates, (ii) equilibration of chromatographic development chamber, (iii) application of sample and standard(s), (iv) development of the chromatogram, (v) detection or location of the solute spots, and (vi) calculation of R_f followed by qualitative and quantitative analyses.

TLC plates of different sizes are used, the most common being 25 cm × 25 cm, 25 cm × 10 cm or 10 cm × 2.5 cm glass plates (microscope slides). The glass plates are cleaned thoroughly and dried. Slurry of the adsorbent in water is coated as a thin layer on the support plate and the water is evaporated by drying at room temperature or at a higher temperature. The drying temperature controls the amount of water held by the adsorbent layer. Other stationary phase particles such as normal phase, reversed phase, ion exchange, and size-exclusion particles may also be used in TLC. Applicators are available commercially to coat the stationary phase particles to a desired thickness. Alumina or silica gel coated on plastic plates or aluminium foils are also available commercially.

The atmosphere within the chromatographic development chamber is presaturated with eluent vapour by allowing sufficient equilibration time of about 20–30 min.

The sample mixture and the standards (0.02–0.1% solutions) are spotted on the start line, about 1–2 cm from the edge, marked with a sharp needle on the TLC plate and the spots are air dried. As in paper chromatography the samples are spotted using a capillary or micropipette and the spots should be compact with diameters less than 5 mm.

New methods of sample preparation and concentration in the case of biological, pharmaceutical, and environmental samples have extended the range of applications of TLC. *Solid phase extraction* (SPE) has been used for concentrating drugs and biological extracts before analysing by TLC. *Ultrasonic extraction* of anionic, cationic, and non-ionic surfactants followed by TLC has been reported to be much more efficient, faster, and less expensive for their determination in laundry wastewater samples compared to traditional shake-flask extraction. *Supercritical fluid extraction* (SCFE) has been used for extracting and concentrating biologically active compounds from plant sources prior to TLC analysis.

The chromatographic development is carried out in the chamber by placing the TLC plate vertically in the chamber avoiding direct contact between the sample spots and the eluent. As the eluent moves past the start line it dissolves the sample components and carries them upward. The sample components get distributed between the stationary and the mobile phases resulting in their different migration and consequent separation. After the elapse of sufficient time of development the plate is taken out and the solvent front is marked. The plate is then dried preferably at room temperature.

The solute spots on the dried plate are located by exposing the plate to iodine vapour by placing it in another chamber containing solid iodine sprinkled on filter paper scrap or cotton wad kept at the bottom of the chamber. The atmosphere within this iodine chamber is saturated with iodine vapour and the iodine vapour gets adsorbed on to solute spots on the dried TLC plate giving a brown or orange colour to the organic solute spots. Alternatively location of the sample spots may be carried out by spraying a solution of iodine or sulphuric acid to produce dark spots of the solutes. The locating reagent may also be specific, for example, ninhydrin reagent for locating amino acids. Yet another method of detection is based on using a stationary phase incorporated with a fluorescent substance. The sample components quench the fluorescence and on viewing the TLC plate under UV light after chromatographic development the whole plate fluoresces exposing the non-fluorescing sample components.

The R_f values of the individual solutes are determined as the ratio of d_S/d_M in a manner similar to that in paper chromatography. Figure 13.31 shows the chromatographic development and determination of R_f values of individual solute components of the sample by TLC.

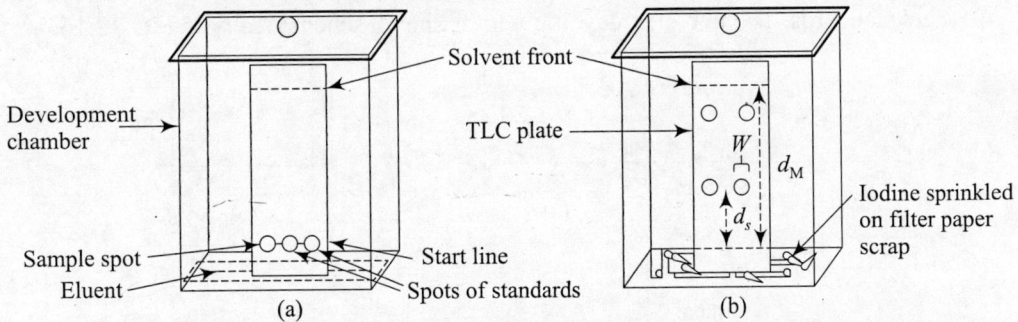

Fig. 13.31 (a) Development of chromatogram in TLC and (b) location of spots in an iodine chamber

The efficiency of the TLC plate is described in terms of the number of theoretical plates and the plate height related to the linear distance d_s of the solute spot from the origin line, and the width w of the spot as

$$N = 16 \, (d_s/w)^2 \tag{13.28}$$

and

$$H = d_s/N \tag{13.29}$$

TLC is more reproducible than PC, and hence, is quite useful for both qualitative and quantitative analysis. Qualitative analysis is carried out by comparing the R_f values with standard or authentic substances under identical conditions. Confirmation of the analysis is carried out by scrapping of the solute spot from the TLC plate, dissolving the solute in a suitable solvent, and subjecting to mass spectroscopy, infrared spectroscopy, or NMR spectroscopy.

Quantitative analysis is usually carried out by applying accurately measured quantities of the sample solution on the TLC plate and measuring the intensity of the orange or brown coloured spots of organic solutes, obtained by exposing to iodine vapour, with a densitometer. Alternatively the sample spots may be scrapped out, dissolved in a suitable solvent and subjected to spectroscopic analysis. TLC can be used even on a preparative scale for handling a few hundred milligrams of samples.

13.14.3 Two-dimensional Planar Chromatography

Two-dimensional paper chromatography (or TLC) can be carried out by adopting the chromatographic development in two steps. In the first step the analyte mixture is separated in the first eluent which should be sufficiently volatile to evaporate at room temperature. After development the paper is dried and then turned through 90 degrees and chromatographic development is carried out as part of the second step using a second eluent. After development with the second eluent, the paper is dried and sprayed with suitable locating agent. From the two-dimensional map the individual solutes are identified by comparing their positions with a map of known compounds obtained under similar conditions. For example, the mixture of amino acids obtained by hydrolysis of a protein may easily be identified using two-dimensional PC/TLC. The first eluent is a mixture of 1-butanol : glacial acetic acid : water (12:3:5). After the development with the first eluent and drying, development in a perpendicular direction with the second eluent of phenol and water mixture (prepared by adding 125 mL water to 500 g of phenol, the mixture being allowed to stand overnight and adding a few drops of ammonia solution to the mixture just before use) is carried out. Ninhydrin solution (0.2 g in 100 mL acetone) is used as the locating reagent and the individual amino acids are identified on the basis of a map prepared using individual amino acid standards. The two-step development is shown schematically in Fig. 12.16.

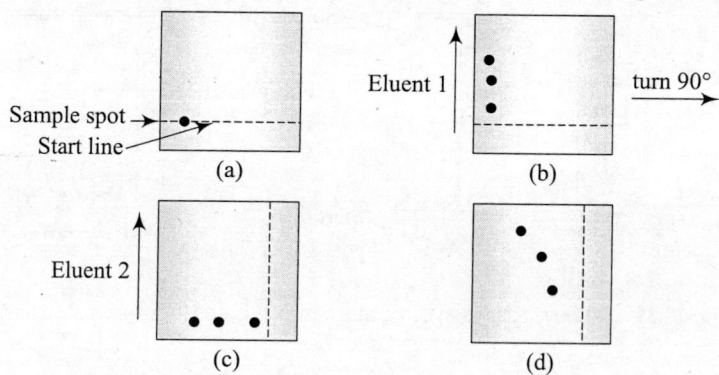

Fig. 13.32   A schematic representation of two-dimensional mapping of a sample mixture

13.14.4   High-Performance Thin Layer Chromatography

The technique has a greater separation efficiency compared to TLC as implied in the name. The stationary phase consists of finely divided adsorbent particles of silica or alumina of uniform size (~5 μm diameter) coated on aluminium support plates to a thickness of about 100 μm. The experimental methodology is the same as in TLC.

HPTLC has several advantages over conventional TLC such as (i) greater separation efficiency due to the smaller particle size of the stationary phase, (ii) shorter analysis time due to the shorter migration distance (about 3–5 cm compared to 15–25 cm in TLC), (iii) greater reliability and reproducibility in qualitative and quantitative analysis, (iv) requirement of less amount of mobile phase for analysis, and (v) suitability for automation in sample spotting (auto sampler) as well as in the detection of solute components (by spectral, fluorescence detectors).

HPTLC plates have an estimated number of theoretical plates of about 4000 on a 3-cm long plate as compared to about 2000 theoretical plates on a 12-cm long TLC plate. Hence, HPTLC provides a sharp separation of solute components requiring a development time of only 10 min in contrast to 25 min required for TLC. However, the sample capacity of HPTLC plates is quite small, and hence, the technique can be used only for analytical purposes. Qualitative and quantitative analyses are usually carried out with a densitometer or fluorescence detector.

13.14.5   Applications of Planar Chromatographic Techniques

TLC offers a greater variety of stationary phases than any other kind of chromatography to provide the required selectivity for a particular separation, including inorganic, organic, adsorption, partition, ion exchange, chiral, mechanically impregnated, polar and non-polar chemically bonded phase, buffered, mixed, and gradient layers. However, the great majority of reported separations and analyses continue to be carried out in the adsorption mode using commercial, precoated NP silica TLC and HPTLC layers. The broad range of applications of TLC and HPTLC include analysis of pharmaceuticals, herbal medicines and dietary supplements, biological and clinical samples, foods and beverages, environmental pollutants, and chemicals. TLC and HPTLC find increasing use in the detection of counterfeit drugs. Another important application is the *quantitative structure activity relationship* (QSAR) studies for the determination of a compound's biological activity for structure-based drug design. For example, lipophilicity is considered to have a dominant role in drug penetration through hydrophobic cell membranes and uptake by target organs or organisms via drug receptor binding. The chromatographic behaviour of polyphenols, such as flavonoids and phenolic acids may be correlated with the compounds' lipophilicity, solubility, plasma–protein binding, and oral absorption.

13.14.6 Developments in Planar Chromatographic Techniques

Microemulsion TLC (ME-TLC) uses a microemulsion as the mobile phase and is claimed to be easier to operate with better resolution and reproducibility. The separation mechanism and retention behaviour are significantly different from those of conventional TLC in many applications. ME-TLC has been used for fingerprinting aqueous extracts of biomolecules, antibiotics, and for the determination of hydrophobicity of compounds.

Ultrathin layer chromatography (UTLC) uses plates with a 10-μm thick monolithic layer grafted onto the glass plate with no binder. The method has been used as a rapid means of chromatographic separation prior to subjecting the sample to *desorption by electrospray ionization mass spectrometry* (DESI)-MS. The rapid analysis of benzodiazepines in urine at picomole concentration was shown to be feasible by combining two-dimensional (2D) UTLC separation followed by *atmospheric pressure matrix-assisted laser desorption/ionization mass spectrometry* (AP-MALDI-MS).

Planar electrochromatography (PEC) involves the use of an electrical field applied across a TLC plate and the resulting electro-osmotic flow drives the mobile phase instead of capillary action as in normal TLC. A new plate for the method, prepared by fusing a mixture of silica gel and glass powder followed by reacting with octadecyltrichlorosilane, was shown to have good mechanical stability and could be regenerated by soaking the chromatogram in acetone. The technique was demonstrated for the separation of five dyes in a PEC chamber with an aqueous acetonitrile containing 25 mM sodium acetate buffer (pH 5), as the mobile phase, and by applying a 2-kV intermittent voltage supply.

Over pressured layer chromatography or *optimum-performance laminar chromatography* (OPLC) is a forced flow method performed in an instrument in which the plate containing an optimized high-performance layer is covered with a flexible, inert polymer sheet under pressure, and the mobile phase is pumped through the absorbent layer. In effect, the layer is analogous to a flat HPLC column. OPLC may be combined with UV densitometry and near-infrared (NIR) spectrometry directly on the layer for qualitative and quantitative pharmaceutical analysis, separation and quantitative determination of aldoses and alditols in hemicellulose hydrolysates, etc.

Thin layer radiochromatography makes use of radiolabelled molecules. The activity of transfer ribonucleic acid (tRNA) modification enzymes was detected using radiolabelled tRNA substrates.

The procedures were based on analysis of pre- or postlabelled nucleotides (with ^{32}P, ^{35}S, ^{14}C, or ^{3}H) generated after complete digestion with selected nucleases of modified tRNA isolated from cells or incubated in vitro with modifying enzyme(s). Nucleotides of the tRNA digests were separated by 2D TLC on cellulose plates to establish the base composition and identification of the nearest neighbour nucleotide of a given modified nucleotide in the tRNA sequence. Maps of 70 modified nucleotides on the TLC plates have been reported. A quantitative method has been evolved for determining compounds labelled with short-lived β-emitting radionucleotides in microdialysates based on TLC with digital photo-stimulated luminescence autoradiography.

Preparative layer chromatography (PLC) has been described for the purification of synthesis products. Samples are applied manually with a syringe as thin 1.5 cm rectangular bands 1.5 cm apart to a 2-mm thickness layer of adsorbent-coated TLC plates. Vertical development followed by detection under UV light and recovery of compounds by scraping and elution has been used for purification and isolation of synthetic products in milligram quantities.

SOLVED PROBLEMS

EXAMPLE 1 *Calculate the retention volumes, capacity factors, and relative retention of two components separated on a column with retention times of 1.8 and 3.2 min respectively on an adsorption column at a flow rate of 1.5 mL/min. The t_o value is 0.8 min.*

SOLUTION

$$V_{R1} = t_{R1} \times F = 1.8 \times 1.5 = 2.7 \text{ mL}$$
$$V_{R2} = 3.2 \times 1.5 = 4.8 \text{ mL}$$
$$V_m = 0.8 \times 1.5 = 1.2 \text{ mL}$$
$$k'_1 = (t_{R1} - t_m)/t_m = (1.8 - 0.8)/0.8 = 1.25$$
$$= (V_{R1} - V_m)/V_m = (2.7 - 1.2)/1.2 = 1.25$$

Similarly, $k'_2 = 3.0$

$$\alpha = t_{R2'}/t_{R1'} = 2.4/1.0 = 2.4$$
$$k'_2/k'_1 = 2.4$$

EXAMPLE 2 *A solute X of a sample mixture is not retained on a GC column and elutes out at 2 min after introduction. Another solute in the same sample mixture Y elutes out at 4.6 min. The mobile phase flow rate is 22 mL/min. Calculate the values of V_m, k' for the solute Y and the time spent by Y in the stationary and mobile phases.*

SOLUTION

$$t_m = 2 \text{ min and } V_m = t_m \times F = 2 \times 22 = 44 \text{ mL}$$
$$k' \text{ for solute } Y = (t_R - t_m)/t_m$$
$$= (4.6 - 2.0)/2.0 = 1.3$$

The time spent by solute Y in the stationary phase $(k' \times t_m) = 2.6$ min

and the time spent by Y in the mobile phase is $(t_R - $ time spent in the stationary phase$)$

$$= 4.6 - 2.6 = 2.0 \text{ min}$$

EXAMPLE 3 *A chromatographic separation of a two-component sample on a 25-cm column gave the retention times in terms of the movement of chart paper for the solutes A and B as 2.2 and 2.6 cm with base widths of the two chromatographic peaks being 0.2 and 0.23 cm respectively. Calculate the (i) number of theoretical plates, (ii) plate height, (iii) resolution of the two peaks, and (iv) the resolution of the two peaks if the column length is changed to 40 cm.*

Solution

(i) Since $n = 16 \, (t_R/W_b)^2$ for the two peaks individually the number of theoretical plates would be

  n for peak $A = 16 \times (2.2/0.2)^2 = 1936$

  n for peak B $= 16 \times (2.6/0.23)^2 = 2044$; average value of $n = 1990$

(ii) Since $h = L/n$

  $h = 25$ cm/1990 plates $= 0.0125$ cm/plate

(iii) Since $R_s = 2 \, [(t_{R2} - t_{R1})/(W_{b1} + W_{b2})]$

  $R_s = 2 \, [(2.6 - 2.2)/(0.2 + 0.23)] = 1.86$

(iv) Since t_R is directly proportional to L (provided the flow rate and capacity factor do not change with the longer column).

  $t_{RA} = 2.2 \times 40/25 = 3.52$ min and $t_{RB} = 2.6 \times 40/25 = 4.16$ min

Since the number of theoretical plates is also directly proportional to L values of n can be calculated for the two peaks on the longer column as

  $n = 1936 \times 40/25 = 3098$ and $1990 \times 40/25 = 3184$

The corresponding base widths of the two peaks are 0.25 and 0.29 cm respectively.

The resolution of the two peaks on the 40 cm long column = 2.37.

Example 4 *In a quantitative analysis of a mixture of three organic compounds by gas chromatography the detector response and the peaks areas respectively of the components were reported as follows: component A – 0.85 and 3.35 cm², component B – 1.0 and 4.2 cm², and component C – 0.95 and 2.8 cm².*

Calculate the percentage composition of the sample mixture by internal normalization.

Solution

The detector response for component A is 0.85.

If it were 1.0 the peak area $= 3.35/0.85 = 3.94$ cm².

Similarly, the peak areas for components B and C are 4.2 cm² and 2.95 cm² respectively.

As per Eq. (13.16)

Percentage A = (peak area of A / total area of all components) $\times$ 100

  $= (3.94/11.09) \times 100 = 35.52$

Percentage $B = (4.2/11.09) \times 100 = 37.87$

Percentage $C = (2.95/11.09) \times 100 = 26.6$

SUMMARY

- Chromatography involves selective distribution of the components of a mixture between the immiscible stationary and mobile phases to bring about their separation by differential migration aided by the mobile phase as it percolates through a bed of the stationary phase.

- Separation of the components of a mixture takes place through any one of the five mechanisms of (i) adsorption, (ii) partition, (iii) ion exchange, (iv) size exclusion, and (v) affinity.

- Retention time and retention volume are useful chromatographic parameters in qualitative identification of the components of a mixture under specified experimental conditions.

- The separation efficiency of a column of stationary phase is related to the number of theoretical plates, the greater the number greater is the separation efficiency. It is related inversely to the height of the theoretical plate, the smaller the plate height greater is the separation efficiency. In GC, the separation efficiency is increased by increasing the number of theoretical plates through the use of longer columns. In contrast, in HPLC the plate height is decreased by efficient high density packing of the column to enhance the separation efficiency.

- The resolving power of a stationary phase column is influenced by the combination of three parameters, namely selectivity factor, capacity factor, and the number of theoretical plates.

- Chromatographic peaks become broad due to a combination of factors such as (i) inhomogeneity in the flow rate of the mobile phase, (ii) longitudinal diffusion of the solute molecules, and (iii) resistance to mass transfer between the two phases. Peak broadening can be minimized and separation efficiency can be attained by optimizing the flow rate of the mobile phase.

- Adsorption chromatography and normal phase partition chromatography make use of a polar stationary phase in conjunction with a non-polar mobile phase to separate polar compounds. In reverse phase chromatography the stationary phase is highly non-polar and the polarity of the mobile phase is tunable by proper selection of a mixture of solvents.

- Ion-exchange chromatography makes use of ion-exchange resins as the stationary phase to separate ions based on their charge.

- In size-exclusion chromatography the separation mechanism is based on the size of the solute molecules and is exclusively used for the separation of biopolymers such as proteins in gel filtration and separation of synthetic polymers in gel permeation chromatography.

- Affinity chromatographic separation is used to separate biological solutes based on their specific interaction between the solute molecule in a mixture and a specially designed stationary phase.

- HPLC is the most versatile liquid chromatographic technique capable of operating on all the five different mechanisms. The technique is useful for qualitative and quantitative analysis of components of a mixture.

- Supercritical fluid chromatography makes use of a supercritical fluid (mostly carbon dioxide) as the mobile phase in conjunction with HPLC stationary phases to separate non-volatile and thermally labile components of a mixture which are not amenable for separation by either HPLC or GC.

- GC is another versatile analytical technique to separate volatile components in a mixture in vapour phase using inert mobile phase or carrier gas in conjunction with a variety of stationary phase columns. The technique is useful for analytical purposes of identification and determination of the amounts of the components of a sample mixture.

- Planar chromatographic techniques of paper chromatography, thin layer chromatography, and HPTLC are essentially analytical techniques for identification and quantitative estimation of components of a mixture.

REVIEW QUESTIONS

1. How are chromatographic methods classified?

2. What are the different mechanisms exploited in chromatography for the separation of mixtures?

3. Describe the experimental set-up for conducting chromatographic separation in the laboratory.

4. Discuss in detail the principle of separation in column chromatographic methods.

5. What is a chromatogram? How is it obtained? What is its use?

6. Give a detailed account on various chromatographic parameters and their significance.

7. Explain the terms retention time, retention volume, partition coefficient, capacity factor, relative retention, resolution, plate height, and number of theoretical plates.

8. What are the factors responsible for broadening of chromatographic peaks? How does the kinetic theory explain band broadening?

9. Write a note on the practice of liquid chromatography.

10. Distinguish between isocratic and gradient elution.

11. Write notes on partition chromatography, hydrophobic interaction chromatography, size-exclusion chromatography, ion-exchange chromatography, and affinity chromatography.

12. How is molecular weight of a protein determined by gel filtration?

13. What is ion chromatography?

14. Give a block diagram of HPLC and describe the components and their functions.

15. Discuss the principle and practice involved in gas chromatography.

16. What is the advantage of temperature programming in GC? How is it carried out?

17. What is a supercritical fluid? What are its characteristics?

18. Explain the principle of supercritical fluid chromatography.

19. What are planar chromatographic techniques?

20. How is a sample mixture of metal ions separated by paper chromatography?

21. Discuss the principle and practice of TLC. How is it useful in the separation and identification of a mixture of amino acids?

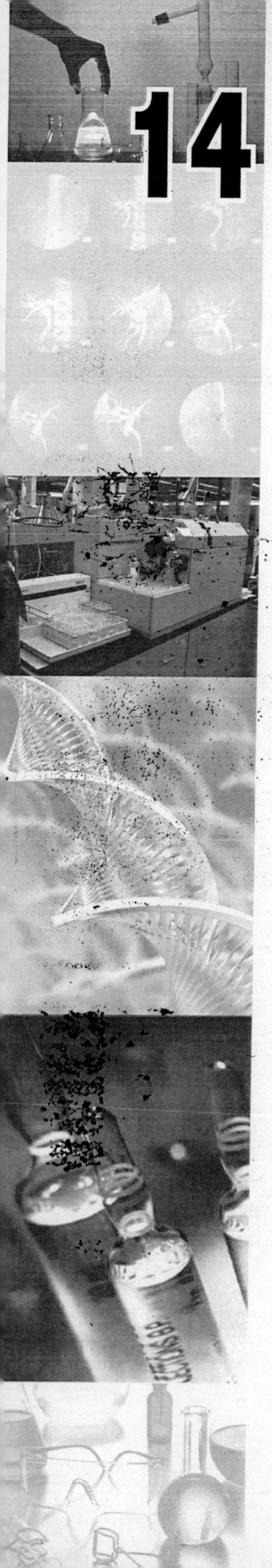

14

Electrophoresis and Related Techniques of Separation

14.1 INTRODUCTION

The different methods of separation of charged species depending on their migration rate when subjected to an applied electric field are broadly grouped as *electro-kinetic methods*. The parameters affecting the migration of charged species include the magnitude and sign of the net charge on the species, surface charge, the strength of the applied electric field, the electrolyte concentration, ionic strength, viscosity, and the pH of the medium, besides temperature. Different methods include electrophoresis and related techniques, isoelectric focusing, and isotachophoresis. *Electrophoresis* monitors the migration of charged species under a constant applied electric field. Isoelectric focusing uses a constant electric field together with a pH gradient so that different charged species get focused at their respective isoelectric points. Isotachophoresis, on the other hand, uses different strengths of electric field in conjunction with a pH gradient to separate the charged species. These techniques find extensive use in the separation and characterization of biomacromolecules such as proteins and nucleic acids.

14.2 ELECTROPHORESIS

The basic principle of electrophoresis involves the separation of charged species due to their differential migration in a buffer solution under the influence of an applied electric field. Electrophoresis can be considered as an incomplete form of electrolysis as the applied electric field is removed before the sample molecules reach the opposite electrodes. The component molecules of a mixture get separated based on the differences in their electrophoretic mobility. Biomolecules, such as amino acids, peptides, proteins, nucleotides, and nucleic acids possess ionizable groups and exist in aqueous media as positive or negative charged species depending on the pH of the medium. Even non-polar substances, such as carbohydrates can be made weakly charged by derivatization as borates or phosphates. Molecules may have similar charge but still may be differentiated on the basis of their charge/mass ratios due to differences in their molecular weights. The combination of these differences results in differential migration of the charged species in an electric field resulting in their separation.

The electrophoretic mobility of an isolated species is its intrinsic property under defined conditions of temperature, buffer concentration, the pH, and composition of the medium. The electrophoretic mobility of a species v' is defined as the migration per unit field strength and given as

$$v' = \frac{ze}{6\pi\eta r} \qquad\qquad (14.1)$$

where z is the valence, e the electron charge (1.602×10^{-19} coulombs), η the viscosity of the liquid medium, and r the radius of the charged species. Viscosity is expressed in Pa s (or Ns m^{-2}) and electrophoretic mobility in m^2 (Vs)$^{-1}$ or cm^2 (Vs)$^{-1}$. The electrophoretic mobility of the charged species, thus, depends on the viscosity of the medium, the size and shape (Stokes' radius), and the charge on the molecule. In the case of macromolecules the electrophoretic mobility depends mainly on the ionizable groups present on the surface of the molecule and the sign and magnitude of the charge carried by the ionizing group. The sign and magnitude of the charge vary depending on the ionic strength and pH of the medium. Separation of molecules can therefore be effected by selecting an appropriate medium.

In an aqueous medium the real mobility of the species depends on three retarding forces, namely, (i) electrophoretic retardation, (ii) relaxation effect, and (iii) electro-osmotic flow of the solvent. Electrophoretic retardation is due to a thin layer of liquid bound to the charged species retarding the mobility of the charged species. The relaxation effect is due to an ion atmosphere formed by oppositely charged species surrounding each charged species. In an applied electric field, the charged species and the surrounding ion atmosphere move in opposite directions resulting in a distorted ion atmosphere and relaxation effect. Electro-osmotic flow, also called electro-endosmotic flow, refers to the flow of solvated water molecules (i.e., bulk solvent water) associated with the cations and anions under the influence of electric field. As cations are more solvated than anions, a net flow of solvated water occurs towards the cathode and due to this electro-osmosis even neutral species migrate towards the cathode, even though they should normally remain at the point of sample introduction.

Different types of electrophoresis have come into vogue depending on whether the electrophoresis is conducted in the presence or absence of a supporting medium. Electrophoresis in the absence of any supporting medium is called the *free solution electrophoresis,* whereas if the technique is carried out using a support medium it is known as *zone electrophoresis.*

14.2.1 Free Solution Electrophoresis

The technique was first introduced by the Swedish biochemist Tiselius for the separation of proteins. It is also called *moving boundary electrophoresis* technique. The sample protein solution is dialysed against a buffer solution and placed in each arm of a U-shaped cell (Fig. 14.1).

Electrophoresis is then carried out by dipping two electrodes connected to a dc source into the sample–buffer mixture in each arm of the tube. The movement of the proteins due to electrolysis is monitored at the boundaries between the sample proteins and the buffer by refractive index detectors placed at the upper ends of the cell. The method was useful in studying the effect of the physical properties of the solution on the mobility of proteins. However, complete separation of proteins could not be achieved due to difficulty in maintaining stable boundaries. Mixing of the boundaries occurs due to diffusion as well as convection.

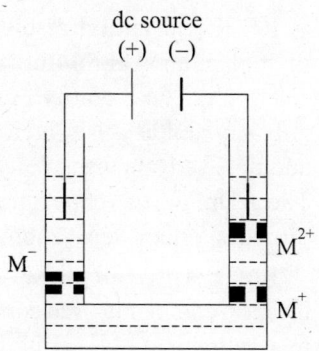

Fig. 14.1 Separation of M$^+$, M^{2+}, and M$^-$ species by moving boundary electrophoresis

14.2.2 Zone Electrophoresis

The technique involves the use of inert and relatively homogeneous supporting medium impregnated with buffer solution for electrophoresis. The sample mixture separates into its components by migration as distinct zones, which can be subsequently detected by suitable analytical techniques. Zone electrophoresis finds extensive use in analytical as well as in preparative work as it has advantages over free solution electrophoresis such as absence of mixing due to convection or diffusion of components and negligible heat generation during electrophoresis. Different supporting media are used in zone electrophoresis and accordingly named depending on the type of support used. *Paper electrophoresis* makes use of a sheet of absorbent paper (cellulose) while *gel electrophoresis* makes use of a gel of cellulose acetate, starch, agarose, or polyacrylamide. Paper electrophoresis has been largely superceded by gel electrophoresis, and hence, rarely used. However, high-voltage paper electrophoresis still finds use for the separation of amino acids and nucleic acids. Cellulose acetate is used as a support for separating low molecular weight bio-organic substances such as amino acids and carbohydrates. Cellulose acetate has several advantages over paper as support medium such as improved resolution, fast separation, less tailing of spots, easy detection of spots by optical methods as the medium is transparent, ease of isolation, and recovery of the separated components by simple dissolution of the cellulose acetate in suitable solvents. Starch gel as the support medium has both chromatographic and electrophoretic properties and the chromatographic properties particularly the ion-exchange capability of starch can hinder the electrophoretic migration of proteins. In addition, starch gel being negatively charged also generates an electro-osmotic flow of the solvent. Polyacrylamide and agarose gels are used as supports for separating large molecular weight substances such as proteins and nucleic acids.

14.2.3 Polyacrylamide.Gel Electrophoresis

Polyacrylamide gel electrophoresis (PAGE) is one of the widely used forms of zone electrophoresis using a gel of polyacrylamide to separate species on the basis of both charge and size. The gel consists of a three-dimensional network of filaments forming pores of differing sizes. The gel can be prepared with controlled pore dimensions suitable for biomolecules of different molecular sizes. Due to the controlled pore dimensions a molecular sieving mechanism aids the separation of molecules in addition to the electrophoretic mobility. In general, large molecules are retarded in the PAGE. In PAGE, the electrophoretic mobility of similarly shaped sample component molecules is related linearly to the logarithm of their molecular weights.

PAGE with its high resolving power is useful for the separation of proteins and oligonucleotides of molecular weights up to 10^6. The technique can separate even molecules of similar electrophoretic mobility provided they differ in their size, and hence, molecular weight. The polyacrylamide gel is inert and does not interact chemically with biomolecules and has the additional advantages of physical stability and capability to handle relatively large sample sizes. The gel minimizes convectional and diffusional movement of the solute components and thereby provides a sharp separation of the sample components into completely resolved bands.

PAGE is used for preparative as well as for analytical separations. For preparative separations, the most commonly used configuration is column PAGE. The monomer acrylamide is polymerized in a cylindrical glass column of ~1 cm diameter and 20–100 cm in length. The experimental set-up is shown in Fig. 14.2.

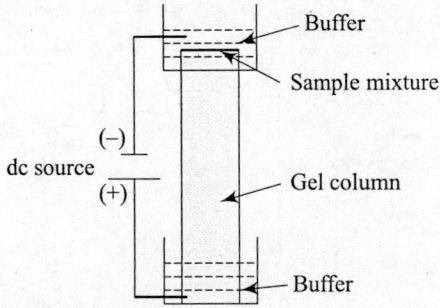

Fig. 14.2 Experimental set-up for column PAGE

During polymerization the gel mixture is covered with layer of water to obtain a flat upper surface after polymerization. After casting the gel, the water layer is removed and the mixture of sample proteins is introduced at the top of the gel. The sample is usually dissolved in concentrated sucrose solution so that its high density prevents mixing of the sample components with the buffer in upper cathodic compartment. The sample mixture also contains a tracking dye either cationic methylene blue or anionic bromophenol blue to monitor the progress of separation during electrophoresis. Initially electrophoresis is run a low current of about 1 mA for about 30 min to allow the sample components to penetrate the gel. Then the current strength is increased to 2–5 mA for normal electrophoresis. Ohmic heating may be reduced by cooling the buffer solutions. Electrophoresis is stopped when the tracking dye reaches the bottom of the column. The gel is extruded from the column and the separated protein bands are fixed by precipitation by soaking the gel in trichloroacetic acid solution. The separated protein bands are then stained for detection.

Analytical gel electrophoresis is mostly carried out in a slab of the gel which allows simultaneous electrophoresis of a number sample mixtures and standards under identical conditions. The different analytical gel electrophoresis techniques include (i) simple or native gel electrophoresis and (ii) sodium dodecyl sulphate–polyacrylamide gel electrophoresis (SDS-PAGE).

14.2.4 Native Gel Electrophoresis

Polyacrylamide gel electrophoresis is mostly practiced as *thin-slab gel electrophoresis* because a thin slab generates less heat during electrophoresis. It has the advantage of analysing several samples simultaneously in a matrix of identical composition. The identification of protein bands with dyes is easier as the dyes (e.g., Coomassie Blue) diffuse rapidly into the thin gel. A vertical slab gel apparatus is shown in Fig. 14.3. It consists of (i) an electrophoresis chamber or gel box, (ii) two buffer reservoirs placed one at the top and the other at the bottom in which electrodes are placed, (iii) a high voltage dc power supply unit, and (iv) two glass plates held parallel by a spacer for casting and holding the slab in vertical position between the reservoirs and a plastic 'comb' (slot former) are also part of apparatus.

The power supply unit is capable of providing constant voltage up to 4 kV and a constant current up to 200 mA. The slab gel of uniform thickness (of 0.5–2.0 mm) and dimensions of 8×10 or 10×10 cm is cast between the vertically held parallel glass plates by polymerization of acrylamide monomer. Acrylamide is mixed with a cross-linking agent (N,N'-methylene-bis-acrylamide) and an initiator-catalyst system such as ammonium persulphate–N,N,N',

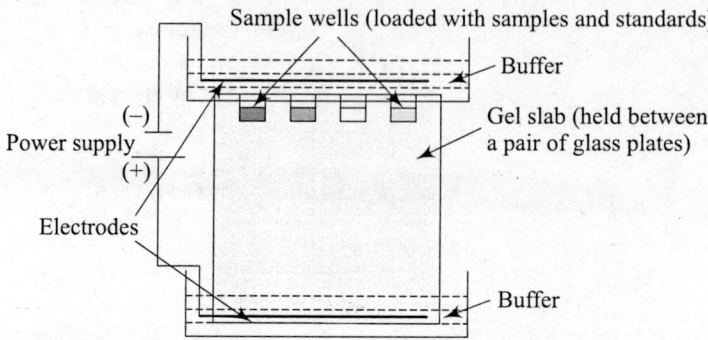

Sample wells (loaded with samples and standards)

Buffer

Power supply (−) (+)

Electrodes

Gel slab (held between a pair of glass plates)

Buffer

Fig. 14.3 Schematic diagram of a vertical slab gel electrophoresis unit

N'-tetramethyl ethylenediamine (TEMED) and polymerized, the polymerization reaction requiring about 30–40 min. The plastic comb is inserted into the top of the slab gel during polymerization to form depressions or indentations which serve as sample wells. Once a slab gel is cast the comb is carefully removed and the sample wells are rinsed with buffer solution to remove any unpolymerized acrylamide and salts. The slab held by the glass plates is clamped into position between the two buffer reservoirs and the electrodes are connected to the power supply unit. The sample proteins are dissolved in a buffer at a pH where the proteins remain stable and in their native conformation. The pH range is usually 8–9 where most of the proteins carry negative charges. The samples are loaded into the sample wells and voltage applied to carry out electrophoresis. The sample proteins move towards the anode placed at the bottom of the vertical polyacrylamide gel. Any basic protein present in the sample remains dissolved in the cathodic buffer at the top of the vertical gel. A buffer with lower pH range may be chosen and the cathode placed at the bottom of the vertical gel to separate basic proteins.

The resolving capability for a given range of molecular weights of proteins depends on the concentrations of the acrylamide monomer and the cross-linking agent bis-acrylamide. Higher concentrations give gels with pores of smaller size which can be used for low molecular weight substances while lower concentrations give gels of bigger pores amenable for high molecular weight compounds. For the separation of proteins in the molecular weight range of 10^5–10^8 a gel of 7.5% polyacrylamide is commonly used together with the cross-linker N,N'-methylene bis-acrylamide with ammonium persulphate (~2 mM) and the free radical scavenger TEMED (~0.1% v/v). The pore size of the gel can be tailored to suit the molecular weights of the sample proteins by altering the concentration of either the monomer acrylamide or the cross-linker. Increasing the concentration of either of the two decreases the pore size and vice versa. For small proteins of about 10 kDa, a gel containing 20% acrylamide and 1% bis-acrylamide may be used while for large proteins up to 1000 kDa, a gel of 4% acrylamide and 0.1% of bis-acrylamide may be used.

The buffer for electrophoresis is carefully selected since some buffer ions react with the compounds under investigation. For example, borate ions in the buffer react with sugars forming complexes. The bulk of the current flowing through the separation channel is carried by the buffer ions, and hence, its composition and ionic strength are of importance. Ions such as Na^+, K^+, Mg^{2+}, F^-, Cl^-, Br^-, SO_4^{2-}, and HPO_4^{2-} are highly mobile and carry much current leading to heating, and hence, may be avoided unless specifically required. In contrast, bulky organic ions have much lower mobility. The composition of the buffer and its ionic strength is chosen on the basis of a compromise between low conductivity (hence, low heating) and a buffer with higher conductivity. The low conducting buffers allow application of a higher voltage but may result in protein–protein interactions. In contrast, the buffer with higher conductivity minimizes protein–protein interaction but a low applied voltage is maintained to avoid heating. The pH of the buffer is chosen so that any chemical change or denaturation of the sample is avoided. Maximum separation of components is obtained generally at the isoelectric point of one of the components.

The buffer systems used for electrophoresis may be continuous or discontinuous systems. In the *continuous buffer system*, the composition of the buffer is the same throughout. The concentration of the buffer in the electrode chambers is higher (~0.2 M) to reduce voltage drop while in the separation channel a lower concentration of the buffer (~0.05 M) is employed. The total ionic strength of the buffer is usually in the range 0.05–0.15 M. The commonly used continuous buffer systems include Tris–borate at pH 8–8.5 without or with a low concentration of EDTA, Tris–Tricine (pH 7.5–8), and imidazole-MOPS (*N*-morpholinopropane-sulphonic acid) (pH 7–7.5).

Discontinuous buffer systems are useful for sharpening the protein bands, particularly in analytical work. A combination of starting buffer anions of higher mobility (e.g., chloride, citrate, phosphate, or EDTAate) and following anions of lower mobility (e.g., borate or glycinate at pH 8–9) is used. The combination of differences in the electrophoretic mobilities of the ions, higher local electric field experienced by the proteins, and the pH facilitate self-sharpening of the protein bands.

The monomer acrylamide is a skin irritant and neurotoxin as well as a cancer suspect agent, and hence, requires careful handling by wearing hand gloves and face mask. Precast acrylamide gels of single concentration or gradient gels provided with a variety of sample well configurations in glass or plastic cassettes are commercially available. Bufferless precast gels of acrylamide containing ion-exchange matrices for sustaining the electric field which do not require any liquid buffers are also commercially available.

14.2.5 Disc Gel Electrophoresis

Disc gel electrophoresis is a modification of the slab gel electrophoresis with relatively greater resolution useful for the separation of proteins and nucleic acid fragments. The method is useful for detecting even microgram quantities of the samples. Two gel layers, an upper *stacking gel* and a lower *resolving gel* are used. The compositions of the two layers and the pH of the buffers used are different such that highly concentrated thin bands of sample proteins or nucleic acids are formed in the stacking gel and these get separated with greater resolution in the resolving gel. The stacking gel has a lower acrylamide concentration (~2–3%) and consequently the pores are bigger in size whereas the pores in the resolving gel are smaller due to greater cross-linking. The sample dissolved in glycine–chloride buffer of pH 8–9 is loaded into the stacking gel. The pH of the stacking gel is 6.9 and the bulk of the glycinate anions are converted to the neutral zwitter ionic form while the sample proteins or nucleic acids still remain as anions. During electrophoresis the applied voltage facilitates the migration of the anionic sample proteins or nucleic acids and chloride while glycine shows negligible electrophoretic mobility. The relative mobilities of the species in the upper stacking gel are chloride > sample proteins or nucleic acids > glycinate. The sample proteins or nucleic acids get accumulated in the form of a thin band sandwiched between the chloride and glycinate ions. On reaching the lower resolving gel whose pH is maintained at

8–9, the order of electrophoretic mobilities changes to chloride > glycinate > sample proteins or nucleic acids. The pore size of the resolving gel is small and the separation of the sample proteins or nucleic acids occurs depending on their unique charge/mass ratio.

14.2.6 Sodium Dodecyl Sulphate–Polyacrylamide Gel Electrophoresis (SDS-PAGE)

This method is useful for the determination of molecular weights of proteins and involves electrophoretic separation of denatured proteins on polyacrylamide gel. The proteins in solution are completely denatured by treatment with the detergent sodium dodecyl sulphate (SDS) and β-mercaptoethanol and boiling the mixture for a few minutes. Addition of β-mercaptoethanol (1% v/v) disrupts disulphide linkages in the protein. SDS binds strongly to the proteins and even oligomeric proteins with polypeptide chains not bound covalently are dispersed as individual subunits. A small amount of SDS (0.1%) is sufficient to saturate the polypeptide chains. Each dodecyl sulphate group carries a negative charge and binds to the hydrophobic regions of the polypeptide at a constant ratio of 1.4 g of SDS per gram of protein (approximately 1 SDS molecule for 2 amino acid residues). The polypeptide chain consequently gets a large negative charge and the charge/size ratio is almost the same for all polypeptide chains. Hence, the separation of the polypeptides occurs based on the size (molecular weight) of the polypeptide brought about by the molecular sieving through the pores of the gel. The electrophoretic mobility is related to the polypeptide size, and hence, it is possible to determine the molecular weight of each component. The electrophoresis is carried out in vertical slab polyacrylamide gel using buffers such as Tris–HCl–glycine and Tris–Tricine buffer for small molecular weight proteins.

The molecular weight of sample polypeptide chains is determined by comparing their mobility with standard polypeptides whose molecular weights are known. Standard polypeptides of molecular weights in the range of 14,000–100,000 and 45,000–200,000 are commercially available. A linear relationship between the mobility and the log of molecular weight of the protein is given by

$$v' = \left(\frac{z}{6\pi\eta r} \right) \left(\frac{A - \log M}{A} \right) \tag{14.2}$$

where A is the log molecular weight of a molecule that has no mobility in the gel and determined by extrapolation of the straight line over the whole range. A graphical plot of the log M versus electrophoretic mobility is useful for the determination of molecular weights of sample proteins (Fig. 14.4).

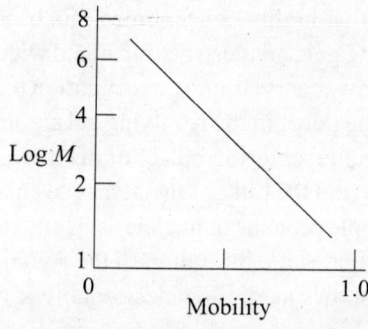

Fig. 14.4 Plot of electrophoretic mobility versus log M of proteins

14.2.7 Agarose Gel Electrophoresis

Agarose obtained from seaweed is a linear polymer of D-galactose and 3,6-anhydrogalactose. A continuous vertical or horizontal slab gel of agarose may be prepared by casting agarose solution in boiling water containing approximately 0.8% for the vertical slab or 0.2% for the horizontal, followed by cooling to room temperature. The concentration of agarose determines the pore size but in general the pore size is much bigger as compared to the pore size of polyacrylamide, and hence, amenable for the separation of large proteins (up to 10^7-10^8 Da) and nucleic acids. Agarose gels contain a small percentage of charged groups such as sulphate and carboxyl groups, and hence, exhibit ion-exchange effects and electro-osmotic flow. Hence, agarose is treated with alkaline solution to hydrolyse these charged groups and improve the sieving characteristics of the gel. During electrophoresis with agarose gels, it is necessary to maintain the temperature critically to prevent changes in viscosity. Agarose gels in general are fragile, and hence, after electrophoresis the gels are examined directly or dried to a thin film before staining.

14.2.8 Parameters Affecting Gel Electrophoretic Separations

The important experimental parameters which affect the separation efficiency of slab gel electrophoresis include (i) electric field strength, (ii) pH and ionic strength of the running buffer, and (iii) temperature during electrophoresis. Usually, the electric field strength in terms of voltage change per unit distance remains constant over the entire gel. Components of a mixture having high mobility separate out easily but species close to the start line suffer from poor resolution. The resolution can be improved by generating an electric field strength gradient across the gel with higher field strength near the start line. The pH of the running buffer determines the charge on the proteins of a mixture, and hence, their mobility. It is possible that two or more proteins may have the same mobility at a given pH, and hence, it is necessary to carry out electrophoresis with running buffers of different pH. The ionic strength of running buffer influences the net charge as well as the mobility of the species and it is preferable to use running buffers of low ionic strength sufficient to dissolve the sample. A temperature change during electrophoresis affects the separation efficiency. Higher temperatures lead to greater diffusion of components, spreading and distortion of zones leading to overlap of adjacent bands. Hence, it is necessary to avoid temperature change due to ohmic heating by proper cooling.

14.2.9 Detection of Proteins and Nucleic Acids in Electrophoresis Gels

After the electrophoretic separation the gel has to be appropriately treated to visualize the separated bands of the biomolecules. The visualization process may be carried by staining the gel with dyes or by blotting techniques.

In the most commonly used visualization method of staining, the gel removed from the electrophoresis set-up is treated with a 10% solution of trichloroacetic acid to fix the proteins by precipitation to prevent their diffusion. The gel is then rinsed with water and immersed in a 0.1–0.25% aqueous solution of the chosen dye. For visualization of protein bands the gel is soaked in an aqueous solution of the Coomassie blue dye (0.25%) for staining. The staining may also be carried out in a solution of the dye in water–methanol–acetic acid (5:5:1). Excess dye is removed (destaining) by soaking the gel or repeated washing in the dye-free solvent to give coloured protein–dye bands in a clear background. Fluorescent staining by reagents such as fluorescamine and anilinonaphthalene sulphonate or silver staining with silver salts such as silver diamine or silver-tungstosilicic acid provides about 10–100 times greater sensitivity in the detection of proteins. Metallic silver is precipitated due to the reduction of Ag^+ by thiol, amine, or tyrosine groups of protein bands and purine bases of nucleic acid bands forming a permanent stain on the gel. For visualization of nucleic acids dyes such as methylene blue or toluidine blue O, methyl green, or ethidium bromide for fluorometric detection may be used. The stained gels may be photographed using visible or ultraviolet light or by scanning with a densitometer to get quantitative information about the separated components.

Enzymes separated by electrophoresis can be detected by enzyme-specific substrate staining (e.g., lactate dydrogenase can be detected by nitroblue tetrazolium salt or catalase by starch-iodide reaction).

Blotting techniques essentially involve the transfer of the separated protein or nucleic acid bands from the electrophoretic gel onto nitrocellulose or nylon membrane for subsequent detection. The *Southern Blot* technique is used to transfer the DNA from agarose gel to a nitrocellulose membrane, while *Northern Blot* technique is used to transfer RNA from agarose gel onto nitrocellulose membrane and *Western Blot* technique is used to transfer proteins from electrophoretic gels on to nitrocellulose membrane. The blotting techniques involve the general steps of treating the gel with alkali to denature the biomolecule, neutralizing the alkali, and placing the gel in close physical contact with the nitrocellulose or nylon membrane. The transfer is effected by a buffer and the transfer process may require as long as 12 h. In the case of Western blot, the transfer of proteins is quickened with minimal diffusional broadening by electrophoretic process.

14.2.10 Pulsed Field Gel Electrophoresis

Pulsed field gel electrophoresis (PFGE) is particularly useful for the separation of very large DNA molecules (>20 kb). DNA molecules exist in solution as ball-like random coils and when applied to agarose gel unravel, elongate, and migrate through the gel (the type of migration is called *reptation*) towards the anode under the influence of an electric field. A molecular sieving effect is brought in as the molecules pass through the pores of the gel. Smaller DNA molecules (<5 kb) can be conveniently separated by conventional agarose gel electrophoresis on the basis of the combined effects of their electrophoretic mobility and molecular size. The small differences in the nucleotide lengths result in large differences in mobility. However, due to the logarithmic relationship between the size and mobility, sensitivity decreases with increasing length. The larger DNA molecules above a size limit migrate at the same rate in a size-independent manner in agarose gel electrophoresis and appear together as a diffuse band. Hence, such large molecules cannot be separated by conventional electrophoretic methods. PFGE was developed to address these problems.

In principle, PFGE involves agarose gel electrophoresis carried out in a manner similar to that in conventional electrophoresis by applying an applied voltage, but the direction of applied voltage is periodically changed. PFGE avoids molecular sieving mechanism through the periodical change in the direction of the applied electric field but exploits the reptation phenomenon to bring about size-dependent electrophoretic mobility in large DNA molecules. The DNA molecules react to the change in the direction of the applied field at different rates depending on their lengths. Smaller lengths of DNA realign themselves quickly and continue their migration. In contrast, the larger DNA molecules are trapped each time the electric field direction changes and start migration only after they reorient themselves along the new field axis. These molecules take more time to realign themselves, and hence, migrate slowly. The change in direction of the applied voltage over a period of time results in the separation of the DNA molecules on the basis on their length or size.

In practice, the applied voltage is periodically switched among three directions, one along the central axis of the gel and the other two directions being at an angle of 120° on either side of the central axis. The pulse time is kept constant and equal in all the three directions which cause a net forward migration of the DNA. The limit of resolution of PFGE has improved to about 5000 kb, and hence, the technique is highly useful in the separation of intact chromosomes facilitating gene mapping and study of evolutionary characteristics of a number of organisms.

14.2.11 Applications of Electrophoresis Techniques

Zone electrophoretic techniques find extensive applications in biochemical analysis which include separation of amino acids, peptides, proteins, and nucleic acids, determination of net charge on the protein, protein subunit composition and molecular weights of the sub units as well as that of the proteins, identification of isoenzymes, sequence analysis of nucleic acids, DNA fingerprinting, and determination of molecular weight of DNA.

14.3 IMMUNOELECTROPHORESIS

Immunoelectrophoresis is useful for the analysis of protein purity and antigenic properties.

This technique involves two sequential steps for the identification and study of the antigenic properties of a complex mixture of proteins (i) separating them by electrophoresis in agar gels and (ii) interacting the separated proteins with specific antibodies to determine the antigenic properties. The introduction of a foreign substance (antigen) into a living system induces the formation of an antibody specific to the antigen. Antibody is a plasma protein belonging to the family of immunoglobulins within the living system. The antigen–antibody interaction is highly specific which results in the formation of an opaque precipitate in the pH range 7–9. The size of the precipitate increases with increasing amount of antigen added to a given amount of antibody till the equivalence point. An excess of antigen beyond the equivalence point causes the precipitate to dissolve. The precipitation reaction is called immunoprecipitin reaction.

Rocket immunoelectrophoresis is a quantitative technique in which the antigen sample is subjected to electrophoretic migration in a gel containing specific antibodies to the antigen. The antigen–antibody complex formed migrates till it reaches the equivalence point. At the equivalence point the complex precipitates and further migration does not occur. This results in the formation of rocket shaped precipitate (hence, the name rocket immunoelectrophoresis). The area of the precipitate is proportional to the concentration of the antigen in the sample.

14.4 CAPILLARY ELECTROPHORESIS

Capillary electrophoresis (CE) is also known as *capillary zone electrophoresis* (CZE), *free solution capillary electrophoresis* (FSCE), or *high-performance capillary electrophoresis* (HPCE). The technique basically involves high voltage electrophoresis carried out in a capillary with online detectors similar to those used in HPLC. The technique requires only micro- to nano-scale samples. The separation efficiency N is related to the electric field strength E, the electrophoretic mobility v', the diffusion coefficient of the migrating species D, and migration distance d as

$$N = \frac{v'Ed}{2D}$$ (14.3)

During electrophoresis the sample mixture is transported through a buffer-filled capillary tube due to a high dc potential applied across the length of the capillary tube. A schematic diagram of the experimental set-up is shown in Fig. 14.5.

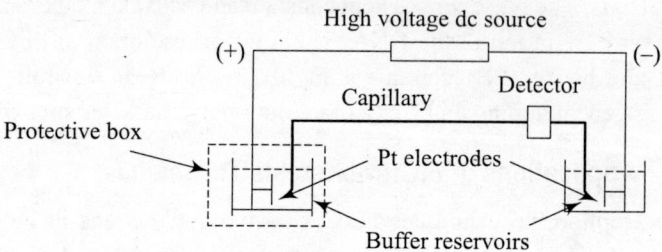

Fig. 14.5 Capillary electrophoresis set-up

The capillary tube is made of fused silica of about 50–100 cm length and an inner diameter of 25–100 μm. The capillary is filled with the chosen buffer and placed between buffer reservoirs which also contain the platinum foil electrodes connected to high voltage dc source. The electrode connected to the high potential side is usually placed inside a protective Plexiglas box as a safety measure against electric shock. A small volume of the sample mixture (a few nanolitres) is injected into the anodic end of the capillary tube hydrodynamically by siphoning or gravity feeding. Alternatively, the sample may be injected electrokinetically by applying a potential. The anodic end of the capillary is removed from the buffer reservoir and placed in a sample tube along with the anode and an injection voltage is applied to bring about an electro-osmotic migration of the sample ions into the capillary tube. A variety of buffers can be used in CE which include phosphate, citrate, acetate, succinate, TRIS, morpholine, etc.

A high voltage of about 20–30 kV is applied across the capillary tube through platinum foil electrodes. Under the influence of the applied electric field the components of the sample mixture migrate towards the negative electrode passing through a detector (e.g., UV spectral detector, laser-induced fluorescence detector, amperometric detector, or mass spectral detector) placed between the end capillary and the buffer reservoir. The sample components get separated depending on their electrophoretic mobilities and the electro-osmotic flow rate of the solvent.

The fused silica surface of the capillary tube develops fixed negative charges due to the dissociation of surface silanol (Si–OH) groups. The fixed negative charges attract positive ions from the buffer giving rise to an electrical double layer. The positively charged species move towards the negative electrode and carry the associated solvent molecules with them resulting in the electro-osmotic flow of the solvent.

Separation of the charged species occurs as positively charged species move through the capillary at a rate greater than the electro-osmotic flow rate of the solvent. Negatively charged species move slowly than the electro-osmotic flow rate of the solvent or even move in the opposite direction as they are repelled by the negative electrode. Neutral species move through the capillary at the electro-osmotic flow rate of the solvent. Thus, the different charged species in the sample mixture get separated under the influence of the applied electric field and reach the detector at different time intervals.

The detector signal recorded as a function of time is displayed in the form of an *electropherogram*, which is similar to HPLC chromatogram with sharp peaks with a very high separation efficiency equivalent to over 200,000 theoretical plates. The technique exhibits very high resolution with the sample peaks appearing as narrow and sharp bands. This is attributed to flat flow profile of the electro-osmotic flow [Fig. 14.6(a)] and negligible joule heating of the solution which minimize longitudinal diffusion and thermally driven convective mixing of the components.

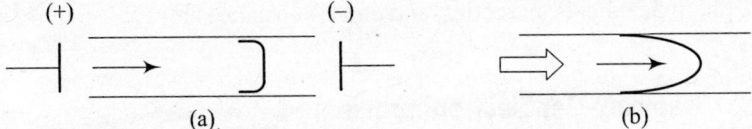

Fig. 14.6 Showing flow pattern in a tube: (a) electro-osmotic flow in CE and (b) hydrodynamic flow under pressure

Capillary electrophoresis finds applications in the analysis of inorganic cations and anions and a large number of biochemicals such as amino acids, peptides, proteins, glycoproteins, metabolites, and drugs in body fluids. The technique is useful for peptide mapping or protein fingerprinting to establish the identity of a protein by peptide sequencing after chemical or enzymatic cleavage of the protein into peptide fragments.

14.4.1 Micellar Electrokinetic Capillary Chromatography

Micellar electrokinetic capillary chromatography (MECC) was introduced in 1984 as a modification of CE to separate even neutral low molecular weight compounds not possible in conventional CE. The technique involves performing capillary electrophoresis in fused silica capillary tube as in CE (Fig. 14.5) with the buffer containing an added surfactant. The surfactant may be anionic (e.g., sodium dodecyl sulphate), cationic (e.g., CTAB), or neutral (e.g., TWEEN).

The surfactant molecule can be described as made of ionic or polar (hydrophilic) head attached to a long non-polar hydrocarbon chain which is hydrophobic. The surfactant molecules on dissolving in aqueous solution at a concentration higher than a level called the *critical micelle concentration* form spherical aggregates called *micelles*. The micelle consists of an aggregate of the surfactant molecules with their hydrophobic hydrocarbon tails facing the interior of the sphere and the polar heads positioned at the surface of the sphere exposed to the surrounding aqueous solution. The hydrocarbon tails within the micelle can dissolve non-polar compounds and retain them while polar compounds prefer the outer surface and the surrounding aqueous medium.

In MECC, the sample components get distributed between the hydrophobic region or phase within the micelle and hydrophilic region on the outer surface. The non-polar components dissolve in the hydrophobic environment within the micelle while polar components dissolve in the aqueous phase. Thus, the micelles constitute an additional stationary phase by partitioning the sample components depending on their polarities similar to that in liquid partition chromatography, and hence, the term chromatography is included in the name of the technique. However, there is one difference as compared to conventional chromatography in that the stationary phase of micelles migrates along the length of the capillary as in capillary electrophoresis. Neutral sample components of greater hydrophobic nature will be retained for a longer time within the hydrophobic micelle, and hence, elute out of the capillary with longer retention times along with the migrating micelle. Neutral hydrophilic components have shorter retention times and elute with the electro-osmotic flow of the buffer. The formation of micelle and interaction between micelle and solute components can be altered by adding electrolytes and organic solvents (e.g., methanol or acetonitrile) or by changing the buffer composition, pH, and temperature to enhance separation efficiency.

MECC finds applications in the analysis of mixtures of polar and non-polar compounds, nucleic acids, pharmaceuticals, narcotics, and chiral substances by adding chiral additives to the mobile phase.

14.4.2 Capillary Gel Electrophoresis

Capillary gel electrophoresis (CGE) involves separation of sample components according to their electrophoretic mobility in conjunction with their molecular size and shape. This technique is, thus, a variation of CE with the capillary tube filled with a polymeric gel (usually cross-linked polyacrylamide or agarose gel) to effect size-based separation or molecular sieving.

The smaller molecules have greater electrophoretic mobility and exit the capillary tube earlier as compared to bigger molecules with lower mobility which exit the capillary tube later. The technique has the advantage of eliminating the electro-osmotic flow as the capillary tube is filled with polymer gel. The polymer gel also prevents band spreading due to the diffusion of solute components and also adsorption on capillary wall. These advantages translate into maximum resolution within the shortest possible migration distance. The technique is useful for the separation of proteins, oligonucleotides, and chiral substances by incorporating cyclodextrins into the polymer gel and for DNA sequencing.

14.4.3 Capillary Electrochromatography

Capillary electrochromatography (CEC) as the name implies is a hybrid of capillary electrophoresis (CE) and chromatography (HPLC) applicable for both charged species and neutral molecules. The separation principle is based on the combination of electrophoretic mobility of charged species and chromatographic sorption of neutral molecules. The apparatus consists of the fused silica capillary tube (~25–50 cm long and <100 μm inner diameter) similar to that in CE. The capillary tube is packed with small-sized (1.5–3 μm diameter) HPLC stationary phase particles of bonded silica and filled with a buffer solution. A variety of stationary phase particles of different polarities, different porosity, and different shapes of particles can be used for packing the capillary thereby facilitating a wide choice of operating modes. The buffer is mostly a phosphate or TRIS or MES buffer of pH > 4. An electro-osmotic flow of the buffer towards the cathode at the end of the capillary tube is generated by an applied electric field. An electrical double layer is formed at the surface of stationary phase particles throughout the length of the capillary and functions as the driving force for the electro-osmotic flow of the buffer. However, there is no pressure drop as in HPLC and the electro-osmotic flow of the buffer has a flat profile similar to that in CE. The flow of the buffer is greater if its pH > 4. A higher electro-osmotic flow facilitates faster separation of the components. The mobile phase is mostly a mixture of methanol or acetonitrile with the buffer solution. The sample (1–20 nL) is introduced by electrokinetic method and the components get separated and move through the detector placed at the end of the capillary as in CE.

Capillary electrochromatography has several advantages over both CE and HPLC in that (i) the flat flow profile of the buffer due to electro-osmotic flow and the presence of a stationary phase with additional selectivity enhance the efficiency of separation by several folds, (ii) the composition of the mobile phase can be varied to improve the selectivity, (iii) solvent consumption is much less compared to that in HPLC, and (iv) CEC can be coupled to mass spectral detector making the technique highly sensitive.

Capillary electrochromatography is gaining importance for the separation and analyses of mixtures of ionic and neutral compounds, e.g., mixtures of polyaromatic hydrocarbons, dyes, peptides, proteins, DNA fragments, pharmaceuticals, etc. Chiral stationary phases can be used to separate mixtures of enantiomers.

14.5 ISOELECTRIC FOCUSING

Isoelectric focusing (IEF) also known as *electrofocusing* is based on electrophoretic separation as a function of pH. The method is highly useful for the separation of amphoteric substances (e.g., amino acids, peptides, and proteins) based on their relative content of acidic and basic residues. The net charge on such a molecule depends on the pH of the medium. When the pH of the surrounding medium is below the isoelectric point (pI) of the amphoteric molecule it is positively charged and moves towards the cathode in an electrophoretic gel. Isoelectric focusing is, thus, a form of gel electrophoresis together with a pH gradient maintained between the cathode and anode, the pH increasing gradually from the anode region to the cathode region. As the positively charged molecule migrates towards the cathode through a gradient of increasing pH, the overall charge on the molecules decreases until the molecule reaches the pH region that corresponds to its pI. When the pH = pI the molecule has no net charge and so its migration ceases as there is no electrical attraction towards either electrode. As a result, the molecule becomes focused into a sharp stationary band. In a mixture of such amphoteric substances the migration of the individual components occurs up to a point in the system where the pH is equal to the isoelectric point of the respective components. The individual components, thus, get focused into sharp bands at points in the pH gradient corresponding to the respective pIs. The technique is capable of very high resolution with proteins differing even by a single charge being fractionated into separate bands.

The principle of isoelectric focusing of a hypothetical sample mixture two proteins A and B having an acidic and basic pI respectively may be explained schematically as shown in Fig. 14.7.

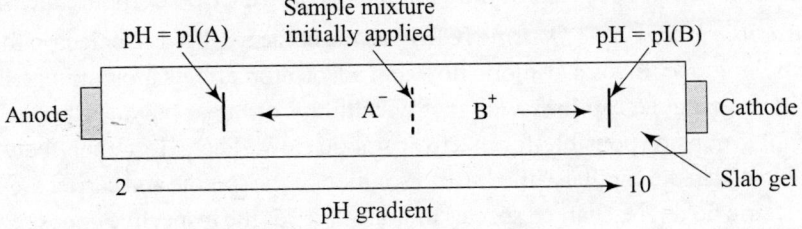

Fig. 14.7 Principle of isoelectric focusing

The sample mixture is applied at the centre of the pH gradient on the slab of gel formed by incorporating the carrier ampholyte buffer. Proteins A and B migrate towards oppositely charged electrodes under the influence of applied electric field. The charge on the proteins decreases as they approach pH values close to their respective pI values. When the pH = pI of the respective proteins, the net charge on the proteins becomes zero, and hence, the proteins do not migrate further. The proteins diffuse away from their pI values due to concentration differences between the regions at pI and away from pI. As they move away from the pI, they acquire charge and are forced to migrate towards the pI thereby bringing in a focusing effect. Thus, they get concentrated or focused as a band at their respective pI values. The resolution or separation of two proteins depends on the difference in their pI values.

Isoelectric focusing is commonly carried out in polyacrylamide gels. The gel mixture containing the acrylamide monomer and the appropriate ampholyte is polymerized in vertical or horizontal columns or slabs. The gelling mixture also contains an electrophoretically inert substance such as sorbitol or glycerol (10%) to increase the osmotic pressure and minimize ripples on the gel surface during the experimental run. For the separation of proteins of molecular weights up to 100,000 Da, a gel containing 7.5% polyacrylamide is normally used. For the separation of larger molecular weight proteins, large pore-sized gel containing lower concentration (2%) of polyacrylamide stabilized by 0.5–1.0% of agarose is used.

A constant power is applied between the electrodes to establish the pH gradient. The formation and stabilization of a continuous pH gradient between the two electrodes is carried out by the use of buffers called carrier ampholytes. Ampholytes are synthetic aliphatic polyamino-polycarboxylic acids available commercially in mixtures covering a wide pH range of 3–10 and in various narrow bands of pH of 5–7, 6–8, etc. The ampholyte mixtures contain large number of both positive and negative charged functional groups with closely spaced pK and pI values. During isoelectric focusing the amphoteric carrier ampholyte species also migrate to their respective isoelectric points. The carrier ampholytes consist of several thousands of amphoteric species differing in their pI values by only 0.05 pH unit and cover the entire pH range. Due to the high buffering capacity of the ampholyte species a small pH plateau is formed around its pI. Since the buffer contains several species with varying pI values, overlapping pH plateaus will be formed and a continuous pH gradient is formed between the two electrodes.

Once the pH gradient is established as indicated by a change in current flowing between the electrodes, the sample with low ionic strength is applied on the slab gel. The experiment is concluded when the focusing of the sample proteins occurs at their respective isoelectric points and a steady state has been reached. The slab gel then is stained with Comassie Blue and finally destained. The pI spectrum of the sample is obtained by determining the pH gradient formed.

Capillary isoelectric focusing (CIEF) as the name implies is isoelectric focusing carried out in a capillary tube. Electro-osmotic flow and adsorption effects are minimized in CIEF by coating or derivatizing the capillary with methylcellulose or linear polyacrylamide to block the surface silanol groups responsible for electro-osmotic flow. The pH gradient between pH 3 and 10 is established in the capillary filled with a solution of the sample and carrier ampholytes by applying an electric field. The charged species migrate towards the respective oppositely charged electrodes till they get focused at pH = pI of the respective species. When all the species get focused the current flow ceases indicating the completion of the experiment. The separated species are then passed sequentially through a detector placed at the end of the capillary by applying pressure at the opposite end of the capillary.

Chromatofocusing involves the separation of proteins by isocratically formed pH gradient on an ion-exchange column. The technique has the favourable attributes of both ion-exchange chromatography and isoelectric focusing. The basic principle depends on the differences in the surface charge of proteins. A majority of the proteins have isoelectric points in the pH range 5–8. However, the net charge distribution varies from protein to protein due to the presence of individual reactive groups and differences in molecular weights and subunit compositions. The surface charge on the protein is influenced by various factors such as nature and concentration of counter ions, presence of water activity modifiers (e.g., urea, ethylene glycol), metal ions, specific ligands, detergents, and temperature. The differences in the net surface charge distribution of different proteins are exploited by a pH gradient developed on the ion-exchange column to facilitate their separation. Proteins have a net negative charge in buffers with pH values above their isoelectric points, and hence, bind to an anionic exchange resin bed in a column equilibrated at a high pH (by an equilibrating buffer). The bound proteins are then eluted by a buffer (focusing buffer) of pH lower than the isoelectric point of the protein of interest. A descending pH gradient develops on the column with the flow of the focusing buffer with a steady drop in the pH. The change in the pH brings about a continuous change in the charge distribution on the protein surface, and hence, the interactions of the proteins with the anion-exchange resin. The proteins elute out of the column close to their isoelectric points. Chromato-focusing is capable of separating proteins differing in their isoelectric points by as little as 0.05 units.

Proteins with a net positive charge can be separated similarly on a cation-exchange resin column by developing an ascending pH gradient through the flow of an appropriate focusing buffer.

Chromato-focusing has analytical as well as preparative applications for the separation and purification of heterogeneous forms and isoforms of proteins, lipoproteins, enzymes, hormones, and other biomolecules and to analyse the heterogeneity of surface charge on molecules.

14.6 TWO-DIMENSIONAL ELECTROPHORESIS

Two-dimensional electrophoresis is used for the analysis of complex mixtures by combining two independent separation techniques, namely, isoelectric focusing and SDS-PAGE. In the first step isoelectric focusing is carried out in a column using carrier ampholytes to generate a pH gradient between 3 and 10 to separate the complex mixture of proteins on the basis of their electrophoretic mobilities and pI values. The gel is extruded from the IEF column and equilibrated in a solution of glycerol, mercaptoethanol, SDS, and buffer for about 30 min. In the second step, discontinuous SDS-PAGE is carried out on a slab of same width as the length of column gel used in IEF, in a perpendicular direction to the direction of migration in column IEF to separate the mixture of proteins on the basis of their molecular weights. The slab gel is finally removed, ● fixed, and stained with Coomassie blue to visualize the separated protein bands.

14.7 ISOTACHOPHORESIS

Isotachophoresis is similar to isoelectric focusing. The name implies that the ions being separated travel at the same speed in an applied electric field. The principle is best explained with a hypothetical sample mixture containing two anions A^- and B^- to be separated on the basis of the differences in their mobilities in an applied electric field (i.e., $m_A^- \neq m_B^-$). The sample anions together with a leading anion L^- (e.g., chloride) with a greater mobility than the sample ions and a trailing or terminating anion T⁻ (e.g., glutamate) with a lower mobility than the sample ions along with a common counter cation (e.g., Tris) constitute the electrolyte solution. In an applied electric field at constant current, the leading anions migrate towards the anode followed by the sample anions according to their mobilities. The terminating anions follow the sample anions. The polarity of the electric field is such that with a homogenous current density all the ions move with same speed at equilibrium and get separated into a number of consecutive zones in immediate contact with each other and arranged in order of their effective mobilities.

Isotachophoresis is carried out in a capillary tube with the sample ions being introduced between the leading and terminating electrolytes and electrophoresis is carried out with an applied voltage of about 30 kV for about 20–30 min. A thermostatically controlled cooling bath surrounds the tube to control the temperature. A low electrical field is created in the leading electrolyte and a high electrical field in the terminating electrolyte. The pH is determined by the counter-ion of the leading electrolyte that migrates in the opposite direction. Initially, the sample constituents separate from each other as they migrate at different speeds. The faster constituents will create a lower electrical field in the leading part of the sample zone and vice versa. Finally the constituents will completely separate from each other and concentrate at an equilibrium concentration, surrounded by sharp electrical field differences. Once equilibrium is achieved the different ions move with the same speed in discreet bands according to their order of mobilities as shown in Fig. 14.8.

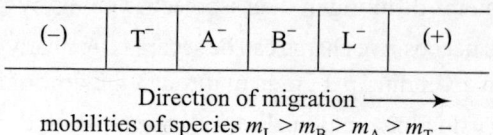

Direction of migration ⟶
mobilities of species $m_L > m_B > m_A > m_T -$

Fig. 14.8 Principle of isotachophoresis

The unique feature of isotachophoresis is that the length of each separated band is directly proportional to the amount of ions, and hence, useful in quantitative estimation of ions.

When the mobilities of the sample ions are similar they can be separated by including synthetic ampholytes called spacer ions which have mobilities intermediate to those of the sample ions. A mixture of cations can similarly be separated into bands by isotachophoresis using a leading cation, terminating cation all the cations having a common counter anion.

Isotachophoresis finds use in the separation of inorganic ions and organic acids, proteins, amino acids, nucleic acids, etc. Applications in industries include quality control in pharmaceutical, beverage and food processing and in pollution control and management.

SUMMARY

- Electrophoresis is an incomplete form of electrolysis in which charged species get separated based on their electrophoretic mobility, that is, differential migration in an applied electric field.

- Zone electrophoresis uses supporting medium in the form of paper or a polymeric gel made of cellulose acetate, agarose or polyacrylamide to separate proteins and nucleic acids based on their electrophoretic mobility.

- PAGE is carried out to separate and identify mixtures of proteins in native gel electrophoresis and for molecular weight determination in the form of SDS–PAGE. In SDS–PAGE the mobility of a polypeptide chain is inversely related to its log molecular weight.

- PFGE involves the separation of large-sized DNA molecules by repetition by applying a voltage periodically changing in its direction to agarose gels. The DNA molecules get separated on agarose gels based on the different lengths or sizes.

- Capillary electrophoresis involves electrophoresis carried out in a capillary tube to separate charged species based on their electrophoretic mobility and diffusion coefficient values. The separation efficiency is very high due to the flat flow profile of the electro-osmotic flow and minimal longitudinal diffusion.

- IEF involves electrophoretic separation of charged species as a function of pH and is particularly suited to amphoteric substances such as amino acids, peptides, and proteins.

- Isotachophoresis involves the separation of ions into bands depending on their effective mobilities between leading and terminating electrolytes in an applied electric field. The technique is useful in quality control and quantitative estimation of ions.

REVIEW QUESTIONS

1. Give an outline of the different electrokinetic methods of separation.

2. Explain the term 'electrophoretic mobility'.

3. Discuss the theoretical principles involved in electrophoretic separation.

4. Describe the various electrophoretic techniques.

5. Give an account of analytical gel electrophoretic techniques and their applications.

6. How is the molecular weight of a protein determined by SDS-PAGE?

7. How are protein and nucleic acid bands detected after their electrophoretic separation?

8. Explain the principle of immunoelectrophoresis.

9. Describe the experimental set-up and procedure of capillary electrophoresis.

10. Write a note on the principle involved in micellar electrokinetic capillary chromatography and its advantages.

11. Describe the techniques of capillary gel electrophoresis and capillary electrochromatography.

12. What is pulsed field gel electrophoresis? How is it useful?

13. Discuss the technique of isoelectric focusing.

14. Explain the principles of capillary isoelectric focusing and chromatofocusing.

15. How are proteins separated by isotachophoresis? Describe the technique.

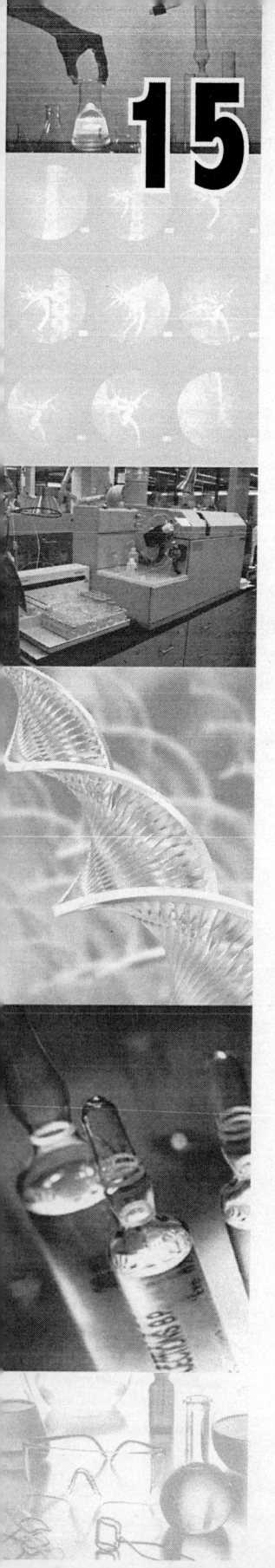

15 Centrifugation

15.1 INTRODUCTION

Centrifugation is a highly useful technique for the separation of materials based on density differences, particularly when gravitational force is insufficient for separation of such materials. Centrifugation is practised for the separation of solid particles from a liquid medium and also for the separation of two immiscible liquids differing in their densities. Centrifugation is applicable to both laboratory scale and industrial scale (or bulk) separations. This chapter focuses on the principles involved and the practice of centrifugation and ultracentrifugation for analytical purposes.

15.2 CENTRIFUGAL FORCE

The centrifugal force F_c acting on a particle during the circular motion of the centrifuge is related to the angular velocity ω (radians s^{-1}) and the radial distance r (cm) of the particle from the centre of rotation as given by

$$F_c = m\omega^2 r \tag{15.1}$$

The equation is similar to the familiar equation $F = ma$. The tangential velocity of the particle in a centrifugal field, v (cm s^{-1}) is given as $v = \omega r$. Equation (15.1) may be rewritten as

$$F_c = mr\left(\frac{v}{r}\right)^2 = \frac{mv^2}{r} \tag{15.2}$$

Since one revolution of the centrifuge is equal to 2π radians (angular velocity in radians, s^{-1}), the rotational speed of a centrifuge (n) may be conveniently expressed in terms of *number of revolutions per minute* (rpm) by substituting the angular velocity ω in Eq. (15.1) by $2\pi n/60$, as given in Eq. (15.3).

$$F_c = m\omega^2 r = m\left(\frac{2\pi n}{60}\right)^2 r \tag{15.3}$$

Since $F_g = mg$ (g is acceleration due to gravity (980 cm s^{-2})), the centrifugal force may be expressed in terms of the *relative centrifugal force* (RCF).

Relative centrifugal force may be defined as the force relative to the gravitational force exerted by the earth or the ratio of the weight of the particle in the centrifugal field to that in the earth's gravitational field as

$$\text{RCF} = \frac{F_c}{F_g} = \frac{m\omega^2 r}{mg} \text{ or } \frac{\omega^2 r}{g} = 1.119 \times 10^{-5} n^2 r \tag{15.4}$$

$$\left(\text{Since } \frac{\omega^2 r}{g} = [(2\pi n/60)]^2/980 = 1.119 \times 10^{-5} n^2\right)$$

Thus, the force developed in a centrifuge is $\omega^2 r/g$ or v^2/rg times as large as the gravity force. RCF is a ratio between forces, and hence, it has no units. However, the numerical value of RCF is commonly expressed as *G number* (number of times the g force) by giving numerical value followed by the symbol *g*. Thus, for example, a particle at an average distance of 6 cm from the rotational axis being centrifuged at 20,000 rpm is subjected to an RCF of about 27,000 times g. (RCF = $1.119 \times 10^{-5} \times (20{,}000)^2 \times 6 = 26{,}856 \times g$. Thus, RCF is dependent on the rotational speed and the distance of the particle from the rotational axis, the latter usually taken as the distance from the centre of the motor drive shaft to the base of the centrifuge tube. However, in the centrifuge tube the distance *r* increases from the tube top to the bottom, and hence, it is usual to define an average g force (g_{av} or RCF_{av}) for a given rotor and speed, a minimum g force (RCF_{min}) at the tube top and a maximum g force (RCF_{max}) at the bottom of the tube.

Manufacturers usually specify the maximum speed of the centrifuge rotor, its radius, and the RCF values. To operate the centrifuge at selected values of RCF below the maximum value it is necessary to choose the rpm, which may be calculated by using formula

$$n = 299 \times \sqrt{\left(\frac{RCF}{r}\right)} \tag{15.5}$$

A *nomogram* (a graphical representation that enables us to read off the value of a dependent variable when the value of two or more independent variables are specified) is useful to directly read the RCF values for the chosen combination of rpm and *r* values as shown in Fig. 15.1.

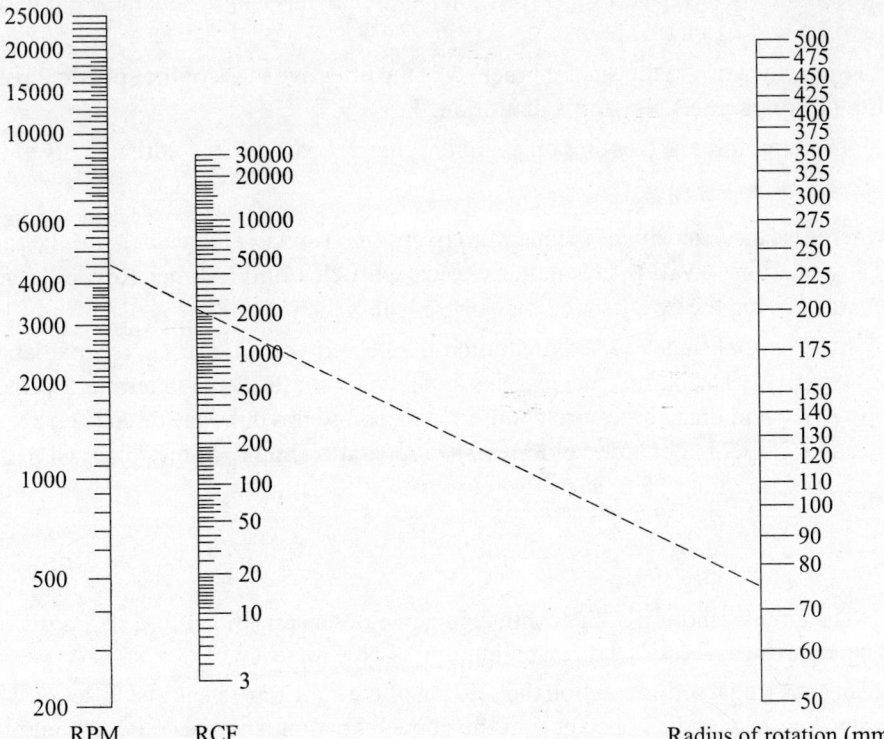

Fig. 15.1 Nomogram for converting rpm into RCF

The RCF for a chosen rpm or the rpm required for a chosen RCF may be directly read off from the nomogram by drawing a line starting from the known value of the radius of the centrifuge as shown in Fig. 15.1. The values may also be obtained by using Eqs. (15.4) and (15.5).

15.3 PRINCIPLES OF CENTRIFUGAL SEDIMENTATION

The solid–liquid separation by centrifugation is similar to gravitation sedimentation as both the methods involve separation depending on the density differences between the solid particles and the surrounding liquid medium aided by gravitational or centrifugal forces.

In the absence of any external force, solid particles suspended in a liquid medium settle down slowly under the influence of gravity. The process is known as sedimentation. Gravitation sedimentation has been known for centuries for separating solid particles on the basis of their density differences.

The velocity of a solid particle moving through a liquid medium during sedimentation depends on the relative magnitude of two opposing forces, the gravitational and drag forces. The particle is accelerated by the gravitational force resulting from the density difference between the particle and the surrounding fluid. Assuming the solid particle to be spherical of diameter d (in cm) and density ρ_s (in g cm^{-3}) the gravitational force F_g, acting on the solid particle is given as

$$F_g = \left(\frac{\pi}{6}\right)\left[d^3\left(\rho_s - \rho\right)\right]g \tag{15.6}$$

where g is gravity and ρ represents the density of the liquid. The above equation is obtained by combining the effects of gravitational and buoyancy forces on the solid particle ($F_g = (\pi/6)$ $[d^3\rho_s]$ g and $F_b = (\pi/6)$ $[d^3\rho]$ g).

Since the term in the square brackets is the effective mass of the spherical particle, the above Eq. (15.6) parallels Newton's definition, $F = ma$.

The drag force, F_d, acting on a single spherical particle in solution is given by Stoke's law

$$F_d = 3\,\pi d\eta v \tag{15.7}$$

where η is the viscosity of the medium (g cm^{-1}s^{-1}) and v is the velocity of the spherical particle. The equation is valid for small spherical particles only so that Reynolds number ($dv\rho/\eta$), characterizing the flow around the sphere is less than one.

In the initial stages of sedimentation the velocity of the particle is small and the drag force is also small. The particle accelerates and reaches its steady state terminal velocity v_g when the buoyancy and drag forces are counterbalanced by gravitational force (i.e., $F_g = F_b + F_d$). The steady state sedimentation velocity or sedimentation rate v_g of the spherical particle is obtained by combining Eqs (15.6) and (15.7) to give

$$v_g = \frac{d^2}{18\eta}(\rho_s - \rho)g \tag{15.8}$$

The circular motion of the centrifuge generates a centrifugal force F_c as given by Eq. (15.1). The centrifugal force is expressed in units of Newton, N.

In centrifugal sedimentation the settling of the solid particle is due to action of the centrifugal force, and hence, the constant g in the above equation is replaced by $\omega^2 r$ and the equation for the sedimentation velocity or sedimentation rate v_c of the particle under centrifugal force is then given as

$$v_c = \frac{d^2}{18\eta} (\rho_s - \rho) \omega^2 r \tag{15.9}$$

Figure 15.2 shows the comparison between sedimentation under gravitational and centrifugal forces.

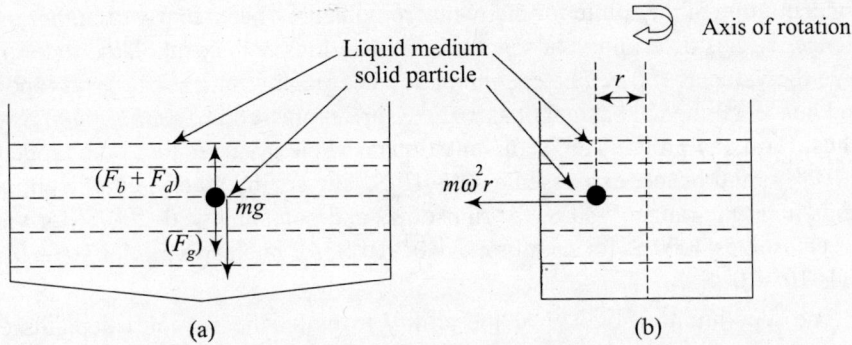

Fig. 15.2 Sedimentation under (a) gravity and (b) centrifugal force

The above equations for sedimentation under centrifugal force are based on Stoke's law which considers the solid particles to be spherical. However, biological solutes are not strictly spherical but spheroidal with one major and one minor axis (ellipsoids of revolution). The frictional force over ellipsoids is greater than that for spherical particles of same volume. Rod-like molecules such as DNA and proteins such as F-actin and myosin exhibit greater friction and such molecules sediment at a slower rate. In such cases a modified form of equation including the frictional ratio f/f_o (where f and f_o are the frictional coefficients of the non-spherical and spherical molecules respectively) is used to determine the terminal velocity. The frictional ratio is close to 1 for spherical molecules

$$v_c = \frac{d^2}{18\eta\left(\dfrac{f}{f_o}\right)} (\rho_s - \rho) \omega^2 r \tag{15.10}$$

The *sedimentation velocity* (v_c) of a particle is also expressed in terms of *sedimentation coefficient s*, which is defined as the sedimentation rate per unit centrifugal field (the ratio of sedimentation velocity to centrifugal force) and given as

$$s = \frac{v_c}{\omega^2 r} \quad \text{or} \quad \frac{dr}{dt} \times \frac{1}{\omega^2 r} \quad \text{or} \quad \frac{d(\ln r)}{dt} \times \frac{1}{\omega^2} \tag{15.11}$$

The value of s is determined from a plot of $\ln r$ versus t. Alternatively integrating r with respect to time between t_1 and t_2 gives

$$s = \frac{2.1 \times 10^2 \log\left(\dfrac{r_2}{r_1}\right)}{n^2 (t_2 - t_1)} \tag{15.12}$$

where n is the rotational speed of the rotor in rpm, r is in cm, and t in seconds. The value of s can be determined by measuring the boundary positions (r_1 and r_2) at t_1 and t_2 for substitution in the above equation.

The sedimentation coefficient determined from sedimentation rate measurements is affected by temperature and solution viscosity and density depending on the solute–solvent system. Hence, the calculated values of s are usually converted to a standard value called *standard sedimentation coefficient* $s_{20, w}$, which is the value that would be obtained in a medium of viscosity and density of water at 20°C. The sedimentation coefficient usually decreases with increase in concentration of the solute for many macromolecules, particularly of higher molecular weights. Hence, $s_{20,w}$ is determined at several concentrations and extrapolated to zero concentration to give the value as $s°_{20,w}$. The sedimentation coefficient values for most biopolymers are small, and hence, taking the basic unit as 10^{-13} S, are expressed in *Svedberg unit S* with $1\ S = 10^{-13}$ s. Thus, for many proteins the sedimentation coefficient values lie in the range 0.1×10^{-13} to 10×10^{-13} s, and hence, expressed as 0.1–10 S. The sedimentation coefficient values for nucleic acids lie in the range 4–80 S, for ribosomes and polysomes 20–200 S, for viruses 50–1000 S, for lysosomes 4000 S, for membranes 10^2–10^5 S, for nuclear particles 10^6–10^7 S, and for whole cells 10^7–10^8 S.

The separation efficiency or the ability to pellet the particles depends on centrifugation conditions represented as the product of time t required for pellet formation by centrifugation in hours and the sedimentation coefficient s of the particles. The pelleting capacity is referred to as the *clearing factor k* given as

$$k = t\,s \times 10^{13} \tag{15.13}$$

where t is in hours. The smaller the value of k greater is the pelleting capacity. The value of k is smaller when the sedimentation path is smaller in the centrifuge tube which is given in terms of the difference between the radius maximum (r_{max}) and minimum (r_{min}) in cm. The value of k can be calculated by the formula

$$k = \frac{2.53 \times 10^{11} \ln\left(\dfrac{r_{max}}{r_{min}}\right)}{n^2} \tag{15.14}$$

where n refers to the speed of rotation in rpm. The value of k is given as a characteristic of the rotor as applied to the sedimentation of a particle in water at 20°C. The actual value of the clearing factor under a given set of centrifugation conditions, k' at lower speeds of the centrifuge may be calculated from the formula

$$k' = k\,[(n_{max})/n]^2 \tag{15.15}$$

The approximate time t required in hours for sedimenting a pellet of a given particle may be calculated if the value of sedimentation coefficient S in Svedberg units is known by using the equation

$$t = k/S \tag{15.16}$$

The above equation gives only the approximate time required for pellet formation because the value of k is based on the assumption that the viscosity and density of the medium are the same as of water.

Centrifugal separation is, thus, primarily based on the differences in densities and sizes of particles. The order of separation of biological macromolecules and cell components based on the decreasing order of density is generally (i) whole cells and cell debris, (ii) nuclei, (iii) chloroplasts, (iv) mitochondria, (v) lysosomes and other microbodies, (vi) microsomes (fragments of endoplasmic reticulum), and (vii) ribosomes. The conditions of centrifugal separation are usually reported specifying the rotational speed, radial dimensions, and time of operation of the centrifuge.

15.4 CENTRIFUGES

Centrifuges are used for analytical and preparative centrifugation. *Analytical centrifuges* are small capacity centrifuges with specially designed rotors and detection systems. They handle only small amounts of pure materials as the primary objective is to study the sedimentation characteristics and molecular structure of macromolecules, particularly of biological macromolecules. *Preparative centrifuges* are used for actual separation, isolation, and purification a variety of products. These centrifuges are commonly used in the separation and harvesting of whole microbial cells from fermentation broths, plant or animal cells from tissue culture, or plasma from blood; fractionation of sub-cellular components, isolation of nucleic acids, lipoproteins, viruses, etc. In industries centrifuges find extensive use for large-scale separations involving three different techniques: (i) *centrifugal sedimentation* to separate fine solids from liquids, (ii) *centrifugal decantation* to separate immiscible liquids with small differences in their densities, and (iii) *centrifugal filtration* in the centrifugation-cum-filtration of solids from liquids. Large-scale centrifugation is advantageous particularly when filtration becomes a cumbersome operation due to the small size of the particles. However, centrifugation produces only a paste of solids or a concentrated suspension whereas filtration produces a dry cake. The suspension obtained by centrifugation requires dewatering to get the dry cake.

Centrifuges consist of a rotor positioned on a motor driven central drive shaft within a closed chamber. The chamber may be armor plated and refrigerated in most high speed centrifuges. The rotors used in centrifuges should be capable of withstanding the stress forces generated during centrifugation. Rotors are made of aluminum alloy for low speed centrifuges and are anodized to protect against corrosion while rotors are made of titanium alloy in high speed centrifuges. The suspension to be centrifuged is usually held in tubes or bottles placed in the rotor. Centrifuge tubes and bottles made of glass, cellulose esters, polyethylene, polypropylene, polycarbonate, nylon, or stainless steel are available in different capacities. A variety of factors such as the type of rotor used, nature and volume of sample, speed of operation and chemical resistance to solvents, transparency, or opaqueness of the tube and the method of harvesting the samples after centrifugation need to be considered in selecting the type and capacity of the tube for centrifugation.

Centrifuges are classified based on their capacity and speed of operation into (i) small bench centrifuges, (ii) large capacity refrigerated centrifuges, (iii) high speed refrigerated centrifuges, and (iv) ultracentrifuges. The *small bench centrifuge* operates at ambient temperatures cooled by the flow of air around the rotor. The centrifuge operates at speeds of 4000–6000 rpm with RCF values in the range of 3000–7000 g. It is used to collect particles which sediment rapidly (e.g., coarse precipitates, yeast cells, erythrocytes, etc.) in small amounts. Centrifuges for handling biological samples called microfuges incorporate refrigeration system to prevent denaturation of samples. They develop RCF of the order of 10,000 g by operating at speeds in the range of 8000–13,000 rpm. Large capacity refrigerated centrifuges operate at 6000 rpm generating RCF of 6500 g. These centrifuges have refrigerated rotor chambers with interchangeable rotors to hold tubes of different capacities in the range of 10–100 mL. These centrifuges are used to collect coarse precipitates, yeast cells, nuclei and chloroplasts and other particles which sediment rapidly. *High speed refrigerated centrifuges* are capable of operating at 25,000 rpm generating RCF of 60,000 g. The rotors are interchangeable. The instrument is used to collect microorganisms, cell debris, large-sized organelle, and proteins precipitated by ammonium sulphate. Ultracentrifuges for analytical and small-scale preparative purposes are available with high speed operation in the range of 60,000–80,000 rpm generating RCF of about 600,000 g.

15.4.1 Rotors

Five different types of rotors are commonly used in centrifuges. These include (i) swinging bucket rotor, (ii) fixed angle rotor, (iii) vertical tube rotor, (iv) zonal rotor, and (v) elutriator. The first three are most widely used in analytical centrifuges.

Swinging bucket rotor holds the sample tube holder or bucket and during acceleration the bucket swings from vertical (gravity force) to the horizontal position aligning with the centrifugal force. The solid particles move radially from the centre of rotation towards the walls of the tube first. The accumulated particles then precipitate in bulk due to convection flow along the length of the centrifuge tube (called the wall effect). The swinging bucket rotor is less efficient for making pellets of particles (Fig. 15.3).

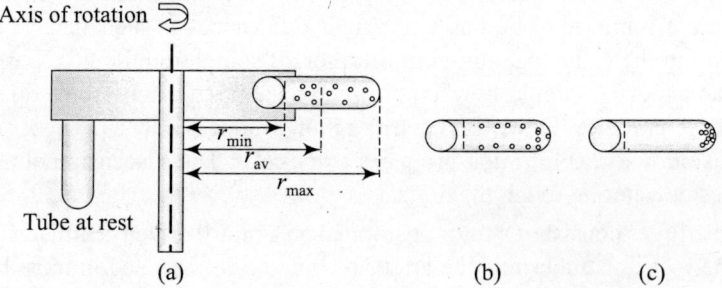

Fig. 15.3   (a) Cross-section of swinging bucket rotor and particle distribution prior to centrifugation, (b) wall effect during centrifugation, and (c) pellet formed after centrifugation

Fixed angle rotor (Fig. 15.4) is a solid rotor and the sample tube fit into the holes in the rotor. The tube angle, thus, remains in a fixed position throughout the loading, centrifugation, and unloading processes. As the centrifugal forces increases during the rotation of the rotor, the solution reorients inside the sample tube. The centrifugation path is shorter in the fixed angle rotor as compared to that in a swinging bucket rotor, and hence, the particles reach the tube wall more rapidly and then slide down the wall to form a pellet. Fixed angle rotors with tube angles in the range 14° to 40° are available. The sedimentation path for the solid particle is shorter when the tube angle is shallow. The fixed angle rotor is more efficient for making pellets.

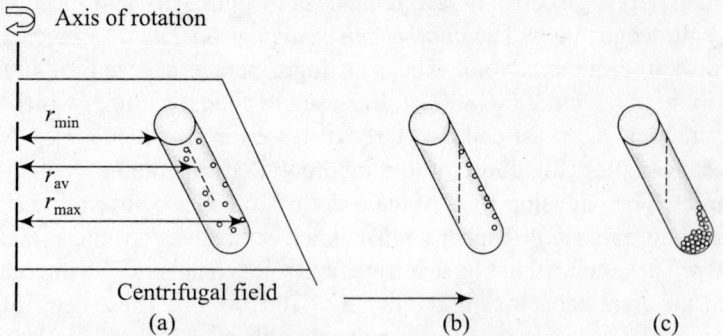

Fig. 15.4   (a) Cross-section of fixed angle rotor and particle distribution prior to centrifugation, (b) wall effect during centrifugation, and (c) pellet formed after centrifugation

Vertical tube rotor (Fig. 15.5) contains sample tubes fixed in a vertical position during centrifugation and the sample reorients towards the centrifugal force. The sedimentation path is the shortest among all types of rotors and since r_{min} is large, a larger minimum centrifugal force is generated and sedimentation occurs faster even at lower speeds. The vertical tube rotor is mostly used for isopycnic and rate zonal centrifugation but not for making pellets as the pellet is formed over the entire tube wall. In addition, the pellet formed detaches easily and falls back into the solution at the end of centrifugation.

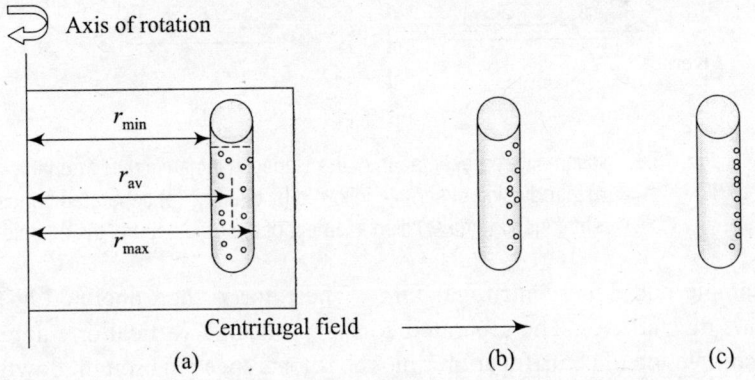

Fig. 15.5   (a) Cross-section of vertical angle rotor and particle distribution prior to centrifugation, (b) wall effect during centrifugation, and (c) pellet formed after centrifugation

Zonal rotor designed to minimize wall effects is a large volume rotor having capacities in the range of 300–1700 mL and used for preparative centrifugation. The zonal rotor consists of a hollow cylindrical bowl provided with a lid. The rotor bowl is divided into four sector-shaped compartments by a vane assembly attached at the centre core of the rotor. The vanes have radial ducts facilitating the flow of gradient from the centre core to the periphery. The rotor core may be loaded or unloaded by dynamic method while the rotor is spinning (called standard core type rotor) or by static method while the rotor is at rest (called reorienting gradient core type rotor).

The most commonly used dynamic mode of loading the standard core-type rotor is carried out while the rotor is revolving at about 2000 rpm. The preformed gradient is pumped from lighter end of the gradient to the denser so that the lighter end of the gradient is placed at the rotor core. After pumping the gradient the rotor is filled with a fluid cushion as dense as or denser than the highest density of the preformed gradient. The sample is introduced at the rotor centre from which it is displaced by pumping an overlay of low density liquid. The rotor bowl is closed and accelerated to the operating speed. Zonal rotors are used for rate zonal or isopycnic centrifugation. After operating the centrifuge for the required time, the gradient, and the separated components are recovered by decelerating the rotor to 2000 rpm and displacing the contents by introducing the fluid cushion at the periphery of the rotor. Figure 15.6 shows the stages of dynamic mode operation of the zonal rotor

In the static method of loading the rotor, a reorienting density gradient is introduced followed by the introduction of the sample as a layer on the top of the density gradient while the rotor is at rest. The rotor is slowly accelerated to about 1000 rpm to allow the reorientation of the gradient

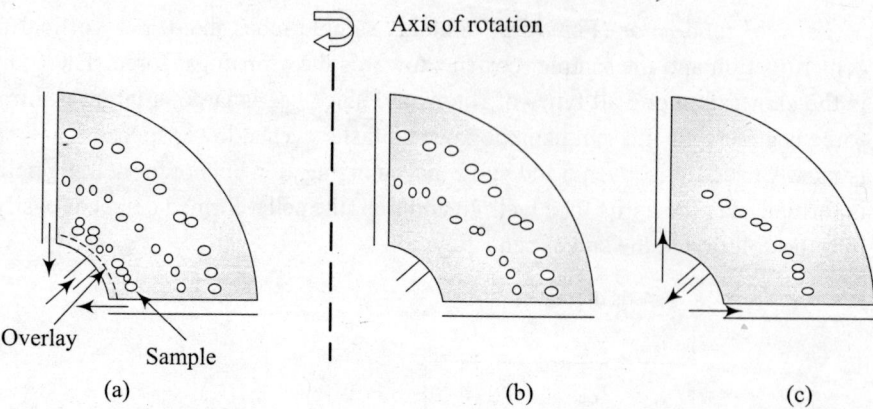

Fig. 15.6 Schematic representation of the stages in dynamic mode zonal rotor centrifugation: (a) introduction of sample followed by overlay, (b) separated bands after high speed centrifugation and (c) displacement of the gradient and the sample components

and sample under the centrifugal force. The rotor is then operated at the required speed to separate the particles. The separated zones approach a vertical orientation at high speeds and after completion of centrifugation time the rotor's speed is brought down first to 1000 rpm and then slowly to rest. The separated zones at rest reorient to horizontal position (Fig. 15.7).

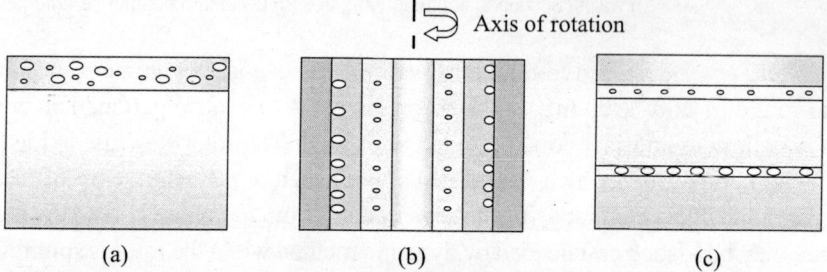

Fig. 15.7 Schematic representation of the stages in static mode zonal rotor centrifugation: (a) sample introduced prior to centrifugation, (b) vertically reoriented gradient and separated bands during high speed centrifugation, and (c) separated sample component bands after centrifugation

Elutriator rotor consists of conical-shaped separation chamber and a counterbalancing bypass chamber at the opposite ends of the rotor. The diameter of the chamber increases towards the axis of rotation (Fig. 15.8). It is a continuous flow rotor with elutriating fluid containing the gradient and the sample being pumped at the periphery of the rotor while the rotor is spinning at a pre-selected rpm in the range of 1000–3000 rpm. The gradient flow velocity decreases towards the axis of rotation (centripetal end) as the diameter of the chamber increases. The tendency of the particles to sediment in the centrifugal field within the separating chamber is balanced by the controlled flow of the elutriating fluid towards the centripetal end. Hence, the particles of different sedimentation rates separate into zones based on their attaining equilibrium position where the opposing centrifugal force and fluid velocity are balanced. The position of equilibrium depends on the shape, density, and size of the particles. The separated bands of particles may be harvested either by stepwise deceleration of the rotor or by stepwise increase in flow rate of the elutriating fluid through the separating chamber.

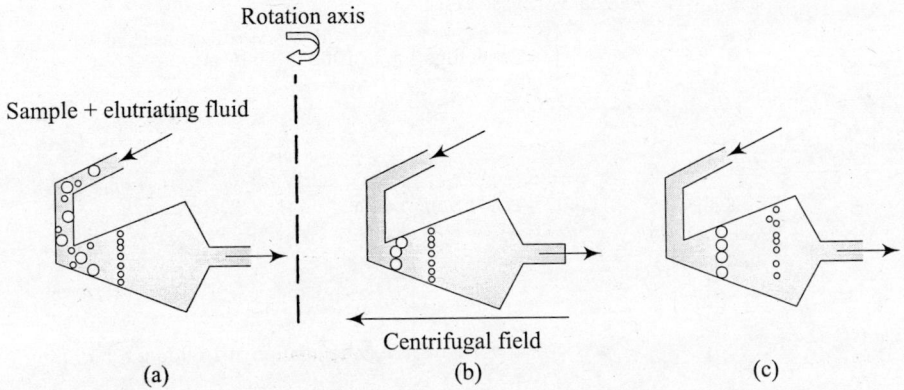

Fig. 15.8 Separation by centrifugal elutriation: (a) sample introduction into the separation chamber, (b) equilibrium positions of the particles, and (c) recovery of separated particles

15.5 CENTRIFUGATION TECHNIQUES

The centrifugation techniques may be used for analytical or preparative separations. Analytical centrifugation is used primarily for characterization of purified samples while preparative techniques aim at purification of the desired material. The specific centrifugation techniques include (i) differential centrifugation, (ii) density gradient centrifugation, and (iii) centrifugal elutriation. Density gradient centrifugation may be carried out by isopycnic centrifugation, equilibrium isodensity centrifugation, or rate-zonal centrifugation.

15.6 DIFFERENTIAL CENTRIFUGATION

Differential centrifugation makes use of the differences in the sedimentation rate of particles of different sizes and densities. The sample mixture such as tissue homogenate or lysed cells to be separated into sub-cellular components is subjected to step-wise increase in the centrifugal field to give a number of fractions. A pellet of particles and a supernatant are formed at each stage for an applied centrifugal field over a given time. The pellet and supernatant are collected separately. The pellet is suspended in the homogenization medium and centrifuged again at the same conditions. This procedure of suspension in homogenization medium, centrifugation under the same conditions, separation into pellet and supernatant is repeated several times to remove any low density particles and to render it free from any cross contamination. Thus, a fairly pure component is obtained in the form of a pellet. The supernatant is then subjected to increased applied centrifugal field in the second step and the entire procedure is repeated several times to isolate relatively pure pellets of different subcomponents. In the first stage of centrifugation the biggest and densest particles sediment out and in each subsequent stage of increasing centrifugal field intermediate-sized particles and densities sediment out, the smallest, and the least dense particles sedimenting out at the higher applied centrifugal field. In the case of particles of same mass but differing in their densities, particles with the highest density sediment at a faster rate as compared to the less dense particles. A summary of a typical protocol of differential centrifugation of a homogenate of lysed cells is shown in Fig. 15.9.

Differential centrifugation is usually the first step in the separation and purification of sub-cellular components as most of them have densities in the range of 1.1–1.3 g/cm^3 in sucrose medium.

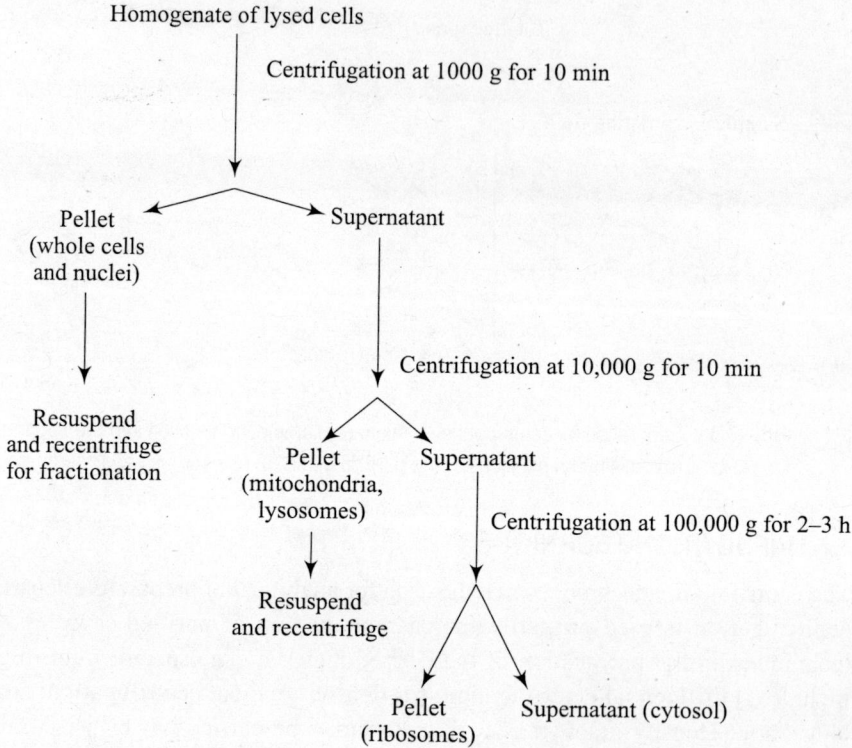

Fig. 15.9 Protocol for differential centrifugation of a homogenate of lysed cells

15.7 DENSITY GRADIENT CENTRIFUGATION

Density gradient centrifugation involves the use of a column of liquid whose density gradually increases from the top to the bottom of the centrifuge tube. Density gradient centrifugation has several advantages such as (i) the sample to be separated can be introduced as a thin layer at the top of the liquid column, (ii) prevention of mixing of the separated particles due to convection, (iii) eliminating convection fluid movement due to thermal or mechanical agitations, and (iv) particles focus into separate bands depending on their densities, thus, facilitating quantitative separation of all the components of a mixture.

The substances used for forming gradients should satisfy certain conditions such as (i) stability in solution, (ii) inertness towards biological samples, (iii) facilitate desired type of separation by covering the density range, (iv) transparent to light at appropriate wavelengths for spectrophotometric monitoring in visible or ultraviolet region, (v) amenable for sterilization, (vi) non-toxic and non-flammable, (vii) have negligible osmotic pressure, and (viii) cause minimum changes in pH, ionic strength, and viscosity. In addition, the substances should be available readily in pure form and should be inexpensive. Satisfying all the above requirements is very difficult, and hence, the substance for forming gradient is chosen on the basis of the nature of particles to be fractionated.

Solutions of cesium chloride (max. density $\rho = 1.91$ at 20°C) and cesium sulphate ($\rho = 2.01$)2 are useful for the separation of DNA and RNA, nucleoproteins and viruses, purification of proteoglycans, and isolation of plasmids. Sucrose solution with a maximum density of 1.32

is useful for the separation of sub-cellular particles, proteins, nucleic acids, and membranes and viruses by rate zonal centrifugation but suffers from the disadvantages of high viscosity even at low concentrations and high osmotic effect. Ficoll ($\rho = 1.17$) a copolymer of sucrose and epichlorohydrin is used instead of sucrose for separation of whole cells and sub-cellular components by isopycnic and rate zonal centrifugation. It has the advantages of low viscosity and inert osmotically at < 20 %w/v concentrations. Other examples of gradient forming substances include dextran ($\rho = 1.13$) for the separation of whole cells and microsomes, bovine serum albumin ($\rho = 1.35$) for the separation of whole cells, colloidal silica, for example, Percoll ($\rho = 1.20$) and Ludox ($\rho = 1.30$) for isolation of whole cells and fractionation of sub-cellular components, and non-ionic iodinated aromatic compounds (e.g., Metrizamide – $\rho = 1.46$, Nycodenz – $\rho = 1.42$) for the isolation of whole cells, fractionation of sub-cellular particles, nucleoproteins, membranes, and viruses.

Density gradients may be formed either by *step* or *discontinuous gradient* technique or *continuous density gradient* technique. The *step density gradient* is formed in the centrifuge tube by carefully adding solutions of decreasing density in successive layers with the help of a pipette. The step gradients usually consist of 4–6 layers. The step gradient can be used as such or transformed into a linear gradient by allowing to stand or gentle stirring to merge the layers. The sample mixture is finally introduced as a thin layer over the lowest density layer at the top of the step gradient and centrifuged to separate the components.

The *continuous density gradient* is formed with the help of a special apparatus called the gradient former or gradient generator. The gradient former consists of two cylinders of identical diameter connected at the base by a tube provided with a control valve (Fig. 15.10). One of the cylinders contains the lighter solution which forms the lowest gradient while the other cylinder contains the more dense solution. The solution levels in the two cylinders must be maintained the same during gradient formation. One of the cylinders called the mixing chamber has an outlet to facilitate the flow of the liquid into the centrifuge tube. When the mixing chamber initially contains the dense solution the centrifuge tube is filled from the top as shown in Fig. 15.10(a). and a continuous linear gradient from dense to light is formed. An equivalent amount of the less dense solution is allowed to flow through the control valve into the mixing chamber to maintain levels in the cylinders the same. The solution in the mixing chamber is maintained homogeneous throughout by constant stirring, and hence, the density of the liquid in the mixing chamber linearly decreases as the centrifuge tube is filled.

Alternatively, if the mixing chamber contains initially the less dense solution the centrifuge tube is filled from the bottom to form a continuous linear gradient from light to dense as shown in Fig. 15.10(b).

Non-linear gradients of convex or concave in concentration as a function of volume may be formed by using cylinders of different geometries or with different volumes. Non-linear gradients may also be formed by using two mechanically driven syringes containing liquids of different densities by operating the syringes at different speeds.

Different types of density gradients are used for different purposes. Step gradients are more suitable for the separation of whole cells and sub-cellular components from plant and animal tissue homogenates. Linear gradients are useful for achieving a higher resolution in the separation of certain viruses and ribosomal subunits. Concave gradients are useful in the separation of lighter particles such as serum lipoproteins by floatation. Long gradient columns and linear–log gradients (the logarithm of the depth of the gradient column is a linear function of the sedimentation coefficient of the particle) are useful in the separation of large particles such as ribosomal subunits, polyribosomes, and certain viruses.

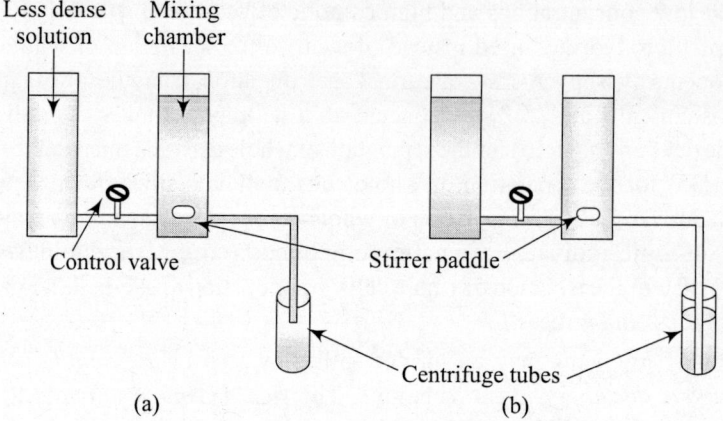

Fig. 15.10 Continuous gradient formation with the help of a gradient generator and filling the centrifuge tube from (a) top and (b) bottom

The gradient formed in the centrifuge before centrifugation adjusts or redistributes itself to the centrifugal forces acting on it. In swinging-bucket rotor the centrifuge tube aligns itself in the direction of the centrifugal field, and hence, the gradient remains without any redistribution. However, in the fixed-angle and vertical tube rotors the gradient and sample redistribute during initial stages of centrifugation, and hence, it is necessary to control the acceleration carefully to maintain the gradient. Similarly deceleration also must be controlled carefully. The redistribution during centrifugation and separation of the components at the end is schematically shown in Fig. 15.11.

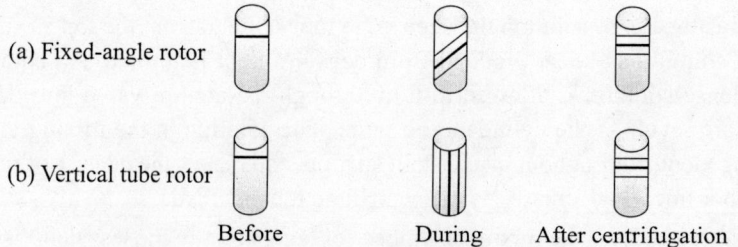

Fig. 15.11 Redistribution of gradient and sample layers

15.7.1 Sample Application and Harvesting Samples from Gradients

Sample volume that can be applied to a gradient depends on the cross-sectional area of the gradient exposed to the sample. For achieving effective separation the sample volume is usually in the range of 0.2–0.5 mL for tubes of 1.0–1.6 cm diameter and about 1 mL for 2.5 cm diameter tubes. A simple guideline is based on the ratio of sample concentration to starting gradient concentration of 1:10 (% w/w). The sample is usually applied to the gradient using a syringe and in the case of fragile samples such as DNA a pipette is used. The sample is allowed to run down the side of the centrifuge tube onto the gradient.

After centrifugation the separated sample components may be recovered by different techniques. If the separated component bands are visible they may be recovered by a hypodermic syringe. Alternatively if the centrifuge tube is made of disposable plastic it may be punctured at the bottom and the drops of gradient containing the separated components recovered in the order from highest density particles to least ones. Another technique for the recovery of the separated components is by displacement of the gradient upward by a dense liquid (e.g., 60% w/v sucrose solution) and collecting the sample components from the top end of the centrifuge tube from lower to higher density parts of the gradient. The individual fractions are then analysed by UV–visible spectrophotometry, refractive index measurement, scintillation counting, or by chemical and enzymatic analysis for characterization.

15.7.2 Density Gradient Centrifugation Techniques

Density gradient centrifugation can be carried out by either of the two methods (i) isopycnic or equal density technique and (ii) rate-zonal technique. Both these techniques are useful for quantitative separation of all the components of a mixture of particles and for determining the buoyant densities as well as sedimentation coefficients of particles.

Isopycnic centrifugation is also known as *equilibrium sedimentation*. The technique depends on buoyant density of the particles and is independent of the shape or size of the particles and also independent of the time of centrifugation. In this technique the sedimenting particles move through a density gradient till they reach their *isopycnic point*. The term *isopycnic point* refers to the condition when the density of the surrounding medium is equal to the density of the particle. Since the density difference between the sedimenting particle and the surrounding medium is zero ($\rho_s - \rho = 0$), further movement of the particle will not occur at this point irrespective of the time of centrifugation. In practice the sample mixture of particles is introduced as a layer at the top of a continuous-preformed density gradient medium covering the entire range of densities of the mixture of particles to be separated. The maximum density of the gradient medium should be greater than the density of the densest particle. On centrifugation the gradient medium establishes itself and the particles separate into bands or zones each at their characteristic buoyant density or isopycnic point as shown in Fig. 15.12.

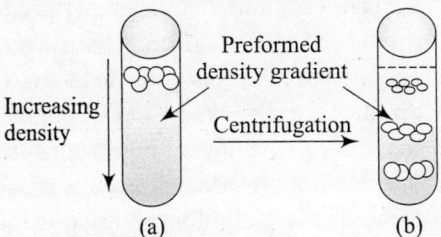

Fig. 15.12 Isopycnic centrifugation (a) prior to centrifugation and (b) after centrifugation

Equilibrium isodensity centrifugation is a modified isopycnic centrifugation method wherein a self-forming gradient is used instead of a preformed gradient. In this method the sample is mixed homogeneously with the gradient medium to form a solution of uniform density. A concentrated solution of sucrose or salts of heavy metals such as rubidium or cesium or a suspension of colloidal silica may be used as gradient medium. Due to centrifugation a concentration gradient, and hence, a density gradient is formed *in situ*. The sample molecules which were distributed uniformly throughout the medium move upward or downward the self-formed gradient during centrifugation and form bands at their isopycnic points.

The isopycnic and equilibrium isodensity centrifugation methods are useful for the determination of buoyant densities of particles and separation of particles based on their densities. For example, sub-cellular components such as Golgi bodies ($\rho_s = 1.11$ g/cm³), mitochondria ($\rho_s = 1.19$ g/cm³), and peroxisomes ($\rho_s = 1.23$ g/cm³) can be effectively separated in a solution of sucrose ($\rho_s = 1.32$ g/cm^3). However, most soluble proteins have the same density as of sucrose solution, and hence, cannot be separated. The methods are also useful to determine the base composition of double stranded DNA and for the separation of linear and circular forms of DNA. The separation efficiency of the different forms of DNA may be enhanced by increasing the density differences. The density differences can be altered by binding with heavy metal ions or dyes or by incorporating heavy isotopes such as ^{15}N during biosynthesis of DNA. However, the methods suffer from the disadvantage of requiring long centrifugation times of about 40–50 h to attain equilibrium.

Rate-zonal centrifugation is also known as *velocity sedimentation technique*. The sample is applied on the top of the preformed density gradient as a thin layer. The gradient medium is of minimal viscosity and density with the highest density of the gradient not exceeding that of the densest particle of the sample. On centrifugation for sufficient time the separation of the sample components occurs on the basis of the sizes of the components or their sedimentation coefficients. After centrifugation the sample components form discrete bands based on their relative velocities of sedimentation. If centrifugation is carried out for too long the separated components form pellets and settle at the bottom of the tube. Hence, centrifugation must be terminated at appropriate time. The rate-zonal method is useful for the separation of proteins, enzymes, hormones, ribosomal sub-units, sub-cellular components, RNA–DNA hybrids, and for determining the size distribution of polysomes and for fractionation of lipoproteins.

15.8 CENTRIFUGAL ELUTRIATION

Centrifugal elutriation is mainly used for the separation as well as purification by washing of a mixture of cells of different tissues or species (e.g., mononuclear leucocytes from human blood, endothelial cells, fat storing cells from liver, etc.). The different types of cells are separated based on their sizes by the action of opposing forces of applied centrifugal field and the centripetal force due to the flow of liquid at controlled rate. A density gradient medium is not required and the medium for suspending cells is chosen so as to allow the sedimentation of the particles. The technique is useful for the fractionation of delicate cells in the size range of 5–50 μm and the separated cells retain their viability. The technique has the advantages of fast separations and good recovery of the products.

15.9 ULTRACENTRIFUGE

Analytical and preparative ultracentrifuges are available commercially which are exclusively used for sedimentation analysis and separation of sub-cellular cell components respectively.

15.9.1 Analytical Ultracentrifuge

Analytical ultracentrifuge is relatively small in size with the capacity of the cells limited to 1 mL or less. Svedberg designed the first analytical ultracentrifuge in 1923 and the present day instruments have basically the same features consisting of a high speed centrifuge and an optical detector system to monitor the sedimentation process continuously. The main components include (i) a rotor housed in a protective armoured chamber, (ii) an optical detector system, (iii) temperature measuring device, and (iv) two cells, an analytical sample cell, and a second counterpoise cell which are positioned in the rotor.

The rotor is kept suspended on a wire connected to the central drive shaft of a high speed motor capable of operating at speeds up to 70,000 rpm (about 500,000 g). The wire suspension of the rotor allows it to find its own axis of rotation. The armoured chamber is refrigerated and evacuated. The optical detector system to monitor the sedimentation of the particles may be a UV–visible spectral detector, a Schlieren optical system for monitoring the changes in refractive index, or a more sensitive Rayleigh or Lebedev interferometric detector. The progress of sedimentation may also be monitored by changes in the UV absorption by the spectral detector. A temperature probe such as a thermistor is mounted on the tip of the rotor. The analytical sample cell is a single sector or double sector centrepiece with a sector angle between 1° and 4° one of about 1 mL capacity, the liquid column having a height of about 10 mm. The walls of the centrepiece will be parallel to the line of centrifugal force when properly aligned in the rotor giving an ideal condition for sedimentation. A schematic diagram of the analytical ultracentrifuge is shown in Fig.15.13.

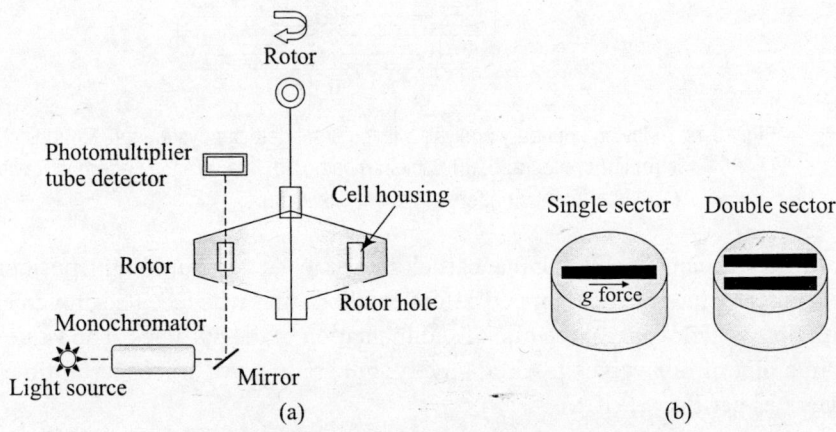

Fig. 15.13   (a) Schematic diagram of an analytical ultracentrifuge with UV–visible spectral detector and (b) single sector and double sector cells

15.9.2   Applications of Analytical Ultracentrifuge

Analytical ultracentrifugation is used for accurate calculation of molecular weights of biopolymers such as polysaccharides, proteins, lipoproteins, nucleic acids, sub-cellular particles, cells, and viruses in their native state. In contrast, other techniques such as gel electrophoresis and size-exclusion chromatography are applicable to denatured samples only. Analytical ultracentrifugation is also useful for the determination of molecular size and shape, density, sedimentation coefficient and frictional coefficient of analyte molecules, study of macromolecular interactions, conformation changes in enzymes, replication of DNA, etc. The technique requires the use of high purity samples.

15.9.3   Determination of Molecular Weight of Macromolecules

Molecular weight of macromolecules can be determined either by (i) sedimentation velocity measurement or (ii) sedimentation equilibrium analysis.

Sedimentation velocity measurement of biopolymers is carried out in analytical ultracentrifuge by operating it at high speeds. The randomly distributed particles of the pure macromolecule suspended in a solvent medium migrate radially outwards from the centre of rotation thereby creating a sharp boundary between the portion of solvent which is clear and the portion of the solvent still containing the particles. The movement of this boundary [Fig. 15.14(a)] is monitored by the optical detector system as a function of time and recorded photographically [Fig. 15.14(b)–(e)]. Hence, the technique is called moving boundary analysis.

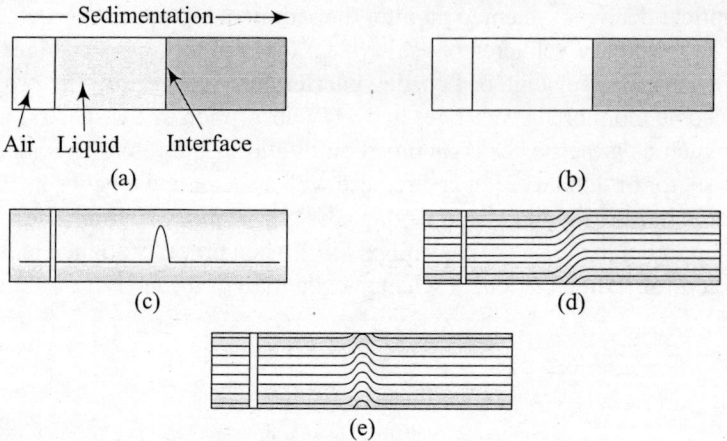

Fig. 15.14 Moving boundary analysis with different detector systems: (a) graphical presentation, (b) UV photograph, (c) Schlieren photo, (d) interference pattern in Rayleigh detector, and (e) interference pattern in Lebedev detector

In this technique the biopolymer particle moves toward an equilibrium position under centrifugal field but the experiment is stopped before equilibrium is reached. The movement of the boundary with time is a measure of the rate of sedimentation of the particles. The value of s is determined from a plot of $\ln r$ versus t. Alternatively, integrating r with respect to time between t_1 and t_2 gives s as per Eq. (15.12).

The molecular weight MW of the biopolymer is calculated from the value of s using the Svedberg relationship

$$MW = s\, RT/(D\,(1 - \upsilon\rho)) \tag{15.17}$$

where R is gas constant, T is absolute temperature, D is diffusion coefficient of the macromolecule, υ (upsilon) is the partial-specific volume, and ρ is the density of the solvent medium. The partial- specific volume is defined as the volume increase when 1 g of solute is added to an infinite volume of solution.

A simple empirical formula which does not require knowledge of the diffusion coefficient of the solute for calculating the molecular weight of macromolecules of globular shape from $s^{\circ}_{20,w}$ value is

$$s^{\circ}_{20,\,w} = 0.00248\ MW^{0.67} \tag{15.18}$$

Sedimentation equilibrium analysis gives more accurate estimation of molecular weights. The ultracentrifuge is operated at different rotor speeds from 800 to about 68,000 rpm giving rise to a large range of centrifugal fields facilitating the estimation of molecular weights of a wide range of molecules from a few hundred to several million. In this technique the particles are allowed to move until they reach an equilibrium position at which there is no net migration of the particles throughout the length of the centrifuge tube. The equilibrium position is brought about due to the opposing forces of sedimentation and diffusion. Once equilibrium position is attained a concentration gradient of the solute as a function of the distance from the rotation axis is set up. Molecular weight of the solute is calculated from the concentration gradient using the relationship

$$MW = 2RT \ln\left(\frac{c_2}{c_1}\right)\Big/[(1 - \upsilon\rho)\,\omega^2\,(r_2^2 - r_1^2)] \qquad (15.19)$$

where c_1 and c_2 are the concentrations of the solute at distances r_1 and r_2 respectively from the rotation axis.

Sedimentation equilibrium analysis is generally carried out at low rotor speeds so that the ratio $c_2/c_1 \approx 4$. The low rotor speed operation is particularly time consuming requiring even several weeks for attaining equilibrium. This disadvantage is overcome using sample cells of short column in the range 1–3 mm.

15.9.4 Determination of Purity of Macromolecules

The purity of samples of nucleic acid, proteins, viruses, etc., is an essential requirement for the accurate estimation of molecular weights by ultracentrifugation. The purity of the sample that is its homogeneity is easily determined by ultracentrifuge by monitoring the sedimentation boundary. Homogeneous samples give rise to a single sharp boundary while impurities in the sample give rise to asymmetry, shoulder, or additional peaks in the display.

15.9.5 Study of Conformation Changes in Macromolecules

Reversible as well as irreversible changes in conformation of macromolecules brought about by exposure of the molecules to agents such as higher temperature, organic solvents, etc., can be investigated by differences in sedimentation rates of the sample under different conditions. Variations in sedimentation rates depend on the frictional resistance of the solute, the more compact molecule showing less frictional resistance, and hence, higher sedimentation rate. In contrast, a more disorganized molecule has greater frictional resistance, and hence, settles down slowly. Changes in the conformation, and hence, changes in sedimentation rates occur when allosteric proteins interact with substrate, activator, or inhibitor and such changes can be easily studied by analytical ultracentrifuge.

15.10 PREPARATIVE ULTRACENTRIFUGE

Preparative ultracentrifuges can produce a relative centrifugal field of about 600,000 g by operating at 80,000 rpm. These centrifuges are provided with relatively more sophisticated control systems for maintaining temperature and avoiding over speeding. The rotor chamber is refrigerated, evacuated, sealed, and enclosed in heavy amour plating. *Airfuge* is an air-driven table-top preparative ultracentrifuge widely used in biochemical and clinical laboratories for handling small volumes. The rotor with provision to accommodate six centrifuge tubes is driven on a virtual friction-free cushion of air in a non-evacuated chamber and is capable of achieving 160,000 g at 100,000 rpm.

● ● ● ● ● ● ● ● ● ● ● ● ● ● ● ● ● ● **SOLVED PROBLEMS**

EXAMPLE 1 *Calculate the RCF$_{min}$, RCF$_{av}$, and RCF$_{max}$ for a centrifuge tube rotating at 15,000 rpm and in which the distance between the rotation axis and the meniscus is 10 cm and that between rotation axis and the bottom of the tube is 18 cm.*

SOLUTION

Using Eq. (15.4) the calculated values of RCF_{min}, RCF_{av}, and RCF_{max} respectively are 25,177 g, 35,248 g, and 45,319 g.

EXAMPLE 2 *Calculate the settling time of spherical particles of 0.1 mm diameter in a column of 50 cm length, the density difference between the solid particles and the liquid being 0.05 g/cm³, and the viscosity of the liquid 1.1 cP.*

SOLUTION

Substituting the given data into Eq. (15.8), the terminal velocity of the particles, v_g is calculated.

$$v_g = (0.01 \text{ cm})^2 (0.05 \text{ g cm}^{-3}) (980 \text{ cm s}^{-2})/(18 (0.011 \text{ g cm}^{-1}\text{s}^{-1}))$$
$$= 0.0247 \text{ cm/s}$$

The time for settling is given by $t = l/v_g$, where l is the length of the column.

$$t = 50 \text{ cm} / (0.0247 \text{ cm s}^{-1}) = 2024.3 \text{ s or } 33.7 \text{ min.}$$

EXAMPLE 3 *Calculate the value of sedimentation coefficient of a particle if it moved a distance from 5 cm to 7.5 cm from the rotation axis on centrifugation at 20,000 rpm during the centrifugation period of 2 h.*

SOLUTION

Using Eq. (15.12) the value of s can be calculated.

$$s = 128.4 \times 10^{-13} \text{ or in Svedberg units} = 128 \text{ S.}$$

EXAMPLE 4 *Calculate the time required to pellet a suspension of particles whose sedimentation coefficient $s_{20, w}$ is 1200×10^{-13} if the particles moved over from a distance of 2 cm to 6 cm from the rotation axis during centrifugation time of 3 h at 30,000 rpm.*

SOLUTION

The value of k is calculated by substituting the given values into Eq. (15.14) and the calculated value of k is substituted into Eq. (15.16) to calculate time required to pellet the suspension.

$$k = 2.53 \times 10^{11} \times \ln (6/2)/(30,000)^2 = 308.8 \text{ and } t = 309/1200 = 0.26 \text{ h.}$$

●   **SUMMARY**

- Centrifugation is a separation technique based on centrifugal force developed by a centrifuge and its effect on substances of different densities. RCF is the centrifugal force of a centrifuge relative to the gravitational force.

- The sedimentation velocity of a particle during centrifugation depends on the particle size, the rotational speed of the centrifuge, and the difference between the densities of the particle and the surrounding medium and is inversely related to the viscosity of the medium.

- Sedimentation coefficient is defined as the sedimentation velocity per unit centrifugal field. The sedimentation coefficient values for most biopolymers are expressed in Svedberg unit, S.

- The separation efficiency is expressed in terms of clearing factor or the pelleting capacity of the centrifugation process and depends on the time required for pellet formation in hours and the sedimentation coefficient of the particle; the smaller the value of the clearing factor greater is the separation efficiency.

- Analytical centrifuges are primarily used to study the sedimentation characteristics of small amounts of pure materials, whereas preparative centrifuges are used for actual separation in biochemical laboratories.

- The different types of preparative centrifuges include (i) small bench centrifuge operating at ambient temperatures at speeds of 4000–6000 rpm with RCF values in the range of 3000–7000 g; (ii) large capacity refrigerated centrifuges operating at 6000 rpm generating RCF of 6500 g; (iii) high speed refrigerated centrifuges operating at 25,000 rpm generating RCF of 60,000 g; and (iv) ultracentrifuges for analytical and small-scale preparative purposes operating at 60,000–80,000 rpm generating RCF of about 60,0000 g.

- Rotors used in centrifuges include (i) swinging bucket rotor, (ii) fixed angle rotor, (iii) vertical tube rotor, (iv) zonal rotor, and (v) elutriator, the first three being more useful for analytical purposes.

- The different centrifugation techniques include (i) differential centrifugation, (ii) density gradient centrifugation, and (iii) centrifugal elutriation.

- Differential centrifugation is based on the differences in the sedimentation rate of particles of different sizes and densities. It is used as the first step in the separation and purification of sub-cellular components.

- Density gradient centrifugation involves the use of a column of liquid whose density gradually increases from the top to the bottom of the centrifuge tube so that all the component particles of different densities in a mixture are quantitatively separated into bands. The different density gradient centrifugation techniques in vogue include (i) isopycnic and (ii) rate-zonal technique.

- Centrifugal elutriation technique is used for the separation as well as purification by washing of a mixture of cells of different tissues or species.

- Analytical ultracentrifuge with a capacity to handle samples of 1 mL or less is capable of operating at speeds up to 70,000 rpm (~500,000 g). It is used for accurate calculation of molecular weights of biopolymers, determination of molecular size and shape, density, sedimentation coefficient and frictional coefficient of molecules, study of macromolecular interactions, conformation changes in enzymes, replication of DNA, etc.

1. Explain the term relative centrifugal force. How is it related to the speed of rotation of a centrifuge?

2. Derive the equation for sedimentation under centrifugal force.

3. What is sedimentation velocity of a particle? How is it calculated? What are its units?

4. Explain the term clearing factor and its significance.

5. Write a note on centrifuges.

6. Explain the principles of operation in different types of rotors.

REVIEW QUESTIONS

7. What is differential centrifugation? How is it performed? What is its application?

8. How are the different density gradients formed?

9. Discuss the different density gradient centrifugation techniques and their uses.

10. Describe the components of an analytical ultracentrifuge with a neat sketch.

11. How is the molecular weight of a biopolymer determined by ultracentrifugation?

12. Give an account on the applications of analytical ultracentrifuge.

16 Electroanalytical Methods

16.1 INTRODUCTION

Electrochemistry deals with the conversion of chemical energy into electrical energy and vice versa and the applications of electrical energy for a wide variety of purposes. The inter-conversion of electrical and chemical energies takes place in an *electrochemical cell*. The electrochemical cell consists of two electrodes immersed in an electrolyte solution kept in a container such as a beaker. The electrochemical cell itself may act as a source of electricity when the two electrodes are suitably connected to a circuit and it functions as a *galvanic cell* or *voltaic cell* transforming spontaneously the chemical energy into electrical energy.

The conversion of the electrical energy supplied by an external source into chemical energy also takes place in an electrochemical cell called the *electrolytic cell*. The two electrodes of the electrochemical cell are connected to an external source of electrical energy such as a battery to allow the flow of current through the cell facilitating oxidation–reduction reactions at the two electrodes thereby utilizing electrical energy to perform chemical reactions. Electroanalytical chemistry basically involves the use of electrical energy for analytical purposes and constitutes an important application dealing with the use of instrumental techniques for analysis of chemical substances. These techniques involve the use of electrical energy and monitoring the responses of the electrical properties of analytes placed in an electrolytic cell.

16.2 CLASSIFICATION OF ELECTROANALYTICAL TECHNIQUES

Electroanalytical techniques may be classified into (i) conductometry, (ii) high-frequency method or oscillometry, (iii) potentiometry and pH-metry, (iv) electrogravimetry, (v) coulometry, and (vi) polarography and amperometry.

Conductometry is based on the measurement of electrical resistance of a solution (mostly aqueous) using two identical inert electrodes made of platinum and expressed in terms of conductance (reciprocal of resistance). The electrical conductance of the solution is due to the presence and migration of charged species (ions) towards the oppositely charged electrodes under the influence of applied electric field. To avoid electrolysis due to red-ox reactions at the electrodes, conductance measurements are taken using alternating current of frequencies in the range of 1000–5000 Hz.

Oscillometry, also known as high-frequency analysis, makes use of high-frequency alternating current (of the order 100–150 MHz) and monitors the changes in conductance, dielectric constant, or both of an electrolyte solution by electrodes which do not come into contact directly with the solution.

Potentiometry involves the measurement of electromotive force (EMF) or potential developed in an electrochemical cell by two non-polarized electrodes using a Wheatstone bridge circuit powered by a direct current source under conditions of zero current flow in the circuit. *pH-metry* is similar to potentiometry and uses the same principles but the measured potentials are usually expressed in pH scale.

Electrogravimetry is carried out under constant applied potential or current using a direct current source to deposit chemical substances on electrodes and determining the weight of the substance deposited.

Coulometry is based on Faraday's law of electrolysis relating to the quantity of electricity passed through an electrolyte solution with the quantity of chemical changes that occur. Coulometry may be carried out under constant applied potential or constant current conditions to monitor the chemical changes.

Polarography and *amperometry* are generally known as *voltammetry* as the techniques involve the study of current–voltage relationship under experimental conditions of diffusion-controlled current flow in an electrolyte solution as a function of applied potential from a direct current source. One or both the electrodes in the electrochemical cell may be polarizable microelectrodes.

16.3 CONDUCTOMETRY

Electrical conductors are broadly classified into two types: (i) electronic conductors, such as metals and alloys and (ii) electrolytic conductors such as solutions of ionic substances and molten or fused salts. Electrochemistry and electroanalytical methods involve the study of electrolytic conductors. Ohm's law applies to both electronic conductors and electrolytic conductors. According to Ohm's law, the current i flowing in a conducting medium is directly proportional to the applied voltage E and inversely proportional to the resistance R of the conducting medium. The units of current, potential difference, and resistance are amperes (A), volts (V), and ohms (Ω), respectively. The resistance is proportional to the length l of a conductor of uniform composition and cross section and inversely proportional to the cross-sectional area A of the conductor. A standard unit of resistance for both metallic and electrolytic conductors is called the *specific resistance* or *specific resistivity*, ρ (in ohm-cm). It is defined as the resistance offered to the flow of current by a conductor of 1 cm-cube in dimensions. The relationship between R and ρ is given by Eq. (16.1) as

$$R = \rho \frac{l}{A} \tag{16.1}$$

Conductance G in mho (ohm^{-1}, Ω^{-1}) and s*pecific conductance* or *specific conductivity*, κ (kappa), in mho cm^{-1} (ohm^{-1}cm^{-1} or S m^{-1}, Siemens per meter) are, respectively, the reciprocals of resistance R and specific resistance ρ. The conductance and specific conductance are related as follows:

$$G = \kappa (A/l) \tag{16.2}$$

Equivalent conductance Λ is a more useful parameter in the case of electrolytic conductors because it takes into account the dependence of specific conductance on the concentration of ionic species present in the electrolytic conductor. The equivalent conductance is defined as the conductance of an electrolyte solution containing 1 g equivalent of solute placed between two parallel electrodes set 1 cm apart and is expressed in units of $ohm^{-1}cm^2$. Equation (16.3) relates the equivalent conductance and specific conductance with the concentration C of the species as

$$\Lambda = 1000 \, (\kappa/C) \tag{16.3}$$

The concentration of the ionic species is expressed as gram-equivalent weight per 1000 cm^3 (obtained by dividing the gram-formula or gram-atomic weight by the charge on the ionic species). When the concentration of the ionic species is expressed in mol/L (gram molecular weight per 1000 cm^3), the equivalent conductance is called the *molar conductance* Λ_M.

16.3.1 Measurement of Conductance

The measurement of electrolytic conductance is carried out under relatively mild conditions of low voltage of about 1–100 V alternating current of frequencies less than 5000 Hz. The magnitude of current flowing through the electrolyte solution depends on the number and types of ions present, their mobility, the nature of the solvent, and the voltage applied. In the case of dilute aqueous solutions of strong electrolytes, the number of ions depends on the concentration, whereas in solutions of weak electrolytes it also depends on the degree of dissociation of the electrolyte and temperature. The measurement of conductance provides information on the concentration of electrolyte solutions, and hence, conductometry finds extensive use in industry both as an analytical tool in quality control and in online process control.

The instrument for measuring the conductance of electrolyte solutions is called the *conductance meter* or *conductance bridge* which is provided with a *conductance cell*. The working principle of the meter is based on Wheatstone bridge circuit to measure the resistance (reciprocal of conductance) of the electrolyte. The electrical circuit of the conductance meter and the block diagram of the conductance cell are shown in Fig. 16.1. The electrolyte solution is taken in the conductance cell and an alternating current of fixed frequency in the range of about 3000–5000 Hz is applied. The resistance R_U of the solution is determined by detecting the point D along the slide wire AB at which no current flows in the circuit as indicated by the galvanometer detector. The ratio of the distances, AD/DB, on the slide wire is equal to the ratio of resistances R_U/R_S where R_S is the resistance of the standard solution (e.g., 0.1 M KCl solution). The reciprocal of the resistance is directly displayed in the conductance meter.

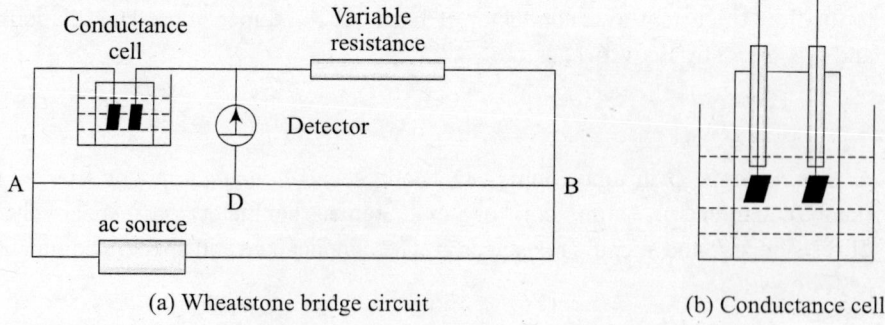

(a) Wheatstone bridge circuit   (b) Conductance cell

Fig. 16.1 Measurement of conductance

The conductance cell has a pair of platinized platinum electrodes welded to platinum wire and sealed rigidly in a heavy Pyrex/Corning glass tube so that the electrodes do not move. The electrodes are held in vertical position to avoid any collection of solids on the surface. The platinum electrodes are platinized by electrodeposition of platinum black on the electrodes by alternating current using an electrolyte solution of $K_2[PtCl_6]$. The surface area of the electrodes is about 1 cm^2 and the electrodes are separated by a fixed distance so that the ratio of the distance between the electrodes and the area of the electrodes (l/A) remains constant. The ratio l/A is called the *cell factor* or *cell constant*. The conductance cell is usually calibrated with solutions of known specific conductivity to determine the cell constant. Potassium chloride solutions of 0.1 M ($\kappa = 0.012856$ ohm^{-1}cm^{-1}) and 0.01 M ($\kappa = 0.00140877$ ohm^{-1}cm^{-1}) at 25°C are used to determine the cell constant. The cell constant is the product of the specific conductance of the solution and its measured resistance, the unit being cm^{-1}.

$$l/A = \kappa R \tag{16.4}$$

Once the cell constant for a given cell is known and the resistance of the solution in the cell is measured, the specific or equivalent conductance of any solution can be calculated using Eqs. (16.2) and (16.3), respectively.

In practice, a conductance cell with a high cell constant (small electrodes and long path length) is used for measuring the conductance of concentrated electrolyte solutions. For solutions of weak electrolytes or dilute solutions, a cell with low cell constant (large electrodes and short spacing) is used.

Kohlrausch observed that specific conductivity decreases with increasing dilution of the electrolyte as the number of charge-carrying species decreases. In contrast, the equivalent conductance increases with increasing dilution of the electrolyte and reaches a limiting value. The limiting equivalent conductance value is called the *equivalent conductance at zero concentration* or *infinite dilution* represented by the symbol Λ_0 or Λ_∞. Kohlrausch found an empirical relationship between the equivalent conductance and concentration C of an electrolyte given as

$$\Lambda = \Lambda_0 - b\sqrt{C} \tag{16.5}$$

where b is a constant.

Theoretical and experimental support for the above empirical relationship came in the form of Debye–Huckel–Onsager equation

$$\Lambda = \Lambda_0 - \left(A + B\Lambda_0\right)\sqrt{C} \tag{16.6}$$

where constant A refers to the *electrophoretic effect* and B to the *relaxation effect* or *asymmetry*. The concentration C is expressed in gram-equivalent per litre. The term *electrophoretic effect* refers to the retardation in the migration of an ion due to its motion against a counter-flow of solvated ionic atmosphere. The relaxation effect refers to the time lag (relaxation time) in the formation of a new ionic atmosphere surrounding the migrating ion replacing the old one. The migration of the ion is retarded by both these effects.

16.3.2 Applications of Conductance Measurements

Applications of conductance measurements are based on *Kohlrausch's law of independent conductance* (or *migration*) *of ions*. According to the law *at infinite dilution, each ion makes a definite contribution to the equivalent conductance of the electrolyte whatever be the nature of the other ion of the electrolyte*. The law is expressed as

$$\Lambda_0 = \lambda_0^{+} + \lambda_0^{-} \tag{16.7}$$

where λ_0^+ and λ_0^- refer to the *ionic conductance* of the cation and anion, respectively, at infinite dilution.

Conductance measurements find use for (i) calculation of the equivalent conductance at infinite dilution (Λ_0), (ii) determination of degree of dissociation of weak electrolytes, (iii) solubility of sparingly soluble salts, (iv) detection of the end points in volumetric titrations, (iv) checking the purity of water for process industries, (v) monitoring pollution of surface waters, (vi) monitoring acidic gases such as SO_2 in ambient air, (vii) determination of moisture content in soil, (viii) monitoring acid strength in a variety of processes, for example, in metal industries (pickling, caustic degreasing, and anodizing), food processing and preservation, pharmaceutical industries, and (ix) detecting the eluted sample components in chromatographic separations.

16.4 CONDUCTANCE TITRATIONS

An important analytical application of conductance measurement is the detection of the end point in volumetric titrations. The principle involves the measurement of conductance of the reaction mixture during the course of a titration and detecting the end point of the titration from a graphical plot of conductance versus volume of titrant. In practice, a known volume of the analyte solution is taken in the conductance cell and its conductance is measured as a function of incremental additions of the titrant from a burette. Figure 16.2 shows the experimental set-up for carrying out the conductance titration.

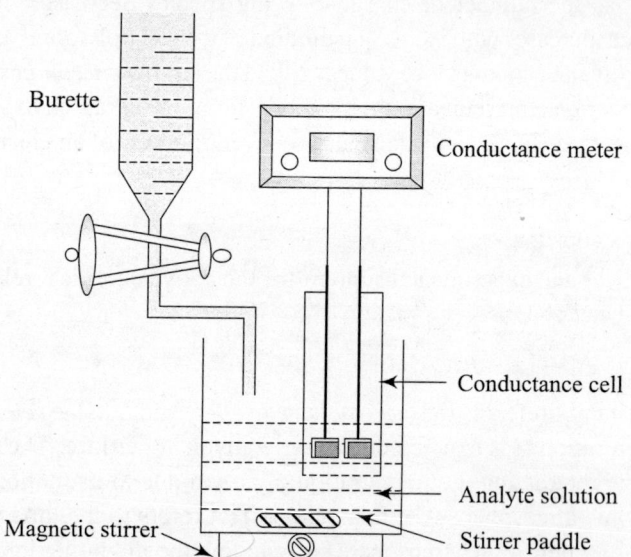

Fig. 16.2 Experimental set-up for conductance titrations

The conductance of the reaction mixture is measured before and also after the end point so as to give sufficient number of data points for the graphical plot. It is not necessary to measure absolute values of conductance as the relative changes in conductance as the reaction progresses are sufficient for preparing a graphical plot. The titration curve consists of two intersecting straight lines, one before the end point and the other after the end point. The end point of the titration is determined by dropping a perpendicular from the intersection point of the straight lines to the x-axis. It is advisable that the titrant is at least 10 times stronger than the analyte in order to minimize the dilution factor.

Conductance titrations have a few advantages in that coloured and even turbid solutions can be handled unlike in volumetric titrations using indicators which require colourless and clear solutions. The level of precision is better than 1% in conductance titrations, and hence, more reliable. Conductance titrations can be carried out both in aqueous as well as in non-aqueous media. However, the disadvantage is that conductance is a non-specific property and high concentrations of other electrolytes cause interference.

The different reactions that can be investigated by conductance titrations include (i) acid–base or neutralization reactions, (ii) displacement reactions, (iii) precipitation reactions, and (iv) complex-formation reactions.

16.4.1 Acid–Base Reactions

Acid–base reactions essentially involve the neutralization of H^+ ions (acid) with OH^- ions (base) to give unionized water.

$$H^+ + OH^- \rightarrow H_2O$$

The ionic conductance values of H^+ and OH^- ions are quite large, whereas that of unionized water is negligibly small. Hence, during the neutralization of an acid with a base, the conductance of the reaction mixture decreases rapidly and linearly till all the acid has been neutralized during a titration and reaches a minimum value. After the end point, the added OH^- ions contribute to the conductance, and hence, there will be a linear increase in the conductance of the reaction mixture. The conductance of the reaction mixture increases somewhat less steeply after the end point as the ionic conductivity of OH^- ions is only half that of H^+. The graphical plot of the measured conductance as a function of the incremental addition of volume of titrant shows two intersecting straight lines and the end point of the titration is determined from the graph.

Titrations involving acid–base reactions may be sub-classified into reactions involving (i) strong acid versus strong base, a typical example being the titration of HCl against NaOH, (ii) strong acid versus weak base (e.g., HCl. vs. ammonia), (iii) weak acid versus strong base (e.g., acetic acid vs. NaOH), (iv) weak acid versus weak base (e.g., acetic acid vs. ammonia), and (v) mixture of strong and weak acids (e.g., $HCl + CH_3COOH$) versus strong base (NaOH). The conductance titration curves for the five types are shown in Fig. 16.3(a–e).

In the titrations involving a strong acid versus a strong base [Fig. 16.3(a)] and a strong acid versus a weak base [Fig. 16.3(b)], the end points can be detected accurately. In the case of weak acid versus strong base [Fig. 16.3(c)], the conductance decreases initially due to the neutralization of H^+ ions but as the titration progresses the conductance increases due to the sodium acetate formed which is completely dissociated. However, sodium acetate undergoes hydrolysis and the conductance of the reaction mixture shows a rounding off at the end point and detection of the end point is difficult particularly at low concentrations of the weak acid. A sharp end point can be obtained only when the concentration of the weak acid taken is greater than 0.005 N. A similar behaviour is observed in the titration of a weak acid versus a weak base and it is necessary to have an initial concentration of the weak acid of 0.005 N to get sharp end point [Fig. 16.3(d)].

The mixture of a strong acid and a weak acid can be titrated with a strong base, provided the dissociation constants of the two acids differ by a factor at least 10^4. The titration gives two end points, the first end point corresponding to the complete neutralization of the strong acid [Fig. 16.3(e)]. The weak acid does not participate in the reaction till all the strong acid has been completely neutralized by the added base. The titration curve shows a steep fall in the conductance till the first end point. After the first end point, the conductance shows a slight increase due to the formation of the salt of the weak acid with the strong base till the second end point due to complete neutralization of the weak acid. Further incremental addition of the strong base after the second end point increases the conductance steeply.

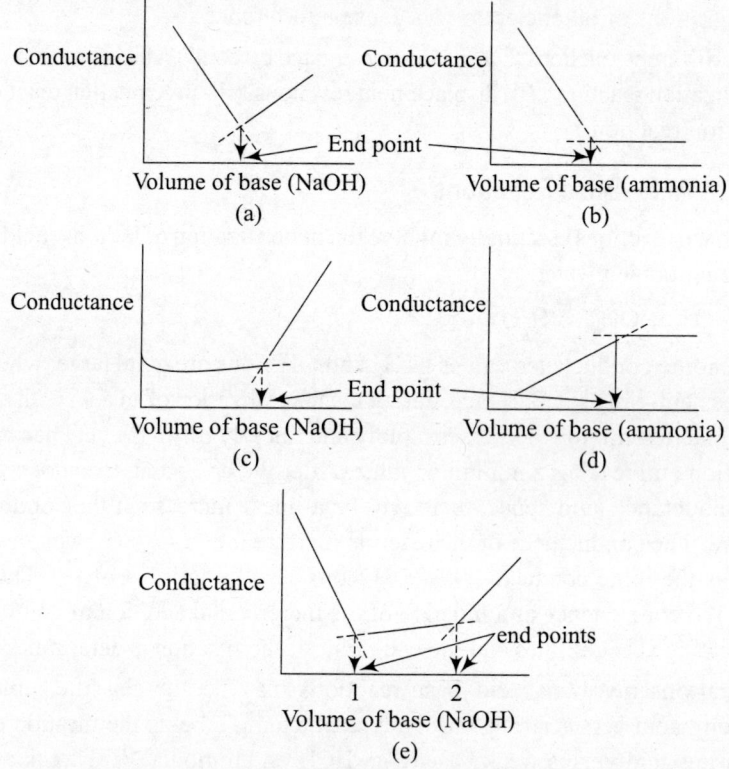

Fig. 16.3 Graphical plots for acid–base titrations by conductometry: (a) HCl versus NaOH (b) HCl versus ammonia (c) acetic acid versus NaOH (d) acetic acid versus ammonia and (e) mixture of HCl and acetic acid versus NaOH

The dicarboxylic acid, oxalic acid, can be considered as a mixture of equivalent amounts of strong and weak acids as the dissociation constants of the two carboxyl groups by about 10^4 (pK_1 and pK_2 are 1.2 and 4.3, respectively). Hence, conductance titration of oxalic acid against a standard solution of sodium hydroxide yields two end points similar to that in Fig.16.3(e).

16.4.2 Displacement Titrations

Displacement titrations are possible because the salt of a weak acid behaves as a Brønsted base, and hence, can be titrated against a strong acid. Thus, the salt of the weak acid, sodium acetate, can be titrated against hydrochloric acid, the strong acid, displacing the weak acid from the salt.

$$NaOAc + HCl \rightarrow HOAc + Na^+ + Cl^-$$

The conductance increases to a small extent till the end point due to greater conductance of chloride ions as compared to the acetate ions and after the end point the excess H^+ of the strong acid added causes a sharp increase in the conductance [Fig. 16.4(a)].

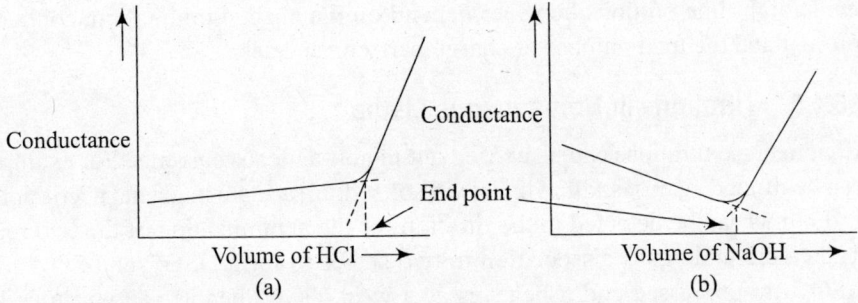

Fig. 16.4 Conductance titration curves for displacement reactions: (a) sodium acetate versus hydrochloric acid and (b) ammonium chloride versus sodium hydroxide

The salt of a weak base (e.g., ammonium chloride) can be titrated against a strong base such as sodium hydroxide to liberate the free base in a similar displacement titration [Fig. 16.4(b)].

16.4.3 Precipitation Titrations

Argentometric titrations in which silver nitrate is titrated against halides (except fluoride) are typical examples of precipitation titrations in which the end points can be determined graphically by conductance measurements. Sharp end points of the titrations can be obtained if the solubility product constant of the precipitate is low and when appropriate concentrations of the salts are employed. The titration curve will round off at the end point if the precipitate is relatively more soluble.

The titration curve for the precipitation reaction of chloride ions titrated against silver nitrate solution is shown in Fig. 16.5. The conductance remains almost constant till the end point as the ionic conductivities of Ag^+ and Na^+ are similar. After the end point, the increase in the conductance is due to the added silver nitrate. A similar titration curve is obtained for the precipitation of sulphate ions with barium hydroxide.

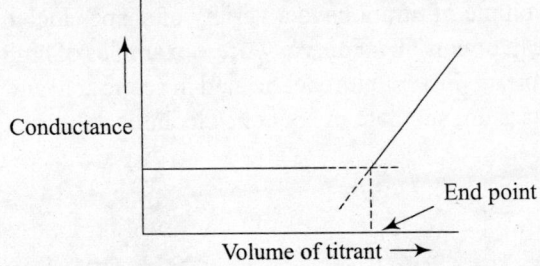

Fig. 16.5 Conductance titration curve for precipitation reactions (e.g., NaCl vs $AgNO_3$)

16.4.4 Complex-formation Reactions

Conductance titration involving complex formation reactions can be performed only when a stable complex is formed. A typical example of complex formation reaction is that of mercury with cyanide.

$$Hg(NO_3)_2 + 2KCN \rightarrow Hg(CN)_2 + 2K^+ + 2\,NO_3^-$$

Initially during the titration, conductance of the reaction mixture varies only slightly as one Hg^{2+} is replaced by $2K^+$ ions. After the end point, addition of potassium cyanide causes the conductance to increase, the titration curve resembling that shown in Fig. 16.5. The slopes of the straight line portions however depend on the change in the ionic conductance of the ions present and the total number of charge-carrying species.

16.4.5 Titrations in Non-aqueous Media

Conductance titrations can be carried out in non-aqueous solvents. For example, sulphuric acid can be titrated against sodium hydroxide or sodium acetate in glacial acetic acid as solvent. Two end points can be detected in the titration for the neutralization of the two replaceable protons as the acid undergoes dissociation in two stages to form HSO_4^- and SO_4^{2-}. The ionization of HSO_4^- is suppressed and it behaves as a weak acid. Phenols can be titrated against bases in non-aqueous solvents of low dielectric constant such as toluene or pyridine. A sharp end point is detected when 2,3,5-trimethylphenol, for example, is titrated against tetrabutyl ammonium hydroxide in toluene or sodium isopropoxide in pyridine.

16.5 OSCILLOMETRIC OR HIGH-FREQUENCY TITRATIONS

Oscillometry uses alternating current in the high-frequency range of 10^6–10^7 Hz as the probe to measure the signal generated by a combination of effects such as the resistance and capacitance of the both the analyte solution and the titration cell (e.g., a glass beaker). The capacitance effect is of importance in solutions of high dielectric constant but of low conductivity, in which molecules absorb and return energy in each frequency cycle owing to the induced polarization and alignment of electrically unsymmetrical molecules. However, if the low dielectric medium contains salts, conductance of ions also contributes to the signal generated. The electrodes are placed in contact with outer walls of the glass vessel containing the analyte solution and connected to an oscillator circuit which supplies the high-frequency alternating current. Since the electrodes are not in contact with the solution no fouling of the electrodes occurs and even corrosive liquids can be handled in the titration cell.

The titration curves obtained by plotting the measured signal for incremental addition of titrant against volume of titrant have a variety of shapes including V-shape, inverted V-shape, intersecting linear, or non-linear curves. A few examples of high-frequency titrations include the titration of glycine against trichloroacetic acid in acetic acid and precipitation titration of barium acetate with potassium sulphate in aqueous alcoholic solution [Figs 16.6(a) and (b)].

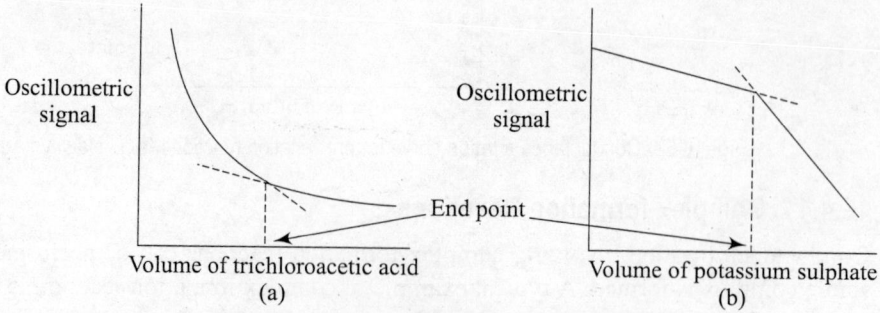

Fig. 16.6 Oscillometric titrations: (a) glycine versus trichloroacetic acid and (b) barium acetate versus potassium sulphate

16.6 PRINCIPLES OF ELECTROGRAVIMETRY AND COULOMETRY

Electrogravimetry and coulometry are typical electroanalytical techniques based on Faraday's laws on electrolysis which established the relationship between the quantity of electricity passed through an aqueous solution of an electrolyte and the amount of chemical substance deposited or dissolved during the electrolysis. Faraday's first law of electrolysis states that *the amount of a substance deposited or dissolved as a result of passage of an electric current is proportional to the quantity of electricity passed*. The law is expressed as

$$w = eit \tag{16.8}$$

where w is the weight (in grams) of the substance deposited or dissolved on the passage of current i (in amperes) for time t (in seconds). The product $i.t$ is expressed as ampere-second (A-s) or as Coulomb (C); and e is a constant called the *electrochemical equivalent* (ECE). The quantity ECE is defined as the weight of the substance deposited or dissolved when 1 A of current is passed for 1 s (i.e., 1 C of electricity or 1 A-s or 3.0×10^9 esu).

Faraday's second law states that *when the same quantity of electricity is passed through different electrolytes, the amounts of different substances deposited or dissolved, are proportional to their chemical equivalent weights* and is given as

$$\frac{w_1}{w_2} = \frac{E_1}{E_2} = \frac{e_1}{e_2} \tag{16.9}$$

where w_1 and w_2 are the amounts deposited or dissolved of the two substances having chemical equivalent weights of E_1 and E_2 on the passage of q coulombs of electricity.

As 1 C of electricity is required to deposit or dissolve one ECE, the quantity of electricity required for 1 g equivalent weight is E/e coulombs or F coulombs ($F = E/e$). F is called the *Faraday* whose value has been found to be 96,450 C approximated to 96,500 C or 26.8 A-h.

The two laws of electrolysis may be combined to give

$$w = i.t\, E/F \tag{16.10}$$

Passage of 1 F of electricity through an electrolyte results in the deposition or dissolution of 1 g equivalent weight (when $i.t = F$, $w = E$).

16.7 ELECTROGRAVIMETRY

Electrogravimetry is one of the oldest and simple electroanalytical techniques based on the principle of plating a metal on an electrode by electrolysis under controlled potential conditions and weighing the deposit. The experimental conditions are maintained such that a smooth and adherent deposit is formed in a short time.

The experimental set-up for electrogravimetry is shown in Fig. 16.7. It consists of an electrolysis cell provided with a platinum cathode of large surface area (gauze form of electrode) and a platinum anode which is relatively small connected to a dc battery through a variable resistance to

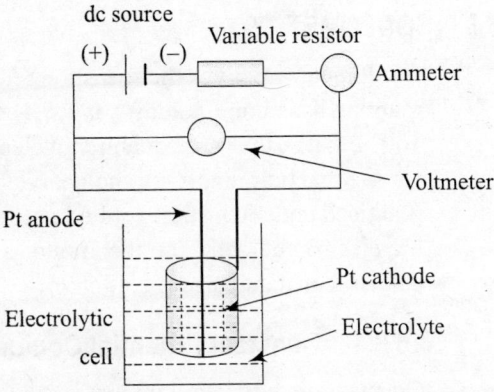

Fig. 16.7 Experimental set-up for electrogravimetry

control or regulate the applied voltage and a milliammeter to measure the current flow. The electrolyte solution consists of the analyte metal of a known volume, buffered to maintain an appropriate pH. The solution also contains complexing agents. Complex species exhibit *throwing power*, the ability to facilitate uniform, smooth, and adherent deposition of the analyte metal on the cathode and avoid tree-like coarse deposits. During electrolysis, the electrolyte solution is heated and stirred to facilitate quicker and uniform deposition of the metal.

Electrogravimetry can be carried out using the above experimental set-up under (i) constant applied potential or (ii) constant cell current. In the *constant applied potential* procedure, the platinum cathode is weighed, immersed in the electrolyte solution, and electrolysis is carried out at a constant cathode potential. As electrolysis progresses and the metal deposits on the cathode, the current flowing in the circuit decreases gradually and when all the metal has deposited, the flow of current becomes zero as indicated by the ammeter reading. The cathode is disconnected, rinsed with water, dried at room temperature, and then weighed to determine the amount of metal deposited from the known volume of the analyte solution. In the *constant current technique* the applied potential is gradually increased to maintain a constant current throughout the electrolysis.

The constant applied potential technique is more widely used as selective deposition of metals from multi-component mixtures is achieved by using a potentiostat, a device which automatically monitors and maintains a cathode potential at a predetermined value. The chosen cathode potential is such that only the metal of interest is deposited without interference from any other metal in the analyte mixture. For example, in an analyte mixture containing copper and nickel, the electrolysis is carried out first at a constant applied potential in the range 2–3 V on a platinum gauze cathode of known weight. The electrolyte is an aqueous solution of a mixture of sulphuric acid and nitric acid containing the analyte. Electrolysis is carried out till the complete deposition of copper on the Pt cathode as indicated by the current decreasing to zero. The cathode is removed and rinsed with water followed by acetone. It is then dried in air and weighed to determine the amount of copper deposited. The electrolyte solution is rendered ammoniacal by the addition of aqueous ammonia. Electrolysis is carried out at an applied potential in the range of 3–4 V on the copper-deposited platinum cathode till the complete deposition of nickel as indicated by the ammeter reading falling to zero. The electrode is removed from the electrolytic cell, rinsed with water followed by acetone, dried and weighed to determine the amount of nickel deposited.

16.8 COULOMETRY

Coulometry involves the measurement of the quantity of electricity used in an electrochemical reaction based on Faraday's laws. The basic requirement is that the current efficiency must be 100%, that is, the entire quantity of electricity passing through the electrolytic cell must be utilized for the reaction involving the analyte species either directly or indirectly. Coulometry may be practiced under constant applied potential conditions or under constant current flowing through the electrolytic cell. The techniques are appropriately called (i) constant potential coulometry and (ii) constant current coulometry.

16.8.1 Constant Potential Coulometry

Constant potential coulometry is similar to constant potential electrogravimetry. The technique makes use of an electrolytic cell with a working electrode (cathode) whose potential is controlled with the help of a potentiostat. Mercury is mostly used as the cathode because many metals and organic compounds can be reduced. The current passing through the electrolytic cell is continuously changing, and hence, it is necessary to introduce an integrating device such as a chemical coulometer or an electronic integrator to determine the quantity of electricity. The chemical coulometer consists of an electrolytic cell in which the total amount of products liberated at the electrodes can be readily measured. For example, the quantity of electricity used can be calculated from the measured amounts of hydrogen and oxygen liberated into gas burettes on the electrolysis of a standard sodium sulphate solution in the chemical coulometer. Examples of analytical determinations include the determination of nickel in the presence of cobalt and determination of trichloroacetic acid in the presence of mono- and di-chloro derivatives.

16.8.2 Constant Current Coulometry

Constant current coulometry or *amperostatic coulometry* is a relatively more versatile technique and is widely used for quantitative analysis. The technique involves electrolysis at constant current till the analytical reaction is complete as indicated by a suitable indicator as in volumetric titrations, and hence, the method is also called *coulometric titration.*

Coulometric titration is an indirect analysis using constant current coulometry. Although the technique is termed as titration, the titrant is not added as such from a burette but generated electrolytically which then reacts stoichiometrically with the analyte. The fundamental requirement in coulometric titration as in constant potential coulometry is that the analyte species should interact with 100% current efficiency, that is, 1 F of electricity brings about a change in the analyte corresponding to 1 mol of electrons. The titration technique involves the electrolysis of a solution of the reagent till the reaction with the analyte present in the same solution is completed. The completion of the reaction is detected by the use of (i) a visual indicator or (ii) instruments such as a pH meter, potentiometer, or spectrophotometer. The amount of reagent generated is evaluated from knowledge of the current passing through the circuit and the generating time. The total quantity of electricity passed (number of coulombs) is derived from the product of current (A) and time (s). The number of moles of electrogenerated titrant is obtained from the number of coulombs of electricity passed since 96,500 C (1 F) will generate 1 g equivalent of the substance. The technique is highly sensitive and it is possible to determine the amounts of analyte at extremely low concentrations. The technique has also the advantages of high degree of precision and accuracy, absence of interferences form other chemical species, non-requirement of standardized solutions, capability to handle even unstable reagents such as chlorine, bromine, etc., and adaptability for quantitative analysis of substances which do not react at an electrode.

The experimental set-up consists of a battery, an accurate timer, and a titration cell as shown in Fig. 16.8. The coulometric titration cell consists of a glass vessel in which a generator electrode to generate the reagent and an auxiliary or second electrode are placed. The generator electrode is usually a platinum strip or wire gauze with a large surface area. If the end point is to be detected by potentiometry or amperometry, a set of indicator and reference (SCE) electrodes are also placed suitably without interfering with the current path between the generator and auxiliary electrodes. The second electrode is isolated by placing it in a glass tube closed with a sintered glass septum to avoid any undesirable effects due to reactions at this electrode. For example, in the anodic generation of an oxidizing agent at the generating electrode, hydrogen gas will evolve from the cathode (second electrode) which must be allowed to escape. The electrolyte solution is stirred and a stream of inert gas (nitrogen) is bubbled through the solution.

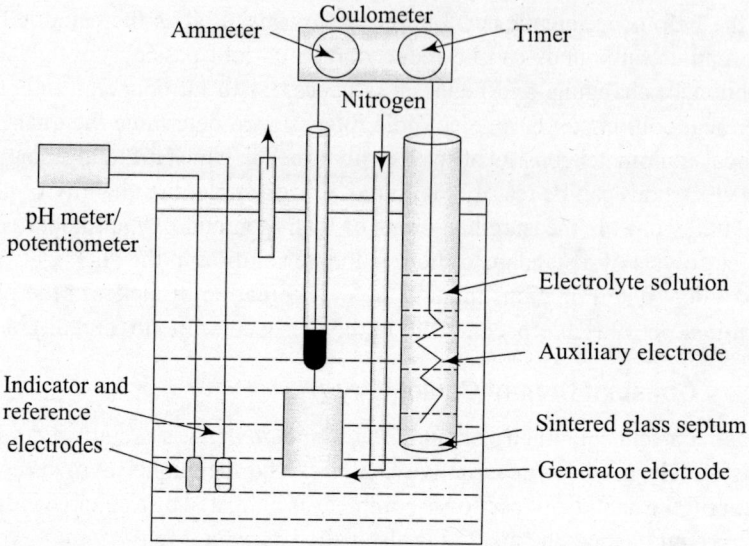

Fig. 16.8 Coulometric titration cell

In practice, the electrolyte solution from which the titrant is generated electrolytically and the analyte solution are taken in the titration cell and constant current is supplied for specified time interval. The progress of the titration is followed periodically by the indicator electrodes connected to a suitable device (pH-meter, potentiometer, etc.). The electrolysis current is usually switched off when the readings of the indicator device are recorded. The end point is detected graphically from a plot of the readings of the indicator device as a function of the quantity of electricity passed. A first derivative plot of the readings gives the location of the end point more accurately.

Coulometric titrations may be used for oxidation–reduction, precipitation, complexation, and acid–base reactions.

Oxidation–reduction reactions

A typical example is the determination of iron(II) by titrating with the oxidant cerium(IV) generated electrolytically from a large excess of cerium(III) solution taken in the titration cell. The titration cell also contains the analyte iron(II) and 1 M sulphuric acid. Cerium(IV) ions are generated in exact stoichiometric amounts at the platinum generator electrode as determined by Faraday's law and the oxidation of iron(II) to iron(III) by the generated cerium(IV) ions proceeds with 100% efficiency. The end point is detected potentiometrically using a platinum indicator electrode in combination with SCE.

A large number of oxidation–reduction titrations may be carried out by constant current coulometry. These include

(i) determination of Ti(III), As(III), and U(IV) with electrolytically generated Ce(IV) from $Ce_2(SO_4)_3$ in 1 M sulphuric acid, the end point being detected potentiometrically,

(ii) titration of As(III), Sb(III), thiosulphate, and sulphide ions and ascorbic acid with electrolytically generated iodine in 0.1 M KI buffered to pH 8 with phosphate buffer, the end point may be detected by using starch indicator or by potentiometric or amperometric methods,

(iii) titration of reducing agents such as Sb(III), U(IV), Tl(I), I^-, SCN^-, NH_3, NH_2OH, and N_2H_4 with electrolytically generated bromine from 0.2 M KBr, the end point being detected by amperometry,

(iv) titration of As(III) and I^- by electrolytically generated chlorine from 2 M HCl using amperometric detector,

(v) determination of V(V), Cr(VI), and MnO_4^- by titrating with Fe(III) generated from 0.3 M ferrous ammonium sulphate in 2 M sulphuric acid and detecting the end point with a potentiometer,

(vi) determination of V(V), Cr(VI), IO_3^-, and bromine with Cu(I) generated from 0.1 M copper sulphate using a potentiometer to detect the end point,

(vii) titrating with Ti(III) generated from 0.6 M titanium(IV) sulphate in 6 M sulphuric acid to determine analytes such as Fe(III), Ce(IV), V(V), and U(VI),

(viii) generating electrolytically Mn(III) from 0.5 M $MnSO_4$ in 2 M sulphuric acid to determine Fe(III), As(III), and oxalic acid using potentiometric detection of the end point,

(ix) titrating As(III), Ce(III), and V(IV) with Ag(II) generated electrolytically from 0.1 M silver nitrate in 5 M nitric acid and detecting the end point potentiometrically.

Precipitation reactions

The titration of chloride, bromide, or iodide by electrolytically generated silver ions is a typical example of a precipitation titration. The coulometric cell (Fig. 16.8) is provided with a silver anode instead of a platinum-generating electrode. The analyte halide is dissolved in 0.5 M potassium nitrate electrolyte and placed in the titration cell. The Ag(I) ions generated by the silver anode react quantitatively with halide ions. The end point of the titration is detected by a potentiometric set-up.

Other examples of precipitation titrations include

(i) determination of chloride or bromide by titrating Hg(I) generated from a mercury anode in 0.5 M perchloric acid and determining the end point by potentiometry,

(ii) determination of zinc by titrating with ferrocyanide ions ($[Fe(CN)_6]^{4-}$) generated by electrolysing 0.2 M potassium ferrocyanide ($K_3Fe(CN)_6$) using the platinum anode. The end point is detected potentiometrically.

Complexation reactions

EDTA may be generated by electrolysing the mercury complex ($[Hg(NH_3)Y]^{2-}$) in 0.1 M ammonium nitrate solution buffered to pH 8.3 with Pt electrodes. The complex $[Hg(NH_3)Y]^{2-}$ is prepared by dissolving mercury(II) nitrate and disodium salt of EDTA in distilled water to get a stock solution of the 1:1 complex of 0.1 M. The stock solution is mixed with 0.1 M ammonium nitrate in 1:3 ratio and the pH of the mixture is adjusted to 8.3 with concentrated ammonia solution. Electrolytically generated EDTA can be titrated against Cu(II), Zn(II), Ca(II), or Pb(II) taken in the titration cell. The end point is detected potentiometrically.

Acid–base titrations

Neutralization reactions can be carried out by generating either H^+ ions or OH^- ions electrolytically using 0.2 M sodium sulphate as the electrolyte and platinum electrodes. Since in acid–base titrations the electrolyte should contain only H^+ or OH^- ions and not any other ions to prevent reduction of such contaminating species at the electrodes, coulometric titration is carried out with an *external generator cell*. The electrolytic cell for external generation of the titrant consists of double arm inverted U-shaped cell (Fig. 16.9) provided with Pt spiral electrodes near the centre of the tube. Glass wool is placed between the electrodes to prevent any turbulent mixing of the ions generated at the electrodes. The electrolyte solution (0.2 M sodium sulphate) for generating H^+ and OH^- ions is fed continuously from the top of the cell. The electrolyte solution flows through the T-joint to the two delivery arms in equal quantities. The electrolysis reactions at the Pt electrodes and the generation of H^+ and OH^- ions may be represented as

$$\text{At Pt cathode, } 2\,H_2O + 2e^- \rightleftarrows H_2 + 2OH^-$$

and

$$\text{At Pt anode, } 2\,H_2O \rightleftarrows O_2 + 4H^+ + 4\,e^-$$

The appropriate delivery arm is connected to the titration cell and the other arm to the drain. Thus, for the titration of acid in the titration cell, the delivery arm containing the Pt cathode is placed suitably to facilitate the flow of OH^- ions into the titration cell.

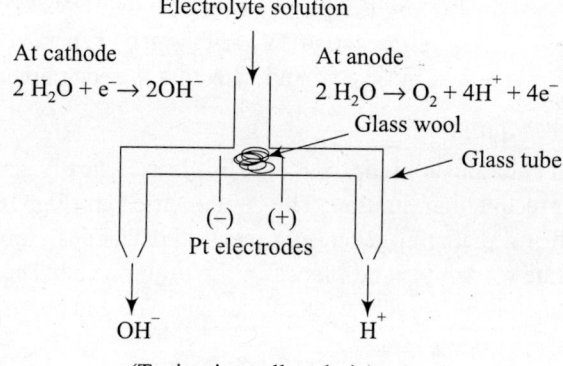

Fig. 16.9 Device for external generation of H^+ and OH^-

External generation of the titrant has the advantage of eliminating any unwanted reactions and is amenable for automation. However, a disadvantage is that the contents of the titration cell get diluted, and hence, it is necessary to regulate the concentration and flow rate of the generator solution.

16.9 POTENTIOMETRY

The principle involved in the use of potentiometry as an analytical tool is based on setting up an electrochemical cell and measuring the electromotive force (emf) of the cell. Hence, it is necessary to understand the working of an electrochemical cell and measurement of emf of the cell. An electrochemical cell converts the chemical energy into electrical energy. In reverse direction, an electrochemical cell transforms the applied electrical energy into chemical energy. It consists of two half-cells or single electrodes, one of which is called the *indicator electrode* and the other the *reference electrode*. At each of the electrodes, reduction or oxidation (red-ox) reactions occur. The indicator and reference electrodes are immersed in appropriate electrolyte solutions kept separately in two separate beakers. The electrical connection between the two electrolyte solutions is made by a salt bridge as shown in Fig. 16.10. The electrodes generate a potential in the cell called the *electromotive force* (emf) represented as E_{cell} and expressed in volts (V). The voltage generated in the cell can be measured by a device called potentiometer. Potentiometer is an accurate voltmeter which measures the E_{cell} under conditions of zero current flow in the circuit.

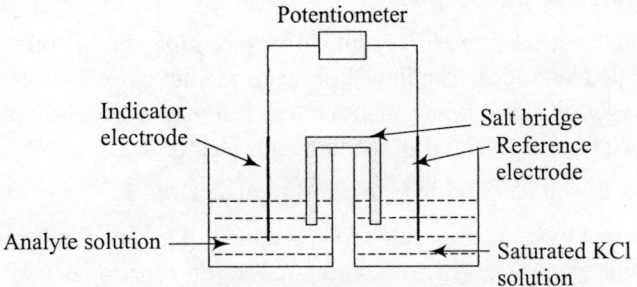

Fig. 16.10 Schematic diagram of an electrochemical cell for measuring the E_{cell}

16.9.1 Thermodynamic Significance of Electrode Potentials

Thermodynamic principles of reversibility govern the energy relationships in electrochemical cells. The driving and opposing forces of the electrochemical reactions of oxidation and reduction are different from each other only by an infinitesimally small value in an electrochemical cell and any change taking place can be reversed by applying a force infinitesimally greater than the one acting. All electrochemical cells are reversible in thermodynamic sense when the current passed or drawn is extremely small. The free energy change ΔG accompanying an electrochemical process at constant temperature and pressure is equal to the reversible work. The electrical work done by a reversible cell having an emf of E_{cell} V on the passage of $n\,F$ coulombs, where n represents the number of electrons involved in the electrode process is given by the product of the emf and the total charge as

$$\text{Electrical work done} = nFE_{cell} \text{ J (or V-C)} \qquad (16.11)$$

The free energy change of the system is given as

$$\Delta G = -nFE_{cell} \qquad (16.12)$$

For a spontaneous process, E should be positive and the change in free energy negative. When the reactants and products are in their standard states where the concentrations are exactly 1 mol each, the change in free energy is called *standard free energy change*, $\Delta G°$ and the cell potential as *standard cell potential*, $E°_{cell}$. The standard cell potential is related to the equilibrium constant for the reaction K as given by the following equation

$$\Delta G° = -nFE°_{cell} = -RT \ln K \qquad (16.13)$$

where R is the gas constant and T the absolute temperature. When the reactants and products are at not in their standard states, the cell potential is given by the following equation known as Nernst equation

$$E_{cell} = E°_{cell} - RT \ln K \qquad (16.14)$$

Substituting the values $R = 8.314 \text{ JK}^{-1}\text{mol}^{-1}$, $F = 96,500 \text{ C mol}^{-1}$, and $T = 298$ K (25°C) and converting the natural logarithm to logarithm to the base 10, Eq. (16.14) may be written as

$$E_{cell} = E°_{cell} - (2.303 \times 8.314 \times 298)/(n \times 96500) \log K \qquad (16.15a)$$

or

$$E_{cell} = E°_{cell} - (0.0591/n) \log ([\text{products}]/\text{reactants}]) \text{ at } 25°C \qquad (16.15b)$$

The standard cell potential $E°_{cell}$ is a constant for a given cell at a constant pressure of 1 atm and varies only with temperature.

16.9.2 Indicator Electrodes

The indicator electrode used for analytical purposes may be any one of the different types of half-cells or single electrodes. The indicator electrode develops a potential or emf called the *single electrode potential E* depending on the concentration to which the indicator electrode is reversible. The single electrode potential of the indicator electrode is given by Nernst equation

$$E = E° - 0.0591/n \log ([\text{product}]/[\text{reactant}]) \text{ at } 25°C \tag{16.16}$$

The different types of single electrodes include (i) metal–metal ion electrode, (ii) metal–metal insoluble salt electrode, (iii) oxidation–reduction electrode, (iv) metal amalgam electrode, and (v) gas electrode. The first three types of single electrodes are more commonly used for electroanalytical work.

The *metal–metal ion electrode* consists of a strip of metal in contact with its own ions, for example, a rod or wire of copper metal in contact with Cu^{2+} ions or zinc metal in contact with Zn^{2+} ions. The electrode may be represented as $M|M^{n+}$ and the red-ox reaction at the electrode as

$$M \rightleftarrows M^{n+} + ne^-$$

The corresponding single electrode potential is given by Nernst equation as

$$E = E° - (0.059/n) \log [M^{n+}]/[M] \tag{16.17}$$

Since the activity of the solid metal M is taken as unity, the equation simplifies to

$$E = E° - (0.059/n) \log [M^{n+}] \tag{16.18}$$

The metal–metal ion electrode is reversible to its own ions, and hence, the potential of the electrode depends on the concentration of the metal ions in solution.

The *metal–metal insoluble salt electrode* consists of a metal in contact with an insoluble (sparingly soluble) salt of the metal and a solution containing the ion present in the salt other than the metal, for example, silver metal in contact with silver chloride and chloride ions or mercury metal in contact with mercurous chloride and chloride ions. The *silver–silver chloride electrode* is represented as $Ag|AgCl|Cl^-$ and can be fabricated by dipping silver wire into a solution containing chloride ions resulting in the deposition of silver chloride on the silver wire. The electrode reaction may be written as

$$Ag(s) + Cl^- \rightleftarrows AgCl(s) + e^-$$

The *calomel electrode* is represented as $Hg|Hg_2Cl_2|Cl^-$ and the electrode reaction as

$$2 Hg (l) + 2 Cl^- \rightleftarrows Hg_2Cl_2 (s) + 2 e^-$$

The silver–silver chloride and the calomel electrodes are reversible to chloride ions, that is, the potentials of the electrodes depend only on the concentration of the chloride ions in solution in contact with the electrodes as given by the following equation

$$E = E° - 0.059 \log (1/[Cl^-]) \tag{16.19}$$

Both these electrodes are used as reference electrodes at saturated concentration of chloride ions as their potentials remain constant over long periods of time and are highly reproducible.

In the *oxidation–reduction electrode* (or red-ox electrode), an inert electronic conductor such as a platinum wire is present. This wire is in contact with a solution containing a substance in two different oxidation states. A typical red-ox electrode is the platinum electrode which is in contact with a solution containing ferrous and ferric ions is represented as $Pt|(Fe^{3+}, Fe^{2+})$. The electrode reaction is represented as

$$Fe^{3+} + e^- \rightleftarrows Fe^{2+}$$

The electrode potential depends on the ratio of the activities of the two ions as given by Eq. (16.20). The electrode is said to be reversible to the red-ox couple.

$$E = E° - 0.059 \log ([Fe^{2+}]/[Fe^{3+}]) \tag{16.20}$$

The *gas electrode* consists of a gas bubbling about a platinum electrode immersed in a solution containing ions to which the gas is reversible. A typical gas electrode is the hydrogen electrode consisting of hydrogen gas bubbling through a solution of H^+ ions in contact with a platinum electrode (Fig. 16.11). The platinum electrode consists of a platinum wire fused to a platinum foil coated with platinum black. The electrode facilitates the establishment of equilibrium between the hydrogen gas and H^+ ions and also provides the electrical contact. The hydrogen electrode is represented as $Pt|H_2|H^+$ and the reaction taking place at the electrode may be represented as

$$\tfrac{1}{2} H_2 (g, Pt) \rightleftarrows H^+(aq) + e^-$$

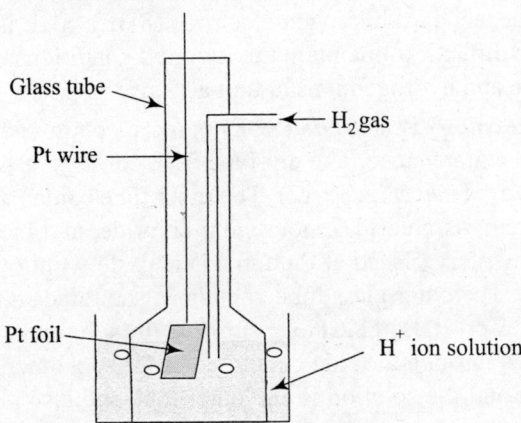

Fig. 16.11 Schematic diagram of a gas electrode (e.g., hydrogen electrode)

The electrode potential is determined both by the activity of hydrogen ions in solution and the pressure of hydrogen gas as given by Nernst equation

$$E = E° - 0.059 \log ([H^+]/[H_2]^{1/2}) \tag{16.21}$$

The hydrogen electrode is called the *standard hydrogen electrode* (SHE) when the hydrogen gas is flowing at exactly 1 atm pressure and the H^+ ion activity in the solution is exactly 1 M. The electrode potential of SHE ($E°$) is arbitrarily fixed as 0.0 V. Since the activity of hydrogen gas in its normal state of existence is unity, the equation simplifies to

$$E = 0 - 0.059 \log ([H^+] \text{ or } E = 0.0591 \text{ pH} \tag{16.22}$$

The *metal amalgam electrode* consists of platinum metal in contact with an amalgam of a reactive metal (e.g., sodium metal dissolved in mercury). The electrode reaction may be represented as

$$M(Hg) \rightleftarrows M^{n+} + n\,e^-$$

where M(Hg) represents the amalgam of the reactive metal and M^{n+} represents the ions of the reactive metal (e.g., sodium amalgam and sodium ions respectively). The electrode potential is given by

$$E = E°_M - (0.059/n) \log ([M^{n+}]/[M(Hg)]) \qquad (16.23)$$

where $E°_M$ is the standard electrode potential of pure metal electrode (e.g., pure sodium) and [M(Hg)] represents the activity of the metal in the amalgam which is not unity.

16.9.3 Reference Electrodes

The requisites for an electrode to be used as a reference electrode in electrochemical cells are as follows.

(i) The electrode should be reversible to oxidation–reduction reactions.

(ii) The electrode potential should remain constant for a long time and should be reproducible.

(iii) Change in its potential with change in temperature should be negligible.

(iv) It should be easily fabricated.

Reference electrodes for use in electrochemical cells include the calomel electrodes (saturated calomel, normal calomel, and decinormal calomel electrodes), silver–silver chloride (Ag/AgCl) electrode, and the standard hydrogen electrode (SHE). SHE is very rarely used as a reference electrode as it is difficult to maintain the required conditions of hydrogen gas flowing exactly at 1 atm pressure and hydrogen ions at unit activity.

The *saturated calomel electrode* (SCE) is more commonly used because of the ease of its construction and maintenance. The dip-type SCE consists of two concentric glass tubes sealed at the top with a rigid cap (Fig. 16.12). The inner tube contains mercury in contact with a paste of calomel (mercurous chloride), potassium chloride, and mercury. A platinum contact wire is immersed in mercury placed at the top of the paste within the inner tube. The inner tube is sealed at the top. The outer glass tube contains a saturated solution of potassium chloride and also contains a few crystals of KCl to ensure that the solution remains saturated at all times. The outer tube is filled through the side opening. Porous septa facilitate contacts between the paste in the inner tube and the solution in the outer tube and also between the solution in the outer tube and sample solution placed in a beaker into which the SCE is dipped. The emf of the SCE is 0.2444 V on the SHE scale at 25°C.

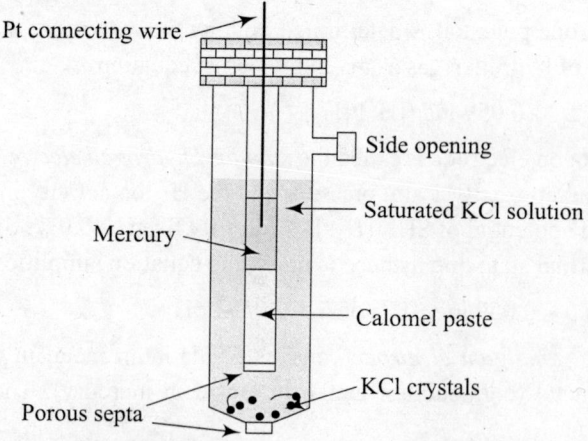

Fig. 16.12 Schematic diagram of saturated calomel electrode

The normal and the decinormal calomel electrodes can be constructed in a similar manner. However, the outer tube contains KCl solution of exactly 1.0 N for the *normal calomel electrode* (NCE) and 0.1 N for the *decinormal calomel electrode*. The electrode potentials are, respectively, 0.2824 and 0.3358 V.

The Ag/AgCl electrode can be fabricated simply by dipping a silver rod or wire in a saturated solution of KCl. The silver chloride precipitate forms a thin film on the silver wire constituting the Ag/AgCl electrode with a potential of 0.1989 V (vs SHE).

16.9.4 EMF Measurement

The emf of a single electrode can be determined by constructing an electrochemical cell in combination with a reference electrode so that oxidation reaction takes place at one of the electrodes and reduction reaction at the other. The emf of the electrochemical cell is the algebraic sum of the two single electrode potentials E_{ind} and E_{REF} constituting the cell and is given as

$$E_{cell} = E_{ind} + E_{REF} \tag{16.24}$$

Only the potential difference E_{cell} between the two electrodes can be measured experimentally by using a potentiometer. Once E_{cell} is determined, the electrode potential of the single electrode can be calculated since E_{REF} is known.

The E_{cell} is usually determined by Poggendorf compensation method using a potentiometer. The potentiometer makes use of a Wheatstone bridge circuit as shown in Fig. 16.13. The circuit consists of a platinum wire AB connected to a dc source (2–4 V). The potentiometer is calibrated with a *standard Weston cadmium cell* (SC) whose emf is accurately known and remains constant at 1.01864 V at 20°C. The emf of the Weston cadmium cell has a negligible (approximately 4×10^{-5} V deg^{-1}) temperature coefficient, and hence, almost independent of temperature at ambient temperature. In addition, the standard cell is a reversible cell and is not subjected to permanent damage due to the passage of current during the short interval of calibration.

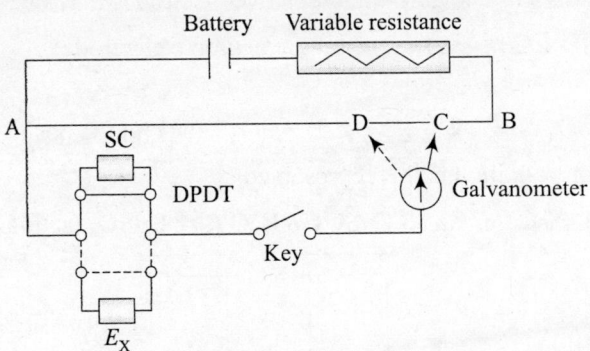

Fig. 16.13 Poggendorf method for measurement of emf

The Weston cell is connected to the end A of the potentiometer wire through its positive terminal, and the negative terminal is connected through a galvanometer to the sliding contact on the wire AB. The circuit is closed by including the SC in the circuit through the double pole-double throw (DPDT) switch and the key. The calibration procedure involves moving the sliding contact along the wire AB till the galvanometer shows zero deflection (zero current flow). The position of the sliding contact is denoted as C. The emf of the standard cell (E_{SC}) is proportional to the length AC as the potential drop across AC is exactly balanced by the E_{SC}.

The electrochemical cell (E_X) consisting of an indicator electrode and a reference electrode whose emf is to be determined is included in the circuit through the DPDT switch and the key while SC is excluded from the circuit. The sliding contact is moved along the wire AB to determine the position D when the galvanometer shows null deflection. The potential drop across AD is proportional to the emf of the unknown cell. The emf of the unknown cell E_X is calculated by the formula

$$E_X = (AD/AC)\, E_{SC} \tag{16.25}$$

The method is called compensation method because the voltage of the cell E_X is compensated by the voltage applied on the potentiometer wire. During the measurement of emf, negligible current is drawn either from the Weston cadmium cell or the unknown cell. The measurement becomes possible only when the positive terminal of the cell is connected to the end A of the potentiometer wire and the negative terminal to the sliding contact. If the terminals are connected wrongly, the balance point cannot be determined, and hence, the polarity of the electrodes is also indicated by the circuit.

16.9.5 Standard Weston Cadmium Cell

The Weston cadmium cell is an H-shaped glass vessel (Fig. 16.14). One of the arms of the vessel contains a short-sealed platinum wire in contact with pure mercury. The mercury is covered with a thick layer of a paste of mercury and mercurous sulphate constituting the active material for the positive electrode. The second arm contains a sealed platinum wire in contact with cadmium amalgam containing 12–14% of cadmium by weight constituting the active material for the negative electrode. Crystals of $CdSO_4.8/3\ H_2O$ are sprinkled over the electrode. Both the arms of the vessel are filled with a saturated solution of cadmium sulphate and the upper ends of the arms are closed with cork and sealed with wax.

The reactions at the negative and positive electrodes may be represented as

$$Cd(s) \rightarrow Cd^{2+}\ (aq) + 2\ e^-\ \text{(at the negative electrode)}$$

and

$$Hg_2SO_4\ (s) + 2\ e^- \rightarrow 2\ Hg\ (l) + SO_4^{2-}\ (aq)\ \text{(at the positive electrode)}$$

The Weston cadmium cell is represented as

$$\text{12–14\% Cd in Hg} \mid 3\ CdSO_4.\ 8\ H_2O\ (s) \parallel CdSO_4\ \text{(saturated solution)}\ Hg_2SO_4 \mid Hg$$

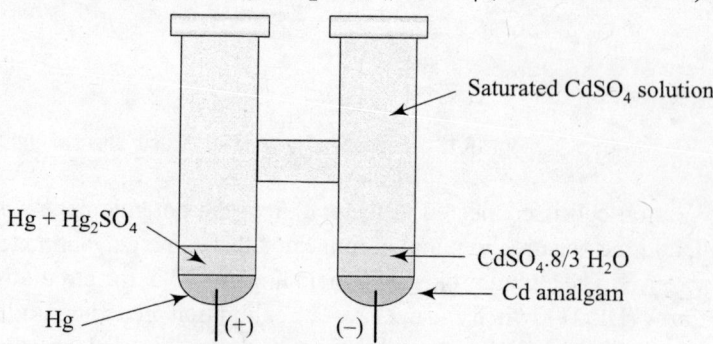

Fig. 16.14   Schematic diagram of standard Weston cadmium cell

16.10 APPLICATIONS OF EMF MEASUREMENTS

The important applications of emf measurements for electroanalytical work include (i) determination of pH, (ii) pH titrations, (iii) potentiometric titrations, and (iv) quantitative analysis by ion-selective electrodes. Other applications include calculation of thermodynamic functions of electrochemical reactions and determination of solubility of sparingly soluble salts.

16.10.1 Determination of pH by Glass Electrode

The pH of an aqueous solution is measured by using a *glass electrode* as an indicator electrode for H^+ ions. The potential of the glass electrode depends on the activity of H^+ ions in solution. The glass electrode is a typical example of an *ion selective electrode*. The glass electrode consists of an electrically conducting glass membrane in the form of a thin bulb fused to a hard glass tube. [Fig. 16.15(a)]. The glass membrane is made of lithium-based glass of composition SiO_2–63%, Li_2O–28%, Cs_2O–2%, BaO–4%, La_2O_3–3% and is mostly used for H^+ ion responsive glass electrode. The bulb contains hydrochloric acid solution of constant concentration (usually 0.1 M) and an internal reference electrode such as Ag–AgCl electrode or a platinum wire in a solution of constant pH. The electrode may be represented as Ag |AgCl(s), HCl (0.1 M) | glass.

Commercially glass electrode is also available as *combination electrode* which consists of the indicator glass electrode (glass membrane bulb) and Ag/AgCl electrode as a reference electrode as a single unit [Fig. 16.15(b)].

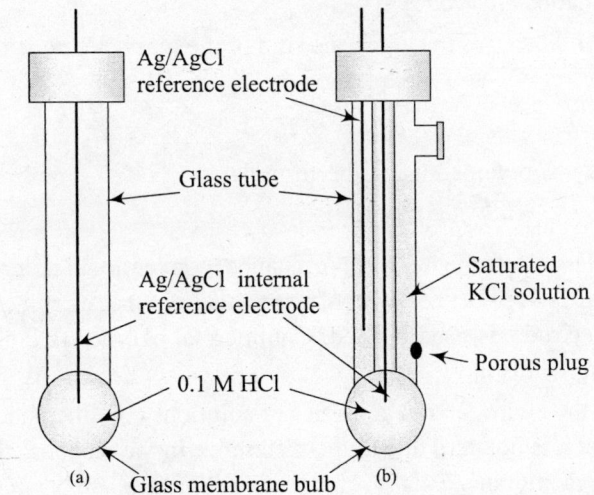

Fig. 16.15   (a) Single glass electrode and (b) combination electrode

The glass electrode develops a potential E_g at the surface of the glass membrane when it is dipped into an aqueous solution having a different pH as compared to that of the solution within the bulb. The potential is related to the transfer of H^+ ions through the glass membrane between solutions of hydrogen ion activities a_1 (external solution) and a_2 (in the internal solution which is a constant) and is given by

$$E_g = -0.059 \log (a_1/a_2) \qquad (16.26)$$

The single glass electrode is combined with a reference electrode (e.g., SCE) and the E_{cell} is measured (Fig. 16.16).

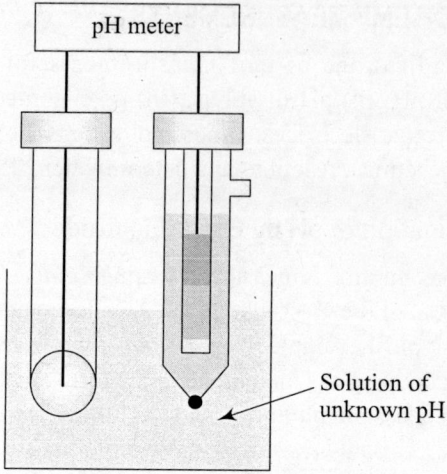

Fig. 16.16 Measurement of pH

The cell may be represented as

indicator electrode   reference electrode

Ag | AgCl(s), HCl (0.1 M) | glass | test solution ‖ (Cl⁻ (satd.) Hg_2Cl_2 | Hg (Or AgCl|Ag in the case of combination electrode)

The value of the E_{cell} is the algebraic sum of the glass electrode potential and the potential of the reference electrode and is expressed as

$$E_{cell} = E_{REF} + E_g - 0.059 \text{ pH at } 25°C \qquad (16.27)$$

and the pH may be obtained using the formula

$$pH = ((E_{REF} + E_g) - E_{cell})/0.059 \qquad (16.28)$$

Since the glass electrode has high resistance, conventional potentiometers cannot be used to determine the E_{cell}. A pH meter, which has a potentiometric circuit capable of measuring the emf of the glass electrode, is used instead. Commercial pH meters indicate the pH of the solution directly in a digital display.

Even when the hydrogen ion activities of solutions on either side of the glass membrane are the same, the emf is not zero due to the difference in the response to pH by the inner and outer surfaces of the membrane. This is referred to as the *asymmetry potential* of the glass electrode. The asymmetry potential changes with time, and hence, periodic calibration of the electrode is necessary for direct measurement of pH. The pH meter along with the electrodes is calibrated using standard buffer solutions before measuring the pH of test solutions. The most commonly used buffers include 0.05 M solution of potassium hydrogen phthalate (pH 4.01 at 27°C) and 0.05 M solution of borax (sodium tetraborate) with a pH of 9.13. The glass electrode is sensitive to ions such as Na^+ in addition to H^+, particularly at pH > 12.

16.10.2 pH Titrations

In the case of acid–base titrations, a pH-meter is used to monitor the pH during the progress of the titration. The experimental set-up for pH titrations is shown in Fig. 16.17. The indicator electrode for an acid–base titration will be a glass electrode. The glass electrode and the reference electrode (usually SCE) are dipped into the analyte solution. Alternatively, the combination electrode can be directly dipped into the analyte solution to be titrated with the two connecting wires of the electrodes connected to a digital pH meter as shown in the Fig. 16.17.

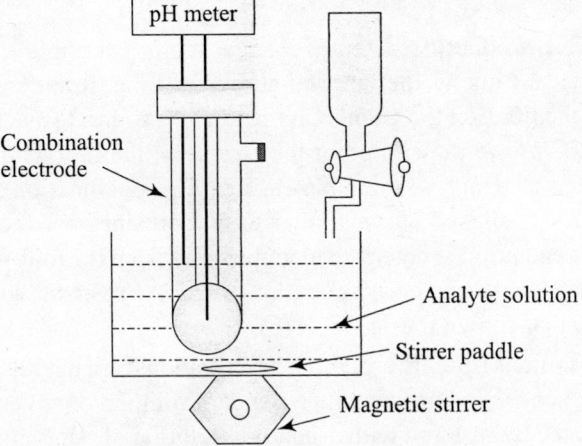

Fig. 16.17 Schematic diagram of pH titration set-up

The pH of the analyte solution placed in the titration cell is noted before the commencement of the titration. During the titration the pH of the solution is continuously monitored as a function of incremental addition of the titrant (base). The plot of pH versus volume of titrant is shown in Fig. 16.18(a). The end point of the titration is detected from the graph by determining the mid point of the vertical straight line portion of the curve. A derivative plot of $d(pH)/dV$ versus volume of the titrant is obtained by monitoring the pH for small (0.1 mL) incremental addition of the titrant closer to the end point in a second titration [Fig. 16.18(b)]. The derivative graph is useful for an accurate detection of the end point.

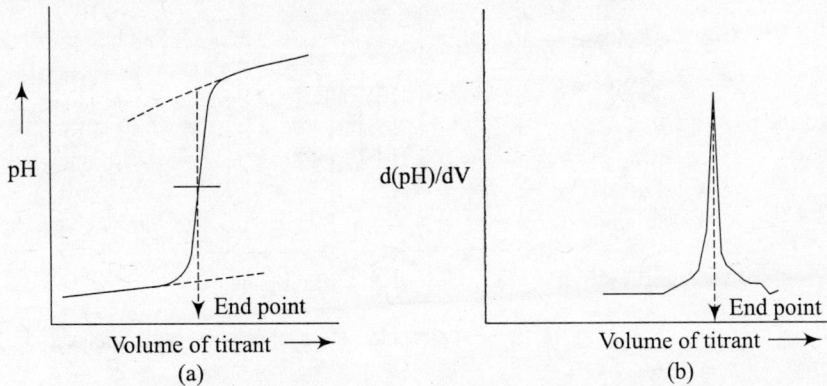

Fig. 16.18 (a) pH titration curve and (b) derivative plot to detect the end point accurately

16.10.3 Potentiometric Titrations

Potentiometric titrations are similar to the pH titrations and are useful for quantitative analysis of analyte undergoing precipitation or oxidation–reduction reactions. The technique involves the measurement of the emf of an electrochemical cell consisting of an indicator electrode reversible to the analyte and a reference electrode immersed in an analyte solution as a function of incremental addition of a standard solution of reagent from the burette. The indicator electrode responds to the concentration of the analyte or the reagent and the measured emf is indicative of the progress of the reaction. The E_{cell} is governed by Nernst equation.

$$E_{cell} = E°_{cell} - (0.059/n) \log \{[product]/[reactant]\} \qquad (16.29)$$

and when the ratio $\{[product]/[reactant]\}$ changes tenfold during the course of the titration, the change in potential ΔE due to incremental addition of the titrant corresponds to $(0.059/n)$. In the beginning of the titration the change in the ratio $\{[product]/[reactant]\}$ is small, and hence, ΔE is also small. Close to the end point the change in the ratio is quite large, and hence, the corresponding ΔE is also large. A plot of measured E_{cell} as a function of the volume of the titrant gives a symmetrical S-shaped curve similar to that obtained for the pH titration as shown in Fig. 16.18(a). The end point is determined graphically from the mid-point of the inflection zone and more accurately from the derivative plot of $(\Delta E/\Delta V)$ versus volume of the titrant which appears similar to that shown in Fig. 16.18(b).

Precipitation titrations (mainly argentometric titrations) are useful for quantitative estimation of halides (Cl^-, Br^- or I^-) by titrating with a standard solution of silver nitrate. Conversely silver ions can be estimated by titrating with a standard solution of potassium chloride. The analyte is placed in the titration cell and the indicator electrode Ag/AgX (formed by dipping silver wire in the halide solution) is reversible to the halide ions in solution. The SCE reference electrode is placed in a separate cell immersed in saturated potassium chloride solution. A salt bridge of ammonium nitrate in agar gel provides the electrical contact between the analyte and the reference cells as shown in Fig.16.19.

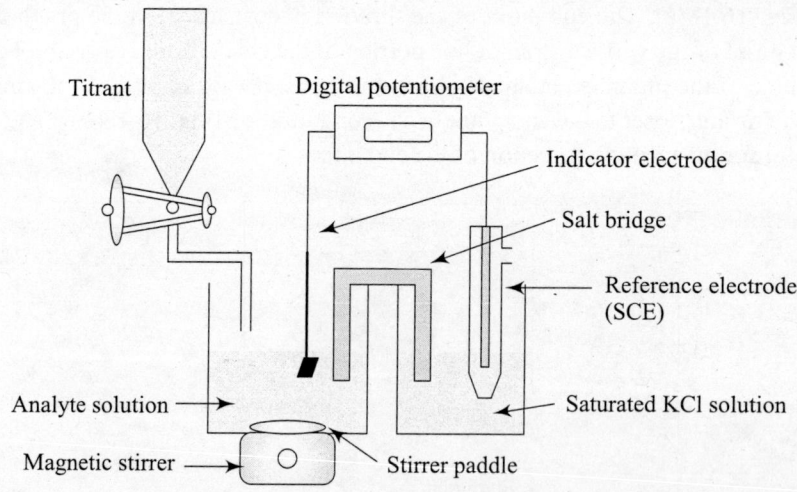

Fig. 16.19 Experimental set-up for potentiometric titrations

Titrations involving oxidation–reduction reactions (e.g., ferrous iron vs. potassium dichromate or potassium permanganate or ceric sulphate) can be conveniently carried out and the end point detected graphically from emf data. The indicator electrode commonly used is the platinum red-ox electrode represented as $Pt|(Fe^{3+}, Fe^{2+})$. The indicator electrode and the reference electrode are dipped into the analyte (ferrous iron solution) placed in the titration cell and connected to a potentiometer. The analyte solution is titrated against a standard solution of potassium permanganate (or potassium dichromate or ceric sulphate). The E_{cell} is measured as a function of added increments of the titrant and the end point is determined from the S-shaped plot of E_{cell} versus volume of titrant and more accurately from the derivative graph [similar to as shown in Figures 16.18(a) and (b)].

16.11 ION SELECTIVE ELECTRODES

Ion selective electrodes consist of specially prepared membranes which are permeable to a single ionic species. The construction is similar to that of the glass electrode for pH measurement. It consists of a tube, one end of which is fused to an electrically conducting membrane. The tube contains a gel or an active ingredient incorporating the ion to which the electrode is sensitive and another inert electrolyte such as potassium chloride. A silver wire in contact with the gel together with the inert electrolyte constitutes the internal silver–silver chloride reference electrode. The ion selective electrode is immersed in the test solution containing the ion to be monitored and the potential developed across the membrane is measured by a potentiometer using an external reference electrode such as SCE. The emf of the cell is related to the activities of the ion of interest in the test solution (a_2) and in the internal gel or medium (a_1) as given by the equation

$$E_{cell} = k - (0.059/n) \log (a_1/a_2) \qquad (16.30)$$

where k is a constant including the internal and external reference electrode potentials.

16.11.1 Different Types of Ion Selective Electrodes

Different types of ion-selective electrodes have come into vogue. These include:

(i) *Glass electrodes* consisting of special glass membranes for the determination of cations in addition to H^+ such as Li^+, Na^+, K^+, NH_4^+, Ag^+, Pb^{2+}, Cd^{2+}, etc.

(ii) *Solid state electrodes* consisting of a compacted disc of a single crystal or microcrystalline material of insoluble salts (e.g., LaF_3 sensitive to F^- ions, doped silver sulphide sensitive to halides (except F^-), sulphide, cyanide, etc.) embedded in a silicone rubber or paraffin matrix moulded in the form of a thin membrane.

(iii) *Liquid ion-exchange membrane electrodes* using liquid ion-exchangers selective to polyvalent cations (Ca^{2+}, Mg^{2+}) or anions (NO_3^-, ClO_4^-, and BF_4^-) and electrodes based on solutions of cyclic polyethers in hydrocarbon solvents with selective complexing ability towards univalent cations.

(iv) *Gas sensing electrodes* which measure the concentrations of gases such as carbon dioxide, ammonia, sulphur dioxide, and nitrogen dioxide in aqueous solutions. The electrode consists of the glass electrode/reference electrode pair inside a plastic tube which is sealed with a thin gas permeable membrane and containing an appropriate electrolyte solution. The CO_2 gas sensing electrode, for example, has sodium hydrogen carbonate solution as the internal electrolyte and the cell reaction may be represented as

$$CO_2 \text{ (sample)} + 2H_2O \rightleftharpoons H_3O^+ + HCO_3^- \text{ (internal electrolyte)}$$

The pH response of the glass electrode/reference electrode pair is a function of dissolved carbon dioxide in the test solution.

ISEs have several advantageous features such as relatively simple to use in a wide range applications and concentrations, amenability for continuous monitoring, useful for cations and anions with better accuracy and precision, unaffected by colour or turbidity of the medium, and capability to operate at temperatures in the range of 0–50°C and in some cases up to 80°C.

Applications of ISEs include the detection and determination of different ions in a variety of fields such as drinking water (fluoride), pollution monitoring (fluoride, chloride, sulphide, nitrate, and cyanide), food processing (nitrate and nitrite in preserved meat products), meat, fish, dairy products and fruit juices (salt content and calcium), and wine (potassium), electroplating and etching baths (fluoride and chloride), clinical laboratories (calcium, potassium, and chloride in body fluids), etc.

16.12 POLAROGRAPHY

Polarography was discovered in 1922 by Heyrovsky. The technique is also known as *voltammetry* as it involves the study of current–voltage relationship in an electrolytic cell where the current flow is entirely controlled by the rate of diffusion of the electroactive species towards a polarizable microelectrode. The technique is useful for the quantitative determination of metal ions as well as electroactive organic compounds at concentrations in the range of 10^{-4}–10^{-8} M as the diffusion current flowing in the electrolytic cell as a function of an applied voltage is directly proportional to the concentration of the electroactive species. The analyte should be electroactive, that is, capable of undergoing oxidation–reduction reactions at the microelectrode.

The basic principle in polarography involves electrolysis of an oxygen-free solution containing the analyte and an inert electrolyte. The applied voltage is increased in increments and the current flowing in the circuit is measured by a microammeter. The dc voltage applied in polarography is usually in the range of –3.0 V to +3.0 V depending on the type of microelectrode. The most commonly used polarizable microelectrode is the *dropping mercury electrode* (DME). The DME consists of pure mercury stored in a reservoir dropping into the electrolyte solution through a fine capillary (~ 0.05 mm inner diameter). A succession of identical drops of mercury is formed and each drop grows in size at the tip of the capillary over a period of about 3–8 s before it falls into the electrolyte solution. In the case of DME the range of applied potential is restricted to +0.25 V to –1.8 V in acidic aqueous media and +0.25 V to –2.3 V in basic aqueous media. At positive potentials higher than +0.25 V mercury dissolves, and hence, cannot be used. The DME has the advantage of providing a fresh cathodic surface in the form of fresh mercury drops emerging out of the capillary, thereby avoiding contamination of the electrode surface. In addition, the high over-voltage for hydrogen formation at the mercury electrode facilitates the reduction of many species in acidic solutions. Instead of the DME a platinum wire can also be used as the microelectrode over the entire range of potentials.

Mercury dropping from the capillary and collecting at the bottom of the electrolytic cell as a pool can itself be used as the non-polarizable anode by connecting it through a platinum wire to the positive terminal of the battery. Alternatively, a reference electrode such as saturated calomel electrode may also be used as the anode. The electrolyte solution consists of a mixture of the analyte and a large excess of an inert or base electrolyte.

During electrolysis three components, namely (i) migration current, (ii) convection current, and (iii) diffusion current contribute to the total current flow in the electrolytic cell. Polarography requires that the contribution of the analyte to the total current should be only due to the diffusion current and not due to the other two components. Hence, the contribution of the analyte to the migration current and convection current is made negligible by suitable experimental conditions. The migration current of the analyte species is made negligible by introducing a large excess (at least 100 times) concentration of an inert electrolyte which does not undergo any reaction at the electrodes. The inert electrolytes commonly used in polarography include potassium chloride, sodium nitrate, or sodium perchlorate. The contribution of the convection current to the total current occurs due to mechanical agitation by stirring and thermal agitation as the temperature of the electrolyte increases during electrolysis. In polarography the electrolyte solution is not stirred during electrolysis and the electrolytic cell is maintained at a constant temperature to avoid thermal agitation of the electrolyte. Hence, the contribution of the convection current to the total current flow in the cell is rendered negligible. Under these conditions the total current flowing in the electrolytic cell as monitored by the microammeter is essentially due to the diffusion current component. The diffusion current arises due to the migration of the analyte species from the bulk of the solution towards the microelectrode and its magnitude is governed by Fick's law of diffusion.

The experimental set-up for carrying out polarography is shown in Fig. 16.20.

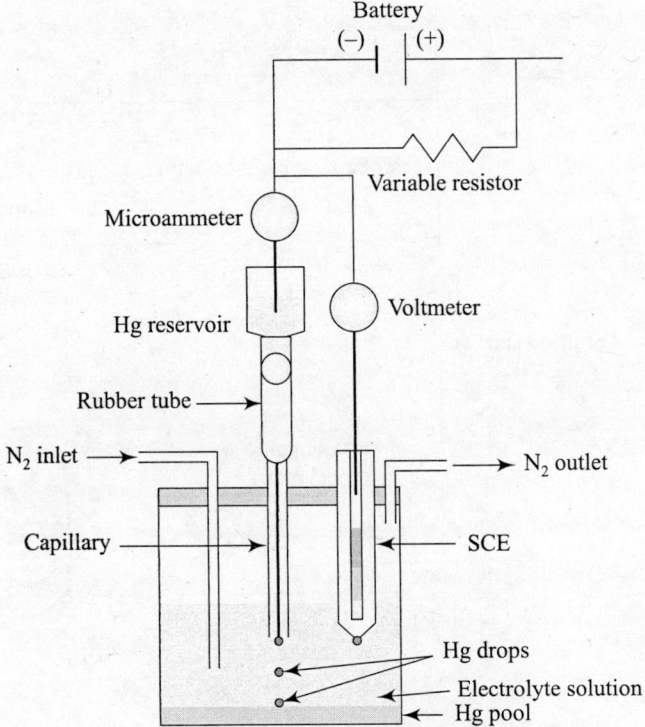

Fig. 16.20 Schematic diagram of polarographic set-up

The experimental procedure involved may conveniently be described with a typical example of the analysis of an analyte such as cadmium in a sample solution. The electrolyte solution consists of the analyte (10^{-3} M) and the inert electrolyte (e.g., KCl) of 10^{-1} M. The electrolyte solution is degassed by bubbling nitrogen gas for about 10–15 min so as to expel the dissolved oxygen which interferes by undergoing reduction at the microcathode.

During electrolysis the solution should not be stirred, and hence, nitrogen gas is passed above the electrolyte solution so as to maintain a blanket of nitrogen atmosphere in the cell. The applied potential at the cathode is gradually varied between 0.0 V to −1.5 V in increments of 0.1 V and the corresponding diffusion current flowing in the cell as indicated by the microammeter reading is recorded. The plot of the diffusion current as a function of applied voltage gives a current–voltage diagram (voltammogram) also called the *polarographic wave* or *polarogram* as shown in Fig. 16.21. The polarogram appears as a saw-toothed curve as shown in Fig. 16.21(a) due to the variations in the condenser current as the mercury drops formed grow in size and detach from the capillary. The smoothed curve taking the average values of the current as a function of applied voltage is shown in Fig. 16.21(b).

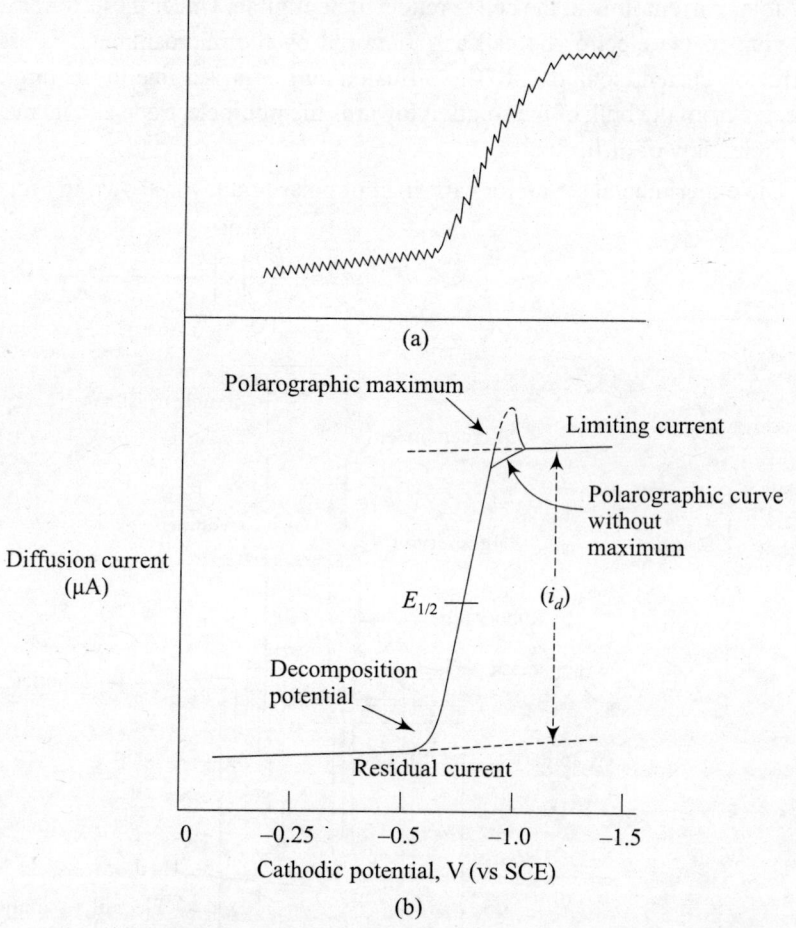

Fig. 16.21 Polarographic wave (a) saw-toothed curve and (b) averaged curve

Initially at small applied voltage the current flow is small and is known as residual current. The current–voltage curve runs close to the x-axis as shown in Fig. 16.21. Once the applied potential just exceeds the decomposition potential of the electroactive species (Cd^{2+}), the reduction of the electroactive species at the DME commences. As cadmium metal gets deposited on the cathode, the concentration of Cd^{2+} in the immediate vicinity of the cathode decreases instantaneously to almost zero thereby causing a concentration polarization at the microcathode. A concentration gradient is also set up between the cathode and the bulk of the solution leading to the diffusion of Cd^{2+} from the bulk of the solution towards the polarized cathode. The rate of diffusion is proportional to the concentration difference of Cd^{2+} between the cathode and the bulk of the solution as governed by Fick's law of diffusion. The resulting current flow i_d, called the *diffusion current* is proportional to the rate of diffusion which in turn depends on the concentration gradient of the electroactive species as given by the equation.

$$i_d = k\,(C - C_o) \tag{16.31}$$

where C and C_o are the concentrations of the electroactive species in the bulk of the solution and at the electrode surface respectively, k is a proportionality constant.

The reduction of the electroactive species at the cathode occurs more rapidly with increasing applied voltage. C_o becomes virtually zero and the concentration gradient reaches a maximum. The rate of diffusion and the rate of deposition increase with increasing applied voltage. Ultimately the two rates become equal and the current flowing in the cell reaches a limiting value. Further increase in the applied voltage does not increase the diffusion current and a plateau corresponding to limiting diffusion current is observed in the polarogram as given by Ilkovic equation

$$i_d = kC \tag{16.32}$$

The constant k is equal to $(607\,n\,D^{1/2}\,m^{2/3}\,t^{1/6})$ where n is the number of electrons involved in the electrode reaction, D is diffusion coefficient of the electroactive species, m is the flow rate of mercury from the capillary, and t is the drop time. The diffusion current given by the above equation is actually the average of the oscillating capacitor current of the mercury drop as it grows and falls.

The polarographic curves obtained by using DME are often distorted due to *polarographic maxima* (also known as *current maxima*) which appear as peaks or as rounded humps as shown in Fig. 16.21. The diffusion current rises sharply with increasing applied voltage and at the limiting current value it increases abnormally until a critical value and then rapidly decreases to the normal limiting current plateau. No theoretical explanation is yet available for the observed maxima. The polarographic maximum is suppressed by the addition of a surface active agent (*maximum suppressor*) at low concentrations (~ 0.005–0.01%) such as gelatin, methyl red, methyl cellulose, or agar in order to determine the true limiting diffusion current plateau.

Ilkovic equation represents the direct proportionality between the concentration of the electroactive species in the bulk of the solution and the diffusion current as given by the height of the polarographic wave. The equation is the basis for quantitative analysis of electroactive species.

The potential corresponding to one-half of the diffusion current indicated by the mid-point of the rising portion of the polarographic wave is called the *half-wave potential*, $E_{1/2}$. The half-wave potential is related to the standard electrode potential of the electroactive species ($E_{1/2} \sim E^\circ$), and hence, is characteristic of the species and useful in qualitative analysis for the identification of the electroactive species. The potential at a polarized electrode is governed by Nernst equation, and hence, the relationship between the potential and current may be expressed by the equation known as the *equation of the polarographic wave*

$$E = E_{1/2} - \frac{0.0591}{n} \log\left(\frac{i_d - i}{i}\right) \tag{16.33}$$

where i is the measured current (after subtracting the residual current) at a given applied potential. For a reversible reduction at the cathode the $E_{1/2}$ is the potential when $i = i_d/2$. A graphical plot of the cathodic potential E versus $\log((i_d - i)/i)$ gives a straight line with a slope of $0.0591/n$ from which the value of number of electrons involved in the electrode process can be determined. The above equation is applicable only for reactions that take place reversibly at the electrode. It indicates that the $E_{1/2}$ is independent of the concentration of the electroactive species.

The half-wave potentials of simple metal ions (metal–aqua complexes) shift to more negative potentials in the presence of complexing ligands. The shift in $E_{1/2}$ is directly proportional to the stability of the complex species, and hence, polarography provides a simple and convenient method to determine the stability constants of metal complexes.

Polarography is a useful analytical technique for qualitative and quantitative analyses of individual electroactive species as well as mixtures. In the case of mixtures, the $E_{1/2}$ values of the different species present in the mixture should be different by at least 0.15–0.2 V for effective separation and identification of the components of the mixture.

The reduction of oxygen at the DME gives rise to two oxygen polarographic waves with $E_{1/2}$ values at -0.05 V and -0.9 V (vs. SCE) respectively.

$$O_2 + 2H^+ + 2e^- \rightarrow H_2O_2 \; ; E_{1/2} = -0.05 \text{ V (vs. SCE)}$$
$$H_2O_2 + 2H^+ + 2e^- \rightarrow 2 H_2O \; ; E_{1/2} = -0.9 \text{ V (vs. SCE)}$$

The reduction waves are useful for quantitative estimation of dissolved oxygen or hydrogen peroxide content of samples.

16.12.1 Quantitative Analysis by Polarography

Quantitative analysis of organic and inorganic analytes by polarography can be carried out by adopting (i) direct evaluation, (ii) calibration method, (iii) standard addition, or (iv) internal standard method.

Direct evaluation is based on comparing the polarograms of the solutions of standard and the test sample under identical conditions of supporting electrolyte composition including that of the maximum suppressor and using Ilkovic equation to evaluate the diffusion current quotient i_d/C of the analyte. The diffusion current of simple ions in neutral or acidic solutions is about 4 μA per milliequivalent of reducible ion. The quotient is used to calculate the concentration of the analyte in the test sample from the height of the polarographic wave. The method does not require the knowledge of the capillary characteristics and accurate control of temperature. Better accuracy can be achieved if the concentration of the analyte in the standard is about the same as that in the test sample.

Calibration method involves determining the wave heights of a series of standard solutions containing different amounts of the analyte and preparing a calibration graph or chart by plotting the wave height as a function of the concentration. The plot is linear over a concentration range of the analyte. The wave height of the analyte in the test sample is determined under identical conditions and the concentration of the analyte is determined from the calibration chart.

Standard addition method is particularly useful when a single analysis is to be performed. The polarogram of the analyte in the test solution is recorded and a known volume of a standard solution of the test ion is added and the polarogram is recorded again to determine the increase in the wave height (diffusion current). The concentration of the analyte C_X is determined using the formula

$$C_X = \frac{-vC_s h}{(hV - H(V + v))} \tag{16.34}$$

where v is the volume (mL) of the standard solution of concentration C_S added to the test sample solution of volume V and h and H refer to the polarographic wave heights obtained for the analyte in the test solution and for the (test sample + standard solution).

Internal standard method also called the *pilot ion method* has the advantage in that it is independent of capillary characteristics as the relative wave heights of two electroactive substances are constant for equal concentrations in a given supporting electrolyte. The concentration of the analyte in the test solution is determined from the wave heights of the analyte and the internal standard (or reference ion) using the formula

$$C_X = \frac{C_S(i_X(I_{dS}))}{i_S I_{dX}} \tag{16.35}$$

where I_{dS} and I_{dX} refer respectively to the diffusion current constants of the standard and the analyte in the given supporting electrolyte and i_S and i_X are the diffusion current values recorded for the standard and the analyte. The method is of limited application because only a small number of ions give well-defined polarographic waves.

16.12.2 Modern Polarographic Techniques

The sensitivity of the dc polarography has been improved by introducing modifications of the technique. The main limitation of the normal dc polarography is due to the variation in the condenser current associated with the charging of each drop of mercury as it forms, grows, and drops. The rhythmic fluctuation of the condenser current gives rise to a saw-toothed current–voltage curve which limits the detection levels of the analyte to about 1×10^{-5} M. Many modifications of the normal dc polarography have come into vogue which eliminate the fluctuating base line current and thereby enhance the sensitivity of the technique. These newer techniques include linear sweep oscillographic polarography, pulse polarography, ac polarography, and stripping voltammetry. The sensitivity of these techniques is about 2–10 times more as compared to normal dc polarography.

Linear sweep oscillographic polarography also called *oscillographic polarography* or *rapid scan polarography* involves rapid scanning of the polarogram within the lifetime of a single drop of mercury. It is carried out by applying a repetitive dc potential sweep of 0.5–1.0 V on each of the growing mercury drop and recording the current–voltage curve on a cathode ray oscilloscope. The current increases with the growing surface area of the mercury drop as in conventional dc polarography, and hence, the potential sweep is timed to synchronize over the last 2 s of the life-time of the drop during which period the changes in the surface area of the drop, and hence, the increase in current is minimal. Exact synchronization is achieved by mechanical tapping of the capillary so that each drop detaches exactly at the same time (usually 6–7 s). The variation of the current for each drop of mercury as it grows as a function of time, the application of the dc potential sweep over the last couple of seconds are shown in Fig. 16.22(a). the variation of the current as a function of applied potential as seen in the oscilloscope is shown in Fig. 16.22(b). The peak current i_p in oscillographic polarogram is due to the rapid potential sweep and is not due to maximum (as in dc polarography). It is proportional to the concentration of the analyte species.

Normal pulse polarography involves maintaining the applied potential to DME at a constant value during most part of the lifetime ($\sim$ 5–6 s) of a single mercury drop and applying a pulse of higher voltage only during the last 50–60 ms of the lifetime of the drop. The current is measured during the last 20 ms of the lifetime of the drop. The potential applied to each successive drop is increased gradually to complete the voltage scan as in the normal dc polarography. The current measured during the last stages of the drop lifetime is proportional to the analyte concentration and the resulting polarogram is similar to the normal polarogram with the saw-toothed pattern being replaced by a stepped curve [as shown in Fig. 16.21(b)].

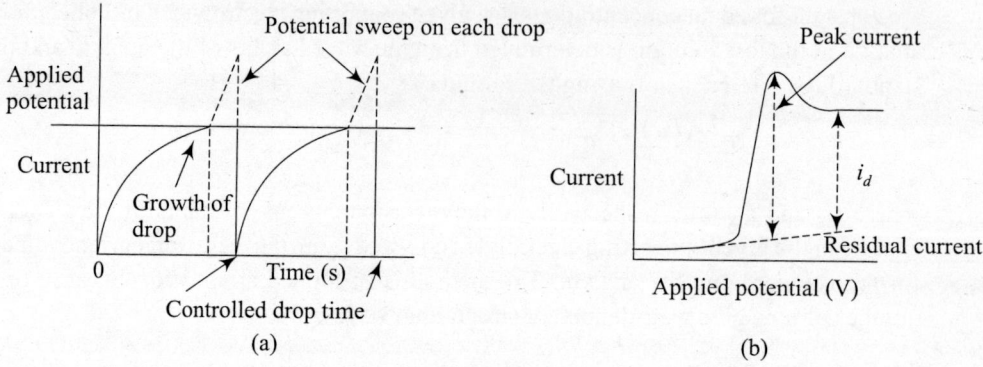

Fig. 16.22   (a) Variation of surface area of the drop and current as a function of time and (b) oscillographic polarogram

Differential pulse polarography (DPP) finds wider application as compared to normal pulse polarography. It involves the application of a 10–100 mV pulse for a short duration of about 40–50 ms superimposed on the increasing dc potential. The pulse is applied during the last quarter of the growth of each mercury drop and synchronization is achieved by mechanical tapping of the capillary to release the drop precisely at a controlled drop time. The diffusion current is measured twice, just before applying the pulse and again during the last 20 ms and the difference between the two measured diffusion current values is plotted as a function of the applied dc voltage. The height of the peak current is directly proportional to the concentration of the electroactive species. The application of the voltage in a pulse for a short duration suppresses the capacitor current of the growing mercury drop. The sensitivity of the technique is enhanced due to the applied pulse voltage as each drop produces a higher diffusion current. In addition, the technique has the advantage in that species with half-wave potentials differing by only 0.05 V can be identified. The superimposition of the voltage pulse on the increasing dc potential as a function of time and the differential pulse polarogram are shown in Figs. 16 23(a) and (b).

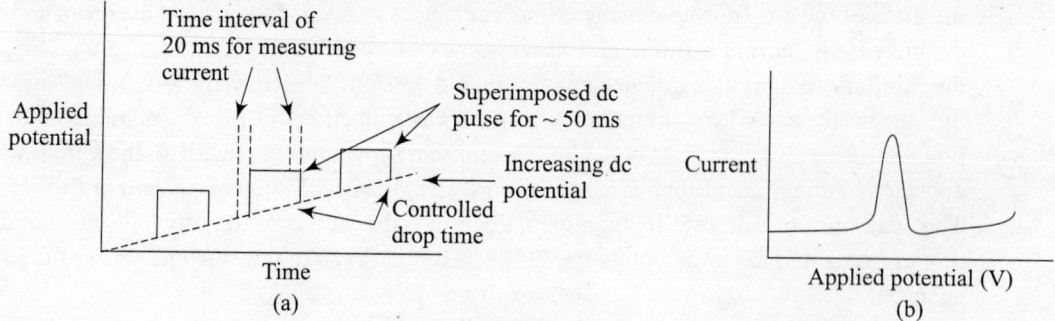

Fig. 16.23   (a) Superimposition of dc pulse potential as a function of time and (b) differential pulse polarogram

AC polarography also called *sinusoidal ac polarography* involves the application of a constant sine wave ac potential of a few mV superimposed on the dc potential. The measured ac current is plotted against the applied dc potential to get an ac polarogram which appears in the form of a peak, the peak potential being the same as the $E_{1/2}$ of the normal dc polarogram and the height of the peak being proportional to the concentration of the analyte species. The technique is used for the separation and identification of electroactive species having close (~ 40 mV) $E_{1/2}$ values which cannot be handled in normal dc polarography. In addition, measurement of peak height is easier as compared to the determination of wave-height, and hence, quantitative analysis is more reliable.

Square-wave polarography is a modification of the ac polarography to eliminate the undesirable condenser current by applying square-wave voltage instead of sinusoidal ac voltage. The potential sweep at a fast scan rate of about 200 mV s^{-1} is applied during the lifetime of a single drop of mercury usually after a few seconds of the birth of the drop to take advantage of the larger surface area of the electrode. The difference in current before and after the application of the pulse is measured and plotted against the scan potential to give a voltammogram in the form of a peak. The height of the peak is proportional to the concentration of the analyte. The technique has the advantage of recording the voltammogram within a few seconds and also a higher sensitivity with detection limits extended to 10^{-7}–10^{-9} M. Because of its speed, the technique finds use as rapid detector in HPLC and also to study electrode kinetics.

Stripping voltammetry also known as *stripping analysis* or *linear-potential sweep stripping chronoamperometry* involves two steps. In the first step, low levels of the analyte is concentrated from the bulk of the solution by electrodeposition onto a cathode or anode of small area. The test solution contains in addition to the analyte metal ions supporting electrolyte just as in the dc polarographic analysis. In the second step, the electrodeposited species is electrolytically stripped from the electrode surface back into solution. Metal ions in trace quantities in the test solution can be deposited by controlled potential reduction for an accurately measured time (ranging from a few minutes to an hour) on a working electrode such as a hanging mercury drop cathode or a mercury-coated inert solid electrode, for example, platinum or carbon. *Anodic stripping* of the metal ions is then carried out by reversing the polarity of the electrode and applying a linear potential sweep to give rise to anodic polarographic waves of the metal ions deposited and concentrated on the working electrode. As the applied potential approaches the oxidation potential of the metal ion, the current increases rapidly and reaches the maximum value at a potential which is approximately equal to the oxidation potential. The polarogram obtained appears as a peak, the diffusion current (height of the polarographic wave) being directly proportional to the amount of metal ion deposited on the working electrode in the first step. A much larger and sharper peak can be obtained by applying a differential pulse of about 2–5 mV s^{-1} and the technique is called *differential pulse anode stripping voltammetry* (DPASV). Figures 16.24(a) and (b) show the polarograms obtained for a metal ion by dc stripping analysis and by DPASV techniques respectively.

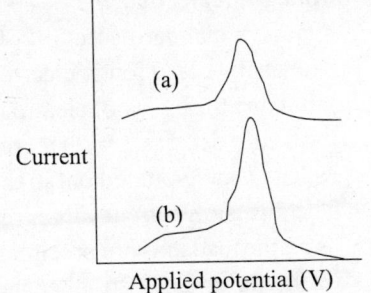

Fig. 16.24 Current–voltage diagrams in stripping analysis: (a) dc stripping and (b) DPASV

16.13 AMPEROMETRIC TITRATIONS

The principle involves measuring the limiting diffusion current flowing in the polarographic cell (Fig. 16.21) at a constant applied potential during a titration of the analyte with a standard solution of the reagent taken in the burette. The requisites of amperometric titrations include (i) one or both the reactants, namely, the analyte in the polarographic cell, the titrant or the product have to undergo reduction at the microcathode which is a polarizable dropping mercury cathode or the platinum microcathode and (ii) the constant applied potential is greater than the $E_{1/2}$ of the electroactive species preferably in the limiting diffusion current region. A known volume of the analyte is taken in the polarographic cell and titrated against a standard solution of the titrant, the titrant being added in incremental volumes and the corresponding flow of diffusion current measured by the microammeter in the polarographic circuit. The graphical plot of the diffusion current as a function of volume of titrant gives two intersecting straight lines and the end point of the titration is determined from the intersection of the straight lines.

Amperometric titrations may be classified into two main types depending whether the polarographic cell contains a single polarizable microindicator electrode or two such electrodes. The latter method is called *biamperometric titration*. Dropping mercury electrode (DME) or a rotating platinum microelectrode at a constant speed of 600 rpm or more may be used as the polarizable electrode. The rotating platinum electrode is preferred for red-ox titrations involving bromine or ferric iron. However, the electrode is useful only in alkaline or weakly acid medium due to its low hydrogen activation overpotential. In addition, the electrode is very sensitive to dissolved oxygen, and hence, requires stringent deaeration of the reagent solutions.

16.13.1 Amperometric Titrations with One Polarizable Indicator Electrode

Amperometric titrations with a single polarizable indicator electrode can be carried out for four different types of titrations depending whether the (i) analyte alone is electroreducible, (ii) titrant alone is electroreducible, (iii) both the analyte and titrant are electroreducible, and (iv) the analyte may give rise to an anodic diffusion current while the titrant gives rise to a cathodic diffusion current.

In a typical example of amperometric titration, lead nitrate solution (analyte) of known volume together with an inert electrolyte of sodium nitrate is placed in the cell and a constant applied voltage greater than the $E_{1/2}$ of lead −0.4 V is applied to the DME. The analyte solution is titrated against a standard solution of the reagent potassium sulphate taken in the burette at −0.8 V. The analyte lead ion is electroactive at the microcathode undergoing reduction at the electrode while the titrant is not electroactive. Initially during the titration the diffusion current in the cell gradually decreases with incremental addition of sulphate ions (as potassium sulphate solution) as lead is precipitated out in the form of lead sulphate. The diffusion current values reaches a minimum value as all the lead ions are precipitated and further addition of the titrant does not alter the magnitude of current. The current flowing in the cell remains constant close to zero as shown in Fig. 16.25. The amperometric titration curve consists of two intersecting straight lines and the end point of the titration is obtained by dropping a perpendicular to the x-axis from the intersection point of the straight lines. A similar titration can be performed by titrating silver (electroactive analyte) against a standard solution of potassium chloride.

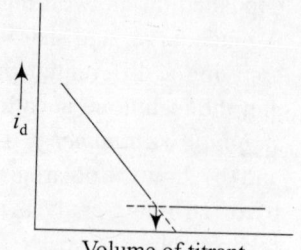

Fig. 16.25 Amperometric titration curve for an electroactive analyte (e.g., Pb^{2+} vs. SO_4^{2-} or Ag^+ vs. Cl^-)

An amperometric titration in which the analyte is not electroactive but the titrant is (e.g., sulphate ions vs. lead or thiosulphate ions vs. iodine) will give rise to a titration curve as shown in Fig. 16.26. Initially during the titration, the diffusion current is small, close to zero, and remains constant till all the analyte is precipitated (in the case of sulphate ions) or oxidized (in the case of thiosulphate ions). After the complete removal of the analyte species, the current flowing in the cell increases linearly with the added titrant as it is electroactive. The end point is determined graphically from the intersection of the two straight lines.

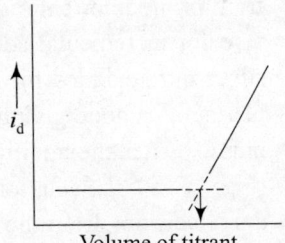

Fig. 16.26 Amperometric titration curve for an electroactive titrant (e.g., SO_4^{2-} vs. Pb^{2+} or $S_2O_3^{2-}$ vs. I_2 or $Cr_2O_7^{2-}$ vs. Pb^{2+} at 0.0 V in acetate buffer of pH 4.2)

When the analyte and the titrant are both electroactive as in the case of lead (lead nitrate) and chromate ions (potassium chromate) the amperometric titration curve is V-shaped as shown in the Fig. 16.27. Initially the diffusion current gradually decreases as the analyte concentration decreases (the product is not electroactive) with increasing volume of titrant and reaches a minimum when all the analyte has reacted. The current once again increases with increasing volume of titrant after the end point. The end point is determined from the intersection point of the two straight lines.

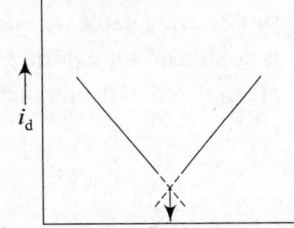

Fig. 16.27 Amperometric titration curve for electroactive analyte and titrant (e.g., Pb^{2+} vs. CrO_4^{2-})

Amperometric titration in which the analyte gives rise to an anodic diffusion current while the titrant gives rise to a cathodic diffusion current or vice versa is also called *titration to zero current*. Typical examples include the titrations of (i) iron(III) with titanium(III) and (ii) copper(II) with tin(II). An applied voltage midway between the half-wave potentials for the cathodic reduction of iron(III) and anodic oxidation of titanium(III) causes a diffusion current for the reduction of iron(III) (and similarly for the reduction of copper (II)) at the start of the titration. The diffusion current decreases with the decrease in the concentration of iron(III) and reaches zero at the end point. After the end point diffusion current due to the oxidation of the added titrant (titanium(III) or tin(II)) is set up which increases with the increasing concentration of the titrant. The cathodic and anodic diffusion currents have different slopes as shown in Fig. 16.28 and the end point is at the intersection point where $i_d = 0$.

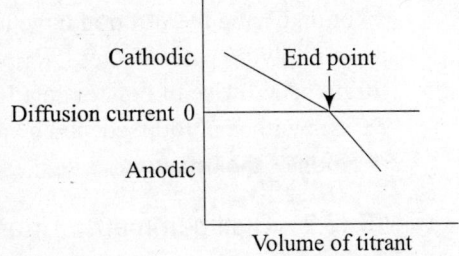

Fig. 16.28 Amperometric titration to zero current

Successive amperometric titrations can be performed in which a mixture of halides (iodide, bromide, and chloride) can be titrated with silver nitrate to get successive end points in a single titration. The amperometric titration is carried out using rotating platinum electrode and a mercury/mercuric iodide/potassium iodide electrode as the reference electrode with a potential of −0.23 V (vs. SCE). Initially the iodide is titrated in ammoniacal medium at pH 9 (0.1–0.3 N ammonia) with silver nitrate. Silver iodide alone will be precipitated without any interference from bromide or chloride. During the titration of iodide the current remains constant close to zero for incremental addition of silver nitrate. When all the iodide is precipitated the excess silver nitrate added forms silver–ammine complex which undergoes reduction at the rotating platinum cathode giving rise to a diffusion current which increases linearly with added silver nitrate. After the addition of three or four increments of silver nitrate beyond the end point (Fig. 16.29), the solution is acidified with dilute nitric acid (0.8–1.0 N). Acidification releases the silver ions from the silver–ammine complex which precipitate bromide ions. The current drops to zero and remains so till all the bromide has been precipitated with the further additions of silver nitrate. The titration is carried out at less negative potential using SCE as the reference electrode. Chloride does not interfere in the titration of bromide because silver chloride particles cause a cathodic current. When all the bromide is precipitated once again the diffusion current increases due to the reduction of silver(I) in the acidic medium indicating the end point for the titration of bromide. After the bromide titration, gelatin (0.1%) is added to the solution to suppress the cathodic current due to silver chloride. The current falls to zero and the titration is continued for chloride. The diffusion current once again rises after all the chloride has been precipitated to the reduction of silver(I) at the rotating platinum cathode.

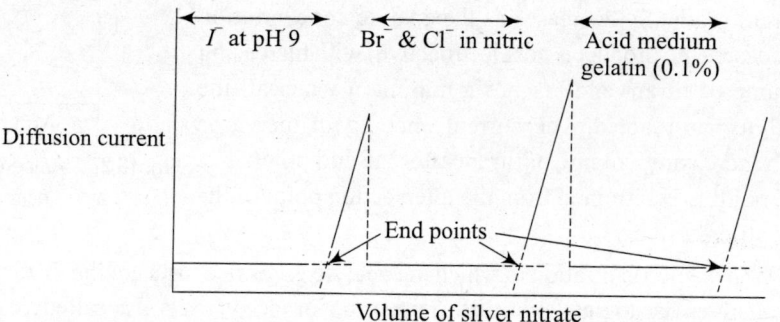

Fig. 16.29   Successive amperometric titrations of a mixture of halides

In amperometric titrations it is preferable to use titrant at 10–20 times more concentrated as compared to that of the analyte to get sharper end points. Otherwise changes in the volume of the solution during the titration have to be taken into account. Current readings must be corrected by multiplying with a factor $[(V + v)/V]$, where V is the initial volume and v is the volume of the titrant added. Use of concentrated titrant has the additional advantage in that less of dissolved oxygen will be introduced, and hence, less time is required for deaeration after each incremental addition of the titrant.

16.13.2   Biamperometric Titrations

Biamperometric titrations involve the use two similar small polarizable Pt electrodes to which a small voltage of about 100 mV is applied (instead of one polarizable microelectrode such as DME or rotating Pt electrode and a reference electrode such as SCE in conventional amperometric titrations). The analyte solution is stirred and titrated against a standard solution of the reagent taken in the burette and the end point is indicated by the sudden appearance or disappearance of current flow in the circuit. A simple apparatus for carrying out the biamperometric titration is shown in Fig. 16.30.

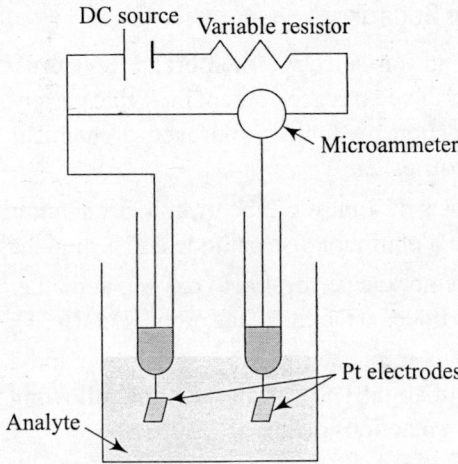

Fig. 16.30 Experimental set-up for biamperometric titrations

A typical example of biamperometric titration is that of a known volume of iodine versus a standard solution of sodium thiosulphate taken in the burette. The iodine–iodide system in the titration cell undergoes reversible red-ox reaction at the electrodes, the amount of iodine undergoing reduction at the cathode being the same as the amount of iodide undergoing oxidation at the anode. With an applied voltage of about 100 mV both the electrodes remain depolarized and an appreciable amount of current flows through the cell. During the titration the added sodium thiosulphate reduces the iodine and current flows through the cell till the end point rapidly decreasing to zero at the end point as one of the electrodes becomes polarized. The end point of the titration is indicated by the sudden drop of current to zero, and hence, is known as *dead-stop end point*. If the titration is carried out by taking the sodium thiosulphate in the titration cell and iodine in the burette, the current flow till the end point will be zero and starts flowing immediately after the end point. The graphical plot of current flow as a function of volume of titrant will be similar to that shown in Fig. 16.25.

Other examples of biamperometric titrations include the titration of iron(II) against cerium sulphate or potassium permanganate in which both the reactants are electroactive. The titration curve will be V-shaped with a minimum in the current flow indicating the end point of the titration.

16.13.3 A Few Important Applications of Amperometry

Amperometric titration for the determination of small amounts of water/moisture in a variety of commercial samples involving Karl Fischer reagent is an important industrial application. The weighed sample is dispersed in the Karl Fischer reagent, sulphur dioxide in anhydrous pyridine-anhydrous methanol, and titrated with a standard solution of iodine in anhydrous methanol amperometrically. One mole of iodine reacts with one mole of water in the sample as per the equation

$$SO_2 + I_2 + H_2O \rightarrow H_2SO_4 + 2\,HI$$

When all the water in the sample is reacted the added iodine may be detected visually or amperometrically to indicate the end point.

Other important applications of amperometry include the determination of dissolved oxygen content in solutions by oxygen sensor and biosensors.

16.13.4 Oxygen Sensor

Oxygen sensor is an ion selective membrane device developed in 1953 by L.C. Clark for determining the dissolved oxygen content in a wide variety samples such as blood, fermentation or biochemical reaction medium in bioreactor, chemical solutions, sewage, sea water, soil, effluents from industries, etc.

The probe consists of a plastic tube or holder containing a saturated solution of potassium chloride into which a platinum disc cathode and a ring-shaped silver anode are immersed.

A thin Teflon membrane permeable to oxygen and other gases covers the electrodes at the end of the probe (Fig. 16.31).

When the probe is immersed in the test solution and a polarizing voltage of about 1.5 V is applied, the following oxidation–reduction reactions occur.

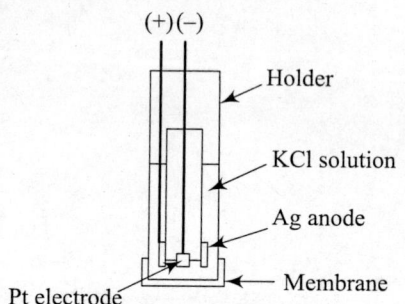

Fig. 16.31 Clark oxygen sensor

At the cathode

$$O_2 + 2\,H_2O + 4\,e^- \rightarrow 4\,OH^-$$

At the anode

$$4\,Ag + 4\,Cl^- \rightarrow 4\,AgCl + 4\,e^-$$

The overall reaction may be represented as

$$4\,Ag + 4\,Cl^- + O_2 + 2\,H_2O \rightarrow 4\,AgCl + 4\,OH^-$$

The reduction of oxygen at the platinum cathode polarizes the electrode with oxygen concentration decreasing virtually to zero within the probe. This results in the diffusion of oxygen from the test solution into the probe through the Teflon membrane and the rate of diffusion depends on the concentration of dissolved oxygen in the test solution. The diffusion current flowing in the electrochemical cell is directly proportional to the amount oxygen diffusing through the membrane. The diffusion current is amplified electronically and displayed in units of % oxygen saturation in the display meter. The oxygen sensor is capable of measurement of dissolved oxygen content at 1 ppm level.

16.13.5 Biosensors

Biosensors are devices similar to ion selective electrodes developed specifically for measuring the concentrations of biologically important molecules such as carbohydrates, urea, enzymes, antibodies, blood gases, blood electrolytes, etc. The main components of a biosensor include (i) a reaction centre, (ii) a transducer, (iii) an amplifier, and (iv) data processing microprocessor or computer which also displays the final report. The reaction centre consists of an immobilized biological material such as an enzyme, antibody or whole cell in a membrane or gel and is the actual sensor through physicochemical changes in absorbance, conductance, mass, concentration, red-ox state, etc. The transducer is in intimate contact with the membrane or gel and converts the physicochemical signal into an electrical signal. The signal is amplified and fed to the data processor. Important examples of biosensors include the oxygen sensor and glucose sensor.

Biochemical oxidation of organic species is catalysed by enzymes known as *oxidases*. Specific oxidases are available naturally for the oxidation of each compound. Oxidases function only in the presence of dissolved oxygen and the extent of depletion of dissolved oxygen is determined by the amount of the oxidizable species. By measuring the dissolved oxygen content by an oxygen sensor facilitates the determination of amount of oxidizable organic species. Glucose sensor is a typical example of a biosensor.

Glucose sensor is similar to the oxygen sensor in construction but the permeable membrane consists of three layers. The outermost layer is made of polycarbonate and is permeable to glucose but not other constituents of blood. The middle layer consists of the glucose sensor enzyme, glucose oxidase, which catalyses the oxidation of glucose to gluconic acid and simultaneously releases hydrogen peroxide.

$$\text{Glucose} + O_2 \xrightarrow{\text{glucose oxidase}} \text{Gluconic acid} + H_2O_2$$

The hydrogen peroxide diffuses through the inner membrane to reach the electrode where it gets oxidized to oxygen.

$$H_2O_2 + 2\ OH^- \rightarrow O_2 + 2H_2O + 2\ e^-$$

The resulting current is directly proportional to glucose concentration in the sample.

SOLVED PROBLEMS

EXAMPLE 1   *Calculate the weight of copper deposited in 2 h by a current of 500 mA. Calculate the weight of copper under the same experimental conditions if the current efficiency is only 95%.*

SOLUTION

No. of Coulombs = 2 h × 60 m × 60 s × 0.5 A = 3600 C

Weight of copper deposited = (63.54 g/2) × 3600 C/96,500 C = 1.1852 g

Weight of copper deposited at 95% current efficiency = 1.1852 × 95/100

$$= 1.1258 \text{ g}$$

EXAMPLE 2   *In a Coulometric titration 25 mL of Ce(IV) was reduced by electrolytically generated Fe(II) at 250 mA current in 15 min. Calculate the concentration of Ce(IV).*

SOLUTION

No. of coulombs of current = $i \times t$ = 15 m × 60 s × 0.25 A = 225 coulombs

One Faraday (96,500 C) will deposit/dissolve 1 g equivalent of substance

Therefore, 225 C = 225/96,500 = 2.332×10^{-3} g equivalent of Fe(II) or Ce(IV) in 25 mL

Concentration of Ce(IV) in the given solution = $2.332 \times 10^{-3} \times 1000/25$

$$= 0.0933 \text{ N}$$

EXAMPLE 3   *Calculate the time required to deposit 0.86 g of silver if the current flow is 0.3 A.*

SOLUTION

107.87 g (1 g equivalent) of silver will be deposited by 96,500 C.

0.86 g of silver will require 96,500 C × 0.86 g/107.87 g = 769.35 C

Time required to deposit 0.86 g of silver at 0.3 a current = 769.35 C/0.3 A

$$= 2564.5 \text{ s} = 42.74 \text{ min}$$

EXAMPLE 4   *In a polarographic experiment a nickel(II) analyte in 0.1 M KCl solution as supporting electrolyte was found to give rise to a diffusion current of 3.8 μA under the experimental conditions of mercury flow m = 26.4 mg. s^{-1} and drop time t = 3.6 s. Calculate the concentration of nickel.*

SOLUTION

Ilkovic equation $i_d = kC$ where $k = (607\ n\ D^{1/2}\ m^{2/3}\ t^{1/6})$ can be simplified to $i_d = k'C$ where $k' = m^{2/3}\ t^{1/6}$ and substituting the given values into the simplified Ilkovic equation concentration of nickel in the solution

$$C = 3.8/(3.56 \times (26.4)^{2/3} \times (3.6)^{1/6}) = 3.8/39.0795 = 0.0972 \text{ M}$$

EXAMPLE 5   *Polarographic curves for standard solutions and test solution of a metal ion were recorded and the following data were obtained. Determine the concentration of the metal ion in the given test sample.*

$[M^{t+}]$ $(10^{-4}$ M)	i_d (μA)
0.5	7.9
1.0	15.2
1.5	22.0
2.0	30.0
2.5	37.8
3.0	45.6
Test sample	32.6

SOLUTION

Plotting the graph with the given i_d values as a function of concentration of the metal ion in the standard solutions gives a straight line from which the concentration of the metal ion in the test sample is determined as 2.2×10^{-4} M.

EXAMPLE 6   *Zinc(II) in a test solution of 50 mL was found to show a galvanometer deflection of 42 mm in a polarographic experiment. The experiment was repeated by adding 5 mL of a 0.0015 M solution of zinc (II) to the test solution and galvanometer deflection was found to be 85 mm. Calculate the concentration of zinc (II) in the test solution by standard addition method.*

SOLUTION

Substituting the given values in the equation

$$C_X = -v \, C_S \, h/(h \, V - H \, (V + v))$$
$$= \frac{-5 \text{ mL} \times 0.0015 \text{ M} \times 42 \text{ mm}}{42 \text{ mm} \times 50 \text{ mL} - 85 \text{ mm} (50 \text{ mL} + 5 \text{ mL})}$$
$$= -0.315/-2575 = 1.22 \times 10^{-4} \text{ M}$$

EXAMPLE 7   *A test solution of 25 mL containing cadmium(II) was found to give a diffusion current of 15.6 μA. An identical test solution of 25 mL containing cadmium of unknown concentration and 3 mL of 0.01 M zinc(II) was found to give a diffusion current value of 12.6 μA. Calculate the concentration of cadmium(II) in the test sample by internal standard method. The diffusion current constants for Cd (II) and Zn (II) in the supporting electrolyte are 3.51 and 3.42 respectively.*

SOLUTION

Substituting the given values in the equation

$$C_X = (C_S \, i_X \, (I_d)_S)/(i_S \, (I_d)_X)$$
$$= (0.01 \text{ M} \times 15.6 \text{ μA} \times 3.42)/(12.6 \text{ μA} \times 3.51)$$
$$= 0.5335/44.26 = 0.012 \text{ M}$$

SUMMARY

- Electroanalytical chemistry makes use of electrical energy for analytical purposes by monitoring the responses of the electrical properties of analytes. The various techniques include (i) conductometry, (ii) high-frequency method or oscillometry, (iii) potentiometry and pH-metry, (iv) electrogravimetry, (v) coulometry, and (vi) polarography and amperometry.

- Conductometry is based on the measurement of electrical resistance of a solution using a Wheatstone bridge circuit and alternating current in the frequency range of 1000–5000 Hz.

- An important analytical application of conductometry is that of conductometric titrations for acid–base, precipitation, and complexation reactions. The end point in such conductance titrations is determined from a graphical plot of conductance versus volume of titrant.

- Oscillometric titrations involve the use of alternating current in the high frequency range of 10^6–10^7 Hz to monitor the signal generated by a combination of resistance and capacitance of both the analyte solution and the titration cell. No fouling of the electrodes occurs as they do not come into contact with the analyte solution and hence even corrosive liquids can be handled in the titration cell.

- Electrogravimetry involves deposition of the analyte metal on the cathode by the application of dc potential either under constant applied potential or under constant applied current conditions and weighing the amount of analyte deposited.

- Coulometry involves the measurement of the quantity of electricity used in an electrochemical reaction at 100% current efficiency either under constant potential conditions or constant current conditions.

- Coulometric titrations involve the electrolytic generation of the titrant in situ in the reaction vessel under constant current conditions and detecting the end point of the titration by visual indictor or pH change, potential change, or amperometrically. A wide variety of titrations involving acid–base, red-ox, complexation and precipitation reactions are amenable for coulometric titrations.

- Potentiometry involves the determination of the Ecell of the electrochemical cell which in turn is related to the concentration of the analyte. Ion-selective electrodes include glass electrode for pH measurement and certain metal ions, gas sensing electrodes, and solid state electrodes.

- Potentiometric titrations are an important analytical application for acid–base, red-ox and precipitation reactions and involve monitoring Ecell as a function of the volume of titrant for detecting end points.

- Polarography monitors the diffusion current generated by the analyte under controlled conditions of suppressed migration and convection currents and absence of dissolved oxygen as a function of applied dc voltage to the DME.

- The half-wave potential and the height of the polarographic wave are useful in qualitative and quantitative analyses respectively.

- Amperometric titrations involve monitoring the limiting diffusion current due to a constant applied potential during a titration of the analyte with a standard solution of the reagent taken in the burette. At least one of the reactants must be electroactive and the end point of the titration is detected graphically from the plot of diffusion current versus volume of titrant.

- Amperometric titrations are performed by using one of the electrodes or both the electrodes under polarized conditions.

- Important examples of applications of amperometry include oxygen sensor and glucose sensor.

REVIEW QUESTIONS

1. Classify the electroanalytical methods.
2. Explain briefly the terms 'specific conductance' and 'equivalent conductance' and their significance.
3. State Kohlrausch law. What are its applications?
4. How is conductance of a solution determined experimentally?
5. What are the different types of titrations that can be performed by conductance measurements?
6. Discuss the different examples of conductance titrations involving acids and bases.
7. How are precipitation titrations carried out conductometrically?
8. Discuss the principle and procedure involved in high frequency titrations.
9. State and explain Faraday's laws of electrolysis.
10. Explain the principle involved in controlled potential electrogravimetry.
11. Discuss the principle involved in coulometry.
12. Give a detailed account of coulometric titrations involving acids and bases.
13. How are the end points determined in precipitation and redox titrations by coulometry? Discuss in detail the experimental procedure with suitable examples.

14. Explain the basic principles involved in potentiometry.

15. What are indicator electrodes? Give a detailed account of the different types of indicator electrodes.

16. Give a neat sketch of the saturated calomel electrode and describe the reactions involved. How is it useful as a reference electrode?

17. How is the EMF of a single electrode determined experimentally?

18. Describe the experimental set-up for conducting potentiometric titrations. Discuss the principle involved in the detection of end point in a red-ox titration by potentiometry.

19. Describe the construction of a pH electrode. How does it function?

20. Write a note on pH titrations.

21. What are ion-selective electrodes? Write a note on the different types of ion selective electrodes and their uses.

22. Explain the basic principle in polarography. How is the technique useful in qualitative and quantitative analyses?

23. Describe the experimental set-up for polarography and the procedure involved for obtaining a polarogram.

24. Explain the terms 'half wave potential' and 'diffusion current'. What is their importance?

25. Discuss the different methods of quantitative analysis in polarography.

26. Give a detailed account of the amperometric titrations and their applications.

27. Explain the principle involved in biosensors. How does glucose sensor function?

28. Describe the construction and function of oxygen sensor.

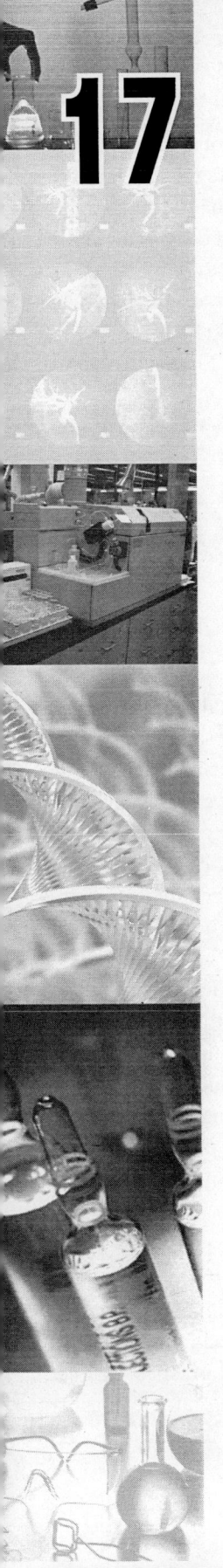

17 Thermal Analytical Methods

17.1 INTRODUCTION

Thermal methods of analysis involve monitoring the changes in some physical property such as weight, temperature, or enthalpy as a function of temperature as the sample is being heated at a controlled heating rate. Other physical properties that may be monitored during the heating of the sample include dimensional changes or mechanical, electrical, acoustic, and optical properties of the sample. Temperature-induced physical changes such as (i) melting, (ii) phase transitions between crystalline phases, (iii) sublimation or volatilization, (iv) glass transition in polymers, (v) chemical reactions such as oxidation, reduction, combustion, decomposition, and combination and (vi) changes in mechanical properties such as compression, expansion, and penetration, can be studied using different thermal analytical methods. Changes in the sample properties as a function of temperature are usually displayed graphically as a *thermal analysis curve* or *thermogram*.

The most commonly used thermal analytical techniques are *thermogravimetry* (TG), *differential thermal analysis* (DTA), *differential scanning calorimetry* (DSC), and *thermomechanical analysis* (TMA), and *dynamic mechanical analysis* (DMA) which monitor changes in mechanical properties; *evolved gas analysis* (EGA) to study the volatile gases/products emitted on heating a sample. Other thermal analytical techniques include *thermodilatometry* for observing changes in dimensions (length or volume), *thermoelectrometry* for changes in electrical properties, *thermomagnetometry* (TM) for changes in magnetic properties, *thermoptometry* or *thermomicroscopy* for changes in optical properties, *thermosonimetry* or *thermoacoustimetry* (TS) for monitoring acoustic properties, and *emanation thermal analysis* (ETA) for monitoring evolution of radioactive gases from samples.

This chapter deals with the important techniques of TG, DTA, DSC, TMA, DMA, and EGA.

17.2 THERMOGRAVIMETRY

In thermogravimetry (TG), the sample is heated at a chosen or specified heating rate and the weight of the sample is continuously monitored as a function of temperature by a sensitive balance. The variation of weight of the sample, as it is being progressively heated to higher temperatures, is displayed as a graphical plot called the thermogravimetric curve or thermogram.

17.2.1 TG Instrument

The instrument used for thermogravimetric analysis is usually called a thermobalance. The main components of the instrument include (i) a sensitive balance which monitors the weight of the sample continuously and (ii) a microprocessor-controlled electrical furnace to heat the sample at a controlled heating rate. The atmosphere within the furnace can be changed at any point of time or temperature using a purge gas. It is possible to enhance or suppress the rates of reactions that the sample may undergo by the use of a purge gas, and thereby, the thermal events that occur in the sample may be controlled. Commonly used purge gases include air; inert gases, such as nitrogen, helium, or argon; or reactive gases such as oxygen, hydrogen, etc. Figure 17.1 shows the schematic diagram of the instrument.

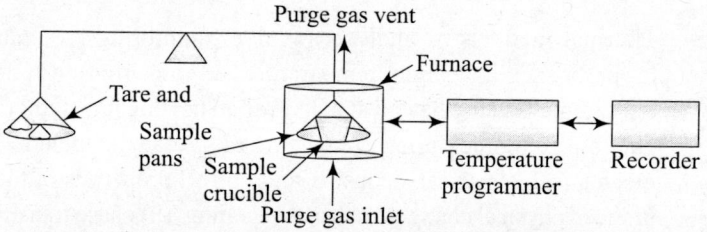

Fig. 17.1 A schematic diagram of thermobalance

The sample pan of the balance is kept suspended inside the electrical furnace whose temperature is accurately controlled by a microprocessor programmer. The sample, usually a solid of 5–50 mg, is placed in a crucible made of alumina, platinum, or gold and placed on the balance pan held in position inside the furnace. The temperature of the furnace (and hence, that of the sample) is increased at a specified heating rate. The temperature range of the furnace can be varied from ambient to about 1500°C at any chosen heating rate. The heating rate can be as low as 1°C min^{-1} to as high as 300–400°C min^{-1}, the optimum heating rate being 10–20°C min^{-1}. Modern instruments are microprocessor/computer-controlled with built-in software that facilitates computation and printing of the graphical plot of sample weight as a function of temperature (thermogram). The software facilitates the computation of the first derivative of the thermogram called the *differential thermogram* (DTG) and plotting the derivative (dw/dT) as a function of furnace temperature. DTG is useful for interpreting the data and more accurate determination of the temperature at which physical or chemical changes occur in the sample, particularly in the case of more complex thermograms or where changes are subtle.

The TG instrument requires temperature calibration for obtaining reproducible thermograms of good quality. Temperature calibration may be carried out by *Curie-point* method. The method is based on the Curie point or temperature where a ferromagnetic material loses its magnetism since the transition temperature is exactly reproducible. Different ferromagnetic metals and alloys having Curie points in the temperature range of 150°C to about 1000°C are available commercially for use as calibration standards. For calibrating the TG instrument, the ferromagnetic calibration standard is placed on the sample pan and a large permanent magnet is placed below the pan. The calibration standard sample being ferromagnetic is attracted towards the permanent magnet showing an increase in weight. As the sample is heated at a specified heating rate, it loses its ferromagnetism at the Curie point with an apparent loss of weight. The temperature experienced by the sample pan can be accurately determined by this method. Figure 17.2 shows the Curie-point calibration of the TG instrument by a series of ferromagnetic materials.

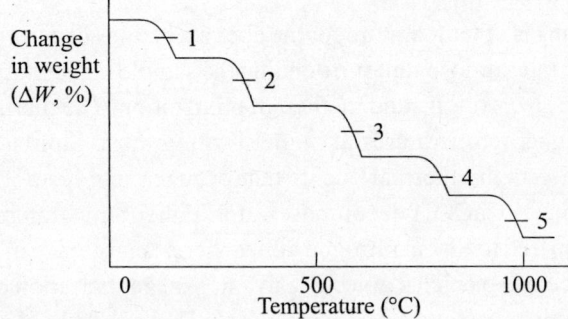

Fig. 17.2 Temperature calibration of TG instrument by Curie-point method: 1. Alumel (163°C), 2. Nickel (354°C), 3. Perkalloy (596°C), 4. Iron (780°C), and 5. Hisat 50 (1,000°C)

17.2.2 Thermogram

The TG thermogram is best explained with the example of the thermal analysis of calcium oxalate monohydrate ($CaC_2O_4.H_2O$), shown in Fig. 17.3. The thermogram is obtained by plotting the mass of the sample as a function of the furnace temperature. The corresponding derivative curve (DTG) curve is also shown as discontinuous line in the same figure.

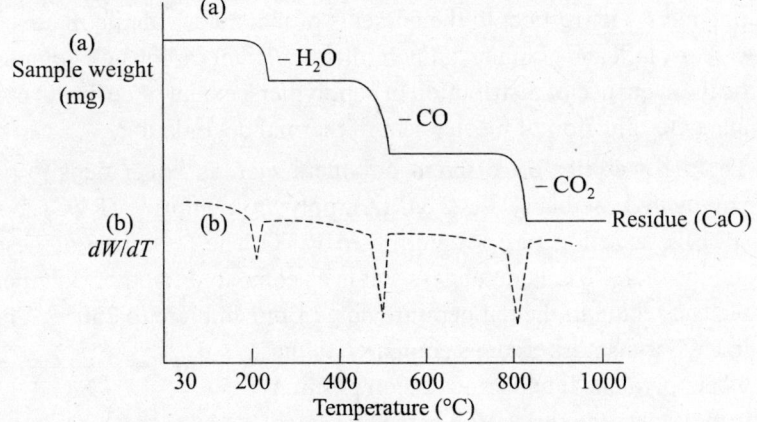

Fig. 17.3 TG analysis of calcium oxalate monohydrate: (a) TG curve and (b) DTG curve

The thermogram may be interpreted as follows. The sample undergoes thermal decomposition in three stages as represented by the following stoichiometric equations:

$$CaC_2O_4.H_2O \xrightarrow{130-190°C} CaC_2O_4 \text{ (anhydrous)} + H_2O \tag{17.1}$$

$$CaC_2O_4 \xrightarrow{400-470°C} CaCO_3 + CO \tag{17.2}$$

$$CaCO_3 \xrightarrow{700-840°C} CaO + CO_2 \tag{17.3}$$

At each stage of decomposition, the weight of the sample decreases due to the evolution of volatile products such as water vapour (H_2O), carbon monoxide (CO), and carbon dioxide (CO_2), respectively. The final residue that remains in the crucible is calcium oxide (CaO). Thus, the TG curve shows four plateau regions and three downward steps. Each plateau indicates the thermal stability of a particular species, and each downward step indicates decrease in mass due to loss of volatile products.

The exact ranges of temperature for the above reactions depend on the experimental conditions such as heating rate, atmosphere surrounding the sample in the furnace, and sample characteristics such as particle size, weight, and method of preparation. The thermal decomposition of a sample will shift to higher temperatures at higher heating rates. Similarly, the atmosphere within the furnace also affects the thermal events that occur during heating of the sample. For example, calcium carbonate ($CaCO_3$) decomposes at a higher temperature in an atmosphere of carbon dioxide as compared to that in nitrogen atmosphere. A large size of the sample as well as a sample in the form of coarse particles gives rise to a non-linear variation of temperature. It is preferable to use small particle size and small sample size. The sample holder or crucible should preferably be a flat plate-shaped to facilitate escape of any evolved gas from the sample.

17.2.3 Applications of Thermogravimetry

The major applications of thermogravimetry include the determination of thermal stability and purity of samples and quantitative determination of composition of mixtures. A few specific case studies are discussed here.

(i) *Thermal characterization of polymers* such as assessment of thermal stability, particularly the high temperature stability and decomposition temperature; composition of filled polymers with respect to the contents of moisture, volatile matter, combustibles such as carbon black and graphite, or inert fillers (calcium carbonate); composition and information on the sequence of distribution in copolymers; extent of curing in condensation polymers; and determination of mechanism of thermal degradation.

Thermal stability of different polymers such as linear density polyethylene (LDPE), polymethyl methacrylate (PMMA), polyvinyl chloride (PVC), polytetrafluoroethylene (PTFE), or Teflon may be determined by TG. The TG curves of these polymers are shown in Fig. 17.4. PVC is much less stable as compared to other polymers as indicated by the plateau region in the temperature range from ambient to 250°C. The degradation pattern in PVC is also different as compared to the other polymers. It undergoes decomposition in two stages to release HCl in the first stage followed by degradation to small fragments in the second stage. The temperatures up to which the polymers are thermally stable are obtained from the plateau regions of the TG curves as 300°C for PMMA, 375°C for LDPE, and 475°C for PTFE. Thus, the thermal stabilities of these polymers decrease in the order PTFE > HDPE > PMMA > PVC.

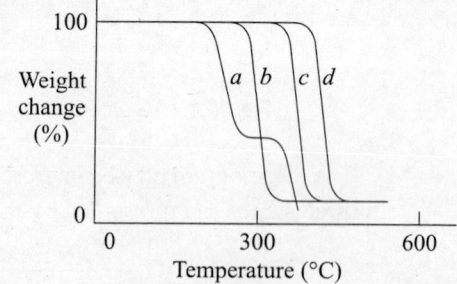

Fig. 17.4 TG curves of polymers indicating thermal stability: (a) PVC, (b) PMMA, (c) LDPE, and (d) PTFE

(ii) *Composition of composite materials* can be determined from TG measurements. For example, the determination of the percentage composition of a commercial elastomer in terms of the polymer, oil extender (plasticizer), and inorganic filler can be carried out by thermogravimetry (Fig. 17.5).

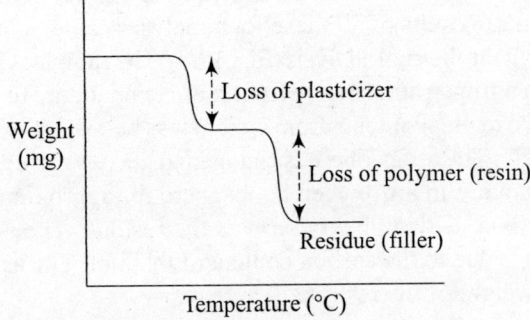

Fig. 17.5 Determination of composition of an elastomer by TG

(iii) Quantitative analysis of a mixture of calcium and strontium in an analyte sample can be carried out by converting them to carbonates and recording the TG of the mixture. Calcium carbonate decomposes quantitatively to yield calcium oxide first in the temperature range 700–850°C, whereas strontium carbonate decomposes only in the temperature range 950–1100°C (Fig. 17.6). The TG curve clearly shows the two distinct thermal decompositions, and from the weight changes in the appropriate temperature ranges, the composition of the mixture can be calculated.

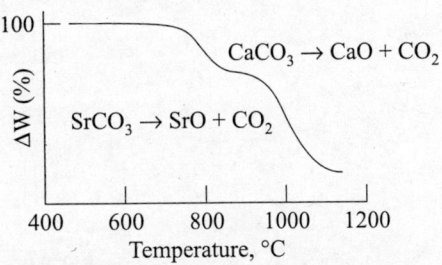

Fig. 17.6 Thermal analysis of mixture of carbonates

(iv) *Moisture content* of a variety of inorganic and organic compounds, industrial raw materials, finished products, coal, foods, pharmaceuticals, etc., can be determined by TG from the loss of weight on subjecting the sample to a programmed heating from room temperature to about 200°C. The presence of different types of water in solids such as moisture, water of hydration or crystallization, and coordinated water can be identified and quantitatively determined. For example, the presence of water of hydration and coordinated water in $CuSO_4.5H_2O$ is shown by the loss of water at different temperatures in the TG thermogram in a stepwise manner until about 220°C. The water of hydration is expelled at relatively lower temperatures below 120°C, whereas the coordinated water molecules are expelled at higher temperatures. The anhydrous copper sulphate loses sulphur dioxide to leave a residue of copper oxide at temperatures above 700°C (Fig. 17.7). The DTG curve shows the corresponding changes more clearly.

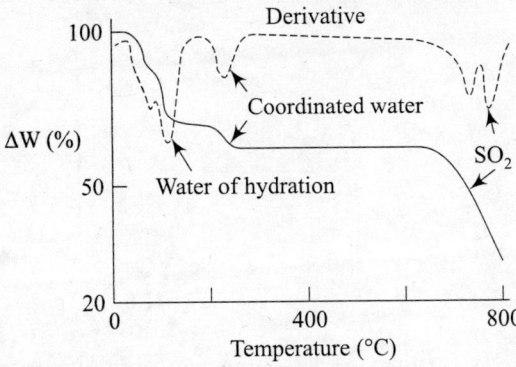

Fig. 17.7 TG and DTG curves of $CuSO_4.5H_2O$

(v) *Proximate analysis of solid fuels* (coal) is useful in evaluating the quality of fuel for use in thermal power plants. The analysis involves the determination of moisture, volatile matter, fixed carbon, and ash contents of solid fuels conventionally carried out in four different experiments. Results of all these four analyses can be obtained in a single experiment by carrying out the thermal analysis (Fig. 17.8). The sample is subjected to programmed heating initially in nitrogen atmosphere from room temperature to 900°C to determine the moisture and volatile matter contents from weight loss below 200°C and between 400°C and 900°C, respectively. The atmosphere is changed to air/oxygen at 900°C, and the sample is held at this temperature in air/oxygen atmosphere for approximately 5 min to completely burn-off the carbon, leaving behind ash as the residue. The weight loss due to heating in air or oxygen is due to the carbon content of the fuel. The ash content of the fuel is obtained from the weight of the residue.

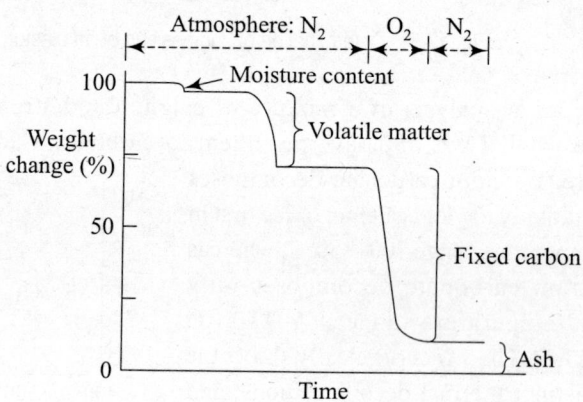

Fig. 17.8 TG curve for proximate analysis of coal

(vi) *Purity of pharmaceuticals* and the contents of the active ingredient and excipient can also be analysed (Fig. 17.9).

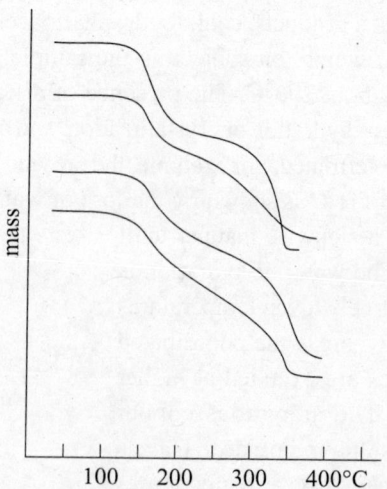

Fig. 17.9 TG analysis of analgesics of different manufacturers

(vii) *Study of temperature-induced reactions* such as thermal degradation reactions, oxidation in the presence of air/oxygen, and reduction in presence of hydrogen can be carried out by

TG. The weight loss or weight gain occurring at different temperatures can be correlated to the reactions and stoichiometric relationships can be deduced.

(viii) *Kinetics of isothermal reactions* can be studied. The rate of isothermal reaction of a substance can be determined by recording the weight loss or weight gain as a function of time at any desired temperature.

17.3 Differential Thermal Analysis

Differential thermal analysis (DTA) involves the measurement of the temperature difference (ΔT) between the test sample and an inert reference substance, such as alumina or glass, as a function of temperature of the furnace as they are heated uniformly at the same heating rate.

17.3.1 DTA Instrument

The DTA instrument consists of the following major components: (i) a furnace for heating both the sample and the reference at the same rate, (ii) thermocouples to measure the temperature of the furnace and that of the reference and sample, (iii) data processor and recorder, and (iv) a device to allow purge gas into the furnace to control the furnace atmosphere. The sample and the reference are placed in crucibles made of ceramic (alumina), and the furnace has provision to hold the crucibles. The furnace is usually microprocessor controlled and can be heated electrically from room temperature to 1500°C at different rates of heating in the range 1–100°C min^{-1}. Temperatures below room temperature even up to –180°C can be attained with cryogenic attachment.

Two types of DTA instruments are available commercially. In the *classical DTA* instrument, the sample and a reference material (α-alumina, silicon carbide, or glass beads) placed in two identical crucible like crevices (sample (S) and reference (R), respectively–Fig. 17.10) on a single heating block which in turn is placed in a furnace and heated at the same rate. The temperatures of the sample (T_S) and the reference (T_R) are measured separately by individual thermocouples placed in contact with the samples. A third thermocouple is used to control the temperature of the furnace. The DTA thermogram is obtained as a plot of the difference ΔT between the test sample and the reference material ($\Delta T = T_R - T_S$) as a function of furnace temperature.

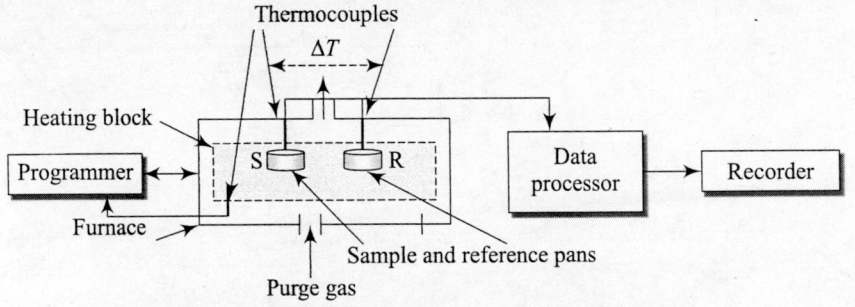

Fig. 17.10 A schematic diagram of classical DTA instrument

The second type of DTA instrument is the *Boersma* DTA or *calorimetric* DTA instrument in which the sample and reference materials are placed on two pans kept on separate heating blocks (Fig. 17.11). The thermocouples are placed in contact with the pans (or the heating blocks) rather than the samples, and a third thermocouple measures the furnace temperature as in the classical instrument.

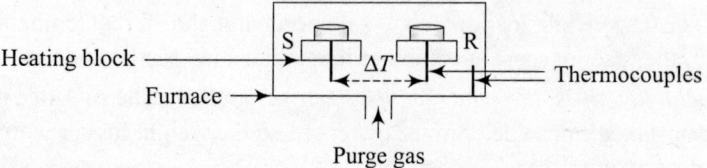

Fig. 17.11 Arrangement of sample holders in calorimetric DTA

The furnace in both the types of DTA is microprocessor controlled. The instrument can operate from ambient to 1500°C with different rates of heating in the range 1–100°C min^{-1}. Low temperatures up to liquid nitrogen temperatures (–180°C) can be reached with suitable cryogenic attachment.

17.3.2 DTA Thermogram

As the sample and the reference are heated in a programmed manner, their temperature increases uniformly. The reference substance does not undergo any physical or chemical changes during heating, and its temperature increases uniformly depending on its specific heat and the rate of heating [Fig. 17.12(a)]. In contrast, the test sample may undergo physical changes such as phase transition, melting, etc. and/or chemical changes such as oxidation, decomposition, etc. These changes may be either endothermic or exothermic. Such changes occurring in the sample lead to a difference in the temperature of the sample as compared to that of the reference even though both are subjected to the same programmed heating. The difference in the temperature ΔT is monitored as a function of the furnace temperature T giving rise to the graphical plot of ΔT versus T as the DTA thermogram as shown in Fig. 17.12(b).

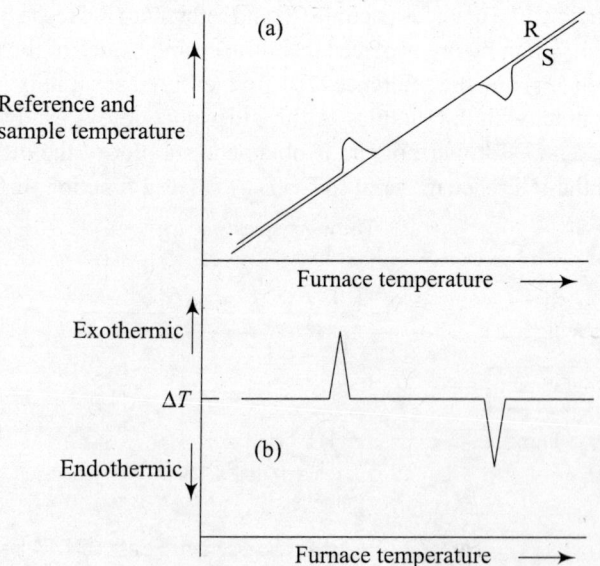

Fig. 17.12 (a) Temperature of reference and sample as a function of furnace temperature and (b) DTA plot of ΔT versus T

DTA thermogram is useful in qualitative and quantitative analyses of a wide variety of materials such as inorganic and organic compounds, pharmaceuticals, foods, polymers, minerals, and biological samples. The pattern of the thermogram can be used as a fingerprint for qualitative identification, whereas the areas under the peaks can be used for quantitative purposes. The thermogram provides information about the nature of energy changes accompanying the physical or chemical transitions that the sample undergoes during heating. The peak temperature indicates the temperature at which such transition occurs, and the peak area is proportional to the energy change (hence, the enthalpy change ΔH) that occurs during the transition. It may be difficult to correlate directly the thermal changes exhibited by the thermogram to the thermal processes that occur in the sample, and in addition, the peak area in the DTA thermogram does not reflect the enthalpy changes exactly, and hence, quantitative analysis requires calibration with appropriate standards.

The area of the peak depends on the amount of material, the heat of reaction, and the heat flow to or from the sample. The relationship between the peak area and the parameters affecting it may be expressed in the form of an equation.

$$A = m\left(\frac{-\Delta H}{gk}\right) \tag{17.4}$$

where m is the amount of the sample in moles, g a constant related to the geometry of the sample, and k a constant related to the thermal conductivity of the sample. The constants g and k can be evaluated provided the heat of reaction of the sample is known. To determine the ΔH of a transition or reaction by DTA, it is necessary to calibrate the instrument and the peak area by using a substance of known ΔH value under controlled conditions of heating. The above equation may then be simplified to

$$A = k'm(-\Delta H) \tag{17.5}$$

where k' is an empirical conversion factor accounting for the geometry and the thermal conductivity of the sample. The above equation can be used to evaluate the ΔH value for any test sample by performing DTA under controlled conditions and determining peak area.

Melting, crystallization, and phase transitions in samples as well as purity estimation from melting characteristics of the sample in comparison with an authentic substance and reaction kinetics of thermal decompositions may be studied conveniently from enthalpy measurements using DTA. Sample sizes required are usually in the range of 5–20 mg.

17.4 DIFFERENTIAL SCANNING CALORIMETRY

Differential scanning calorimetry is similar to DTA except that the sample and reference are heated by separate electrical heaters programmed to heat, both exactly at the same rate. The principle involves the measurement of the difference in the heat flow or flux to the sample in comparison to the reference or alternatively heat gain (or heat loss) by the sample in comparison to the reference. The measured energy difference between the sample and the reference is directly equal to the enthalpy change (ΔH) of the sample as it undergoes changes such as melting, phase transition, or chemical reaction during heating. The DSC thermogram is a plot of ΔH versus T and similar to DTA thermogram in appearance [Fig. 17.12(b)] except that the y-axis is ΔH.

17.4.1 DSC Instrument

The instrument consists of the following main components: (i) a furnace with provision to hold the sample and reference pans and provided with heaters and thermal sensors for the sample and reference, (ii) data processor and recorder, and (iii) facility for atmospheric control. The sample is placed in aluminum pan and an empty aluminum pan is used as the reference. The furnace is controlled by a microprocessor and has a temperature range from room temperature to 700°C and can be programmed to provide different heating rates. The temperature range of the instrument can be extended to liquid nitrogen (77 K) temperatures.

Commercially two different types of DSC instruments are available: (i) *heat flux DSC* and (ii) *power compensation DSC*.

Heat flux DSC is based on the quantitative measurement of the differential heat flow or flux to the sample and the reference as they are heated by the same source. The sample pan containing the sample and the empty reference pan (both made of aluminum) are placed on raised platforms on a constantan thermoelectric disc placed inside a heating block. As heat is transferred to the pans from the electrically heated disc, the differential heat flow to sample and reference is monitored by chromel/constantan area thermocouples. The thermocouples formed by the junction of the constantan disc and chromel discs or wafers attached to the underside of the constantan disc are connected in series to measure the differential heat flow. The temperature of the sample is determined by chromel/ alumel thermocouple attached to the chromel wafers. The heat flux is directly related to enthalpy changes in the sample. The instrument has facility for introducing purge gas to control the atmosphere within the dynamic sample chamber (Fig. 17.13).

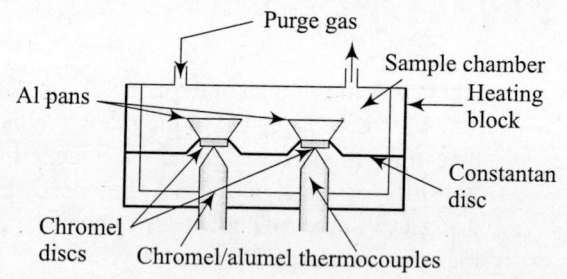

Fig. 17.13   Schematic diagram of the furnace in heat flux DSC

Power compensation DSC makes a direct measurement of the enthalpy change by making use of a different instrument design. The sample and reference pans are heated by separate heating blocks in the furnace by separate electrical heaters (Fig. 17.14) programmed such that the sample and the reference are heated exactly at the same rate. Whenever the sample undergoes a phase transition (endothermic) such as simple melting or glass transition in a polymer or exothermic changes (e.g., oxidation or crystallization), it results in a change in the sample temperature in comparison to the temperature of the reference and is monitored by the individual Pt sensors. The power to the sample or the reference heater is modified so that the temperature difference (ΔT) between the sample and reference is maintained as zero ($\Delta T = 0$). The difference in power supplied to the sample and reference heaters is accurately monitored and represents the energy change or enthalpy change (ΔH) in the sample.

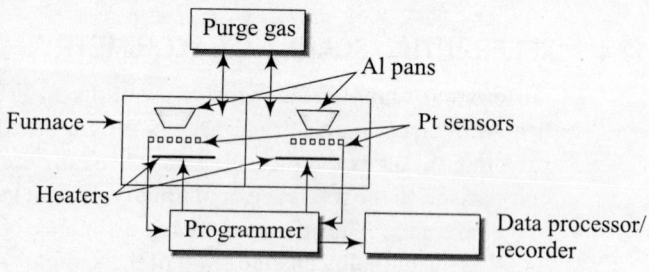

Fig. 17.14   A schematic diagram of power compensated DSC

17.4.2   Applications of DTA and DSC

The DSC and DTA thermograms are similar and provide almost the same information except that DSC provides the direct measurement of enthalpy changes (ΔH) of thermal events, whereas the peak area in DTA is only proportional to enthalpy change, and hence, requires calibration with standard substances. The techniques are useful in the study of phase transitions and temperature-induced changes that occur even at very low temperatures. Both the techniques find applications in the study of a wide variety of inorganic, organic, food, pharmaceutical, polymer, and biological samples. The techniques are useful in solving many manufacturing/processing problems and for the determination of per cent purity of samples. The techniques find use in safety analysis in automobile, aviation, and electronic industries.

A few typical applications of DTA to inorganic samples include (i) study of phase transitions and (ii) deducing the stoichiometric reactions that are illustrated here. Figure 17.15 shows the DTA thermogram of sulphur exhibiting four distinct endothermic peaks. The peak at 113°C is attributed to the phase transition of rhombic sulphur to monoclinic, whereas the peak at 124°C corresponds to the melting of the element. Liquid sulphur is known to exist in three different forms and the peak at 179°C is assigned to one of these transitions. The large endothermic peak at 446°C corresponds to the boiling point of sulphur.

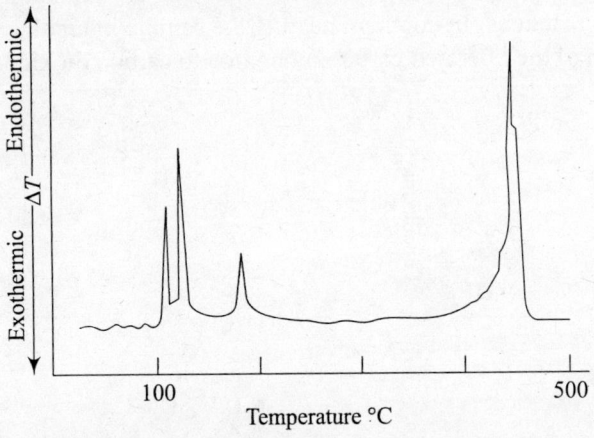

Fig. 17.15 Differential thermogram of elemental sulphur

DTA thermogram of manganese phosphinate monohydrate (Mn $(PH_2O_2)_2 \cdot H_2O$) shows both exothermic and endothermic peaks (Fig. 17.16)

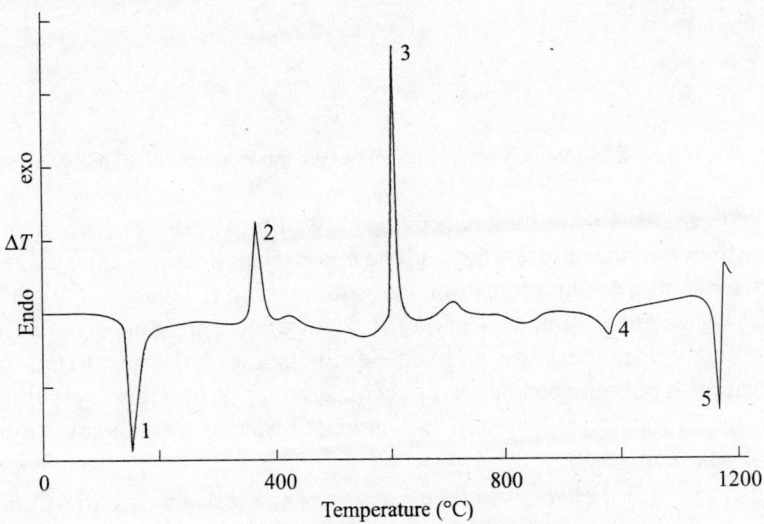

Fig. 17.16 DTA thermogram of Mn $(PH_2O_2)_2 \cdot H_2O$

The stoichiometric equations of the thermal transitions may be expressed as follows:

(i) $Mn(PH_2O_2)_2 \cdot H_2O \xrightarrow{170°C} Mn(PH_2O_2)_2 + H_2O$ (endothermic)

(ii) $Mn(PH_2O_2)_2 \xrightarrow{380°C} MnHPO_4(\alpha) + PH_3$ (exothermic)

(iii) $MnHPO_4(\alpha) \xrightarrow{580°C} MnHPO_4(\beta)$ (exothermic)

(iv) $2\,MnHPO_4(\beta) \xrightarrow{950°C} Mn_2P_2O_7 + H_2O$ (endothermic)

(v) $Mn_2P_2O_7 \xrightarrow{1180°C}$ melting (endothermic)

The effects of the change in the atmosphere as the sample is undergoing programmed heating is nicely demonstrated in DTA thermograms of calcium oxalate monohydrate in nitrogen and air atmospheres as shown in Fig. 17.17. Heating the sample in nitrogen atmosphere gives rise to a thermogram indicating the loss of water, carbon monoxide, and carbon dioxide due to endothermic reactions. In contrast, heating the sample in air shows an exothermic peak due to the oxidation of the liberated carbon monoxide to carbon dioxide.

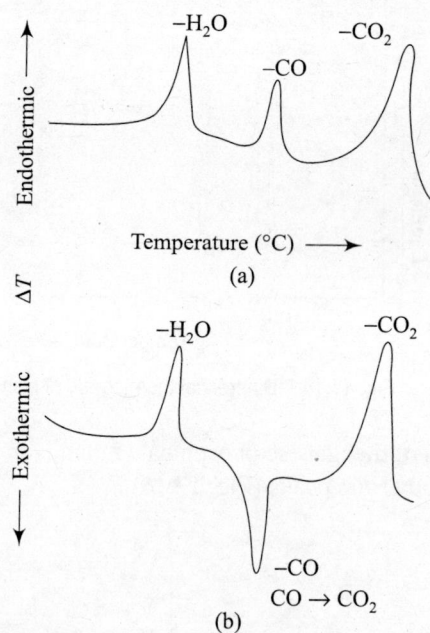

Fig. 17.17 DTA thermogram of calcium oxalate monohydrate in (a) nitrogen and (b) air

DTA /DSC analysis of polymers provides a wealth of information on several parameters of importance in polymer processing and the end use applications. The parameters include (i) softening temperature or glass transition temperature (T_g), (ii) melting temperature, (iii) heat of melting, (iv) degree of cure (or cross-linking), (v) temperature and time of curing, (vi) residual cure, (vii) crystallization temperature and time, (viii) isothermal crystallization characteristics, (ix) composition of polymer blends, (x) effect of additives on polymer crystallinity, (xi) oxidative induction time or long-term stability, (xii) impact resistance, (xiii) enthalpy of transitions or reactions of polymers, etc.

Polymers show major changes in their properties at one of the two transition points, namely, melting point and glass transition temperature. Crystalline polymers show characteristic melting point, whereas amorphous polymers exhibit glass transition temperature.

The DTA/DSC thermogram of a typical polymer sample is shown in Fig. 17.18. The glass transition of a polymer is usually seen as a step rather than as a peak as it involves change in the heat capacity of the polymer sample.

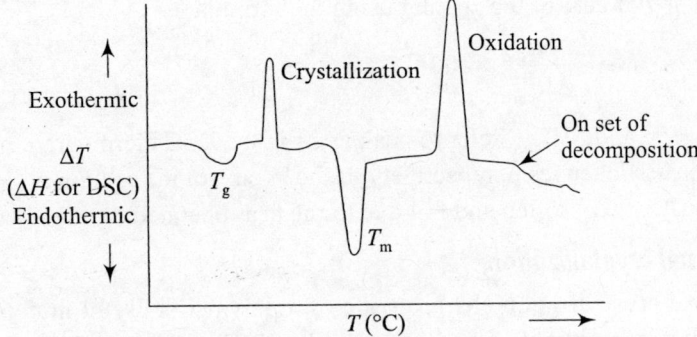

Fig. 17.18   DTA/DSC thermogram of a typical polymer sample

Few of the above-mentioned parameters are explained in the following text.

Glass transition temperature

The glass transition temperature (T_g) is an important characteristic of polymers as it indicates the lower end use temperature. All amorphous (non-crystalline or semi-crystalline) materials will exhibit a T_g during heating which is the transformation temperature of the amorphous phase. The glass transition event occurs when a hard, solid, amorphous material, or component undergoes its transformation to a soft, rubbery, liquid phase. The T_g is observed as an endothermic stepwise change in the DSC.

T_g is a valuable characterization parameter associated with a material and can provide very useful information regarding the end use performance of a product. A polymeric material cannot be processed or worked below its T_g. T_g is also related to the end use characteristics of a wide range of polymeric materials. For example, the epoxy thermosetting adhesive resin with a T_g below 25°C is a liquid at room temperature. It undergoes curing on the addition of a cross-linking agent and its T_g increases as it undergoes curing and setting. The T_g of the polymers used in electrical wire coatings and motor windings should be well above the operating temperature to prevent any failure of the protective polymer coating.

The crispness of a food such as a cookie can be evaluated on the basis of T_g of the biopolymer starch portion of the cookie. Prolonged or improper storage of the food results in the absorption of moisture by the food resulting in the loss of crispness. Gradually the food becomes stale. The loss of crispness of the food is indicated by a decrease of T_g of the starch component well below room temperature.

The setting in of the crease during ironing of the garment is related to T_g of the polyester in the garment. During ironing, the temperature of the amorphous component of the polyester exceeds its T_g and a crease can be formed in the garment. The crease sets in when the iron is removed and the temperature drops below the T_g.

Degree of cure

The degree of cure or extent of cross-linking is an important parameter in determining the effectiveness of polymer coatings. For example, when a thermoset is coated on wires or motor windings the polymer should be highly cross-linked so that it does not soften due to heat liberated during the end use. In general, the higher the degree of cure or cross-linking, the higher the glass transition temperature or softening temperature. The percentage of cure can be calculated from the enthalpy of cure of the coating using the formula

$$\% \text{ cure} = \frac{\Delta H_{cure}}{\Delta H_{cure}^0} \times 100 \tag{17.6}$$

where ΔH_{cure} and ΔH_{cure}^0 refer to peak areas in the DSC thermogram of the test sample and the pure uncross-linked resin, repectively. If the % cure of the coating is low, the T_g will be low and such a coating will soften and fail due to the heat liberated during operation of the motor.

Isothermal crystallization

Isothermal crystallization characteristic of a polymer is useful in determining the percentage crystallinity of thermoplastic polymers such as PET, nylon, etc. The % crystallinity is important as it is related to stiffness, ability to resist creep, impact resistance or toughness, barrier resistance, and optical clarity of the final product. The % crystallinity of a polymer is calculated from the DSC thermogram using the following formula:

$$\% \text{ crstallinity} = \frac{\Delta H_M - \Delta H_c}{\Delta H_m^0} \times 100 \tag{17.7}$$

where ΔH_m and ΔH_c refer to the enthalpies of melting and crystallization of the test sample, respectively, and ΔH_m^0 refers to the heat evolved from a reference thermoplastic resin such as PE ($\Delta H_m^0 = 286.1 \text{ J g}^{-1}$).

Oxidative degradation

Oxidative degradation of materials such as printed circuit boards made of epoxy resin and non-woven cloth or semiconductor diodes is indicated by the appearance of an exothermic peak in DTA/DSC thermogram. Similarly, oxidative degradation tests can be carried out by DSC to evaluate the storage stability of food products.

17.5 THERMOMECHANICAL ANALYSIS

Thermomechanical analysis (TMA) involves the measurement of the changes in the mechanical properties such as dimensional changes (length or volume), density, etc., of materials as a function of temperature when the sample specimens of specific geometries are subjected to heating at a specified heating rate. The data are displayed in the form of TMA thermograms.

17.5.1 TMA Instrument

The instrument consists of (i) a microprocessor-controlled sample furnace, (ii) a quartz probe containing a thermocouple, (iii) a transducer coupled to the quartz probe, (iv) a heat sink, and (v) data processor and recorder. The sample furnace is electrically heated and can be programmed for specific heating rates in the range of about $-100°C$ to approximately $1000°C$. The thermocouple attached to the quartz probe measures the temperature of the sample while the probe itself monitors the movement of the sample. The sample movement is translated into a movement of the transformer core within the transducer, the linear variable differential transformer (LVDT). The electrical signal generated is proportional to the displacement of the quartz probe and the sign of the signal indicates the direction of the movement. Different probes are used for measuring different mechanical properties such as expansion/compression, penetration, tension, volume change, etc. Schematic diagrams of the basic components of TMA analyser and the different types of probes are shown in Fig. 17.19. The sample rests on a quartz stage surrounded by the furnace for the expansion and penetration tests. The sample in the form of film or fibre is held between a pair of stationary and movable hooks made of silica in the probe for studying the sample under tension. Volume expansion of the sample is measured by placing it in a cylinder-piston type quartz probe or immersing the sample in silicon oil or alumina powder so that the volume change is translated into a linear motion of the piston.

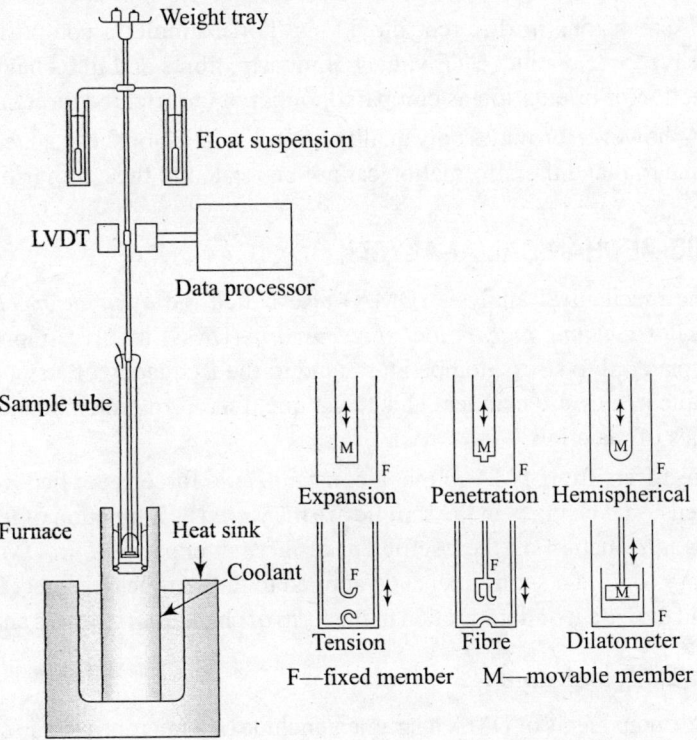

Fig. 17.19 (a) TMA analyser and (b) probe configurations

17.5.2 Applications of TMA

TMA is an important tool in material science and technology, providing information required by design and process engineers on the behaviour of materials at different temperatures. It has been used extensively in the study of polymers and composites with regards to their viscoelastic characteristics, aging, penetration by solvents, and mechanical properties. For example, the glass transition temperature T_g of a polymer corresponds to the expansion of the free volume allowing greater chain mobility above the transition temperature. The coefficient of thermal expansion (CTE) changes abruptly at the transition temperature as indicated in the TMA curve of a polymer (Fig. 17.20).

When the product is heterogeneous and consists of different materials in contact with each other, large differences in the CTE of the different materials can lead to a variety of failures of the product during operation. The materials of the product may be anisotropic, i.e., have different

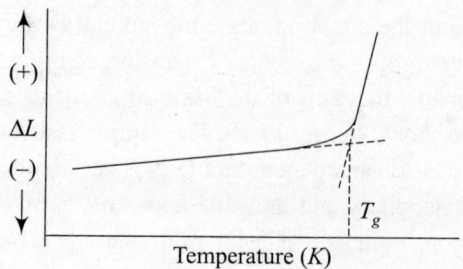

Fig. 17.20   TMA curve showing glass transition region of a polymer

thermal expansions in different directions. For example, a composite of graphite fibres and epoxy have three distinct CTE values. Similarly, fibres and films have different CTE values in the direction of orientation as compared to that in un-oriented direction.

TMA, however, provides only qualitative information on the changes in mechanical properties of the material and the information cannot be related to the composition of the sample.

17.6   DYNAMIC MECHANICAL ANALYSIS

Dynamic mechanical analysis (DMA) also called the *dynamic mechanical thermal analysis* (DMTA) or *dynamic mechanical spectroscopy* (DMS) involves monitoring the response of a sample material to stress, temperature, and to the frequency of a small deformation applied to the sample in a cyclic manner. This technique, therefore, links the thermal and the mechanical properties of materials.

DMA differs from TMA in that a constant static force is applied to the sample in TMA and the dimensional changes in the sample are measured as a function of temperature or time. TMA provides information on the coefficient of thermal expansion. In DMA, on the other hand, an oscillatory force at a set frequency is applied to the sample to subject it to a controlled stress or strain and the extent of deformation in the form of changes in stiffness and damping are monitored.

17.6.1   DMA Instrument

The main components of DMA instrument include (i) a microprocessor-controlled sample furnace, (ii) sample clamps to hold the sample, (iii) a drive shaft and shaft guidance system connected to a drive motor for providing the load for the applied force, and (iv) LVDT which measures the change in voltage as the probe moves through the magnetic core (Fig. 17.21).

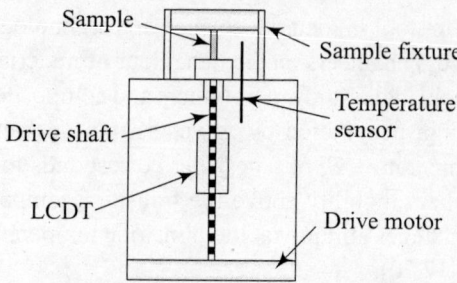

Fig. 17.21   Schematic diagram of DMA

The sample is prepared in specified dimensions for specific measurements.

17.6.2 DMA Applications

DMA finds extensive use in the study of viscoelastic properties of polymers. Materials can be classified as elastic solids or viscous fluids based on their mechanical characteristics. These two extremes of mechanical behaviour of a material are evaluated by determining the stress–strain curve of a material. Polymers are composed of long molecular chains and have unique viscoelastic properties, a combination of the characteristics of an elastic solid and a viscous fluid. In DMA, the viscoelastic properties are determined by applying a sinusoidal force or stress and monitoring the resultant strain in the form of displacement. For a perfectly elastic solid, the stress and the strain will be perfectly in phase, whereas for purely viscous fluids there will be a 90° phase displacement of the strain with respect to stress. Polymers exhibit a phase lag between stress and strain by an angle δ, and thus, the characteristics are in between that of the elastic solids and viscous fluids.

DMA data provide information on elastic (Young's) modulus, storage modulus, loss modulus, and damping of a material. Modulus is the ratio of stress to strain, whereas Young's modulus is the ratio of strain when an increasing stress is applied to the sample. Young's modulus is calculated from the slope of the initial part of the stress–strain curve of a material and indicates the modulus of elasticity under tension or the ability of a material to undergo elastic deformation. DMA provides information on three important parameters, namely, storage modulus (different from Young's modulus), loss modulus, and damping. The storage modulus (E') is a measure of the elastic behaviour of a material under conditions of temperature, load, and the frequency of an applied sinusoidal force, and thus, is a measure of the stored energy in the material and represents the elastic portion of material. The loss modulus is a measure of energy dissipated as heat and represents the viscous portion of a material. Damping is the ratio of loss modulus to storage modulus. Damping is dimensionless property, usually expressed as tan δ and is a measure how well the sample can dissipate energy or absorb energy as it oscillates. DMA also provides information on glass transition temperature T_g and melting temperature T_m of polymer samples.

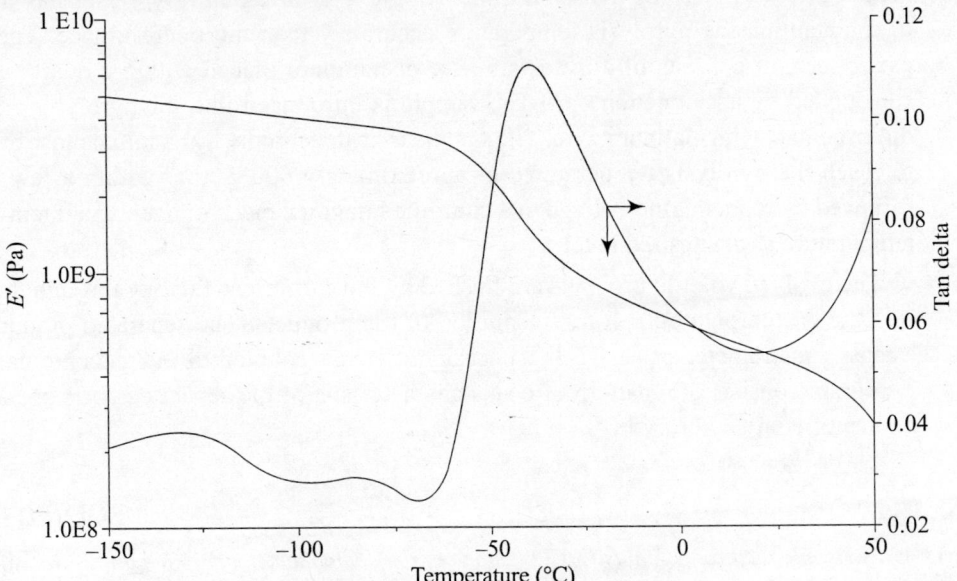

Fig. 17.22 DMA spectrum of polycaprolactone

DMA spectrum of a polymer sample is shown in Fig. 17.22. The drop in storage modulus between $-60°C$ and $-30°C$ is attributed to glass transition of the amorphous region in the semi-crystalline polymer (polycaprolactone). The polymer loses its mechanical integrity above 50°C as it begins to melt and flow. The damping (tan δ) curve shows a bigger peak at approximately $-40°C$ indicating the T_g (known as α transition) and smaller peaks due to local motions within the polymer chain (called β transitions) between $-80°C$ and $-130°C$. The β transitions are not observed in DSC but clearly seen in DMA and are important in evaluating the impact resistance of the polymer.

DMA can be used to evaluate the humidity and immersion effects on materials. The physical changes a material undergoes when exposed to water as, e.g., when pasta is immersed in warm water it penetrates and softens the starch or when gelatin is undergoing dissolution, can be studied by immersion tests using DMA. Similarly, the paper used for wrapping materials can be evaluated for its humidity characteristics.

17.7 EVOLVED GAS ANALYSIS

Evolved gas analysis (EGA) involves the study of volatile breakdown products or gases evolved on heating a sample. By identifying the emitted products, the sample itself can be characterized. One of the common methods of EGA is *pyrolysis gas chromatography* (Py-GC). The technique involves gas chromatographic analysis of the evolved volatile products by the pyrolysis (thermal dissociation) of the sample under controlled conditions in an inert atmosphere or vacuum. The pyrogram obtained is useful for characterizing the sample.

17.7.1 Pyrolysis Gas Chromatograph Instrument

The main components of the instrument consist of (i) a pyrolyser and (ii) a gas chromatograph coupled to mass spectrometer or FT-IR spectrophotometer. The pyrolyser unit is an electrically heated microfurnace for pyrolysing the sample. A Curie-point pyrolyser has the advantage of maintaining the pyrolysis temperature accurately than the microfurnace. The Curie-point pyrolyser unit consists of a pyrolysis wire of platinum placed within a quartz tube which is surrounded by an induction coil. The sample is introduced into the pyrolyser and is put into direct contact with platinum wire. The sample is heated rapidly by an initial pulse of high voltage to reach the pyrolysis temperatures of approximately 600–800°C within a few milliseconds followed by reducing the voltage to a controlled maintenance voltage to maintain the pyrolysis temperature at the desired level.

The evolved volatile products are flushed by the carrier gas flowing through the quartz tube into the gas chromatograph for separation of the products. The separated products are led to a mass spectrometer or an FT-IR spectrophotometer coupled to the gas chromatograph. The pyrogram consists of a pattern of peaks characteristic of the separated components which can be identified by a library fit.

● ● ● ● ● ● ● ● ● ● ● ● ● ● ● ● ● **SOLVED PROBLEMS**

EXAMPLE 1 *A sample mixture contains CaCO$_3$ as one of the components and 15.6 mg of it was subjected to TG analysis between 500°C and 850°C. The weight loss was found to be 4.8 mg. Calculate the CaCO$_3$ of the given sample mixture.*

SOLUTION

The only reaction observed between 500°C and 850°C may be formulated as

$$CaCO_3 \rightarrow CaO + CO_2$$

Since 1 mol of CO_2 (mol. wt. = 44) is lost from 1 mol of $CaCO_3$ (mol. wt. =100), 4.8 mg of CO_2 will be lost from

$(100/44) \times 4.8 = 10.91$ mg of $CaCO_3$

% $CaCO_3$ in the sample mixture = $(10.91/15.6) \times 100 = 69.9$

EXAMPLE 2 *Proximate analysis of a coal sample was carried out using 22 mg and the following weight changes were observed at different temperature ranges : 105–115°C (– 0.5 mg), 500–800°C in nitrogen atmosphere (– 5 mg), and above 900°C in air (– 15 mg). Interpret the observed data.*

SOLUTION

The observed weight loss of 0.5 mg corresponds to loss of moisture in the temperature range of 105–115°C, 5 mg between 500°C and 800°C is due to loss of volatile matter in the sample coal, and 15 mg loss above 900°C is due to the loss of carbon dioxide formed by complete oxidation (combustion) of carbon in the sample. The residue is ash obtained by the combustion process. Hence,

Moisture content of the sample coal = $(0.5 \text{ mg}/22 \text{ mg}) \times 100 = 2.27\%$

Volatile matter content = $(5 \text{ mg}/22 \text{ mg}) \times 100 = 22.7\%$

Fixed carbon content = $(15 \text{ mg}/22 \text{ mg}) \times 100 = 68.2\%$

Ash content = $(22 – 0.5 – 5 – 15 = 1.5 \text{ mg}) = (1.5 \text{ mg}/22 \text{ mg}) \times 100 = 6.82\%$

EXAMPLE 3 *A pure compound having a stoichiometric formula of $CrCl_3.6H_2O$ (mol. wt. = 266.35) was subjected to TG analysis. The sample weight was 12.5 mg. A weight loss of 1.7 mg was observed at 105–115°C and another weight loss of 3.4 mg was observed at 180–200°C. Interpret the observed data.*

SOLUTION

The observed weight losses may be attributed to the loss of water of hydration and coordinated water molecules, respectively.

Loss due to expulsion of water of hydration = $(1.7 \text{ mg}/12.5 \text{ mg}) \times 100 = 13.6\%$

Loss due to expulsion of coordinated water = $(3.4 \text{ mg}/12.5 \text{ mg}) \times 100 = 27.2\%$

The amount of water molecules in the compound = $[(6 \times 18)/266.35] \times 100 = 40.55\%$

Weight loss at 105–115°C corresponds to loss of two molecules of water of hydration.

Weight loss at 180–200°C corresponds to loss of four molecules of coordinated water.

Hence, the molecular formula of the compound is written as $[Cr (H_2O)_4 Cl_2] Cl.2H_2O$

EXAMPLE 4 *DTA thermogram of 20 mg sample of a pure compound A of molecular weight 120 and enthalpy of fusion 1.58 kcal mol^{-1} showed the melting temperature to be 150°C and the peak area was calculated to be 52 cm^2. A sample of 15 mg of compound B of molecular weight 85 having melting temperature of 145°C gave a DTA peak area of 48 cm^2 under identical experimental conditions. Calculate the enthalpy of fusion of the compound B.*

SOLUTION

Assuming the geometries and the thermal conductivities of the two samples to be nearly the same, the enthalpy of fusion of compound B can be evaluated using Eq. (17.5).

ΔH of compound B = ΔH_A (peak area of B/peak area of A) (moles of A/moles of B)

$= 1.58$ kcal mol^{-1} (52 cm^2/48 cm^2) [(20 mg $\times$ 120)/(15 mg $\times$ 85)]

$= 3.22$ kcal mol^{-1}

SUMMARY

- Thermal analytical methods involve monitoring temperature-induced changes in physical properties of substances as they undergo heating at a controlled heating rate.

- TGA involves monitoring changes in the weight of the sample continuously as a function of temperature. The on-line plot of weight change as a function of increasing temperature is called the TG thermogram and is useful to understand the nature of temperature-induced changes in the sample. The first derivative of the thermogram is known as DTG.

- The major components of TGA include a balance and a furnace controlled by a programmer.

- TGA is useful in determining the thermal stability and the composition of materials as the sample undergoes thermal decomposition reactions.

- DTA involves determining the difference in temperature between the reference material (e.g., alumina) and the sample as both of them are subjected to heating at the same heating rate. The DTA thermogram is a plot of ΔT versus T and is helpful in detecting and quantitatively analysing the physical changes such as melting, glass transition, and phase changes and chemical changes such as exothermic and endothermic reactions that occur in the sample. The peak temperature in the DTA thermogram is indicative of the transition temperature and the peak area is proportional to the enthalpy change.

- DSC is similar to DTA except that the sample and the empty crucible are heated at the same heating rate. The heat flow between the sample and the furnace or vice versa is monitored as a function of the temperature of the furnace to give a plot of ΔH versus T called the DSC thermogram. The information generated by DSC is similar to that obtained by DTA, the peak area in the DSC thermogram being equal to the enthalpy change.

- DTA and DSC are extensively used in the study of polymeric materials.

- TMA involves monitoring the changes in dimensions of the sample as a function of temperature.

- DMA is useful to study the relationship between the thermal and mechanical properties of materials as it monitors the response of the sample to changes in temperature or stress.

REVIEW QUESTIONS

1. Give an outline of the different types of thermal analytical methods.

2. Explain the principle involved in thermogravimetry with a suitable example.

3. What is a thermogram? How is it obtained?

4. What is DTG? What is its use?

5. Give a neat sketch of the block diagram of thermobalance and describe the functions of its components.

6. How is TG useful in the analysis of solid fuels?

7. Write a note on the applications of TG in the study of polymers.

8. Explain the principle of DTA.

9. Sketch the block diagram of DTA instrument.

10. How are endothermic and exothermic transitions identified by DTA?

11. Write a note on atmospheric control in TG and DTA.

12. How is DSC different from DTA?

13. Discuss the principles and practice involved in DSC analysis of polymers.

14. Explain the principle involved in TMA. What are its applications?

15. What is DMA? Give a brief description of the instrument.

16. How is DMA useful in the study of polymers and composites?

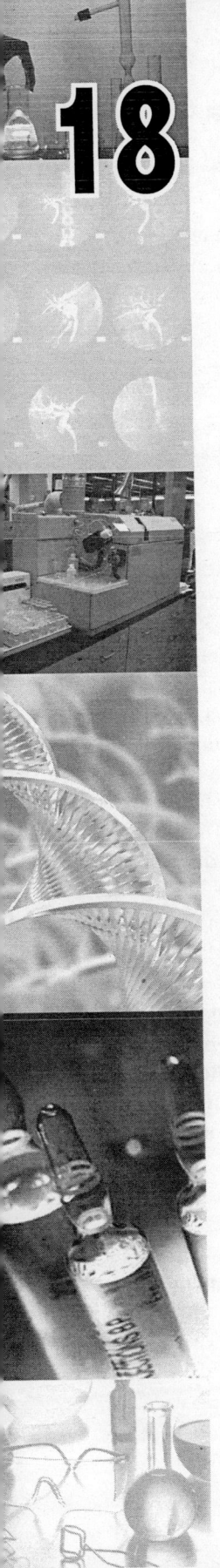

18 Radiochemical Methods of Analysis

18.1 INTRODUCTION

Radiochemical methods are based on the use of radioactive isotopes which emit ionizing radiation spontaneously in the form of alpha (α), beta (β), and gamma (γ) particles or radiations. The nature of emitted particle and its intensity are analysed in different radiochemical methods for qualitative and quantitative analyses in a variety of samples. The methods include radioisotope dilution analysis, neutron activation analysis, autoradiography, and radioimmunoassay. All the methods basically use radioisotopes for labelling the chemical species taking advantage of the facts that (i) normal and radioisotopes are chemically identical (except in the case of the radioisotope of hydrogen, tritium which gives rise to isotope effect), (ii) the radioactive characteristics of an isotope do not undergo any change as it undergoes chemical reactions, and (iii) availability of very highly sensitive detectors for the detection of the emitted particles with detection limits at picogram (10^{-12}) level.

18.2 ORIGIN OF RADIOACTIVITY

Radioactivity is a spontaneous disintegration of certain nuclei which are unstable. Such unstable nuclei are called *radioisotopes*. It is well known that elements beyond atomic number 83 (bismuth) are radioactive. The nucleus of an atom with an extremely small radius in the order of femtometer (10^{-15} m) consists of an assembly of neutrons and protons, collectively called *nucleons*. The number of protons increases linearly with atomic number but the number of neutrons added to the nucleus for each proton added varies. The neutron–proton ratio is one till atomic number 20 (calcium) and it increases gradually in higher atomic number elements to about 1.5 at atomic number 82 (lead) and 83 (bismuth). The stability of the nucleus seems to depend on both the total number of nucleons as well as the neutron–proton ratio. When the atomic mass (sum of neutrons and protons) exceeds the value of 209 with the simultaneous increase in the neutron–proton ratio beyond 1.5, the nucleus becomes unstable.

The neutron–proton ratio in the range 1–1.5 seems to accord maximum stability to the nucleus. As the number of protons increases electrostatic repulsive forces increase proportionate to Z^2/R where Z is the atomic number and R is the radius of the nucleus. The radius of the nucleus is proportional to $A^{1/3}$ where A is the mass number, and hence, the electrostatic repulsion is proportional to $Z^2/A^{1/3}$. The calculated electrostatic repulsion is 117 for calcium whereas for bismuth it is 1155. Thus, for an approximate five times increase in the atomic weight the electrostatic repulsion increases by about ten times. In order to stabilize the nucleus the neutron–proton ratio has to increase with increasing atomic number beyond calcium to about 1.5 in bismuth.

18.2.1 Decay Modes of Radioactive Isotopes

Nuclei having neutron–proton ratio outside the range of 1–1.5 become unstable and decay by converting neutron to protons so as to achieve the stable ratio. An unstable nucleus achieves stability through any one or more pathways which include (i) alpha-particle emission, (ii) β-decay involving either emission of negatron (electron) or positron, (iii) K electron capture, (iv) internal conversion, and (v) gamma-ray emission.

Alpha-decay or *alpha-particle emission* is more common with nuclei of masses >209 which undergo decay through emission of α particles. An excited state daughter nucleus is formed which relaxes to the ground state by emitting γ-rays. The α-decay of a nucleus, for example, naturally occurring radioactive elements of thorium and uranium may be represented as

$$_Z^A X \rightarrow \, _{(Z-2)}^{(A-4)} Y + \, _2^4 He^{2+} + \gamma \tag{18.1}$$

The alpha particles have high energy of the order of 10 MeV and penetrate through 5–7 cm of air. Because of their low penetrating power they can be stopped by thin sheets of solids. The ionizing power of alpha particles is quite high because of their high energy. A large number of ion pairs (of the order 25,000 per cm) are formed in the linear path of alpha particles as they travel through air or any other medium. The discrete energy of the alpha particles is characteristic of the emitting nucleus and appears as a sharp peak in the energy spectrum. Because of their high ionizing power the alpha particles can be distinguished from beta and gamma radiation. Ionization chamber is the commonly used detector for detecting alpha particles. Alpha particles can be detected by detector even in the presence of beta and gamma radiation and the radioactivity due to alpha particles alone can be calculated by subtracting the residual activity of other ionizing particles determined after filtering off the alpha particles from the total activity due to all types of radiation. Examples of alpha particle emitting isotopes include ^{238}U, ^{234}U, ^{230}Th, ^{226}Ra, ^{210}Po, etc.

β-decay can occur either through the loss of negatron (electron) or through the emission of positron depending on the neutron–proton ratio in the nucleus. Negatron emitters include ^{234}Pa, ^{214}Pb, ^{207}Tl, etc.

Negatron decay (loss of electron) is a characteristic decay process of nuclei with a high neutron–proton ratio (neutron-rich nucleus). Such neutron-rich nuclei are formed by neutron capture reaction. The decay is initiated by converting the neutrons into protons by emitting negatrons (electrons) $β^-$. Beta decay is usually accompanied by the emission of γ-photons. The emission of an additional particle, namely, the antineutrino (ν), has been postulated to account for the energy balance during negatron decay.

$$_Z^A X \rightarrow \, _{(Z+1)}^{(A)} Y + \beta^- + \bar{v} \tag{18.2}$$

The beta particles are energy-rich electrons which are emitted from a given radioisotope with continuous range of energy. The energy the emitted electron depends on the angle between the path of the electron and that of the antineutrino. The angle varies from atom to atom in any given radioactive element, and hence, the energy of the emitted electrons also varies continuously over a wide range. The energy spectrum of the emitted electrons is characterized by an upper limit E_{max}, which corresponds to the transition energy of the decay process and average or mean energy E_{mean} which is unique for the particular isotope, and hence, useful in the identification of the isotope.

The beta particles of energy greater than 0.5 MeV can be relatively easily detected on the basis of their ionization capacity by a scintillation detector. A beta particle with energy of 0.5 MeV penetrates through 1 m of air and produces about 60 ion pairs per cm.

Positron emission occurs with nuclei having low neutron–proton ratio (proton rich nuclei) which decay by converting the protons to neutrons and emit positrons, β^+. In addition the decay process is accompanied by emission of γ-rays. The emission of positron is postulated to be accompanied by the emission of neutrino (ν) to satisfy the energy balance. The positron has a very short life and is quickly annihilated as it collides with an electron. The annihilation process generates γ-photons of characteristic energy of 0.51 MeV.

K-electron capture is yet another pathway by which an unstable nucleus with a low neutron–proton ratio stabilizes itself in which the electron from the K shell is captured by the nucleus and thereby a proton is converted into a neutron. The vacancy in K shell is filled by an electron by a higher energy level and the excess energy is released in the form of X-rays. Positron emission and K-electron capture may occur in different atoms of the same radioisotope as both these arise due to low neutron–proton ratio of the unstable nuclei.

Internal conversion involves the transfer of energy of an unstable nucleus to an extra-nuclear orbital electron which is expelled from the atom. The expelled electron has a relatively low energy as compared to energy of the electron (negatron) expelled from the nucleus. Internal conversion is accompanied by the emission of low energy γ-photons.

Gamma decay involves the emission of γ-rays in most naturally occurring radioactive decay processes as the excited state daughter nuclei relax to the ground state. The gamma-rays are mono-energetic high energy photons. The γ-ray spectrum consists of discrete lines characteristic of the nuclear energy levels of the isotope, and hence, the spectrum is highly useful for identifying the radioisotope and its determination.

18.2.2 Kinetics of Radioactive Decay Process

The radioactive decay process of radioisotope A to form a stable daughter nucleus B is a first-order process and the rate of the decay process depends only on the number of atoms of N and the disintegration rate (dN/dt) referred to as *activity* is given by the equation

$$-\frac{dN}{dt} = \lambda N \tag{18.3}$$

where λ is called the decay constant. The number N_t of remaining nuclei (from an initial number N_0) at any given time t is given as

$$N_t = N_0 \exp(-\lambda t) \text{ or } \log N_t = \log N_0 - 0.4343\ \lambda t \tag{18.4}$$

A plot of $\log N_t$ against time t is a straight line with a slope $= -0.4343\ \lambda$. Rutherford introduced a more practical way of expressing the rate in terms of *half-life period*, $t_{1/2}$, defined as the time taken for the initial number N_0 of nuclei to reduce to half the initial number ($N_t/N_0 = 0.5$) given as

$$t_{1/2} = 0.693/\lambda \tag{18.5}$$

The value $t_{1/2}$ is characteristic of a particular radioisotope, and hence, useful in the identification of the isotope. The *average life* of an isotope is $1/\lambda$ which varies for natural radioisotopes in the range of 10^{-6} s to 10^{10} years.

The rate equation becomes complex when the daughter element B itself is not stable and undergoes further decay. Many of the natural radioisotopes undergo multistep decay, and hence, the mathematical treatment is not simple for quantitative analysis. Hence, detectors are usually calibrated with suitable standards for measuring the activity of the isotope.

18.2.3   Units of Radioactivity

The unit of radioactivity was originally based on one disintegration per second and named as *becquerel* (Bq) in honour of Becquerel who discovered the phenomenon of radioactivity in naturally occurring uranium ore of pitchblende. Radioactive units are expressed in larger units as mega becquerel (MBq = 10^6 Bq) and tera becquerel (TBq = 10^{12} Bq). Another unit of radioactivity commonly used is the *curie* (Ci), one curie being the amount of radioactive material which emits particles at the rate of 3.7×10^{10} disintegrations per second (dps) or 2.2×10^{12} disintegrations per minute (dpm). Hence, 1 curie = 3.7×10^{10} Bq.

The millicurie (mCi – 2.2×10^9 dpm) and microcurie (μCi – 2.2×10^6 dpm) are used as standard units in most radioactive measurements.

The number of disintegrations emitted by a radioactive source depends on its purity, that is, the number of radioactive atoms and the decay constant λ. The radioactivity is also expressed in terms of *specific activity* defined as the disintegrations per unit mass of radioactive atoms, the units being mCi mmole^{-1} and μCi μmole^{-1}.

In practice the data obtained from the radiation detectors are in *counts per minute* (cpm) as they detect and count only a small fraction of the emitted particles. Usually the detector is calibrated to determine its counting efficiency using a standard substance of known radioactivity. The counting efficiency is the ratio of detected activity in cpm to the actual activity of the standard in dpm.

$$\% \text{ efficiency} = \frac{\text{observed cpm of the standard}}{\text{dpm of 1 } \mu\text{Ci of the standard}} \times 100 = \frac{\text{observed cpm}}{2.2 \times 10^6 \text{ dpm}} \times 100 \qquad (18.6)$$

18.3   MEASUREMENT OF RADIOACTIVITY

Radioactive emissions are detected and measured by radiation detectors. The materials of construction of the detector are chosen so as to have minimum absorption of radiation. The radioactive emissions generate a series of electrical pulses in the detector. The number of pulses generated in a given time is called the *count* which is a measure of the radioactivity. The detector is calibrated with standard radiation sources and the proportionality between the activity of the source and the count is established. In certain cases the size of the pulse or the pulse height is directly proportional to the energy of the incident particles or photons. *Pulse height analysis* is useful to distinguish between the different types of radioactive emissions from different sources. In radiochemical methods of analysis the radiations most commonly used include the β-rays, γ-rays, and α-particles are rarely used.

Figure 18.1 shows the general layout of the components of radiation detection and counting systems.

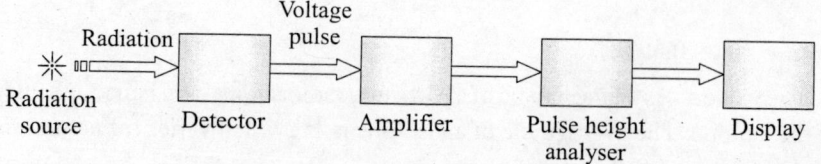

Fig. 18.1   General layout of radiation detection and measurement system

Radiation detectors may broadly be classified into three types based on the effects produced by radioactive emanations in the detector: (i) detectors based on ionization of gases (or liquids or solids) as in ionization chamber, proportional counter and Geiger–Muller tube detectors, (ii) detectors based on photo effect as in scintillation detector, and (iii) detectors based on chemical reactions as in photographic emulsions.

18.3.1 Detectors Based on Ionization

Becquerel discovered that gases and vapours exposed to γ-rays became electrical conductors due to ionization and this property was made use of in several detectors to determine radioactivity quantitatively. The intensity of ionization produced by a radioactive particle moving in a gas or vapour is called *specific ionization* which is *equal to the number of ion pairs formed per centimeter of the path*. For particles of the same mass and energy, the intensity increases with the charge and for particles of the same charge and energy, the intensity increases with the mass or with decreasing speed. Thus, for a given charge, the slower moving particles bring about greater ionization. The α-particles being the heaviest among the particles, produce about 50–100 thousand ion pairs in air, whereas β-particles are very fast, and hence, do not produce more than a few hundred ion pairs. The γ-rays also ionize atoms but in an indirect way, by knocking out electrons from a few atoms. The knocked out electrons have sufficient kinetic energy to bring about ionization of other atoms.

The detector basically consists of a sealed metal cylinder filled with a gas (e.g., argon) and provided with a thin axial metal rod or wire. The metal cylinder functions as the cathode and the metal rod or wire functions as the anode and suitable voltage is applied to the electrodes. The electrons produced by ionization of the gas generate electrical pulses when they are collected by the anode. The number of pulses generated in a given time is called the *count* and is a measure of radioactivity. When the pulse size of current is plotted against the applied voltage to the detector a characteristic curve is obtained as shown in Fig. 18.2

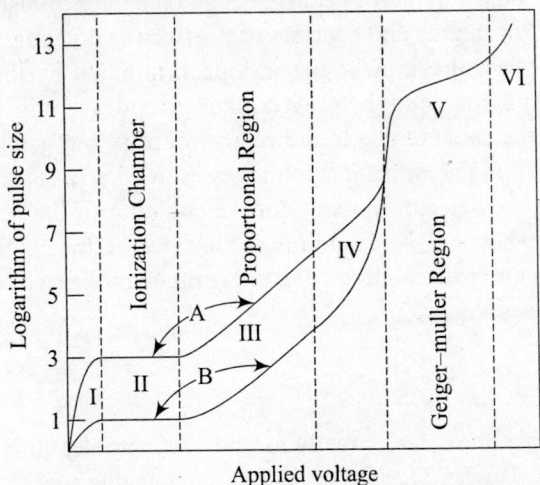

Fig. 18.2 Pulse size as a function of applied voltage

The curve shows six distinct regions (I–VI). In region I at less than 150 V of applied voltage the potential gradient between the electrodes is shallow and the primary electrons produced by ionization move slowly towards the anode. Most of them, however, combine with argon ions before being discharged at the electrode. As the potential increases the electrons are collected more efficiently by the anode and the pulse height increases with increasing applied voltage.

With increasing applied voltage a plateau region (region II) is obtained as the pulse becomes independent of the applied voltage in the range 150–300 V because the number of ions produced is exactly equal to the number discharged at the electrode. The region II of the curve is used for the *ionization chamber*.

In region III as the applied voltage increases, electrons are accelerated towards the anode and the pulse becomes once again proportional to the increasing voltage in the range 300–600 V and the radiation detector used in this region is called *proportional counter*. In addition electrons are sufficiently energetic to knock out electrons from the surrounding filling gas by collision. This *secondary ionization* results in the multiplication of ion pairs known as Townsend's avalanche or cascade. The number of secondary ion pairs produced by a single primary ion pair is known as *gas phase amplification* or *gas phase multiplication*. The pulse size is proportional to the total number of secondary electrons in regions I–IV which in turn is proportional to the number of primary ion pairs produced by the original ionizing particle.

The gas phase amplification due to secondary ionization of the filling gas ultimately reaches a maximum when a single ionizing particle leads to complete ionization of the filling gas thereby saturating the detector. There is no possibility for further ionization of the filling gas and a plateau known as the *Geiger region* or *Geiger–Müller region* (region V) results in which the number of ion pairs formed (pulse size) is independent of the applied voltage at > 800 V. This region is used for the most popular radiation counter, *Geiger–Müller detector*.

All the gas ionization detectors are characterized by a *paralysis time* or *dead time* during which time the detector is not functional. The ionization initiated by an ionizing particle at a given applied voltage causes the electrons to accelerate towards to the anode and get discharged quickly. However, the positive ions of the filling gas are sluggish in their progress towards the wall (cathode) of the cylinder. The difference in the rates of discharge results in the build up of a charge around the anode and decreases the effective potential difference thereby terminating the discharge. The voltage pulse that has been initiated by the ionizing particle persists for several hundred microseconds before it diminishes and is terminated. The discharge of secondary electrons from the cathode due to the reduction of filling gas ions also takes time. Thus, the avalanche started by the ionization continues even when a subsequent ionizing particle arrives thereby making the detector unready for the counting of the newly arriving ionizing particle. During this period, called the dead time of the detector, the counts recorded by the detector C_M will be less than the 'true count' C_T. The magnitude of the two counts depends on the length of the dead time t as given by

$$C_T = \frac{C_M}{1 - C_M t} \qquad (18.7)$$

In order to make the detector ready for the next particle, the dead time must be shortened by quenching the avalanche. The introduction of a quenching agent or electron attracting substances (e.g., ethyl alcohol, bromine, or carbon dioxide) at low pressure into the detector quenches the voltage by absorbing the secondary electrons. The quenching agent has relatively a lower ionization potential as compared to the filling gas, and hence, the filling gas ions transfer the energy to the quenching substance forming new positive ions. The positive ions of the quenching gas have lower energy content, and hence, their discharge at the cathode does not produce photoelectric emission of the electrons. The dead time of the detector can also be shortened by reducing the anode potential immediately after the pulse is initiated. The combination of these methods reduces the dead time of the gas ionization detectors to a large extent.

Ionization chamber is useful for the detection of α, β, γ, and X-rays and with suitable adaptations even neutrons. The chamber consists of a metal cylinder functioning as the cathode and an axial metal rod as the anode (Fig. 18.3).

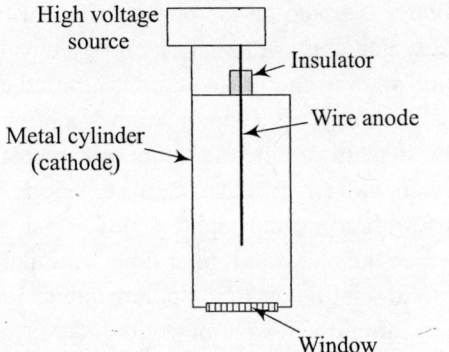

Fig. 18.3 Schematic diagram of ionization chamber

The cylinder is filled with a filling gas (air for alpha particles, xenon or krypton for γ and X-rays). The applied voltage of 150–400 V corresponds to the second region of the curve shown in Fig. 18.2 and the magnitude is such that the recombination of ion pairs is minimized without any gas amplification. The radioactive sample is placed just outside the window and the intensity of radiation can be counted in the form of electronic pulse which is proportional to the number of electrons released by the ionizing radiation by *pulse ion chamber detector* type. Alternatively, *current ion chamber detector* type integrates the signal and provides a dc current, the magnitude of which indicates the count. Ionization chamber detector is an accurate and quick-acting detector useful for detection of even very weak radiations.

Proportional counter is similar in construction to the ionization chamber. The applied voltage is much higher as compared to that of ionization chamber, and hence, an inherent gas phase amplification occurs. The total number of secondary electrons is proportional to the number of primary ion pairs produced by the ionizing particle. The deat time of the detector is very short (~0.25 µs), and hence, the counter is capable of high counting rates of the order of 50,000–200,000 counts s^{-1}. However, the signal generated is quite small and requires amplification and it is fed to a scaler.

Geiger–Müller counter (GM counter or tube) or Geiger counter consists of a cylindrical cathode made of brass or glass which has been silvered and a tungsten wire anode at the centre of the cathode cylinder and is similar in construction to the ionization chamber (Fig. 18.3). The two electrodes are enclosed in a gas-tight envelope. The envelope is filled with a mixture of ionizing argon gas (~ 80 mm) and a quenching vapour of methane (20 mm) or ethanol or 0.1% chlorine. Incident radiation or ionizing particle is allowed into the envelope through a mica window. A potential in the range of 800–2500 V is applied between the electrodes. The counting rate is limited to about 15,000 counts min^{-1} because of the relatively long dead time of about 200 µs. The GM tube exhibits high sensitivity for the detection of beta particles but is less sensitive to X-rays and gamma-rays.

18.3.2 Detectors Based on Photo Effect

In 1899, Becquerel discovered that radiation could produce luminescence in certain inorganic compounds such as zinc sulphide, barium platinocyanide, and diamond. Rutherford and Geiger established that the phenomenon was due to α-activity. Visual observation of the luminescence was eye-tiring and the use of luminescence phenomenon for detection and quantification of radioactivity had to wait till the development of photomultiplier tube during World War II.

Scintillation counter is based on the phenomenon of luminescence exhibited by chemicals called *scintillators*, which convert radiation energy into light energy. The scintillator absorbs the ionizing radiation and the energy acquired is emitted as a pulse of light in the UV or visible region. The scintillation counter is a combination of a scintillator and a photomultiplier tube. The nature of scintillator depends on the radiation to be measured. Solid scintillators, for example, zinc sulphide activated with silver coated on the envelope of the photomultiplier tube detector is used for α-particles. Anthracene and naphthalene (emission wavelengths 445.0 nm and 410.0 nm respectively) coated on the photomultiplier tube are suitable for beta particles of moderate and high energy. Other solid scintillators (e.g., sodium iodide, lithium iodide, anthracene, naphthalene, and loaded polymers) find use for the detection of γ-rays and X-rays of which sodium iodide activated with 1% thallium iodide is extensively used as it has several advantageous features. The scintillator has a large photoelectric cross-section and high probability of absorption of incident radiation because of its high density. In addition because of its high transparency to its own optical emission line of thallium, thicker slices of the solid material can be used for absorbing incident radiation thereby enhancing its efficiency of detection. The incident radiation excites the iodine atom and the excited iodine atom relaxes to its ground state by emitting a light pulse in the ultraviolet region which is promptly absorbed the thallium atom which in turn emits fluorescent light at 410.0 nm. The thallium fluorescence emission is detected by the photomultiplier tube detector. The crystal is usually protected from extraneous light and atmospheric moisture by enclosing it in aluminum foil which also serves as an internal reflector. The NaI(Tl) scintillator has a high and uniform quantum efficiency over the wavelength region of 0.3–2.5 Å and can be used up to 4 Å. The most commonly used NaI(Tl) 'well type' scintillation counter in which a well is drilled in the crystal of 5–10 cm diameter for placing the sample as shown in Fig. 18.4.

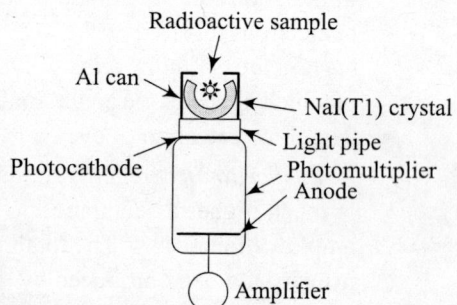

Fig. 18.4 Well type NaI (Tl) scintillation detector

Organic liquid scintillators are conjugated aromatic molecules (e.g., 2,5-diphenyloxazole -PPO and 1,4-bis-2-(5-phenyloxazolyl)-benzene–POPOP) find use for the detection of negatrons (electrons) and low energy (< 1 MeV) beta emitting radionucleids such as ^{3}H, ^{14}C, ^{32}P, and ^{35}S and especially ^{3}H (0.018 MeV) and ^{14}C (0.016 MeV) because of their shorter decay times. The liquid scintillator and the sample are dissolved in a suitable solvent (e.g., toluene or dioxan or dioxan–water (20%) mixture) to ensure intimate contact between the sample and the scintillator. If the sample is insoluble a suspension of the sample stabilized with silica gel is used for measurement. The solvent absorbs the energy of the incident particle of the radioactive sample and transfers it to the scintillator. The scintillator emits a pulse of UV radiation which is detected by the photomultiplier tube. A secondary scintillator (e.g., POPOP) may be used in conjunction with the primary liquid scintillator to shift the emission to visible region where the photomultiplier exhibits greater sensitivity.

Semiconductor detector is based on the generation of electron-hole pairs in a semiconductor material by the ionizing radiation from a radioactive sample. The semiconductor detector consists of an intrinsic semiconductor crystal such as germanium and silicon coated with a thin layer of gold plated on each end of the crystal. Electrodes are attached to the gold layers and a potential of about 3–5 kV is applied. The semiconductor crystal on exposure to radiation forms free electrons and holes in numbers proportional to the energies of the incident radiation which migrate to oppositely charged electrodes and generate electrical pulses. The pulse size is proportional to the energy of the incident radiation.

Two types of semiconductor detectors are in use. *Surface barrier detector* (or intrinsic *detector*) consists of a *p-n* junction formed at the surface of wafer of high purity silicon. *Lithium drifted detector* is formed by doping first an intrinsic germanium or silicon crystal with *p*-type dopants (e.g., B, Al, Ga, or In) to give a *p*-type semiconductor followed by 'drifting' the strongly *n*-type element lithium into the *p*-type crystal. The crystal is cooled to liquid nitrogen temperature to form a permanent *p-n* junction in the crystal. The *p-n* junction is called the depletion region and its thickness increases on application of a reverse bias. On exposure to ionizing radiation electron-hole pairs are formed in the depletion layer which migrate to oppositely charged electrodes and generate electrical pulses. The lithium drift detector has to be maintained at liquid nitrogen temperature at all times to prevent precipitation of lithium and also to reduce thermal noise. The detector is also placed within a high vacuum jacket to minimize background current.

18.3.3 Detector Based on Chemical Reaction

Detection of radiation through chemical reaction essentially involves the use of photographic film coated with a silver halide emulsion sensitive to the radiation. The silver halide undergoes reduction to metallic silver on exposure to radiation and forms a latent image which is developed and fixed on the film as a black metallic deposit. A linear relationship exists between the intensity of the radiation and the extent of blackening of the film.

18.4 AMPLIFIERS AND OTHER ELECTRONIC EQUIPMENT

Detectors used in *X*-ray spectroscopy and radiochemical methods require auxiliary electronic equipment, such as amplifier, scaler, and coincidence and anti-coincidence units. The signal generated by the detector is quite small and requires a primary or pre-amplification and a second stage amplification for further processing by a scaler or recorder. The pre-amplifier is usually located just after the detector within the detector housing. After a second amplification the detector signal in the form of electric pulse is fed to an electronic divider called scaler which records every tenth pulse (decade scaler) thereby counting or registering the radioactivity in terms of pulses as a function of time. Electronic coincidence unit eliminates the background noise due to pulses generated, particularly by photomultiplier in scintillation detector. Anti-coincidence unit is used to minimize the effects due to external ionizing radiation such as cosmic rays.

18.5 PULSE HEIGHT ANALYSER

Pulse height analyser unit distinguishes between different types of radiation based on the proportionality between the energy of the radiation and the size or height of the pulse generated.

Pulse height analysis, thus, distinguishes the ionizing radiation of the radioactive sample from noise pulses and cosmic radiation and is particularly useful in the analysis of a complex mixture of signals from X-rays, gamma-rays, etc. A *single channel pulse height analyser* has an electronic gate of about 0.1 V wide and allows only pulses above certain amplitude so that it functions as a baseline discriminator to eliminate weak pulses due to scattered radiation. *Multichannel pulse height analyser* consists of a large number of electronic gates which allow pulses having amplitudes between the preset lower and upper limits and are controlled by a computer with built-in software to provide a spectrum and also other added features such as peak search, identification, and integration.

18.6 ANALYTICAL APPLICATIONS OF RADIOISOTOPES

Analytical applications of radioisotopes include isotope dilution analysis, activation analysis mainly using neutrons, autoradiography, and radioimmuno assay in biological and clinical diagnosis and environmental monitoring of radioactive pollutants.

18.6.1 Isotope Dilution Method

The isotope dilution method is based on measuring the change in isotopic ratio when radioactive and non-active isotopes are mixed. The method is particularly useful when quantitative isolation of an analyte element from a given sample is not known or is difficult. The method involves mixing thoroughly a known weight W_1, of the radioisotope of known activity (A_1) with the sample containing the analyte element. A part of the analyte element is isolated from the sample and the activity of the isolated material is determined (A_2). The isolation of the total amount of analyte element need not be quantitative. It is just necessary to isolate a small fraction of the analyte from the total amount, sufficient enough for quantitative weighing or measuring the activity accurately. The amount of the analyte (W_2) in the sample is calculated from the extent of dilution of the radioactive tracer using the formula

Total activity added to the sample $= A_1 W_1 = A_2 (W_1 + W_2)$

$$W_2 = W_1 \left(\frac{A_1}{A_2} - 1 \right) \tag{18.8}$$

The method is simple and has the advantages of non-requirement of quantitative separation of the analyte from the sample and high sensitivity, and hence, useful in trace analysis of elements, particularly in complex materials, for example, radiocarbon dating of archaeological samples, anthropological specimens, and determination of the amount of hydrogen adsorbed in transition metal lattice. Analysis of complex biochemical mixtures, can be conveniently carried out, for example, determination of vitamin B_{12} in biological samples using the radioisotope ^{60}Co. A classical example of this method was the study of reaction mechanism of ester hydrolysis by Polanyi in the 1930s. Amyl acetate was hydrolysed by water containing the isotope ^{18}O which was found in acetic acid after the completion of the reaction resulting in the identification of the chemical pathway as

$$CH_3-\overset{\overset{\text{O}}{\|}}{C}-[OC_5H_9 + H]-O^*H \rightarrow CH_3-\overset{\overset{\text{O}}{\|}}{C}-O^*H + C_5H_9OH$$

and not as

$$\underset{\substack{\|\\ CH_3-C-O}}{O} \;[C_5H_9 + HO^*]-H \rightarrow \underset{\substack{\|\\ CH_3-C-OH}}{O} + C_5H_9O^*H$$

The main disadvantage of this method is the determination of the chemical yield of the analyte that is being isolated. In order to overcome this limitation a method called *substoichiometric yield determination* is used. The method involves adding a known substoichiometric amount of a complexing agent and conditions maintained so that it reacts quantitatively with the analyte. The amount of analyte isolated is determined from the amount of the complexing agent used. The method can be used to determine nanogram quantities of analyte element (e.g., Cupferron is used as the complexing agent for extracting iron into chloroform using ^{59}Fe as tracer).

Reverse isotope dilution method involves the isolation of a radioactive isotope from a sample and dilute it with its inactive isotope.

18.6.2 Activation Analysis

Activation analysis refers to *neutron activation analysis* (NAA), a sensitive multi-element non-destructive analytical technique applicable to a variety of elements at different concentration levels up to trace quantities. The basic principle of the method involves bombarding the sample with a flux of low energy (of the order of a few MeV) thermal neutrons for sufficient length of time to form radioisotopes of elements in the sample and carry out qualitative as well as quantitative analyses based on the known radioactive emissions (mostly γrays) and decay paths for each element. Most of the radioisotopes produced by neutron activation decay by emitting gamma rays. Since the method is non-destructive it finds applications in the study of works of art and historical artifacts.

During neutron bombardment the rate of production of the radioisotopes is given by the relationship

$$\frac{dN^*}{dt} = \Phi\sigma N - \lambda N^* \tag{18.9}$$

where N refers to the number of target nuclei from which N^* number of radioactive atoms are produced, Φ is the flux of bombarding particles (expressed in $cm^{-2}\,s^{-1}$), σ is the reaction cross-section (expressed as barns; 1 barn $= 10^{-24}$ cm^2/target atom), and λ is the decay constant ($0.693/t_{1/2}$).

The number of target nuclei depends on the weight of the sample (w) and its atomic weight (M) and the fractional abundance of the target nucleus f as given by the relationship

$$N = w\,N_A\,f/M \tag{18.10}$$

where N_A is the Avogadro number (6.02×10^{23} nuclei mol^{-1}).

Integrating Eq. (18.9) over time t of irradiation gives the number of radioactive nuclei formed at the end of irradiation as

$$N^* = \frac{\Phi\sigma N}{\lambda}(1 - e^{-\lambda t}) \tag{18.11}$$

The amount of radioactivity A_0 at the end of irradiation time is equal to the product λN^*. For a given radioisotope produced the A_0 can be calculated by substituting for λ in Eq. (18.11)

$$A_0 = \Phi\sigma N\left[1 - \exp\left(\frac{-0.693t}{t_{1/2}}\right)\right] \tag{18.12}$$

The exponential term within the brackets of Eq. (18.12) is the *saturation factor*, S and represents the ratio of the amount of activity produced during the time of irradiation to that produced in infinite time. The time of irradiation required to achieve *saturation activity* with a saturation factor > 0.98 is usually six times the $t_{1/2}$ for a given radionuclide.

In order to avoid problems due to inhomogeneity of the neutron flux a standard is also irradiated along with the sample. Assuming that identical specific activities for the analyte element are induced in the sample and the standard the amount of analyte element in the sample w_2 is calculated from the known values of the weight of the standard w_1 and the activities a_1 and a_2 respectively of the standard and the sample using the formula

$$w_2 = w_1 \, \frac{a_1}{a_2} \tag{18.13}$$

In practice the sample and the standard are encapsulated in separate quartz or polyethylene vials and irradiated in a suitable reactor at a constant and known neutron flux. The most commonly used neutron sources for neutron activation analysis include (i) a nuclear reactor, (ii) californium-252 isotope, and (iii) an alpha-particle source such as radium or americium mixed with beryllium. Uranium fission reactor provides a neutron flux of thermal neutrons in the order of 10^{12}–10^{14} neutrons $cm^{-2} s^{-1}$. Californium-252 source provides a neutron flux of about 10^{12} neutrons $cm^{-2} s^{-1}$. A portable but weaker neutron source is based on the reaction $^9Be \, (\alpha, n)^{12}C$. It contains a mixture of an alpha emitter such as ^{226}Ra, ^{210}Po, or ^{239}Pu with beryllium and provides a neutron flux of 10^7 neutrons $cm^{-2} s^{-1}$.

The thermal neutrons are captured by the target nucleus (analyte element) resulting in the formation of a compound nucleus which undergoes transmutation almost instantaneously to a stable configuration through the emission of prompt particles or prompt gamma photons. In most cases the stable configuration yields a radioactive nucleus which in turn decays by the emission of both particles and one or more characteristic delayed gamma photons. The decay process of the radioactive nucleus is relatively a slow process and is dependent on the unique half-life of the radioactive nucleus. After irradiation the sample is left for a specific decay period and then placed in a radiation detector (e.g., thallium doped sodium iodide scintillation detector or a semiconductor detector) to detect the emanations.

Neutron activation analysis can be applied to 74 elements depending on the experimental procedure, the minimum detection limits being in the range of picograms to micrograms g^{-1} of sample. For example, Dy, Eu, Lu, Mn, and In can be detected at 1–10 picograms while elements such as Au, Ho, Ir, Re, Sm, and W can be detected at 10–100 picogram levels. A large number of elements can be detected only at nanograms or microgram levels. Heavier elements with larger nuclei in general have higher neutron capture cross-sections.

Different techniques are possible in neutron activation analysis depending on the experimental parameters employed. *Epithermal neutron activation analysis* (ENAA) basically involves activation of the sample by slow neutrons having kinetic energies < 0.5 MeV obtained by moderation within the reactor. *Fast neutron activation analysis* (FNAA) involves the use of primary fission neutrons (fast neutrons) having kinetic energies > 0.5 MeV without moderation. FNAA technique is particularly suitable for elements such as N, O, F, Al, Si, P, Cr, Nb, Fe, Cu, Y, Ni, Mo, and Pb. Determination of oxygen at levels of 1 μg/g of sample is conveniently carried out in less than a minute by FNAA by irradiating the sample with 14 MeV neutrons to form

^{16}N isotope via the ^{16}O (n, p) ^{16}N reaction. The product ^{16}N isotope emits gamma radiation of 6.13 MeV (and a small percentage of 7.13 MeV) which can be detected without any interference.

If the decay products of activation (mostly gamma-rays) are analysed almost immediately after activation the analysis is known as *Prompt gamma neutron activation analysis* (PGNAA) while *Delayed gamma neutron activation analysis* (DGNAA) refers to analysis of gamma-rays sometime after irradiation. PGNAA is mostly suitable for elements having high neutron capture cross-sections, elements that form only stable isotopes or elements that emit gamma-rays of weak intensity. PGNAA is characterized by short irradiation time and short decay time for the products often in the range of seconds to a few minutes. DGNAA on the other hand is used for most elements that form artificial radioisotopes with relatively long decay times. DGNAA is characterized by long irradiation time and long decay times lasting over hours, days, and even weeks.

Instrumental neutron activation analysis (INAA) refers to analysis directly on the irradiated sample. The sample containing the analyte is exposed to a flux of neutrons. Neutron capture by the analyte nucleus produces a neutron-rich radioisotope which then undergoes decay through beta emission with a unique half-life mostly accompanied by gamma emission. A high resolution gamma-ray spectrometer is used to detect the delayed gamma-rays for qualitative and quantitative analyses.

The method has found immense use in quality control for the production of polysilicon used in solar cells. Though the purity requirements of polysilicon for solar cells is relatively less (about five-seven 9s purity) as compared to nine purity for semiconductor industry, the power conversion efficiency of the solar cell depends on the concentration of the impurities in raw material. INAA has the advantage of providing accurate data for large-sized or bulk samples without the necessity of dissolving or digesting the sample. However, it suffers from the disadvantages of the requirement of a high-flux neutron source which can only be provided by a reactor, long time of the order of a few weeks for analysis and inability to analyse lighter elements particularly B, C, O, and Al which are of importance in semiconductor industry.

18.6.3　Radioimmunoassay

Radioimmunoassay (RIA) is based on the use isotope dilution principle to immunoassay techniques for analyses of biological compounds at picogram (10^{-12} g) levels in clinical laboratories. Immunoassay is a technique practiced in biochemistry based on a competitive reaction between an analyte biomolecule called the antigen (or immunogen) and a labelled or tagged standard antigen for the limited binding sites available on a specific biomolecule called the *antibody* also known as *immunoglobulin* (Ig). The tag or label may be a radioactive tracer as in RIA, or an enzyme as in ELISA (enzyme-linked immunosorbent assay) or a fluorophore as in fluorescence-based immunoassay.

An *antigen* (*anti*body *gen*erator) is defined as potentially a harmful substance or molecule which on entering the blood stream of animals or human beings triggers the production of an *antibody* by the immune system of the recipient and reacts with the antibody to form an *antigen–antibody complex*. The antigen is a large-sized protein or polysaccharide produced by invading organisms such as viruses, bacteria, and other microorganisms or the parts, coats, flagella, cells, etc., of microorganisms. *Immunogen* is a specific type of antigen which provokes an immune response. Vaccines are immunogenic antigens intentionally administered to induce an acquired immunity in the recipient. Self-produced antigens are usually tolerated by the immune system and when they are not tolerated an *autoimmune disorder* is said to occur.

The immune system consisting of a white blood cell called *plasma cell* is activated by the invading antigen and responds through the formation of specific antibodies or *immunoglobulins*. The terms *antibody* and *immunoglobulin* are interchangeably used. The antibody is also a high molecular weight protein (~150 kDa) which recognizes and neutralizes the foreign molecule (antigen) by binding it the through stereospecific interaction to form the antigen–antibody complex. Five different types or classes called *isotypes* of immunoglobulins have been identified in human blood, namely, IgG, IgA, IgD, IgE, and IgM, each having different biological properties, functional location and ability to bind different antigens. IgG is the most abundant among all the immunoglobulins.

The structure of immunoglobulin is represented as Y-shaped molecule (Fig. 18.5) consisting of four polypeptide chains, two identical lighter chains containing about 211–217 amino acid residues and two heavy chains made of about 450–550 amino acid residues connected through disulphide bridges.

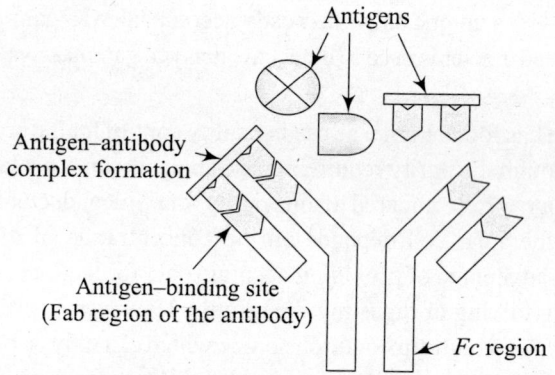

Fig. 18.5 Antibody molecule

Each heavy chain consists of two regions, a constant domain or region and a variable domain (~110 amino acid long). The light chain also has a constant domain and a variable domain.

Hydrolysis of the immunoglobulin with the enzyme papain produces three fragments, each with a molecular weight of about 50,000. Two fragments are identical and are capable of binding the antigen, hence, referred to as *Fab* (*fragment, antigen binding*) which are located on each arm of the Y. The compositions of the two Fab fragments vary depending on the nature of the antigen and the two regions bind to the specific antigen with great precision leaving out other types of antigens. The mechanism of the binding is such that the antibody can tag an invading microbe or an infected cell so that other parts of the immune system can neutralize the invading organism directly.

The third fragment is incapable of binding the antigen by itself but is crystallizable from solution, hence, called Fc (*fragment, crystallizable*) which is located at the base of the Y. The composition of Fc is almost constant and the region is involved in modulating the immune cell activity.

In practice the RIA consists of several steps. (i) In the first step the antibody specific for the analyte antigen is obtained by injecting the antigen into the blood stream of an animal to trigger the natural defense mechanism and produce the immunoglobulins in the animal body in sufficient concentration over a period of 6–12 weeks. The blood is drawn from the animal and the serum containing the specific antibodies for the antigen called the *antiserum* is separated for use in RIA. (ii) In the second step a known amount of the antigen is tagged with a gamma emitting radioisotope of iodine attached to tyrosine (or fluorophore or an enzyme depending on the immunoassay method). The radiolabelled antigen is mixed with a known amount of the antiserum and incubated for about 2 h at 4°C to allow the formation of antigen–antibody complex. (iii) In the third step a sample of serum from a patient containing the same antigen whose amount is to be determined (assayed) is added to antigen-antibody complex formed in the second step. The unlabelled antigen ('cold antigen' from the serum of the patient) competes with the radiolabelled antigen ('hot antigen') for the specific binding sites of the antibody and displaces some of the radiolabelled antigen. As the concentration of the cold antigen increases it displaces more of the hot antigen from antigen–antibody complex thereby reducing the ratio of the antibody-bound hot antigen to the free hot antigen. (iv) In the fourth step the antibody-bound antigens (both cold and hot) are separated from the free or unbound ones by adsorption on charcoal or precipitation. The supernatant solution contains the free antigens of both cold and hot type. (v) In the final step the radioactivity of bound antigens as well as that of the supernatant is measured using a gamma counter. The amount of the antigen in the patient's serum is determined by a calibration chart called binding curve prepared using known standards. The binding curve and the graphical determination of the antigen in the test sample are shown in Fig. 18.6.

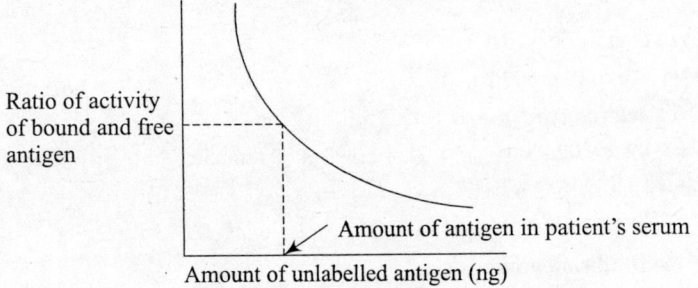

Fig. 18.6 Binding curve for an antigen–antibody system

18.6.4 Autoradiography

Autoradiography is a simple process to locate the position of a radionuclide used as label within a solid specimen. The prefix *auto* in autoradiography is used to indicate the presence of the radioisotope within the specimen in contrast to microradiography in which the specimen is exposed to X-rays from an external source.

The procedure in autoradiography involves placing the specimen in intimate contact with a photographic film or plate coated with special emulsions sensitive to nuclear radiations or X-rays. The emulsion consists of silver halide crystals suspended in gelatin and coated on a film or plate. The beta or gamma particles emitted by the radionuclide reduce the silver halide to silver metal in the form of a latent image which is converted to visible image by chemical development. The unexposed silver halide is removed by fixing agent and the autoradiographic image (autoradiogram) clearly shows the distribution of the radionuclide within the specimen. Direct autoradiography uses radioisotopes which emit relatively weak radiation (e.g., ^{14}C or ^{35}S) while indirect autoradiography makes use of radioisotopes (e.g., ^{32}P or ^{125}I) emitting high energy particles and the emitted particles are detected by a liquid scintillator which in turn generates light photons for detection by photographic film.

Autoradiography finds use in biology, for example, to study the absorption of minerals and their circulation in the plant body, to determine the tissue localization of a radioactive substance introduced via a metabolic pathway or bound to a receptor or enzyme or hybridized to a nucleic acid.

SOLVED PROBLEMS

EXAMPLE 1 *A crude sample of an organic compound X was analysed by the technique of isotope dilution. Radiocarbon labelled pure organic compound X (50 mg) having a specific activity of 30,000 cpm g^{-1} was added to the crude sample and equilibrated. The organic compound was separated and 100 mg of the pure sample showed an activity of 25 cpm. Calculate the weight of the organic compound X in the crude sample.*

SOLUTION

Equation (18.8) is used for calculating the weight of organic compound X.

$$W_1 = 50 \text{ mg}; A_1 = 30,000 \text{ cpm g}^{-1}; A_2 = 25 \text{ cpm per} 100 \text{ mg} = 250 \text{ cpm g}^{-1}$$

Weight of compound X in the crude sample $W_2 = 50 \times ((30,000/250) - 1) = 5950$ mg.

EXAMPLE 2 *A 10-mg sample of steel containing 0.2% of Mn was irradiated for 60 min at a neutron flux of 2×10^{12} neutrons $cm^{-2} s^{-1}$. The half-life period of ^{56}Mn isotope is 2.6 h and the cross-section for neutron absorption is 13.3 barns for ^{55}Mn. Calculate the activity of the sample.*

SOLUTION

Substituting the given data in Eq. (18.12)

$$A_0 = \Phi \sigma N [1 - \exp(-0.693 \, t/t_{1/2})]$$
$$\Phi = 2 \times 10^{12} \text{ neutron cm}^{-2} \text{ s}^{-1}; \sigma = 13.3 \times 10^{-24} \text{ cm}^{-2}$$
$$N = 0.01 \times 0.02 \times 6.023 \times 10^{23}/54.94 = 2.193 \times 10^{18} \text{ nuclei}$$
$$t = 60 \text{ min} = 1 \text{ h}; t_{1/2} = 2.6 \text{ h}$$
$$A_0 = 2 \times 10^{12} \times 13.3 \times 10^{-24} \times 2.193 \times 10^{18} \times (1 - \exp(-0.693 \times 1/2.6))$$
$$= 1.364 \times 10^7 \text{ disintegrations s}^{-1}$$

SUMMARY

- Radioactivity is a spontaneous disintegration of unstable nuclei called radioisotopes, the instability being attributed to neutron–proton ratio >1.5. The radioisotopes emit ionizing radiation of α, β, or γ particles.

- Alpha-decay is more common with nuclei of masses >209. β-decay involves emission of negatron (electron) nuclei with a high neutron–proton ratio (neutron-rich nuclei) or positron decay with nuclei having low neutron–proton ratio (proton-rich nuclei). Gamma-rays are emitted as the excited state daughter nuclei relax to the ground state.

- The radioactive decay process follows the first-order kinetics, the rate being expressed in terms of half-life period, which is useful in the identification of the isotope.

- The commonly used unit of radioactivity is curie which is the amount of radioactive material that emits particles at the rate of 3.7×10^{10} disintegrations per second (dps). Specific activity is defined as the disintegrations per unit mass of radioactive atoms. Radiation detectors detect and count the emitted particles and express as counts per minute (cpm).

- Radiation detectors are based on (i) ionization of gases (or liquids or solids) (e.g., ionization chamber, proportional counter and Geiger-Müller tube detectors), (ii) photo effect (e.g., in scintillation detector), and (iii) chemical reaction on photographic emulsion.

- Pulse height analysis involves distinguishing different types of radiation based on the proportionality between the energy of the radiation and the size or height of the pulse generated by the radioactive sample or from cosmic rays which contributes to the background noise.
- Isotope dilution analysis, neutron activation analysis, autoradiography, and radioimmunoassay are the main analytical applications of radioisotopes.
- Isotope dilution analysis involves measuring the change in isotopic ratio when radioactive and non-active isotopes are mixed. The method is adopted when quantitative isolation of the analyte element is difficult.
- Neutron activation analysis is a non-destructive analytical technique applicable to about 74 elements at different concentration levels up to trace quantities. The method involves bombarding the sample with a flux of low energy thermal neutrons for sufficient length of time to form radioisotopes of elements in the sample and carry out qualitative as well as quantitative analyses based on the known γ-ray emissions and decay paths of each element.
- Radioimmunoassay involves the use of isotope dilution principle to immunoassay techniques for analyses of antigens in a sample at picogram $(10^{-12}$ g) levels in clinical laboratories.
- Autoradiography involves location of a radionuclide used as label within a solid specimen.

REVIEW QUESTIONS

1. Write a note on the origin of radioactivity.
2. Give an account of the different modes of decay of an unstable nucleus.
3. What are the different units of measurement of radioactivity?
4. Give a general outline of measurement of radioactivity.
5. Write a note on the different detectors used in the measurement of radioactivity.
6. What is pulse height analysis? What is its use?
7. Discuss the principles and practice of isotope dilution analysis.
8. How is neutron activation analysis carried out? What are its salient features?
9. What is radioimmuno assay? How is it carried out?

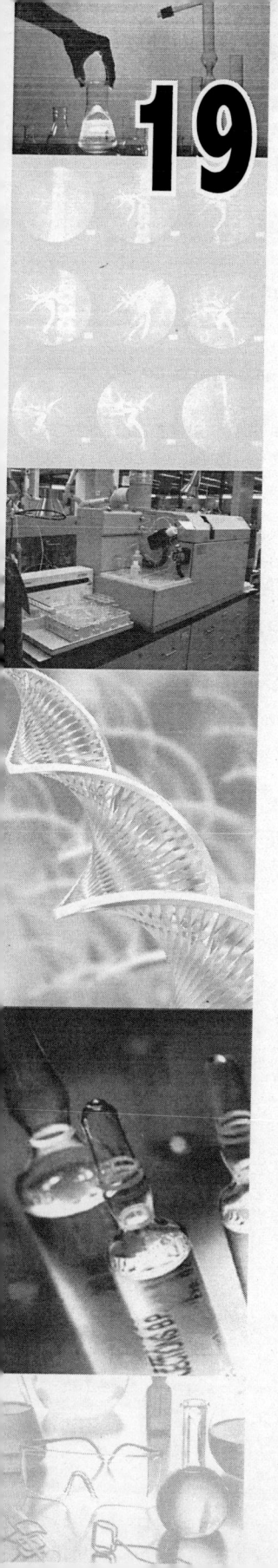

19 Surface Analytical Methods

19.1 INTRODUCTION

The surface of a solid may be considered theoretically as an infinitesimally thin layer separating the two phases, namely the bulk solid and the gas (air) phase surrounding the surface. The surface layer may extend between atomic dimensions to a thickness of 10 nm or more. A concentration gradient of surface-adsorbed species may exist extending to the bulk in a direction perpendicular to the surface. The study of surfaces involves the compositional changes that occur in the first 20 atomic layers. Surface chemistry and the characteristics of surface are quite important in a number of areas of industrial importance such as semiconductor manufacture, thin films and coatings, adhesion of layers, corrosion, and heterogeneous catalysis. Similarly, morphological studies are of importance in the characterization of cell, tissues and microorganisms from the perspectives of industrial practice, health care, and clinical diagnosis.

Study of surface characteristics poses the unique problem of surface contamination because any surface in contact with the surrounding atmosphere adsorbs various components of the atmosphere such as moisture, oxygen, and carbon dioxide depending on the nature of the surface and the relative concentrations of the atmospheric components. Adsorption occurs even when the surface is placed in highly evacuated conditions. For example, even at a low pressure of 10^{-5}–10^{-6} torr the surface gets contaminated due to adsorption of residual gases and vacuum pump oil vapours within a few seconds and the surface remains clean for about an hour or so under ultra-high vacuum conditions of about 10^{-10} torr. Hence, high vacuum systems are essential auxiliary equipment required in the study of surfaces.

Clean surface is obtained by specialized techniques such as baking the sample at high temperatures, mechanical scraping, and polishing using a variety of abrasives, ultrasonic cleaning, sputtering the sample surface with a beam of argon ions, and scribing with a carbide tip. Depth profiling is yet another technique adopted in the study of surfaces which involves bombarding the surface of the sample with a beam of argon ions or xenon ions to sputter and etch away the surface layer and simultaneously bombarding the sputter-etched surface with a beam of X-rays (in X-ray photoelectron spectroscopy) or a beam of electrons (in Auger electron spectroscopy).

19.2 CLASSIFICATION OF SURFACE ANALYTICAL METHODS

A variety of analytical methods find extensive use in the study of the surface characteristics of solids. The different surface analytical methods can broadly be classified into five major groups: (i) microscopic methods based on visual examination of surfaces of biological as well as industrial materials, (ii) methods based on adsorption–desorption of probe molecules, (iii) spectroscopic methods, (iv) X-ray methods, and (v) thermal methods of analysis.

Microscopic methods for visual examination of biological materials along with the basic principles electron microscopic methods, such as TEM, SEM, STEM, and AFM have been dealt with exclusively in Chapter 5. The basic principles of the remaining methods listed above and their applications are discussed in the following sections.

19.3 METHODS BASED ON ADSORPTION–DESORPTION OF PROBE MOLECULES

·A variety of molecules can be used as probes to investigate the surface of a solid. The molecules get adsorbed through physical or chemical interaction on the surface of the solid.

19.3.1 Physisorption

The term *physisorption* or *physical adsorption* refers to the surface phenomenon of adsorption of fluids, mostly gases, onto the surface of a solid and held by relatively weak physical forces such as van der Waals' forces. Physisorption of gases on solid surfaces occurs at low temperatures close to the boiling point of the gas adsorbed (adsorbate) and in general decreases with increase in temperature. Higher pressures facilitate physisorption onto the surface of a solid (adsorbent). Multilayer adsorption consisting of several molecular layers of adsorbate gas is quite common in physisorption. The heat or enthalpy of physisorption is usually low and lies in the range of 10–40 kJ mol^{-1}. The process is reversible in that the physisorbed gas can easily be desorbed as the activation energy for desorption of the adsorbate gas is low.

Physisorption of the probe molecule forms the basis of characterization of the surface of a solid and provides information on the nature of the surface in terms of adsorption isotherm; surface area and porous nature of the solid in terms of pore volume and pore size distribution.

Adsorption isotherm refers to the relationship between the amount of gas adsorbed by a solid and the equilibrium pressure at a given temperature and is obtained by plotting the amount of gas adsorbed as function of partial pressure of the gas at a constant temperature. Experimental studies have shown that the

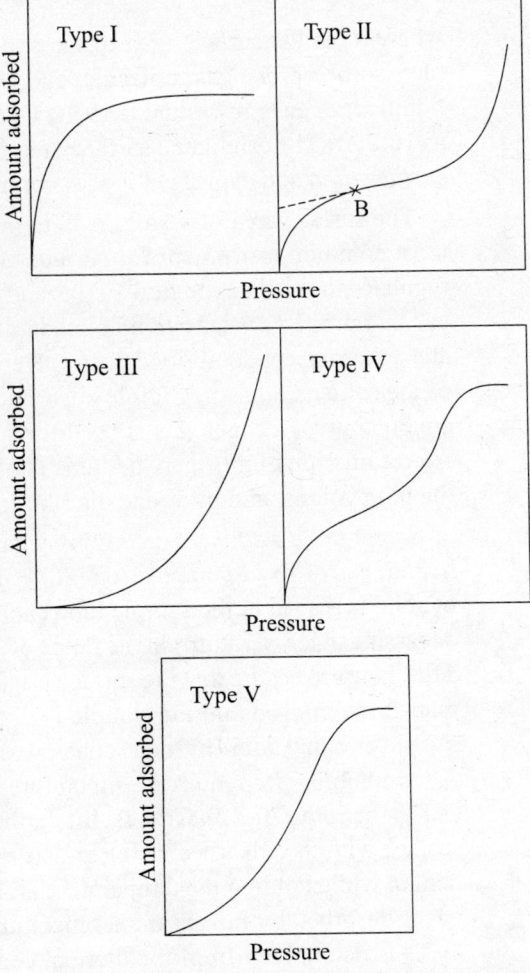

Fig. 19.1   Five different types of adsorption isotherms

gas–solid interaction may be studied in terms of five types of adsorption isotherms designated as Types I, II, III, IV, and V as shown in Fig. 19.1

Type I adsorption isotherm corresponds to monolayer adsorption of the gas on the surface of the solid as postulated by Langmuir theory of adsorption. The surface is saturated and further adsorption does not occur as indicated by the plateau. Type I adsorption isotherm exhibited by a gas–solid system indicates the possibility of chemisorption of the gas due to its specific affinity to the solid surface. Types II and III adsorption isotherms indicate multi-layer adsorption of the gas on the solid surface. Types IV and V adsorption isotherms indicate condensation of the gas in capillary pores of the solid indicating the solid surface to be porous.

Surface area of a solid is usually determined by either *single point method* or *multipoint method* based on Brunauer, Emmett, and Teller (BET) equation. According to BET theory the volume of gas V_m necessary to form a monolayer of adsorption on the surface of a solid corresponds to point B in the Type II adsorption isotherm observed for most solids. The surface area of the solid is calculated using the equation

$$A_S = \frac{V_m N A_{GAS}}{M_V} \tag{19.1}$$

where A_S is the surface area of the solid, N is the Avogadro number, A_{GAS} is the area of one physisorbed probe gas molecule, and M_V is the molar volume. Nitrogen gas is physisorbed at liquid nitrogen temperature ($-195.8°C$) and the area occupied one molecule of nitrogen is taken as $16.2 Å^2$. The calculated surface area in $Å^2$ is divided by the weight of the solid to get *specific surface area* and expressed in units of $m^2 g^{-1}$.

The surface area of a solid is determined by using a surface area analyser instrument. The most common instrument for surface area determination consists of a manifold and pressure regulator to regulate the flow of gas and also vary the composition of the gas mixture of carrier gas and probe gas, sample tube made of quartz, electrically heated mantle to heat the sample and a thermal conductivity detector. Helium gas is used as a carrier gas for cleaning the surface of the solid as it has negligible adsorption characteristics and also because of its high thermal conductivity. Nitrogen gas is used for surface area determination as it alone gets physisorbed from a mixture of nitrogen–helium. The surface area analysers are also capable of determining the pore volume and pore size distribution on the surface of the solid.

Single-point method of determining the surface area involves the use of a mixture of nitrogen–helium gas of fixed composition (30% nitrogen). In practice a weighed quantity (~0.1 to 1 g) of solid is placed in the sample tube enclosed in a heating mantle and the surface is cleaned by degassing at temperatures in the range of 200–300°C and simultaneously passing helium gas for a few hours to ensure that the surface is clean and does not contain any adsorbed gas. The heating mantle is removed and the sample is cooled to liquid nitrogen temperature by immersing the sample tube into liquid nitrogen contained in a Dewar flask while passing helium gas continuously. After attaining the required temperature the solid sample is exposed to a mixture of nitrogen (30%)–helium (70%) gas to facilitate the physisorption of nitrogen gas. After saturating the surface with the adsorbed nitrogen, any excess and condensed gas is removed by flowing pure helium while holding the temperature at liquid nitrogen temperature. The physisorbed nitrogen gas is desorbed by raising the temperature and the volume of the desorbed gas corresponding to V_m is determined from the thermal conductivity detector. The detector is calibrated to read directly the volume of desorbed gas using a known volume of nitrogen gas.

Multipoint method of determining the surface area is based on BET equation

$$\frac{P}{V(P^0 - P)} = \frac{1}{V_m c} + \frac{c-1}{V_m c}\left(\frac{P}{P^0}\right) \tag{19.2}$$

where V and V_m represent the total volume adsorbed and the volume adsorbed when the surface of the solid is covered completely by a monolayer of gas at the pressure P and temperature T respectively. The saturated vapour pressure of the gas is represented as P^0 at the same temperature. The constant c depends on the nature of the gas at the given temperature T and is given as

$$c = \frac{\exp\left(E_1 - E_L\right)}{RT} \tag{19.3}$$

where E_1 is the energy of adsorption in the first layer assumed to be constant and E_L is the energy of adsorption of the succeeding layers equal to the energy (heat) of liquefaction of the gas. The volume of gas required for monolayer adsorption V_m is a constant for a given adsorbate gas-adsorbent system and c is also a constant for the given gas.

In practice the composition of nitrogen–helium mixture is varied between 5% and 30% of nitrogen with an increase in nitrogen by 5% for each incremental step. The BET plot of P/P^0 $[V(1-(P/P^0))]$ as a function of P/P^0 is a straight line (Fig. 19.2) and the value of V_m is obtained from the slope $(c-1)/V_m c$ and the intercept $1/V_m c$ in the equation.

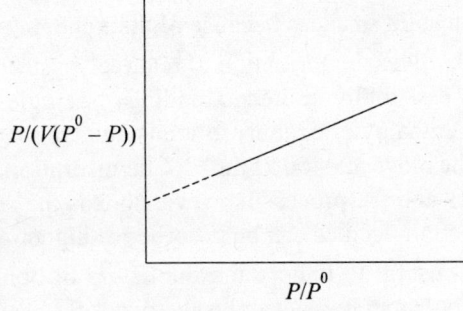

Fig. 19.2   BET plot

Total pore volume refers to the volume of pores available or accessible to a gas (usually nitrogen) on the surface of the solid. Pore volume is determined from the volume of nitrogen gas which condenses as a liquid within the pores present in the surface of a catalyst from a 95% to 5% nitrogen–helium gas mixture at liquid nitrogen temperature using the formula

$$V_T = \frac{273.2}{\text{room temperature}} \times \frac{\text{Atmospheric pressure}}{760} \times 0.00155 \times 100 \times V \tag{19.4}$$

where V is the gas volume and the factor 0.00155 is the ratio of the molar volume of liquid nitrogen (34.67 cm^3) and the gaseous nitrogen (22, 414 cm^3).

Pore size and pore size distribution are important in deciding the end use of a solid for adsorption and catalysis applications because these parameters determine the accessibility of the pores within the solid to adsorbate molecules. Solids are classified based on the pore dimensions as predominantly microporous, mesoporous, or macroporous materials. Solids with pore dimensions of < 2 nm are classified as microporous and those with pore dimensions in the range 2–50 nm as mesoporous. Macroporous solids have large-sized pores with diameters >50 nm.

The distribution of pores of different volumes (or sizes) in porous materials is calculated from adsorption–desorption isotherms. The adsorption–desorption isotherm is constructed from data collected on the volume of nitrogen adsorbed or desorbed from nitrogen–helium gas mixture of different compositions over the range of 0–90% nitrogen. The method of calculation is based on Barrett, Joyner, and Halenda (BJH) method and its modification assuming that the pore can be considered as a cylindrical pore or as a pore bound by two parallel plates. Both the cylindrical pore model and the parallel plate model use Kelvin equation as the basic equation in the theory of capillary condensation given as

$$r_{\mathrm{p}} = \frac{-2\gamma\rho\, M \cos\theta}{RT \ln(P/P_0)} \tag{19.5}$$

where r_{p} represents the radius of the largest cylindrical capillary for the cylindrical pore model or the distance between the plates of the largest parallel plates for the parallel plate pore model, γ is the surface tension of the adsorbate (nitrogen), ρ is the density of the adsorbate, M is its molecular weight, θ is the contact angle (usually assumed to be zero), R is the gas constant, T is absolute temperature, and P/P_0 is the ratio of the actual pressure to the vapour pressure of the bulk adsorbate.

19.3.2 Chemisorption

The term *chemisorption* or *chemical adsorption* refers to the adsorption of certain substances (gases or liquids) on solid surfaces because of their chemical affinity for the solid.

Chemisorption involves the formation of relatively a strong bond between the adsorbate and the adsorbent and is essentially limited to binding of a single layer (monolayer) of the adsorbate on the surface. The enthalpy of chemisorption is quite high and comparable to the heat of bond formation lying in the range 50–400 kJ mol^{-1}. Chemisorption is favoured by higher temperatures and higher pressures and the process is irreversible as the adsorbate is strongly held on to the surface. Chemisorbed molecules can be desorbed at higher temperatures.

Chemisorption is useful to understand the nature of bonding between the catalyst surface and a reactant molecule and how a catalyst activates the reactant molecule. Another important application of chemisorption is to study the nature of active sites, their distribution, strength, and density of active sites as well as the homogeneity or heterogeneity of the catalyst surface through the use of appropriate probe molecules.

Many of the industrial catalysts are acidic in nature (e.g., aluminosilicates such as clay minerals and zeolites, metal containing phosphotungstates, etc.) and the active sites in such catalysts are either Brønsted or Lewis acid sites. It is possible to determine the nature of acidic sites, their density, strength and distribution through chemisorption of probe molecules such as ammonia or pyridine or butylamine. The interaction of the chemisorbed probe molecule with the solid surface is investigated by IR spectroscopy (see Section 19.4.1) or thermal desorption methods (e.g., TPD) discussed later in Section 19.7.1

19.4 VIBRATIONAL SPECTROSCOPIC TECHNIQUES FOR SURFACE STUDIES

Spectroscopic methods useful for studying the vibrational motions of species adsorbed on a surface include the more common techniques such as IR spectroscopy and specially developed vibrational spectroscopic techniques for the study of surfaces such as electron energy loss spectroscopy (EELS) and reflection–absorption infrared spectroscopy (RAIRS).

19.4.1   IR Spectroscopy

Infrared spectroscopy finds extensive use for the study of surface-adsorbed species on solid samples (e.g., heterogeneous catalysts). The different techniques include transmission IR spectroscopy for the study of supported metal catalysts provided the concentration of the surface-adsorbed species is quite high and the solid sample is transparent to IR radiation. Diffuse reflectance FTIR spectroscopy (DRIFTS) technique is a modified technique useful for the study of high surface area catalyst samples which are not sufficiently transparent to IR radiation involving the analysis of the diffusely scattered or reflected IR radiation from the sample. Attenuated total reflection (ATR) spectroscopy (also known as multiple internal reflection spectroscopy—MIR) is yet another modification amenable for the study of surfaces. These techniques have been introduced in Chapter 8.

One of the classic examples of the application of vibrational spectroscopy in the study of surface-adsorbed species is that of adsorption of carbon monoxide on metallic surfaces. The C–O stretching frequency of the free molecule at 2143 cm^{-1} shifts to lower to frequencies and the extent shift depends on the nature of bonding between CO and metal surface. Carbon monoxide can bind to a metal surface as a terminal group or as bridging group between two metal atoms (doubly bridging) or three metal atoms (triply bridging). Terminal CO group shows IR absorption band due to C–O stretching vibration in the range 2100–1920 cm^{-1} while doubly bridged CO group shows the absorption band in the region 1920–1800 cm^{-1}. Triply bridged CO shows the C–O stretching frequency at <1800 cm^{-1}. A bonding model envisages sigma bonding between the carbon and the metal synergistically strengthened by pi-bonding through electron density flow from the filled metal orbitals to the vacant antibonding pi-molecular orbitals of CO.

Another important example of the application of IR spectroscopy involves the study of surface acidity of solid acid catalysts. Acidity of solid surface is characterized through infrared spectroscopy by using pyridine or ammonia as probe molecules. Pyridine being a weaker base as compared to ammonia is a more selective indicator of surface acidity. Chemisorption of pyridine can distinguish between Lewis and Brønsted acid sites depending on the nature of interaction between pyridine and the acidic surface. Pyridine interacts with a Lewis acid site to form a coordination complex whereas it reacts with Brønsted acid site to form a pyridinium ion. If the acidic site on the surface is relatively weak pyridine can bind to the site through hydrogen bonding. All the three types of acidic sites can be distinguished from IR spectral data. The presence of strong IR absorption bands at 1445–1460 cm^{-1} and at 1476–1490 cm^{-1} indicate the presence of pyridine bound to Lewis acid sites as well as weak acid sites through hydrogen bonding. Pyridine adsorbed on Brønsted acid sites gives rise to an exclusive strong IR absorption band at 1540–1550 cm^{-1}. The presence of Brønsted as well as Lewis acid sites can be identified from the overlapping IR absorption bands in the region 1600–1640 cm^{-1}.

Ammonia adsorbs onto Brønsted acid sites of a catalyst surface by interacting through the lone pair of electrons of the nitrogen with surface hydroxyl groups resulting in the formation of adsorbed ammonium ions which shows an IR absorption band at 1450 cm^{-1}. It also gets adsorbed as coordinated ammonia onto Lewis acid sites which are electron pair acceptors giving rise to absorption band at 1640–1620 cm^{-1}. In addition chemisorption of ammonia can take place in undissociated, dissociated, and ionized (NH_2^-) forms. IR absorption bands at 3400 cm^{-1} and 3320–3200 cm^{-1} have been assigned to N–H asymmetric stretch and N–H symmetric stretch respectively of chemisorbed undissociated ammonia. Ammonia chemisorbed in the dissociated form shows IR absorption bands at 3526 cm^{-1} and 3446 cm^{-1} for the N–H asymmetric stretch and the N–H symmetric stretch respectively. Ammonia chemisorbed as NH_2^- group shows IR absorption bands at 3386 cm^{-1}, 3335 cm^{-1}, and 1410 cm^{-1} assigned to asymmetric stretching, symmetric stretching, and the bending modes of the group respectively.

19.4.2 Electron Energy Loss Spectroscopy

Electron energy loss spectroscopy (EELS) also referred to as *high-resolution electron energy loss spectroscopy* (HREELS) or *vibrational energy loss spectroscopy* (VELS) involves focusing a beam of electrons of low but definite energy (in the range 1–10 eV) on the surface of a sample and monitoring the energy of the scattered electrons. When the incident beam of electrons is focused on the surface at a particular incident angle a majority of the electrons ($\sim$ 99.9%) are scattered elastically, that is, do not gain or lose energy on interacting with the molecules adsorbed on the surface. The elastically scattered electrons are reflected from the surface at the same angle as the incident angle and the reflection is called *specular reflection*. A small percentage ($\sim$0.1%) of electrons of the incident beam undergoes inelastic scattering as they gain or lose energy on interacting with the molecules on the surface. These electrons are reflected at any angle including the incident angle.

The inelastic scattering of electrons is due to the long-range interaction between the electric fields of incident electrons and the vibrating dipoles of the adsorbed molecule and is known as *dipole scattering*. The adsorbed molecule may have molecular vibrations either parallel to the adsorbent surface or in a perpendicular direction. If the adsorbent is a metal the interaction between the conduction band electrons of the metal and the oscillating bonds of the adsorbed molecule is such that there will be a net cancellation of the dipole when the molecular vibration is parallel to the surface. On the other hand, if the molecular vibration is perpendicular to the surface of the metal the dipole becomes twice in magnitude. At other intermediate angles the magnitude of the dynamic dipole will be intermediate. According to quantum mechanical selection rule only those vibrational modes of the adsorbed species exhibiting dynamic dipoles perpendicular to the surface interact with the incident of electrons. The adsorbed molecule absorbs energy from the incident beam of electrons when its molecular vibration is perpendicular to the surface of the adsorbent. The energy change corresponds to the change in the vibrational energy of the adsorbed molecule. Hence, by monitoring the energy of the inelastically scattered electrons gives rise to electron energy loss spectrum of the adsorbed molecule which provides information on the adsorbate geometry.

For example EELS spectrum of acetylene adsorbed on the [111] plane of copper metal shows peaks at 2920, 1307, and 920 cm^{-1} assigned to C–H stretch, C–C stretch, and C–H deformation modes respectively of the adsorbed molecule from which it is concluded that bond formation occurs between the metal and C≡C bond (π bonding) and also between the C and Cu atoms through σ bonding. Since the C–C stretch at 1307 cm^{-1} is in between the stretching frequencies of the C–C and C=C bonds of ethane and ethylene (990 and 1620 cm^{-1} respectively) the adsorbed molecule has C–C bond characteristics between those of the C–C single and C=C double bonds indicating that π bonding between acetylene and the metal is extensive and the molecule is strongly adsorbed on the metal surface.

The EELS spectrometer consists of an electron gun consisting of white hot tungsten filament for emitting electrons, monochromator systems consisting of electrostatic lenses, sample holder, electron energy analyser, and electron multiplier detector. The incident beam energy is usually in the range of 1–10 eV. The resolving power of the instrument is about 1 MeV of electron energy corresponding to about 8 cm^{-1}, which is poor as compared to the resolving power of about 1 cm^{-1} of IR spectrophotometers.

19.4.3 Reflection–absorption Infrared Spectroscopy

Reflection–absorption infrared spectroscopy (RAIRS) is a convenient technique for the study of surface-adsorbed species and together with EELS becomes a powerful technique for the determination of surface structure. The basic principle of RAIRS involves the absorption of IR radiation by surface-adsorbed molecules as they exhibit their characteristic vibrations. Absorption of incident IR radiation occurs at a specific frequency only when a vibration possesses a continuously varying dipole. Based on the surface dipole selection rule, only vibrations which are perpendicular to the adsorbent surface give rise to an oscillating dipole, and hence, are IR active thereby giving rise to an observable absorption band. In contrast, vibrations of the adsorbed molecule parallel to the adsorbent surface are IR inactive.

In general the amount of surface-adsorbed species will be low in the range of nanomoles in most cases and dispersive IR spectrophotometers may not be sensitive to detect very weak signals. However, FTIR spectrophotometers are inherently capable of detecting such weak signals and also exhibit high resolution of about 1 cm^{-1}, and hence, RAIRS studies are best carried out using FTIR instruments. The range of frequencies in RAIRS is usually 3600–600 cm^{-1} and it is not possible to observe metal–adsorbate bond vibrations which occur at frequencies < 600 cm^{-1}.

An example of the application of RAIRS involves the study of ethylene adsorbed on the [111] surface of platinum. The RAIRS spectrum shows a strong band at 1339 cm^{-1} and two other weak bands at 2884 cm^{-1} and 1118 cm^{-1}. The spectrum has been interpreted on the basis that on adsorption ethylene rearranges to give an ethylidyne fragment, C–CH$_3$, which gives rise to the absorption band at 2884 cm^{-1} and the C–C singe bond at 1118 cm^{-1}. The interpretation is based on the fact that the symmetric CH stretch of the methyl groups is observed at 2870 cm^{-1} in ethane (close to 2884 cm^{-1} seen in RAIRS spectrum). Further the symmetric deformation vibration of pure ethane gives rise to a band at 1380 cm^{-1} which correlated well with the strong band at 1339 cm^{-1} seen in the RAIRS spectrum.

RAIRS can be used as a simple structural characterization technique particularly for the determination of adsorbent–adsorbate bonding characteristics, identification of surface intermediates, and elucidation of reaction pathways in catalytic reactions.

19.5 ELECTRONIC SPECTROSCOPIC METHODS

Spectroscopic methods for the study of surfaces involve the use of a probe beam of energetic particles such as UV or *X*-ray photons, electron, or ions which impinges on the target surface. A variety of radiations or particles are emitted from the target surface after the interaction with the probe beam. The different types of probe beams and emissions giving rise to different types of spectroscopic techniques are summarized in Fig. 19.3.

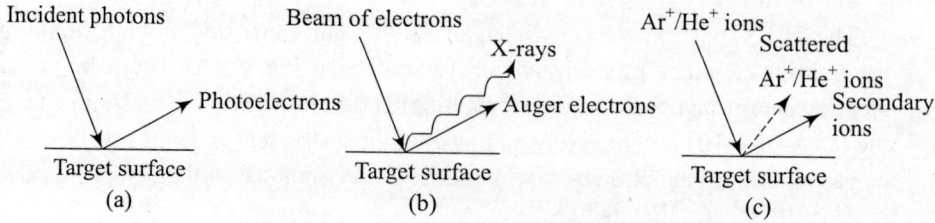

Fig. 19.3 Interaction between probe beam and the target surface in (a) ESCA, (b) EPMA, XRF, and AES, and (c) SIMS and ISS

The probe beam of particles may be electromagnetic radiation such as UV light or *X*-rays giving rise to *electron spectroscopy for chemical analysis* (ESCA). A beam of electrons is used in *Auger spectroscopy* or *Auger electron spectroscopy* (AES), *electron microprobe analysis*, *and X-ray fluorescence analysis*. A beam of helium or argon ions is used in *secondary ion mass spectrometry* (SIMS) and *ion scattering spectroscopy* (ISS)

19.5.1 Electron Spectroscopy for Chemical Analysis

Electron spectroscopy for chemical analysis (ESCA) is actually *photoelectron spectroscopy* based on photoelectric effect enunciated by Einstein. Experimental studies indicated that when a metal is exposed to an incident beam of light there is a threshold frequency (energy) of the light below which it does not eject electrons from the surface regardless of its intensity. Einstein postulated that light consists of photons of energy $E = hv$ (where h is Planck's constant and v is the frequency) and since an electron in an atom is held by an energy called the binding energy E_b depending on the quantized energy level it occupies, the threshold frequency is the minimum frequency required to eject the electron from the metal to empty space. The threshold frequency corresponds to the minimum energy and is called the *work function* φ. At frequencies greater than the threshold frequency for a particular metal the incident beam of photons can eject the electron which escapes to vacuum with a kinetic energy E_k. The kinetic energies of the ejected photoelectron is analysed by a detector called the energy analyser in photoelectron spectroscopy. The resulting kinetic energy spectrum provides information on the energy levels of the metal, and hence, direct imaging of the electronic states of atoms or molecules becomes possible through photoelectron spectroscopy. The electron binding energy, E_b, relative to the Fermi level of solid can be calculated using the relationship

$$E_b = hv - E_k - \varphi \tag{19.6}$$

Thus, the photoelectric effect involves irradiation of a sample with photons of sufficient energy to facilitate the emission of electrons from the sample. The number of photoelectrons emitted depends on the intensity of the incident beam of light and the energy of the photoelectrons in turn depends on the wavelength of light. Depending on type of incident beam of radiation ESCA can be subdivided into two techniques (i) *ultraviolet photoelectron spectroscopy* (UPS) and (ii) *X-ray photoelectron spectroscopy* (XPS).

Ultraviolet photoelectron spectroscopy was originally developed as molecular photoelectron spectroscopy to study the photoelectron spectra of free gaseous molecules using helium discharge lamp as the source of photons of wavelength 58.4 nm corresponding to an energy of 21.2 eV. The technique was useful in identifying the bonding, non-bonding, and antibonding molecular orbitals based on the peaks observed in the spectrum corresponding to the valence region molecular orbitals and the fine structure due to vibrational levels. The technique was later extended to identify the surface-adsorbed species on solids and for the determination of work function of the material based on the emission of photoelectrons.

X-ray photoelectron spectroscopy has emerged as an important technique in the study of surfaces. The basic principle of XPS involves irradiating the sample with a beam of *X*-rays of relatively low energy (1–2 keV) under high vacuum conditions to knock off photoelectrons from the atoms of the sample through photoelectric effect as shown in Fig. 19.4.

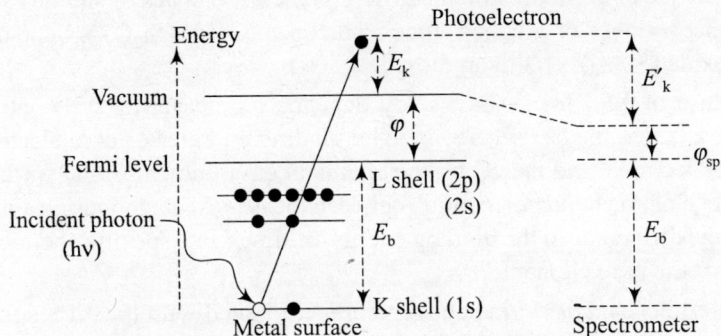

Fig. 19.4 Principle of *X*-ray photoelectron spectroscopy

The kinetic energy and the number of electrons that escape from the top surface layers of the material are analysed by a spectrometer. The information is generated in the form of XPS spectrum which is plot of number of photoelectrons ($N(E)$) emitted with energy E, against E_k or often against E_b (the binding energy of escaping electrons).

In the spectrometer the ejected electron is subjected to a contact potential between the sample and the spectrometer which changes the kinetic energy of the electron to E'_k. The binding energy of the electron E_b is obtained using the relationship

$$E_b = h\nu - E'_k - \varphi_{sp} \tag{19.7}$$

where φ_{sp} is an instrument constant called the *spectrometer work function*. The value of φ_{sp} is about 4 eV and varies with the type of instrument. The Fermi level of the solid surface is taken as reference energy (and taken as zero). It is less compared to the binding energy of the electron in the same quantized energy level in a gaseous atom. Thus, the binding energy of 1*s* electron of gaseous carbon atom is 288 eV while the binding energy of the same electron in the solid carbon sample is $(288 - \varphi_{sp})$ eV. The value of 284.8 eV is used as the reference value for the 1*s* electron of carbon in hydrocarbons attached to metals.

The XPS spectrometer consists of the major components of *X*-ray generator to produce a highly focused beam of monochromatic *X*-rays of aluminium K_α or magnesium K_α, an ultra-high vacuum system, electron collector lens, a hemispherical electron energy analyser, electron detector, and sample stage. The aluminium K_α radiation of wavelength 8.3386 Å corresponds to photon energy of 1.487 keV, while the K_α *X*-rays of magnesium source have a wavelength of 9.89 Å corresponding to photon energy of 1.254 keV. The resolving power of XPS depends on the line width of *X*-ray source which is about 1 eV and can be reduced to about 0.3 eV by using a quartz crystal monochromator to achieve better resolution. The monochromator system also prevents heat and secondary electrons generated by the *X*-ray source from reaching the sample. The instrument is operated under ultra-high vacuum conditions at 10^{-10}–10^{-9} torr.

XPS is essentially a surface analytical technique which provides quantitative information on the elemental composition, empirical formula, chemical, and electronic states of elements in a material. XPS can detect all the elements other than hydrogen and helium with detection limits in the range of parts per thousand (ppt) and provides unique information on the surface of a material to a depth of 1–10 nm. For example, XPS analysis of a polymer surface reveals the presence or absence of the different chemical states of carbon known as carbide (C^{2-}), hydrocarbon (C–C), alcohol (C–OH), ketone (C=O), organic ester (COOR), carbonate (CO_3), fluorohydrocarbon (CF_2–CH_2), and trifluorocarbon (CF_3). Similarly, the surface of a silicon wafer provides information on the different chemical states of silicon by revealing the presence or absence n-doped or p-doped silicon, silicon suboxide (Si_2O), silicon monoxide (SiO), silicon sesquioxide (Si_2O_3), or silicon dioxide (SiO_2).

The use of XPS for surface study depends on measuring the relative binding energies of electrons, called the *chemical shift*. The binding energies of core electrons are affected by the valence electrons, and hence, by the chemical environment of the surface atom, that is, on the degree of electron bond polarization between the nearest neighbouring atoms. A *specific chemical shift* is the difference in the binding energy values of one specific chemical state and the binding energy of the pure element.

The term *chemical shift* is conveniently explained with the XPS spectrum of the carbon 1s electron in ethyl trifluoro acetate as shown in Fig. 19.5.

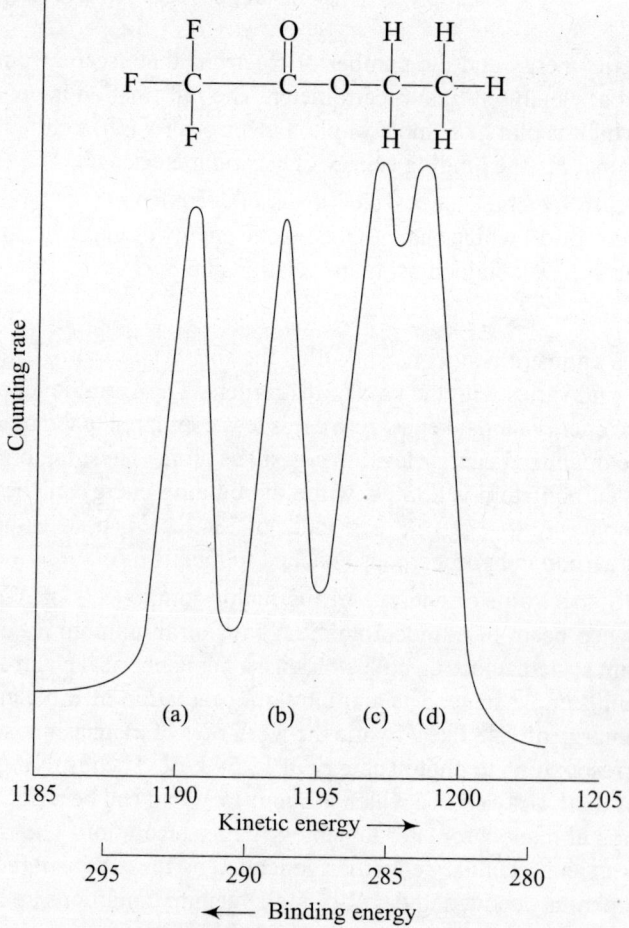

Fig. 19.5 XPS spectrum of ethyl trifluoro acetate

The spectrum shows four distinct peaks representing the $1s$ electron-binding energies of the four carbon atoms located in four different chemical environments. The $1s$ electron of CF_3 group has the highest binding energy because the three F atoms withdraw electron density from the carbon atom and consequently the $1s$ electron is bound tightly, and hence, requires higher energy to eject it. The next higher binding energy is shown by the $1s$ electron of the carbon atom in the COO group.

XPS finds use in the analyses of a wide variety of materials such as metals and alloys, semiconductors, catalysts, ceramics, glasses, paints, polymers, biomaterials and implants, oils, papers, inks, plants and plant parts, wood, etc. XPS is widely used in the study of catalysts. The technique provides information on (i) elemental composition, (ii) oxidation states of elements, and (iii) in certain cases the dispersion of one phase over another.

Analysis of sample composition is based on the characteristic binding energies of elements. Almost all the photoelectrons emitted in XPS have kinetic energies in the range of 0.2–1.5 keV and the probing depth of XPS is about 1.5–6 nm depending on the kinetic energy of the photoelectron. The intensity of the XPS peak (area) is proportional to the number of atoms of particular element present in the sample.

The presence of different oxidation states of an element in a sample surface can be detected by XPS. The XPS spectra of copper metal and its oxides Cu_2O and CuO are shown in Fig. 19.6.

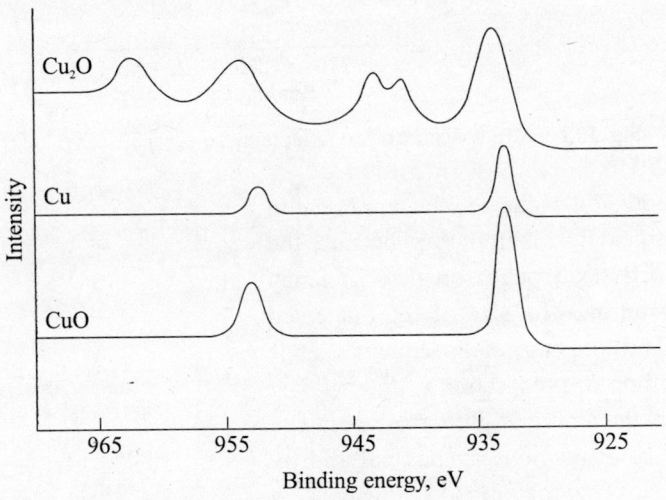

Fig. 19.6 XPS spectra of copper metal and its oxides

It is easy to distinguish between copper metal (0) and its oxidation state of +2 in the form of oxide CuO (Fig. 19.6) but not between copper metal and Cu_2O. The two oxides can be distinguished from the $1s$ electron-binding energies of oxygen (530.8 eV for Cu_2O and 530.1 eV for CuO).

Similarly, the presence of Mo(IV), Mo(V), and Mo(VI) in the Mo/Al_2O_3 catalyst prepared by impregnating $(NH_4)_2MoO_7$ on Al_2O_3 followed by reduction at different temperatures is shown in the XPS spectrum. The sample calcined at 400°C shows XPS peak of Mo $3d_{5/2}$ electron at 238.2 eV due to Mo(VI). The XPS spectrum of the sample on reduction at different temperatures shows peaks due to Mo(V) and Mo(IV) with Mo(IV) concentration increasing with increasing reduction temperature as shown in Fig. 19.7.

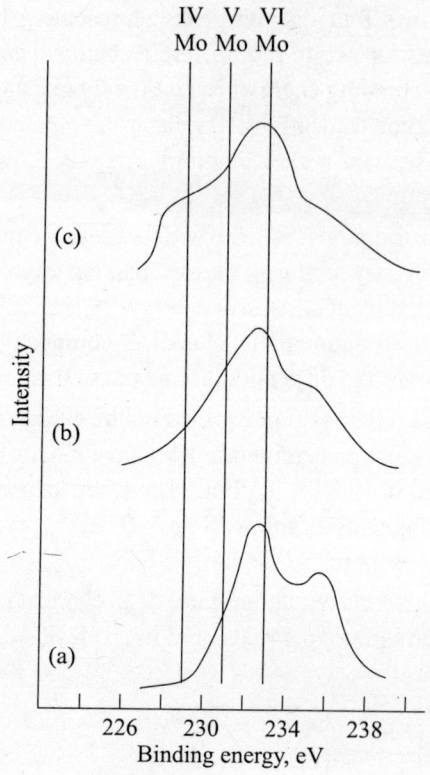

Fig. 19.7 XPS spectrum of Mo/Al$_2$O$_3$ reduced at (a) 300°C, (b) 400°C, and (c) 500°C

The dispersion of the active component in a supported catalyst is important because the catalytic activity depends on the efficiency of dispersion and the size of the dispersed particles. The efficiency of dispersion depends on the method of preparation of the catalyst and also on the nature of precursor used. The efficiency of dispersion can be analysed by XPS by measuring the ratio (I_p/I_s) where I_p is the intensity of the XPS due to the dispersed particle and I_s is the intensity of the peak due to the support. Figure 19.8 shows the XPS spectra of ZrO$_2$/SiO$_2$ (ZrO$_2$ dispersed on SiO$_2$ support) prepared from zirconium nitrate solution and from zirconium ethoxide. The spectra indicate that the dispersion of ZrO$_2$ is much better for the catalyst prepared from zirconium ethoxide as indicated by the greater intensity of Zr 3d peak.

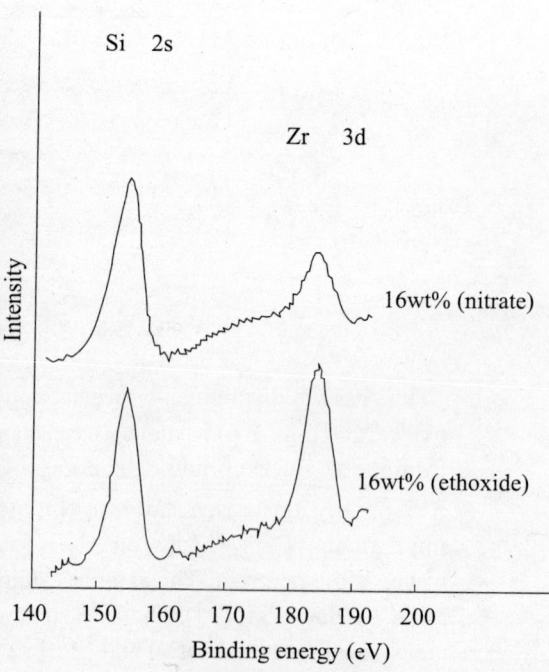

Fig. 19.8 XPS spectra of ZrO$_2$/SiO$_2$ catalysts after calcining at 700°C

The effect of calcination temperature on the dispersion of NiO on Al_2O_3 can be seen in XPS spectra with peaks shown for Ni (2p 3/2), O (1s), and Al (2p) + Ni (3p) electrons in Fig. 19.9. The dispersion of NiO is better when calcined at 1050°C.

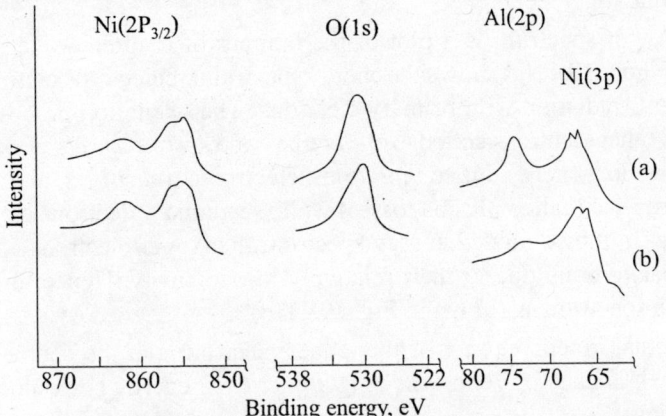

Fig. 19.9 XPS spectra of NiO/Al_2O_3. (a) calcined at 1050°C and extracted with HCl, (b) calcined at 1050°C, (c) calcined at 850°C, (d) $NiAl_2O_4$, and (e) NiO

19.5.2 Auger Electron Spectroscopy

Auger electron spectroscopy (AES) is based on the Auger process or Auger effect in which a beam of primary electrons having kinetic energy in the range of 1–10 keV impinge on the sample surface and excite the atom by creating holes in the core levels. The excited atom relaxes by filling the core hole with an electron from a higher shell with the simultaneous release of energy in the form of an *X*-ray photon (*X*-ray fluorescence). Alternatively the excited atom may relax through Auger transition which involves the emission of a second electron called the Auger electron from a higher energy level as shown in Fig. 19.10. The kinetic energy of the Auger electron depends only on the nature of energy levels involved in the Auger process or transition and independent on the energy of the primary electrons.

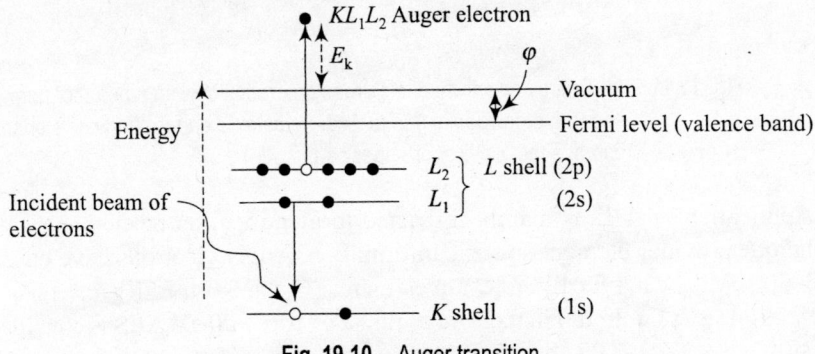

Fig. 19.10 Auger transition

The Auger transition is designated as KL_1L_2 transition indicating that the initial core hole was created in the *K* shell by the impingement of the primary electrons which was filled by an electron from L_1 level and the Auger electron was emitted from the L_2 level.

AES instrument consists of an electron gun which emits a beam of electrons onto the sample and the emitted electrons are deflected to a cylindrical mirror analyser and then directed to an electron multiplier detector. The system is operated under ultra-high vacuum conditions. The emitted electrons have energies in the range of 50 eV to about 3 keV, the escape depth of the electrons is limited to a few nanometers from the top surface of the sample. Hence, AES is a surface analytical technique.

The Auger spectrum is a plot of the number of emitted Auger electrons as a function of energy. Figure 19.11(a) shows the energy spectrum of electrons coming off the surface of a solid sample on irradiation with primary electrons. The excitation process yields along with Auger electrons other electrons called *loss electrons* and *secondary electrons*. Loss electrons are those which have lost energy due to vibrations, electronic transitions, and to collective excitations of the electron sea (called plasma losses) while secondary electrons are generated due to inelastic processes. In the combined energy spectrum the Auger electrons appear as small peaks in an intense background due to their relatively low intensity. Hence, the AES is presented as the derivative spectrum as shown in Fig. 19.11(b).

The actual kinetic energy of the Auger electron is at the centre of the integral peak (E_0). However, it is usual to report the Auger energy as energy corresponding to the negative peak (E_a) as shown.

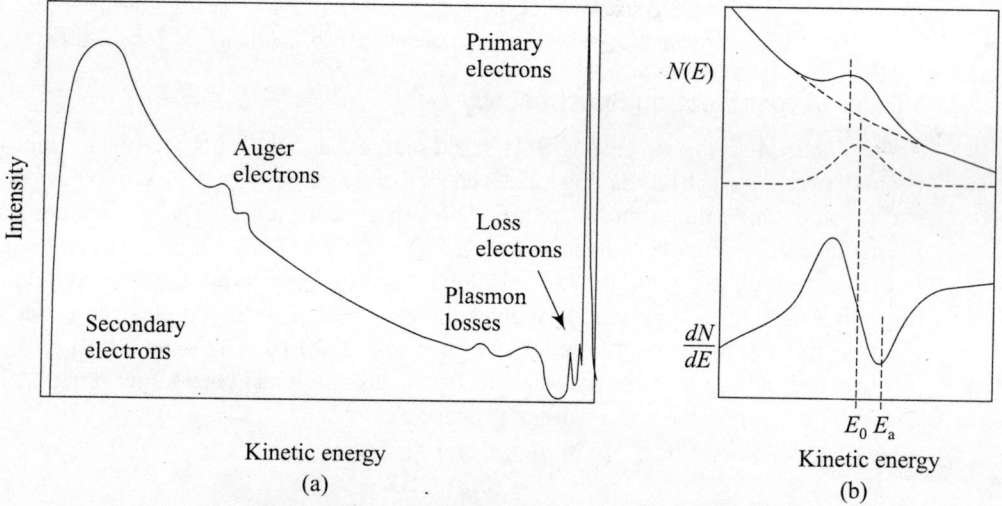

Fig. 19.11   (a) Energy spectrum of electrons on irradiation with a beam of primary electrons showing small peaks due to Auger electrons and (b) derivative spectrum to improve the visibility of Auger peaks

Application of AES is mainly restricted to identification of elements on the surface as the technique provides element-specific information. Auger electrons have energies in the range of 100–300 eV for catalytically relevant elements (C, Cl, S, Pt, Ir, Rh, etc.). As the mean-free path of electrons is at minimum in the energy range of 100–300 eV AES is considerably more surface sensitive than XPS. XPS and AES provide similar information but the spatial resolution in XPS is poor as compared to AES because the incident beam of X-rays cannot be made smaller than 150 microns whereas in AES the electron beam can be made still smaller, and hence, provides better resolution.

A typical example of the application of AES is the study of the synthesis of methanol over a copper single crystal by reacting a mixture of gases of CO_2, CO, and H_2 on Cu [100] at temperatures 500–550 K at pressure of 1 bar. The spectrum exhibiting the characteristic LMM peaks of Cu between 700 and 900 eV is shown in Fig. 19.12. The peaks due to KVV (V = valence shell) transitions of C at 275–280 eV and of O at 510 eV are also shown.

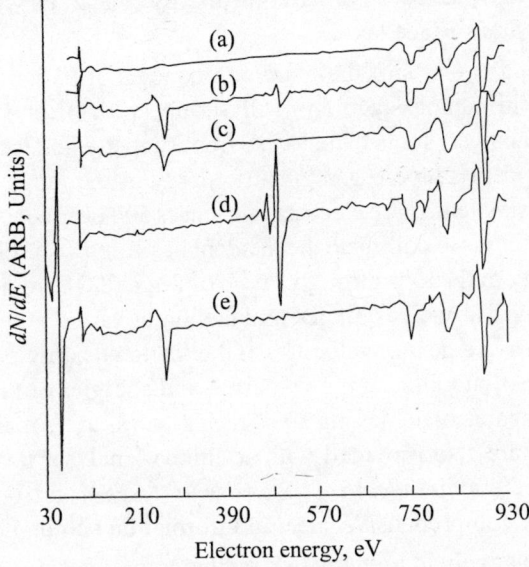

Fig. 19.12   AES of Copper catalyst (a) clean Cu[100], (b) after methanol synthesis, (c) after heating to reaction temperature of 500–550 K, (d) after oxidation at 555 K, and (e) after subsequent methanol synthesis

The O KVV peak at 505 eV is seen predominantly in curve (d). The oxidized copper surface is twice as active as reduced copper in methanol synthesis. The oxidized copper surface contains more carbon than the initially reduced Cu surface after methanol synthesis and subsequent heating (compare curves (e) and (c)). However, it must be mentioned that since AES is only element sensitive it is not clear whether the carbon and oxygen are present as CO, CO_2, CH_2OH, or CHO, etc., or as atomic species. Only IR can help identifying the nature of the species on the surface.

19.5.3   Ion Scattering Spectrometry

ISS includes a group of surface analytical techniques such as *low-energy ion scattering spectroscopy* (LEIS), *medium-energy ion scattering* (MEIS), and *high-energy ion scattering* (HEIS) also known as *Rutherford backscattering spectroscopy* (RBS). The basic principle in all the three techniques involves bombarding the surface of a target material with noble gas ions ($^3He^+$, $^4He^+$, $^{20}Ne^+$, or $^{40}Ar^+$) and observing the positions, velocities, and energies of the elastically scattered ions. The techniques differ only in the energies of the incident beam of particles. The energy of the ions in LEIS is the range of 100 eV–10 keV while it is in the range of 100–200 keV for MEIS and of the order of several MeV in HEIS. The surface analytical techniques are useful in the characterization of the chemical and structural aspects of the surface of the target material, particularly LEIS is quite sensitive and is one of the techniques capable of observing directly hydrogen atoms on the surface of a material.

The interaction between the incident beam of ions and the target surface leads to one or more of events such as scattering, adsorption, sputtering or recoiling, displacement, electron emission and photon emission. Each event is characterized by a specific interaction for a given incident ion-target surface system and each event can be monitored to give rise to individual spectroscopic techniques. ISS and particularly LEIS is primarily concerned with scattering of the incident ions by atoms at the first surface layer.

The instrument consists of the main components of (i) ion source, (ii) focusing lenses and chopper lenses, (iii) sample stage, (iv) TOF set-up, (v) detector, (vi) ultra-high vacuum pump, and (vii) additional analyser tools. The ion source is mostly an electron impact ionization chamber in which monopositive noble gas ions of energy in the range 100 eV–10 KeV (in LEIS) are generated. Focusing lenses are electrostatic lenses and beam chopping lenses have either plate or cylindrical geometries to collimate the incident beam and also selectively filter the beam based on mass and velocity of the ions through a time of flight (TOF) region. The sample stage allows the sample to be moved as well as rotated and facilitates varying the incident angle. The TOF set-up facilitates the measurement of velocities of the scattered particles. The detector is an electrostatic analyser capable of monitoring the velocities and energies of the scattered particles. Ultra-high vacuum pump capable of achieving 10^{-10} torr is necessary to maintain the surface clean. Usually ISS instruments are also provided with additional analyser tools such as *low-energy electron diffraction* (LEED), *Auger electron spectroscopy* (AES), and *X-ray photoelectron spectroscopy* (XPS) to provide comprehensive spectral information simultaneously.

In LEIS the relatively low-energy incident ions are scattered as they undergo elastic collision with surface atoms of the first layer. The energy of the scattered ions depends on the incident beam energy, mass of the surface atom, and the scattering angle. The energy of the scattered ion at a particular angle (90°) depends only on the mass the surface atom and the energy of the incident beam of ions as given by the expression.

$$\frac{E_S}{E_I} = \frac{M_{Sur} - M_I}{M_{Sur} - M_I}$$

(19.8)

where E_S and E_I are the energies of the scattered ion and the incident noble gas ion, M_I, and M_{Sur} are the masses of the incident noble gas ion and the surface atom respectively.

Thus, by measuring the energy spectrum of the scattered ions it is possible to get information on the mass of the target atom and the nature of atoms on the surface. The surface arrangement or distribution of different types of atoms on the surface may also be determined by rotating the surface and monitor the energy at different scattering angles. The graphical plot of the intensity of the scattered ions as a function of kinetic energy appears as peaks, the intensity being proportional to the number of atoms of the particular kind.

LEIS finds use in determining the surface composition of heterogeneous catalysts, thin film coatings, adhesive layers, etc., and to provide information on the arrangement of surface atoms.

19.5.4 Secondary Ion Mass Spectrometry

Secondary ion mass spectrometry (SIMS) involves *sputtering* or bombarding the target solid surface with a primary beam of ions and analysing the ejected secondary ions from the surface by a mass spectrometer.

The term *sputtering* refers to the three effects brought about by the interaction of the primary ions with the surface of a sample: (i) mixing up of the upper layers of sample surface resulting in amorphization of the surface, (ii) implanting of the primary ions on the surface, and (iii) ejection of atoms or small molecules from the surface generally called secondary ions. The molecules may undergo dissociation to generate neutral species and positively as well as negatively charged fragments ions. The secondary ions are extracted from the sputtering area by the application of an electric field between the sample and an extraction lens. The extraction lens focuses the secondary ions into a secondary ion beam which is passed on to a mass spectral analyser which separates the ions on the basis of mass and energy. The count rates of the different secondary ions detected by the ion detector provide information on the nature of the surface with respect to its composition. The elemental, isotopic, and molecular composition of the surface or thin film can be studied as the technique is sensitive with detection limits of elements being at ppb levels. All elements including hydrogen can be detected and quantitatively determined. Only noble gases are difficult to handle as they do not ionize easily. SIMS can determine the isotopic composition because the secondary ions of different masses can be analysed. The sputtered area is determined by the diameter of the primary ion beam which is in the order of micrometers SIMS is capable of high lateral resolution. In additions SIMS is also capable of depth profiling and three-dimensional measurements because the sample is eroded during sputtering.

The main components of the secondary ion mass spectrometer include (i) primary ion source, (ii) accelerating and focusing elements, (iii) high vacuum chamber for holding the sample and the secondary ion extraction lens, (iv) mass analyser, and (v) ion detector.

Three basic types of primary ion sources are commonly used in SIMS. These include (i) electron ionization source or duoplasmatron used for generating noble gas ions (Ar^+ or Xe^+), oxygen ions; (O^- or O_2^+), and ionized molecules such as C_{60}^+ or SF_5^+ from SF_6; (ii) surface ionization source consisting of a porous tungsten plug through which Cs atoms vapourize and generate Cs^+ ions, and (iii) liquid metal ion source (LMIG) for generating metal or alloy ions from metal and alloys which have low melting temperatures (e.g., gallium metal and indium or bismuth alloys). The LMIG consists of a liquid metal-covered tungsten tip which emits ions under the influence of intense electric field. The primary ion source is selected depending on the requirements of continuous or pulsed operation, beam dimensions, and the sample to be analysed. Oxygen ions are used to investigate electropositive elements in the sample while cesium ions are used to investigate electronegative elements. LMIG is used mainly for generation of pulsed primary ion beam.

The electrostatic accelerating and focusing lenses focus the primary ion beam onto the surface of the sample. A high vacuum pump is an essential component providing vacuum in the order of 10^{-6} torr in the sample chamber to prevent surface contamination by adsorption of gases and also to prevent the collision of secondary ions with the background gases.

The different types of mass analysers used in SIMS to separate the secondary ions include (i) sector field mass spectrometer with electrostatic and magnetic analyser components, (ii) quadrupole mass spectrometer, and (iii) time of flight (TOF) mass spectrometer. The TOF mass spectrometer separates the secondary ions according to their kinetic energy as they pass through a field-free region and requires a pulsed primary ion gun or a pulsed secondary ion extraction unit. Electron multiplier detector or a Faraday cup or a microchannel plate detector coupled with a fluorescent screen is commonly used as detector.

SIMS is a non-destructive technique useful in the analysis of trace elements in samples as small as pollen grains. Because of its high sensitivity in the range of ppb, analytical applications include elemental and isotopic composition in minerals, semiconductors, microfossils, meteorites, interplanetary dust, etc.

19.6 X-RAY METHODS

X-ray methods applicable to the analysis of surfaces include X-ray absorption spectroscopy, X-ray fluorescence spectroscopy, and electron probe microanalysis which have been introduced in Chapter 11.

Extended X-ray absorption fine structure (EXAFS) and *X-ray absorption near-edge structure* (XANES) are techniques associated with X-ray absorption spectroscopy (XAS) (see Chapter 11). When the energy of the incident beam of X-rays matches the binding energy of an atom in a sample, the intensity of X-rays transmitted decreases sharply exhibiting an absorption edge which is element specific in the absorption spectrum. The X-ray absorption spectrum is a graphical plot of the absorption coefficient the sample as a function of energy of the incident X-ray beam. The XAS spectrum of a solid (or liquid and molecular gases) consists of several larger and smaller crests and troughs and is generally known as the *X-ray absorption fine structure* (XAFS). XAFS provides information on the local structure around the absorbing atom. XAFS cannot be observed in the case of free atoms, for example, noble gases.

XAS can be conveniently analysed in the form of regions (i) *pre-edge* region where the energy of the incident beam of X-rays in insufficient to bring about ionization of the atom, but sufficient to bring out the transition to higher energy orbitals, (ii) *edge* region where the incident X-ray photon energy $E < E^0 + 10$ eV (where E^0 is the ionization energy) where XANES is observable, (iii) *near-edge* region in the energy range $E^0 + 10$ eV $< E < E^0 + 50$ eV where *near-edge X-ray absorption fine structure* (NEXAFS) is observable, and (iv) *EXAFS* region where the X-ray photon energy $E > E^0 + 50$ eV.

Surface-extended X-ray absorption fine structure (SEXAFS) is EXAFS extended to study of surfaces and involves detecting and measuring the intensity of the Auger electrons as a function of the energy of the incident beam of X-ray photons. The technique provides information on the inter-atomic distances of adsorbates and their coordination environment.

19.7 THERMAL METHODS

Thermal methods of surface characterization involve heating the sample at a controlled heating rate in a controlled atmosphere while monitoring the evolved gas. The most commonly used method is temperature-programmed desorption (TPD). Temperature-programmed reduction (TPR) is yet another technique which depends only on the reducibility of the solid.

19.7.1 Temperature-Programmed Desorption

Temperature-programmed desorption (TPD) is also known as *thermal desorption spectroscopy* (TDS) and involves, as the name implies, temperature-induced desorption of a previously adsorbed substance. The technique is one of closely related techniques in which a chemical process such as oxidation, reduction, or sulphidation that takes place on a solid surface is monitored as the temperature of the solid sample increases linearly with time. In these temperature-programmed techniques, time and temperature are related by the expression

$$T(t) = T_0 + \beta t \qquad (19.9)$$

where T and T_0 refer respectively to the temperature of the solid sample at a given time t (in min) and when $t = 0$ (initial temperature) and β is the heating rate (in °C min^{-1}) which is usually in the range of 0.1°C min^{-1} to a maximum of about 20°C min^{-1}. The concentration profile of the species undergoing the reaction is monitored as a function of temperature by a detector placed suitably. The most commonly used detectors are thermal conductivity detector (TCD) and mass spectral detector (MSD). The experimental data is present in the form of a graphical plot of the concentration of the species as a function of temperature to give thermal reaction spectrum.

TPD is concerned only with desorption of an adsorbed species from the solid surface. A block diagram of the TPD instrument is shown in Fig. 19.13.

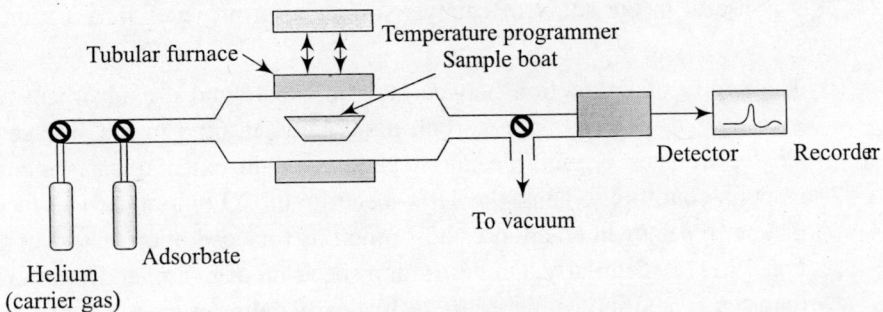

Fig. 19.13   Schematic diagram of TPD set-up

The solid sample (e.g., a heterogeneous catalyst) is subjected to first an adsorption process of a chosen gas (adsorbate) followed by desorption of the adsorbed gas by a programmed heating. The heating rate is usually in the range of $0.1°C \, min^{-1}$ to about $20°C \, min^{-1}$. The desorbed gas is monitored with the help of a suitable detector. The detector used may be a TCD or a MSD. The graphical plot of the amount of desorbed gas as a function of increasing temperature is called the *thermal desorption spectrum* or *temperature-programmed desorption spectrum*.

The steps involved in recoding the TPD spectrum of a catalyst sample include the following: (i) The surface of the catalyst sample is cleaned by evacuating the sample to low pressures of the order of 10^{-3}–10^{-4} torr while maintaining a flow of the pure carrier gas (He) and a temperature of about 200°C, (ii) the catalyst sample is maintained at a required temperature (usually room temperature) and a small amount of the adsorbate gas is introduced into the flowing carrier gas to facilitate the adsorption process, (iii) any excess adsorbate gas which gets condensed on the catalyst surface is removed by flowing the pure carrier gas so that all the physisorbed adsorbate is completely removed as indicated by the base line in the TPD spectrum, and (iv) the catalyst sample is now subjected to a programmed heating at a specified heating rate and the concentration of the desorbing species is monitored by the online detector (either TCD or MSD).

A typical spectrum is shown in Fig. 19.14.

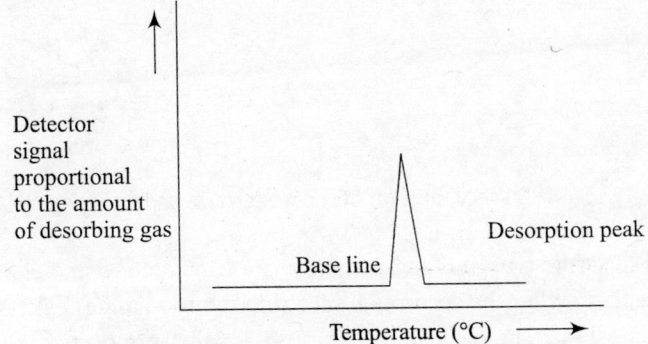

Fig. 19.14   A hypothetical TPD spectrum

The TPD spectrum provides information on the temperature range in which desorption occurs. In a simple spectrum with one desorption peak, the area under the peak is proportional to the amount of gas desorbing (or the amount of gas adsorbed prior to desorption). The appearance of a single peak indicates the homogeneity of the active sites on the surface. In contrast, if the catalyst surface has different types of active sites multiple peaks can be observed.

TPD is useful in the study of catalyst surface as it provides information on a variety of aspects:

(i) The nature of interaction between the adsorbate and the adsorbent (catalyst) can be investigated. For example, carbon monoxide can bind to the surface of a catalyst in single, two sites, or multiple sites depending on the extent of surface coverage and other reaction conditions. Thus, the TPD spectrum of CO on tungsten surface indicates three distinct types of interaction or adsorption as (or desorption states) of CO as shown in Fig. 19.15(a). Similarly, the desorption spectrum of hydrogen from the [100] surface of tungsten crystal shown in Fig. 19.15(b) clearly indicates the effect of surface coverage on the desorption profile. The low temperature peak at about 430 K appears only at higher surface coverage.

(ii) The extent of surface coverage by the adsorbate prior to desorption n_0, can be determined from the TPD spectrum as given by Eq. (19.10).

$$n_0 = \int_0^{\inf} \frac{-dn}{dt}\, dt = \int_0^{\inf} c\, P\, dt \tag{19.10}$$

where c is the proportionality factor relating the partial pressure of the desorbed gas to its desorption rate. When c is constant, the area of the desorption peak is proportional to the surface coverage.

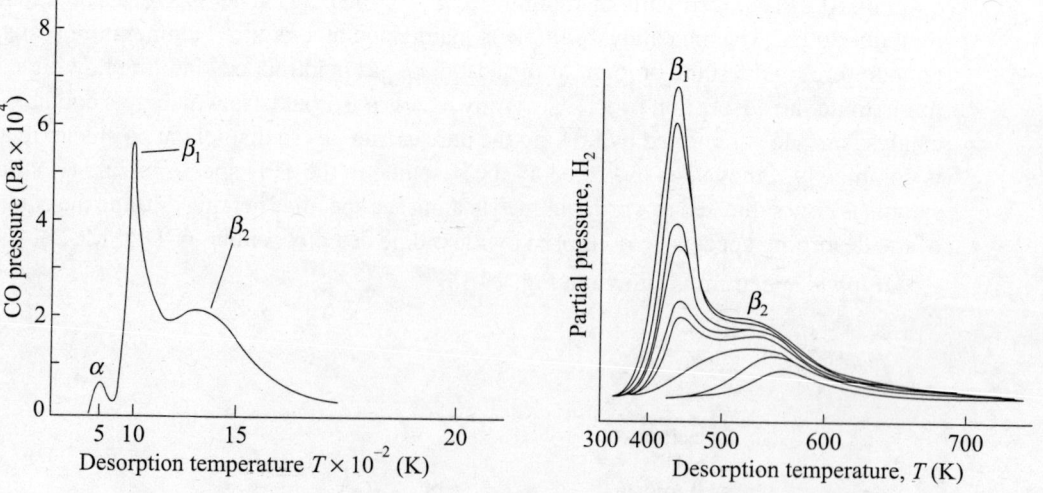

Fig. 19.15 TPD spectrum of (a) CO from W surface and (b) H_2 from [100] surface of W crystal

(iii) TPD spectrum is useful for the determination of kinetic parameters for the desorption process, particularly when the desorption spectrum shows a single peak with a coverage- independent rate constant. The peak in the desorption spectrum occurs when the desorption rate is a maximum and this condition can be found by setting $d^2n/dt^2 = 0$. Equations (19.11) and (19.12) are applicable to the first-order desorption ($x = 1$) and the second-order desorption ($x = 2$) respectively.

$$\frac{E}{RT_m^2} = \frac{A}{\beta} \exp\left(\frac{-E}{RT_m}\right) \qquad \text{for} \quad x = 1 \tag{19.11}$$

$$\frac{E}{RT_m^2} = \frac{(2n_m A)}{\beta} \exp\left(\frac{-E}{RT_m}\right) \qquad \text{for} \quad x = 2 \tag{19.12}$$

where E is the activation energy for desorption, R is gas constant, T_m is temperature maximum in K or peak temperature of desorption occurring at maximum surface coverage n_m, x is the order of desorption, and β is heating rate (K/min). Since the second-order desorption curve is mostly symmetric about T_m, the initial surface coverage $n_0 = 2n_m$. Hence, Eq. (19.12) can be rewritten as

$$\frac{E}{RT_m^2} = \frac{(n_0 A)}{\beta} \exp\left(\frac{-E}{RT_m}\right) \qquad \text{for} \quad x = 2 \tag{19.13}$$

Thus, as given in Eq. (19.13) the peak maximum temperature depends on the initial coverage n_0 only for the second-order desorption. Thus, if the peak shifts to lower temperatures with increasing surface coverage the order of desorption is 2. In contrast, if the peak temperature is independent of surface coverage the order of desorption is 1. In Fig. 19.15(b) the desorption peak at higher temperatures depends on the surface coverage and shifts to lower temperatures with increasing coverage indicating that it is a second-order desorption. The first peak in the same figure shows a peak temperature independent of surface coverage, and hence, it represents a first-order desorption process.

For a constant initial surface coverage n_0 of the first-order or the second-order desorption process differentiation of Eq. (19.11) or (19.13) gives

$$\frac{d \ln (T_m^2/\beta)}{d (1/T_m)} = \frac{E}{R} \tag{19.14}$$

The activation energy E of the desorption can be determined from the slope of the plot of $d \ln (T_m^2/\beta)$ versus $1/T_m$. The pre-exponential factor A (frequency factor) can then be determined from Eq. (19.15)

$$-\frac{dn}{dt} = k_d n^x = A \exp\left(\frac{-E}{RT}\right) n^x \tag{19.15}$$

where n is the surface coverage (concentration of the desorbing species per unit area) and k_d is the rate constant for desorption.

(iv) TPD spectrum is useful for determining the surface acidity in terms of the nature of acid sites and their densities and the strengths of sites. Probe molecules such as ammonia, pyridine or alkylamines can be used in such studies. The presence of multiple peaks in the TPD spectrum indicates the heterogeneity of the acidic surface. The peak temperatures are indicative of the strength of acidic sites while the areas under the peaks provide information on the density of acidic sites.

19.7.2 Temperature-Programmed Reduction

Temperature-programmed reduction (TPR) is highly sensitive for characterizing solids which undergo reduction is the presence of reducing agent, usually hydrogen gas. The experimental set-up and methodology is similar to that of TPD, the only difference being the reducing atmosphere provided by the hydrogen gas flowing during programmed heating of the sample. The amount of hydrogen consumed by the sample as a function of temperature is determined by monitoring the change either in the pressure of the flowing gas or in the weight of the sample. The TPR spectrum is a plot of amount of hydrogen consumed as a function of the sample temperature.

The technique is useful in the determination of the effect of catalyst precursors and the type of interactions between the active precursor and the support. For example, Fig. 19.16 shows the TPR profile of the reduction of Ni $(OH)_2$ prepared by urea hydrolysis. The TPR shows double peaks corresponding to the reduction of the rhombohedral and cubic forms of NiO. The reduction profile of NiO (cubic) is also shown in Fig. 19.16.

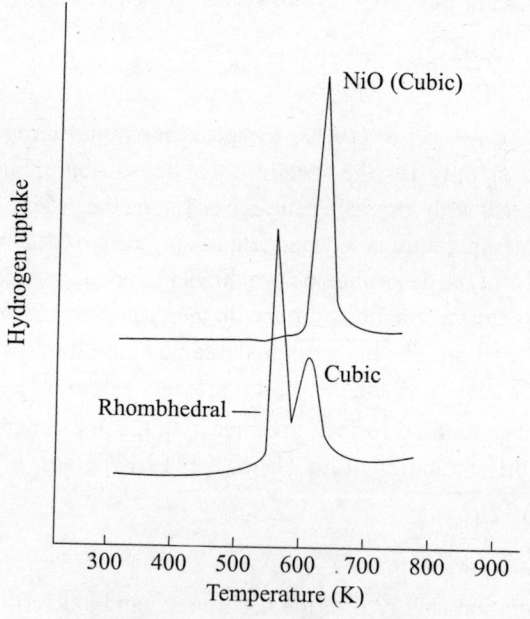

Fig. 19.16 TPR profile of Ni $(OH)_2$

Temperature-programmed reduction is useful to probe the nature of interaction between a dispersed metal oxide and its support. For example, in the preparation of Ni/SiO$_2$ catalyst the reduction of NiO to Ni occurs at four different temperatures as shown in Fig. 19.17.

The four reduction peaks are assigned to (i) reduction of Ni^{3+} at 503 K, (ii) reduction of bulk NiO at 588 K, (iii) reduction of NiO at 658 K is indicative of interaction between NiO and the support SiO_2, and (iv) NiO reduction at 823 K indicative of strong interaction between NiO and SiO_2.

The effect of calcination temperature on the reduction and dispersion of the Ni metal during the preparation of Ni/SiO$_2$ catalyst is analysed by TPR (see Fig. 19.18). Reduction becomes more difficult with increasing calcination temperature. In calcination above 773 K water vapour decreases the formation of surface Ni and also causes inhomogeneity and low dispersion of the metal. Hence, lower calcination temperature is preferred.

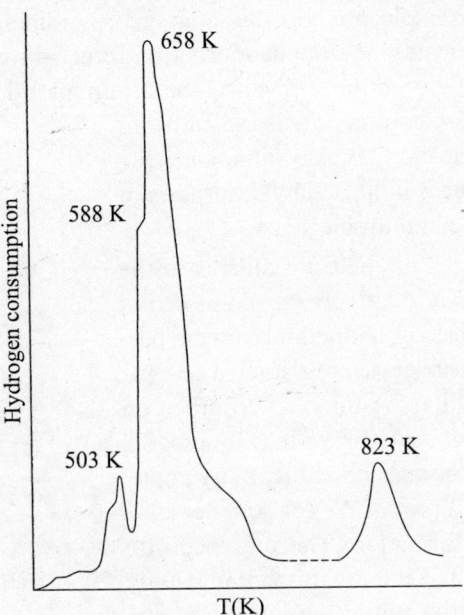

Fig. 19.17 TPR profile of Ni/SiO$_2$ calcined at 593 K for 30 min

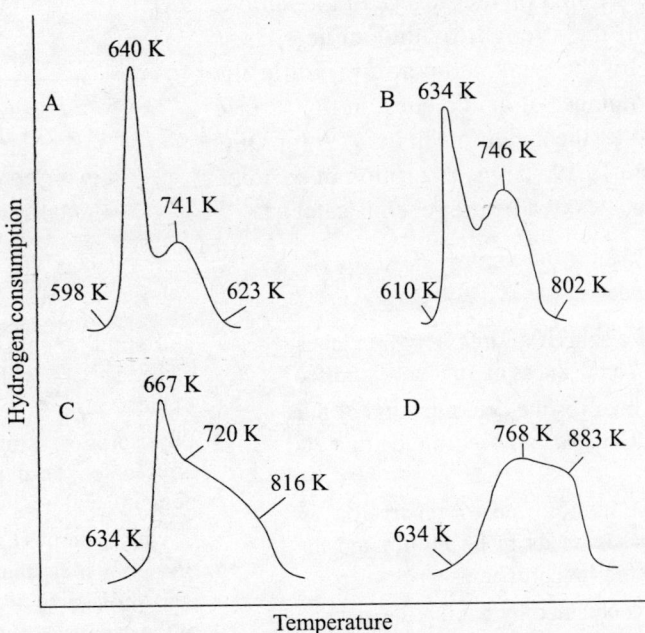

Fig. 19.18 TPR profile showing effect of calcination temperature on Ni/SiO$_2$
(A) 623 K, (B) 673 K, (C) 773 K, (D) 973 K

19.7.3 Desorption Studies by TG, DTA, and DSC

Thermal analytical instruments such as thermogravimetric analyser, differential thermal analyser, and differential scanning calorimeter can also be used for TPD. The pre-adsorbed sample is subjected to programmed heating in TGA and the weight loss is monitored as a function of temperature. For example, pre-adsorbed ammonia, pyridine, or *n*-butylamine can be desorbed and the weight loss indicates the amount of probe molecule adsorbed on a given weight of the sample from which the acidity of the surface can be determined. The temperature at which desorption of the adsorbed base commences is indicative of the strength of the acidic sites. Any inhomogeneity in the acidic surface is indicated by desorption of the probe molecule in multiple steps.

In DTA and DSC the detector signals in the form of ΔT and ΔH respectively are monitored as a function of increasing temperature as the pre-adsorbed sample is subject to programmed heating. For example, pyridine or ammonia adsorbed on the solid sample can be desorbed in DTA or DSC for characterizing the surface acidity of a sample. The DTA (or DSC) plot of ΔT (or ΔH) versus T provides useful information. The presence of a single desorption peak indicates homogeneity of the surface acidity while the appearance of multiple peaks indicates the presence of different types of acidic sites on the surface of the sample. In addition, the strength of the acidic sites is indicated by the peak temperature while the density or number of acidic sites on the surface is indicated by the area of peak in DTA (or DSC plot). Figure 19.19 shows desorption of pyridine from acidic sites of a heterogeneous catalyst.

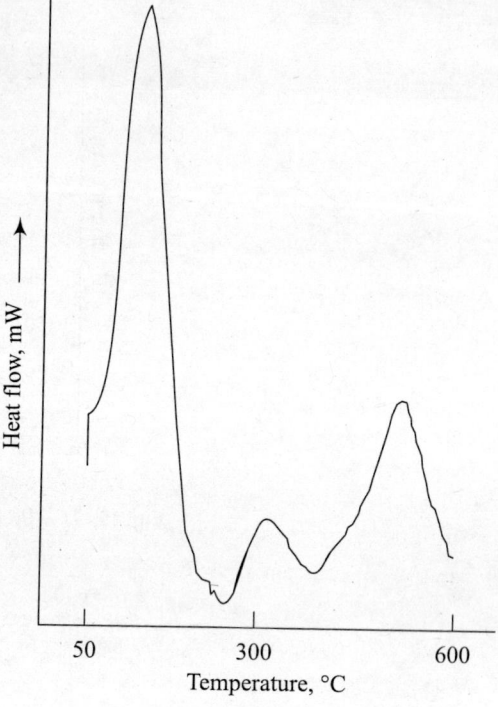

Fig. 19.19 DSC thermogram showing pyridine desorption from acidic sites

SUMMARY

- The surface of a solid is an infinitely thin layer extending up to a thickness of 10 nm or more.
- Surface study involves the determination of the compositional changes that occur in the first 20 atomic layers.
- Surface contamination due to adsorption of surrounding gases needs to be eliminated by evacuating to ultra-low pressures.
- Clean surface is obtained by baking the sample at high temperatures, mechanical scraping and polishing using a variety of abrasives, ultrasonic cleaning, sputtering the sample surface with a beam of argon ions, and scribing with a carbide tip.
- Depth profiling involves bombarding the surface of the sample with a beam of argon ions or xenon ions to sputter and etch away the surface layer

and simultaneously bombarding the sputter-etched surface with a beam of *x*-rays or a beam of electrons.
- Physisorption using nitrogen as probe molecule is used for the determination of surface area, porosity, and pore size distribution of a solid.
- Chemisorption is useful to understand the nature of active sites, their distribution, strength and density of active sites as well as any heterogeneity of the catalyst surface, and the nature of binding forces involved between the catalyst surface and a reactant molecule and how a catalyst activates the reactant molecule.
- Chemisorption using probe molecules of ammonia or pyridine is useful in the determination of surface acidity and to distinguish between Brønsted and Lewis acid sites, through IR spectroscopy and thermal desorption methods.

- IR spectroscopy of chemisorbed probe molecule carbon monoxide reveals the type of bonding in meta carbonyls.

- Electron energy loss spectroscopy involves focusing a beam of electrons of low energy in the range 1–10 eV on the surface of a sample and monitoring the energy of the inelastically scattered electrons to get information on the nature of bonding between the solid and the adsorbate and also on adsorbate geometry.

- Reflection–absorption infrared spectroscopy together with EELS is useful for the study of surface-adsorbed species and surface structure.

- Electron spectroscopy for chemical analysis is based on photoelectric effect and involves irradiation of a sample with X-ray photons or UV photons to facilitate the emission of electrons from the sample. The kinetic energies of the ejected photoelectrons are analysed by a detector to provide information on the energy levels of the metal.

- XPS provides quantitative information on the elemental composition, empirical formula, chemical, and electronic states of elements on the surface of a material to a depth of 1–10 nm. XPS can detect all the elements other than hydrogen and helium with detection limits in the range of parts per thousand.

- Auger electron spectroscopy involves bombarding a beam of primary electrons of 1–10 keV energy on the sample surface. The bombardment creates holes in the core levels of the surface atom. The excited atom relaxes by filling the core hole with an electron from a higher shell with the simultaneous release of energy in the form of X-ray fluorescence or emission of an Auger electron. The kinetic energy of the Auger electron provides information on the nature of energy levels of the sample and is useful for the identification of elements on the surface.

- Ion scattering spectroscopy involves bombarding the surface of a target material with noble gas ions and observing the positions, velocities, and energies of the elastically scattered ions. The technique is sensitive and capable of observing directly hydrogen atoms on the surface of a material and is useful in the study of surface composition of heterogeneous catalysts, thin film coatings, and adhesive layers and also to provide information on the arrangement of surface atoms.

- Secondary ion mass spectrometry involves bombarding the target solid surface with a primary beam of ions and analysing the ejected secondary ions from the surface by a mass spectrometer.

- The non-destructive technique is highly sensitive and provides information on the elemental and isotopic composition on surfaces at ppb levels.

- X-ray absorption spectroscopy along with extended X-ray absorption fine-structure and X-ray absorption near-edge structure provide information on the local structure around the absorbing atom.

- Surface-extended X-ray absorption fine-structure technique provides information on the inter-atomic distances of adsorbates and their coordination environment.

- Temperature-programmed desorption involves temperature-induced desorption of a previously adsorbed substance and provides information on the nature and energy of interaction between the surface of a solid and the adsorbate.

- Temperature-programmed reduction is used for characterizing solids which undergo reduction in the presence of hydrogen gas and is particularly suited to study interaction between the active catalyst and its support.

REVIEW QUESTIONS

1. What are the characteristics of the surface of a solid?
2. Give a brief outline of the classification of the surface analytical methods.
3. Describe the use of probe molecules in the study of the surface of a solid by adsorption.
4. How is surface area of a solid determined?
5. How is chemisorption of pyridine useful in the study of surface acidity?
6. Give an account of the use of IR spectroscopy as a surface analytical tool.

7. Discuss the principles involved in EELS and RAIRS techniques and their applications.

8. Give an account of the use of different probe beams used in electronic spectroscopic methods.

9. What is ESCA? Discuss in detail the principle and practice involved in ESCA. What are its applications?

10. Write a note on Auger spectroscopy and its applications.

11. Discuss the principles involved in ISS and SIMS.

12. Write a note on the different techniques associated with XAS.

13. Discuss in detail the technique of TPD and its applications in the study of surfaces.

14. What is TPR? How is it carried out? How is it useful in the study of catalysts?

Bibliography

Christian, G., 2004, *Analytical Chemistry*, 6th edn, Wiley.

Dean, J.A., 1995, *Analytical Chemistry Handbook*, McGraw-Hill.

Evans, A., 1987, *Potentiometry and Ion Selective Electrodes*, Wiley.

Haines, P.J., 1995, *Thermal Methods of Analysis: Principles, Applications and Problems,* Blackie A & P.

Harvey, D., 2000, *Modern Analytical Chemistry*, McGraw-Hill.

Mendham, J., R.C. Denney, J.D. Barnes, M. Thomas, and B.Sivasankar, 2009, *Vogel's Textbook of Quantitative Chemical Analysis*, 6th edn, Pearson.

Metcalfe, E., 1987, *Atomic Absorption and Emission Spectroscopy*, Wiley.

Mikkelsen, S.R. and E Corton, 2004, *Bioanalytical Chemistry*, Wiley Interscience.

Pavia, D.L., G.M. Lampman, and G.S. Kriz Jr. 1979, *Introduction to Spectroscopy*, Saunders.

Riley, T. and C. Tomlinson, 1987, *Principles of Electroanalytical Methods*, Wiley.

Skoog, D.A., D.M.West, F.J.Holler, and S.R.Crouch, 2004, *Fundamentals of Analytical Chemistry*, 8th edn, Thomson /Brooks/Cole.

Skoog, D.A., F.J. Holler, and T.A. Nieman, 1998, *Principles of Instrumental Analysis*, 5th edn, Brooks/Cole.

Wendlandt, W.W., 1986, *Thermal Analysis*, 3rd edn, Wiley.

Willard, H.H., L.L. Merritt, J.A. Dean, and F.A. Settle, 1988, *Instrumental Methods of Analysis*, 7th edn, Wadsworth Pub. Co.

Williams, D.H. and I. Fleming, 1989, *Spectroscopic Methods in Organic Chemsitry*, 4th edn, McGraw-Hill.

Index

Related Titles

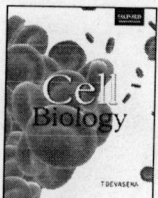

Cell Biology | 9780198075516

Devasena T., Centre for Nanoscience and Technology, Anna University, Chennai.

Cell Biology is a textbook for students of biotechnology and allied disciplines to understand the functioning of a cell—the fundamental unit of life.

Key Features

- Includes self-explanatory line diagrams and flow charts for easy understanding of concepts
- Provides explanations to important topics such as cellular metabolic pathways, haematopoiesis, and stem cell biology

Enzymology | 9780198064435

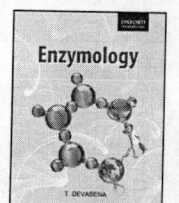

Devasena T., Centre for Nanoscience and Technology, Anna University, Chennai

Enzymology is designed as a textbook for the students of biotechnology and chemical engineering. It will also serve as an invaluable reference for undergraduate and postgraduate students of biotechnology and life science programmes.

Key Features

- Addresses topics on mechanics of enzyme action, microbiological enzymes, and purification and characterization of enzymes that are of interest to a life science student
- Discusses latest topics such as biosensors and biochips

Immunology | 9780198068266

Raj Khanna, Former Professor, University of Lucknow

The book has been written primarily for undergraduate biotechnology, biochemistry, and medical courses; it would also be useful to postgraduate students of biochemistry, biotechnology, and microbiology.

Key Features

- Presents recent research areas such as regulatory T cells and their manipulation for clinical use and cancer immunotherapy including cancer vaccines
- Includes major immunological methods, immunoassays, and experimental systems in the appendix

Other Related Titles

- 9780195697810 Pal and Ghaskadbi: Fundamentals of Molecular Biology
- 9780195676884 Chakravarty: Immunology and Immunotechnology
- 9780195699609 Nallari: Medical Biotechnology
- 9780195687828 Bhattacharyya and Banerjee: Environmental Biotechnology
- 9780195692303 Ghosh and Mallick: Bioinformatics Principles and Applications
- 9780195676839 Bosu and Thukral: Bioinformatics Experiments, Databases, Tools, and Algorithms
- 9780195696578 Rastogi and Pathak: Genetic Engineering